Student Study Guide and Solutions Manual

to accompany

CHEMISTRY
THE MOLECULAR SCIENCE
Second Edition

Olmsted & Williams

Wayne Tikkanen
California State University, Los Angeles

John Olmsted III
California State University, Fullerton

WCB Wm. C. Brown Publishers

Dubuque, IA Bogotá Buenos Aires Caracas Chicago Guilford, CT London
Madrid Mexico City Seoul Singapore Sydney Taipei Tokyo Toronto

Copyright © 1997 by The McGraw-Hill Companies, Inc. All rights reserved

ISBN 0-8151-5515-8

No part of this publication may be reproduced, stored in a retrieval system, or transmitted, in any form or by any means, electronic, mechanical, photocopying, recording, or otherwise, without the prior written permission of the publisher.

Printed in the United States of America

10 9 8 7 6 5 4 3 2 1

Contents

Chapter 1	**The Science of Chemistry**	1
	Solutions to Chapter Problems	26
Chapter 2	**The Atomic Nature of Matter**	33
	Solutions to Chapter Problems	52
Chapter 3	**The Composition of Molecules**	61
	Solutions to Chapter Problems	99
Chapter 4	**Chemical Reactions and Stoichiometry**	113
	Solutions to Chapter Problems	142
Chapter 5	**The Behavior of Gases**	161
	Solutions to Chapter Problems	187
Chapter 6	**Atoms and Light**	201
	Solutions to Chapter Problems	218
Chapter 7	**Atomic Structure and Periodicity**	227
	Solutions to Chapter Problems	256
Chapter 8	**Fundamentals of Chemical Bonding**	267
	Solutions to Chapter Problems	295
Chapter 9	**Chemical Bonding: Multiple Bonds**	311
	Solutions to Chapter Problems	336
Chapter 10	**Effects of Intermolecular Forces**	353
	Solutions to Chapter Problems	383
Chapter 11	**Macromolecules**	393
	Solutions to Chapter Problems	417
Chapter 12	**Chemical Energetics**	429
	Solutions to Chapter Problems	454
Chapter 13	**Spontaneity of Chemical Processes**	465
	Solutions to Chapter Problems	492
Chapter 14	**Mechanisms of Chemical Reactions**	503
	Solutions to Chapter Problems	540
Chapter 15	**Principles of Chemical Equilibrium**	553
	Solutions to Chapter Problems	580
Chapter 16	**Aqueous Equilibria**	599
	Solutions to Chapter Problems	629
Chapter 17	**Electron Transfer Reactions**	651
	Solutions to Chapter Problems	677
Chapter 18	**The Transition Metals**	693
	Solutions to Chapter Problems	708
Chapter 19	**The Main Group Elements**	717
	Solutions to Chapter Problems	726
Chapter 20	**Nuclear Chemistry and Radiochemistry**	735
	Solutions to Chapter Problems	755

Chapter 1 The Science of Chemistry

1.1 What Is Chemistry?

QUESTIONS TO ANSWER, SKILLS TO LEARN
1. **What is chemistry?**
2. **What do chemists study?**
3. **How do scientists develop explanations for phenomena?**

Chemistry is the study of matter; it attempts to understand the properties of the substances in our world and to predict how these substances will behave under certain conditions. Chemistry has made possible many of the technological advances we enjoy and offers the only hope to repair environmental damage from improper storage of hazardous waste and oil spills. Chemists seek to understand how and why matter undergoes changes, and how matter may be used to create new materials for engineering or new medicines. Here are some of the questions chemists often ask about matter:

- How and why does matter undergo change from one form to another?
- What types of substances and how much energy are required for the change?
- How many different substances are present in a mixture?
- What is the composition of matter (relative amounts of different elements)?

Chemists make (**synthesize**) the substances in a sample of interest to see if they can duplicate the properties of the sample . They deliberately introduce small changes in molecules to see the effect that change has on the properties of that substance. Chemists are also molecular architects, using the building blocks of matter, the elements, to make new types of materials.

Exercise 1.1 A plant in the Amazon rain forest has proved to be effective in curing several diseases. How can chemists help find ways to use this property?

Steps to Solution: Chemists can help find uses for this property by analyzing the different substances in the plant to find the one responsible for the desired property. They may then try to synthesize one or several of the compounds found in the plant and compounds related to the naturally occurring ones to see if the desirable property can be duplicated or improved.

Exercise 1.2 List several reasons why chemistry is important (see text for some examples and make up your own).

An Experimental Science

Chemistry is an experimental science in that chemists make discoveries by experimental measurements or observations which are then recorded. Chemists make explanations for these observations and test them by experiments.

Experiments *measure properties of matter (there are many that can be measured). Experiments are designed to test explanations which predict physical or chemical properties or the occurrence of chemical and physical changes.*

Properties *are specific attributes of substances/matter that can be measured in experiments. Some more specific definitions of properties will be provided later in this chapter.*

Theories *are tentative general explanations that have been tested and explain the properties of a variety of substances. Theories are tested with new experiments to see if the theory correctly predicts the behavior in the new experiment. If a theory passes a large number of experimental tests, we can be confident that it is a reasonably good predictor of nature.*

Principles *or* *natural laws* *are observations that have never been violated An example is the principle of conservation of energy.*

Exercise 1.3 Superconductivity is described in the text in Section 1.1. Is superconductivity a property, principle or theory?

Steps to Solution: Superconductivity is a phenomenon that is observed under certain conditions for certain materials. Unlike a theory, it is not an explanation of properties nor a broad statement regarding physical behavior. Therefore superconductivity is a property.

1.2 The Molecular Nature of Chemistry

QUESTIONS TO ANSWER, SKILLS TO LEARN
1. **What are atoms and molecules?**
2. **How do you describe atoms and molecules with symbols?**

An *atom* is the smallest particle of matter that has chemical identity as an element. An isolated atom can be envisioned as a sphere. Atoms have:
- very small mass (0.00000000000000000000000002 kg for a carbon atom)
- very small size (0.0000000003 m for a C atom)

A sample of matter that you can detect through your senses (a **macroscopic** sample) contains millions of billions of atoms.

Molecules are comprised of two or more atoms that are bound together in some definite ordered shape. In molecules, certain atoms are found next to other atoms. If the arrangement of atoms is changed, a different molecule is obtained. In molecules with only two atoms, the two atoms have only one possible distinct structure: the atoms are next to each other. The following sketches show several stable diatomic molecules. Different shades of gray and the atomic symbols are used to distinguish the different atoms in these molecules: the only unlabeled element is carbon.

Consider two different arrangements of two carbon (C) atoms, one oxygen (O) atom and four hydrogen (H) atoms (rules for obtaining these two arrangements will be developed later):

Ethylene oxide is a stable substance that can be stored. Among other things, this reactive gas is used for sterilizing surgical equipment. In contrast, ethenol is a substance that cannot be prepared in usable quantities. Thus, although these two molecules have the same number of atoms of the same type, the molecules' different structures give them different properties.

Exercise 1.4 Sketch diagrams of the molecules (a) hydrogen chloride (HCl) and (b) nitrogen oxide (NO).

Steps to Solution: Consider the number of ways that the atoms can be arranged. Draw the atoms as spheres in contact with one another to show how they are attached to each other.
(a) HCl. Because HCl is a diatomic molecule, there is only one arrangement of the atoms: side by side.

(b) NO (see part (a) of this question).

Chemical symbols are a shorthand notation for naming elements. Two or three letters are used to represent each element in a chemical formula. Because not all of the symbols correspond to abbreviations of the names, it is best to memorize the names and symbols. The elements' symbols sometimes, but not always, are similar to the beginning of the name. Before completing this chapter, you should memorize the names and symbols of the first 36 elements. You may find it helpful to make flash cards with the symbol of the element on one side and its name on the other.

Exercise 1.5 Give the names of the elements with the following symbols:
(a) K (b) S (c) Ar (d) B (e) Nb, (f) Na (g) Ne (h) Br

Steps to Solution: This is an exercise in helping you to learn the names of the elements. Look at the periodic table and memorize the first 36 elements!
(a) potassium (b) sulfur (c) argon (d) boron (e) niobium (f) sodium (g) neon (h) bromine

Exercise 1.6 What are the symbols for the following elements?
(a) phosphorus (b) beryllium (c) nitrogen (d) lead (e) lanthanum

Steps to Solution: This is an exercise in helping you to learn the names of the elements. Look at the periodic table and memorize the first 36 elements!
(a) P (b) Be (c) N (d) Pb (e) La

Chemical compounds contain more than one atom and nearly always more than one element. A given compound always has the same relative numbers of the different types of atoms present. Sodium chloride, the primary substance in table salt, has one atom of sodium for each atom of chlorine. A more convenient way to describe the elemental composition of sodium chloride (or any chemical compound) is *chemical formulas*. A chemical formula is written by listing the symbols for each element consecutively, with the number of atoms of any particular element subscripted after the element's symbol. If there is only one atom of an element in the compound, no subscript is written. The lack of a subscript after an element's symbol indicates only one atom of that element is present in the formula.

If we want to write the formula for sodium chloride, we first find the symbols for each element (Na for sodium, Cl for chlorine). There is one sodium atom for each chlorine atom, so the subscripts would be 1. However, subscripts of 1 are not written; thus the chemical formula for sodium chloride is NaCl. At this point, we have not developed the rules for the order of the elements' symbols in the formulas. Do not be concerned if at this point the order of the symbols in your formula is different. However, you should be concerned if your subscripts are different from those in the exercises.

Exercise 1.7 What are the chemical formulas for the molecular models shown in:

(a) hydrazine (b) formaldehyde (c) ethylene oxide

Steps to Solution: The formulas are found by first finding the different type of atoms in the molecule and then the number of each type of atom present. The elements' symbols are then listed and the subscripts noting the different numbers of atoms placed after the appropriate symbol.

(a) We see that hydrazine contains two nitrogen (N) atoms and four hydrogen (H) atoms. Thus the N in the formula will have a subscript of 2 and the H will have a subscript of 4. The chemical formula is $\boxed{N_2H_4}$.

(b) Formaldehyde has one carbon (C) atom, two hydrogen (H) atoms and one oxygen (O) atom. C and O will have no subscript, indicating that only one atom of each of these elements is present and H will have a subscript of 2. The chemical formula is $\boxed{CH_2O}$.

(c) The formula of ethylene oxide is $\boxed{C_2H_4O}$.

1.3 The Periodic Table of the Elements

QUESTIONS TO ANSWER, SKILLS TO LEARN
1. What is the periodic table?
2. What are general types of matter formed by the elements?
3. What kind of similarities can be found using the periodic table?

Arrangement of the Periodic Table

The periodic table is arranged in rows and columns with two rows that are removed from the main body of the periodic table. Moving from left to right in a particular row, the masses of the atoms increase and the chemical properties change relatively dramatically. Moving down a column, the mass of the elements increases, but the chemical properties (or reactivity) of the elements are remarkably similar. Sometimes the elements in columns are called groups or families to emphasize this similarity in properties.

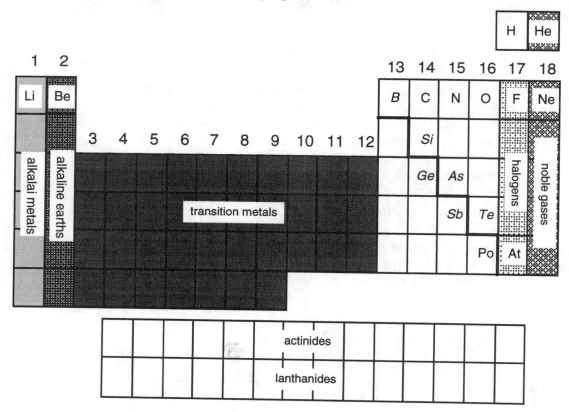

Chemists describe the elements in terms of three general classes: *metals*, *nonmetals*, and *metalloids*. Metals are characterized by their ability to conduct electricity well, to be hammered into thin sheets (malleability), and to be drawn into wires (ductility). Metals also often have a shiny appearance. Only one metal (Hg, a liquid) is not a solid at room temperature. Most elements are metals, except for those in the upper right of the periodic table. Of these elements that are not metals, most of them are nonmetals that have decidedly nonmetallic properties. They can be liquids, solids, or gases at room temperature and are poor conductors of heat and electricity. Six elements (B, Si, Ge, As, Sb, Te) have moderate electrical conductivity and are called the metalloids or sometimes, semiconductors. In the periodic table, a heavy zig-zag line divides the non-metals and the metals. The metals fall to the left of the line and the nonmetals to the right. The metalloids, which have properties intermediate between metals and nonmetals, fall along this line and their symbols are shown in italic print.

Exercise 1.8 What elements would you expect to have chemical properties similar to those of sodium?

Steps to Solution: Elements in a column have similar chemical properties. Therefore, we expect that elements that are in the same column would have similar chemical properties.

Since sodium is in the first column or group of the periodic table, other elements in the first group would have similar chemical properties. Therefore, we would expect that potassium, lithium, and the other first-group elements would have similar properties.

Exercise 1.9 Classify the following elements as metals, nonmetals or metalloids.
(a) fluorine (b) silicon (c) Al (d) Rb (e) carbon (f) Ge

Steps to Solution: The periodic table is used to classify the elements in these categories. Use the complete periodic table in your text to locate the symbol of the element in question. Its location is used to classify that element as a metal, non-metal or metalloid.

(a) Fluorine is F. It is found in the upper right section of the periodic table in the nonmetals.
F is a nonmetal .

(b) Silicon is Si. It falls in the category between metals and nonmetals. Silicon is a metalloid .

(c) Al is the symbol for aluminum. It is found to the left of the line dividing metals and non-metals. Aluminum is a metal .

(d) Ge is the symbol for germanium. It is found in the category between metals and nonmetals. Germanium is a metalloid .

The elements in given columns of the periodic table have similar chemical properties. They repeat regularly and are called the ***periodic properties***. The elements in a column are called a ***group***. The general properties of elements in groups proceeding from the leftmost group to the right are:

Chapter 1: The Science of Chemistry

- **Column 1**: the alkali metals. These are soft shiny metals that react with the halogens (X = F, CL, Br, I, At) to give compounds of the formula MX where M is an alkali metal. Also referred to as the Group 1 elements.

- **Column 2**: the alkaline earths. These are metals that react with halogens to give compounds of formula MX_2 and with oxygen to give compounds of formula MO. Also referred to as the Group 2 elements.

- **Columns 3-12**: the transition metals. These are metals of differing reactivity and properties.

- **Column 13-16**: the main group elements. Metals, non-metals and metalloids are found in these 4 columns.

- **Column 17**: the halogens. These elements are found as X_2, but they are quite reactive, forming compounds with metals (as described above) and nonmetals.

- **Column 18**: the noble gases. These elements are found as atomic gases and are quite unreactive, although some compounds have been made with them.

Exercise 1.10 Predict the formulas of the compounds formed in the reaction between: (a) sodium and iodine (b) Ra and Cl_2 (c) Cs and F_2 (d) strontium and oxygen

Steps to Solution: Determine the groups to which both substances belong and use the general descriptions of chemical reactivity listed above to predict the formulas.

(a) Sodium is Na and is one of the alkali metals. Iodine has the symbol I and is one of the halogens, found as I_2. The alkali metals form compounds MX with halogens; therefore, we expect sodium and iodine to form the compound $\boxed{NaI}$.

(b) Ra is one of the Group 2 elements and Cl is a halogen. Since alkaline earths form complexes of the formula MX_2 with halogens, we expect that the compound will have the formula $\boxed{RaCl_2}$.

(c) Cs is a Group 1 element (an alkali metal) that forms compounds of the formula MX with halogens, of which fluorine is a member. The expected formula is $\boxed{CsF}$.

(d) Strontium is an alkaline earth, one of the Group 2 elements, and forms compounds MO with oxygen. The expected formula is $\boxed{SrO}$.

1.4 Characteristics of Matter

QUESTIONS TO ANSWER, SKILLS TO LEARN
1. What is matter?
2. How do chemists categorize matter?
3. How do chemists classify events that change matter?

Matter is anything that occupies space and has mass. Matter can be composed of one compound or many compounds. It is useful to categorize matter (1) in terms of the number of substances of which it is composed and (2) if several compounds are present, how they are distributed throughout that particular sample of matter.

There are several ways to categorize matter. We can classify matter as either pure substances or mixtures. In the diagram below, we have a sample and the arrows point to two different places which we sample. If both samples show the same composition (relative numbers of atoms or molecules), we can say that the sample is **homogeneous**. If the samples have different compositions (different numbers or different types of atoms or molecules), we say that the sample is **heterogeneous** which means it must be a mixture of two or more compounds.

sample 1 ←→ sample 2

Looking at a homogeneous sample at the molecular level, a further distinction emerges. *If only one type of molecule or atom is present, we have a pure substance. If more than one molecule is present, we have a homogeneous mixture*, commonly called a **solution**.

Pure substances can be broken down into two categories. If the substance contains only one element it is an *elemental substance*. If it is composed of *more than one element, it is a chemical compound*.

Exercise 1.11 As a review, define: (a) pure substances (b) elemental substance (c) compounds (d) mixtures (e) homogeneous mixtures (f) heterogeneous mixtures.

Steps to Solution: Review the definitions given above.

(a) Pure substances have only one kind of molecule, or if an atomic substance, one kind of atom present. Examples include neon gas (Ne) and nitrous oxide, N_2O.

(b) Elemental substances have only one element present. Examples include oxygen (O_2) and ozone (O_3).

(c) Compounds contain a single type of molecule with more than one element present in that molecule, such as methane (CH_4) and nitric oxide (NO_2).

(d) *Mixtures contain more than one chemical substance, such as coffee or air.*

(e) *In homogeneous mixtures the different substances are found in the same proportions throughout the sample. Examples include filtered coffee, steel and filtered air.*

(f) *In heterogeneous mixtures the different substances are found in different proportions in different locations in the sample. Examples include a sandwich and a chocolate bar with nuts.*

Exercise 1.12 A new class of compounds, the fullerenes, has been recently discovered. They are made by vaporizing carbon and collecting the resulting solid. C_{60}, C_{70} and other substances are obtained from the solid. Is the solid a mixture, a pure substance, or a compound?

Steps to Solution: We must determine the composition of the solid from the supplied information and use the provided definitions to determine what category the substances belong in. The solid is a mixture because it contains a number of pure substances, including C_{60} and C_{70} as stated in the problem.

Beyond its composition, matter can be further classified by its volume and shape (phase). The ***phases of matter*** that are most commonly encountered are the solid, liquid and gas phases.

- ***Solids*** have a definite shape and volume.

- ***Liquids*** have a definite volume but not a definite shape. Liquids take the shape of their container.

Solids and liquids are called *condensed* phases because of their nearly constant volume

- ***Gases*** have neither a specific volume nor shape. They take the shape of their container and change their volume to fill the container.

Matter can be converted from one phase to another by changing temperature and/or pressure. Changes of phases are one class of ***transformations of matter.*** A transformation of matter causes the properties of the matter to change. Converting water to steam is a transformation of matter. The property that changes is the phase: steam is a gas and the water was a liquid. But the gas and the liquid are both water. Because the chemical substance is unchanged, this a ***physical transformation***. Another transformation of matter is the burning of wood in a fireplace. From one solid and oxygen, we obtain a new solid (the ashes) and various gases and heat. The generation of new chemical substances is characteristic of a ***chemical transformation***.

Chapter 1: The Science of Chemistry

Physical transformations lead to no change in the molecules and/or atoms participating in the change. *No new molecules are produced in a physical transformation.* The more common physical transformations of matter arecalled changes of phase. A common phase change is the melting of ice, in which water is transformed from the solid to the liquid phase. In the beginning, water molecules are in the solid phase; afterwards they are liquid. However, the chemical substance is still water after the transformation. The melting of water is a physical transformation because the chemical structure and identity of the substance involved have not changed.

In a *chemical transformation*, the substances present at the end of the transformation have changed. *Different molecules are present at the end of the transformation than were present at the beginning.* The arrangement of atoms has changed in the molecules, or atoms may have been transferred from one molecule to another during the process.

Exercise 1.13 Classify the following as physical or chemical transformations: (a) eggs being fried; (b) gasoline evaporating from an open can; (c) charcoal briquettes turning into powder in the bag; (d) the rearrangement shown in the following molecular picture.

Steps to Solution: We must determine if the substances involved are different after the given transformation. We can determine this from the observed properties of the substances or from the structure of the molecules if provided.

(a) This is a chemical transformation. The properties of the eggs have been changed from fluid to stiff. The white of the egg has changed from transparent to opaque. Also, the change cannot be reversed.

(b) This is a physical change. The gasoline is changing from the liquid to the gas phase, but no chemical change is occurring.

(c) This is a physical change. The charcoal is being broken up into smaller pieces, but is not undergoing any chemical change. The powder has the same properties as the briquettes: it is black and will burn like the briquettes.

(d) This is a chemical change because a different molecule results from the process. The atoms in the left-hand molecule are arranged differently than the atoms in the right-hand molecule. Different arrangements of atoms correspond to different molecules.

Chapter 1: The Science of Chemistry

1.5 Measurements in Chemistry

QUESTIONS TO ANSWER, SKILLS TO LEARN
1. Why are measurements important?
2. What quantities are measured?
3. What are standards and conversion units ?
4. Determining significant figures and using scientific notation
5. How do measurements limit the precision of an answer?

In order to determine if changes have occurred in matter, we must compare properties of the sample. Changes are determined by measurement of the **physical and chemical properties** of the sample. Chemists and other scientists make many types of measurements. *Chemical properties* are usually descriptions of the types of chemical reactions that a substance will undergo.

An important physical property that we can measure is the **amount of space** an object occupies. In one dimension we call this measurement *length*. In two dimensions, it is called *area* and in three dimensions, *volume*. Length is measured with a ruler, measuring tape and other devices whose length is known. Area (length times length, or length^2) and volume (length times length times length, or length^3) can be measured indirectly or through several measurements of length.

The amount of matter an object possesses is reflected in another physical property we can measure: its **mass**. The more mass an object possesses, the more strongly it is pulled to the center of the earth (it weighs more) and the harder it is to move. Mass measurements are invaluable in chemistry because the quantity of matter is an important property, and mass measurements are relatively easy to perform.

Time and **temperature** are other quantities often measured.

Magnitude and Units

Measurements have several attributes that give them meaning: (1) *units*, which relate the number to the actual size of the object (a sense of scale) ; (2) *magnitude*, the numerical relation between the size of the object and the units; and (3) *precision*, determined by the limits of the measuring instrument and the repeatability of the measurement. The table below shows some measurements to illustrate these attributes.

Chapter 1: The Science of Chemistry

Measured Quantity	Magnitude	Units	Precision
length of a size 9 shoe	0.290	m	± 0.001 m
volume of standard can of soda	0.355	L	± 0.001 L
length of average sitcom, including ads	1800	s	± 10 s
mass of average chocolate bar	0.0100	kg	±0.0001 kg

In the first example, the length of the shoe, the magnitude, 0.290, tells us how long the shoe is with respect to the standard unit, the meter (m). The precision, (± 0.001 m) tells us how precise the measuring instrument was.

In an effort to have all workers use the same units of measurement, the SI (short for Systeme Internationale) units of measurement were developed. The base units are shown in the table below.

Base SI Units of Measurements

Quantity	Unit	Symbol
mass	kilogram	kg
length	meter	m
time	second	s
temperature	Kelvin	K
amount	mole	mol
electric current	ampere	A
luminous intensity	candela	cd

Some units are obtained from these base units. For example, volume has units of length cubed (length^3). Because the SI unit of length is m, the SI unit of volume is m^3.

Exercise 1.14 Give the SI and more common metric units for (a) length (b) temperature (c) time and (d) volume.

Steps to Solution: Refer to the table above or in your text for SI units.

(a) The SI unit is the meter, m. Other common metric units are the centimeter and the millimeter.

(b) The SI unit for temperature is Kelvin, K. The metric unit is degrees Celsius, °C.

(c) The SI unit of time is the second, s. Other common units are minutes and hours.

(d) The SI unit of volume is the cubic meter, m³. The common metric units are the liter, L, and milliliters, mL.

Chapter 1: The Science of Chemistry

The SI system has not been completely adopted by all societies nor even by all scientists. Although scientists use the metric system, chemists find the SI volume unit, m³, very large and most laboratory chemists rarely use such amounts of materials. Therefore, they often convert one set of units to another.

Types of Unit Conversions

Several types of conversions are used to solve problems involving different units:

(1) One system of measurement to another system of measurement such as: m to yards

To convert units we recognize their equality. Looking at the first conversion unit shown above:

$$0.9144 \text{ m} = 1 \text{ yard}$$

We can rearrange this expression to give:

$$1 = \frac{0.9144 \text{ m}}{1 \text{ yard}}$$

Multiplying $\frac{0.9144 \text{ m}}{1 \text{ yard}}$ times a length in yards gives the length in m.

$$1 = \frac{1 \text{ yard}}{0.9144 \text{ m}} = 1.094 \frac{\text{yard}}{\text{m}}$$

Multiplying $1.094 \frac{\text{yard}}{\text{m}}$ times a length in m gives the length in yards.

These ratios are conversion factors. Both ratios are equal to 1; but which one should we use to convert yards to meters in converting the length of a 50.0 m pool to yards? We use a technique called dimensional analysis (sometimes called the factor label method) to determine the correct conversion factor. The following rule is useful:

The correct conversion ratio is the one that leads to cancellation of the unwanted units

$$\text{length (yards)} = 50 \text{ m} \cdot 1.094 \frac{\text{yard}}{\text{m}} = \boxed{54.70} \text{ yards}$$

If we use the other conversion factor, the units are quite different and are not the desired units.

$$\text{length (yards)} = 50 \text{ m} \cdot \frac{0.9144 \text{ m}}{1 \text{ yard}} = 45.72 \frac{\text{m}^2}{\text{yard}}$$

The correct result is that the pool is 54.70 yards long.

Chapter 1: The Science of Chemistry

> **COMMON PITFALL: Using reciprocal conversions.**
> This example shows one of the common pitfalls in the use of conversion factors: using the reciprocal of the appropriate conversion factor. You can avoid this error by including the units and checking the units of the answer to see that they are correct.

This answer is reasonable because a meter is a bit longer than a yard and so one would expect that the length of the pool in yards should be a greater number of yards than of meters. Other conversion factors may be found in your textbook or other sources.

(2) *Between different decimally related units (different prefixes on the units) such as*: cm to m. These conversions require use of the decimal conversions summarized in the following table.

Commonly used prefixes and decimal conversion factors

Prefix name	pico	nano	micro	milli	centi	kilo	mega
Abbreviation	p	n	μ	m	c	k	M
Decimal conversion	10^{-12}	10^{-9}	10^{-6}	10^{-3}	10^{-2}	10^{3}	10^{6}

Exercise 1.15 Convert (a) 50.72 cm to m (b) 0.0932 kg to g

Steps to Solution: This problem is solved by using the decimal conversions in a fashion similar to the unit conversions described above. You should memorize the decimal conversions in the above table.

(a) 50.72 cm to m
The conversion factor between cm to m is: 10^{-2} m/cm. We can also write: 10^{2} cm/m

$$100 \text{ cm} = 1 \text{ m} \quad \text{or} \quad 1 = \frac{1 \text{ m}}{100 \text{ cm}}$$

The right conversion factor will lead to cancellation of the units of cm: therefore we choose:

$$50.72 \text{ cm} \cdot \frac{1 \text{ m}}{100 \text{ cm}} = \boxed{0.5072 \text{ m}}$$

(b) 0.0932 kg to g
The conversion factor between kg to g is: 10^{3} g/kg. We can also write: 10^{-3} kg/g

$$1 = \frac{10^{-3} \text{ kg}}{\text{g}} \quad \text{or} \quad 1 = \frac{10^{3} \text{ g}}{\text{kg}}$$

The right conversion factor will lead to cancellation of the units of kg: therefore we choose:

$$0.0932 \text{ kg} \cdot \frac{10^3 \text{ g}}{\text{kg}} = \boxed{93.2 \text{ g}}$$

(3) *Between fundamental to derived units such as*: cm³ to m³ as illustrated in Exercise 1.16.

Exercise 1.16 Conversion of 152.67 cm³ to m³

Steps to Solution: This problem is solved by using the decimal conversions to the SI to non-SI units. Then we must take the non-SI conversion through the arithmetic operations used to obtain the derived units. First, we find the conversion between cm and m.

$$\frac{1 \text{ m}}{100 \text{ cm}}$$

The units are volume, length cubed, so the ratio must be raised to the third power

$$\left(1 = \frac{1 \text{ m}}{100 \text{ cm}}\right)^3 = \frac{1 \text{ m}^3}{100^3 \text{ cm}^3} = 10^{-6} \frac{\text{m}^3}{\text{cm}^3}$$

Now we can proceed with the conversion:

$$152.67 \text{ cm}^3 \cdot 10^{-6} \frac{\text{m}^3}{\text{cm}^3} = \boxed{1.5267 \times 10^{-4} \text{ m}^3}$$

1.6 Calculations in Chemistry

QUESTIONS TO ANSWER, SKILLS TO LEARN
1. **What is density?**
2. **Converting volume to mass and mass to volume**
3. **What is the difference between intensive and extensive properties?**
4. **What is the difference between precision and accuracy?**

Density is a quantity that relates the mass and the volume of an object. The density of any matter is its mass divided by its volume:

$$\text{density} = \frac{\text{mass}}{\text{volume}} \qquad \rho = \frac{m}{V}$$

If we multiply the volume by the density of the matter, then we obtain the mass of the matter. Conversely, if we divide the mass of matter by its density, we obtain its volume. However, be careful with the units! *The units of volume and mass must be the same as in the density.*

Density is a property of matter that depends only on the type of matter involved and not on the quantity of the matter. The density of a substance also depends slightly on temperature, so you will often see a temperature specified for precise values of densities.

Exercise 1.17 Gold has a density of 19.282 g/cm³. What is the mass (in kg) of a sphere of solid gold with a radius of 6.0 cm? (The volume of a sphere is $V = \frac{4}{3}\pi r^3$ where r is the radius.)

Steps to Solution: Density relates the volume and mass. In this problem we are asked to determine the mass of an object. We are given data (the radius) which can be used to find the volume. The volume and density can be used to calculate the mass of the object. The mass can be calculated by rearranging the density expression:

$$\rho = \frac{m}{V} \quad \text{gives} \quad m = \rho \cdot V$$

We calculate the volume:
$$V = = 9.0 \times 10^2 \text{ cm}^3$$

We know the density (given) and can use the rearranged density equation to find the mass in g:
$$m = \rho \cdot V = 9.0 \times 10^2 \text{ cm}^3 \cdot 19.282 = 1.7 \times 10^4 \text{ g}$$

(note the units of cm³ cancel out)

Because the question asks for the mass in kg, we must convert g to kg.

$$m = 1.7 \times 10^4 \text{ g} \cdot \frac{1 \text{ kg}}{10^3 \text{ g}} = \boxed{1.7 \times 10^1 \text{ kg}} \text{ (note the units of g cancel out)}$$

Because it is a characteristic of the type of matter, density can be used to identify substances. A property of matter that is independent of the amount of matter is called an ***intensive property***. The converse, a property of matter which depends upon the amount of matter present, is called an ***extensive property***.

Precision and Significant Figures in Calculations

Accuracy and precision are two very important concepts in science. ***Accuracy*** describes how well a measurement or result agrees with the "true" value. ***Precision*** depends upon the type of measuring instrument used and how reproducible the measurement is. The precision of *the measurement is reflected in the number of significant figures in the value.* In the following examples, the significant figures are underlined.

- 7.345 has four significant figures (one significant figure per digit).

- 0.004560 has four significant figures; the leading zeros are not significant because no non-zero numbers precede them. The final zero is significant because it is to the right of the decimal point.

- 10.002 has five significant figures. The zeros to the right of the decimal point are significant because there are non-zero digits to their left.

The number of significant figures can be ambiguous when very large or very small numbers are being dealt with. Scientific notation removes this ambiguity by expressing the number as a decimal figure multiplied by 10 raised to the appropriate power. It also makes the numbers more manageable. In the numbers below, the large numbers with zeros to the left of the decimal point do not have an unambiguous number of significant figures. The significant figures have been underlined.

Measurement	Number of significant figures	Number expressed in scientific notation
4.6433 kg	5	4.6433 kg
356.001 m	6	3.56001×10^2 m
0.000432 L	3	4.32×10^{-4} L
10010 g or 10010 g	4 or 5 (ambiguous due to last 0)	1.001×10^4 g or 1.0010×10^4 g
10 eggs	infinite (a count of items is an integer quantity)	

A measurement can be precise but not accurate. For example, a digital thermometer may show a temperature of 95.02°C in a sample of pure boiling water at conditions where we know the temperature should be 100.00°. The different temperature value indicates a precise but inaccurate reading; one often tests precision instruments by measuring a quantity whose value is known (a standard) to calibrate the instrument.

Significant figures in calculations

In a calculation, there often are several values with different numbers of significant figures to be added, multiplied, etc. Your calculator will give you more figures than are significant, and *you must be able to determine how many significant figures there are in your answer*.

In general, the least precise measurement (least significant figures) determines the precision (significant figures) in a result.

We will apply this rule in two problems to see how it works. We proceed through the calculation and remove the extra, non-significant figures only at the end of the calculation. Removing the nonsignificant figures is sometimes called *rounding off* the answer. Two rules suffice for most situations:

- If the first digit to be removed is 5 or greater, round the remaining digit upward by one digit. For example, 14.5362 rounds to 14.54

- If the first digit to be removed is less than 5 leave the last significant digit as it is. For example, 14.5322 rounds to 14.5

Exercise 1.18 Gold has a density of 19.282 g/cm³. What is the mass (in kg) of a sphere of solid gold with a radius of 6.0 cm? (This is the same calculation as in Exercise 1.17, but now we will find the number of significant figures.)

Steps to Solution: We perform the calculation retaining many nonsignificant figures, and rounding only at the end after we determine the number of significant figures that the answer can have. You may want to read up on using the memory function in your calculator to keep the values of intermediate numbers of calculations.
We calculate the volume retaining nonsignificant figures:

$$V = \left(\frac{4}{3}(3.14)(6.0 \text{ cm})^3\right) = 904.778684 \text{ cm}^3$$

Then the mass is calculated:

$$m = 19.282 \frac{g}{cm^3} \cdot 904.778684 \text{ cm}^3 = 17445.9426 \text{ g}$$

To determine the number of significant figures, we examine all the data used:

- $r = 6.0$ cm   • $\pi = 3.14$ • $\rho = 19.282$ g/cm³

The radius has only two significant figures; therefore the answer must have only two significant figures.

$$m = \underline{17}445.9426 \text{ g where the significant figures are underlined}$$

Because the digit to the right of the last significant figure is less than five, the last significant digit is left as is:

Chapter 1: The Science of Chemistry

$m = 17000$ g: converted to scientific notation gives: $m = 1.7 \times 10^4$ g

Because the question asks for the mass in kg, we must convert g to kg.

$$m = \boxed{1.7 \times 10^1 \text{ kg}}$$

> **Common Pitfall: Rounding too soon.**
> You should always use all the non-significant figures possible in calculations and remove them (round off) only at the last step of the calculation. Small errors can accumulate in the course of a calculation that can effect the magnitude of your final result

Exercise 1.19 A flask with a mass of 160.342 g is filled with 22° C water. The mass of the flask filled with water is found to be 310.5 g. Given that the density of water at 22°C is 0.99780 g/cm³, what is the volume of the flask?

Steps to Solution: The problem asks the volume of the flask. The data supplied are: the mass of the flask, the mass of flask filled with water and the density of water.

We can determine the mass of the water, m_{H_2O} in the flask, and using the density, obtain the volume occupied by that mass of water.

First we rearrange the density equation to solve for volume:

$$\rho = \frac{m}{V} \quad \text{can be rearranged to give}$$

$$V = \frac{m_{H_2O}}{\rho}$$

The mass of the water, m_{H_2O}, is the difference between the mass of flask filled with water and the empty flask:

$$m_{H_2O} = 310.5 - 160.342 \text{ g} = 150.158 \text{ g}$$

$$V = \frac{50.158 \text{ g}}{0.99780 \text{ g/cm}^3} = 150.489076 \text{ cm}^3$$

How many significant figures will the answer have? How many significant figures are in the data?

- ¥ 310.5 g   4 significant figures
- ¥ 160.342 g   6 significant figures
- ¥ 0.99780 g/cm³   5 significant figures

However, the quantity used in the equation is the difference of the two masses:

```
   310.5   g
  -160.342 g
   150.158 g
```

There are four significant figures in the difference. In the 310.5, there is no precision in the second place to the right of the decimal point. Therefore, in an addition or subtraction operation, the significant figures in the answer will stop at that decimal place. Thus, the answer will have 4 significant figures (underlined).

V = 1<u>50.4</u>89076 cm³ = $\boxed{150.5 \text{ cm}^3}$ (after rounding)

The number of significant figures in an addition or subtraction operation is determined from examining the numbers involved with decimal points aligned.

1.7 Chemical Problem Solving

QUESTIONS TO ANSWER, SKILLS TO LEARN
 1. **Developing a strategy for solving chemistry word problems.**

The seven steps described below constitute a general strategy for solving word problems in chemistry.

1. Identify the type of problem. What is the question asking you to find, calculate, describe, etc.? Identify the data and the goal.

2. If the problem involves molecules, think molecules! Develop a mental picture (which you may draw as a diagram or sketch) and visualize what is happening.

3. Are there equations or relationships that can be applied to the problem? Write them down.

4. Organize the data you identified in step 1.

5. Algebraically manipulate the equations from step 3 to give a form that will give the desired quantity. Alternatively, break a large problem up into a series of smaller problems. Both of these procedures will provide a plan for you to execute in the next step.

6. Execute the plan. Substitute the numerical values into the rearranged equation and solve. Make sure you keep track of all units and significant figures.

7. Ask yourself "Is the answer reasonable?" Here is where it is important to think molecules and remember the relative sizes of objects. Questions you should ask yourself about your result are:

- Does your result answer the question completely?
- Are the units consistent with the quantity you are solving for?
- Is the magnitude right?
- If a chemistry question, does it make chemical sense?
- Is the answer clearly identified and labeled (including the correct units)?

Exercise 1.19 In an adventure movie, the hero wants to remove a spherical solid gold object (density = 19.282 g/cm³) with a radius of 3.0 inches and replace it with an equal mass of sand (density = 1.6 g/cm³) using a 10. inch diameter cylindrical bag to hold it. The empty bag weighs 50. g. Assuming the sand takes the shape of a cylinder ($V_{cylinder} = 2\pi L r^2$), how long will the bag of sand need to be to weigh as much as the gold ball?

Steps to Solution: *Apply the seven-step method.*

Step 1. We need to find the length of a 10 inch diameter cylinder of sand in a 50 g bag that will have the same mass as the gold ball.

The data provided are:

- radius of the ball: 3.0 inches
- density of gold: 19.282 g/cm³
- diameter of the cylindrical sandbag: 10. in (2 significant figures)
- mass of the empty sandbag: 50. g (2 significant figures)
- density of sand: 1.6 g/cm³

Step 2. Picture the situation; perhaps draw a sketch. Include references to the values that are in the equations.

mass (gold) = mass (sand) + mass(bag)

We need to fill the bag with enough sand to give equal masses. Since the densitites are different, the volumes will be different.

Chapter 1: The Science of Chemistry

Step 3. This problem involves masses, volumes and density. We should remember the definition of density and the relation between mass and volume:

$$\rho = \frac{mass}{volume} \quad \text{and} \quad mass = density(volume)$$

$$V_{cylinder} = 2\pi L r^2 \qquad V_{sphere} = \frac{4}{3}\pi r^3$$

Step 4. Let's organize the data by the object it relates to:

gold ball	sand bag
radius of the ball: 3.0 in.	diameter of the sand bag: 10. in
density of gold: 19.282 g/cm³	density of sand: 1.6 g/cm³
	mass of the sand bag: 50. g

Step 5. Since the mass of the gold (mass(gold) and mass of the sand (mass(sand)+ mass of bag (mass(bag)) need to be the same we can write:

$$mass(gold) = mass(sand) + mass(bag)$$

We know from the equations in step 3 and the fact that the only data we have involve density and volume that:

$$mass(gold) = density(gold)(volume(gold))$$

$$mass(sand) = density(sand)(volume(sand))$$

$$mass(bag) = 50 \text{ g}$$

At this point we can pursue two different tactics: breaking the problem down into smaller pieces or performing some algebra to obtain a single expression that can be solved. The first of these methods will be illustrated.

This is one way we might break up the problem. Other ways might be just as acceptable. We first solve for the mass of the gold ball:

$$m_{gold} = \text{density(gold)(volume)} = \text{density(gold)}\frac{4}{3}\pi r_{ball}^3$$

The next step is to find the mass of sand required. The mass of sand will be the mass of the gold minus the mass of the bag:

$$\text{mass(sand)} = \text{mass(gold)} - \text{mass(bag)}$$

Using the density of sand we can solve for the volume of sand required:

$$V(\text{sand}) = \frac{\text{mass(sand)}}{\text{density(sand)}}$$

The data are in mixed units: length in inches and density in g/mL. We will need the appropriate conversion factor in the solution: 1 in = 2.54 cm
Using the known radius of the bag and the expression for the volume of a cylinder, we can solve for the length of the bag:

$$V(\text{sand}) = 2\pi r^2 L_{cyl} \quad \text{gives} \quad L_{cyl} = \frac{V(\text{sand})}{2\pi r^2}$$

by dividing both sides of the equation by $2\pi r^2$.

Step 6. Now we execute the plan.

We solve for numerical answers for each section of the problem.

$$m_{gold} = \text{density(gold)}\frac{4}{3}\pi r_{ball}^3$$

$$m(\text{gold}) = 19.282 \text{ g/cm}^3 \, (4/3)(3.14)(3.0 \text{ in}(2.54 \text{ cm/in}))^3 = 2.5205 \times 10^4 \text{ g}$$

We retain nonsignificant figures until we obtain the final answer. We next solve for the mass of sand required:

$$\text{mass(sand)} = \text{mass(gold)} - \text{mass(bag)}$$

$$\text{mass(sand)} = 2.5205 \times 10^4 \text{ g} - 50 \text{ g} = 2.5155 \times 10^4 \text{ g}$$

We now solve for the volume of the sand:

$$V(\text{sand}) = \frac{\text{mass(sand)}}{\text{density(sand)}} = \frac{2.5155 \times 10^4 \text{ g}}{1.6 \text{ g/cm}^3} = 21978 \text{ cm}^3$$

Now we solve for the length of the cylinder; here we need to convert the radius in inches to cm.

Chapter 1: The Science of Chemistry

$$L_{cyl} = \frac{V(sand)}{2\pi r^2} = \frac{21978 \text{cm}^3}{2(3.14)(5.0\text{in}(2.54\text{cm/in}))^2} = 29.3858 \text{ cm}$$

Rounding to two significant figures gives:

$$L_{cyl} = \boxed{2.9 \times 10^1 \text{ cm}}$$

Step 7. Check the answer.

1. Does the result answer the question completely? **Yes**; we are asked to find the length of the cylinder and that is the quantity we have found.

2. Are the units consistent with the quantity we are solving for? **Yes**; a length should have units of length. The units are cm, units of length.

3. Is the magnitude right? **Yes**, the answer seems large (about a third of a meter), but a gold ball is heavy; it is reasonable for a bag of sand to be bigger than a gold ball.

4. If a chemistry question, does it make chemical sense? No chemistry here, so this question does not apply.

5. Is the answer clearly identified and labeled (including the correct units)? **Yes**; the answer is identified as L_{cyl} and the units are written down. On an examination, it is wise to circle or otherwise identify the answer so that it is readily found.

Test Yourself

1. In each of the following there is a property underlined. Classify the property as intensive or extensive and as a chemical or a physical property.
    (a) <u>One pound</u> of margarine
    (b) Sodium and chlorine <u>react to form rock salt</u>.
    (c) The density of aluminum is <u>2.70 g/mL.</u>
    (d) Salt is <u>soluble in water</u>.

2. Cadmium is frequently found as an impurity in minerals containing zinc. Explain this fact and give the symbol for cadmium.

3. The speed of sound in air at sea level is 340 m/s. Convert this speed to miles per hour.

4. A piece of metal with mass of 134.4 g is placed in a graduated cylinder that contains 17.08 mL of water. The volume of the water and metal sample is found to be 24.13 mL. What is the density of the metal?

Answers

1. (a) Extensive physical; (b) intensive chemical; (c) intensive physical; (d) intensive physical.

2. Cadmium (symbol Cd) is in the same group as Zn, and thus forms many compounds similar to Zn.

3. 761 miles/hour.

4. 19.1 g/mL.

Solutions to Chapter Problems

1.1 Think about political issues that involve chemical reactions. Examples include the following:

(a) What are appropriate standards for "clean" air and water?
(b) What should be the regulations for testing new drugs before they are approved for human use?
(c) How can toxic waste sites best be cleaned up?
(d) Should chlorofluorocarbons be banned because of potential damage to the ozone layer?

1.3 Think about what a pharmacist does that involves chemistry. A pharmacist deals in chemical substances, most of which are toxic when incorrectly administered. Among the chemistry-related skills that a pharmacist uses are the following:

(a) Weighing, volume measurements and unit conversions;
(b) Knowing chemical compatibility of different drugs;
(c) Identifying similar or chemically equivalent drugs; and
(d) Protecting drugs from degradation due to exposure to adverse conditions.

1.5 Your criticism should be based on the experimental nature of chemistry. When experimental results fail to match theoretical expectations, the experiment may be flawed or the theory may be incorrect. The chemist should redo the experiment, adjusting conditions if necessary, to determine whether or not the results are correct. If repeated measurements show that the results consistently differ from what theory predicts, then the chemist should examine how to revise the theory to accommodate the results.

1.7 You must commit to memory the correspondence between names and symbols of various elements, but remembering them is simplified by the fact that most English names and elemental symbols are related.

(a) H;   (b) He;   (c) Hf;   (d) N;   (e) Ne;   and   (f) Nb.

Chapter 1: The Science of Chemistry

1.9 Associating an element's name with its symbol requires memorization of both names and symbols. The examples in this problem all begin with the letter "a."

   (a) arsenic; (b) argon; (c) aluminum; (d) americium; (e) silver; (f) gold; (g) astatine; and (h) actinium.

1.11 In writing a chemical formula, remember to use elemental symbols and subscripts for the number of atoms: CCl_4.

1.13 To convert a molecular picture into a molecular formula, count atoms of each type and consult (or recall) the color scheme used for the elements. See p. 5 of your textbook for the color scheme used in this and many other texts.

   (a) Br_2;   (b) HCl;   (c) C_2H_5I;   (d) PCl_3;   (e) SF_4; and   (f) N_2O_4.

1.15 The next element after the end of a row is the first element in the succeeding row: cesium.

1.17 Elements in the same column of the periodic table share similar chemical properties. For sulfur, these are its vertical neighbors, oxygen (O) and selenium (Se). The horizontal neighbors are phosphorus (P) and chlorine (Cl).

1.19 Metals from group 1 react in 1:1 ratio with Br from Group 17:   Li, Na, K, Rb, Cs.

1.21 Consult the periodic table: lithium, Li; beryllium, Be; boron, B; carbon, C; oxygen, O; fluorine, F; and neon, Ne.

1.23 Remember that a pure substance contains a single chemical compound and a solution is a homogeneous mixture of substances.

   (a) Pure substance;
   (b) Solution of various substances in water;
   (c) Heterogeneous mixture (indicated by its opacity);
   (d) Solution of nitrogen, oxygen, and other gases;
   (e) Heterogeneous mixture of gases and particulate matter; and
   (f) Heterogeneous (indicated by the presence of grains).

1.25 Remember that gases and liquids both flow, but gases are much less dense than liquids.
   (a) liquid;   (b) liquid;   (c) solid;   and   (d) solid.

1.27 In a chemical transformation, one substance is converted into another.

   (a) Water vapor deposits as ice, a physical transformation;
   (b) Liquid water evaporates, a physical transformation; and
   (c) A flame is a chemical transformation.

Chapter 1: The Science of Chemistry

1.29 A mixture contains more than one substance, while a compound is a single substance containing more than one element.
(a) A mixture of water and dissolved substances; (b) A pure compound, H_2O;
(c) A mixture of water and one or more solids; (d) A single element (He);
(e) Isopropanol, a pure compound; and (f) A mixture of a solvent and various pigments.

1.31 Power of ten notation expresses any number as a value between 1 and 10 times a power of ten. Trailing zeros are retained only when they are significant.
(a) 1.00000×10^5; (b) 1.00×10^4; (c) 4.00×10^{-4}; (d) 3×10^{-4}; and (e) 2.753×10^2.

1.33 To do unit conversions, multiply by a ratio that cancels the unwanted unit(s).
(a) $(430 \text{ kg})(10^3 \text{ g/kg}) = 4.30 \times 10^5$ g;
(b) $(1.35 \text{ mm})(10^{-3} \text{ m/mm}) = 1.35 \times 10^{-3}$ m;
(c) $(624 \text{ ps})(10^{-12} \text{ s/ps}) = 6.24 \times 10^{-10}$ s;
(d) $(1024 \text{ ng})(10^{-9} \text{ g/ng})(10^{-3} \text{ kg/g}) = 1.024 \times 10^{-9}$ kg;
(e) $(93,000 \text{ km})(10^3 \text{ m/km}) = 9.3000 \times 10^7$ m;
(f) $(1 \text{ day})(24 \text{ hr/day})(60 \text{ min/hr})(60 \text{ s/min}) = 8.6400 \times 10^4$ s (assuming exactly 1 day);
and (g) $(0.0426 \text{ in})(2.54 \text{ cm/in})(10^{-2} \text{ m/cm}) = 1.08 \times 10^{-3}$ m.

1.35 This is a unit conversion problem involving summation and unusual units. First convert all masses into SI units, then put into the same power of ten and add the masses.
$(5.0 \times 10^{-1} \text{ carat})(3.168 \text{ grains/carat})(1 \text{g}/15.4 \text{ grains})(10^{-3} \text{ kg/g}) = 1.0 \times 10^{-4}$ kg;
$(7.00 \text{ g})(10^{-3} \text{ kg/g}) = 7.00 \times 10^{-3}$ kg;
total mass = $(7.00 \times 10^{-3} \text{ kg}) + (0.10 \times 10^{-3} \text{ kg}) = 7.10 \times 10^{-3}$ kg.

1.37 Because the quart is a volume measure, density must be used to convert from volume to mass. The density of water is 1.00 g/cm^3 (Table 1-5). Assume exactly one quart.
$(1 \text{ quart})(1 \text{ L}/1.057 \text{ quart})(10^3 \text{ mL/L})(1 \text{ cm}^3/\text{mL})(1.00 \text{ g/cm}^3) = 9.46 \times 10^2$ g.

1.39 Density is mass divided by volume, and the volume of a block is $V = l\,w\,h$.
$V = (15.5 \text{ cm})(4.6 \text{ cm})(1.75 \text{ cm}) = 1.25 \times 10^2$ cm^3;
$\rho = m/V = (98.456 \text{ g}) \div (1.25 \times 10^2 \text{ cm}^3) = 0.79$ g/cm^3.
This result has two significant figures because one dimension is known to only two significant figures.

1.41 To convert from mass to volume, divide by density. See Table 1-5 for densities.
$V = m/\rho = (15.4 \text{ g}) \div (2.70 \text{ g cm}^{-3}) = 5.70$ cm^3.

1.43 The question asks for the density of water expressed in SI units. Density = m/V, the volume of water is the volume of the cylindrical container, and the mass of water is the difference between the masses of the filled and empty container.
$m = 270.064 \text{ g} - 93.054 \text{ g} = 177.010$ g; $V = \pi r^2 h$;
$r = (0.875 \text{ in})(2.54 \text{ cm/in}) = 2.2225$ cm; $h = (4.500 \text{ in})(2.54 \text{ cm/in}) = 11.43$ cm;
$V = \pi(2.2225)^2(11.43) = 177.37$ cm^3; $\rho = m/V = (177.010 \text{ g}) \div (177.37 \text{ cm}^3) =$ 0.998 g cm^{-3};
This result is rounded to three significant figures because the radius of the cylinder is known to only three significant figures.

1.45 The question asks for the length of a wire in meters. The data provided are the diameter of the wire, its mass, and its density. Think of the wire as a very long, thin cylinder. The volume of the wire is $V = \pi r^2 h$, where h is the length of the wire. The volume of the wire is not provided, but it can be calculated from the mass and density of the wire.
$V = m/\rho = (5.00 \text{ lb})(454 \text{ g/lb}) \div (8.92 \text{ g cm}^{-3}) = 2.54 \times 10^2 \text{ cm}^3$;
$r = \text{diameter}/2 = (0.0508 \text{ in})(2.54 \text{ cm/in}) \div 2 = 6.45 \times 10^{-2} \text{ cm}$;
$h = V/\pi r^2 = (2.54 \times 10^2 \text{ cm}^3) \div \pi (6.45 \times 10^{-2} \text{ cm})^2 = 1.94 \times 10^4 \text{ cm}$;
Multiply by 10^{-2} m/cm to convert to meters: 1.94×10^2 m.

1.47 In a molecular picture, each atom is shown as a sphere. A ring of eight sulfur atoms looks like a necklace.

1.49 Consult the periodic table to answer questions about elemental symbols. Eight elements have symbols beginning with "T": Ti (titanium), Tc (technetium), Te (tellurium), Ta (tantalum), Tl (thallium), Tb (terbium), Tm (thulium), and Th (thorium).

1.51 Volume and mass are related through density: $V = m/\rho$.

$V = (36.5 \text{ g}) \div (3.12 \text{ g/mL}) = 11.7 \text{ mL}$.

1.53 "Do not match" means that the letters in the symbol are not related to the English name of the element. See Table 1-1 in your textbook.
Lead, Pb.

1.55 This is a unit conversion problem. Multiply by dimensional ratios to convert units.
$(100 \text{ yr})(365.24 \text{ day/yr})(24 \text{ hr/day})(60 \text{ min/hr})(60 \text{ s/min}) = 3.1557 \times 10^9$ s.

The number of days in a year is not exact, so the result has the same number of significant figures as the conversion factor for days/yr.

1.57 The periodic table is organized with elements that have similar chemical properties in the same column. Gold shares Group 11 with silver and copper. All three are shiny, non-reactive, soft, and good conductors of heat and electricity.

1.59 Your list might include questions about which you have a special interest. Our list includes pollution questions such as how to reverse the processes that form the ozone hole; health questions such as how to design new drugs to combat bacterial infections and how important for our health are trace elements in our diets; biochemical questions such as what molecular processes take place when we smell an odor; and technological questions such as how to design materials that are high-temperature superconductors.

1.61 The key to this question is to recognize how solids, liquids, and gases are organized at the molecular level.

A solid has a rigid, constant shape because its molecules are in fixed, regular arrangements. Thus B matches D.
A liquid is dense yet able to flow easily, because its molecules, though close together, are not in fixed positions. Thus C matches E.
A gas has low density and flows easily because its molecules are separated by much empty space. Thus A matches F.

1.63 Kelvin-Celsius conversions involve addition/subtraction rather than multiplication/division, because the two scales have the same unit size but different zero points.
K = °C + 273.15;
$T = -11.5$ °C + 273.15 = 261.7 K.

1.65 The significant figures in a result of multiplication/division is determined by the number containing the least number of significant figures.

(a) 4.52×10^{34};   (b) 1×10^{-4}.

1.67 The value of an extensive property changes with the amount of the substance, while the value of an intensive property is independent of the amount of the substance.
Intensive: (a), (c), and (e); Extensive: (b) and (d).

1.69 Determine a molecular formula from a model by counting the atoms of each kind and referring to the color code in use for the elements.

(a) $CHClF_2$; (b) CH_2O_2; (c) BrF_3; and (d) C_4H_{10}.

1.71 Do unit conversions by multiplying by the appropriate ratios.
(a) $(340 \text{ in}^3)(2.54 \text{ cm/in})^3 (10^{-2} \text{ m/cm})^3 = 5.57 \times 10^{-3} \text{ m}^3$.
(b) $(35 \text{ mi/hr})(1.609344 \text{ km/mi})(10^3 \text{ m/km})(1 \text{ hr}/60 \text{ min})(1 \text{ min}/60 \text{ s}) = 16 \text{ m s}^{-1}$.
(c) $[(6 \text{ ft})(12 \text{ in/ft}) + 10 \text{ in}](2.54 \text{ cm/in})(10^{-2} \text{ m/cm}) = 2.1 \text{ m}$.
(d) $(220 \text{ lb})(1 \text{ kg}/2.2 \text{ lb}) = 1.00 \times 10^2 \text{ kg}$.

1.73 Time is related to distance through time = distance/speed.

time = $(2.37 \times 10^5 \text{ mi})(1.609344 \text{ km/mi})(10^3 \text{ m/km})/(3.00 \times 10^8 \text{ m/s}) = 1.27 \text{ s}$.

1.75 To draw molecular models, make use of the color-coded, scaled atoms shown in Figure 1-4 in your textbook. In the figures below, w = white, B = black, b = blue, and r = red.

HCN: H₂O: CO: NNO:

Chapter 1: The Science of Chemistry

1.77 Distance is related to speed through the equation, speed = distance/time. Thus,
distance = (speed)(time)
distance traveled in one year =
  $(1.86 \times 10^5$ mi/s$)(1.61$ km/mi$)(60$ s/min$)(60$ min/hr$)(24$ hr/day$)(365.24$ day/yr$)$
distance traveled in one year = 9.44×10^{12} km.

1.79 There are 11 ammonia molecules, each made up of one N atom and three H atoms. Thus, the figure contains 11 N atoms and 33 H atoms.

1.81 Physical properties do not relate to other substances, while chemical properties do.
  Physical properties: appearance, melting point, softness, density.
  Chemical properties: reaction with chlorine and reaction with water.

1.83 Consult the periodic table to find examples of various classes of compounds.
  (a) Halogens are found in Group 17: fluorine (F), chlorine (Cl), bromine (Br), iodine (I).
  (b) Alkaline earth metals are found in Group 2: beryllium (Be), magnesium (Mg), calcium (Ca), strontium (Sr), barium (Ba).
  (c) Actinides are those elements in the $n = 5$ f-block, for example actinium (Ac), uranium (U), and berkelium (Bk).
  (d) Noble gases are in Group 18: helium (He), neon (Ne), argon (Ar), krypton (Kr), etc.

1.85 Each atom contributes a distance equal to its diameter, so the number of atoms is the total distance divided by the diameter of an atom. A unit conversion is required between inches and picometers.
Number = $(1$ in$)(2.54$ cm/in$)(1$ m/10^2 cm$)(10^{12}$ pm/m$)/200$ pm/atom = 1.3×10^8 atoms.

1.87 Similar chemical properties are displayed by elements in the same column of the periodic table (In addition, lanthanides are similar to one another, as are actinides).
  Same-column pairs: Be and Ba; Pd and Pt; Ag and Au; S and Se; Si and Sn; Pr and Pa.

1.89 First convert the mixed units (ft, in) to inches, then make the metric conversion.

height = $[(7$ ft$)(12$ in/ft$) + 1$ in$](2.54$ cm/in$)(10$ mm/cm$) = 2.159 \times 10^3$ mm;

The precision of the height is $\pm (0.25$ in$)(2.54$ cm/in$)(10$ mm/cm$) = \pm 6.4$ mm;
thus the result should be reported to the nearest 10 mm: 2.16×10^3 mm.

1.91 (a) Copper, Cu; sulfur, S; and oxygen, O; (b) $CuSO_4$.

1.93 A liquid of lower density floats on a liquid of higher density. In the B + C photo, B floats on C, so C is denser than B. In the A + C photo, C floats on A, so A is denser than C. The ranking is B (ether) < C (water) < A (vegetable oil).

1.95 An element could only be named in antiquity if it was known in pure form in those times. Several metals (gold, silver, and copper) occur naturally in pure form. Several others (iron, tin, antimony, tungsten, mercury, and lead) can be obtained in pure form by heating their ores.

1.97 The appearance of this sample makes it clear that it is a heterogeneous mixture. The gemstone has an entirely different color, structure, and texture than the ore in which it is embedded.

1.99 The question asks about the material making up the crown. To answer this, we need some property of the crown that depends on composition. Density is one such property. To calculate density, we need mass and volume. The mass of the crown is measured to be 2.65 kg. When placed in a basin full of water, the crown displaces 145 cm^3 of water, so the volume of the crown is 145 cm^3.

$\rho = m/V = (2.65 \text{ kg})(10^3 \text{ g/kg}) \div (145 \text{ cm}^3) = 18.3 \text{ g cm}^{-3}$;

Compare this with the densities of metals: Au, 19.3 g cm^{-3}; Ag, 10.50 g cm^{-3};

From the fact that the crown was less dense than gold, Archimedes concluded that it must be an alloy of gold and some less dense metal, probably silver.

Chapter 2 The Atomic Nature of Matter

2.1 Atomic Theory

QUESTIONS TO ANSWER, SKILLS TO LEARN
1. What is the composition of matter?
2. What are the characteristics of atoms?
3. Drawing molecular pictures
4. Understanding dynamic equilibrium

Matter, Atoms and the Atomic Theory

Matter is comprised of atoms that combine into larger aggregates we call molecules. An abundance of experimental evidence supports this notion: technology now allows us to discern features of matter of the size we expect atoms to be, and we "see" features that are consistent with the presence of atoms and molecules. The idea that there are limits to the smallness of the pieces into which matter can be broken down goes back to the ancient Greeks, who called the smallest indivisible bits of matter *atomos*.

Fundamentals of Atomic Theory

- All matter is composed of tiny particles called atoms.

- All atoms of a given element have identical chemical properties
  We will describe chemical properties in more detail later. For now, you should know that the element we call oxygen sustains the chemical reactions necessary to life regardless of its source. Oxygen atoms have the same properties, whether they are obtained from our atmosphere or from rocks from the moon. Chemical reactivity is one way of recognizing elements.

- Atoms of each different element have distinct properties.
  Different elements have different properties. For example, oxygen is essential to sustain life, whereas chlorine is poisonous and will terminate life rather than sustain it. In addition, chlorine gas has a pale yellow color, whereas oxygen gas is colorless.

- Atoms form chemical compounds by combining in whole number ratios. All samples of a given pure compound contain the same number of elements and the same numbers of atoms of those elements.
  Atoms form compounds only as whole atoms, because atoms cannot be divided and retain the properties of that atom. For molecules of a substance to have the same properties, different molecules of that substance must contain the same number of atoms of the same types connected in the same way.

- In chemical reactions, there are changes in the way the atoms are joined in the substances, but atoms are neither created nor destroyed.

This facet of the atomic theory is a recognition of the law of conservation of matter. This law would be violated if matter were destroyed or created. Because chemical processes make new types of matter from other matter, each process (called a reaction) must involve transfer of atoms (or electrons) between the molecules to produce the new molecules.

Drawing Molecular Pictures

Chemical processes involve the transfer of atoms between molecules. To better understand the nature of chemical processes, it is useful to visualize the process that is occurring. For example, sulfur atoms react with oxygen to give sulfur dioxide, a gas that is a pollutant but also a substance that is transformed into sulfuric acid, the industrial chemical produced in the largest quantity. Sulfur in the gas phase can be present as atoms of sulfur, which have a spherical shape. Oxygen occurs in molecules in which two oxygen atoms are combined. The product of the reaction between sulfur and oxygen, sulfur dioxide, is a bent molecule. Three ways to represent such a process are shown below.

One sulfur atom and one oxygen molecule combine to form one molecule of sulfur dioxide

$$S + O_2 \rightarrow SO_2$$

The first method of describing the process uses **molecular pictures**. The second way describes the process in words. The third method is a **chemical equation of reaction**. It is important to learn to draw molecular pictures. In Chapter 3 you will learn how to identify molecules by names and how to name molecules: the rules of chemical nomenclature. In Chapter 4 you will learn about chemical equations of reaction.

Certain features of these representations should be emphasized. An arrow separates two sets of molecules or atoms. The arrow is chemists' shorthand for "react to form" or "are transformed to." In the molecular picture and in the equation, the **starting materials** or **reactants** are placed to the left of an arrow pointing right. To the right of the arrow are placed the molecule(s) that result from the reaction or process. These are called the **products** of the reaction.

Finally we should note a third, less obvious feature. In the molecular picture and the equation of reaction, the number of atoms on the left-hand side of the arrow is the same as the number of atoms on the right-hand side of the arrow. Any chemical equation of reaction that has the same number of atoms of the elements present on the reactant and product sides of the arrow is a ***balanced* chemical equation of reaction**. Such an equation conforms to the natural law that mass cannot be destroyed or created even at the atomic level.

Exercise 2.1 An important step in the manufacture of sulfuric acid is the reaction of two molecules of sulfur dioxide, pictured above, with oxygen, a molecule containing two atoms of oxygen, to form sulfur trioxide, a molecule that has a sulfur atom to which are attached three oxygen atoms. Represent this process using molecular pictures.

Steps to Solution: To represent this reaction, we must draw the molecules of reactants and products. The reactants are sulfur dioxide and oxygen, because the exercise states that they react to form sulfur trioxide. Sulfur dioxide contains two atoms of oxygen attached to a sulfur atom (as described above). Oxygen is described as containing two atoms of oxygen, so we draw two circular atoms attached to one another.

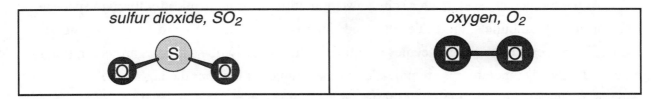

Because these two molecules are the reactants, they must appear to the left of the arrow in the reaction.

Sulfur trioxide is described as a molecule with three oxygen atoms attached to a sulfur atom. The drawing must show this. Because there is no information to indicate that the oxygen atoms are attached to one another, we will not show any oxygen-oxygen contact (bonds).

The problem asks for two molecules of sulfur dioxide and one molecule of oxygen to give sulfur trioxide. Drawing these we have:

We count up atoms to make sure that the numbers of atoms are the same on both products and reactants and discover that there are too few sulfur and oxygen atoms in the product molecule. There must be two molecules of sulfur trioxide in the products so that atoms are not "destroyed" in the reaction.

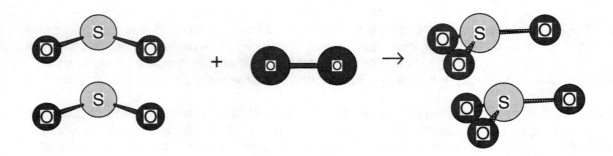

Dynamic Equilibrium

At the molecular level, matter is in constant motion. Gas molecules fly through space as fast as or faster than bullets; molecules in liquids bump around like people in a crowd trying to go different places; atoms and molecules in solids move restlessly like a kindergartner waiting to go out and play. Temperature is an indication of the amount of motion: the higher the temperature, the greater the amount of motion at the molecular and atomic level.

Despite this molecular energy, there are many instances where no change can be observed at the macroscopic level. In many instances, there is change occurring at the molecular level, but two processes are occurring at equal speeds, one undoing the effects of the other. This is **dynamic equilibrium**. Equilibrium is a situation where no observable change is occurring, but no matter is being supplied.

$$A \rightarrow B \quad \text{is just as fast as} \quad B \rightarrow A$$

Dynamic equilibrium is change at the molecular level occurring at equal speeds in both directions

Therefore a burner on a gas stove is not at equilibrium because gas (fuel) is being constantly supplied to it and it produces waste gases constantly. Only one process is occurring (combustion of gas) and its reverse is not occurring at an appreciable speed. A closed bottle of water is at equilibrium, because both evaporation of water and condensation of water are occurring, but the amount of water evaporating is the same as the amount of water condensing.

Exercise 2.2 (a) Consider a bathtub where the amount of water coming in from the faucet is the same as the amount leaving by the drain. Is this an example of dynamic equilibrium? Give reasons for your answer.
(b) A capped bottle of soda is under pressure caused by the carbon dioxide used to make the "fizz." When capped, the bottle is an example of dynamic equilibrium. Explain the processes that are occurring . (Hint: consider what happens when you open a bottle of soda.)

Steps to Solution: *We must remember the defining characteristics of a dynamic equilibrium: there is no apparent change, and processes must be occurring at equal speeds, one counteracting the other.*

(a) *In the bathtub example, although no net change is occurring in the level of the water, water is being supplied from outside the tub and being disposed of outside the tub. The amount of water in the drain and the amount of water in the supply pipes are not the same and at least in the short run, water from the drain is not being transferred to the faucet. Because matter is being transferred in and out of the tub, this is not an example of dynamic equilibrium.*

(b) *Two pieces of information are provided in this problem: (1) the bottle is under carbon dioxide gas pressure and (2) a hint asking you to think what happens when you open a bottle of soda. When you open the bottle, you see bubbles forming, sometimes quite violently. When you leave the bottle open for a while, the soda goes "flat" or loses its carbonation. This happens because the carbon dioxide leaves the soda when the bottle is opened. The same process must take place in the capped bottle at the same temperature, but in the closed bottle, the carbon dioxide cannot escape. Two processes are occurring: carbon dioxide is leaving the soda to enter the space above the liquid and then carbon dioxide above the liquid enters the liquid. Because we see no changes in the capped soda, these two processes must occur at the same speed; therefore this is a case of dynamic equilibrium.*

2.2 Atomic Architecture: Electrons and Nuclei

QUESTIONS TO ANSWER, SKILLS TO LEARN
1. **Why do atoms stick together or fly apart?**
2. **What are atoms made of?**

In chemistry, we are concerned about matter and we tend to think of it from the microscopic point of view. Matter is made up of molecules, but how do such small units become observable pieces of matter? You have learned that matter is made up of atoms bound to each other in the form of molecules, but what holds atoms together into molecules, or molecules together in what we call liquids and solids? Are atoms themselves composed of smaller particles and if so, what are they?

For objects to be brought closer to other objects (subatomic particles, atoms, molecules, planets), they require a force to push or pull them closer together. In the following pages, we will discuss the different types of natural forces. To understand chemistry, the most important force is that which acts between electrically charged particles; we will briefly discuss the other forces as well.

Forces in Nature

Gravitational Force. To be influenced by gravitational force, an object need only have mass and be near another object that has mass. The magnitude of the force depends upon the masses of the two objects and the reciprocal of the square of the distance between them:

$$F_{gravitational} = G\frac{m_1 m_2}{r^2}$$

In this equation, m_1 and m_2 are the masses of the two objects, r is the distance between them and G is a constant. Gravitational force is very small between small objects such as atoms and molecules.

Electrical Force. To be influenced by electrical forces, an object must have an electrical charge. The magnitude of the force between two charged objects depends upon the size of the two electrical charges and the square of the reciprocal of the distance between them, Coulomb's Law:

$$\text{Coulomb's Law:} \qquad F_{electrical} = k\frac{q_1 q_2}{r^2}$$

In this equation, q_1 and q_2 are the sizes of the two charges, r is the distance between them and k is a constant. Because there are two types of electrical charge (positive and negative) several combinations are possible. However, you need only to remember that like charges repel and unlike charges attract. Also, as the size of a charge increases, the magnitude of the force increases in direct proportion to the increase in the size of the charge.

The Atom: What Are Its Parts

Electrons

The electron is the lightest particle in the atom and has a negative charge. Often the letter "e" is used to represent the electron. The mass and charge of the electron were deduced from the complementary experiments of Thomson and Millikan. Thomson determined the charge to mass ratio, q/m of the electron, and Millikan determined the charge on an electron through his oil drop experiment (see Figure 2-20 in text). In the oil drop experiment, Millikan determined the charge on negatively charged drops of oil floating between two electrically charged plates that would be necessary so that the electrical force pulling the drops up was exactly as strong as the gravitational force pulling them down.

$$|F_{el}| = |F_{gravity}|$$

The charge Millikan computed was always some integer multiple of 1.6×10^{-19} coulombs

Chapter 2: The Atomic Nature of Matter

($n \cdot (-1.6 \times 10^{-19}$ coulombs) , $n = 1,2,3 \ldots$), so the smallest unit of charge that was present was -1.6×10^{-19} coulomb. Combining this data with Thomson's charge to mass ratio gave the mass of the electron, m_e:

$$m_e = 9.1 \times 10^{-31} \text{ kg} \qquad \text{electron charge} = -1.6 \times 10^{-19} \text{ coulomb}$$

Nuclei

Thomson's and Millikan's experiments determined that atoms contained negative electrical charges: because atoms are neutral (possess no net electrical charge), there must be an equal amount of positive charge in the atom as well. In addition, the mass of the electron was so small that most of the mass of atoms must be in particles other than the electrons. Rutherford's work showed that, except for the electrons, all the mass and all the positive charge of the atom were in a small volume, about 1/10,000 of the radius of the atom itself. The massive core was called the **nucleus**, and the description of an atom now placed most of the mass in the nucleus which is surrounded by a swarm of electrons. The nucleus is not an indivisible unit; further experiments showed that it contained two other types of particles, both with masses nearly 2000 times the mass of the electron. The **proton** has a positive charge equal in magnitude but opposite in charge to that of the electron. The **neutron** has no electrical charge. The electrical charge and masses of the electron, proton and neutron are listed below.

Particle name	Symbol	Electrical charge	Mass
electron	e	-1.6022×10^{-19} C	9.1094×10^{-31} kg
proton	p	$+1.6022 \times 10^{-19}$ C	1.6726×10^{-27} kg
neutron	n	0	1.6749×10^{-27} kg

The following picture of the atom emerges from these results. An atom is built from a positively charged nucleus about 1/10,000 the diameter of the atom which contains nearly all the mass of the atom. The electrons are attracted to the nucleus, but are fighting the electrical repulsions between themselves. The next section describes how elements can be identified through counting the numbers of the three particles in an atom.

Exercise 2. 3 Calculate the mass of 6.022×10^{23} protons + 6.022×10^{23} electrons.

Steps to Solution: We are asked to calculate the mass of a large number of protons and electrons. The above table supplies the mass of one proton and one electron. Therefore the mass of the group is found by multiplication:

$$m = 6.022 \times 10^{23} \ (9.1094 \times 10^{-31} \ \text{kg} + 1.6726 \times 10^{-27} \ \text{kg})$$

$$m = \boxed{1.0078 \times 10^{-3} \ \text{kg}}$$

2.3 Atomic Diversity: The Elements

QUESTIONS TO ANSWER, SKILLS TO LEARN
1. **What particles establish chemical identity?**
2. **What are isotopes?**
3. **Drawing the mass spectrum knowing the number of isotopes**

In the previous section we described the types of particles found in atoms and some of their properties. It is perhaps astounding to think that all the positive charge in an atom is found in the small confines of the nucleus; the electrical repulsions between protons must be large! However, the particles in the nucleus are held together by the very strong nuclear forces described above which dwarf electrical forces. Nuclei, in general, do not undergo changes because they are held together by such strong forces. The nuclei hold the ingredients of chemical identity because they do not undergo changes during chemical reactions.

Nuclear Charge = Chemical Identity

Atoms are composed of a nucleus, protons and neutrons, and electrons. The complete representation of a particular neutral atom is given by the number of protons and the number of neutrons. We summarize this information in the representation

$$^{A}_{Z}X$$

where X is the symbol for the particular element. The quantity Z is the number of protons and is called the **atomic number**. Another important quantity is the sum of the numbers of protons and neutrons, called the **mass number**, and is identified with the symbol A. The atomic number identifies an element and since it is a count of the number of protons, it is always an integer. In most periodic tables, one finds the atomic number above the symbol of the element as shown below for sulfur, S.

The atomic number defines an element since it fixes several atomic characteristics:
   (a) the number of electrons in a neutral atom
   (b) how strongly the electrons are bound to the nucleus

Exercise 2.4 What element has 15 protons? Give both the symbol and the name.

Steps to Solution: *The number of protons is the atomic number, so we must look for the element with Z = 15 on the periodic table. This element has the symbol P, which is phosphorus.*

Exercise 2.5 How many protons do atoms of the following elements contain? Also name the element.
(a) Fr (b) Ne (c) Br

Steps to Solution: *We need only to look up the elemental symbol in the periodic table and then find the atomic number. We will find the name of the element in your text.*

(a) *Fr is francium with Z = 87; thus francium atoms contain 87 protons.*

(b) *Ne is neon with Z = 10; thus neon atoms contain 10 protons*

(c) *Br is bromine with Z = 35; thus bromine atoms contain 35 protons.*

Isotopes

Atomic identity is fixed by the number of protons in the nucleus, but the mass of a given atom depends on the sum of all its particles, including neutrons and electrons. The mass of the nucleus depends on the number of protons and neutrons. Because the number of neutrons does not affect the charge of the nucleus, the mass number has little effect upon the chemical properties of an element. We distinguish nuclei with different mass numbers by calling them **isotopes**, and label each isotope by a superscript preceding the element's symbol: ^AX.

The isotope of uranium used as fuel for nuclear reactors has a mass number of 235; we identify this isotope as ^{235}U. The number of neutrons in this isotope is the difference between the mass number and the atomic number: 235 - 92 = 143 neutrons. Another isotope of uranium, with mass number 238, is not suitable for use as a nuclear fuel, but is used as ballast or in applications where high density is required. We designate this isotope as ^{238}U, and it has 146 neutrons. For neutral atoms, the number of electrons is the same as the number of protons: 92. In chemical processes, both isotopes undergo essentially identical reactions.

Exercise 2.6 Carbon has several different isotopes, with mass numbers 12, 13 and 14. Write the symbolic representation of each isotope and indicate the number of neutrons present in each isotope.

Steps to Solution: Carbon has atomic number $Z = 6$, so each carbon nucleus contains 6 protons. The number of neutrons present in each nucleus is the mass number minus the atomic number: $A - Z$.

^{12}C: number of neutrons $= A - Z = 12 - 6 = 6$ neutrons
^{13}C: number of neutrons $= 13 - 6 = 7$ neutrons
^{14}C: number of neutrons $= 14 - 6 = 8$ neutrons

Determining the Numbers and Masses of Isotopes: Mass Spectroscopy

The existence of ions is established by measuring the mass to charge ratio, m/q, with an instrument called the mass spectrometer. The mass spectrometer (shown schematically in the text) takes atoms (or molecules) and removes electrons by one of several methods. The charged species are then accelerated by an electrical charge and the stream of charged particles passes through a magnetic field which bends the path of the ions. The amount of bending depends on the mass to charge ratio, where larger mass fragments are deflected less than lighter ones. Therefore, if a mass spectrometer is used to examine a sample of gaseous atoms, the isotopes can be separated and the masses and the relative numbers of the different isotopes determined. The information is usually displayed in a format similar to a bar graph, where the mass number is the "x" axis, and the relative number of the isotope is the height of the bar or "peak."

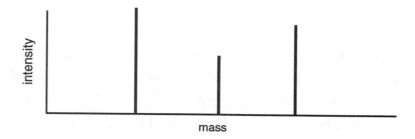

In a sample of an element obtained from naturally occurring substances, the element has a characteristic number of isotopes, each of which constitutes a percentage of the total number of atoms of that element.

> *% natural abundance*: **the number of atoms of a particular isotope per 100 atoms of that element obtained from naturally occurring sources.**

Exercise 2.7 The mass spectrum of chlorine atoms has two peaks, one for $A = 35$ (intensity = 1.00) and the other at $A = 37$ (intensity = 0.32). Write the isotopic symbols for both of isotopes and calculate their natural abundances.

Steps to Solution: Since we are given the mass numbers, we can write the isotopic symbols by placing the mass number as a superscript preceding the element's symbol, Cl. Therefore we have a mixture of ^{35}Cl and ^{37}Cl. The natural abundances are a bit more complicated to determine. The number of Cl atoms per peak is proportional to the intensity of the peak.

$$\text{number of } ^{35}Cl \text{ atoms} = k\,(1.00)$$

$$\text{number of } ^{37}Cl \text{ atoms} = k\,(0.32)$$

where k is a constant converting signal intensity to the number of atoms.

The total number of Cl atoms is proportional to the sum of intensities.

$$\text{Total Cl atoms} = k(1.00 + 0.32) = 1.32\,k$$

The percentage of ^{35}Cl atoms is the abundance:

$$\%\,^{35}Cl = \frac{\text{number of }^{35}Cl \text{ atoms}}{\text{Total Cl atoms}} 100\% = \frac{k \bullet 1.00}{k \bullet 1.32}\% = 76.0\%$$

The abundance of ^{37}Cl is:

$$\%\,^{37}Cl = 100\% - \%\,^{35}Cl = 100 - 76.0\% = \boxed{24.0\,\%}$$

2.4 Charged Atoms: Ions

QUESTIONS TO ANSWER, SKILLS TO LEARN
1. **What are cations and anions?**
2. **What are the differences between ions and atoms?**

Ions are atoms or groups of atoms (polyatomic ions) in which the number of electrons is not equal to the number of protons. Ions have considerably different properties than the elements from which they are formed. Therefore, it is imperative that charges be written to identify a species as an ion. For example, the metal K does not have the same properties as the ion K^+. Ions that have excess positive charge (fewer electrons than protons) are called **cations**; ions with excess negative charge (more electrons than protons) are called **anions**.

- **Cation**: fewer electrons than protons; have a positive (+n) charge (n = Z - [number of electrons]).

- **Anion**: more electrons than protons; have a negative charge of -n.

Because like electrical charges repel each other, significant numbers of ions of one charge or another cannot be stored as free ions; we find cations in combination with anions so that the total charge is zero. Common sources of ions for use in reactions are **ionic solids** and **ionic solutions**.

Ionic solids are composed of cations and anions spaced at regular, repeating intervals so that the cations are closer to anions rather than other cations. In this way, the attractive electrical forces are maximized. The structure of an ionic solid is sometimes referred to as the **lattice**. In the ionic solid, the total positive charge is equal to the total negative charge so no electrical charge is present in an ionic solid. When describing ions in symbols, we write the symbol of the element from which the ion is formed and then write the charge as a following superscript. The charge is expressed in the number of electrons lost or gained to form the ion from the element.

Exercise 2.10 Strontium forms an ion by losing 2 electrons and bromine forms ions in which the bromine atoms each gain 1 electron.
(a) Write the symbols for the ions formed by strontium and bromine.
(b) What is the formula of ionic solid formed by the ions described above?

Steps to Solution: *In part (a) we must write the charge as a superscript following the symbol of the element. For part (b), we must remember that the formula will have the lowest possible number of cations and anions so that the sum of the positive charges is equal to the sum of the negative charges*

(a) *Because the strontium atom loses two electrons, it will have twice the positive charge of an electron. In the symbol, the charge is shown in multiples of the electron charge; therefore, we write Sr^{2+}. For the ion formed from bromine, there is one extra electron, so we write Br^- (the 1 is understood).*

(b) *To determine the formula for the ionic compound the charge on the compound must be zero. The total charge on the cations is the same as the total charge on anions. Since the charge on the cation is +2, and the anion has a charge of -1, there must be two anions per cation to have a neutral (uncharged) compound. Therefore, the formula is $SrBr_2$.*

Ionic Solutions

When an ionic solid dissolves in water it gives separate cations and anions, which move independently of each other. A schematic of the conversion of an ionic lattice to an ionic solution is shown below.

ionic solid lattice → Separate cations and anions in solution

The highly ordered ions in the solid now are scattered about when dissolved in water. In a well mixed solution, the distribution of Na⁺ and Cl⁻ ions is uniform. You will note that in the drawing above, the number of cations is the same as the number of anions. One property of ionic solutions is that they are good conductors of electricity. For electrical current to flow in a solid or liquid, electrical charges must move. When two wires that are connected to a source of electrical power are placed into an ionic solution, positive ions migrate toward the negatively charged wire and negative ions migrate toward the positively charged wire. Because electrical charge is carried by ions moving through the solution, electric current flows. Without ions to carry charge, no current flows in water.

Exercise 2.11 Draw a molecular picture that illustrates the reaction between Na metal and Br_2 gas.

Steps to Solution: The compound formed by the reaction is NaBr because Na (sodium) is a Group 1 metal that forms cations of charge +1 and Br_2 (bromine) is a halogen that forms anions of charge -1. Sodium is a metal where all the atoms are in contact with several others, and bromine is found as Br_2 molecules. The product is an ionic solid that has alternating cations and anions.

4 Na   2Br₂   4NaBr

Exercise 2.12. What are the numbers of protons and electrons in the ions in KBr? Write the symbols for those ions.

Steps to Solution: The atomic number of K (potassium) is 19 and that of Br (bromine) is 35. These are the numbers of protons in those elements. Because potassium has a +1 charge, it has one less electron than proton: 19 - 1 = 18 electrons. Bromine has a -1 charge; therefore it has one more electron than proton: 35 + 1 = 36 electrons. We write these charges as superscripts after the symbols: K⁺ and Br⁻.

2.5 Conservation Laws

QUESTIONS TO ANSWER, SKILLS TO LEARN
1. What does conservation of physical quantities mean?
2. What are potential and kinetic energy?
3. Calculating kinetic energy

In studying the natural world around us, scientists have found that certain quantities do not change during the course of physical and chemical processes. For example, in the examples shown above, the number of atoms remain the same through the process. The fact that a quantity is constant and does not change is stated by saying that quantity is **conserved**.

Conservation of Mass

Perhaps the most familiar conservation law is the law of conservation of mass:

Mass is neither created nor destroyed during physical and chemical transformations.

Although the conservation of mass is an approximation, for the purposes of most of this course (except for transformations of nuclei in nuclear "chemistry"), the amount of matter at the end of a process is the same as the amount of matter at the beginning of the process. We can observe this in many physical processes; when water boils to form steam, we can retrieve at least some of the water by holding a cooled surface over the boiling water, and with a large enough surface we could collect all the water. The law of conservation of mass applies in chemical processes as well: although we may not observe the source of the reactant (as in iron rusting) or where the product goes (as when burning a piece of wood), mass is still conserved.

Exercise 2.12 When the space shuttle is launched, it uses a variety of fuels, among them hydrogen and oxygen which burn to give water. If 1.54×10^4 kg of hydrogen and 1.22×10^5 oxygen are burned in a few seconds of boost, how much water and unreacted starting material are produced?

Steps to Solution: We are asked to determine the mass of products and unreacted starting materials from the masses of the original amount of fuel present on the shuttle. Because mass is conserved we expect that the total mass of products will be the mass of the reactants:

mass = mass hydrogen + mass oxygen = 1.54×10^4 kg + 1.22×10^5 kg = $\boxed{1.38 \times 10^5 \text{ kg}}$

Chapter 2: The Atomic Nature of Matter

Conservation of Atoms

In all transformations, the law of conservation of mass applies. Because matter is the source of mass and matter is made of atoms, the number of atoms must be conserved in a chemical process. The law of conservation of atoms states:

The number of atoms of each element is conserved in chemical and physical processes.

For example, the reaction of bromine with sodium is shown below where there are eight atoms of sodium and four atoms of bromine giving NaBr. Because there is extra sodium, the extra sodium atoms remain unreacted:

    8 Na   2 Br$_2$ 4 NaBr   4 Na

The numbers of atoms of sodium and bromine are the same before and after the transformation. We will use the laws of conservation of mass and atoms later in the discussion of stoichiometry, the calculation of the amounts of reactants consumed or required in a given chemical process and the amounts of products obtained from that process.

We will now discuss a second law of conservation derived from the conservation of mass, the conservation of electrons (or the conservation of electrical charge). Because electrons have mass (although it is small) they too are conserved in chemical transformations.

There is no change in the total electrical charge in a chemical process.

Electrons are somewhat more confusing to account for; it is easier to keep track of the total charge on the reactants and the products.

Exercise 2.13 One of the two processes that occur in the production of chlorine and sodium hydroxide is a reaction of the Cl$^-$ anion. Deduce the product X in the reaction below using the above conservation laws (X is one molecule):

$$2\ Cl^- \rightarrow X + 2e^-$$

Steps to Solution: To solve this problem we must apply the principles of conservation of atoms and conservation of electrons. Therefore, the strategy will be to tally the numbers of atoms of each type in the products and reactants and to compare the total charge on the reactants to that on the products.

(a) There are two chlorine atoms on the left-hand side of the arrow, but none in the products. Therefore X will contain two chlorine atoms, since we are told that X is a single molecule.

The charge on the reactants is found next: on the left are two Cl⁻ ions giving a net charge of 2(1-) = -2. The charged species in the products are 2 electrons, each with a -1 charge giving a total charge of 2(-1) = -2. Since the charge on the known species in the products is equal to the charge on the products, the product X will have no charge.

From conservation of atoms, X contains 2 Cl atoms and from conservation of electrons, X is neutral. Therefore $\boxed{X = Cl_2}$

Conservation of Energy

The final law of conservation we will discuss is the law of conservation of energy:

Energy is neither created nor destroyed in *any* process, but is transferred from one body to another or converted from one form to another.

Energy is an abstract notion: it is defined by what it can do rather than by what it is.

Energy is the ability to do work.

Energy is a property of molecules and atoms that can be reflected in the motion of the molecules or in the different positions of atoms within a molecule. When we heat a substance, we increase the motion of the molecules or atoms, increasing their energy by transferring energy from the heat source to the molecules. Energy differences due to different arrangements of atoms are related to an ambiguous concept, stability of the molecule; these energy differences find application in understanding fuels. These two types of energy are some of the many forms that energy takes. Among the forms of energy, we will discuss kinetic energy, potential energy, thermal energy and radiant energy.

Kinetic Energy

Kinetic energy is energy possessed by an object that is in motion. Kinetic energy depends on the mass of the object, m, and how rapidly it is moving: its velocity, u. The mathematical relationship is:

$$\text{kinetic energy} = kE = \frac{1}{2}mu^2$$

From this equation we find that the units of energy are mass $\cdot \dfrac{\text{distance}^2}{\text{time}^2}$: in SI units, kg·m²/s².

The SI unit for energy, no matter what form it takes, is the Joule, J, where

Chapter 2: The Atomic Nature of Matter

$$1 \text{ J} = 1 \text{kg} \cdot \text{m}^2/\text{s}^2$$

the kinetic energy of any moving piece of matter can be calculated using the above formula. It is important to use appropriate unit conversions when finding the energy in J.

Exercise 2.14 At what speed would a 1.65 x 10² pound person have to move to have kinetic energy of 2.145 x 10³ J?

Steps to Solution: To solve this problem we need to find the velocity of a body of known mass with a given kinetic energy. This will require rearrangement of the kinetic energy equation and because the mass is supplied in non-SI units, a conversion must also be made. The kinetic energy equation will have to be rearranged to find the velocity of the person:

$$kE = \frac{1}{2} mu^2 \text{ is rearranged to give}$$

$$u^2 = \frac{2(kE)}{m} \quad \text{so} \quad u = \sqrt{\frac{2(kE)}{m}}$$

Now we find the speed at which the person must move to have a kinetic energy of 2.145 x 10³ J. The units of mass must be converted from pounds to kg; the velocity will be obtained in m/s:

$$m = \frac{165 \text{ lb}}{2.2 \text{ lb/kg}} = 75.0 \text{ kg}$$

$$u = \sqrt{\frac{2(kE)}{m}} = \sqrt{\frac{2(2.145 \times 10^3 \text{ kg} \cdot \text{m}^2/\text{s}^2)}{75 \text{kg}}} = 7.56 \text{ m/s}$$

Is the answer reasonable? It is difficult to tell: the quantitative nature of energy is complicated by the square of the velocity term. However, more work will show that this is a reasonable answer (7.56 m/s is about 17 miles per hour; you should verify this conversion for yourself).

Potential Energy

Potential energy is a function of the position of an object. Potential energy is stored energy, energy that may be released when conditions are correct.

Gravitational potential energy is one case where energy because of position is easy to understand. An object in a high place falls down to earth (or whatever celestial body you are on). As the object falls, its speed increases as the potential energy is converted to kinetic energy (because energy is conserved).

Chemical potential energy is energy due to the attractive forces between atoms in molecules. When gasoline is burned in an automobile engine, the positions of the carbon, hydrogen and oxygen atoms in the molecules of reactants are different than the positions of the atoms in the products. This energy difference is used to make heat to move the car. The attractive forces between atoms that form molecules are generated by a third type of potential energy.

Electrical potential energy results from the forces between electrically charged particles. We will consider this type of potential energy to help understand the forces between the nuclei and electrons in atoms and molecules.

Thermal Energy.

When the space shuttle reenters the earth's atmosphere, it is moving at speeds of greater than 25,000 km/hour. When the craft lands it is moving at speeds of 350-450 km/hour, much slower than when it started its trip back to earth. Because the craft weighs many thousands of kg, much of the kinetic energy as well as the gravitational potential energy the shuttle lost as it fell to earth has been converted to thermal energy: portions of the spaceship reach temperatures of around 3000°C during the reentry process. Thermal energy is related to temperature: an increase in temperature indicates addition of thermal energy.

Radiant Energy

Another type of energy that will be important in understanding chemistry is **radiant energy**, which is observed when you sunbathe: you absorb sunlight and get warm. Radiant energy is being converted to thermal energy. Sunlight has no mass or electrical charge; therefore there is no conversion of kinetic energy or gravitational or electrical potential energy to thermal energy. We will discuss radiant energy later on in more detail.

Exercise 2.15. Explain how the conservation laws discussed above are obeyed in the following situations:

(a) You stay out in the sun too long and get a sunburn.
(b) A bicyclist pedals up a steep hill and then coasts down. He then notices that he is quite hungry.

Steps to Solution:

(a) *The radiant energy of the sunlight has changed the properties of your skin. This change is a change in chemical potential. Thus radiant energy has been transformed into chemical energy.*

(b) *To pedal up the hill, the cyclist requires the input of energy to increase the gravitational potential energy. The source of this energy is chemical energy, which he obtains from his food. Coasting down the hill converts the gravitational potential energy into kinetic energy. Finally, the cyclist notices the output of all the chemical energy he used to go up the hill. He is hungry, nature's call to refill his energy stores.*

Test Yourself

1. Draw a molecular picture illustrating the reaction of two molecules of hydrogen with one molecule of oxygen to give two molecules of water.

2. A helium nucleus has a velocity of 4.91×10^6 m/s. What is its kinetic energy?

3. Write the symbol that describes each of the following isotopes:
    (a) an atom that contains eight protons and eight neutrons.
    (b) an atom that contains forty-five protons and fifty-eight neutrons.
    (c) an ion that contains nineteen protons, twenty neutrons and eighteen electrons

4. When a bullet hits a hard surface it can get hot enough to melt. Explain this in terms of the law of conservation of energy.

Answers

1.

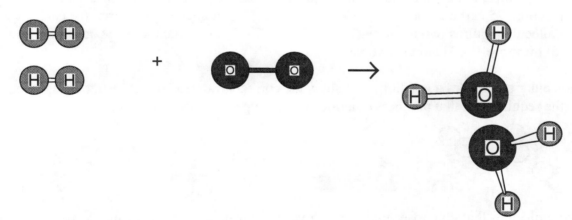

2. 3.20×10^{13} J

3. (a) ^{16}O (b) ^{103}Rh (c) $^{39}K^+$

4. The kinetic energy of the bullet is transformed into thermal energy.

Solutions to Chapter Problems

2.1 Molecular pictures must show the structures of individual molecules and the differences between phases. All these "molecules" have monatomic units. A gas is mostly empty space, a liquid is tightly packed but not entirely regular, a solid has a regular repeating structure.

(a) gaseous helium   (b) solid tungsten   (c) liquid mercury

2.3 See your textbook for the five features of atomic theory.
(1) All starting materials and products are made from atoms of carbon and oxygen.
(2) All oxygen atoms are in diatomic molecules in the starting materials and CO molecules in the products; likewise, C atoms behave all in the same way.
(3) Carbon atoms and oxygen atoms do different things in the reaction.
(4) C and O combine in 1:1 atomic ratio to give CO molecules.
(5) Atoms of each type are conserved: 4 O atoms and 13 C atoms.

2.5 In any chemical reaction, the amount of atoms of each element must be conserved. In this reaction, 25 carbon atoms are consumed. (a) 25 CH_4 molecules will form; (b) each CH_4 molecule requires two hydrogen molecules, so 50 H_2 molecules will be consumed; and (c) there are 100 H atoms involved.

2.7 Molecular pictures of a reaction must show the correct arrangements of the atoms and obey the requirement that numbers of atoms of each type be conserved.

2.9 The molecules that give roses their aroma evaporate from the surface of the flower. Once in the gas phase, they collide countless times with other gas molecules, moving slowly away from the rose until, when they reach a nose, they are sensed by the olfactory sensors.

2.11 "Dynamic equilibrium" refers to the fact that molecules continue to be transformed from one phase or form to another even though no net change is taking place. In this example, iodine molecules sublime from the crystals at the bottom of the flask into the gas phase, where their presence imparts a pale violet color to the gas. Some of these molecules then condense on the surfaces of the flask, forming crystals. At any time, there is a constant number of molecules in the gas phase, but some molecules are subliming into the gas from the crystals while equal numbers of molecules are condensing onto the crystals.

Chapter 2: The Atomic Nature of Matter

2.13 The force between two charged bodies is given by Coulomb's law, $F = k\, q_1 q_2 / r^2$.
(a) the force increases by a factor of three; (b) the force increases by a factor of $3 \times 3 = 9$; and (c) the force decreases by a factor of $2^2 = 4$.

2.15 In the Millikan experiment, gravitational force was balanced by electrical force for some of the charged particles. Changing the polarity would reverse the direction of the electrical force for all the charged particles. Negatively-charged particles would be attracted downward, but now positively-charged particles would be attracted upward. Under these conditions, the proper electrical field would cause some of the positively-charged particles to be suspended in space. Because each of these particles had lost one or more electrons, the charge on the electron could be determined from these observations.

2.17 The charge and mass of an electron are fixed (see Table 2-1). Divide total charge by electronic charge to find the number of electrons, then multiply by the mass per electron to find the total mass:
$(-1.00 \times 10^{-6}\ \text{C}) \div (-1.6022 \times 10^{-19}\ \text{C/electron}) = 6.24 \times 10^{12}$ electrons;
$(6.24 \times 10^{12}\ \text{electrons})(9.1094 \times 10^{-31}\ \text{kg/electron}) = 5.69 \times 10^{-18}$ kg.

2.19 Use the properties of electrons in Table 2-1 to answer this question.
(a) To achieve neutrality, the two protons must be balanced by two electrons.
(b) This can be calculated two ways:
Using proton and neutron masses, $m_{total} = 2(1.6726 \times 10^{-27}\ \text{kg}) + 2(1.6749 \times 10^{-27}\ \text{kg}) + 2(9.1094 \times 10^{-31}\ \text{kg}) = 6.6968 \times 10^{-27}$ kg;
electron mass fraction $= 2(9.1094 \times 10^{-31}\ \text{kg}) \div (6.6968 \times 10^{-27}\ \text{kg}) = 2.7205 \times 10^{-4}$.
Using the atomic mass of He,
$m_{total} = (4.003\ \text{g/mol})(10^{-3}\ \text{kg/g}) \div (6.022 \times 10^{23}\ \text{atom/mol}) = 6.649 \times 10^{-27}$ kg;
electron mass fraction $= 2(9.1094 \times 10^{-31}\ \text{kg}) \div (6.649 \times 10^{-27}\ \text{kg}) = 2.741 \times 10^{-4}$.
These two results differ because, as described in Chapter 20 of your textbook, the mass of a nucleus is not exactly equal to the sum of the masses of its constituent protons and neutrons.

2.21 Consult a table of element names and symbols: Ta, tantalum, 73; Th, thorium, 90; Ti, titanium, 22; Tl, thallium, 81; Tm, thulium, 69; Te, tellurium, 52; and Tc, technetium, 43.

2.23 An isotopic symbol has the elemental symbol prefaced by a superscript giving the mass number and a subscript giving the atomic number.
(a) $^{56}_{26}\text{Fe}$;   (b) $^{38}_{18}\text{Ar}$;   (c) $^{236}_{92}\text{U}$;   and   (d) $^{19}_{9}\text{F}$.

2.25 There are two isotopes, so the mass spectrum has two peaks. The mass-11 peak is 4 times as intense as the mass-10 peak.

2.27 There are two isotopes of chlorine, so Cl_2 can be $^{35}Cl-^{35}Cl$, $^{35}Cl-^{37}Cl$, $^{37}Cl-^{35}Cl$, or $^{37}Cl-^{37}Cl$. These have mass 70, 72, 72, and 74, so there are three peaks in the mass spectrum. To find the relative intensities of the three peaks, multiply the fractional isotopic abundances: mass 70, (.7577)(.7577) = .5741; mass 72, (.7577)(.2423) + (.2423)(.7577) = .3672; and mass 74, (.2423)(.2423) = .0587.
Mass 70 is most intense and mass 74 is least intense.

2.29 Cations are positively charged, anions are negatively charged, and neutral species carry no charge.
cations: Cl_2^+, CO^+, Cr^{3+}; anions: Cl^-, $Cr_2O_7^{2-}$; and neutrals: C, CCl_4, CO_2.

2.31 The chemical formula for a charged species has a right superscript charge.

(a) $OH^- + H^+ = H_2O$; (b) $Na - e^- = Na^+$; (c) $HCl - H^+ = Cl^-$; and (d) $O + 2e^- = O^{2-}$.

2.33 A molecular picture of a solution of a salt contains cations, anions, and water molecules. The total positive charge on the cations should be equal to the total negative charge on the anions. Thus there should be equal numbers of Li^+ and I^- ions.

2.35 Elements in Groups 1 and 2 easily lose one and two electrons, respectively. Those in Groups 16 and 17 easily gain two and one electrons, respectively.
(a) Rb^+; (b) F^-; and (c) Ba^{2+}.

2.37 Cations and anions combine to form neutral compounds. In any neutral compound, the total amount of positive charge must match the total amount of negative charge.
RbF and BaF_2.

2.39 (a) Conservation of mass applies to an entire system, not any individual item in that system. The marathon runner loses mass through perspiring, but the water that evaporates in this process is added to the mass of the atmosphere.
(b) Total energy, which is conserved, is the sum of potential, kinetic, and thermal energy. On the tree, an apple possesses some gravitational potential energy. As an apple falls, that potential energy is converted into kinetic energy of motion.
(c) When an apple hits the ground, the impact transfers energy to molecules in the earth and in the apple. As a result, there is a slight increase in temperature; the kinetic energy of the apple has been converted into thermal energy.

Chapter 2: The Atomic Nature of Matter

2.41 Speed and kinetic energy are related through the equation, $kE = mu^2/2$. The joule has units kg m^2/s^2, so mass must be expressed in kg and u has units m/s.
m (Table 2-1) = (1.6749 × 10^{-27} kg);
$u = [2\,kE/m]^{1/2} = [2(3.75 \times 10^{-23}\,kg\,m^2/s^2) \div (1.6749 \times 10^{-27}\,kg)]^{1/2} = 2.12 \times 10^2$ m/s.

2.43 Positive ions are generated in a mass spectrometer by removal of electrons
(a) CHCl$_3$ → CHCl$_3^+$ + e^-;
(b) CHCl$_3^+$ → CCl$_2^+$ + H + Cl.

2.45 Energy comes in various forms: radiant, kinetic, potential, and thermal.
(a) Radiant energy is consumed and thermal energy is produced.
(b) Gravitational potential energy is consumed and kinetic energy is produced.
(c) Chemical potential energy is consumed and thermal energy is produced.

2.47 Consult a table of elemental symbols and names.
Mg, magnesium, 12; Mn, manganese, 25; Cu, copper, 29; Co, cobalt, 27; P, phosphorus, 15; and Pb, lead, 82.

2.49 Conservation of charge requires that the net charge remain the same during any process.
(a) The products have six negative charges, so six electrons must be added to the neutral N$_2$ molecule.
(b) An oxalate ion has two negative charges, so two electrons must be removed.
(c) The products have equal numbers of positive and negative charge, so charge is conserved.

2.51 Energy and speed are related through $kE = mu^2/2$. The numerical calculation can be simplified by first solving the problem algebraically.

Because kE (proton) = kE (electron), $m_p u_p^2 = m_e u_e^2$ and $u_p^2 = m_e u_e^2/m_p$;
thus, $u_p = u_e(m_e/m_p)^{1/2}$.
$u_p = (1.55 \times 10^6$ m/s$)[(9.1094 \times 10^{-31}$ kg$) \div (1.6726 \times 10^{-27}$ kg$)]^{1/2} = 3.62 \times 10^4$ m/s.

2.53 Elements in Group 2 of the periodic table form cations with +2 charge and ionic compounds whose formulas are XF$_2$.

Be, Mg, Ca, Sr, Ba, and Ra.
(A number of transition metals and f-block elements also form compounds with this formula, but you are not expected to know them at this stage).

2.55 The left superscript in an isotopic symbol gives the sum of protons and neutrons, the atomic number is the number of protons, and a right superscript indicates (protons – electrons), when this quantity is non-zero.
(a) 8 protons, 8 neutrons, 10 electrons; (b) 7 p, 8 n, 7 e; (c) 25 p, 30 n, 22 e;
(d) 17 p, 18 n, 18 e; and (e) 17 p, 20 n, 16 e.

2.57 An atomic mass spectrum contains a peak at each isotopic mass, and the peak heights are proportional to the isotopic abundances. The pie chart shows five isotopes.

2.59 The number in a symbol such as Co-60 represents the sum of protons and neutrons. The number of protons can be deduced from the elemental symbol and a periodic table.
(a) 27 protons, 33 neutrons; (b) 6 p, 8 n; (c) 92 p, 143 n; and (d) 92 p, 146 n.

2.61 Kinetic energy is given by $kE = m\, u^2/2$, but to obtain energy in joules, we must convert to SI units.
(2250 lb)(0.453592 kg/lb) = 1.021×10^3 kg;
(57.5 mi/hr)(1.609344 km/mi)(10^3 m/km)(1 hr/60 min)(1 min/60 s) = 25.70 m/s;
kE = (1.021×10^3 kg)(25.70 m/s)^2/2 = 3.37×10^5 kg m^2/s^2 = 3.37×10^5 J.

2.63 A salt is composed of cations and anions, which exist as individual ions in solution.

2.65 Each of these isobars has mass number 40, so the upper left superscript is 40.
$^{40}_{17}$Cl, $^{40}_{18}$Ar, $^{40}_{19}$K, $^{40}_{20}$Ca, and $^{40}_{21}$Sc.

2.67 The law of conservation of mass applies to an entire system, not to any individual item within that system. When magnesium burns in air, magnesium atoms combine with oxygen from the air to form magnesium oxide. The mass of the solid increases, but the mass of solid plus gas remains constant.

2.69 The atomic number of an element, which equals the number of protons in its nucleus, can be found from the atomic symbol and a periodic table. The right-hand superscript in an ionic symbol represents (protons – electrons).

(a) 11 protons, 10 electrons; (b) 7 p, 10 e; (c) 22 p, 18 e; and (d) 53 p, 54 e.

2.71 Charge must be conserved, and every chemical compound must be electrically neutral. Thus, two chlorine atoms gain an electron for each calcium atom that loses two electrons.

(a) Ca^{2+}, Cl^-; (b) $CaCl_2$;
(c)

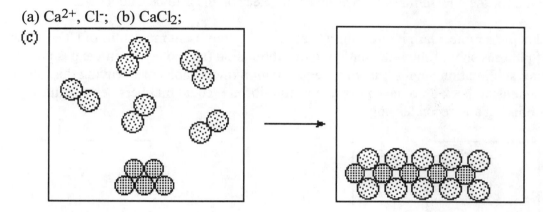

2.73 There are only two isotopes that meet these criteria.
$^{14}_{7}N$ and $^{14}_{6}C$. ($^{14}_{5}B$ has a ratio of 1.6 to 1).

2.75 Consult the periodic table to determine the atomic number of technetium.
^{98}Tc has 43 protons, 55 neutrons, and 43 electrons; ^{99}Tc has one more neutron, 56.

2.77 The question asks for a comparison of macroscopic and molecular views. Figure 2-12 B shows a homogeneous solution, where the solute molecules are evenly distributed in the solution. The right photo in Figure 2-11 is closest to this uniform distribution. Figure 2-12 A shows a solution before much mixing, with dye molecules concentrated in one region. The left photo in Figure 2-11 matches this distribution.

2.79 Molecular pictures of reactions must show that atoms of each element are conserved. There must be equal numbers of atoms of each element on the two sides of the arrow. The question asks you to adjust the left-hand side to make the figure correct. On the right-hand side there are eight atoms of each element, so the left-hand side must show four molecules of each type:

2.81 Any ionic compound must be electrically neutral. The charges on the ions are Al = +3, O = -2, and F = -1. Two Al ions match the charge on three O ions and three F ions match the charge on one Al ion. Aluminum oxide is Al_2O_3; aluminum fluoride is AlF_3.

2.83 Millikan's experiment allowed the mass of the electron to be determined indirectly. The oil droplets that he observed had one or more excess electrons attached to them, but their masses were changed insignificantly by these extra electrons. Analysis of the Millikan experiment gave a value for the charge on the electron. This value along with Thomson's value for the charge/mass ratio allowed the electron mass to be calculated.

2.85 To keep the molecular picture simple, use circles to represent molecules of CO_2. A series of pictures should show the initiation of sublimation (no molecules in the gas phase), after sublimation begins but before equilibration (more molecules leaving the solid than condensing back onto the solid), and at equilibrium (equal numbers of molecules subliming and recondensing).

Initial

In progress

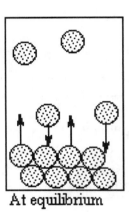
At equilibrium

2.87 In a crystalline solid, the ions lie in a regular array, in this case cubic. In the liquid phase, the ions are still close together but the arrangement is no longer entirely regular.

Solid

Liquid

Chapter 2: The Atomic Nature of Matter

2.89 An electric discharge is capable of stripping electrons off of molecules to create positively-charged ions. As mass spectrometry shows, molecules can also fragment under these conditions. Finally, electrons can be captured by neutral species to generate negative ions. Multiply-charged positive atomic ions may form when more than one electron is removed from an atom (Multiply-charged negative ions are also possible in principle, but in the gas phase these species are not stable. Multiply-charged positive molecular ions also are generally unstable and break apart to give atomic ions).

Nitrogen can form N_2^+, N_2^-, N^+, N^-, N^{2+}, etc.; Oxygen can form O_2^+, O_2^-, O^+, O^-, O^{2+}, etc.

Problem 2.4 described how NO can form during lightning strikes. Thus ions of NO are also possible: NO^+ and NO^-.

2.91 The conservation laws described in Chapter 2 deal with mass, atoms, energy, and charge.
The total mass remains the same in any physical or chemical transformation.
The number of atoms of each type (element) remains the same in any physical or chemical transformation.
Energy can be transformed from one form to another, but the total amount of energy remains the same in any process.
The total amount of charge, taken as the sum of the positive charges minus the sum of the negative charges, remains unchanged in any process.

2.93 Assuming that the vitamin C from both sources is pure, the compound made in the laboratory has identical value as that extracted from oranges. According to atomic theory, every sample of a particular substance contains identical atoms of the various elements use to construct that substance. Orange trees use atoms of carbon, oxygen, and hydrogen that are identical to those used by a scientist in the laboratory. (Oranges may be superior to pills, because they supply compounds besides vitamin C that can be beneficial to health.)

Chapter 3 The Composition of Molecules

3.1 Representing Molecules

QUESTIONS TO ANSWER, SKILLS TO LEARN
1. **Describing molecules through:**
    - a list of symbols of the elements: chemical formulas
    - pictures
        - structural formulas
        - three-dimensional models
        - line structures
2. **Using chemical formulas**

A chemical formula lists the elements in a substance and the number of atoms of each element by using each elements' symbols and a numerical subscript that designates the number of atoms in that substance (except when only one atom is present; then no subscript is used). To prevent confusion, rules are used for the order in which elements are listed in chemical formulas:

1. The element farther to the left in the periodic table appears first *except* for hydrogen.

2. If hydrogen is present, it appears last *except* when the other element is from the oxygen family or the halogens (Groups 16 and 17).

3. If both elements are from the same column of the periodic table, the element in the lower period (row) is listed first.

When three or more elements are present in a compound, the order of symbols depends on whether the compound contains ions (ionic compounds are described in the next section). Multi-element compounds that are not ionic often contain carbon and follow this rule.

- For non- ionic compounds that contain carbon, their formulas are written by listing carbon first and the remaining elements are listed alphabetically.

Exercise 3.1. Write the correct chemical formulas for the compounds whose molecular structures are pictured below.

Steps to Solution: We must first determine what elements are contained in each molecule and the number of atoms of each element. Then we apply the rules listed above in writing the chemical formula.

(a) This molecule contains four atoms of phosphorus. The formula is P_4.

(b) This molecule contains one nitrogen atom and three hydrogen atoms. Using Rule 2, H will follow the N; therefore the formula is NH_3.

(c) This molecule contains carbon and thus follows the rule for naming non-ionic carbon-containing compounds. It contains one carbon atom, four hydrogen atoms and one oxygen atom. The formula is CH_4O.

(d) This molecule contains two elements from the same family, chlorine and bromine. From Rule 3 the element from the lower group is listed first. The formula is $BrCl$.

Compounds that contain more than one carbon atom often have several arrangements of atoms that have the same chemical formulas, or **isomers**. Three isomers for the formula C_2H_4O are shown below:

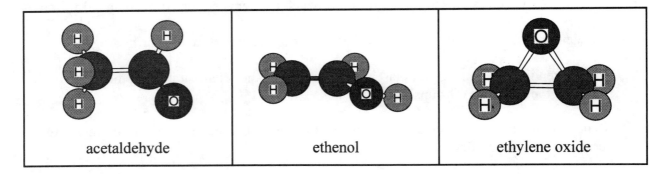

acetaldehyde   ethenol   ethylene oxide

There are several ways to distinguish among these three molecules. One way is to write the formula in terms of structural subunits instead of the total numbers of atoms. The formulas are often written with the atoms grouped by substructure as shown below.

$CH_3(CO)H$ $CH_2CH(OH)$ CH_2CH_2O

The first formula shows that acetaldehyde contains a CH_3 group attached to a $(CO)H$ group. The second formula shows that a CH_2 unit is attached to a $CH(OH)$ unit. The third formula is somewhat ambiguous, but less so than C_2H_4O. Ambiguity in structures is greatly reduced by the use of **structural formulas**, which show not only the numbers of atoms, but also the connections from atom to atom. The diagrams used up to this point are examples of structural formulas.

Chapter 3: The Composition of Molecules

Structural Formulas and Three-Dimensional Representations

In the diagrams used up to this point molecular structures have been represented by circles attached to other circles. How do the atoms maintain this arrangement? The atoms are held together by the electrical attractions between the positively charged nuclei and negatively charged electrons. These attractions are called **chemical bonds**, a topic that will be explored in Chapters 8 and 9. For now, we can say that a chemical bond is formed by two atoms that share a pair of electrons. In a structural formula, a chemical bond is represented as a line connecting the symbols of the elements joined by the bond. If two atoms share more than one pair of electrons, there are multiple bonds between the atoms. Four shared electrons create a double bond between two atoms and six shared electrons, a triple bond. The following exercise shows some compounds which contain carbon, hydrogen and nitrogen with different types of C-N bonds.

Exercise 3.2 Write chemical formulas for the following compounds whose structural formulas are shown.

Steps to Solution: To write chemical formulas, we need to know the elements present and the number of atoms of each element in the molecule. Therefore, we must count the atoms of each element and write the formula consistent with the guidelines.

(a) This compound is ethylamine, which has a fishy smell. The molecule contains two C atoms, seven H atoms and one N atom. A chemical formula that conveys no structural information is C_2H_7N. Writing it using the subunits present conveys structural information: $CH_3CH_2NH_2$.

(b) This molecule, methylimine, contains two C atoms, five H atoms, one N atom and a CN double bond. The chemical formula of C_2H_5N is not as informative as writing the subunits: CH_3CHNH.

(c) This molecule, acetonitrile or methyl cyanide, contains two C atoms, three H atoms and one N atom. The chemical formula can be written as C_2H_3N or CH_3CN.

Remember that these diagrams are two-dimensional representations of three-dimensional objects: molecules. The structures of molecules are best rendered by using three-dimensional molecular models. Two types are common: **ball and stick models** and **space filling models** shown below for acetonitrile.

ball and stick model

space filling model

Line Structures

Another two-dimensional way to represent the structure of a molecule is a **line structure** which is similar to a structural formula, but is more useful for representing complicated structures. Line formulas are less cluttered than structural formulas because they do not show all the bonds and atom labels explicitly. They are constructed using the following guidelines:

- All bonds except C-H bonds are shown as lines.

- C-H bonds are not shown.

- Single bonds are shown as a single line; double bonds are shown as two parallel lines; triple bonds are shown as three parallel lines.

- All atoms except carbon and hydrogen are labeled with their elemental symbols.

- Hydrogen atoms are labeled when they are attached to any atom other than carbon.

Exercise 3.3. Draw a line structure for the compounds whose structural formulas are shown below.

(a)	(b)	(c)
H, H, H, C-C-H, H—N, H, H	H, H, C=C, H—N, H, H	H, N≡C-C-H, H

<u>Steps to Solution</u>: *To draw a line structure from a structural formula we apply the guidelines listed above. We will work part (a) in stages:*

We first remove the C labels and H labels from hydrogens attached to carbon atoms:

Next we remove the C-H bonds:

To make the central C atom noticeable, we make sure that the C-C bond and the C-N bond meet at an angle:

(b) *We first remove the atom labels for carbon and hydrogen when bonded to carbon giving:*

Removing the C-H bonds gives the line structure:

(c) *The line structure is:*

Note we do not have to make the C-C and C-N bonds meet at an angle because one bond is a single bond and the other is a triple bond, making the C atom's position known.

Structural Formulas from Line Structures

The line structure is a convenient way to represent formulas of molecules. However, we must be able to convert a line structure back to a structural formula so that we can obtain the chemical formula. The conversion relies on the following principle:

In all neutral compounds that contain carbon, each carbon atom has four chemical bonds.

This principle is readily verified. If you examine all of the carbon-containing structural formulas in this chapter, you will observe that all have four bonds.

Because the only bonds not drawn in a line structure are C-H bonds, if any carbon atom has fewer than four bonds, we add C-H bonds until that atom has a total of four bonds. A good strategy for conversion of line structures to structural formulas is: (1) place a C at the intersection of bonds and at the ends of the lines; (2) add C-H bonds to each C atom until the total number of bonds is four. Having obtained the structural formula, the chemical formula is

Chapter 3: The Composition of Molecules

readily obtained. This method is summarized schematically below and illustrated in the following exercise.

$$\text{line structure} \longleftrightarrow \text{structural formula} \longrightarrow \text{chemical formula}$$

The right-hand arrow only points right because one can have more than one structural formula for a given chemical formula.

Exercise 3.4 Obtain the chemical formula for the following line structures.

(a)	(b)	(c)	(d)

Steps to Solution: The strategy that should be followed is to: (1) find the structural formula and (2) use the structural formula to obtain the chemical formula:

(a) First we place a C atom at all the unmarked intersections and line ends:

Now we add C-H bonds until there are 4 bonds to each C atom. There are 2 C atoms in this molecule. One has only 1 bond to the other C atom; it will need 3 C-H bonds. The other C atom has 3 bonds: 1 to the C atom, a double bond. It will need one C-H bond to have 4 bonds, giving the completed line structure.

Counting the atoms gives the chemical formula, C_2H_4O.

(b) First a C atom is placed at all the unmarked intersections and line ends (i):

Then (ii) we add C-H bonds until there are 4 bonds to each C atom. There are 2 C atoms in this molecule, both with 2 bonds, one to C and the other to O. We must add 2 C-H bonds to each C atom to obtain 4 bonds to each C atom to give the completed structural formula.

Counting the atoms gives the chemical formula, C_2H_4O .

(c) First a C atom is placed at all the unmarked intersections and line ends (i).

Then C-H bonds are added until there are 4 bonds to each C atom (ii). There are 6 C atoms in this molecule. Three of them have 2 bonds to other C atoms and require 2 C-H bonds. One C atom has 1 bond to another C atom and needs 3 C-H bonds. Another already has 4 bonds (2 to O [double bond] and 2 to C atoms) and requires no C-H bonds. One has 3 bonds to neighboring C atoms and needs 1 C-H bond.

The chemical formula is $C_6H_{10}O$.

(d) Use the techniques demonstrated above. The chemical formula is C_4H_8S .

3.2 Naming Chemical Compounds

QUESTIONS TO ANSWER, SKILLS TO LEARN
    1. **Describing molecules with names: chemical nomenclature**

In the preceding sections we saw that compounds with different structures may have the same chemical formula. Each different compound may also have several names. There are **common names**, which are historical in origin and rarely carry any information about the structure. There are **systematic names** from which we can deduce the chemical formula and structure. The systematic names you will learn follow the guidelines of **IUPAC** (International Union for Pure and Applied Chemistry). However, you need to know some common names because some are used more often than IUPAC names.

Naming Binary Compounds

    Binary compounds are those which contain only two elements. In this section we will review the rules for naming binary compounds which do not contain metals. Use the guidelines listed in Section 3.1 for finding the order of elements in the formula. The order in which the elements appear in the name is the same as in the formula. In writing the name of a compound:

    1. The element that appears first is identified by its full name.

2. The second element is named by a root word for that element that ends with the suffix *-ide*. (Table 3.1 in the text lists the common roots.)

3. When there is more than one atom of an element(s) in the formula, the number of atoms is given by a prefix to the element name (or root). If the prefix ends with the letter "*o*" or "*a*" *and* the name of the element begins with a vowel, the *last* letter of the prefix is dropped. (Table 3.2 in the text gives the common prefixes.)

Exercise 3.5 Give the names of the following compounds using the IUPAC system:
(a) CCl_4 (b) IF_5 (c) BBr_3 (d) NF_3 (e) OF_2 (f) P_4O_{10}

Steps to Solution: Apply the guidelines for naming binary compounds.

(a) C is carbon. Because it is named first in the formula, the name starts with carbon. The root for Cl, chlorine, is chlor-. Adding the suffix, ide, we obtain chloride. The prefix tetra must be added to chloride to indicate 4 chlorine atoms (tetra + chlor + ide). The name is carbon tetrachloride.

(b) I is iodine. There are five fluorine atoms (penta + fluor + ide); thus the name is iodine pentafluoride.

(c) B is boron. There are three bromine atoms (tri + brom + ide); thus the name is boron tribromide.

(d) N is nitrogen). There are three fluorine atoms (tri + fluor + ide); thus the name is nitrogen trifluoride.

(e) O is oxygen. There are two fluorine atoms (di + fluor + ide); thus the name is oxygen difluoride.

(f) Use the above method to show that the name is tetraphosphorus decoxide.

Exercise 3.6. Write the chemical formulas for the following compounds.
(a) sulfur trioxide (b) antimony pentachloride (c) sulfur hexafluoride (d) boron tri-iodide

Steps to Solution: The information necessary to write the formula is present ; the elements' names, roots, and prefixes tell us the number of each element.

(a) Sulfur has the symbol S. Oxide indicates the other element is oxygen, and the prefix tri fixes the number of oxygen atoms at three. The chemical formula is $\boxed{SO_3}$.

(b) Antimony has the symbol Sb. Chloride contains the root for chlorine (chlor-), and the prefix penta tells us that there are five chlorine atoms present in the molecule. The chemical formula is $\boxed{SbCl_5}$.

(c) Sulfur has the symbol S. Fluoride contains the root for fluorine, <u>fluor</u>-. The prefix <u>hexa</u> indicates that there are six fluorine atoms in the molecule, thus, the chemical formula is $\boxed{SF_6}$.

(d) Use the above methods to show the formula is $\boxed{BI_3}$.

Binary Compounds with Hydrogen

Hydrogen is unusual in that it can come first in formulas (and therefore names) of compounds. Names of some of these compounds follow the IUPAC systematic rules (hydrogen fluoride and lithium hydride), but many of them do not. Refer to the text for specifics.

Carbon-Based (Organic) Compounds

Carbon forms many stable binary compounds with hydrogen. These compounds are classified into three broad categories by the types of carbon-carbon bonds present.

- **Alkanes** contain only C-C single bonds.

- **Alkenes** contain one or more C-C double bonds.

- **Alkynes** contain one or more C-C triple bonds.

Number of C atoms	Root	Alkane	Alkene	Alkyne
1	*meth*	CH_4, methane	none (no C-C bonds)	none (no C-C bonds)
2	*eth*	C_2H_6, ethane	C_2H_4, ethene	C_2H_2, ethyne (acetylene)
3	*prop*	C_3H_8, propane $H_3C-CH_2-CH_3$	C_3H_6, propene	C_3H_4, propyne $H_3C-C\equiv CH$

More Than Two Elements: Functional Groups

The chemical properties of organic compounds depend largely on the presence of structural subunits called **functional groups**. An organic molecule that contains the two atom unit, -OH, is called an **alcohol**.

Chapter 3: The Composition of Molecules

Exercise 3.7 Give the names and chemical formulas of the following molecules:

(a)	(b)	(c)
OH structure (2-butanol)	OH structure (1-butanol)	Br structure (2-bromopropane)

Steps to Solution: We do not have to answer the two questions in order. It will be easier to draw the structural formula, obtain the chemical formula, and then name the compound. In this exercise, we will draw the structural formulas and then obtain the names and chemical formulas. The C-H bonds will be represented by writing the H's after the C atom trailing the appropriate subscript, for example, CH_2. Then the formulas and names can be obtained.

(a) We convert the line structure to the structural formula:

The chemical formula is $C_4H_{10}O$ and the presence of the -OH group identifies the molecule as an alcohol. A 4-carbon chain has the root butan- and an alcohol has the suffix -ol. The -OH group is on the number 2 carbon, therefore, the compound is <u>2-butanol.</u>

(b) We convert the line structure the structural formula:

The chemical formula is $C_4H_{10}O$ and the presence of the -OH group identifies the molecule as an alcohol. A 4 carbon chain has the root butan- and an alcohol has the suffix -ol. The -OH group is on the number1 carbon, therefore, the compound is <u>1-butanol.</u>

(c) Use the methods illustrated above. The name is <u>2-bromopropane</u>.

COMMON PITFALL: Numbering groups.
When identifying the position of a functional group, make sure that the number indicating the position of the functional group is as small as possible.

In Exercise 3.7 (a), one could name the molecule 2-butanol or 3-butanol. Because 2 is smaller than 3, we named it 2-butanol.

Chapter 3: The Composition of Molecules

3.3 Ionic Compounds

QUESTIONS TO ANSWER, SKILLS TO LEARN
1. **Describing ionic compounds using names of monatomic and polyatomic cations and anions**
2. **Learning the rules for naming ionic compounds**

Naming Atomic Ions

Atomic ions are formed from a single atom that has gained or lost electrons to form respectively, an **anion** or a **cation**. The alkali metals (the lithium family) form atomic cations with charge +1 and the alkaline earths (the beryllium family) form cations with charge +2. Most of the compounds that contain metals of these two families are ionic. The transition metals form atomic ions with variable charges. There are six elements that form monatomic anions: the halogens form anions of -1 charge, and oxygen and sulfur form anions of -2 charge.

Naming Polyatomic Ions

Ions that contain groups of atoms with an electrical charge are called molecular ions because the group of atoms is bound together by sharing electrons as in a molecule, but they have an electrical charge. They are also called **polyatomic ions**. You should memorize the names and formulas of the common polyatomic ions as they are often referred to by name. To help you in this task, at the end of this chapter you will find a set of flash-cards that have the names of common ions on one side and their formulas on the other.

Ionic Compounds: Nomenclature Rules and Deduction of Formulas from Names

Ionic compounds contain anions and cations in the appropriate numbers so that there is no net electrical charge:

The negative charge of all the anions balances the positive charge of all the cations.

Because ions have specific charges and chemical formulas, we can determine the ratio of cations to anions if we know the charges on the anion and cation. Knowing this ratio allows us to determine the chemical formula of the ionic compound. The following guidelines are used in writing the formulas of ionic compounds:

- The cation is listed first.

- The formula of a polyatomic ion is written as a unit.

- The formulas of polyatomic ions are written in parentheses with a following subscript indicating the number of that ion in the formula if more than one is present.

Some cations, especially the transition metals and some of the lower main group metals can form several stable ions. When naming compounds which contain elements that can form several different charged ions, the charge must be specified by a Roman numeral in parentheses after the name of the element. For example, copper (I) chloride is CuCl, but copper (II) chloride is $CuCl_2$.

Hydrates

Many ionic compounds formed in the presence of water are obtained with water molecules incorporated in the substance, called waters of hydration. Ionic substances with incorporated water are called **hydrates**. Because the water is present as molecules, the chemical formula shows the water molecules separated from the ions by a dot. We name hydrates by naming the ionic compound and indicate the number of waters of hydration by the Greek prefix followed by the word *hydrate*.

$$\text{Copper (II) sulfate hexahydrate} = CuSO_4 \cdot 6\ H_2O$$

The number of waters of hydration is not predictable and heating can change the number of waters of hydration. For example, blue copper sulfate hexahydrate loses water when heated to become white copper (II) sulfate, $CuSO_4$.

Exercise 3.8 Give the formula for each of the following compounds.
 (a) sodium phosphate b) ammonium carbonate (c) sodium hydrogenphosphate
 (d) rhodium (III) chloride trihydrate

Steps to Solution: Identify the ions and their charges. Then determine the combination of cations and anions that leads to a neutral composition (no electrical charge) with the smallest whole number of cations and anions. The number of cations will be the subscript for the cation and the number of anions will be the subscript for the anion.

(a) *Phosphate is the name for the PO_4^{3-} ion. To obtain a compound with no electrical charge three sodium ions are required. The cation is listed first, giving the formula* $\boxed{Na_3PO_4}$.

(b) *Ammonium is the name for the NH_4^+ ion, and carbonate is the name for the CO_3^{2-} ion. Two ammonium ions are required to cancel the 2- charge of the carbonate anion. The formula is* $\boxed{(NH_4)_2CO_3}$.

(c) *Sodium is the name for the Na^+ ion. Hydrogen phosphate indicates that one hydrogen ion is bound to the phosphate ion, giving the HPO_4^{2-} ion. Two sodium ions are needed to cancel the 2- charge, so the formula is* $\boxed{Na_2HPO_4}$.

(d) *Use the methods above to show that the compound's formula is* $\boxed{RhCl_3 \cdot 3\ H_2O}$.

Exercise 3.9 Write the name of each of the following compounds: (a) $NaHCO_3$ (b) $FeBr_2$ (c) KH_2PO_4 (d) $CaCl_2 \cdot 2H_2O$

Steps to Solution: We must find the names of the polyatomic ions (if you haven't memorized them) and then list the cation name and then the anion name.

(a) Na is the sodium ion and HCO_3^- is the hydrogen carbonate ion. The name is sodiumhydrogencarbonate .

(b) Fe is iron, and there also are 2 Br⁻ ions present (because Br belongs to Group 17). The charge on iron must be +2 to cancel out the 2(-1) charges on the two bromides. We indicate the +2 charge with the Roman numeral after the name for iron: iron(II)bromide .

(c) K is the potassium ion with a charge of +1. The $H_2PO_4^-$ ion is the dihydrogenphosphate ion. The compound is potassiumdihydrogenphosphate .

(d) Use the methods above to show that the compound's name is \X(calciumchloride dihydrate.)

Recognizing Ionic Compounds

The vast majority of ionic compounds have one or both of these characteristics:
- a metal ion from the alkali metals (Group 1) or alkaline earths (Group 2) is present
- a polyatomic ion is present

Exercise 3.10 Deduce which of the following compounds are ionic: (a)$HSiCl_3$ (b)$BaSO_4$ (c)C_6H_6 (d) $(NH_4)_2SO_4$ (e)$PbSO_4$ (f) OsO_4

Steps to Solution: To identify ionic compounds on the basis of their formula, we must see if they contain either a Group 1 or 2 metal or a polyatomic ion. If so, the compound is an ionic substance.

(a) $HSiCl_3$ contains neither a Group 1 or 2 metal ion or a polyatomic ion. Therefore it is not an ionic compound.

(b) $BaSO_4$ contains a Group 2 metal, barium, and a polyatomic ion, sulfate. Therefore, it is an ionic compound.

(c) C_6H_6 contains no metal ions or polyatomic ions. Therefore C_6H_6 is not an ionic compound.

(d) $(NH_4)_2SO_4$ contains the polyatomic ions ammonium and sulfate. Therefore ammonium sulfate is an ionic compound.

(e) $PbSO_4$ contains the polyatomic ion sulfate. Lead sulfate is an ionic compound.

(f) OsO₄ does not contain either a polyatomic ion or a Group 1 or 2 metal. It is not an ionic compound.

3.4 The Mole

QUESTIONS TO ANSWER, SKILLS TO LEARN
1. What is the chemist's collective term for numbers of molecules?
2. What is the difference between atomic mass, elemental molar mass, isotopic molar mass?
3. Converting mass of substance to number of molecules or moles of substance.

Chemists are interested in the ways in which molecules interact and/or react with each other. Molecules have very little mass and are very small; therefore, any observable amount of a substance has a huge number of molecules. When dealing with large numbers of items, we often use names that stand for a certain number of items. For example, you might refer to 144 paper clips as a gross or 12 eggs as a dozen; the words *gross* and *dozen* are **collectives**. Chemists have their own collective term because they routinely deal with very large numbers of molecules: this term is the **mole** (abbreviated as mol).

One mol is the number of atoms in exactly 12 g of the isotope carbon-12.

This number is important enough to be a fundamental unit. It is the SI unit for the amount of matter present in terms of the number of particles. It is called **Avogadro's number** and has the symbol N, N_A, or N_O. In this text we will use N_A for Avogadro's number and its value to 4 significant figures is:

$$N_A = 6.022 \times 10^{23} \text{ 1/mol}$$

Note that the units are 1/mol; N_A is the conversion factor for changing the number of items to the moles of items. The mol is a very large number: one mol of paper stacked flat would be more than 5000 light years tall; this length is 1000 times longer than the distance of the nearest star to Earth. However, a charcoal briquette (about 35 g) contains about 3 moles of carbon atoms. The mol is a unit that is highly suitable for very small objects like molecules and atoms.

In Chapter 2, we discussed the existence of isotopes in elements. To summarize: (a) each isotope has a somewhat different mass, but all isotopes of an element have the same chemical properties; (b) elements occur naturally as several isotopes; the numbers of atoms of the different isotopes in different naturally occurring samples are the same. A sample of sulfur obtained in Texas has the same number of isotopes and the same number of atoms of each isotope as a sample of sulfur obtained from Saudi Arabia. The mass numbers of the sulfur isotopes, their percent abundance and the mass of the isotope in g/mol are shown below.

Mass Number of Sulfur Isotope	% Abundance	Mass of Isotope(g/mol)
32	95.00	31.97207
33	0.76	32.97146
34	4.22	33.96786
36	0.014	35.96709

There are several ways to define molar mass:

- The **elemental molar mass** is the mass of one mol of an element containing the naturally occurring isotopes in their natural abundances. This value is sometimes called the atomic weight.
- The **isotopic molar mass** is the mass of one mol of a particular isotope of an element.
- The **atomic mass** is the mass of one atom of an element.
- The **molar mass** refers to the mass of one mol of a substance.

The elemental **molar mass** is the quantity you will most often use. It is the mass of a mole of a naturally occurring element and depends on the number of atoms of each isotope and the mass of each isotope. If we have 1 million (1×10^6) atoms of sulfur, 95% of them will be the isotope of mass number 32, 0.76% will be the isotope of mass number 33, and so on. The molar mass of sulfur is found by calculating the weighted average:

$$\text{molar mass} = \Sigma(\text{fractional abundance})(\text{isotopic molar mass})$$

$$\text{where the fractional abundance} = \frac{\% \text{ abundance}}{100}$$

The molar mass of sulfur is given by:

molar mass = 0.9500(31.97207 g/mol) + 0.0076(32.97146 g/mol) + 0.0422(33.96786 g/mol)
                   + 0.00014(35.96709 g/mol)

molar mass = (30.37 + 0.25 + 1.43 + 0.0050) g/mol = 32.06 g/mol

(note: only significant figures shown)

The elemental molar mass is the value we find in the periodic table or in the tables of atomic weights. We rarely have to use the isotopic molar masses and percent abundances as shown above, although the method is sometimes applied to other problems.

Exercise 3.11 Rhenium is a metal that is used in the conversion of crude oil to gasoline in "catalytic cracking" processes. Rhenium has two naturally occurring isotopes, ^{185}Re (37.500%) and ^{187}Re (62.500%), with isotopic molar masses of 184.9530 and 186.9560 g/mol, respectively. What is the elemental molar mass of rhenium?

Steps to Solution: We will approach this problem using the seven-step strategy outlined in Chapter 1.

(1) Identify the problem. *The problem asks for the elemental molar mass, which is the weighted average of the isotopic molar masses.*

(2) Develop a mental picture of what is happening. *There are atoms of the same element which have different masses.*

(3) What equations/formulas are relevant? *The equation:*

molar mass = $\sum$(fractional abundance)(isotopic molar mass)

relates the desired quantity, elemental molar mass, to quantities that are readily obtained.

(4) Organize the data provided and find other data if needed. *We are given the percent abundances and isotopic molar masses (underlined in the problem).*

(5) Rearrange or combine the equations from step 3 or break the problem into smaller pieces. *The equation does not need to be manipulated or rearranged.*

(6) Execute the plan. *Now we substitute in the data, making sure that the appropriate units are used:*

molar mass = (0.37500)(184.9530 g/mol) + (0.62500)(186.9560 g/mol)

molar mass = $\boxed{186.20 \text{ g/mole}}$

(7) Is the answer reasonable? *The answer is more than the isotopic mass of the lighter isotope and less than the isotopic mass of the heavier isotope as we would expect for a mixture of the two. It is also closer to the mass of the more abundant isotope. Therefore, the answer is reasonable.*

Chapter 3: The Composition of Molecules

Molar Masses and Conversion Between Mass and Moles

We now have the tools to determine the number of atoms in a sample. We can use Avogadro's number and the elemental molar mass to convert mass to moles to numbers of atoms.

$$\text{moles} = \frac{\text{mass}}{\text{molar mass}}$$

mass = moles(molar mass) and number of atoms = moles (N_A)

Exercise 3.12 Methanol is produced using a catalyst that contains 6.8% chromium. How many moles of chromium and how many atoms of chromium are in 100 g of the catalyst?

Steps to Solution: *We will approach this problem using the seven-step strategy outlined in Chapter 1.*

(1) Identify the problem. *The problem asks for the number of moles of Cr and atoms of Cr in 100 g of the catalyst.*

(2) Develop a mental picture of what is happening. *The atoms are part of a mixture whose composition is partially known (we know the % Cr).*

(3) What equations/formulas are relevant?

$$\text{moles} = \frac{\text{mass}}{\text{molar mass}} \quad \text{and} \quad \text{number of atoms} = \text{moles}\,(N_A)$$

(4) Organize the data provided and find other data if needed. *We are supplied with the % Cr in the catalyst (6.8 %), and the mass of catalyst (100 g). You can find the molar mass of Cr in the periodic table in your text (51.996 g/mol).*

(5) Rearrange or combine the equations from step 3 or break the problem into smaller pieces. *The equations do not need to be manipulated or rearranged to give the desired quantities. We do need to determine the mass of chromium.*

(6) Execute the plan. *Now we substitute in the data, making sure that the appropriate units are used:*

$$\text{mass Cr} = 6.8\%(100\text{ g}) \cdot \frac{1}{100\%} = 6.8\text{ g}$$

$$\text{moles Cr} = 6.8\text{g} \cdot \frac{1\text{ mol Cr}}{51.996\text{ g}} = 0.13\text{ mol Cr}$$

$$\text{atoms Cr} = 0.13 \text{ mol} \cdot 6.022 \times 10^{23} \frac{1}{\text{mol}} = 7.9 \times 10^{22} \text{ atoms Cr}$$

Note that the units of mol cancel out. Also note that the units for N_A are written as atoms/mol. The units of Avogadro's number are items per mol; we can use whatever items we are counting as the numerator in N_A.

(7) Is the answer reasonable? *Yes, because the number of moles is less than 1 and the mass of chromium is less than the molar mass. The number of atoms is large, but it is less than a mol.*

3.5 Mass-Mole-Number Conversions

QUESTIONS TO ANSWER, SKILLS TO LEARN
1. **Determining the molar mass of a substance from the chemical formula**
2. **Determining the number of molecules or moles of a substance from the mass of a substance**
3. **Determining the mass percent composition of each element in a substance**

A molecule is a collection of atoms that is held together by forces called chemical bonds. Therefore one mol of a substance contains one mol of its chemical formula unit.

Exercise 3.13 Find the number of moles of atoms in one mol of each of the following substances:
(a) ammonia, NH_3; (b) formaldehyde, CH_2O ; (c) cadmium nitrate, $Cd(NO_3)_2$.

Steps to Solution: Remember that the number of atoms per molecule or formula unit is found from the subscript in the formula.

(a) *Ammonia, NH_3, contains one mol of nitrogen atoms and three moles of hydrogen atoms.*

(b) *Formaldehyde, CH_2O, contains one mol of carbon atoms, one mol of oxygen atoms and two moles of hydrogen atoms.*

(c) *Cadmium nitrate, $Cd(NO_3)_2$, contains one mol of Cd^{2+} ions and two moles of NO_3^- ions. Each mol of nitrate contains one mol of nitrogen atoms and three moles of oxygen atoms. In two moles of nitrate ions there are 2(1) = 2 moles of N atoms and 2(3) = 6 moles of O atoms. One mol of cadmium nitrate contains one mol of Cd, two moles of N and six moles of O.*

Chapter 3: The Composition of Molecules

> **COMMON PITFALL: Parentheses with subscripts.**
> It's easy to make mistakes when ions or other groups are in parentheses with subscripts. Make sure you interpret the parentheses correctly in reading formulas and multiply the subscript of the parentheses times the subscripts of the atoms inside the parentheses.

The mass of a molecule is the sum of the masses of all its parts. Therefore, the molar mass (*MM*) of a substance is the sum of the elemental molar masses of all the atoms in its chemical formula.

Exercise 3.14 Calculate the molar mass of each of the substances in Exercise 3.13.

(a) *The molar mass of ammonia is the sum of the masses of one mol of nitrogen and three moles of hydrogen:*

$$MM \text{ of } NH_3 = \frac{1 \text{ mol N}}{\text{mol } NH_3} \cdot 14.007 \text{ g/mol N} + \frac{3 \text{ mol H}}{\text{mol } NH_3} \cdot 1.0079 \text{ g/mol H}$$

$$MM \text{ of } NH_3 = 17.030 \text{ g/mol } NH_3$$

(b) *The molar mass of CH_2O is the mass of one mole of carbon, one mole of oxygen and two moles of hydrogen:*

$$MM \text{ of } CH_2O = \frac{1 \text{ mol C}}{\text{mol } CH_2O} \cdot 12.011 \text{ g/mol C} + \frac{2 \text{ mol H}}{CH_2O} \cdot 1.0079 \text{ g/mol H} + \frac{1 \text{ mol O}}{CH_2O} \cdot 16.00 \text{ g/mol O}$$

$$MM \text{ of } CH_2O = 30.026 \text{ g/mol } CH_2O$$

(c) *The molar mass of $Cd(NO_3)_2$ is the mass of one mole of cadmium, two moles of nitrogen and six moles of oxygen:* Show that the

$$MM \text{ of } Cd(NO_3)_2 = 236.42 \text{ g/mol } Cd(NO_3)_2$$

Exercise 3.15 Calculate the molar mass of testosterone, a hormone, whose line structure is shown below in the left figure.

Steps to Solution: Since the chemical formula is not provided, the task of determining the chemical formula via converting the line formula to a structural formula will have to be performed. The structural formula is used to find the chemical formula, from which the molar mass is calculated.

Using the method demonstrated earlier we obtain the structure on the right.

Counting the atoms in the structural formula gives us the chemical formula $C_{19}H_{28}O_2$. Now the molar mass of testosterone can be calculated. In previous exercises, the units have been written out completely to insure that the correct units are obtained. The more compact notation demonstrated below is faster to use:

MM (testosterone) = 19 mol C(12.011 g/mol) + 28 H(1.0079 g/mol) + 2 O (15.9994g/mol)

MM (testosterone) = 288.434 g/mol

Given the chemical formula, we can determine the molar mass, the number of moles (or atoms) and the mass of each element in the substance. The schematic diagram that follows shows how you can convert between these quantities:

mass substance ⟷ (chemical formula / molar mass) ⟷ moles substance ⟷ (chemical formula) ⟷ moles elements ⟷ (elemental molar mass) ⟷ mass elements

The following exercise gives some examples of the conversions among the three leftmost quantities, mass of substance to moles of atoms and to numbers of atoms.

Exercise 3.16 Sulfuric acid is the industrial chemical produced in the greatest mass. In 1991, 86.62 x 10^9 pounds of sulfuric acid, H_2SO_4, were manufactured. How many sulfur atoms are present in this amount of sulfuric acid?

Steps to Solution: We will use the seven-step method.

(1) Identify the problem. We are asked to find the number of sulfur atoms in 86.62 x 10^9 pounds of sulfuric acid.

(2) Develop a mental picture of what is happening. Sulfuric acid is a compound which contains one sulfur atom, four oxygen atoms and two hydrogen atoms per molecule. Therefore, the number of molecules of sulfuric acid is the same as the number of atoms of sulfur:

moles S = moles H_2SO_4

(3) What equations/formulas are relevant?

Chapter 3: The Composition of Molecules

$$\text{moles} = \frac{\text{mass}}{\text{molar mass}} \quad \text{and} \quad \text{number of atoms} = \text{moles}\,(N_A)$$

(4) Organize the data provided and find other data if needed. We are given the chemical formula for sulfuric acid and the mass of sulfuric acid. The molar mass of sulfuric acid is readily determined. We must make the conversion between pounds and g (because molar mass is in g/mol).

(5) Rearrange or combine the equations from step 3 or break the problem into smaller pieces. The equations do not need rearrangement. One way to solve this problem is to:
(a) find the molar mass of sulfuric acid, (b) find the moles of sulfuric acid in 86.62×10^9 pounds of sulfuric acid (this will require a units conversion, 454 g/pound) and (c) use Avogadro's number to convert the number of moles of sulfur atoms to the number of sulfur atoms.

(6) Execute the plan.

(a)

$$MM\,(H_2SO_4) = 1\,S\,(32.06\text{ g/mol}) + 4\,O\,(15.9994\text{ g/mol}) + 2\,H\,(1.00797\text{ g/mol})$$

$$MM\,(H_2SO_4) = 98.07 \text{ g/mol } H_2SO_4$$

(b)

$$\text{mols S} = \frac{1 \text{ mol S}}{\text{mol } H_2SO_4} \bullet \text{mols } H_2SO_4 = \frac{86.62 \times 10^9 \text{ pounds}(454 \text{ g/pound})}{98.07 \text{ g/mol} H_2SO_4}$$

$$\text{moles S} = 4.01 \times 10^{11} \text{ molS}$$

(c)

atoms S = mol S(N_A) = 4.01×10^{11} mol S $(6.022 \times 10^{23}$ 1/mol) = $\boxed{2.42 \times 10^{35} \text{ atoms S}}$

(7) Is the answer reasonable? Yes. The number of atoms is large, but so is the amount of substance being described.

3.6 Determining Chemical Formulas

QUESTIONS TO ANSWER, SKILLS TO LEARN
1. **Calculating elemental mass percent (percent composition) from chemical formulas**
2. **Converting percent composition to empirical formulas**
3. **Using molar masses and empirical formulas to obtain molecular formulas**

Chemical formulas contain a large amount of useful information. In this section we learn how to confirm a proposed chemical formula from experimental data on the percent mass of the elements in the substance, and propose a chemical formula of a compound from the percent mass of the elements it contains. Both processes involve converting between chemical formulas and elemental percent composition (also called elemental analysis or mass percent composition).

The elemental percent composition of a substance is the mass of each element in grams found in 100 grams of the substance.

The following exercise illustrates the calculation of elemental percent composition.

Exercise 3.17. The solvent, 1,2-dichoroethane, $C_2H_4Cl_2$, is a major industrial chemical used in the manufacture of polyvinylchloride (PVC). What is the elemental percent composition of 1,2-dichlororethane?

Steps to Solution: First the molar mass is calculated, and then the percent of each element is calculated.

$$MM = 2\,C\,(12.011\text{ g/mol}) + 4\,H\,(1.0079\text{ g/mol}) + 2\,Cl\,(35.453\text{ g/mol}) = 98.960\text{ g/mol}$$

The % of each element is the fraction of the mass of that element divided by the mass of a formula unit times 100%:

$$\%C = \frac{g\,C\text{ per chemical formula}}{MM}100\% = \frac{2\,C(12.011\text{ g/mol})}{98.960\text{ g/mol}}(100\%) = 24.27\,\%$$

$$\%H = \frac{g\,H\text{ per chemical formula}}{MM}100\% = \frac{4\,H(1.0079\text{ g/mol})}{98.960\text{ g/mol}}(100\%) = 4.07\,\%$$

$$\%Cl = \frac{g\,Cl\text{ per chemical formula}}{MM}100\% = \frac{2\,Cl(35.453\text{ g/mol})}{98.960\text{ g/mol}}(100\%) = 71.65\,\%$$

To check the accuracy of the calculations, we take the sum of the percent compositions of all the elements; it should be 100%:

$$24.27\% + 4.07\,\% + 71.65\,\% = 99.99\,\% \text{ (which rounds to 100.0\%)}$$

Therefore the percent composition of dichloroethane is:

$$\boxed{24.27\% \text{ C}; 4.07 \% \text{ H}; 71.65\% \text{ Cl}}.$$

The elemental percent composition can be also used to confirm the identity of a compound. To do this, we have the elemental composition determined by a commercial laboratory and compare those experimental figures with the composition calculated for the chemical formula that we believe the compound to be. If the agreement is good, then the compound most likely has the chemical formula used in the calculation. If not, the compound has a different chemical formula or is impure.

Exercise 3.18 The compound fructose (fruit sugar) has the following structure:

A sample of a sugar thought to be fructose is tested by elemental analysis and the results are:
%C = 40.1%; % H = 6.67 %; %O = 53.2%. Is the percent composition consistent with the sample being fructose?

Steps to Solution: *We must compare the percent composition found by experiment with that calculated for the chemical formula of fructose. First we determine the chemical formula. Next, we calculate the elemental percent composition. Finally we compare the calculated and the experimental values for the composition*

After we convert the line structure to a structural formula, the chemical formula is found to $C_6H_{12}O_6$. The molar mass is 180.16 g/mol.

We calculate the elemental percent composition:

$$\% \text{ C} = \frac{6 \text{ C}(12.011 \text{ g/mol})}{180.16 \text{ g/mol}} 100\% = 40.00 \%$$

$$\% \text{ H} = \frac{12 \text{ H}(1.0079 \text{ g/mol})}{180.16 \text{ g/mol}} 100 \% = 6.71 \%$$

$$\% \text{ O} = \frac{6\text{O}(15.994 \text{ g/mol})}{180.16 \text{ g/mol}} 100 \% = 53.28 \%$$

Finally we compare the values for calculated and experimental percent composition:

Chapter 3: The Composition of Molecules

Element	Calculated %	Experimental %
C	40.00	40.1
H	6.71	6.67
O	53.28	53.2

Because the experimental and calculated values are quite similar (the differences are about that of the precision of the measurement ± 1 unit of the last decimal place shown) the elemental analysis shows that the composition is consistent with the sample having the chemical formula of fructose.

Empirical Formulas

   The elemental percent composition of a compound can be used to confirm the chemical formula of a compound because the mass of the various elements is fixed by the ratio of the elements to each other. We can also deduce the ratio of elements with respect to each other from the elemental percent composition. Consider the example of fructose. Given the elemental percent composition (the calculated ones in the table) we can calculate the moles of carbon, hydrogen and oxygen in 100 g of sample (100 g is chosen because the mass of an element in 100 g is equal to the percent of that element). Next, the molar ratios of the elements to the least abundant element are calculated. For the first step:

$$\text{mol C in 100 g} = \frac{40.00 \text{ g}}{12.011 \text{ g/molC}} = 3.330 \text{ mol C}$$

$$\text{mol H in 100 g} = \frac{6.71 \text{ g}}{1.0079 \text{ g/mol H}} = 6.66 \text{ mol H}$$

$$\text{mol O in 100 g} = \frac{53.28 \text{ g}}{15.9994 \text{ g/mol O}} = 3.330 \text{ mol O}$$

Since the moles of carbon and oxygen are equal, we can find either the moles of hydrogen per mole carbon or the moles of hydrogen per mole oxygen:

$$\frac{\text{mol H}}{\text{mol C}} = \frac{6.66 \text{ mol H}}{3.33 \text{ mol C}} = 2.00 \text{ mol H/mol C}$$

The number of moles of O per mol of C is found similarly:

Chapter 3: The Composition of Molecules

$$\frac{\text{mol O}}{\text{mol C}} = \frac{6.66 \text{ mol H}}{3.330 \text{ mol C}} = 1.00 \text{ mol O/mol C}$$

We will use these ratios to determine the subscripts in a chemical formula. **We must divide by the *smallest number of moles* because subscripts in a chemical formula cannot be less than one; dividing by the smallest number of moles ensures that all the ratios will be 1 or larger**.

This result tells us that the ratio of hydrogen to oxygen to carbon is 2 to 1 to 1. The chemical formula that corresponds to these ratios is CH_2O. This is not the chemical formula for fructose; it is the **empirical formula**. Any compound which contains the elements H, O and C in a 2:1:1 ratio will give the same elemental analysis and the same empirical formula.

The empirical formula, CH_2O, is related to the chemical formula, $C_6H_{12}O_6$, because its subscripts are one sixth those of fructose. The chemical formula often differs from the empirical formula because the chemical formula is some whole number of empirical formula units. For example, fructose contains six empirical formula units.

To determine the chemical formula from the empirical formula we need the molar mass to find the number of empirical formula units in each molecule.

If an experiment showed the molar mass of fructose to be 180 g/mol, we could use this datum with the empirical formula to determine the chemical formula for a molecule of fructose. The empirical formula is CH_2O, which has a molar mass (sometimes called empirical molar mass or empirical formula weight) of

$$MM \text{ of } CH_2O = 30.03 \text{ g/mol}$$

The number of CH_2O units in fructose, N, is found by the ratio of the molar mass to the empirical molar mass:

$$N = \text{\# of empirical formula units/molecule} = \frac{180 \text{ g/mol fructose}}{30.03 \text{ g/mol empirical formula}}$$

$$N = 6.0 \frac{\text{mol empirical formula}}{\text{mol fructose}}$$

So the chemical formula of fructose contains six empirical formula units. The chemical formula is found by multiplying the subscripts of the empirical formula by 6:

$$\text{chemical formula} = C_N H_{2(N)} O_N = C_6 H_{12} O_6$$

Chapter 3: The Composition of Molecules

Exercise 3.19 A compound containing only phosphorus and oxygen is found to have a molar mass of 284 g/mol and to be 43.62% P by mass. What is the chemical formula of this substance?

Steps to Solution: The seven-step method will be used.

(1) We need to determine the chemical formula. To accomplish this task, we need to determine the empirical formula and then find how many empirical formulas are in the chemical formula.

(2) We are told that the compound contains only phosphorus and oxygen. Therefore, given the percent of P in the compound, we should realize that the rest of the mass of the molecule is oxygen.

$$\%O = 100 - \%P$$

(3) The formula we need to convert mass of elements to moles via atomic mass is:

$$\text{moles} = \frac{\text{mass}}{\text{molar mass}}$$

(4) The data given are:

the molar mass of the compound (284 g/mol) and

% P = 43.62 and indirectly, % O = 100 - 43.62 = 56.38 %

(5) The plan:
- *(a) Calculate the numbers of moles of P and O.*
- *(b) Divide the numbers of moles of P and O by the smaller number of moles.*
- *(c) If necessary, adjust mol ratios to whole numbers by multiplying both by a common factor to obtain the empirical formula.*
- *(d) Calculate the empirical formula mass and take the ratio of*

$$\frac{\text{molar mass}}{\text{empirical formula mass}}$$

- *(e) Multiply the subscripts of empirical formula by the ratio obtained in (d) to obtain the chemical formula.*

(6) Execute plan: (a)

$$\text{mol P} = \frac{43.62 \text{ g}}{30.97 \text{ g/mol}} = 1.408 \text{ mol P} \quad \text{mol O} = \frac{56.38 \text{ g}}{16.00 \text{ g/mol}} = 3.524 \text{ mol O}$$

(b) Since phosphorus has the smaller number of mols, divide the mols O by the mols P:

$$\frac{\text{mol O}}{\text{mol P}} = \frac{3.524 \text{ mol O}}{1.408 \text{ mol P}} = 2.503 \text{ mol O/mol P}$$

(c) $PO_{2.5}$ is not an acceptable empirical formula because subscripts must be whole numbers.

Multiply both subscripts by 2 to give : P_2O_5

(d) The empirical molar mass = 141.94 g/mol

The number of empirical formula units per molecules, N, is:

$$N = \frac{284 \text{ g/mol}}{141.94 \text{ g/mol}} = 2$$

(e) The chemical formula is $P_{2(2)}O_{5(2)}$ or $\boxed{P_4O_{10}}$

7. The answer is reasonable. The molar mass calculated from the chemical formula is the same (within limits of precision) as that given in the problem.

Combustion Analysis

Combustion analysis is one method for determining the carbon and hydrogen content of a compound. The sample is burned in an atmosphere of oxygen to convert all carbon to carbon dioxide and all hydrogen to water. Under the conditions of the experiment, there is more than enough oxygen present to effect this transformation completely.

$$C \rightarrow CO_2 \quad \text{and} \quad 2H \rightarrow H_2O$$

For every mol of CO_2 obtained in combustion analysis, one mol of C was present.

For every mol of H_2O obtained in combustion analysis, two moles of H were present.

The moles of C and H are determined as shown below:

$$\text{mass } CO_2 \rightarrow \text{mol } CO_2 = \text{mol C} \rightarrow \text{mass C}$$

$$\text{mass } H_2O \rightarrow \text{mol } H_2O = \frac{1}{2} \text{ mol H} \rightarrow \text{mass H}$$

The masses of carbon and hydrogen are used to determine the percentages of carbon and hydrogen in the sample. By the law of conservation of mass, the percent mass of other elements is the difference between 100% and the percentages of C and H:

$$\text{\% other elements} = 100\% - (\% C + \% H)$$

Chapter 3: The Composition of Molecules

The percentages of other elements may be determined by other methods. The methods illustrated above can then be applied to determine empirical and (with appropriate data) chemical formulas.

Exercise 3.20 Bisphenol A is an important industrial chemical used in the manufacture of epoxy glues and polycarbonate plastics.

$$HO-\text{C}_6H_4-C(CH_3)_2-\text{C}_6H_4-OH$$

A 5.35 mg sample containing only C, H and O and believed to bisphenol A undergoes combustion analysis. The products of combustion are 15.49 mg CO_2 and 3.37 mg H_2O. In a separate experiment the molar mass is found to be 228 g/mol. Find the empirical and molecular formulas for this sample. Are they consistent with the sample being bisphenol A?

Steps to Solution: The seven step method will be applied to this problem.

(1) The question asks us to determine the empirical and molecular formulas, given the combustion analysis and molar mass. We then compare the determined molecular formula with that of bis-phenol A to see if they are the same.

(2) The carbon and hydrogen in the sample are converted to CO_2 and H_2O respectively. Oxygen is determined by difference; it is the only other element present besides C and H.

(3) For this problem we will convert between mass and moles using the molar masses of the substances involved.

(4) The data are <u>underlined</u> above.

(5) Following the flow chart in the text, the first quantities we find are (a) the moles of carbon and hydrogen and then (b) the masses of carbon, hydrogen and oxygen. The next step (c) is to determine the empirical formula and then (d) the chemical formula. Finally (e), we will compare this result with the chemical formula determined from the line structure supplied for bisphenol A.

(6) (a) Convert mass of CO_2 and H_2O to moles C and H:

$$\text{mol } CO_2 = \frac{15.49 \times 10^{-3} g}{44.01 \text{ g/mol}} = 3.520 \times 10^{-4} \text{ mol } CO_2 = \text{mol C}$$

$$\text{mol } H_2O = \frac{3.37 \times 10^{-3} g}{18.01 \text{ g/mol}} = 1.871 \times 10^{-4} \text{ mol } H_2O = \frac{1}{2} \text{ mol H}$$

$$\text{mol H} = \frac{2 \text{ mol H}}{1 \text{ mol H}_2\text{O}}(\text{mol H}_2\text{O}) = 2 \cdot 1.871 \times 10^{-4} \text{ mol H}_2\text{O} = 3.741 \times 10^{-4} \text{ mol H}$$

(b)

$$\text{mass C} = \text{mol C}(12.011 \text{ g/mol C}) = 3.520 \times 10^{-4} \text{ mol C } (12.011 \text{ g/mol C})$$

$$\text{mass C} = 4.23 \times 10^{-3} \text{ g C}$$

$$\text{mass H} = \text{mol H}(1.0079 \text{ g/mol H}) = 3.741 \times 10^{-4} \text{ mol H } (1.0079 \text{ g/mol H})$$

$$\text{mass H} = 3.77 \times 10^{-4} \text{ g H}$$

$$\text{mass O} = \text{mass sample} - (\text{mass C} + \text{mass H})$$

$$\text{mass O} = 5.35 \times 10^{-3} \text{ g} - (4.23 \times 10^{-3} \text{ g C} + 3.77 \times 10^{-4} \text{ g H}) = 7.4 \times 10^{-4} \text{ g O}$$

$$\text{mol O} = \frac{7.4 \times 10^{-4} \text{ g O}}{15.994 \text{ g/mol O}} = 4.7 \times 10^{-5} \text{ mol O}$$

(c) The empirical formula is determined by calculating the mol ratios of the elements to the element with the smallest number of mols. Oxygen has the smallest number of mols present in the sample.

$$\frac{\text{mol H}}{\text{mol O}} = \frac{3.741 \times 10^{-4} \text{ mol H}}{4.7 \times 10^{-5} \text{ mol O}} = 8.0 \text{ mol H/mol O}$$

$$\frac{\text{mol C}}{\text{mol O}} = \frac{3.520 \times 10^{-4} \text{ mol C}}{4.7 \times 10^{-5} \text{ mol O}} = 7.55 \text{ mol C/mol O}$$

The mol ratio of C : H : O is 7.55 : 8 : 1.

To obtain integer coefficients, we need to multiply each of these by 2. The empirical formula obtained is $C_{15}H_{16}O_2$.

(d) The empirical formula weight is found to be 228.29 g/mol, which is the same (within limits of precision) as that found for the substance. Therefore the chemical formula is $\boxed{C_{15}H_{16}O_2}$.

(e) Converting the line structure of bisphenol A to a structural formula gives a chemical formula of $C_{15}H_{16}O_2$; therefore the results are consistent with the sample being bisphenol A.

(7) Yes, the answer is reasonable.

3.7 Aqueous Solutions

QUESTIONS TO ANSWER, SKILLS TO LEARN
1. **What are the common concentration units for solutions?**
2. **What are the different types of aqueous solutions?**
3. **Converting volume to mol**
4. **Calculations involved in the preparation of solutions**

To this point, we have focused attention on pure substances and on using their mass to determine the number of moles of substance present. However, liquid solutions form an important part of chemistry because chemical reactions generally occur more rapidly in solutions than as solids. A **solution** is a homogeneous mixture of two or more components. We call the major component the **solvent**; it dissolves the minor components, which are called the **solutes**. **Aqueous** solutions are those where water is the solvent. An attractive feature of solutions is that one can easily measure volumes of solutions that contain known numbers of solute molecules or ions. The converting factor between volume and amount of solute is called **concentration**. The concentration unit we will be discussing now is **molarity**, which relates volume of solution to moles of solute.

Molarity

Molarity relates the volume of *solution* to the number of moles of solute:

$$\text{molarity}, M = \frac{\text{mol solute}}{\text{total volume solution}} = \frac{\text{mol solute}}{\text{L solution}}$$

Molarity converts the volume of solution of known molarity to moles of solute:

$$\text{mol solute} = M\,(\text{mol/L}) \cdot V(\text{L}) = \text{mol}$$

A solution of known molarity is prepared by dissolving a known amount of solute in the solvent and then adding water to give a known volume.

Calculating molarity

Molarity is calculated from the ratio of the moles of solute to the volume of solution.

Chapter 3: The Composition of Molecules

Exercise 3.21 A solution is prepared by the addition of 18.0 g fructose, $C_6H_{12}O_6$, to water, allowing it to dissolve and then adding more water until the total volume is 0.350 L. What is the molarity of fructose in the solution?

Steps to Solution: We are asked to find the molarity of the solution. We must (a) find the number of moles of solute from the mass of solute given, and (b) divide it by the total volume of the solution.

(a)
$$\text{mol } C_6H_{12}O_6 = \frac{18.0 \text{ g}}{180.16 \text{ g/mol}} = 9.99 \times 10^{-2} \text{ mol } C_6H_{12}O_6$$

(b)
$$M(C_6H_{12}O_6) = \frac{9.99 \times 10^{-2} \text{ mol } C_6H_{12}O_6}{0.350 \text{ L}} = 0.285 \text{ mol } C_6H_{12}O_6/L = 0.285 \text{ M } C_6H_{12}O_6$$

Because molar concentration is a common way to express the amount of solute per unit of volume, the concentration of a solute is often shown by enclosing the solute in square brackets. For example, the concentration of $C_6H_{12}O_6$ would be represented by $[C_6H_{12}O_6]$. We write $[C_6H_{12}O_6] = 0.285$ M rather than "the concentration of $C_6H_{12}O_6 = 0.285$ M."

We broadly categorize aqueous solutions by their ability to conduct electricity. Pure water is a poor conductor of electricity, but when ions are present, it is a much better conductor. When a substance is dissolved by a solvent, the molecules of the solvent surround the solute. If the solution is a poor conductor, then we expect that the solute dissolves as molecules, and the solution contains solute molecules surrounded by solvent molecules. Compounds that we have learned to identify as ionic dissolve to form solutions that conduct electricity well. In these solutions, we conclude that the solute is present as ions, surrounded by water molecules.

Ionic solids dissolve in water to give a solution of cations and anions

If we dissolves 1.0 mol of $SrCl_2$ in water to give 1.0 L of solution, the solution will contain Sr^{2+} ions and Cl^- ions. The fact that an ion is in aqueous solution is usually stated by labeling the elemental symbol with an "*aq*" in parentheses.

$$1.0 \text{ mol } SrCl_2 \text{ (solid)} \xrightarrow{\text{water to give 1.0L}} 1.0 \text{ mol } Sr^{2+}(aq) + 2.0 \text{ mol } Cl^-$$

Since matter is conserved, there will be 1.0 moles of of Sr^{2+} ions in solution and 2.0 moles of Cl^- ions in solution.

$$[Sr^{2+}] = \frac{1.0 \text{ mol } Sr^{2+}}{1.0 L} = 1.0 \text{ M} \qquad [Cl^-] = \frac{2.0 \text{ mol } Cl^-}{1.0 L} = 2.0 \text{ M}$$

The solution may be described as a 1.0 M $SrCl_2$ solution, but for ionic compounds, it is important to remember the solution contains ions dissolved in water, which may or may not have the same concentration as the named concentration of the solution.

Solution Preparation

In preparing a solution of known molarity, one needs to first determine the number of moles of solute required to obtain the desired concentration. This amount of solute is dissolved in a small amount of solvent and then solvent is added to give the final, desired volume.

Exercise 3.21 We wish to make 250 mL of a solution 0.150 M in Ca^{2+} using calcium chloride dihydrate as the source of calcium. Describe how this would be done.

Steps to Solution: The question asks for the procedure to prepare a solution. We need first to determine the number of moles of solute required and then the number of grams required.

$$\text{mol solute} = M(V) = 0.150 \frac{\text{mol Ca}^{2+}}{L}(250\text{mL})\frac{1L}{1000\text{mL}} = 3.75 \times 10^{-2} \text{ mol Ca}^{2+}$$

Calcium chloride dihydrate is the source of calcium; the formula is $CaCl_2 \cdot 2H_2O$. The molar mass of this ionic compound is found to be 147.017 g/mol. Because there is one Ca^{2+} ion per $CaCl_2 \cdot 2H_2O$, the number of moles of Ca^{2+} and $CaCl_2 \cdot 2H_2O$ are the same.

$$\text{g } CaCl_2 \cdot 2H_2O = 3.75 \times 10^{-2} \text{ mol Ca}^{2+} (147.017 \text{ g/mol}) = 5.51 \text{ g } CaCl_2 \cdot 2H_2O$$

Having determined the mass of solute required, we need to weigh it out and then prepare the solution. We dissolve the 5.51 g $CaCl_2 \cdot 2H_2O$ in water (much than 250 mL) and then add water until the total volume of the solution is 250 mL using a volumetric flask (as described in the text).

COMMON PITFALL: Adding too much solvent.
We do not prepare a solution of known molarity by adding the target volume of solvent to the solute. For example, to prepare an 0.10 M solution of fructose, we do not add 0.10 mol fructose to 1.0 L of water because the final volume would be greater than 1.0 L. We dissolve the solid in a smaller amount of water and then add water to give a total volume of 1.0 L.

Dilutions

Solution of high concentrations are often described as **concentrated**; solutions of lower concentrations are called **dilute**. Commercially, solution reagents are sold and shipped at the highest concentration feasible. However, these concentrations are too high for most laboratory uses and these solutions must be converted to lower concentration solutions. The process of obtaining lower concentration solutions from more concentrated solutions is called **dilution**.

Chapter 3: The Composition of Molecules

In preparing a dilute solution, one has a desired molarity, M_D, and volume, V_D, of solution to prepare. The number of mols of solute contained in this solution is:

$$\text{mol solute} = M_D V_D$$

The number of moles of solute contained in the amount of concentrated solution to be used must be the same. The concentration of the concentrated solution, M_C, and the volume of the concentrated solution, V_C, must also be equal to the number of mols of solute desired in the solution:

$$\text{mol solute} = M_C V_C$$

The number of mols of solute is the same for both solutions; the **dilution formula** is:

$$\text{mol solute} = M_C V_C = M_D V_D \quad \text{or} \quad M_C V_C = M_D V_D$$

The use of this formula is illustrated in the next exercise.

Exercise 3.22 Nitric acid is a corrosive acid capable of dissolving many metals, with the evolution of brown nitrogen dioxide gas as a byproduct. It is often sold as concentrated nitric acid which has [HNO$_3$] = 16. M. A common concentration used in the laboratory is 6.0 M HNO$_3$. If you need 20. L of 6.0 M HNO$_3$, what volume of concentrated nitric acid is required?

Steps to Solution: We will solve this problem in two ways. (i) First, we will explicitly show the equality of the number of mols of solute. (ii) Second, we will use a shorter method which does not explicitly emphasize the equal number of mols of solute.

(i) We need to calculate the number of mols of HNO$_3$ contained in the dilute solution, because that number of moles will be contained in the volume of concentrated acid.

$$\text{mol HNO}_3 = M(V) = 6.0 \text{ mol/L } (20.\text{ L}) = 120 \text{ mol HNO}_3$$

The volume of concentrated solution required must also contain 120 mol HNO$_3$, so

$$120 \text{ mol HNO}_3 = M_C V_C$$

Rearranging the equation to solve for V_C we obtain

$$V_C = \frac{120 \text{ mol HNO}_3}{M_C} = \frac{120 \text{ mol HNO}_3}{16.\text{ mol /L}} = \boxed{7.5 \text{ L}}$$

(ii) Alternatively, we can rearrange the dilution equation to solve directly for the volume of concentrated acid required:

Chapter 3: The Composition of Molecules

$$M_C V_C = M_D V_D$$

$$V_C = \frac{M_D V_D}{M_C} = \frac{6.0\ M\ (20.\ L)}{16.\ M} = \boxed{7.5\ L}$$

The volume of concentrated nitric acid required is 7.5 L. This answer is reasonable because it is smaller than the volume of dilute acid required.

In preparing a solution from concentrated acid, you never add water to the concentrated acid. Diluting acids releases enough heat to boil water and cause splashing or spraying of the acid solution in preparation. You add the concentrated acid to the water (less than the total volume desired), mix and then add water to the desired volume. This mnemonic reminds you of the safe way to dilute acids:

Do like you oughter: add acid to water

Precipitation Analysis

Some elements cannot be analyzed by combustion analysis because they do not give a single product or because of other problems. One alternative route for analysis is to convert the element to an ion that will react with another ion of opposite charge to give an insoluble product. The insoluble product can be separated from the solution, dried to remove the solvent and then its mass determined. The mass of the product can then be converted to moles which, with the use of the chemical formula, will tell us the number of moles and subsequently the mass of the ion we precipitated, as illustrated by the flow chart shown below.

$$\text{compound} \rightarrow \text{ions} + \text{other ions} \rightarrow \text{ionic solid}$$

$$\text{mass ionic solid} \rightarrow \text{moles ionic solid} \rightarrow \text{moles ion being analyzed} \rightarrow \text{mass ion being analyzed}$$

The following exercise illustrates this method.

Exercise 3.23 A 1.345 g sample of solid residue from a hazardous waste dump is being analyzed for barium. The sample is dissolved and then sodium sulfate is added. The insoluble barium sulfate is dried, and a total of 73.8 mg of $BaSO_4$ is collected. What percent of the sample is barium?

Steps to Solution: The percent of barium in the sample is given by the formula:

$$\%\ Ba = \frac{g\ Ba}{g\ sample}\ 100\%$$

Given the g of the sample we must find the g of Ba. We know the mass of $BaSO_4$ collected; therefore we can determine the mass of barium collected:

$$g\ BaSO_4 \rightarrow mol\ BaSO_4 \rightarrow mol\ Ba \rightarrow g\ Ba$$

$$g\ Ba = mol\ Ba\ (MM\ Ba)$$

$$mol\ Ba = (mol\ BaSO_4)\frac{1\ mole\ Ba}{1\ mole\ BaSO_4} \quad and \quad mol\ BaSO_4 = \frac{g\ BaSO_4}{MM\ BaSO_4}$$

Combining these expressions,

$$g\ Ba = \frac{g\ BaSO_4}{MM\ BaSO_4}(MM\ Ba) = \frac{73.8\ mg}{233.39\ g/mol}\frac{(1g)}{(1000mg)}(137.33\ g/mol)$$

$$g\ Ba = 4.34 \times 10^{-2}\ g\ Ba$$

$$\%\ Ba = \frac{4.34 \times 10^{-2}\ g\ Ba}{1.345\ g}\ 100\% = \boxed{3.23\ \%\ Ba}$$

Test Yourself

1. A compound containing only sulfur and fluorine with a molar mass of 146.0 g/mol has an elemental composition of % F = 78.05 % and % S = 21.95 %. What are the empirical and chemical formulas and the name of this compound?

2. How many moles of (a) Sr^{2+} ions and (b) Cl^- ions are contained in 51.5 mL of 0.23 M $SrCl_2$?

3. Determine the chemical formula and elemental % composition of the following compound:

4. How many mL of 0.75 M NaCl would be required to prepare 1.5 L of 2.0×10^{-3} M NaCl? How would this solution be prepared?

Answers

1. The empirical and molecular formulas are the same: SF_6. Its name is sulfur hexafluoride.

2. (a) 1.18×10^{-2} mol Sr^{2+}; (b) 2.37×10^{-2} mol Cl^-.

3. $C_{11}H_{17}NO_3$; % H = 8.11, % C = 62.54 %, % N = 6.63 %, % O = 22.72 %.

4. Measure 4.0 mL of 0.75 M NaCl and add water to give 1.5 L of solution.

sulfate	ammonium	chloride	cyanide
carbonate	nitrite	hydronium	nitrate
chlorate	acetate	hydroxide	phosphate
potassium	sodium	sulfite	permanganate
aluminum	dihydrogen-phosphate	hydrogen-carbonate	calcium

Chapter 3: The Composition of Molecules

CN^-	Cl^-	NH_4^+	SO_4^{2-}
NO_3^-	H_3O^+	NO_2^-	CO_3^{2-}
PO_4^{3-}	OH^-	$CH_3CO_2^-$	ClO_3^-
MnO_4^-	SO_3^-	Na^+	K^+
Ca^{2+}	HCO_3^-	$H_2PO_4^-$	Al^{3+}

Chapter 3: The Composition of Molecules

Solutions to Chapter Problems

3.1 Determine the chemical formula from a ball-and-stick model by counting balls of each color and consulting the color code for the elements (Figure 1-4 of your textbook).

(a) CH_4; (b) C_2H_4; (c) C_2H_6O; (d) HBr; (e) PCl_3; (f) CH_4N_2O; (g) C_2H_5I; and (h) N_2H_4.

3.3 Structural formulas look like ball-and-stick structures but with lines in place of sticks and elemental symbols in place of balls.

3.5 To convert a line structure into a structural formula, place a C atom at the end of each line and at each line intersection, then add enough –H connections to give each C atom 4 connections.

3.7 To convert a structural formula into a line structure, remove all –H connections to carbon atoms and remove all C atom labels.

3.9 Determine a chemical formula from a space-filling model by counting atoms of each color and using the color code (Figure 1-4 of your textbook) to identify the elements. Name binary compounds using the rules given in Section 3-2 of your textbook.
(a) CO_2, carbon dioxide; (b) HCl, hydrogen chloride; and (c) CCl_4, carbon tetrachloride.

3.11 Determine a chemical formula from a ball-and-stick model by counting atoms of each color and using the color code (Figure 1-4 of your textbook) to identify the elements. Name binary compounds using the rules given in Section 3-2 of your textbook.
(a) CH_4, methane; (b) O_2, oxygen; and (c) PCl_3, phosphorus trichloride.

3.13 Identify the compound as binary or carbon-based. Then apply the rules given in Section 3.2 of your textbook.
(a) CH_4; (b) HF; (c) CaH_2; (d) PCl_3; (e) N_2O_5; (f) SF_6; (g) $C_5H_{12}O$ (note: in the structural formula, an –OH group is attached to the middle C atom); and (h) BF_3.

3.15 Identify the compound as binary or carbon-based. Then apply the rules given in Section 3.2 of your textbook. (a) disulfur dichloride; (b) iodine heptafluoride; (c) hydrogen bromide; (d) dinitrogen trioxide; (e) silicon carbide; (f) methanol.

3.17 Ionic compounds can be identified from the presence of either metals from Column 1 or Column 2 of the periodic table or polyatomic ions. Any compound is neutral, so the formula for an ionic compound must contain equal numbers of positive and negative charges.
(a) not ionic, HF; (b) Column 2 metal (+2) and Column 17 element (-1), ionic, CaF_2, ; (c) polyatomic anion (-2) and +3 metal, ionic, $Al_2(SO_4)_3$; (d) ammonium cation (+1) and Column 16 element (-2), ionic, $(NH_4)_2S$; (e) not ionic, SO_2; and (f) not ionic, CCl_4.

3.19 Identify the compound as ionic, binary or carbon-based. Then apply the rules given in Section 3.2 of your textbook. Ionic compounds can be identified from the presence of either metals from Column 1 or Column 2 of the periodic table or polyatomic ions.
(a) not ionic, carbon-based, dichloromethane; (b) not ionic, binary, carbon dioxide; (c) Column 2 metal, ionic, calcium oxide; (d) polyatomic anion, ionic, potassium carbonate; (e) not ionic, binary, phosphorus tribromide; (f) not ionic, binary, hydrogen fluoride; and
(g) polyatomic anion, ionic, sodium hydrogenphosphate.

Chapter 3: The Composition of Molecules

3.21 To determine a chemical formula from the name of a compound requires knowledge of the polyatomic anions and charges. The compound must be electrically neutral.

(a) +1 cation and -2 anion, Na_2SO_4; (b) +1 cation and -2 anion, K_2S;
(c) +1 cation and -1 anion, KH_2PO_4; (d) +2 cation, -1 anion, $CaF_2 \cdot 4H_2O$;
(e) +4 metal and -2 anion, PbO_2; (f) +1 cation and -1 anion, $NaHCO_3$; and
(g) +1 cation and -1 anion, $LiBrO_4$.

3.23 Follow the rules for naming ionic compounds (cation first, then anion), using Roman numerals for cation charge only when multiple possibilities exist.

(a) calcium chloride hexahydrate; (b) iron(II) ammonium sulfate; (c) potassium carbonate; (d) tin(II) chloride; (e) sodium hypochlorite; (f) silver sulfate; (g) copper(II) sulfate;
(h) potassium dihydrogenphosphate; (i) sodium nitrate; (j) calcium sulfite; and
(k) potassium permanganate.

3.25 Mass-mole-number conversions require the use of masses in grams, molar masses, and/or Avogadro's number.

(a) $n = m/MM = (7.85 \text{ g}) \div (55.85 \text{ g/mol}) = 0.141$ mol;
(b) $n = \#/N_A = (6.55 \times 10^{13} \text{ atoms}) \div (6.022 \times 10^{23} \text{ atoms/mol}) = 1.09 \times 10^{-10}$ mol;
(c) $n = m/MM = (4.68 \text{ μg})(10^{-6} \text{ g/μg}) \div (28.09 \text{ g/mol}) = 1.67 \times 10^{-7}$ mol; and
(d) $n = m/MM = (1.46 \text{ ton})(10^3 \text{ kg/ton})(10^3 \text{ g/kg}) \div (26.98 \text{ g/mol}) = 5.41 \times 10^4$ mol.

3.27 The molar mass of a naturally occurring element can be calculated by summing the product of the fractional abundance of each isotope times its isotopic molar mass.

^{36}Ar: $(0.337 \%)(10^{-2}/\%)(35.96755 \text{ g/mol}) = 0.121$ g/mol;
^{38}Ar: $(0.063 \%)(10^{-2}/\%)(37.96272 \text{ g/mol}) = 0.024$ g/mol;
^{40}Ar: $(99.60 \%)(10^{-2}/\%)(39.9624 \text{ g/mol}) = 39.803$ g/mol;
Elemental molar mass = $0.121 + 0.024 + 39.804 = 39.948$ g/mol.

3.29 To calculate the molar mass of a compound, multiply each elemental molar mass by the number of atoms in the formula and sum over the elements.
(a) CCl_4, $MM = (12.01 \text{ g/mol}) + 4(35.45 \text{ g/mol}) = 153.8$ g/mol;
(b) K_2S, $MM = 2(39.10 \text{ g/mol}) + 32.07 \text{ g/mol} = 110.27$ g/mol;
(c) O_3, $MM = 3(16.00 \text{ g/mol}) = 48.00$ g/mol;
(d) LiBr, $MM = 6.94 \text{ g/mol} + 79.90 \text{ g/mol} = 86.84$ g/mol;
(e) GaAs, $MM = 69.72 \text{ g/mol} + 74.92 \text{ g/mol} = 144.64$ g/mol; and
(f) $AgNO_3$, $MM = 107.87 \text{ g/mol} + 14.01 \text{ g/mol} + 3(16.00 \text{ g/mol}) = 169.88$ g/mol.

Chapter 3: The Composition of Molecules

3.31 Determine the molecular formula from the line drawing, taking into account the "missing" carbon atoms at the ends and vertices of lines and the "missing" hydrogen atoms attached to carbon atoms. To calculate MM, multiply each elemental MM by the number of atoms in the formula and sum over the elements.

Tyrosine: 9 C atoms, 7 missing H atoms, 4 shown H atoms, 1 N atom, 3 O atoms;
$C_9H_{11}NO_3$, $MM = 9(12.01) + 11(1.008) + 1(14.01) + 3(16.00) = 181.19$ g/mol;
Lysine: 6 C atoms, 9 missing H atoms, 5 shown H atoms, 2 N atoms, 2 O atoms;
$C_6H_{14}N_2O_2$, $MM = 6(12.01) + 14(1.008) + 2(14.01) + 2(16.00) = 146.19$ g/mol;
Glutamic acid: 5 C atoms, 5 missing H atoms, 4 shown H atoms, 1 N atom, 4 O atoms;
$C_5H_9NO_4$, $MM = 5(12.01) + 9(1.008) + 1(14.01) + 4(16.00) = 147.13$ g/mol;
Tryptophan: 10 C atoms, 6 missing H atoms, 4 shown H atoms, 2 N atoms, 2 O atoms;
$C_{10}H_{10}N_2O_2$, $MM = 10(12.01) + 10(1.008) + 2(14.01) + 2(16.00) = 190.20$ g/mol.

3.33 To calculate the number of atoms in a mass, convert the mass to grams, divide by molar mass to obtain moles and multiply by Avogadro's number to obtain number.

(a) # = $(5.86\ \mu g)(10^{-6}\ g/\mu g)(6.022 \times 10^{23}\ atoms/mol) \div (9.01\ g/mol) = 3.92 \times 10^{17}$ atoms;
(b) # = $(5.86\ \mu g)(10^{-6}\ g/\mu g)(6.022 \times 10^{23}\ atoms/mol) \div (30.97\ g/mol) = 1.14 \times 10^{17}$ atoms
(c) # = $(5.86\ \mu g)(10^{-6}\ g/\mu g)(6.022 \times 10^{23}\ atoms/mol) \div (91.22\ g/mol) = 3.87 \times 10^{16}$ atoms;
(d) # = $(5.86\ \mu g)(10^{-6}\ g/\mu g)(6.022 \times 10^{23}\ atoms/mol) \div (238.0\ g/mol) = 1.48 \times 10^{17}$ atoms

3.35 To calculate the mass of some number of molecules of a substance, divide by Avogadro's number to obtain moles and multiply by molar mass to obtain grams.

(a) CH_4, $MM = 16.04$ g/mol;
$m = (3.75 \times 10^5\ molecules)(16.04\ g/mol) \div (6.022 \times 10^{23}\ molec/mol) = 9.99 \times 10^{-18}$ g;
(b) $C_9H_{13}NO_3$, $MM = 183.2$ g/mol;
$m = (2.5 \times 10^9\ molecules)(183.2\ g/mol) \div (6.022 \times 10^{23}\ molec/mol) = 7.6 \times 10^{-13}$ g;
(c) $C_{55}H_{72}MgN_4O_5$, $MM = 893.5$ g/mol;
$m = (1\ molecule)(893.5\ g/mol) \div (6.022 \times 10^{23}\ molec/mol) = 1.484 \times 10^{-21}$ g;

3.37 Moles and mass in grams are related through the equation, $n = m/MM$. When masses are not given in grams, unit conversions must be made.

$MM (H_3PO_4) = 98.00$ g/mol; 1 lb = 453.6 g;
$n = (26.19 \times 10^7\ lb)(453.6\ g/lb) \div (98.00\ g/mol) = 1.212 \times 10^9$ mol;
The chemical formula shows that there is one mole of P in every mole of H_3PO_4. The problem states that 15% of the annual production of H_3PO_4 comes from elemental P:
$n(P) = (15\%)(10^{-2}/\%)(1.212 \times 10^9\ mol) = 1.818 \times 10^8$ mol;
$m(P) = n\ MM = (1.818 \times 10^8\ mol)(30.97\ g/mol)(10^{-3}\ kg/g) = 5.6 \times 10^6$ kg

Chapter 3: The Composition of Molecules

3.39 To calculate mass percent composition, which involves mass ratios, work with one mole of substance. Take the ratio of the mass of each element that one mole contains to the mass of one mole (molar mass).
CaO: MM = 40.08 g/mol + 16.00 g/mol = 56.08 g/mol;
% Ca = (100%)(40.08) ÷ (56.08) = 71.47 %;
% O = (100%)(16.00) ÷ (56.08) = 28.53 %;

SiO_2: MM = 28.09 g/mol + 2(16.00 g/mol) = 60.09 g/mol;
% Si = (100%)(28.09) ÷ (60.09) = 46.75 %;
% O = (100%)(2)(16.00) ÷ (60.09) = 53.25 %;

Al_2O_3: MM = 2(26.98 g/mol) + 3(16.00 g/mol) = 101.96 g/mol;
% Al = (100%)(2)(28.09) ÷ (101.96) = 52.92 %;
% O = (100%)(3)(16.00) ÷ (101.96) = 47.08 %;

Fe_2O_3: MM = 2(55.85 g/mol) + 3(16.00 g/mol) = 159.70 g/mol;
% Fe = (100%)(2)(55.85) ÷ (159.70) = 69.94 %;
% O = (100%)(3)(16.00) ÷ (159.70) = 30.06 %.

3.41 When working with mass percent compositions, it is convenient to consider 100 g of substance. Divide the mass of each element by its elemental molar mass to obtain relative amounts of each element in the substance. Then divide each amount by the smallest among them and round to whole numbers to obtain the empirical formula. To convert an empirical formula to a molecular formula, multiply the numbers in the empirical formula by the ratio of the compound's molar mass to its empirical formula mass.

C: (74.0 g/100 g) ÷ (12.01 g/mol) = 6.1615 mol C/100 g;
H: (8.65 g/100 g) ÷ (1.008 g/mol) = 8.5813 mol H/100 g;
N: (17.35 g/100 g) ÷ (14.01 g/mol) = 1.2384 mol N/100 g;

Divide each by the smallest among them, 1.2384 mol N/100 g:
C: (6.1615 mol C/100 g) ÷ (1.2384 mol N/100 g) = 4.975 mol C/mol N, round to 5 C/N;
H: (8.5813 mol H/100 g) ÷ (1.2384 mol N/100 g) = 6.929 mol H/mol N, round to 7 H/N;

Empirical formula: C_5H_7N, empirical MM = 81.12 g/mol;
MM/empirical MM = (162 g/mol) ÷ (81.12 g/mol) = 2; Molecular formula: $C_{10}H_{14}N_2$.

3.43 This is an empirical formula problem involving combustion analysis. Do a mass analysis of the combustion data to obtain the mass percent composition of the compound. Determine the mass percent oxygen by difference. Once the mass percentages are known, use elemental molar masses to determine relative amounts of the elements.
% N = 3.8 % (stated in the problem);
m C = (97.46 mg CO_2)(12.01 g/mol C) ÷ (44.01 g/mol CO_2) = 26.60 mg;
% C = (100 %)(26.60 mg) ÷ (38.7 mg) = 68.7 %;
m H = (20.81 mg H_2O)(2)(1.008 g/mol H) ÷ (18.02 g/mol H_2O) = 2.329 mg;
% H = (100 %)(2.329 mg) ÷ (38.7 mg) = 6.02 %;

% O = 100 % − (3.8 % + 68.7 % + 6.02 %) = 21.5 %;
C: (68.7 g) ÷ (12.01 g/mol) = 5.72 mol C; H: (6.02 g) ÷ (1.008 g/mol) = 5.97 mol H;
N: (3.8 g) ÷ (14.01 g/mol) = 0.271 mol N; O: (21.5 g) ÷ (16.00 g/mol) = 1.34 mol O;
Divide each by the smallest among them, 0.271 mol N:
(5.72 mol C) ÷ (0.271 mol N) = 21.11 C/N, round to 21;
(5.97 mol H) ÷ (0.271 mol N) = 22.0 H/N, round to 22;
(1.34 mol O) ÷ (0.271 mol N) = 4.94 O/N, round to 5;
These values match those for heroin, $C_{21}H_{22}O_5N$.

3.45 The solution process is $KOH_{(s)} \rightarrow K^+_{(aq)} + OH^-_{(aq)}$. Each mole of solid generates one mole of each ion.

(a) Molarity is found using the equation, $M = n/V$.
$M = n/V$ = (4.75 g) ÷ (56.11 g/mol)(275 mL)(10^{-3} L/mL) = 0.308 M = [K^+] = [OH^-];

(b) In a dilution, the number of moles of solute remains constant while volume increases, so $M_1V_1 = M_2V_2$. The starting volume is 25.00 mL and the final volume is 100.00 mL
$M_2 = M_1V_1/V_2$ = (0.308 M)(25.00 mL) ÷ (100.00 mL) = 0.077 M = [K^+] = [OH^-];

(c) Molecular pictures of solutions must illustrate the relative numbers of ions of each type present in the solution. Because of the four-fold dilution, the solution in (b) is 1/4 as concentrated as the solution in (a). We omit solvent molecules for clarity, but remember that there are many more molecules of water than of either of the ions.

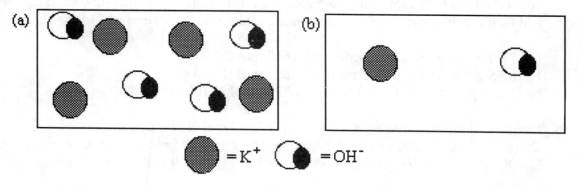

3.47 The solution process is $MgCl_{2(s)} \rightarrow Mg^{2+}_{(aq)} + 2\,Cl^-_{(aq)}$. Each mole of solid generates one mole of magnesium cations and two moles of chloride anions.

(a) Molarity is found using the equation, $M = n/V$. MM_{MgCl_2} = 95.24 g/mol.
$M_{Mg^{2+}}$ = (4.68 g) ÷ (95.24 g/mol)(150 mL)(10^{-3} L/mL) = 0.328 M;
$M_{Cl^-} = 2\,M_{Mg^{2+}})$ = 0.656 M.

(b) Molecular pictures of solutions must illustrate the relative numbers of ions of each type present in the solution. Your picture must show twice as many chloride ions as magnesium ions. There are many more molecules of water than of either of the ions.

3.49 When working empirical formula problems that involve multiple analyses, it is convenient to determine the mass percent composition from the analyses and then determine the empirical formula from the mass percent composition. Each analysis has its own mass of starting sample.
m C = (4.34 g CO_2)(12.01 g/mol C) ÷ (44.01 g/mol CO_2) = 1.184 g;
% C = (100 %)(1.184 g) ÷ (4.00 g) = 29.61 %;
m Cl = (0.334 g AgCl)(35.45 g/mol Cl) ÷ (143.32 g/mol AgCl) = 0.08267 g;
% Cl = (100 %)(0.08267 g) ÷ (0.125 g) = 66.09 %;
% H = 100 % − (29.61 % + 66.09 %) = 4.30 %;

C: (29.61 g) ÷ (12.01 g/mol) = 2.47 mol C; H: (4.30 g) ÷ (1.008 g/mol) = 4.27 mol H;
Cl: (66.09 g) ÷ (35.45 g/mol) = 1.86 mol Cl;

Divide each by the smallest among them, 1.86 mol Cl:
(2.47 mol C) ÷ (1.86 mol Cl) = 1.33 C/Cl;
(4.27 mol H) ÷ (1.86 mol Cl) = 2.30 H/Cl;
Multiply all values by 3 to get round numbers: 3(1.33) = 4; 3(2.30) = 6.9, round to 7;
The empirical formula is $C_4H_7Cl_3$.

3.51 Use Avogadro's number to convert from moles to number of atoms. Multiply the total number of atoms by the atomic diameter to find the total length.
l = (1 mol)(6.022 x 10^{23} atoms/mol)(127.8 pm/atom)(10^{-12} m/pm) = 7.696 x 10^{13} m.

3.53 Use nomenclature rules to determine a name from a chemical formula.

(a) ammonium chloride; (b) xenon tetrafluoride; (c) iron(III) oxide; (d) sulfur dioxide; and
(e) potassium perchlorate.

3.55 To calculate mass percent composition, which involves mass ratios, work with one mole of substance. Take the ratio of the mass of each element that one mole contains to the mass of one mole (molar mass).
Fe_2SiO_4: MM = 2(55.85 g/mol) + 28.09 g/mol + 4(16.00 g/mol) = 203.79 g/mol;
% Fe = (100%)(2)(55.85) ÷ (203.79) = 54.81 %;
% Si = (100%)(28.09) ÷ (203.79) = 13.78 %;

%O = (100%)(4)(16.00) ÷ (203.79) = 31.40 %;

NaAlSi$_3$O$_8$: MM = 22.99 + 26.98 + 3(28.09) + 8(16.00) = 262.24 g/mol;
% Na = (100%)(22.99) ÷ (262.24) = 8.77 %;
% Al = (100%)(26.98) ÷ (262.24) = 10.29 %;
% Si = (100%)(3)(28.09) ÷ (262.24) = 32.13 %;
% O = (100%)(4)(16.00) ÷ (262.24) = 24.41 %;

Al$_2$Si$_2$O$_5$(OH)$_4$: MM = 2(26.98) + 2(28.09) + 5(16.00) + 4(16.00 + 1.008) = 275.18 g/mol;
% Al = (100)(2)(26.98) ÷ (275.18) = 19.61 %;
% Si = (100)(2)(28.09) ÷ (275.18) = 20.42 %;
% O = (100)(9)(16.00) ÷ (275.18) = 52.33 %;
% H = (100)(5)(1.008) ÷ (275.18) = 1.83 %;

MgSi$_4$O$_{10}$(OH)$_8$: MM = 24.31 + 4(28.09) + 10(16.00) + 8(16.00 + 1.008) = 432.73 g/mol;
% Mg = (100)(24.31) ÷ (432.73) = 5.62 %;
% Si = (100)(4)(28.09) ÷ (432.73) = 25.97 %;
% O = (100)(18)(16.00) ÷ (432.73) = 66.55 %;
% H = (100)(8)(1.008) ÷ (432.73) = 1.86 %;

3.57 The question asks about mole-mass-number quantities. Although there is much interesting biomedical information provided, the only relevant data are the chemical formula of Verapamil, C$_{27}$H$_{38}$O$_4$N$_2$, and the amount of Verapamil per tablet, 120.0 mg
(a) MM = 27(12.01) + 38(1.008) + 4(16.00) + 2(14.01) = 454.59 g/mol;
(b) $n = m/MM$ = (120.0 mg)(10^{-3} g/mg) ÷ (454.59 g/mol) = 2.640 x 10^{-4} mol;
(c) # atoms = n N_A (atoms/molecule)
= (2.640 x 10^{-4} mol)(6.022 x 10^{23}/mol)(2 atoms N/molecule) = 3.180 x 10^{20} atoms N.

3.59 The statement of this question identifies it as an empirical formula problem. Use the decomposition data to determine the mass percent composition of the compound, then use elemental molar masses to find the empirical formula.
% Hg = (100 %)(0.302 g Hg/0.350 g sample) = 86.3 %;
m S = (0.0964 g SO$_2$)(32.07 g/mol S) ÷ (64.07 g/mol SO$_2$) = 0.04825 g;
% S = (100 %)(0.04825 g S/0.350 g sample) = 13.8 %;
86.3 % + 13.8 % = 100.1 %, so the compound contains no oxygen;
Hg: (86.3 g) ÷ (200.6 g/mol) = 0.430 mol Hg; S: (13.8 g) ÷ (32.07 g/mol) = 0.430 mol S;
The compound contains equal numbers of moles of each element, so its formula is HgS.

3.61 To write correct chemical formulas, you must know the compositions and charges of polyatomic ions. The overall formula must be electrically neutral.
(a) NaNO$_2$ and NaNO$_3$; (b) K$_2$CO$_3$ and KHCO$_3$; (c) FeO and Fe$_2$O$_3$; and (d) I$_2$ and I$^-$.

3.63 The chemical formula of a substance provides all the information needed to compute its molar characteristics.
(a) $MM = 2(26.98) + 3[32.07 + 4(16.00)] = 342.17$ g/mol;
(b) $n = m/MM = (25.0$ g$) \div (342.17$ g/mol$) = 7.31 \times 10^{-2}$ mol;
(c) % Al $= (100)(2)(26.98) \div (342.17) = 15.77$ %;
% S $= (100)(3)(32.07) \div (342.17) = 28.12$ %; % O $= (100)(12)(16.00) \div (342.17) = 56.11$ %;
(d) There are 12 O atoms per formula unit, so $m = MM \div 12 = 28.5$ g.

3.65 To convert a line structure into a structural formula, place a C atom at the end of each line and at each line intersection, then add enough –H connections to give each C atom four connections. From the structural formula, determine the chemical formula by counting and calculate the molar mass from numbers of atoms and atomic molar masses.

benzaldehyde, C_7H_6O, $MM = 106.05$ g/mol;
methylbutyl acetate, $C_7H_{14}O_2$, $MM = 130.18$ g/mol;
jasmone, $C_{11}H_{16}O$, $MM = 164.24$ g/mol; limonene, $C_{10}H_{16}$, $MM = 136.23$ g/mol; and vanillin, $C_8H_8O_3$, $MM = 152.14$ g/mol.

Chapter 3: The Composition of Molecules

3.67 A six-carbon straight chain has three different kinds of carbon atoms: ones on the end (1-position), ones next to the end (2-position), and ones nearest the middle (3-position). Attaching an –OH group to each position in turn gives three different isomers.

1-hexanol   2-hexanol   3-hexanol

3.69 This problem poses a set of mole-mass-number conversion questions. From Sample Problem 3-13, Sevin is $C_{12}H_{11}NO_2$, MM = 201.2 g/mol.
(a) n_C = (C atoms/molecule) m/MM = (12)(8.3 g) ÷ (201.2 g/mol) = 0.50 mol;
(b) m_O = ($m_{compound}$)(% O)(10^{-2}/%) = (4.5 g)(2)(16.00) ÷ (201.2) = 0.72 g;
(c) n = (75 mL)(1.00 g/mL)(0.010 g/100 g) ÷ (201.2 g/mol) = 3.7 x 10^{-5} mol;
= $n\,N_A$ = (3.7 x 10^{-5} mol)(6.022 x 10^{23} molecules/mol) = 2.2 x 10^{19} molecules;
(d) 15 gallons of spray requires 15 mL of insecticide. The insecticide contains 3.7 x 10^{-5} mol/75 mL, so n = (15 mL)(3.7 x 10^{-5} mol/75 mL) = 7.5 x 10^{-6} mol.

3.71 To write correct chemical formulas, you must know the compositions and charges of polyatomic ions. The overall formula must be electrically neutral.
(a) $CaBr_2$; (b) $NaBrO_3$; (c) $NaBrO_2$; and (d) $NaHCO_3$.

3.73 The statement of this question identifies it as an empirical formula problem. Use the combustion data to determine the mass percent composition of the compound, then use elemental molar masses to find the empirical formula.

m C = (11.8 g CO_2)(12.01 g/mol C) ÷ (44.01 g/mol CO_2) = 3.22 g;
% C = (100 %)(3.22 g) ÷ (5.00 g) = 64.4 %;
m H = (2.42 g H_2O)(2)(1.008 g/mol H) ÷ (18.02 g/mol H_2O) = 0.2707 g;
% H = (100 %)(0.2707 g) ÷ (5.00 g) = 5.41 %;
% Fe = 100 % – (64.4 % + 5.41 %) = 30.2 %;

C: (64.4 g) ÷ (12.01 g/mol) = 5.36 mol C; H: (5.41 g) ÷ (1.008 g/mol) = 5.37 mol H;
Fe: (30.2 g) ÷ (55.85 g/mol) = 0.541 mol Fe;

Divide each by the smallest among them, 0.5.41 mol Fe:
(5.36 mol C) ÷ (0.541 mol Fe) = 9.91 C/Fe, round to 10;
(5.37 mol H) ÷ (0.541 mol Fe) = 9.93 H/Fe, round to 10;
The empirical formula is $C_{10}H_{10}Fe$.

3.75 Chemical names follow the standard rules for chemical nomenclature.

(a) carbon dioxide; (b) potassium nitrate; (c) sodium chloride;
(d) sodium hydrogencarbonate; (e) sodium carbonate; (f) sodium hydroxide;
(g) calcium oxide; and (h) magnesium hydroxide.

Chapter 3: The Composition of Molecules

3.77 There is much interesting information provided about allicin, but the question asks you to determine its molecular formula, for which the essential data are the combustion analysis and the approximate molar mass.
m C = (8.13 mg CO_2)(12.01 g/mol C) ÷ (44.01 g/mol CO_2) = 2.219 mg;
% C = (100 %)(2.219 mg) ÷ (5.00 mg) = 44.37 %;
m H = (2.76 mg H_2O)(2)(1.008 g/mol H) ÷ (18.02 g/mol H_2O) = 0.3088 mg;
% H = (100 %)(0.3088 mg) ÷ (5.00 mg) = 6.18 %;
m S = (3.95 mg SO_2)(32.07 g/mol S) ÷ (64.07 g/mol SO_2) = 1.977 mg;
% S = (100 %)(1.977 mg) ÷ (5.00 mg) = 39.54 %;
% O = 100 % − (44.37 % + 6.18 % + 39.54 %) = 9.91 %;

C: (44.37 g) ÷ (12.01 g/mol) = 3.69 mol C; H: (6.18 g) ÷ (1.008 g/mol) = 6.13 mol H;
S: (39.54 g) ÷ (32.07 g/mol) = 1.23 mol S; O: (9.91 g) ÷ (16.00 g/mol) = 0.619 mol O
Divide each by the smallest among them, 0.619 mol O:
(3.69 mol C) ÷ (0.619 mol O) = 5.96 C/O, round to 6;
(6.13 mol H) ÷ (0.619 mol O) = 9.90 H/O, round to 10;
(1.23 mol S) ÷ (0.619 mol O) = 1.99 S/O, round to 2;
The empirical formula is $C_6H_{10}S_2O$, with an empirical molar mass of 162 g/mol. This is nearly equal to the experimental molar mass, so the molecular formula is the same as the empirical formula.

3.79 When salts dissolve in water, they dissociate into cations and anions. Molecular substances retain their structure in solution, and H_2O is always present as a species in aqueous solutions.
(a) H_2O, NH_4^+, SO_4^{2-}; (b) H_2O, CO_2; (c) H_2O, Na^+, F^-; (d) H_2O, K^+, CO_3^{2-};
(e) H_2O, Na^+, HSO_4^-; and (f) H_2O, Cl_2.

3.81 This question asks about solution concentration. The solution in question is prepared in two stages: First a solid is dissolved in water, then a sample of the resulting solution is diluted with more water. Calculate the concentration in two steps:

First solution: $M = n/V$ and $n = m/MM$.
For $KMnO_4$, MM = 39.098 + 54.938 + 4(15.999) = 158.033 g/mol.
n = (474.1 g) ÷ (158.033 g/mol) = 3.000 mol;
M = (3.000 mol) ÷ (2.000 L) = 1.500 mol/L.
Second solution: $M_2V_2 = M_1V_1$, so $M_2 = M_1V_1/V_2$.
M_2 = (1.500 mol/L)(50.00 mL)(10^{-3} L/mL) ÷ (1.500 L) = 5.000 × 10^{-2} mol/L.

3.83 Use the definition of mass percentage composition to compute the molar mass:
% Co = (100 %)(atoms Co/molecule)(MM_{Co}) ÷ ($MM_{molecule}$).
MM = (100 %)(1)(58.93 g/mol) ÷ (4.34 %) = 1.36 × 10^3 g/mol.

3.85 Convert from mass to moles to determine molarity.
$M = n/V = m/MM\ V$.
M = (8.3 mg/L)(10^{-3} g/mg) ÷ (32.00 g/mol) = 2.6 × 10^{-4} mol/L.

Chapter 3: The Composition of Molecules

3.87 Calculate mass fractions using molar masses and the chemical formula.
f_C = (atoms C/molecule)(MM_C) ÷ ($MM_{compound}$).

(a) C_3H_8: f_C = 3(12.01) ÷ [3(12.01) + 8(1.008)] = 0.8172;
(b) C_8H_{18}: f_C = 8(12.01) ÷ [8(12.01) + 18(1.008)] = 0.8412;
(c) C_2H_5OH: f_C = 2(12.01) ÷ [2(12.01) + 6(1.008) + 16.00] = 0.5214;
(d) CH_4: f_C = 1(12.01) ÷ [12.01 + 4(1.008)] = 0.7487.

3.89 Calculate mass percentages in order to answer this question.

% N = (100 %)(atoms N/formula unit)(MM_N) ÷ ($MM_{compound}$)
$m_{compound}$ = (kg$_N$)(100 %) ÷ (% N).
NH_4NO_3: MM = 80.05 g/mol; % N = (100 %)(2)(14.01) ÷ (80.05) = 35.00 %;
m = (1 kg)(100 %) ÷ (35.00 %) = 2.857 kg;
$(NH_4)_2SO_4$: MM = 132.15 g/mol; % N = (100 %)(2)(14.01) ÷ (132.15) = 21.20 %;
m = (1 kg)(100 %) ÷ (21.20 %) = 4.716 kg;
$(NH_2)_2CO$: MM = 60.06 g/mol; % N = (100 %)(2)(14.01) ÷ (60.06) = 46.65 %;
m = (1 kg)(100 %) ÷ (46.65 %) = 2.143 kg;
$(NH_4)_2HPO_4$: MM = 132.06 g/mol; % N = (100 %)(2)(14.01) ÷ (132.06) = 21.22 %;
m = (1 kg)(100 %) ÷ (21.22 %) = 4.713 kg;

3.91 You should memorize the names and chemical formulas, including charges, of common polyatomic ions.

(a) IO_3^-; (b) IO_4^-; (c) PO_4^{3-}; (d) NO_3^-; (e) SO_4^{2-}; (f) $C_2H_3O_2^-$; and (g) HCO_3^-.

3.93 This is a mole-mass-number conversion problem.

ions = $M V N_A$ = (2.2 x 10^{-9} mol/L)(1 L)(6.022 x 10^{23} ions/mol) = 1.3 x 10^{15} ions;
Convert mass to mol, then use molarity to find the volume containing this amount of moles:
n = m/MM = (1.00 kg)(10^3 g/kg) ÷ (85.47 g/mol) = 11.70 mol;
V = n/M = (11.70 mol) ÷ (2.2 x 10^{-9} mol/L) = 5.3 x 10^9 L.

3.95 The question asks about mole-mass-number quantities. Although there is much interesting biomedical information provided, the only relevant data are the chemical formula of FK-506, $C_{44}H_{69}O_{12}N$, and the mass of the sample, 5.0 mg.

(a) MM = 44(12.01) + 69(1.008) + 12(16.00) + 1(14.01) = 804.0 g/mol;
(b) n = m/MM = (5 mg)(10^{-3} g/mg) ÷ (804.0 g/mol) = 6.2 x 10^{-6} mol;
(c) # atoms = $n N_A$ (atoms/molecule)
= (6.2 x 10^{-6} mol)(6.022 x 10^{23}/mol)(12 atoms O/molecule) = 4.5 x 10^{19} atoms O.

Chapter 3: The Composition of Molecules

3.97 To convert a structural formula into a line structure, remove all –H connections to carbon atoms and remove all C atom labels.

[Structures shown: 3-methylbutane thiol, trimethyl amine, cadaverine, pyridine]

3.99 To determine a molecular formula, mass percent composition and an approximate molar mass must be known. The data given are insufficient to calculate mass percent composition. To calculate the mass percentages of carbon and hydrogen, the mass of compound that was burned must be known. Unless the percentages of C and H together total 100 %, further information about the other elements present in the compound is also needed.

3.101 This is a "real-life" example of empirical formula analysis. Do a mass analysis of the combustion data to obtain the mass percent composition of the compound. Determine the mass percent oxygen by difference. Once the mass percentages are known, use elemental molar masses to determine relative amounts of the elements.

m C = (54.9140 – 54.4375 g CO_2)(12.01 g/mol C) ÷ (44.01 g/mol CO_2) = 0.1300 g;
% C = (100 %)(0.1300 g) ÷ (2.8954 – 2.7534 g) = 91.57 %;
m H = (47.9961 – 47.8845 g H_2O)(2)(1.008 g/mol H) ÷ (18.02 g/mol H_2O) = 0.01249 g;
% H = (100 %)(0.01249) ÷ (2.8954 – 2.7534 g) = 8.79 %;
% O = 100 % – (91.57 % + 8.79 %) = – 0.36 %, so the compound contains no O;
C: (91.57 g) ÷ (12.01 g/mol) = 7.62 mol C; H: (8.79 g) ÷ (1.008 g/mol) = 8.72 mol H;
Divide H by C: 8.72/7.62 = 1.144 multiply by 7 to obtain whole numbers:
7(1.144) = 8. The empirical formula is C_7H_8.

3.103 To prepare a solution of low concentration from a solution of higher concentration, a dilution is made using pure solvent. In this problem, we know the final concentrations and volume and the initial concentrations, so we need to calculate the initial volumes:
$M_2V_2 = M_1V_1$, so $V_1 = M_2V_2/M_1$
V_1(sodium acetate) = (0.30 M)(1.5 L) ÷ (5.0 M) = 0.090 L
V_1(acetic acid) = (0.15 M)(1.5 L) ÷ (5.0 M) = 0.045 L

To prepare the solution, the worker should mix 0.090 L of sodium acetate solution, 0.045 L of acetic acid solution, and enough water to make the final volume 1.5 L.

Chapter 4 Chemical Reactions and Stoichiometry

4.1 Writing Chemical Equations

QUESTIONS TO ANSWER, SKILLS TO LEARN
1. **The structure of chemical equations of reaction**
2. **Balancing chemical equations of reaction**

A compact notation for describing the starting materials and products in chemical changes is the *chemical equation of reaction*. The general structure of a chemical equation of reaction is shown below:

$$X \text{ molecules } A + Y \text{ molecules } B \rightarrow Z \text{ molecules } C + U \text{ molecules } D$$

- The chemical formulas of the starting substance(s), the *reactants*, are listed on the left of the arrow, separated with + signs.

- The substance(s) resulting from the transformation, the *products*, represented by the chemical equation are listed to the right of the arrow; if there is more than one product, they are all listed and separated by + signs.

- The physical state of the substance may be indicated in the equation of reaction by a letter in parentheses following the chemical formula: $_{(g)}$ for a gas; $_{(l)}$ for a liquid; $_{(aq)}$ for an aqueous solution.

- The arrow indicates the direction of the change; other information about the conditions of the change may be listed above or below the arrow.

For a chemical equation of reaction to be complete and accurate, the number of atoms of each element on the reactant side must be the same as the number of atoms of that element on the product side, because the number of atoms is conserved in chemical changes. The numbers of molecules of reactants and products are represented by a numerical coefficient preceding the chemical formula of each substance. The coefficients must be whole numbers, and they must be the smallest whole numbers that will satisfy the conservation of atoms. A chemical equation of reaction that satisfies these criteria is said to be **balanced**.

Let's examine the following reaction which is used in the manufacture of sulfuric acid:

sulfur dioxide   plus   oxygen   forms   sulfur trioxide:

The chemical equation of reaction is:

$$2\ SO_{2\ (g)} + O_{2(g)} \xrightarrow[400\text{-}500\ °C]{V_2O_5\ \text{catalyst}} 2\ SO_{3(g)}$$

The equation contains more information than the molecular pictures described in Chapter 3 and is much more compact. It states that two gaseous molecules of sulfur dioxide combine or react with one molecule of oxygen to give two molecules of sulfur trioxide as molecules in the gas phase. Because a mol is just a large number of molecules, the equation of reaction also states that the reaction of two moles of SO_2 and one mol of O_2 will give two moles of SO_3. The information above and below the arrow informs us that the conversion requires V_2O_5 and high temperatures (400 to 500 °C). However, this information is optional and is often omitted.

Balancing Equations of Reaction

This equation of reaction is also balanced. One way to check is to write a table that compares the number of atoms in each reactant to the number of atoms in each product.

$$2\ SO_{2\ (g)} + O_{2(g)} \rightarrow 2\ SO_{3(g)}$$

atoms S	2	2
atoms O	2(2) + 2 = 6	2(3) = 6

To balance chemical equations, we can only change the coefficients of the molecules.

COMMON PITFALL: CHANGING SUBSCRIPTS

When balancing chemical equations of reaction, NEVER change the subscripts or superscripts of a chemical formula. Doing so changes the chemical reaction itself because one or more of the compounds have been changed.

Exercise 4.1 Balance the following chemical reactions, which illustrate the synthesis of three important industrial chemicals. In these examples, tables of the atoms in starting materials and products will be written. With practice, you will find more compact ways of keeping track of the numbers of atoms.

(a) <u>Urea: $NH_2(CO)NH_2$</u>

$$NH_3 + CO_2 \rightarrow NH_2(CO)NH_2 + H_2O$$

(b) <u>Hydrogen chloride: HCl</u>

$$NaCl_{(s)} + H_2SO_4\,_{(aq)} \rightarrow Na_2SO_4\,_{(s)} + HCl_{(g)}$$

(c) <u>Aluminum sulfate: $Al_2(SO_4)_3$</u>

$$Al_2O_3 \cdot 2H_2O_{(s)} + H_2SO_4\,_{(aq)} \rightarrow Al_2(SO_4)_3\,_{(s)} + H_2O_{(l)}$$

Steps to Solution: Tabulate the elements and the number of atoms of each element present in the starting materials and the products. Then adjust the coefficients of reactants and products until the equation is balanced.

(a) <u>Urea: $NH_2(CO)NH_2$</u>

$$NH_3 + CO_2 \rightarrow NH_2(CO)NH_2 + H_2O$$

Elements	Atoms in starting materials	Atoms in products	Difference
N	1	2	-1
H	3	6	-3
C	1	1	0
O	2	2	0

Examining the table, we see that C is balanced (one atom each in products and reactants), O is balanced (two atoms in products and reactants), but N and H are not. If we change the coefficient of NH_3 to 2, N will be balanced. The additional molecule of ammonia balances the hydrogen H. The elements C and O need not be considered because they are already balanced and are not affected by addition of NH_3.

$$2\,NH_3 + CO_2 \rightarrow NH_2(CO)NH_2 + H_2O$$

(b) <u>Hydrogen chloride: HCl</u>

$$NaCl_{(s)} + H_2SO_4\,_{(aq)} \rightarrow Na_2SO_4\,_{(s)} + HCl_{(g)}$$

Elements	Atoms in starting materials	Atoms in products	Difference
Na	1	2	-1
Cl	1	1	0
H	2	1	1
S	1	1	0
O	4	4	0

From the table, it is clear that S and O are balanced, whereas H, Cl, and Na are not. If two NaCl units are used, then Na is balanced but H and Cl are not balanced as seen below.

$$2\ NaCl_{(s)} + H_2SO_{4\ (aq)} \rightarrow Na_2SO_{4\ (s)} + HCl_{(g)}$$

Elements	Atoms in starting materials	Atoms in products	Difference
Na	2	2	0
Cl	2	1	1
H	2	1	1

Placing a coefficient of 2 in front of the HCl in the products will balance H and Cl to give the balanced chemical equation:

$$2\ NaCl_{(s)} + H_2SO_{4\ (aq)} \rightarrow Na_2SO_{4\ (s)} + 2\ HCl_{(g)}$$

(c) <u>Aluminum sulfate: $Al_2(SO_4)_3$</u>
Use the methods shown above to find the balanced equation of reaction:

$$Al_2O_3 \cdot 2H_2O_{(s)} + 3\ H_2SO_{4(aq)} \rightarrow Al_2(SO_4)_{3\ (s)} + 5\ H_2O_{(l)}$$

The balanced chemical equation of reaction gives the molecule to molecule (or mol to mol) relationship between reactants and products. In practice we directly measure quantities such as mass that are subsequently converted to moles.

4.2 The Stoichiometry of Chemical Reactions

QUESTIONS TO ANSWER, SKILLS TO LEARN
1. **Using mol ratios of reactants (or products) to find out how much of a substance will be required (or produced) in a chemical reaction**
2. **Determining masses of reactants required for a reaction**
3. **Determining the mass of product expected from a reaction**

In chemical reactions, transformations occur between molecules, atoms, or ions to produce different molecules or ions. The balanced chemical equation contains information about the relative numbers of molecules/atoms/ions involved in the reaction; the molar mass allows conversion of mass to moles and vice versa. The amounts of products and reactants involved in a chemical reaction are calculated in the **stoichiometry** of the reaction. The synthesis of urea will serve as an example.

The balanced chemical equation for the reaction of ammonia and carbon dioxide to give urea is:

$$2\, NH_3 + CO_2 \rightarrow NH_2(CO)NH_2 + H_2O$$

The equation tells us that two molecules of ammonia are required for each molecule of carbon dioxide to produce one molecule of urea and one molecule of water. To illustrate some strategies in calculating masses of products and reactants, we will find the mass of ammonia and the mass of CO_2 required to produce 1.0×10^3 kg of urea. Remember that the coefficients indicate the numbers of molecules (or moles) of the substances involved in the reaction. We must convert the quantities we are given to moles before determining the other quantities.

Reactions take place between molecules/ions;
moles are the "currency" of stoichiometry problems

In this example we are given the number of kg of product desired. The moles of reactant required per mole of urea are found from the equation.

There are five steps in every stoichiometry problem:

<u>Step 1:</u> *Have a balanced chemical reaction; otherwise all subsequent calculations may be meaningless.*

In this example, the equation is already balanced.

<u>Step 2:</u> *Determine the relationship of moles of reactants to products. These are the stoichiometric ratios. Here you "think molecules"!*

For each mole of $CO(NH_2)_2$, two moles of NH_3 are required. This mole relationship can be written as a ratio:

mole relationship: 1 mol $CO(NH_2)_2$ reacts with 2 mols NH_3 or

stoichiometric ratios $\dfrac{2 \text{ mol } NH_3}{1 \text{ mol } CO(NH_2)_2}$ (converts mols $CO(NH_2)_2$ to mols NH_3)

and $\dfrac{1 \text{ mol } CO(NH_2)}{2 \text{ mol } NH_3}$ (converts mols NH_3 to mols $CO(NH_2)_2$)

For each mole of $CO(NH_2)_2$, one mole of CO_2 is required. Similarly, the moles of CO_2 equals the moles of urea:

1 mol $CO(NH_2)_2$ = 1 mol CO_2 stoichiometric ratio : $\dfrac{1 \text{ mol } CO_2}{1 \text{ mol } CO(NH_2)_2}$

Step 3: *Determine the number of moles of reactant supplied or product desired.*
We are given the required mass of urea and its chemical formula, so we find the number of moles of urea:

$$\text{mol } CO(NH_2)_2 = \dfrac{1.0 \times 10^3 \text{ kg} \times 10^3 \text{ g/kg}}{60.06 \text{ g/mol}} = 1.67 \times 10^4 \text{ mol } CO(NH_2)_2$$

(Note that one nonsignificant figure is being kept throughout the calculation and will be discarded at the end of the problem.)

Step 4: *Determine the moles and then masses of the other substances in the equation.*

$$\text{mass } NH_3 = MM(NH_3)\,(\text{mol } NH_3)$$

The moles of NH_3 is found from the moles of $CO(NH_2)_2$ using the stoichiometric ratio between urea and NH_3 found in step 2.

$$\text{mol } NH_3 = \text{mol } CO(NH_2)_2 \cdot \dfrac{2 \text{ mol } NH_3}{1 \text{ mol } CO(NH_2)_2}$$

$$\text{mol } NH_3 = 1.67 \times 10^4 \text{ mol } CO(NH_2)_2 \cdot \dfrac{2 \text{ mol } NH_3}{1 \text{ mol } CO(NH_2)_2} = 3.34 \times 10^4 \text{ mol } NH_3$$

$$\text{mass } NH_3 = 3.34 \times 10^4 \text{ mol } NH_3 \cdot 17.03 \text{ g/mol } NH_3$$

$$\text{mass NH}_3 = 5.7 \times 10^5 \text{ g NH}_3 \text{ or } 5.7 \times 10^2 \text{ kg NH}_3$$

We can find the mass of CO_2 by a similar calculation:

$$\text{mass CO}_2 = \text{mol CO}_2 \cdot MM\,CO_2 = \text{mol CO(NH}_2)_2 \cdot \frac{1 \text{ mol CO}_2}{1 \text{ mol CO(NH}_2)_2} \cdot MM\,CO_2$$

$$\text{mass CO}_2 = 1.67 \times 10^4 \text{ mol CO(NH}_2)_2 \cdot \frac{1 \text{ mol CO}_2}{1 \text{ mol CO(NH}_2)_2} \cdot 44.01 \text{ g/mol CO}_2$$

$$\text{mass CO}_2 = 7.3 \times 10^5 \text{ g CO}_2 \text{ or } 7.3 \times 10^2 \text{ kg CO}_2$$

Step 5: Check the answer.

The answer makes sense. Even though we need more moles of ammonia, the required mass is less than that of CO_2 because the molar mass of ammonia is less than half that of carbon dioxide.

Exercise 4.2 Benzene is used in the manufacture of nylon. One of the intermediate molecules is cyclohexane, which is produced by the reaction of benzene with hydrogen as shown in the unbalanced reaction below.

benzene   cyclohexane

How much H_2 (in kg) would be required to convert 2.00×10^3 L of benzene to cyclohexane (ρ[benzene] = 0.8765 g/mL) ?

Steps to Solution:

- First, balance the reaction. The line structures must be converted to chemical formulas. Doing so gives:

$$C_6H_6 + H_2 \rightarrow C_6H_{12}$$

Balancing the equation gives:

$$C_6H_6 + 3\,H_2 \rightarrow C_6H_{12}$$

- Second, determine the relevant stoichiometric ratios. We need to know how much hydrogen will react with the benzene. Therefore, we need to know how many molecules of H_2 will react with each C_6H_6 molecule. From the chemical equation of reaction:

1 mol C_6H_6 reacts with 3 mol H_2: the stoichiometric ratio is $\dfrac{3 \text{ mol } H_2}{1 \text{ mol } C_6H_6}$

- *Third, determine the mols of appropriate substances, in this case C_6H_6. We must convert the volume in L to mass in g:*

$$\text{mol } C_6H_6 = 2.00 \times 10^3 \text{ L} \cdot \dfrac{10^3 \text{ mL}}{\text{L}} \cdot 0.8765 \dfrac{\text{g}}{\text{mL}} \cdot \dfrac{1 \text{ mol } C_6H_6}{78.11 \text{g}} = 2.24 \times 10^4 \text{ mol } C_6H_6$$

- *Fourth, convert mols of benzene to mols of hydrogen and then to mass of hydrogen.*

$$\text{mol } H_2 = \text{mol } C_6H_6 \cdot \dfrac{3 \text{ mol } H_2}{1 \text{ mol } C_6H_6} = 2.24 \times 10^4 \text{ mol } C_6H_6 \cdot \dfrac{3 \text{ mol } H_2}{1 \text{ mol } C_6H_6}$$

$$\text{mol } H_2 = 6.72 \times 10^4 \text{ mol } H_2$$

$$\text{mass } H_2 = 6.72 \times 10^4 \text{ mol } H_2 \cdot 2.016 \dfrac{\text{g}}{\text{mol } H_2} = \boxed{1.36 \times 10^5 \text{ g } H_2 = 1.36 \times 10^2 \text{ kg } H_2}$$

- *Fifth, does the answer make sense? Yes. The amount of H_2 needed is large, but so is the starting amount of benzene.*

4.3 Yields of Chemical Reactions

QUESTIONS TO ANSWER, SKILLS TO LEARN
1. **What makes actual yields smaller than theoretical yields?**
2. **Calculating theoretical and actual yields**

In the calculations in the preceding section, we assumed implicitly that all of the reactants combine and yield the products listed in the equation of reaction with 100% efficiency: every molecule of reactant ultimately would be incorporated into a molecule of product. In the real world, this is rarely observed because a number of factors reduce the amount of product formed, among them:

- incomplete consumption of starting materials

- contamination that consumes some of the starting materials, leading to the formation of other products through other reactions (competing reactions)

Chapter 4: Chemical Reactions and Stoichiometry

- formation of products other than those specified in the equation of reaction
- loss of product in purification and collection steps.

The amount of product obtained from a chemical reaction is called the **yield**. The amount actually collected is called the **actual yield**, whereas the amount predicted by stoichiometry, assuming complete conversion of the given amount of starting materials, is called the **theoretical yield**. The efficiency of the reaction in forming the product is described by the **percent yield (% yield)**, which is defined in terms of the actual and theoretical yields:

$$\% \text{ yield} = \frac{\text{actual yield}}{\text{theoretical yield}} \cdot 100\%$$

The percent yield has no units of mass or moles because it is a ratio. Therefore the yields may be expressed in units of mass or moles *as long as both the theoretical and actual yields are expressed in the same units.*

Exercise 4.3 In Exercise 4.1(a), which describes the synthesis of urea, the amounts of reactants were calculated for a theoretical yield of 1.0×10^3 kg. If 9.7×10^2 kg was obtained, what was the percent yield of urea?

Steps to Solution: The relationship between the percent yield and the actual and theoretical yield is:

$$\text{percent yield} = \frac{\text{actual yield}}{\text{theoretical yield}} \cdot 100\%$$

To solve this problem, we must find the values for the actual and theoretical yields and then substitute them into the equation. The theoretical yield is 1.0×10^3 kg and the actual yield is 9.7×10^2 kg. Therefore:

$$\text{percent yield} = \frac{9.7 \times 10^2 \text{ kg}}{1.0 \times 10^3 \text{ kg}} \cdot 100\% = 97\%$$

Exercise 4.4 In Exercise 4.2, 2.0×10^3 L of benzene and 1.36×10^2 kg of H_2 react to give 1.80×10^3 kg of cyclohexane, C_6H_{12}. What is the percent yield of cyclohexane?

Steps to Solution: The formula is the same as that shown above. The actual yield is given; however, the theoretical yield must be calculated from the stoichiometry of the reaction. From the equation, 1 mole cyclohexane is obtained for each mole of benzene reacted. Therefore:

$$\text{mol } C_6H_{12} = \text{mol } C_6H_6 = 2.24 \times 10^4 \text{ mol (from Exercise 4.2)}$$

$$\text{mass } C_6H_{12} = 2.24 \times 10^4 \text{ mol} \cdot (84.16 \text{ g/mol}) = 1.89 \times 10^3 \text{ kg}$$

$$\text{percent yield} = \frac{1.80 \times 10^3 \text{ kg}}{1.89 \times 10^3 \text{ kg}} \cdot 100\% = \boxed{95\%} \quad \text{(2 significant figures in } 2.0 \times 10^3\text{)}$$

4.4 The Limiting Reagent

QUESTIONS TO ANSWER, SKILLS TO LEARN
1. **What is a limiting reagent?**
2. **Calculating theoretical yields when a limiting reagent is present**
3. **Preparing a table of amounts**

In the reactions shown above, the calculations assume that we have placed the exact stoichiometric amounts of reactants in the container in which the reaction is going to occur. However, in many cases a reaction is run under conditions such that one reactant is consumed before the other$_{(s)}$. The reactant that is consumed first will limit the amount of product that is formed and therefore is called the **limiting reagent**. The limiting reagent determines the theoretical yield because when it runs out, no more product can form. It is important to realize that mass can be deceptive in determining the limiting reagent, because the number of molecules depends on both the mass *and* the molar mass of the reagent. In addition, stoichiometric ratios of required reactants may cause the limiting reagent to be a reactant that does not have the smallest number of moles.

The limiting reagent is identified by comparing the number of moles of product that would result from each of the starting materials.

The starting material that will give the smallest number of moles of product is the limiting reagent.

The stoichiometric ratios of reactant to product must be used in conjunction with the number of moles of the starting materials.

Exercise 4.5 Quicklime, CaO, is converted to slaked lime, calcium hydroxide, by reaction with water:

$$CaO_{(s)} + H_2O_{(l)} \rightarrow Ca(OH)_{2\ (s)}$$

If 5.10 kg of CaO is allowed to react with 2.0 kg of H_2O, what is the theoretical yield of $Ca(OH)_2$?

Steps to Solution: This a limiting reagent problem. To solve it, we must determine the number of moles of product that would be formed from both reactants. The reagent that will give the least product is the limiting reagent.

Reactant	Moles	$\left(\dfrac{\text{mol reactant}}{\text{mol Ca(OH)}_2}\right)$	Moles product
CaO	90.9	1	90.9
H$_2$O	1.11 x 10^2	1	1.11 x 10^2

Because the moles of Ca(OH)$_2$ based on the amount of CaO are the smallest, CaO is the limiting reagent even though it is the reactant present in greater mass. Therefore the theoretical yield of Ca(OH)$_2$ is calculated from the number of moles of CaO:

$$\text{g Ca(OH)}_2 = \text{mol CaO} \cdot 74.09 \text{ g/mol} = 90.9 \text{ mol} \cdot (74.09 \text{ g/mol})$$

$$\text{g Ca(OH)}_2 = \boxed{6.74 \times 10^3 \text{ g or 6.74 kg}}$$

COMMON PITFALL: USING MASSES TO FIND THE LIMITING REAGENT
The limiting reagent is not always the reagent with the smallest mass!
Work with moles and stoichiometric ratios!!

In cases where the ratios are not 1:1, even more care must be taken!

One way to organize and simplify limiting reagent problems is with **tables of amounts**. The format of a table of amounts is illustrated in the solution of the following problem. The table contains one column for each reactant and product and five rows:

<u>Row 1</u> lists the substances, preferably by chemical formula.

<u>Row 2</u> lists the ratio of the moles of starting material to its stoichiometric coefficient, which allows the determination of the limiting reagent (after some practice, the second row may be omitted).

<u>Row 3</u> lists the moles of each substance before any reaction begins. The reactant with the smallest ratio will be the limiting reagent.

<u>Row 4</u> contains the change in the number of moles of each substance. Because reactants are consumed, the amount will decrease. All changes in the amounts of reactants are negative because they are being consumed. Conversely, changes in the amounts of products are positive because the amounts of products are increasing.

Row 5 is the sum of the entries in rows 3 and 4. In a limiting reagent problem, one of these entries must be zero. If any entry is less than zero, an error has been made either in choosing the limiting reagent or in arithmetic.

Exercise 4.6. Sulfur hexafluoride, a very good electrical insulator, is prepared by the reaction of sulfur with elemental fluorine:

$$S_{(s)} + 3\ F_{2(g)} \rightarrow SF_{6(g)}$$

How much SF_6 could be produced from the reaction of 1.50×10^2 kg of S with 6.50×10^2 kg of F_2?

Steps to Solution: First we balance the equation and enter the formulas into the first row. The next step is to calculate the moles of the starting materials:

$$\text{mol S} = \frac{1.50 \times 10^2 \text{ kg} \cdot 10^3 \text{g/kg}}{32.07 \text{g/mol}} = 4.68 \times 10^3 \text{ mol}$$

$$\text{mol F}_2 = \frac{6.50 \times 10^2 \text{ kg} \cdot 10^3 \text{g/kg}}{38.00 \text{g/mol}} = 1.71 \times 10^4 \text{ mol}$$

These values are entered into the third row.

The ratio of moles of reactant to stoichiometric coefficient is calculated in the second row of the table. In this case, the ratio is smallest for S; therefore, sulfur is the limiting reagent. We now calculate the entries for the fourth row. Starting with sulfur, the change in each amount of substance is:

$$\text{mol S consumed} = -4.68 \times 10^3 \text{ mol SF}_6 \cdot \frac{1 \text{ mol S}}{\text{mol SF}_6} = -4.68 \times 10^3 \text{ mol S}$$

$$\text{mol F}_2 \text{ consumed} = -4.68 \times 10^3 \text{ mol SF}_6 \cdot \frac{3 \text{ mol F}_2}{\text{mol SF}_6} = -1.40 \times 10^3 \text{ mol F}_2$$

$$\text{mol SF}_6 \text{ produced} = +4.68 \times 10^3 \text{ mol S} \cdot \frac{1 \text{ mol SF}_6}{\text{mol S}} = 4.68 \times 10^3 \text{ mol SF}_6$$

Adding the quantities in the third and fourth rows gives the fifth row to complete the table.

Table of Amounts

Quantity/substance	S	F$_2$	SF$_6$
$\dfrac{\text{mol reactant}}{\text{coefficient}}$	$\dfrac{4.68 \times 10^3}{1}$ $= 4.68 \times 10^3$	$\dfrac{1.71 \times 10^4}{3}$ $= 5.70 \times 10^3$	----
Starting amount	4.68×10^3 mol	1.71×10^4 mol	0
Change in amount	-4.68×10^3 mol	-1.40×10^4 mol	4.68×10^3 mol
Final amount	0	3.07×10^3 mol	4.68×10^3 mol

From the moles of SF$_6$ produced, we can calculate the mass of SF$_6$:

$$\text{mass SF}_6 = 4.68 \times 10^3 \text{ mol} \cdot 146.06 \text{ g/mol SF}_6 \cdot \frac{1 \text{ kg}}{1000 \text{ g}} = \boxed{6.83 \times 10^2 \text{ kg}}$$

If desired, the remaining mass of F$_2$ could be readily calculated using the number of moles of F$_2$ in row 5.

4.5 Precipitation Reactions

QUESTIONS TO ANSWER, SKILLS TO LEARN
1. **What happens in a precipitation reaction?**
2. **What is a spectator ion?**
3. **Writing net ionic equations of reaction**
4. **Which ionic compounds are soluble?**
5. **Synthesis through precipitation**

In precipitation reactions, two different ionic compounds are mixed in a solvent (in our case, water). A solid forms, which is called the **precipitate**. The remaining solution contains the ions that have not precipitated and is called the **supernatant liquid**. The key event in precipitation reactions is the formation of a product that is a solid and can be readily separated from the solution by processes such as filtration. To understand precipitation reactions (and other reactions involving ions in solution), it is helpful to identify the chemical species that are in the solution and their role in the reaction of interest.

Mixing a potassium fluoride solution and a calcium chloride solution leads to the formation of a white precipitate. Some combination of ions has led to the formation of an insoluble salt or salts. Drawing pictures is useful when "thinking molecules" to understand a problem, so remember that KF and CaCl$_2$ dissolve in water to give solutions of the ions:

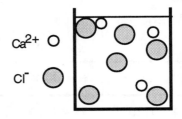

$$KF_{(s)} \rightarrow K^+_{(aq)} + F^-_{(aq)} \qquad\qquad CaCl_{2(s)} \rightarrow Ca^{2+}_{(aq)} + 2\,Cl^-_{(aq)}$$

Immediately upon mixing these two solutions, a solution containing K$^+$, F$^-$, Cl$^-$, and Ca^{2+} ions is formed. These ions can now undergo reaction to give a precipitate.

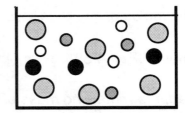

The precipitate is formed from some combination of cations and anions in the solution. The four possible reactions to give salts are listed below:

<u>possible reactions with Ca^{2+}</u>   <u>possible reactions with K$^+$</u>

$$Ca^{2+}_{(aq)} + 2\,Cl^-_{(aq)} \rightarrow CaCl_{2(s)} \qquad\qquad K^+_{(aq)} + F^-_{(aq)} \rightarrow KF_{(s)}$$

$$Ca^{2+}_{(aq)} + 2\,F^-_{(aq)} \rightarrow CaF_{2(s)} \qquad\qquad K^+_{(aq)} + Cl^-_{(aq)} \rightarrow KCl_{(s)}$$

The upper two reactions do not give insoluble salts; they are the reverse of the reactions that gave the solutions of starting materials. The remaining two salts that might be formed are KCl and CaF$_2$. Mixing aqueous NaCl and KNO$_3$ solutions results in no precipitate; however, mixing NaF and Ca(NO$_3$)$_2$ solutions does result in a precipitate. The first experiment shows that mixing K$^+$ and Cl$^-$ ions does not result in a precipitate; the second shows that mixing Ca^{2+} and F$^-$ ions does result in a precipitate. Therefore the solid is CaF$_2$. The solution that remains contains the potassium ions and chloride ions that did not react (a very small concentration of calcium and fluoride ions is present because CaF$_2$ does have a very small solubility).

$$Ca^{2+}_{(aq)} + 2\,Cl^-_{(aq)} + 2\,K^+_{(aq)} + 2\,F^-_{(aq)} \rightarrow CaF_{2(s)} + 2\,K^+_{(aq)} + 2\,Cl^-_{(aq)} \quad \textbf{complete equation}$$

Net Ionic Equations and Spectator Ions

What role is played by the K$^+$ and Cl$^-$ ions? They do not actually participate in the reaction; they start off as aqueous ions and end up as the same aqueous ions after the precipitation has occurred. Because these two ions do not participate in the precipitation reaction, they are often called **spectator ions**. The chemistry that occurs can be written listing only the reacting ions and the products:

$$Ca^{2+}{}_{(aq)} + 2\ F^{-}{}_{(aq)} \rightarrow CaF_2{}_{(s)} \quad \textbf{net ionic equation}$$

This is the **net ionic equation**. *The net ionic equation lists only the ions that participate in the reaction.* Spectator ions are omitted.

If one does not have a laboratory at hand to test the reactions of various ions, how can one predict the formation of a precipitate from the ions present in the mixture? The following guidelines are useful in predicting precipitation of ionic solids:

Guidelines for Predicting Solubility of Ionic Salts
1. Salts that contain the following cations are soluble: NH_4^+ and Group 1 metal cations.
2. Salts that contain the following anions are soluble: NO_3^-, Cl^-, Br^-, I^-, SO_4^{2-}, HSO_4^-, $H_3C_2O_2^-$ (acetate), and ClO_4^- (perchlorate)
3. Any salt not included in guidelines 1 and 2 is **insoluble**.
4. Exceptions to guideline 2: AgX, PbX_2, Hg_2X_2 (X = Cl, Br, I), Ag_2SO_4, $BaSO_4$, and $PbSO_4$ are **insoluble**.
5. The following compounds are an exception to guidelines 1 and 2: $Ba(OH)_2$, MgS, CaS, and BaS are **soluble**.

The use of these rules is demonstrated in the following exercise. When writing the formula for the ionic salt, remember that the solid will have no net electrical charge: the sum of the charges on the anion$_{(s)}$ and cation$_{(s)}$ must be 0.

Exercise 4.7 Predict the product and write the net ionic equation of reaction that occurs upon mixing solutions of the reactants listed below.

Steps to Solution: First identify the ions that are present in the solution. Next use the solubility guidelines to determine what salt, if any, will precipitate. Write the equation of reaction. Identify spectator ions and then write the net ionic equation of reaction.

(a) $NH_4Cl_{(aq)} + AgNO_{3(aq)} \rightarrow$
The solution that results will contain four ions: NH_4^+, Cl^-, Ag^+, and NO_3^-. The two products that might result (ignoring the starting materials) are $AgCl$ and NH_4NO_3. Ammonium nitrate will be soluble by guidelines 1 and 2; silver chloride will be insoluble by guideline 3. The reaction can be written in full:

$$NH_4Cl_{(aq)} + AgNO_{3(aq)} \rightarrow AgCl_{(s)} + NH_4NO_{3\ (aq)}$$

or as the net ionic equation:

$$Ag^+_{(aq)} + Cl^-_{(aq)} \rightarrow AgCl_{(s)}$$

(b) $MgCl_{2\,(aq)}$ + NaOH $_{(aq)}$ →
The resulting solution will contain four ions before any precipitation occurs: Mg^{2+}, Cl^-, Na^+, and OH^-. The two salts that could form that are not starting materials are $Mg(OH)_2$ and NaCl. NaCl is soluble because it contains the soluble cation, Na^+ (guideline 1); $Mg(OH)_2$ is expected to be insoluble by guideline 3. Therefore the reaction is:

$$MgCl_{2\,(aq)} + 2\,NaOH_{(aq)} \rightarrow Mg(OH)_{2\,(s)} + 2\,NaCl_{(aq)}$$

and the net ionic equation is:

$$Mg^{2+}_{(aq)} + 2\,OH^-_{(aq)} \rightarrow Mg(OH)_{2(s)}$$

(c) $MnBr_{2\,(aq)}$ + $Na_2S_{(aq)}$ →

Use the methods outlined above to show the reaction is:

$$MnBr_{2\,(aq)} + Na_2S_{(aq)} \rightarrow MnS_{(s)} + 2\,NaBr_{(aq)}$$

and the net ionic equation is:

$$Mn^{2+}_{(aq)} + S^{2-}_{(aq)} \rightarrow MnS_{(s)}$$

Precipitation Stoichiometry

Precipitation reactions may be treated quantitatively to determine the amount of precipitate formed and the number of ions remaining in the solution. After we convert molarity and volume into moles, knowledge of the stoichiometry of the reaction will allow us to determine the limiting reagent and construct a table of amounts. Then we can determine the amount of precipitated product and ions remaining in solution.

Exercise 4.8 Phosphoric acid is manufactured from phosphorus sources that contain variable amounts of arsenic. Phosphoric acid that is to be used in food products (check the label on your soda) must have this arsenic removed; the removal is effected by the addition of sodium sulfide to give the very insoluble arsenic sulfide. The unbalanced reaction may be summarized as:

$$As^{3+} + S^{2-} \rightarrow As_2S_{3\,(s)}$$

Treatment of a 5.00×10^2 L sample of phosphoric acid with excess sodium sulfide gives 8.45 g of solid As_2S_3.

(a) How many g of As were removed by the treatment of the sample?

(b) What was the molar concentration of As, and how many mg of arsenic were removed from each liter of phosphoric acid solution by the sodium sulfide treatment?

(c) What volume of a 6.0 M sodium sulfide solution would be needed to react with this amount of arsenic?

Steps to Solution: To answer parts (a) and (b) we need to determine the amount of arsenic present in the solution in terms of total mass and conversion of that mass to concentration. This information will allow us to answer part (c).

(a) This part of the problem can be solved by calculating the moles of As from the moles of As_2S_3.

$$g\ As = mol\ As(74.92\ g/mol)$$

$$mol\ As = \frac{2\ mol\ As}{mol\ As_2S_3}\ mol\ As_2S_3 = \frac{2\ mol\ As}{mol\ As_2S_3} \cdot \frac{8.45g}{246.02g/mol\ As_2S_3} = 6.87 \times 10^{-2}\ mol\ As$$

$$g\ As = 6.87 \times 10^{-2}\ mol(74.92\ g/mol) = \boxed{5.15\ g\ As}$$

(b) The molar concentration of arsenic (regardless of its chemical form) in the solution is determined by the definition of molarity:

$$M = \frac{mol\ substance}{liter\ solution}$$

The moles of As were determined in part (a) (6.87×10^{-2} mol As), and the volume of the solution is known from the data in the problem to be 5.00×10^2 L. Substitution gives:

$$M = \frac{6.87 \times 10^{-2}\ mol\ As}{5.00 \times 10^2\ L} = 1.37 \times 10^{-4}\ mol/L$$

The concentration in mg/L is readily found through conversion of units from *M* to mg/L:

$$concentration\ (mg/L) = \frac{mole}{L} \cdot \frac{g}{mol} \cdot \frac{1000\ mg}{g}$$

$$mg/L = 1.37 \times 10^{-4}\ \frac{mol}{L} \cdot 74.9216\ \frac{g}{mol} \cdot \frac{1000\ mg}{g} = \boxed{10.3\ mg/L}$$

(c) The first step in the solution to this problem is to balance the equation of reaction so that the molar relationship between reactants and products is known.

$$2\ As^{3+} + 3\ S^{2-} \rightarrow As_2S_3\ (s)$$

From the balanced equation of reaction, the relationship between moles of arsenic and sulfide is :

$$3 \text{ mol } S^{2-} = 2 \text{ mol } As^{3+} \text{ gives } \frac{2 \text{ mol } As^{3+}}{3 \text{ mol } S^{2-}} \text{ or } \frac{3 \text{ mol } S^{2-}}{2 \text{ mol } As^{3+}}$$

Because we want to convert moles of As to moles of S^{2-}, the second conversion factor is the desired one:

$$\text{mol } S^{2-} = \text{mol } As^{3+} \cdot \frac{3 \text{ mol } S^{2-}}{2 \text{ mol } As^{3+}} = \text{mol } S^{2-}$$

$$\text{mol } S^{2-} = 6.87 \times 10^{-2} \text{ mol As} \cdot \frac{3 \text{ mol } S^{2-}}{2 \text{ mol } As^{3+}} = 1.03 \times 10^{-1} \text{ mol } S^{2-}$$

$$V \text{ Na}_2\text{S solution} = \frac{\text{mol } S^{2-}}{M} = \frac{1.03 \times 10^{-1} \text{ mol } S^{2-}}{6.0 \text{ mole/L}} = \boxed{1.72 \times 10^{-2} \text{ L} = 17.2 \text{ mL}}$$

4.6 Acid-Base Reactions

QUESTIONS TO ANSWER, SKILLS TO LEARN
1. What are acids and bases?
2. What happens in an acid-base reaction?
3. What is the difference between strong and weak acids?
4. Mol calculations through volumetric analysis: titration

Another important class of chemical reactions involves the transfer of protons, H^+, from one substance to another:

$$HA + B \rightarrow BH^+ + A^-$$

Reactions in which the hydrogen cation, H^+, is transferred are called acid-base reactions. In the schematic reaction above, the acid, HA, donates a proton to the base, B. For now, the following definitions apply:

- An **acid** is a substance that donates protons to other chemical substances.
- A **base** is a substance that accepts protons.

Exercise 4.9 Identify the acid and the base in the following acid-base reactions.

Steps to Solution: An acid is a proton donor and a base is a proton acceptor. You must examine the reactants to see which one gives up a proton (the acid) and which accepts a proton (the base).

(a) $H_3O^+_{(aq)} + OH^-_{(aq)} \rightarrow 2\, H_2O_{(l)}$

In this reaction, the acid is the hydronium ion, H_3O^+, and the base is the hydroxide ion, OH^-.

(b) $NH_{3\,(aq)} + H_3O^+_{(aq)} \rightarrow NH_4^+_{(aq)} + H_2O_{(l)}$

The acid in this reaction is the hydronium ion, H_3O^+, and the base is ammonia, NH_3.

(c) $CH_3CO_2H_{(aq)} + NH_{3\,(aq)} \rightarrow NH_4^+_{(aq)} + CH_3CO_2^-_{(aq)}$

Here the compound CH_3CO_2H, acetic acid, is the acid, transferring a proton to the base, ammonia, NH_3.

(d) $NH_4^+_{(aq)} + OH^- \rightarrow NH_{3\,(aq)} + H_2O_{(l)}$

The ammonium ion transfers a proton to the hydroxide ion; therefore the ammonium ion is the acid and hydroxide is the base.

In each of these reactions, the common event is transfer of hydrogen ions, which is the most common acid-base reaction we will encounter.

Acid and Base Strength

Chemists often speak about *strong* and *weak* acids. The notion of acid strength is often confused with how corrosive the acid is. In fact, acid strength depends on another property, the ability of the acid to behave as an ionic substance when dissolved in a solvent, which, for our purposes, will be water. When 1.0 mol of NaCl is dissolved in water to give 1.0 L of solution, the solution contains 1.0 M Na^+ ions and 1.0 M Cl^- ions. NaCl is called a strong electrolyte because it dissolves to give a strongly conducting solution.

When 1.0 mol of HNO_3 is dissolved in water to give 1.0 L of solution, the solution contains 1.0 M H_3O^+ ions and 1.0 M NO_3^- ions. This is shown below to left. Because *the acid completely dissociates to give ions by transferring a proton to water*, we call nitric acid a **strong acid**. A strong acid is a strong electrolyte. Common strong acids include sulfuric acid, hydrochloric acid, and nitric acid.

Strong acids: highly dissociated into ions   Weak acids: mostly undissociated

Some acids, on the other hand, dissolve in water without completely dissociating into ions. An example is acetic acid, CH_3COOH, which gives vinegar its sour taste. When dissolved in 1.0 L water, 1.0 mol of acetic acid does not give a solution containing 1.0 M H_3O^+ and 1.0 M acetate (CH_3COO^-) ions; rather it produces a mixture containing the molecule CH_3COOH, and small quantities of acetate and hydronium ions as illustrated above to right.

This solution contains relatively few ions through the transfer of protons from acetic acid to water compared to strong acids; therefore acetic acid is called a **weak acid**. *Weak acids dissolve in water with incomplete transfer of protons to the solvent*, resulting in the presence of molecules of the acid as well as ions. Common weak acids include acetic acid and citric acid.

Similarly, there are strong and weak bases. Strong bases dissolve in water to give stoichiometric amounts of hydroxide ions. Potassium hydroxide and barium hydroxide are examples of strong bases:

$$KOH_{(s)} \rightarrow K^+_{(aq)} + OH^-_{(aq)}$$

$$Ba(OH)_{2\,(s)} \rightarrow Ba^{2+}_{(aq)} + 2\,OH^-_{(aq)}$$

Dissolving 1.0 mol of $Ba(OH)_2$ in water results in the formation of 1.0 mol of Ba^{2+} ions and 2.0 moles of OH^- ions. Dissolving BaO in water also leads to the formation of stoichiometric quantities of barium ions and hydroxide ions:

$$BaO_{(s)} + H_2O_{(l)} \rightarrow Ba^{2+}_{(aq)} + 2\,OH^-_{(aq)}$$

Therefore barium oxide is also a strong base. Other bases do not show a stoichiometric relationship between the number of molecules of base dissolved and the number of moles of hydroxide ions produced. These bases are weak bases. Ammonia (NH_3) and pyridine (C_5H_5N) are examples of weak bases.

The following guidelines dependably predict how completely an acid-base reaction will proceed.

- A **strong acid** will react completely with a **strong or weak base**.

- A **strong base** will react completely with a **strong or weak acid**.

- A **weak acid** will not react completely with a **weak base** and vice versa. The reactions of weak acids with weak bases will be discussed in more detail in later chapters.

Exercise 4.10 A factory uses HCl to clean metal products before painting them. One step that is performed before disposal of the waste acid solution is neutralization of the excess acid, ideally with an inexpensive base such as $Ca(OH)_2$. How many kg of $Ca(OH)_2$ would be required to neutralize the acid in a tank containing 450 L of a solution that is 4.73 M HCl?

Steps to Solution: The question asks us to determine the amount of $Ca(OH)_2$ needed to neutralize an amount of acid characterized by the volume and molarity. From the data supplied, we can calculate the moles of HCl. This value must be converted to moles of $Ca(OH)_2$.

To determine the amount of calcium hydroxide required, we must determine the mol relationship between the acid and the base: what ions are present and in what ratio will they react? Because HCl is a strong acid and $Ca(OH)_2$ is a strong base, they will completely dissociate to give the following ions in the reaction mixture:

$$H_3O^+; Cl^-; Ca^{2+}; OH^-$$

The net ionic reaction for a neutralization reaction involving hydrogen ion and hydroxide ion is:

$$H_3O^+_{(aq)} + OH^-_{(aq)} \rightarrow 2\ H_2O_{(aq)}$$

However, this does not really indicate the stoichiometry between $Ca(OH)_2$ and HCl as well as the full equation:

$$2\ HCl_{(aq)} + Ca(OH)_{2\ (s)} \rightarrow 2\ H_2O_{(l)} + CaCl_{2\ (aq)}$$

From this equation, it is clear that :

1 mol Ca(OH)$_2$ reacts with 2 mol HCl; the stoichiometric ratio is $\dfrac{2 \text{ mol HCl}}{\text{mol Ca(OH)}_2}$

With this information and the number of moles of HCl, we can determine the required amount of calcium hydroxide:

$$\text{kg Ca(OH)}_2 = \dfrac{\text{g Ca(OH)}_2}{1000 \text{ g/kg}} = MM(\text{Ca(OH)}_2) \cdot \dfrac{1 \text{ mol Ca(OH)}_2}{2 \text{ mol HCl}} \cdot \text{mol HCl} \cdot \dfrac{1 \text{ kg}}{1000 \text{ g}}$$

$$\text{kg Ca(OH)}_2 = \dfrac{74.095 \text{ g}}{\text{mole Ca(OH)}_2} \cdot \dfrac{1 \text{ mol Ca(OH)}_2}{2 \text{ mol HCl}} \cdot 4.73 \text{ mol/L } (450 \text{ L}) \cdot \dfrac{1 \text{ kg}}{1000 \text{ g}}$$

$$\text{kg Ca(OH)}_2 = \boxed{78.8 \text{ kg Ca(OH)}_2}$$

Is the answer reasonable? It seems large, but the amount of solution being neutralized is also large. The relative sizes are similar: kg of base and hundreds of liters of acid. Therefore the answer is reasonable.

Titrations and Volumetric Analysis

It is fairly easy to measure volumes of liquids, so one method to determine the amount of a substance present in a sample is through reaction with a solution of known concentration of a reagent using apparatus that allows precise measurement of volume, such as a burette. This technique is called **titration**. The number of moles of the reagent is calculated from the relationship:

$$\text{mol} = M \cdot V$$

From the stoichiometry of the reaction between the reagent whose volume is being measured and the substance being analyzed, we can determine the amount of that substance. The substance being analyzed may be a solid or a solution. For accurate results, we must know to a high degree of precision the concentration of the reagent whose volume is being measured. The concentration of the reagent is often determined through a process called **standardization**, titrating a known amount of compound of known composition whose reaction with the titrant is well characterized. The calculations involved in titration and standardization are presented in the next exercise.

Exercise 4.11 A standard for determining the concentrations of acid solutions is sodium carbonate, Na_2CO_3. In the acid-base titration of an 0.2610 g sample of pure Na_2CO_3 with a solution of HCl, a total of 37.32 mL were required to reach the stoichiometric point. What is the molar concentration of HCl in the solution?

Chapter 4: Chemical Reactions and Stoichiometry

Steps to Solution: In this problem we are to calculate the concentration of the HCl titrant, $M(HCl)$. The data supplied are the mass of sodium carbonate and the volume of HCl solution used. Thinking in terms of molecules, the number of moles of sodium carbonate is related to the number of moles of HCl (and molarity) via the volume of HCl at the stoichiometric point by the equation of reaction:

$$2\ HCl_{(aq)} + Na_2CO_{3(aq)} \rightarrow H_2CO_{3\,(aq)} + 2\ NaCl_{(aq)} + H_2O_{(l)}$$

From the chemical equation, the mol ratio of HCl to Na_2CO_3 is:

$$2\ mol\ HCl\ reacts\ with\ 1\ mol\ Na_2CO_3 \quad or \quad mol\ HCl = \frac{mol\ Na_2CO_3}{2}$$

The moles of Na_2CO_3 can be found from its mass:

$$mol\ Na_2CO_3 = \frac{g\ Na_2CO_3}{MM\ Na_2CO_3} = \frac{0.2610\,g}{105.989\ g/mol\ Na_2CO_3} = 2.463 \times 10^{-3}\ mol\ Na_2CO_3$$

The moles of HCl in the solution are found from the molarity and the volume of HCl:

$$mol\ HCl = M(HCl) \cdot V(HCl) = (mol\ Na_2CO_3) \cdot \frac{2\ mol\ HCl}{mol\ Na_2CO_3}$$

$$mol\ HCl = 2 \cdot 2.463 \times 10^{-3}\ mol\ Na_2CO_3 = 4.926 \times 10^{-3}\ mol$$

Substituting for the moles of sodium carbonate gives:

$$M(HCl) \cdot V(HCl) = mol\ HCl$$

Converting the volume of HCl into L gives:

$$37.32\ mL = 0.03732\ L$$

Solving for $M(HCl)$ gives:

$$M(HCl) = \frac{mol\ HCl}{V(HCl)} = \frac{4.926 \times 10^{-3}\,mol}{0.03732\ L} = 0.1320\ mol/L$$

4.7 Oxidation-Reduction Reactions

QUESTIONS TO ANSWER, SKILLS TO LEARN
1. What are redox reactions?
2. What happens in a redox reaction?
3. What are half reactions?
4. Using the activity series to predict the products of metal displacement reactions

The third class of reactions to be discussed in this chapter involves the transfer of not only atoms, but also electrons between molecules, atoms, or ions. Placing a piece of aluminum into a solution of acid results in the formation of aluminum ions and hydrogen gas:

$$2\ Al_{(s)} + 6\ H_3O^+_{(aq)} \rightarrow 2\ Al^{3+}_{(aq)} + 3\ H_{2\ (g)} + 6\ H_2O_{(l)}$$

The aluminum in the reactants is converted to the aluminum ion in the products through the loss of three electrons per aluminum atom. Because two aluminum atoms are consumed in the reaction, there must be six electrons produced through the electron loss process. The reaction, which involves only electron loss, is written as:

$$2\ Al_{(s)} \rightarrow 2\ Al^{3+}_{(aq)} + 6\ e^-$$

where the electrons are written as a product of this process. This equation is balanced in both atoms and electrons.

However, because electrons are conserved, they must be present somewhere in the products. The hydronium ions have been converted to water and hydrogen gas, showing a loss of positive charge. Writing an equation for the part of the reaction not involving the aluminum, it is clear that without six electrons combining with some reactant$_{(s)}$, the equation will not be balanced in electrons. In the next equation, the reaction of H_3O^+ with electrons is shown:

$$6\ H_3O^+_{(aq)} + 6\ e^- \rightarrow 3\ H_{2\ (g)} + 6\ H_2O_{(l)}$$

Because these reactions describe different but complementary processes (electron loss by one substance and electron gain by another), they are called **half reactions**. The electron loss process is called **oxidation**, and the process in which substances gain electrons is called **reduction**. The two reactions describing electron loss and electron gain when added together give the balanced equation of the reaction.

oxidation half reaction	$2\ Al_{(s)} \rightarrow 2\ Al^{3+}_{(aq)} + 6\ e^-$
reduction half reaction	$6\ H_3O^+_{(aq)} + 6\ e^- \rightarrow 3\ H_{2\ (g)} + 6\ H_2O_{(l)}$
net *or* **redox** reaction	$2\ Al_{(s)} + 6\ H_3O^+_{(aq)} \rightarrow 2\ Al^{3+}_{(aq)} + 3\ H_{2\ (g)} + 6\ H_2O_{(l)}$

Oxidation and reduction processes must occur together: electrons must be transferred in equal numbers from the substance undergoing oxidation (being oxidized) to the substance undergoing reduction (being reduced); otherwise, electrons must be either destroyed or created, which cannot occur because electrons are conserved. Therefore, reactions where electrons are transferred are called reduction-oxidation reactions or **redox** reactions. The two broad classes of redox reactions to be discussed in this section are **metal displacement** and **oxidation by oxygen**.

Metal Displacement and the Activity Series

In metal displacement reactions, a metal ion, M^{n+} (or other ion), reacts with another metal, M', to form a cation of M' and the metal, M:

$$M^{n+}_{(aq)} + M' \rightarrow M'^{m+}_{(aq)} + M$$

The metal ion M^{n+} is said to be displaced from the solution. In metal displacement reactions, one metal (M) is reduced and another metal (M') is oxidized. The relative reactivity of metals and their ions toward displacement is summarized in the activity series. A partial listing of the activity series is shown below.

Relative Reactivity of Metals

most reactive (easily oxidized)			increasing stability of metal $\longrightarrow$						least reactive (difficult to oxidize)			
react with water			react with acid (H_3O^+)						unreactive metals			
metal	K	Ca	Na	Mg	Al	Zn	Fe	Ni	Pb	H_2	Cu	Ag	Au
cation	K^+	Ca^{2+}	Na^+	Mg^{2+}	Al^{3+}	Zn^{2+}	Fe^{2+}	Ni^{2+}	Pb^{2+}	H_3O^+	Cu^{2+}	Ag^+	Au^{3+}
ions difficult to displace (difficult to reduce)			increasing ease of reduction of ion $\longrightarrow$						ions readily displaced (easily reduced)			

It is important to note that ions that are difficult to reduce are derived from reactive (easily oxidized) metals. Conversely, the unreactive metals have ions that are easily displaced (easily reduced). The activity series is useful in predicting whether redox reactions between a metal and a metal ion will occur. The following guidelines are used in applying the activity series:

- Metals that are more reactive than H_2 will be displaced only by metal ions that are more reactive than H_3O^+.

- Metal ions are displaced only by metals that are less reactive than the metal from which the ion is derived.

Chapter 4: Chemical Reactions and Stoichiometry

- The greater the difference in reactivity of the metal and the metal ion (the distance between them in the activity series), the more vigorous the reaction.

The following exercise will illustrate the application of these guidelines.

Exercise 4.12 Describe the results of the following experiments and write the balanced equation of reaction for that reaction.

(a) A piece of iron metal is dipped into a solution containing copper(II) ions.
(b) Silver metal is placed into a solution of Zn^{2+}.
(c) Sodium metal is dropped into water.
(d) A solution containing Au^{3+} has a piece of zinc metal added to it.

Steps to Solution: Examine the positions of the substances on the activity series and use the above guidelines to predict the reactions that will occur. Write the oxidation and reduction half reactions and make sure the same number of electrons must be transferred in the reduction and oxidation half reactions. If necessary, multiply one of the half reactions by an integer factor. Finally, add the half reactions to obtain the net reaction.

(a) *Iron has higher activity than copper. Therefore, iron will displace copper (II) ions from solution. Iron will be oxidized and copper (II) will be reduced.*

oxidation	$Fe \rightarrow Fe^{2+} + 2\ e^-$
reduction	$Cu^{2+} + 2\ e^- \rightarrow Cu$
sum (redox reaction)	$\boxed{Cu^{2+} + Fe \rightarrow Cu + Fe^{2+}}$

(b) *Silver has lower activity than zinc; therefore, silver cannot displace zinc ions. Therefore no reaction will occur.*

(c) *Sodium is one of the most active metal. Examination of the activity series shows that sodium is one of the metals that react with water. Sodium will react with water to give hydrogen and sodium ions.*

oxidation	$Na \rightarrow Na^+ + e^-$
reduction	$2\ H_2O + 2\ e^- \rightarrow H_2 + 2\ OH^-$

To balance this reaction, the same number of electrons must be transferred in the oxidation and reduction half reactions. To have the same number of electrons transferred, the number of electrons transferred in the oxidation process must be doubled. This is accomplished by multiplying the oxidation reaction by 2 and then adding the oxidation and reduction half reactions.

oxidation	$2\ Na \rightarrow 2\ Na^+ + 2\ e^-$
reduction	$2\ H_2O + 2\ e^- \rightarrow H_2 + 2\ OH^-$
sum (redox reaction)	$\boxed{2\ Na + 2\ H_2O \rightarrow 2\ Na^+ + H_2 + 2\ OH^-}$

(d) The redox half ractions are:

oxidation $Zn \rightarrow Zn^{2+} + 2\ e^-$
reduction $Au^{3+} + 3\ e^- \rightarrow Au$

The balanced redox reaction is found by the methods above:

$$\boxed{3\ Zn + 2\ Au^{3+} \rightarrow 3\ Zn^{2+} + 2\ Au}$$

Oxidation by Oxygen

Molecular oxygen is quite reactive and reacts with all but the least active metals to form oxides. When oxygen reacts to form oxides, the oxidation half reaction can be written as:

$$O_2 + 4\ e^- \rightarrow 2\ O^{2-}$$

The highly reactive metals in the activity series oxidize readily-; only copper, silver, and gold do not form oxides. The formulas of metal oxides can be predicted given the charge on the metal cation, because the oxide anion has the characteristic -2 charge.

Exercise 4.13 What is the chemical formula of the following metal oxides and the balanced equation of reaction for its formation?

(a) Barium oxide, used to remove water from reagents
(b) Niobium oxide, where the oxidation number of Nb is +5

Steps to Solution: To predict the product of the reaction, we must find the characteristic charge on the metal and then supply enough oxygen to form that compound.

(a) Barium is a Group 2 metal and forms only ions of +2 charge. Because the oxide ion has a charge of -2, uncharged barium oxide would have the formula BaO. The equation of reaction can be readily balanced by inspection. The unbalanced equation of reaction is:

$$Ba + O_2 \rightarrow BaO$$

Balancing the oxygen gives $2\ Ba_{(s)} + O_{2(g)} \longrightarrow 2\ BaO_{(s)}$, which is balanced.

(b) Niobium has a charge of +5; because the oxide ion has a charge of -2, the formula must be Nb_2O_5 to obtain an electrically neutral compound. The unbalanced equation of reaction is:

$$Nb_{(s)} + O_{2(g)} \rightarrow Nb_2O_{5\ (s)}$$

Balancing O first $Nb_{(s)} + 5\ O_{2\ (g)} \rightarrow 2\ Nb_2O_{5\ (s)}$
Balancing Nb $4Nb_{(s)} + 5\ O_{2(g)} \longrightarrow Nb_2O_{5(s)}$, the balanced equation.

Exercise 4.14 Gold is mined by forming a solution containing the Au^{3+} ion and then displacing the gold ion with an active metal such as zinc. A 4.00×10^4 L pool of such a solution has a density of 1.08 g/mL and contains 0.0051% gold by mass. Treatment with zinc gives a 98% yield of the dissolved gold.

(a) What is the stoichiometric mass of zinc (in kg) required to remove the gold in this solution?
(b) What mass of gold will be recovered by this treatment?

Steps to Solution: This is a multi-step problem. First we must determine the amount of zinc required to reduce the gold in the solution, which requires determining the amount of gold that is present.

(a) The minimum amount of zinc required will be the stoichiometric amount. The balanced chemical equation for the reaction was found in Exercise 4.13(d).

$$3\ Zn + 2\ Au^{3+} \rightarrow 3\ Zn^{2+} + 2\ Au$$

The equation supplies the stoichiometric information needed for the problem: three moles of Zn are required to displace two moles of Au^{3+}.

The next step is to determine the moles of gold present in the solution. From the supplied volume and density of the solution, we can determine the total mass of the solution:

$$\text{mass} = \rho \bullet V = 1.08 \text{ g/mL } (4.00 \times 10^4 \text{ L} \bullet 1000 \text{ mL/L}) = 4.32 \times 10^7 \text{ g solution}$$

The mass of gold contained in the solution is determined using the mass percent gold in the solution supplied: 0.0051%.

$$\text{mass Au} = (4.32 \times 10^7 \text{ g solution}) \bullet \frac{0.0051\%}{100\%} = 2.20 \times 10^3 \text{ g Au}$$

$$\text{mol Au} = \frac{2.20 \times 10^3 \text{ g Au}}{196.97 \text{ g/mol}} = 11.2 \text{ mol Au}$$

The mol of Zn are determined from the mol Au and the stoichiometric ratio found above:

$$\text{mol Zn} = \text{mol Au} \bullet \frac{3 \text{ mol Zn}}{2 \text{ mol Au}}$$

$$\text{mass Zn} = \text{mol Zn}(65.38 \text{ g/mol Zn}) = \text{mol Au} \bullet \frac{3 \text{ mol Zn}}{2 \text{ mol Au}} \bullet (65.38 \text{ g/mol Zn})$$

$$\text{mass Zn} = 1.10 \times 10^3 \text{ g} = \boxed{1.10 \text{ kg}}$$

(b) *The mass of gold recovered by this treatment will be determined by the mass present and the percent yield of the process. Because a total of 2.20 kg of Au were present in the solution(the theoretical yield) and the percent yield is 98%, the amount of gold recovered is:*

$$\text{mass Au} = 2.20 \text{ kg} \frac{98\%}{100\%} = \boxed{2.16 \text{ kg}}$$

Test Yourself

1. How many g of $CaCO_3$ (a popular calcium supplement) would be required to supply the calcium in 1.00 g of $Ca_5(OH)(PO_4)_3$ (the form of calcium found in bones)?

2. Pieces of copper, aluminum and iron are placed into a solution containing Ni^{2+} ions. Which metal is most likely to react?

3. A 0.327 g sample of impure sodium bicarbonate is titrated with 0.124 M H_2SO_4. A total of 14.91 mL is required to reach the stoichiometric point. What is the percent $NaHCO_3$ in the sample? (The reaction that occurs is:
$$2 \text{ NaHCO}_3 + H_2SO_4 \rightarrow Na_2SO_4 + H_2O + CO_2.)$$

4. Hydrogen cyanide, HCN, is manufactured by the following reaction:

$$2 \text{ NH}_{3(g)} + 3 \text{ O}_{2\,(g)} + 2 \text{ CH}_{4\,(g)} \rightarrow 2 \text{ HCN}_{(g)} + 6 \text{ H}_2\text{O}_{(g)}$$

If 250.0 kg each of NH_3, O_2 and CH_4 react, what would be the theoretical yield of HCN?

Answers

1. 1.00 g
2. Aluminum
3. 95.0%
4. 140.8 kg

Solutions to Chapter Problems

4.1 A balanced chemical equation must have equal numbers of atoms of each element on each side of the arrow. Balance each element in turn, beginning with those that appear in only one reactant and product, by adjusting stoichiometric coefficients. Generally, H and O are balanced last.
(a) N is already balanced. Give H_2O a coefficient of 2 to balance H. Then O is also balanced: $NH_4NO_3 \rightarrow N_2O + 2\ H_2O$;
(b) Give H_3PO_4 a coefficient of 4 to balance P. Now there are 12 H on the right; give H_2O a coefficient of 6 to balance H. Then O is also balanced: $P_4O_{10} + 6\ H_2O \rightarrow 4\ H_3PO_4$;
(c) Give HIO_3 a coefficient of 2 to balance I; then H and O are also balanced:
$2\ HIO_3 \rightarrow I_2O_5 + H_2O$;
(d) Give Cl_2 a coefficient of 5/2 to balance Cl. Then multiply through by 2 to clear fractions:
$2\ As + 5\ Cl_2 \rightarrow 2\ AsCl_5$.

4.3 A balanced chemical equation must have equal numbers of atoms of each element on each side of the arrow. Balance each element in turn, beginning with those that appear in only one reactant and product, by adjusting stoichiometric coefficients. Generally, H and O are balanced last.
(a) The reaction is $N_2 + H_2 \rightarrow NH_3$. Give NH_3 a coefficient of 2 to balance N; then give H_2 a coefficient of 3 to balance H: $N_2 + 3\ H_2 \rightarrow 2\ NH_3$;
(b) The reaction is $H_2 + CO \rightarrow CH_3OH$. C and O are balanced. Give H_2 a coefficient of 2 to balance H: $2\ H_2 + CO \rightarrow CH_3OH$;
(c) Give C a coefficient of 3 to balance C; then all elements are balanced:
$CaO + 3\ C \rightarrow CO + CaC_2$;
(d) Give HCl a coefficient of 2 to balance Cl; then H is also balanced. Give O_2 a coefficient of 1/2 to balance O; then multiply through by 2 to clear fractions:
$2\ C_2H_4 + O_2 + 4\ HCl \rightarrow 2\ C_2H_4Cl_2 + 2\ H_2O$.

4.5 Molecular pictures of reactions must show the correct number of molecules undergoing reaction. In 4.3a, one molecule of N_2 reacts with three of H_2 to give two of NH_3:

Chapter 4: Chemical Reactions and Stoichiometry

4.7 A balanced chemical equation has equal numbers of atoms of each element on each side of the arrow. Balance each element in turn, beginning with those that appear in only one reactant and product, by adjusting stoichiometric coefficients. Generally, H and O are balanced last.
(a) Give $Ca(OH)_2$ a coefficient of 3 and H_3PO_4 a coefficient of 2 to balance Ca and P. There are now 12 H atoms on the left, so give H_2O a coefficient of 6 to balance H, whereupon O is also balanced (14 on each side):
$$3\ Ca(OH)_2 + 2\ H_3PO_4 \rightarrow 6\ H_2O + Ca_3(PO_4)_2;$$
(b) Give NaOH a coefficient of 2 to balance Na; then give H_2O a coefficient of 2 to balance H, whereupon O is also balanced (4 on each side):
$$Na_2O_2 + 2\ H_2O \rightarrow 2\ NaOH + H_2O_2;$$

(c) Give HF a coefficient of 3 to balance F; then give H_2O a coefficient of 3 to balance H, whereupon O is also balanced (3 on each side):
$$BF_3 + 3\ H_2O \rightarrow 3\ HF + H_3BO_3;$$

(d) Give NH_3 a coefficient of 2 to balance N; then give H_2O a coefficient of 3 to balance H. Now give CuO a coefficient of 3 to balance O and Cu a coefficient of 3 to balance Cu:
$$2\ NH_3 + 3\ CuO \rightarrow 3\ Cu + N_2 + 3\ H_2O.$$

4.9 Remember that calculations of amounts in chemistry always center on the mole. Convert mass of the first starting material into moles by dividing by molar mass, multiply by the stoichiometric ratio to obtain moles of the second starting material, then convert moles to mass by multiplying by molar mass:

(a) $\left(\dfrac{5.00\ g\ N_2}{28.02\ g/mol}\right)\left(\dfrac{3\ mol\ H_2}{1\ mol\ N_2}\right)(2.016\ g/mol) = 1.08\ g\ H_2;$

(b) $\left(\dfrac{5.00\ g\ H_2}{2.016\ g/mol}\right)\left(\dfrac{1\ mol\ CO}{2\ mol\ H_2}\right)(28.01\ g/mol) = 34.7\ g\ CO;$

(c) $\left(\dfrac{5.00\ g\ CaO}{56.08\ g/mol}\right)\left(\dfrac{3\ mol\ C}{1\ mol\ CaO}\right)(12.01\ g/mol) = 3.21\ g\ C;$ and

(d) $\left(\dfrac{5.00\ g\ C_2H_4}{28.05\ g/mol}\right)\left(\dfrac{1\ mol\ O_2}{2\ mol\ C_2H_4}\right)(32.00\ g/mol) = 2.85\ g\ O_2.$

Chapter 4: Chemical Reactions and Stoichiometry

4.11 To calculate masses for a chemical reaction, balance the equation, then do appropriate mass-mole-mass conversions.
The reaction is $CCl_4 + HF \rightarrow CCl_2F_2 + HCl$. C is balanced. Balance F by giving HF a coefficient of 2; then balance H by giving HCl a coefficient of 2, whereupon Cl is also balanced (4 on each side): $CCl_4 + 2\,HF \rightarrow CCl_2F_2 + 2\,HCl$.
Mass of HF required:

$$\left(\frac{(175 \text{ kg } CCl_4)(10^3 \text{ g/kg})}{153.8 \text{ g/mol}}\right)\left(\frac{2 \text{ mol HF}}{1 \text{ mol } CCl_4}\right)(20.01 \text{ g/mol})(10^{-3} \text{ kg/g}) = 45.5 \text{ kg HF};$$

Masses of products formed:

$$\left(\frac{(175 \text{ kg } CCl_4)(10^3 \text{ g/kg})}{153.8 \text{ g/mol}}\right)\left(\frac{1 \text{ mol } CCl_2F_2}{1 \text{ mol } CCl_4}\right)(120.9 \text{ g/mol})(10^{-3} \text{ kg/g}) = 138 \text{ kg } CCl_2F_2;$$

$$\left(\frac{(175 \text{ kg } CCl_4)(10^3 \text{ g/kg})}{153.8 \text{ g/mol}}\right)\left(\frac{2 \text{ mol HCl}}{1 \text{ mol } CCl_4}\right)(36.46 \text{ g/mol})(10^{-3} \text{ kg/g}) = 83.0 \text{ kg HCl}.$$

4.13 You are asked to calculate the mass of a chemical reactant. Calculations of amounts in chemistry always center on the mole. Convert mass of the product into moles by dividing by molar mass, multiply by the stoichiometric ratio to obtain moles of the starting material, then convert moles to mass by multiplying by molar mass:

$$\left(\frac{(1.50 \text{ kg } I_2)(10^3 \text{ g/kg})}{253.8 \text{ g/mol}}\right)\left(\frac{2 \text{ mol NaI}}{1 \text{ mol } I_2}\right)(149.9 \text{ g/mol}) = 177 \text{ g NaI}$$

4.15 You are asked to calculate the mass of a chemical product. Calculations of amounts in chemistry always center on the mole. Convert mass (in grams) of the reactant into moles by dividing by molar mass, multiply by the stoichiometric ratio to obtain moles of the product, then convert moles to mass by multiplying by molar mass.
The molar mass of sugar ($C_6H_{12}O_6$) is 180.2 g/mol.

$$\left(\frac{(1.00 \text{ kg sugar})(10^3 \text{ g/kg})}{180.2 \text{ g/mol}}\right)\left(\frac{2 \text{ mol } C_2H_5OH}{1 \text{ mol sugar}}\right)(46.07 \text{ g/mol}) = 511 \text{ g } C_2H_5OH$$

4.17 You are asked to calculate moles for a chemical reaction. To do this requires the use of mole ratios, which in turn requires a balanced chemical equation. Balance the equation, then do appropriate mass-mole-mass conversions:

The reaction is $C_5H_5N + O_2 \rightarrow H_2O + CO_2 + N_2$. Balance N by giving C_5H_5N a coefficient of 2. Then CO_2 requires a coefficient of 10 and H_2O requires a coefficient of 5. This gives 25 O atoms on the right, so give O_2 a coefficient of 25/2 and multiply through by 2 to clear fractions:
$4\,C_5H_5N + 25\,O_2 \rightarrow 10\,H_2O + 20\,CO_2 + 2\,N_2$.

Use the density equation to calculate the mass of pyridine: $m = \rho V$

$$\left(\frac{(0.982 \text{ g/mL})(2.95 \text{ mL } C_5H_5N)}{79.1 \text{ g/mol}}\right)\left(\frac{10 \text{ mol } H_2O}{4 \text{ mol } C_5H_5N}\right) = 9.16 \times 10^2 \text{ mol } H_2O$$

4.19 In a yield problem, we compare actual amounts with theoretical amounts. Carry out standard mole-mass conversions to find the theoretical amount (FA = fluoroapatite).

The reaction is $Ca_3(PO_4)_3F + 5\ H_2SO_4 + 10\ H_2O \rightarrow 3\ H_3PO_4 + 5\ CaSO_4 \cdot 2H_2O + HF$. Use stoichiometry to calculate the theoretical yield. $MM_{FA} = 504.3$ g/mol:

$$\left(\frac{(1.00\ \text{kg FA})(10^3\ \text{g/kg})}{504.3\ \text{g/mol}}\right)\left(\frac{3\ \text{mol H}_3\text{PO}_4}{1\ \text{mol FA}}\right)(97.99\ \text{g/mol}) = 5.83 \times 10^2\ \text{g H}_3\text{PO}_4$$

Now use the actual amount to determine the percent yield:

$$\%\ \text{yield} = (100\%)\left(\frac{\text{actual amt}}{\text{theor amt}}\right) = (100\%)\left(\frac{4.00 \times 10^2\ \text{g}}{5.83 \times 10^2\ \text{g}}\right) = 68.6\%.$$

4.21 In a multiple-step synthesis, the yield of each step must be multiplied to determine the overall yield. In this case, there are eight steps, each with a yield of 88%, so the overall fractional yield is $(0.88)(0.88)(0.88)(0.88)(0.88)(0.88)(0.88)(0.88) = (0.88)^8 = 0.36$ Use this value and the desired amount of phenobarbitone in moles to determine how many moles of toluene are required. Do the usual mole-mass conversions. For phenobarbitone, $MM = 232.2$ g/mol.

$$\text{Actual yield} = \frac{(25\ \text{kg})(10^3\ \text{g/kg})}{232.2\ \text{g/mol}} = 107.7\ \text{mol};$$

$$\text{Theoretical yield} = \text{Moles required} = \frac{107.7\ \text{mol}}{0.36} = 299\ \text{mol};$$

Mass required = $(299\ \text{mol})(92.13\ \text{g/mol})(10^{-3}\ \text{kg/g}) = 28$ kg.

4.23 The calculation of a yield requires a theoretical amount and an actual amount. The theoretical amount is calculated as in Problem 4.11:

$$\left(\frac{(175\ \text{kg CCl}_4)(10^3\ \text{g/kg})}{153.8\ \text{g/mol}}\right)\left(\frac{1\ \text{mol CCl}_2\text{F}_2}{1\ \text{mol CCl}_4}\right)(120.9\ \text{g/mol})(10^{-3}\ \text{kg/g}) = 137.6\ \text{kg CCl}_2\text{F}_2;$$

(carry an additional significant figure until the calculation is complete)
The actual amount (given) is 105 kg. The yield is $(100\%)(105\ \text{kg}/137.6\ \text{kg}) = 76.3\%$.

To find how much of each reactant is required, divide the desired amount by the percent yield to obtain the theoretical yield and then do the usual stoichiometric calculations:

$$\text{Theoretical yield} = \frac{155\ \text{kg}}{(10^{-2}/\%)(76.3\%)} = 203\ \text{kg};$$

$$\left(\frac{(203\ \text{kg CCl}_2\text{F}_2)(10^3\ \text{g/kg})}{120.9\ \text{g/mol}}\right)\left(\frac{1\ \text{mol CCl}_4}{1\ \text{mol CCl}_2\text{F}_2}\right)(153.8\ \text{g/mol})(10^{-3}\ \text{kg/g}) = 258\ \text{kg CCl}_4;$$

$$\left(\frac{(203\ \text{kg CCl}_2\text{F}_2)(10^3\ \text{g/kg})}{120.9\ \text{g/mol}}\right)\left(\frac{2\ \text{mol HF}}{1\ \text{mol CCl}_2\text{F}_2}\right)(20.01\ \text{g/mol})(10^{-3}\ \text{kg/g}) = 67.4\ \text{kg HF}.$$

Chapter 4: Chemical Reactions and Stoichiometry

4.25 To determine which ingredient will run out first, calculate how many cheeseburgers could be made with each ingredient:

INGREDIENT	INVENTORY	AMT/BURGER	# BURGERS
roll	(12 dozen)(12/dozen)	1	144
beef	40 lb	2(1/4 lb)	80
cheese	(2 pkg)(65/pkg)	1	130
tomato	40	1/4	160
lettuce	(1 kg)(10^3 g/kg)	15 g	66.7

The lettuce will run out first, after 66 burgers have been made.

4.27 When amounts of all reactants are specified, one reactant will be limiting. Use amount tables and divide starting amount by coefficient to find the limiting reagent (LR). Starting amounts are in metric tons (1 metric ton = 10^6 g), so it is convenient to work with 10^6 mol amounts.

(a) Reaction	N_2 +	$3 H_2$ →	$2 NH_3$
MM (g/mol)	28.02	2.016	17.03
Amt (10^6 mol)	(1.00/28.02) = .0357	(1.00/2.016) = .496	0
10^6 mol/coeff	.0357 (LR)	.165	
Change (10^6 mol)	− 0.0357	−3(0.0357)	+2(0.0357)
Final (10^6 mol)	0	.389	.0714

The mass that could be produced is (0.0714 x 10^6 mol)(17.03 g/mol) = 1.22 metric ton

(b) Reaction	$2 H_2$ +	CO →	CH_3OH
MM (g/mol)	2.016	28.01	32.03
Amt (10^6 mol)	(1.00/2.016) = .496	(1.00/28.01) = .0357	0
10^6 mol/coeff	.248	.0357 (LR)	
Change (10^6 mol)	− 2(0.0357)	−0.0357	+0.0357
Final (10^6 mol)	.176	0	0.0357

The mass that could be produced is (0.0357 x 10^6 mol)(32.03 g/mol) = 1.14 metric ton

(c) Reaction	CaO +	3 C →	CO +	CaC_2
MM (g/mol)	56.08	12.01	28.01	64.10
Amt (10^6 mol)	.0178	.0833	0	0
10^6 mol/coeff	.0178 (LR)	.0278		
Change (10^6 mol)	−.0178	−3(.0178)	+.0178	+.0178
Final (10^6 mol)	0	.0299	+.0178	+.0178

The masses that could be produced are:
CO: (0.0178 x 10^6 mol)(28.01 g/mol) = 0.499 metric ton;
CaC_2: (0.0178 x 10^6 mol)(64.10 g/mol) = 1.14 metric ton

(d) Reaction	2 C$_2$H$_4$ +	O$_2$ +	4 HCl →	2 C$_2$H$_4$Cl$_2$	+ 2 H$_2$O
MM (g/mol)	28.05	32.00	36.46	98.97	18.01
Amt (10^6 mol)	.0357	.0313	.0274	0	0
10^6 mol/coeff	.0179	.0313	.00685 (LR)		
Change (10^6 mol)	- (2/4)(.0274)	-(.0274/4)	-.0274	+.(2/4)(.0274)	+.0137
Final (10^6 mol)	.022	.0245	0	+.0137	+.0137

The masses that could be produced are:
C$_2$H$_4$Cl$_2$: (0.0137 x 10^6 mol)(98.97 g/mol) = 1.36 metric ton;
H$_2$O: (0.0137 x 10^6 mol)(18.01 g/mol) = 0.247 metric ton

4.29 When amounts of all reactants are specified, one reactant will be limiting, so these are limiting reagent problems. Use an amount table and divide starting amount by coefficient to determine which is the limiting reagent (LR).

Reaction	SiO$_2$ +	3 C →	SiC +	2 CO$_2$
MM (g/mol)	60.086	12.01	40.10	44.01
Amt (mol)	.832	4.16	0	0
mol/coeff	.832 (LR)	1.39		
Change (mol)	-.832	-3(.832)	+.832	+(2)(.832)
Final (mol)	0	1.66	+.832	+1.66

The mass that could be produced is: (0.832 mol)(40.10 g/mol) = 33.4 g.

4.31 When amounts of all reactants are specified, one reactant will be limiting, so these are limiting reagent problems. Use an amount table and divide starting amount by coefficient to determine which is the limiting reagent (LR).

Reaction	P$_4$ +	5 O$_2$ →	P$_4$O$_{10}$
MM (g/mol)	123.9	32.00	283.9
Amt (mol)	(3.75/123.9) = .0303	(6.55/32.00) = .205	0
mol/coeff	.0303 (LR)	.0410	
Change (mol)	- .0303	-(5)(.0303)	+0.0303
Final (mol)	0	0.0535	0.0303

The mass that could be produced is (0.0303 mol)(283.9 g/mol) = 8.60 g.
There would be (0.0535 mol)(32.00 g/mol) = 1.70 g of O$_2$ left unreacted.

4.33 Identify species based on the type of substance present: Ions for salts and strong acids, molecules for other substances. H$_2$O is always a major species in aqueous solution.
(a) Salt: NH$_4^+$, Cl$^-$, H$_2$O; (b) Salt: Fe^{2+}, ClO$_4^-$, H$_2$O; (c) Salt: Na$^+$, SO$_4^{2-}$, H$_2$O;
(d) Not a salt: Br$_2$, H$_2$O; (e) Salt: K$^+$, Br$^-$, H$_2$O.

4.35 Identify species based on the type of substance present: Ions for salts and strong acids, molecules for other substances. H$_2$O is always a major species in aqueous solution.
(a) Salt: K$^+$, HPO$_4^-$, H$_2$O; (b) Weak acid: CH$_3$CO$_2$H, H$_2$O; (c) Salt: Na$^+$, CH$_3$CO$_2^-$, H$_2$O; (d) Weak base: NH$_3$, H$_2$O; (e) Salt: NH$_4^+$, Cl$^-$, H$_2$O.

Chapter 4: Chemical Reactions and Stoichiometry

4.37 To determine whether a precipitate will form, consider the solubility guidelines. If any combination of cations and anions is insoluble, that salt will precipitate.
(a) $AgNO_3$: All nitrates are soluble, so a precipitate forms only if the silver salt is insoluble.   (a) AgCl precipitates; (b, c, and d) No precipitate; (e) AgCl precipitates.

(b) Na_2CO_3: All sodium salts are soluble, so a precipitate forms only if the carbonate salt is insoluble.   (a) No precipitate; (b) $FeCO_3$ precipitates; (c, d, and e) No precipitate.

(c) $Ba(OH)_2$: Both ions form insoluble salts unless their partners convey solubility.
    (a, d, and e) No precipitate; (b) $Fe(OH)_2$ precipitates; (c) $BaSO_4$ precipitates.

4.39 A net ionic equation shows which ions combine to give new products. Spectator ions do not appear in the net equation, and charges must be balanced.

(a) $Ag^+ + OH^- \rightarrow AgOH_{(s)}$;
(b) $2\ Fe^{3+} + 3\ C_2O_4^{2-} \rightarrow Fe_2(C_2O_4)_{3(s)}$;
(c) No reaction;
(d) $Ni^{2+} + 2\ OH^- \rightarrow Ni(OH)_{2(s)}$.

4.41 This is a limiting reagent problem involving solutions of ionic substances. Start with the balanced net ionic reaction and set up an amounts table. Calculate amounts using the equation $n = MV$:

Reaction	$2\ Ag^+\ +$	$CO_3^{2-}\ \rightarrow$	$Ag_2CO_{3(s)}$
Start (10^{-3} mol)	2.75	3.33	0
Change (10^{-3} mol)	-2.75	-2.75/2	+2.75/2
Final (10^{-3} mol)	0	1.95	1.38

The mass of solid that forms is $(1.38 \times 10^{-3}\ mol)(275.7\ g/mol) = 0.381$ g. The ions remaining in solution are the excess CO_3^{2-} and the spectators, NO_3^- and K^+.

4.43 Acids generate hydronium ions, and bases generate hydroxide ions, in aqueous solution. You must memorize those that are strong; all others are weak.
(a) Nitric acid (strong acid); (b) acetic acid (weak acid); (c) phosphoric acid (weak acid); (d) calcium oxide (strong base, generates OH^- in water); and (e) acetone (not acid or base).

4.45 This is a stoichiometry problem involving an acid-base reaction. Determine the net ionic reaction and then use mole-mass conversions to find the mass:
The balanced reaction is $Al(OH)_{3(s)} + 3\ H_3O^+ \rightarrow Al^{3+}_{(aq)} + 6\ H_2O_{(l)}$.
$n_{H_3O^+} = MV = (0.175\ mol/L)(155\ mL)(10^{-3}\ L/mL) = 2.71 \times 10^{-2}$ mol;

$$m_{Al(OH)_3} = \left(\frac{1\ mol\ Al(OH)_3}{3\ mol\ H_3O^+}\right)(mol\ H_3O^+)(MM_{Al(OH)_3})$$

$m_{Al(OH)_3} = (1/3)(2.71 \times 10^{-2}\ mol)(78.0\ g/mol) = 0.704$ g.

Chapter 4: Chemical Reactions and Stoichiometry

4.47 This is a stoichiometry problem involving an acid-base reaction. Information about both reactants are given, so it is a limiting reagent problem, and an amounts table is called for. First, the net reaction must be identified. The species present are H_2O, Ba^{2+}, OH^-, H_3O^+, and Cl^-. Of these, only OH^- and H_3O^+ react:

Reaction	H_3O^+ +	OH^- →	$2 H_2O$
Start (10^{-3} mol)	5.00	6.00	0
Change (10^{-3} mol)	-5.00	-5.00	+ 10.00
Final (10^{-3} mol)	0	1.00	10.00

The final volume is the sum of volumes of the two solutions, 2.50×10^2 mL = 0.250 L. Do dilution calculations to find the final concentrations of the spectator ions.
$[H_3O^+] = 0$; $[OH^-] = (1.00 \times 10^{-3}$ mol$) \div (0.250$ L$) = 4.00 \times 10^{-3}$ M.
$[Ba^{2+}] = (2.00 \times 10^{-2}$ M$)(1.50 \times 10^2$ mL$) \div (2.50 \times 10^2$ mL$) = 1.20 \times 10^{-2}$ M;
$[Cl^-] = (5.00 \times 10^{-2}$ M$)(1.00 \times 10^2$ mL$) \div (2.50 \times 10^2$ mL$) = 2.00 \times 10^{-2}$ M.

4.49 This is a stoichiometry problem involving a chemical reaction and formation of a solution. Concentration is expressed in molarity, moles of substance divided by volume of solution. First determine the balanced chemical equation, then do standard mass-mole conversions to calculate the amount of phosphoric acid formed; finally, divide by volume to find molarity. The unbalanced reaction is $P_4O_{10} + H_2O \rightarrow H_3PO_4$. Give H_3PO_4 a coefficient of 4 to balance P and H_2O a coefficient of 6 to balance H; then O is also balanced (16 on each side): $P_4O_{10} + 6 H_2O \rightarrow 4 H_3PO_4$.
$n_{P_4O_{10}} = m/MM = (1.00 \times 10^2$ g$) \div (283.9$ g/mol$) = 0.3522$ mol;
$n_{H_3PO_4} = (4)(0.3522) = 1.409$ mol;
$[H_3PO_4] = n/V = (1.409$ mol$) \div (1.55$ L$) = 0.909$ M.

4.51 A redox reaction between a metal and a cation occurs when the metal is higher in the activity series than the cation. The activity series appears as Table 4-2 of your textbook.

(a and b) Cu is below H_2 and Mg in the activity series, so no reaction occurs;
(c) Cu is above Ag in the activity series, so a displacement reaction occurs:
$$Cu_{(s)} + 2 Ag^+_{(aq)} \rightarrow Cu^{2+}_{(aq)} + 2 Ag_{(s)};$$
(d) K is above H_2 in the activity series, so a displacement reaction occurs:
$$2 K_{(s)} + 2 H_2O \rightarrow 2 K^+_{(aq)} + H_{2(g)} + 2 OH^-_{(aq)}$$

4.53 Formation reactions are easily balanced once the chemical formula of the oxide is known. Remember that O carries a charge of –2 and that the compound must be electrically neutral.

(a) Sr is in Group 2, so it forms a +2 cation and its oxide is SrO:
$$2 Sr + O_2 \rightarrow 2 SrO;$$
(b) Cr(III) has +3 charge and its oxide is Cr_2O_3:
$$4 Cr + 3 O_2 \rightarrow 2 Cr_2O_3;$$
(c) Sn(IV) has +4 charge and its oxide is SnO_2:
$$Sn + O_2 \rightarrow SnO_2.$$

4.55 This is a stoichiometry problem involving a redox reaction. Information is provided about both reactants, so it is a limiting reagent problem and a table of amounts is called for. To determine the net reaction, remember that Al forms Al^{3+} cations and that charge must be balanced: $\quad 6 H_3O^+ + 2 Al \rightarrow 2 Al^{3+} + 3 H_2 + 3 H_2O$.

Because the problem asks only about the mass of H_2 formed, we do not need to do calculations on the other products.

Reaction	$6 H_3O^+ +$	$2 Al \rightarrow$	$3 H_2$
Amt (10^{-2} mol)	4.8	1.316	0
10^{-2} mol/coeff	0.8	0.658 (LR)	
Change (10^{-2} mol)	-3(1.316)	-1.316	+(3/2)(1.316)
Final (10^{-2} mol)	0.852	0	1.974
Mass (g)			3.97×10^{-2} g

4.57 Examine the molecular picture to see what chemical reaction occurs. The reactants are atomic X and atomic Y and the product of the reaction is YX_2 (there are atoms of X left over). Thus, the reaction is (d) $Y + 2 X \rightarrow YX_2$.

4.59 The first two parts of this problem ask about an ionic reaction. Identify the species present in the solution and use the products to find the net reaction and the spectator ions. The species present are H_2O, NH_4^+, HCO_3^-, Na^+, and Cl^-. The solid product is $NaHCO_3$.

(a) The net reaction is $Na^+ + HCO_3^- \rightarrow NaHCO_{3(s)}$;
(b) NH_4^+ and Cl^- are spectator ions;
The second two parts of this problem involve stoichiometric calculations. This is a limiting reagent problem as well as a yield problem, requiring a table of amounts to determine the theoretical yield.

(c) Reaction	$Na^+ +$	$HCO_3^- \rightarrow$	$NaHCO_3$
Start (10^2 mol)	30.0	7.50	0
Change (10^2 mol)	-7.50	-7.50	+7.50
Final (10^2 mol)	22.5	0	7.50

The theoretical yield is 7.50×10^2 mol.
The actual yield is (35.0 kg)(10^3 g/kg) ÷ (84.0 g/mol) = 4.17×10^2 mol;
The percent yield is (100%)(4.17×10^2 mol) ÷ (7.50×10^2 mol) = 56%
(Low because $NaHCO_3$ is relatively soluble, so many ions remain in solution)..

(d) To find concentrations in the final solution, first determine how many moles of each ion remain in solution, then divide by the final volume of the solution, which is 10.00×10^2 L. The amounts of the two spectator ions are unaffected by the reaction:
$[NH_4^+]$ = (7.50×10^2 mol) ÷ (10.00×10^2 L) = 0.750 M;
$[Cl^-]$ = (30.0×10^2 mol) ÷ (10.00×10^2 L) = 3.00 M;
The amounts of the reactant ions are reduced by the amount that precipitates:
$[Na^+]$ = (30.0×10^2 mol − 4.17×10^2 mol) ÷ (10.00×10^2 L) = 2.58 M;
$[HCO_3^-]$ = (7.50×10^2 mol − 4.17×10^2 mol) ÷ (10.00×10^2 L) = 0.33 M.

Chapter 4: Chemical Reactions and Stoichiometry

4.61 The formation reactions for oxides are easily balanced by inspection. The number of electrons lost by the metal atom can be determined by assigning two additional electrons to each oxygen atom in the oxide.
$4 Ru + 3 O_2 \rightarrow 2 Ru_2O_3$. Three O atoms gain six electrons, so each Ru loses three.
$Ru + O_2 \rightarrow RuO_2$. Two O atoms gain four electrons, so each Ru loses four.
$2 Ru + 3 O_2 \rightarrow 2 RuO_3$. Three O atoms gain six electrons, so each Ru loses six.
$Ru + 2 O_2 \rightarrow RuO_4$. Four O atoms gain eight electrons, so each Ru loses eight.

4.63 The problem asks for a yield, so this is a yield problem; it is specified that excess fluorine is present, so the stoichiometric calculation is a straightforward mole-mass conversion.

The reaction is $Xe + 2 F_2 \rightarrow XeF_4$.
Theoretical yield = $\left(\dfrac{5.00 \text{ g Xe}}{131.3 \text{ g/mol}}\right)\left(\dfrac{1 \text{ mol } XeF_4}{1 \text{ mol Xe}}\right) = 3.81 \times 10^{-2}$ mol;
Actual yield = (4.00 g) ÷ (207.3 g/mol) = 1.93×10^{-2} mol;
% yield = (100%)(1.93×10^{-2} mol) ÷ (3.81×10^{-2} mol) = 50.6%;
Amount unreacted = (3.81×10^{-2} mol – 1.93×10^{-2} mol)(131.3 g/mol) = 2.47 g.

4.65 In this problem, we must calculate how much ilmenite is in the ore, then how much titanium is in the ilmenite, and finally how much of the titanium will be extracted.
$m_{ilmenite}$ = (15%)(10^{-2}/%)(1.00×10^3 ton) = 1.5×10^2 ton;

m_{Ti} = ($m_{ilmenite}$) ($MM_{Ti}/MM_{ilmenite}$) = (1.5×10^2 ton)$\left(\dfrac{47.88 \text{ g/mol}}{151.7 \text{ g/mol}}\right)$ = 47.3 ton;

Mass recovered = (95%)(10^{-2}/%)(47.3 ton) = 45 metric tons.

4.67 When determining reaction products, first identify the type of substances present to determine what kind of reaction can occur.

(a) $NaHSO_4$ is a salt, but HSO_4^- is a weak acid and KOH is a strong base:
$HSO_4^- + OH^- \rightarrow SO_4^{2-} + H_2O$ (acid-base reaction);
(b) Oxygen can oxidize metals to metal oxides:
$2 Al + 3 O_2 \rightarrow 2 Al_2O_3$ (redox reaction);
(c) Oxygen can oxidize hydrocarbons to carbon dioxide and water:
$C_3H_8 + 5 O_2 \rightarrow 3 CO_2 + 4 H_2O$ (redox reaction);
(d) Mg displaces hydrogen gas from strong acids such as HBr:
$Mg + 2 H_3O^+ \rightarrow Mg^{2+} + H_2 + 2 H_2O$ (redox reaction);
(e) Na_2O is a soluble salt which generates the O^{2-} ion. This is a strong base that reacts with water: $Na_2O + H_2O \rightarrow 2 Na^+ + 2 OH^-$ (acid-base reaction);
(f) Both compounds are salts, so a precipitate will form if any combination is insoluble:
$Pb^{2+} + S^{2-} \rightarrow PbS_{(s)}$ (precipitation reaction).

Chapter 4: Chemical Reactions and Stoichiometry

4.69 The first step in working an analysis problem is to determine what chemical reactions are occurring. Here, the possible reactants are Zn, Cu, H_3O^+, Cl^-, and H_2O, and one product is H_2. The product alerts us that a redox reaction has occurred, and the hint tells us what that redox reaction must be: $Zn + 2\, H_3O^+ \rightarrow H_2 + Zn^{2+} + 2\, H_2O$.

Knowing the chemistry, we can now perform standard mole-mass conversions to determine how much Zn reacted from the amount of H_2 produced:
$n_{Zn} = n_{H_2} = m/MM = (21.3 \times 10^{-3}\ g) \div (2.016\ g/mol) = 1.057 \times 10^{-2}\ mol$;
$m_{Zn} = n\, MM = (1.057 \times 10^{-2}\ mol)(65.39\ g/mol) = 0.690\ g$;

% Zn = $(100\%)(m_{Zn}/m_{total}) = (100\%)(0.690\ g) \div (5.73\ g) = 12.0\%$;
% Cu = $100\% - 12.0\% = 88.0\%$.

4.71 A balanced chemical equation has equal numbers of atoms of each element on each side of the arrow. Adjust stoichiometric coefficients to balance each element in turn, beginning with those that appear in only one reactant and product. Generally, balance H and O last.

(a) N and O are already balanced. Give H_2 a coefficient of 5/2 to balance H, then multiply through by 2 to clear fractions:
$$5\, H_2 + 2\, NO \rightarrow 2\, NH_3 + 2\, H_2O;$$
(b) C is already balanced. Give NO a coefficient of 2 to balance N; then O is also balanced:
$$CO + 2\, NO \rightarrow N_2O + CO_2;$$
(c) Give NH_3 a coefficient of 2 to balance N and H_2O a coefficient of 3 to balance H; then there are 4 O on the right, so give O_2 a coefficient of 2 to balance O:
$$2\, NH_3 + 2\, O_2 \rightarrow N_2O + 3\, H_2O;$$
(d) The reaction is $NO + NH_3 \rightarrow N_2 + H_2O$. N appears in three reagents, so balance it last. Give NH_3 a coefficient of 2 and H_2O a coefficient of 3 to balance H, then give NO a coefficient of 3 to balance O. Now there are 5 N on the left, so give N_2 a coefficient of 5/2 to balance N and multiply through by 2 to clear fractions:
$$6\, NO + 4\, NH_3 \rightarrow 5\, N_2 + 6\, H_2O;$$
(e) The reaction is $H_2O + NO \rightarrow O_2 + NH_3$. Give NH_3 a coefficient of 2 and H_2O a coefficient of 3 to balance H, then give NO a coefficient of 2 to balance N. Now there are 5 O on the left, so give O_2 a coefficient of 5/2 to balance O and multiply through by 2 to clear fractions:
$$6\, H_2O + 4\, NO \rightarrow 5\, O_2 + 4\, NH_3;$$
(f) The reaction is $H_2 + O_2 \rightarrow H_2O$. Give H_2O a coefficient of 2 to balance O, then give H_2 a coefficient of 2 to balance H:
$$2\, H_2 + O_2 \rightarrow 2\, H_2O.$$

4.73 A convenient synthesis of ionic salts starts with solutions of soluble salts of the cation and anion which, when mixed, form a precipitate of the desired product. Sodium salts are inexpensive soluble sources of anions and chlorides or nitrates are inexpensive soluble sources of cations. Do mole-mass conversions to compute the required masses.

(a) Mix solutions of Na_3PO_4 and $FeCl_3$. The net reaction is $Fe^{3+} + PO_4^{3-} \rightarrow FePO_{4(s)}$;
$n_{FePO_4} = m/MM = (2.50 \text{ kg})(10^3 \text{ g/kg}) \div (150.77 \text{ g/mol}) = 16.6$ mol;
$n_{FeCl_3} = n_{Na_3PO_4} = n_{FePO_4}$;
$m_{FeCl_3} = n\ MM = (16.6 \text{ mol})(162.2 \text{ g/mol}) = 2.69 \times 10^3$ g;
$m_{Na_3PO_4} = n\ MM = (16.6 \text{ mol})(163.94 \text{ g/mol}) = 2.72 \times 10^3$ g;

(b) Mix solutions of NaOH and $ZnCl_2$. The net reaction is $Zn^{2+} + 2\ OH^- \rightarrow Zn(OH)_{2(s)}$;
$n_{Zn(OH)_2} = m/MM = (2.50 \text{ kg})(10^3 \text{ g/kg}) \div (99.41 \text{ g/mol}) = 25.1$ mol;
$n_{ZnCl_2} = n_{Zn(OH)_2}$; $n_{NaOH} = 2\ n_{Zn(OH)_2}$;
$m_{ZnCl_2} = n\ MM = (25.1 \text{ mol})(111.4 \text{ g/mol}) = 2.80 \times 10^3$ g;
$m_{NaOH} = n\ MM = 2\ (25.1 \text{ mol})(40.01 \text{ g/mol}) = 2.01 \times 10^3$ g;

(c) Mix solutions of Na_2CO_3 and $NiCl_2$. The net reaction is
$Ni^{2+} + CO_3^{2-} \rightarrow NiCO_{3(s)}$;
$n_{NiCO_3} = m/MM = (2.50 \text{ kg})(10^3 \text{ g/kg}) \div (118.7 \text{ g/mol}) = 21.1$ mol;
$n_{NiCl_2} = n_{Na_2CO_3} = n_{NiCO_3}$;
$m_{NiCl_2} = n\ MM = (21.1 \text{ mol})(129.59 \text{ g/mol}) = 2.73 \times 10^3$ g;
$m_{Na_2CO_3} = n\ MM = (21.1 \text{ mol})(105.99 \text{ g/mol}) = 2.24 \times 10^3$ g.

4.75 The quantity asked for is molarity of a solution of base, and the analysis involves a titration. Titration calculations are based on the mole relationship at the stoichiometric point. For an acid-base titration, $n_{acid} = n_{base}$. The acid is a solid, for which $n = m/MM$, and the base is a solution, for which $n = MV$.

$n_{base} = n_{acid} = (0.7996 \text{ g}) \div (204.2 \text{ g/mol}) = 3.916 \times 10^{-3}$ mol;
$V_{base} = (43.75 \text{ mL} - 0.15 \text{ mL})(10^{-3} \text{ L/mL}) = 4.360 \times 10^{-2}$ L;
$M = n_{base}/V = (3.916 \times 10^{-3} \text{ mol}) \div (4.360 \times 10^{-2} \text{ L}) = 8.982 \times 10^{-2}$ M.

4.77 Molecular pictures and limiting reagent calculations require a balanced chemical equation for the reaction under consideration. For this process, the balanced equation is
$N_2 + 3\ H_2 \rightarrow 2\ NH_3$.

(a) There are 15 H_2 molecules and 6 N_2 molecules in this picture. Divide number of molecules by coefficient to see which reactant is limiting: 15/3 = 5 for H_2, 6/1 = 6 for N_2. The smaller number identifies the limiting reagent, H_2.

(b) The new molecular picture must show that all the H_2 has been consumed and NH_3 has been produced, but the number of atoms of each element must be the same as before: 30

atoms of H and 12 atoms of N. Fifteen H_2 molecules react with five N_2 molecules to produce 10 molecules of NH_3:

(c) Construct an amounts table for the stoichiometric calculations:

Reaction	N_2 +	3 H_2 →	2 NH_3
MM (g/mol)	28.02	2.016	17.03
Amt (mol)	6	15	0
mol/coeff	6	5 (LR)	
Change (mol)	-15/3	-15	+(2/3)(15)
Final (mol)	1	0	10
Mass (g)	28.02	0	170.3

4.79 Identify species based on the type of substance present: Ions for salts and strong acids, molecules for other substances. H_2O is always a major species in aqueous solution. Reactions that can occur included precipitation, proton transfer (acid-base), and redox.

(a) NH_3 is a weak base, so the species present are NH_3 and H_2O; HCl is a strong base, so the species present are H_2O, H_3O^+ and Cl^-; an acid-base reaction occurs:
$$NH_3 + H_3O^+ \to NH_4^+ + H_2O;$$

(b) Both substances are salts, so the species present are the appropriate cations and anions: H_2O, Ca^{2+}, Cl^-, Na^+, and SO_4^{2-}; sulfates are generally insoluble, so a precipitation reaction occurs:
$$Ca^{2+} + SO_4^{2-} \to CaSO_{4(s)};$$

(c) KOH is a strong base, so the species present are H_2O, K^+, and OH^-; HBr is a strong acid, so the species present are H_2O, H_3O^+, and Br^-; an acid-base reaction occurs:
$$H_3O^+ + OH^- \to 2 H_2O;$$

(d) HNO_2 is a weak acid, so the species present are H_2O and HNO_2; KOH is a strong base, so the species present are H_2O, K^+, and OH^-; an acid-base reaction occurs:
$$HNO_2 + OH^- \to NO_2^- + H_2O.$$

4.81 Use standard rules for acid nomenclature to determine the names of acids:
(a) sulfuric acid; (b) carbonic acid; (c) hydrobromic acid; (d) perchloric acid;
(e) sulfurous acid; and (f) acetic acid.

Chapter 4: Chemical Reactions and Stoichiometry

4.83 Because amounts of each reactant are provided, this is a limiting reagent problem, for which it is convenient to use an amounts table.

(a) The reaction is $CaCO_3 + H_3O^+ \rightarrow Ca^{2+} + CO_2 + H_2O$. Carbon and calcium are already balanced. Give H_3O^+ a coefficient of 2 to balance the charges and H_2O a coefficient of 3 to balance H. Then O is also balanced (5 on each side):
$$CaCO_3 + 2\ H_3O^+ \rightarrow Ca^{2+} + CO_2 + 3\ H_2O.$$

(b and c) Reaction	$CaCO_3$ +	$2\ H_3O^+$ $\rightarrow$	Ca^{2+} +	CO_2
Amt (mol)	0.050	0.050	0	0
mol/coeff	0.050	0.025 (LR)		
Change (mol)	-0.025	-0.050	+0.025	+0.025
Final (mol)	0.025	0	0.025	0.025

(b) $m = n\ MM = (0.025\ \text{mol})(44.01\ \text{g/mol}) = 1.1$ g;

(c) The spectator ion concentration remains unchanged: $[Cl^-] = 0.10$ M; the other ionic species present in solution is Ca^{2+}; $[Ca^{2+}] = (0.025\ \text{mol}) \div (0.100\ \text{L}) = 0.050$ M.

4.85 This problem asks about the stoichiometry of a mixture of two salts, $CaSO_4$ and $Ca(H_2PO_4)_2$.

(a) The 2:1 mole ratio indicates the coefficients for the balanced equation. There is no water present during this reaction, so we need not ask about species present:
$$2\ H_2SO_4 + Ca_3(PO_4)_2 \rightarrow 2\ CaSO_4 + Ca(H_2PO_4)_2$$

(b) Because the mixture has fixed stoichiometric proportions of 2:1, it has a combined molar mass of $2\ MM_{CaSO_4} + MM_{Ca(H_2PO_4)_2}$:
$MM_{combination} = 2(136.1\ \text{g/mol}) + 234.1\ \text{g/mol} = 506.3$ g/mol;
$n_{combination} = (50.0\ \text{kg})(10^3\ \text{g/kg}) \div (506.3\ \text{g/mol}) = 98.76$ mol;
$n_{acid} = 2\ n_{combination}$; $n_{Ca_3(PO_4)_2} = n_{combination}$;
$m_{acid} = 2(98.76\ \text{mol})(98.09\ \text{g/mol}) = 19.4 \times 10^3$ g;
$m_{Ca_3(PO_4)_2} = (98.76\ \text{mol})(310.2\ \text{g/mol}) = 30.6 \times 10^3$ g.

(c) Each mole of $Ca_3(PO_4)_2$ yields 2 mol of phosphate ion, so the amount of phosphate ion available is $2\ n_{combination} = 2(98.76\ \text{mol}) = 198$ mol.

4.87 There is much information about propylene oxide provided in this problem, but all that is needed for the calculations is the balanced chemical equation, the amount of starting material, and molar masses. The stoichiometry of this reaction is 1:1 in all reagents.

(a) $m_{\text{propylene oxide}} = m_{\text{hydroperoxide}} (MM_{\text{propylene oxide}}/MM_{\text{hydroperoxide}})$;
$m_{\text{propylene oxide}} = (75\ \text{kg})(58.1\ \text{g/mol}) \div (90.1\ \text{g/mol}) = 48$ kg;

(b) $m_{\text{propene}} = m_{\text{hydroperoxide}} (MM_{\text{propene}}/MM_{\text{hydroperoxide}})$;
$m_{\text{propene}} = (75\ \text{kg})(42.1\ \text{g/mol}) \div (90.1\ \text{g/mol}) = 35$ kg.

4.89 A molecular picture must show the stoichiometry of the reaction and conserve atoms of each element. The balanced reaction is 2 Mg + O$_2$ → 2 MgO. The desired picture starts with six Mg atoms and four O$_2$ molecules. The six Mg atoms react with three O$_2$ molecules to produce six MgO molecules, leaving one O$_2$ molecule unreacted. Your figure should show the Mg-containing species as solids:

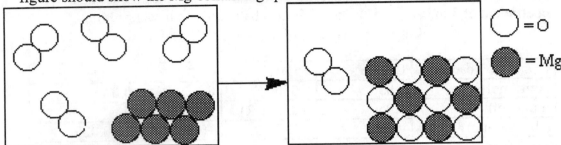

4.91 The quantity asked for is the amount of SO$_2$ in the sample. The data include a titration from which the amount of acid can be determined. The chemical reaction indicates that one mole of sulfuric acid forms from one mole of SO$_2$.
$n_{SO_2} = n_{H_2SO_4} = 2n_{OH^-} = 2(5.70$ mL$)(10^{-3}$ L/mL$)(5.96 \times 10^{-3}$ mol/L$) = 6.79 \times 10^{-5}$ mol;
$m_{SO_2} = n\ MM = (6.79 \times 10^{-5}$ mol$)(64.1$ g/mol$) = 1.09 \times 10^{-3}$ g.

4.93 There is much interesting information in this problem, but the essential features are the balanced chemical equation and the amounts of starting materials. This is a limiting reagent problem, which is conveniently solved using an amounts table. The question asks only about starting materials, so we need not complete the products part of the table:

Reaction	B$_{10}$H$_{14}$ +	11 O$_2$ →	5 B$_2$O$_3$ +	7 H$_2$O
MM (g/mol)	122.22	32.00		
Amt (10^2 mol)	3.27	25.0		
10^2 mol/coeff	3.27	2.27 (LR)		
Change (10^2 mol)	-(1/11)(25.0)	-25.0		
Final (10^2 mol)	0.997	0		

Liquid oxygen is used up first;
There is $(0.997 \times 10^2$ mol$)(122.22$ g/mol$) = 122 \times 10^2$ kg of decaborane left over.

4.95 Because amounts of both starting materials are given, this is a limiting reagent problem, for which it is convenient to use an amounts table.
The chemical reaction is As$_2$O$_3$ + C → As + CO. Give As a coefficient of 2 to balance As, give CO a coefficient of 3 to balance O, and give C a coefficient of 3 to balance C:
As$_2$O$_3$ + 3 C → 2 As + 3 CO

Reaction	As$_2$O$_3$ +	3 C →	2 As +	3 CO
MM (g/mol)	197.8	12.01	74.922	28.01
Amt (mol)	0.250	0.600	0	0
mol/coeff	0.250	0.200 (LR)		
Change (mol)	-(1/3)(0.600)	-0.600	+(2/3)(0.600)	+0.600
Final (mol)	0.050	0	0.400	0.600

$m_{As} = (0.400$ mol$)(74.922$ g/mol$) = 30.0$ g; $m_{CO} = (0.600$ mol$)(28.01$ g/mol$) = 16.8$ g.

Chapter 4: Chemical Reactions and Stoichiometry

4.97 This problem asks about yields in an economic context. The reaction stoichiometry indicates mole ratios of 1:1 for each reactant and the product. (1 Mkg = 10^6 kg)

(a) Calculate moles of product, use the yield to determine moles of reactant required, then convert to mass:
$n_{product}$ = (3.40 x 10^2 Mkg)(10^6 kg/Mkg)(10^3 g/kg) ÷ (88.15 g/mol) = 3.860 x 10^9 mol;
$m_{isobutene}$ = $n_{product}$ (1 mol isobutene/1 mol product)($MM_{isobutene}$) ÷ (% yield)(10^{-2}/%)
$m_{isobutene}$ = (3.860 x 10^9 mol)(56.1 g/mol) ÷ (86%)(10^{-2}/%) = 2.52 x 10^{11} g.

(b) Use the new yield to calculate moles of product, then convert to mass:
$$m_{product} = (2.52 \times 10^{11} \text{ g}) \left(\frac{88.15 \text{ g/mol}}{56.1 \text{ g/mol}} \right)(93\%)(10^{-2}/\%)(1 \text{ kg}/10^{-3} \text{ g})(10^{-6} \text{ Mkg/kg})$$
$= 3.67 \times 10^2$ Mkg.

(c) The improvement in yield means more product is produced at no increase in production cost, so the revenues from the extra product produced represent increased profits:
Profit = (mass)(price) = (367 – 340 Mkg)(10^6 kg/Mkg)(22¢/kg)(10^{-2} \$/¢) = \$ 5.9 x 10^6.

4.99 The quantity asked for is a volume ratio, which can be related to a mass ratio using densities and to a mole ratio using molar masses. First balance the chemical reaction, leaving H, which appears in both reactants, for last. N is already balanced; give H_2O a coefficient of 2 to balance O. Now there are six H on the left and four on the right; multiply the coefficients of both H_2O_2 and H_2O by 2 to give 8 on each side:

$N_2H_{4\,(l)}$ + 2 $H_2O_{2\,(l)}$ → $N_{2\,(g)}$ + 4 $H_2O_{\,(g)}$. The two reactants should be present in a mole ratio of 1 N_2H_4:2 H_2O_2. Now use $m = n\,MM$ and $V = m/\rho$, giving $V = n\,MM/\rho$:

$$\text{Vol. ratio} = \left(\frac{(1 \text{ mol } N_2H_4)(32.05 \text{ g/mol})}{(1.44 \text{ g/mL})} \right) : \left(\frac{(2 \text{ mol } H_2O_2)(34.02 \text{ g/mol})}{(1.01 \text{ g/mL})} \right)$$

Vol. ratio = 22.3 : 67.4 or 1:3.02.

4.101 The statement of the problem indicates what reactions take place:

$H_2S + O_2 \rightarrow SO_2 + H_2O$. H and S are balanced; give O a coefficient of 3/2 to balance O, then multiply through by 2 to clear fractions:
$$2\,H_2S + 3\,O_2 \rightarrow 2\,SO_2 + 2\,H_2O;$$

$SO_2 + H_2S \rightarrow S + H_2O$. H is already balanced; give H_2S and H_2O coefficients of 2 to balance O and S a coefficient of 3 to balance S:
$$SO_2 + 2\,H_2S \rightarrow 3\,S + 2\,H_2O.$$

4.103 A molecular picture must show the stoichiometry of the reaction and conserve atoms of each element. The net ionic reactions are $H_3O^+ + OH^- \rightarrow 2\, H_2O$ and $OH^- + HSO_4^- \rightarrow H_2O + SO_4^{2-}$. The first of these reactions occurs until all the hydronium ions are consumed:
(a) The first 3 OH^- ions react with the 3 H_3O^+ ions, generating 6 water molecules:

(b) An additional 3 OH^- ions react with the 3 HSO_4^- ions, generating 3 more H_2O molecules and three sulfate ions:

(c) There are no acid molecules present to react with additional hydroxide ions:

4.105 Section 4.3 of your textbook describes the reasons why yields are less than 100%: Incomplete reaction, competing reactions, multiple reactions, losses during collection and purification.

4.107 The combustion of a mixture of hydrocarbons involves several reactions occurring simultaneously. The combustion reactions are as follows:
$$CH_4 + 2\, O_2 \rightarrow CO_2 + 2\, H_2O; \qquad 2\, C_2H_6 + 7\, O_2 \rightarrow 4\, CO_2 + 6\, H_2O;$$
$$C_3H_8 + 5\, O_2 \rightarrow 3\, CO_2 + 4\, H_2O;$$
To determine the total output of CO_2, we must calculate how much results from each of the individual reactions. The amount of each starting material is found from the percentage composition of the mixture:
$n_{CH_4} = (750\text{ g})(74\%)10^{-2}/\% \div (16.0 \text{ g/mol}) = 34.7 \text{ mol};$
$n_{C_2H_6} = (750\text{ g})(18\%)10^{-2}/\% \div (30.0 \text{ g/mol}) = 4.5 \text{ mol};$
$n_{C_3H_8} = (750\text{ g})(8\%)10^{-2}/\% \div (44.0 \text{ g/mol}) = 1.36 \text{ mol};$
$n_{CO_2} = n_{CH_4} + n_{C_2H_6} (4 \text{ mol } CO_2/2 \text{ mol } C_2H_6) + n_{C_3H_8} (3 \text{ mol } CO_2/1 \text{ mol } C_3H_8)$
$n_{CO_2} = (34.7 \text{ mol}) + 2(4.5 \text{ mol}) + 3(1.36 \text{ mol}) = 47.8 \text{ mol}.$

Chapter 4: Chemical Reactions and Stoichiometry

4.109 Identify species based on the type of substance present: Ions for salts and strong acids, molecules for other substances. H_2O is always a major species in aqueous solution.
(a) Strong acid: H_3O^+, NO_3^-, and H_2O; (b) Salt: NH_4^+, SO_4^{2-}, and H_2O; (c) Salt: K^+, HCO_3^-, and H_2O; (d) Not a salt: CO_2 and H_2O; (e) Salt: Na^+, $CH_3CO_2^-$, and H_2O; (f) Salt: NH_4^+, $H_2PO_4^-$, and H_2O; (g) Weak base: NH_3 and H_2O.

4.111 The problem asks about the amount of reactant needed to react completely with a given amount of another reactant. The reaction is $CO_{2\,(g)} + 2\,KOH_{(s)} \rightarrow K_2CO_{3(s)} + H_2O_{(l)}$.
Use mole amounts and stoichiometric ratios:
n_{CO_2} = (1.0 kg/astronaut day)(5 astronaut)(6 day)(10^3 g/mol) ÷ (44.0 g/mol) = 6.8 x 10^2 mol;
m_{KOH} = (6.8 x 10^2 mol)(2 mol KOH/1 mol CO_2)(56.1 g/mol) = 76 kg KOH.

Chapter 5 The Behavior of Gases

5.1 Molecules in Motion

QUESTIONS TO ANSWER, SKILLS TO LEARN
1. **What does a gas look like at the molecular level?**
2. **What is a "velocity profile"?**
3. **What is an energy distribution profile?**
4. **Calculating average molecular kinetic energies**
5. **Calculating relative rates of effusion**

Gases differ from liquids and solids in two observable properties: first, they change volume and shape to fit their container; second, they are much more easily compressed than liquids or solids. The first property results from the very weak forces attracting the molecules to each other. The second property is because much of the space occupied by a gas is empty, and compressing the sample simply decreases the space available to each molecule.

The diagram at left shows two flasks separated by a valve, one filled with a gas and the other under vacuum. When the valve is opened, the gas rapidly expands until the number of gas molecules per volume in both containers is the same, as illustrated at right.

If molecules in a gas will fill their container when its shape changes, they must be in constant motion. Otherwise, they would not go to other parts of the container when the valve opens. Experiments that measure the velocity of molecules in a sample containing only one type of molecule show that the molecules in that sample have different velocities. The number of molecules with a velocity such as 200 m/s is not the same as the number of molecules with velocity 300 m/s. If the number of molecules with a given velocity is plotted versus velocity, the resulting graph is called a **velocity profile** or **distribution** such as that shown below. For the particular sample whose velocity distribution is shown below, more molecules have velocity of about 190 m/s than any other velocity. This is the maximum in the distribution and is the most probable velocity. The number of molecules with velocity of about 300 m/s is about 1/3 the number of molecules with a velocity of about 190 m/s. There are many molecules in the group whose velocities are less than the most probable velocity and many whose velocities are greater. Measurements of the velocity distributions of other molecules at the same temperature show similar distributions, but the most probable velocity is different : heavier molecules move more slowly on average than lighter molecules at a given temperature.

A velocity profile. Data for a measurement of the number of molecules of menthol acetate (198.31 g/mol) at different velocities (T= 287 K).

The kinetic energy (kE) is related to the velocity, u, through:

$$kE = \frac{1}{2} mu^2$$

The kinetic energy of a molecule can be calculated from this relationship given the mass of the molecule and its velocity. The kinetic energy of a menthol molecule traveling at 190 m/s, is calculated using the mass of one molecule:

$$m = \frac{MM}{N_A} = \frac{198.31 \text{ g/mole}}{6.022 \times 10^{23}/\text{mole}} \cdot 10^{-3} \text{ kg/g} = 3.293 \times 10^{-25} \text{ kg}$$

COMMON PITFALL: FAILING TO CONVERT MASSES TO KG
Be sure to convert to mass per molecule from molar mass, otherwise very large (and wrong) kinetic energies result!

At a velocity of 190 m/s, the kinetic energy of a menthol molecule is:

$$kE = \frac{1}{2} mu^2 = \frac{1}{2} (3.293 \times 10^{-25} \text{ kg})(190 \text{ m/s})^2$$

$$kE = 5.944 \times 10^{-21} \text{ kg} \cdot \text{m}^2/\text{s}^2 = \boxed{5.944 \times 10^{-21} \text{ J}} \quad (\text{remember: } 1 \text{ J} = 1 \text{ kg} \cdot \text{m}^2/\text{s}^2)$$

Repeating this calculation for a variety of molecules with different molar masses at the same temperature leads to the observation that:

Chapter 5: The Behavior of Gases

Although molecules of different molar masses have different distributions of velocities, the distribution of kinetic energies is the same for all gases at a given temperature.

Average Kinetic Energy

The number of molecules that have the most probable kinetic energy is quite small compared to the total number of molecules. A more meaningful quantity for describing the attributes of a large group is the average. The average kinetic energy of a gas molecule is related to the temperature by the equation:

$$kE_{average} = \frac{3RT}{2N_A}$$

(T is temperature in Kelvins; N_A is Avogadro's number; R is a constant)

The new constant, R, is called the gas constant and has units of energy per mol per Kelvin:

$$R = 8.314 \text{ J/mol·K}$$

At 298 K the average kinetic energy of a gas molecule is:

$$kE_{average} = \frac{3 \cdot (8.314 \text{ J/mole·K}) \cdot (298 \text{ K})}{2 \cdot 6.022 \times 10^{23} \text{ 1/mole}} = 6.17 \times 10^{-21} \text{ J}$$

The kinetic energy of a group of molecules is obtained by multiplying the average kinetic energy by the number of molecules.

$$kE_{total} = N(kE_{average}) = N \cdot \frac{3RT}{2N_A}$$

For a mol of a gaseous substance, $N = N_A$ and the equation is:

$$kE_{mole} = N_A(kE_{average}) = N_A \cdot \frac{3RT}{2N_A} = \frac{3}{2} RT$$

Exercise 5.1 ClO has been shown to be active in the depletion of the ozone layer. Calculate the average kinetic energy of a molecule of ClO at 250 K and the total kinetic energy of one mole of ClO at 250 K.

Steps to Solution: We are asked to calculate the kinetic energy of a molecule and a mole of the substance, ClO. However, because the kinetic energy of gas molecules at a given temperature is independent of molar mass, we need not calculate the molar mass. *We find the average kinetic energy per molecule using the relationship:*

$$kE_{average} = \frac{3RT}{2N_A} = \frac{3(8.314 \text{J/mol} \cdot \text{K})(250K)}{26.022 \times 10^{23} 1/\text{mol}} = \boxed{5.18 \times 10^{-21} \text{ J}}$$

The total kinetic energy of a mol of ClO is Avogadro's number times the kinetic energy per molecule:

$$kE_{mol} = \frac{3}{2} RT = \frac{3}{2}(8.314 \text{J/mol} \cdot \text{K})(250K) = 3.11 \times 10^3 \text{ J/mol}$$

Confirm that at 298 K the kinetic energy of a mol of gas molecules is 3.72 x 10⁸ J. The answers to this exercise are consistent with the idea that at a lower temperature, the kinetic energy of the gas molecules is lower.

How fast do gaseous molecules move? : Effusion and Diffusion

In the preceding material, we have seen the mathematical relationship between the kinetic energy and (1) velocity and mass and (2) temperature. For a mol of gas,

$$\frac{1}{2} mu^2_{average} = kE_{average} = \frac{3}{2} RT$$

Because the mass of one molecule, m, is related to the molar mass and N_A $\left(m = \frac{MM}{N_A}\right)$, substitution and rearrangement gives an equation for the average velocity of a gas molecule:

$$u_{average} = \sqrt{\frac{3RT}{MM}}$$

In terms of this equation, consider how the average velocity changes with changes in molar mass and temperature. When the temperature increases, the velocity increases (this makes sense because the kinetic energy is increasing). The average velocity of a heavy molecule is slower that of a lighter molecule at the same temperature (this is because both molecules have the same average kinetic energy). We will discuss two phenomena involving the movement of gas molecules, *diffusion* and *effusion*.

Diffusion *is the movement of a gas from a region of high concentration to region of lower concentration.* Diffusion is the phenomenon that occurs when you smell someone's cologne from across the room; the molecules of scent have traveled across the room to your nose.

Effusion *is the movement of molecules through small openings.* Effusion is the phenomenon responsible when a latex rubber helium balloon loses gas overnight and falls from the ceiling to the floor. The helium escaped through microscopic holes in the balloon. The speed at which both effusion and diffusion occur depends on the average velocity of the molecules involved. Effusion is more easily quantified by equations and will be be the focus of this discussion. The amount of gas that effuses in a given time is called the **rate of effusion** and is measured in units that reflect the number of molecules and the time involved; typical units are mL/minute, mL/s or molecules/s. We will compare the effusion or diffusion of two different molecules under the same conditions of temperature and pressure. Under these conditions, the rate is proportional to the average velocity of the molecules. The ratio of the rates of effusion of two substances, A and B, depends only on the molar masses of the substances:

$$\frac{\text{rate(A)}}{\text{rate(B)}} = \sqrt{\frac{MM(B)}{MM(A)}}$$

Lighter molecules effuse or diffuse more rapidly than heavier molecules because lighter molecules, on average, have greater velocity. This expression is known as Graham's Law.

Exercise 5.2 Three identical latex rubber balloons (Latex rubber has microscopic pores or holes in it) are filled to the same size with different gases. One is filled with helium, the next with hydrogen, $^1H_2$, and the last balloon is filled with deuterium, $^2H_2$. What will happen overnight and will there be any difference in the sizes of the balloons?

Steps to Solution: We are asked to explain what will happen to the balloons and whether there will be differences in the sizes of the balloons. Your experience should tell you that overnight all the balloons will deflate to some extent. However, they will deflate at a rate governed by Graham's Law, because the gas escapes by effusion through the microscopic pores in each balloon. To determine the relative rates, we must compare the molar masses of the gases: $^1H_2$(2.016 g/mol); He (4.0026 g/mol) ; $^2H_2$ (4.0 g/mol). The gas with the smallest molar mass will effuse the most rapidly, so the hydrogen balloon will deflate the most rapidly. The deuterium and helium balloons will deflate at almost the same rates because the gases have almost the same molar mass.

Exercise 5.3 Boron has two naturally occurring isotopes, ^{11}B (*MM* = 11.009 g/mol) and ^{10}B (*MM* = 10.013 g/mol) with natural abundances of 80.1% and 19.9%, respectively. Boron-10 is of interest because it efficiently captures neutrons and might be used in cancer treatment to destroy tumors. One way to separate substances of different molar masses is through effusion. How much faster will $^{10}BF_3$ effuse than $^{11}BF_3$?

Steps to Solution: This question is asking us to find how much faster the lighter isotope will effuse, a direct application of Graham's Law. We place the rate of ^{10}B in the numerator of the expression because the rate of boron-10 effusion will be expressed relative to the rate of boron-11 effusion:

$$\frac{\text{rate}(^{10}BF_3)}{\text{rate}(^{11}BF_3)} = \sqrt{\frac{MM(^{11}BF_3)}{MM(^{10}BF_3)}}$$

Rearrangement of this equation to solve for rate of $^{10}BF_3$ gives:

$$\text{rate}(^{10}BF_3) = \text{rate}(^{11}BF_3) \sqrt{\frac{MM(^{11}BF_3)}{MM(^{10}BF_3)}}$$

Calculating the molar masses of BF_3 with the two different isotopes and substituting gives:

$$\text{rate}(^{10}BF_3) = \text{rate}(^{11}BF_3) \sqrt{\frac{68.0042 \text{ g/mole}}{67.0082 \text{ g/mole}}} = 1.0074 \text{ rate}(^{11}BF_3)$$

The gas effusing from a 1:1 mixture of the isotopic types of boron trifluoride would contain more of the lighter isotope of boron; sometimes the effused gas mixture would be described as being enriched in the lighter isotope.

Note that we used the isotopic molar masses in the calculation of the molar masses of the borontrifluoride compounds. Also, the rate of effusion could be expressed in mL/s or molecules/s. The rate of $^{10}BF_3$ effusion calculated by this formula will be given in the same units as the $^{11}BF_3$ rate.

5.2 The Ideal Gas Equation

QUESTIONS TO ANSWER, SKILLS TO LEARN
1. **How do gases exert pressure?**
2. **What is an ideal gas?**
3. **What is the ideal gas law?**

Important properties of a gas that are readily measured include **pressure, temperature, volume,** and the **moles** of molecules that make up the sample. A useful mathematical description (equation) would relate these four quantities and allow us to predict how changing one of these quantities would affect the other properties. The equation should also be consistent with the molecular nature of chemistry so that we can use this model to make qualitative predictions.

As we saw in the preceding section, temperature is a measure of the average kinetic energy of a gas sample. We can determine the number of moles by knowing the mass of the gas and molar masses of the components. The volume is the size of the vessel containing the gas. Pressure is a less straightforward quantity; pressure is related to the force exerted by molecules

Chapter 5: The Behavior of Gases

when they collide with any surface, whether it be the rubber skin of a balloon or your skin. The force exerted by a gas depends on the number of times molecules collide with a surface. The force exerted by each collision depends on the kinetic energy of the molecules or atoms of the gas. A simple relationship between the pressure, P, volume, V, number of moles of gas, n, and absolute (Kelvin) temperature is obtained if the following assumptions are made.

1. The volume occupied by the molecules of the gas is insignificant; that is, it is much smaller than that of the volume of the container.

2. Forces acting between the gas molecules are very weak.

Gases which have these properties are described as **ideal**.

Ideal gas behavior is *not* observed when the assumptions of the ideal gas law are not valid: (1) when the volume occupied by the gas molecules is not negligible or (2) when significant attractive forces between molecules are present. These conditions occur at high pressures and temperatures near the boiling point of the substance.

The ideal gas law relates the four variables listed above through this equation:

$$PV = nRT$$

where R is a constant, which has different values depending on the units we choose for R. The two most common values of R are:

$$R = 0.0821 \text{ L·atm/mol·K and } R = 8.314 \text{ J/mol·K}$$

Rearranging the ideal gas law to solve for pressure shows how pressure depends on the number of molecules, the volume of the container and the temperature:

$$P = \frac{n}{V} RT$$

The pressure depends on the number and the energy of collisions of molecules with the walls of the container; therefore, it is not surprising that the pressure is proportional to the temperature, which is proportional to the average kinetic energy of the molecules. The pressure also depends on the number of collisions of the molecules with the walls of the container. Two factors can affect the number of collisions. One is the absolute number of molecules in the container. Increasing the number of molecules in the container will increase the number of collisions with the walls. However, if we reduce the size of the container of a gas, the number of collisions will increase, also leading to an increase in pressure.

Decreasing the volume of the container without decreasing the number of molecules inside it increases the number of collisions with the walls of the container.

A quantity that will account for both of these effects is the **molecular density** of the gas, (sometimes number density):

$$\text{molecular density} = \frac{n}{V} \quad (\text{mol}/V \text{ or mol/L})$$

As the number of molecules per liter increases, the number of collisions with the walls of the container increases and so does the pressure. Conversely, decreasing the molecular density (increasing the volume of the container or decreasing the moles of gas) lowers the pressure exerted by the gas.

Exercise 5.4 Consider an ideal gas and the concept, molecular density, we have just discussed:
(a) Under what conditions will a gas exhibit high molecular density?
(b) Will the fraction of volume occupied by the gas molecules be larger or smaller at higher molecular density?

Steps to Solution: The molecular density is defined as $\frac{n}{V}$. For the molecular density to be large, the number of mols must be large compared to the volume. To describe the conditions in terms of the other measurable properties of a gas, the following form of the ideal gas law is useful:

$$P = \frac{n}{V} RT$$

(a) If the molecular density is large, then the pressure may be large because of the large number of energetic collisions the walls of the container experience. Another possibility is low temperature, where the energy of the each collision is sufficiently low so that the pressure is not very high even though the number of collisions is high.

(b) The fraction of volume occupied by the gas molecules is $\frac{V_{molecules}}{V_{gas}}$. As molecular density increases, the number of molecules per unit volume will increase. As a result the volume of the molecules, $V_{molecules}$, will increase, but the volume of the gas will remain constant. Therefore, the fraction of volume occupied will increase as the molecular density increases (with either increasing pressure, decreasing temperature or both).

Exercise 5.5 For a sample of ideal gas, describe the changes at the molecular level and the effect on the observable property, pressure for each of the following changes in conditions. All other properties of the gas are assumed to not change.
(a) The volume is decreased so that the final volume is one third of the original volume.
(b) One half of the gas molecules are removed from the container.

Steps to Solution:
(a) If the volume is decreased to give a final volume one third that of the original, then the molecular density will have tripled. Because there are now three times as many molecules per unit volume, the number of collisions with the walls of the container is tripled and the pressure is expected to triple. Use of the ideal gas law solved for the pressure gives the same result:

$$P_1 = \frac{n}{V_1} RT_1 \quad \text{and} \quad P_2 = \frac{n}{V_2} RT_1$$

Because $V_2 = \frac{V_1}{3}$, $P_2 = \frac{3n}{V_1} RT_1$, the ratio $\frac{P_2}{P_1}$ is: $\frac{P_2}{P_1} = \frac{\frac{3n}{V_1} RT_1}{\frac{n}{V_1} RT_1} = 3$; thus, $P_2 = 3P_1$

(b) If one half the molecules are removed from the container, then the pressure will be decreased by a factor of 2, since there will only be half the number of collisions with the walls of the container. The pressure would be expected to be one half the initial pressure.

$$P_1 = \frac{n_1}{V_1} RT_1, P_2 = \frac{n_2}{V_1 RT_1}, \text{ and } n_2 = \frac{n_1}{2}$$

$$\frac{P_2}{P_1} = \frac{\frac{n_1}{2V_1} RT_1}{\frac{n_1}{V_1} RT_1} = \frac{1}{2}; \text{ thus, } P_2 = \frac{P_1}{2}$$

5.3 Pressure

QUESTIONS TO ANSWER, SKILLS TO LEARN
1. What is pressure?
2. What are common units of pressure?
3. Converting pressure units and choosing the correct units for *R*

Pressure is defined as force per unit area:

$$P = \frac{F}{A}$$

The pressure exerted by solids and liquids depends on their mass and the area on which their mass "sits". In contrast, the pressure exerted by a gas measures the force of all the collisions that the gas molecules make with the container or surface. In a mercury barometer, the end of the column of mercury at the top of the tube has very few gas molecules hitting it to exert any force whereas the other end of the tube is immersed in a pool of mercury exposed to the atmosphere, which is undergoing many collisions with the gas molecules in the air. At higher elevations there are fewer molecules per liter of air; therefore, there are fewer collisions, and the pressure is lower.

The common units of pressure are defined relative to the pressure exerted by the earth's atmosphere. At sea level the atmosphere will support a column of mercury about 760 mm Hg tall.

Pressure Unit Name	Dimensions	Units/ Atm
atmosphere	atm	1.00
mm Hg	mm Hg	760 mm Hg/atm
torr	torr	760 torr/atm
Pascal	N/m²	1.01325 x 10^5 Pa/atm

The common use of the mercury barometer has lead to the use of a unit called the **torr** or **mm Hg**, the pressure that will support a 1 mm Hg column. The SI unit for pressure is the Pascal (Pa). The above table gives the different units of pressure and their relationships with each other. The Pascal is cumbersome, so most pressures are reported in units of atmospheres, torr or mm Hg.

Exercise 5.6 The pressure in an underwater dwelling is 9.52 atm. What is the pressure in torr and Pascals?

Chapter 5: The Behavior of Gases

Steps to Solution: This problem involves the conversion of units using the conversion factors listed in the table above. The rule to remember is that the correct conversion factor is the one that gives the correct units.

We convert the pressure to torr as follows:

$$P(\text{torr}) = 9.52 \text{ atm} \left(\frac{760 \text{ torr}}{\text{atm}} \right) = \boxed{7.24 \times 10^3 \text{ torr}}$$

We use a similar procedure to find the pressure in Pascals:

$$P(\text{torr}) = 9.52 \text{ atm} \left(\frac{1.01325 \times 10^5 \text{ Pa}}{\text{atm}} \right) = \boxed{9.65 \times 10^5 \text{ Pa}}$$

Exercise 5.7 A steel cylinder 1.2 m tall and 20 cm in diameter is filled with nitrogen at a pressure of 2200 pounds per square inch (psi) (14.7 pounds per square inch = 1 atm) at a temperature of 24°C. How many kg of nitrogen does the cylinder contain?

Steps to Solution: This problem requires determining the amount of nitrogen in the cylinder. First, the number of moles of nitrogen can be determined using the ideal gas law, because the pressure, volume, and temperature of the gas are given. Second, the number of moles of gas is related to the mass of the nitrogen through its molar mass.

Before we can substitute the data into the ideal gas law, we must convert the units of volume, pressure and temperature to those of R (0.0821 L • atm/mol•K). First the volume must be calculated, using length in dm, because 1 dm^3 = 1 L.

$$V_{\text{cylinder}} = \pi r^2 L = \pi (20 \text{ cm} \frac{1 \text{ dm}}{10 \text{ cm}})^2 \cdot 1.2 \text{ m} \cdot \frac{10 \text{ dm}}{1 \text{ m}} = 1.51 \times 10^2 \text{ dm}^3 = 1.51 \times 10^2 \text{ L}$$

We must convert the pressure to atm before use in the ideal gas equation:

$$P = 2200 \text{ psi} \cdot \frac{1 \text{ atm}}{14.7 \text{ psi}} = 1.50 \times 10^2 \text{ atm}$$

The final conversion takes the temperature from Celsius to Kelvins:

$$T = 24 \text{ °C} + 273 = \boxed{297 \text{ K}}$$

> **COMMON PITFALL: Mismatched units.**
> The most common problem experienced in use of the ideal gas law is unit conversion, especially the conversion of temperature to Kelvins. Make sure that the units you use are consistent with the value of *R* you use.

The ideal gas law must be rearranged to solve for the number of moles of gas:

$$PV = nRT \rightarrow n = \frac{PV}{RT}$$

$$n = \frac{PV}{RT} = \frac{1.50 \times 10^2 \text{ atm}(1.51 \times 10^2 \text{ L})}{0.0821 \text{ L} \cdot \text{atm/mole} \cdot \text{K}(297 \text{ K})} = 9.26 \times 10^2 \text{ mol}$$

The cylinder will contain this many moles of any ideal gas under the conditions specified above. The mass of nitrogen, N_2, contained in the cylinder is found using the molar mass:

$$m = 9.26 \times 10^2 \text{ mol } (28.0134 \text{ g/mol}) = 2.59 \times 10^4 \text{ g} = 25.9 \text{ kg}$$

Rounding to two significant figures (remember that the volume was calculated using measurements with only two significant figures), the mass of nitrogen contained in the cylinder is $\boxed{26 \text{ kg}}$.

5.5 Applying the Ideal Gas Equation

QUESTIONS TO ANSWER, SKILLS TO LEARN
1. **How is the ideal gas law used to predict the behavior of a gas?**
2. **Calculating changes in pressure, volume or temperature after changing one of these properties**
3. **Calculation of molar mass from vapor density**

Pressure-Volume Relationships

The mathematical relationship between the pressure and volume of a unchanged number of gas molecules at a constant temperature was important in the development of the ideal gas law. Robert Boyle performed a series of experiments in which the pressure of a sample of gas could be varied and the volume of the gas then measured. The results of that work led to the expression called Boyle's Law:

Boyle's Law: For a fixed amount of a gas at constant temperature: PV = constant

This result is what we expect from the molecular point of view. If we halve the volume of a gas, the molecules have half the volume to move about in and they collide with the walls of the container more often. Because the average kinetic energy of the molecules is still the same, the force exerted on surfaces is greater and the pressure is greater.

Boyle's Law is a special case of the ideal gas law: n is constant and T is constant, so the product, nRT, is constant.

$$PV = nRT = \text{constant}$$

If we change the pressure on a gas, we can use this relationship to calculate the new volume. Labeling the initial conditions with the letter "i" (for initial) and the new pressure and volume with the letter "f" (for final), this useful expression results in:

$$P_iV_i = nRT = P_fV_f \quad \text{or} \quad P_iV_i = P_fV_f$$

This expression's power is that we need not know either the amount or the temperature of gas as long as the amount and temperature of the gas are the same at the initial and final conditions.

Exercise 5.8 A spherical weather balloon is constructed so that the gas inside can expand as the balloon ascends to higher altitudes where the pressure is lower. If the radius of the spherical balloon is 2.5 m at sea level where the pressure is 753 torr, what will be the radius at an altitude of about 10 km where the pressure of the gas is 210 torr if the temperature has not changed?

Steps to Solution: *We are asked for the radius of the balloon at a new pressure. The data supplied are the initial pressure (753 torr) and final pressure (210 torr) and the radius of the balloon at the initial pressure (2.5 m). The radius is related to the volume of the balloon because the volume of a sphere can be calculated using the radius. The volume of the balloon is then calculated at the new pressure using Boyle's Law.*

$$V_{\text{sphere}} = \frac{4}{3}\pi r^3 = \frac{4}{3}\pi (2.5 \text{ m})^3 = 65.4 \text{ m}^3 = V_i$$

We use Boyle's Law to solve for the final volume (note that because both pressures are in torr, no conversion is necessary):

$$V_f = \frac{P_iV_i}{P_f} = \frac{753 \text{ torr} \cdot 65.4 \text{ m}^3}{210 \text{ torr}} = 235 \text{ m}^3$$

Note that the pressure and volume do not have to be in any particular units; it is only necessary that the units of pressure be the same so that the final volume has the same units as the initial volume.

The problem is nearly solved; calculation of the radius of the balloon at the new volume is the remaining step.

$$r = \sqrt[3]{\frac{3V}{4\pi}} = \sqrt[3]{\frac{3 \cdot 235 \text{ m}^3}{4\pi}} = 3.83 \text{ m}$$

Rounding this answer to two significant figures, the new radius is $\boxed{3.8 \text{ m}}$.

Chapter 5: The Behavior of Gases

Temperature-Volume Relationships

The quantitative relationship between the temperature and volume of a fixed amount of gas at constant pressure was discovered by Jacques Charles. He found that the volume of a fixed amount of gas increased with increasing temperature in a linear fashion, giving an expression known as Charles' Law:

Charles' Law: For a fixed amount of gas at constant pressure, $V = \text{constant} \cdot T$.

Charles' Law is a special case of the ideal gas law.

$$V = \frac{nR}{P} T = \text{constant} \cdot T \quad or \quad \frac{V}{T} = \text{constant}$$

Charles' Law can be used to predict the volume change of a sample of gas that undergoes a temperature change at constant pressure provided the temperature is expressed in Kelvins. Using the subscripts "i" and "f" for the initial and final states of the gas, Charles' Law can be written:

$$\frac{V_i}{T_i} = \text{constant} = \frac{V_f}{T_f} \quad or \quad \frac{V_i}{T_i} = \frac{V_f}{T_f}$$

The temperature must be expressed in K.

Exercise 5.9 A 1.56 L sample of nitrogen gas (1.21 x 10⁻² g) is heated from 24 °C to 350 °C. Compare the densities of the nitrogen at the different temperatures.

Steps to Solution: *The question asks us to compare the densities of the gas at two different temperatures. Density is defined as m/V; because the mass is supplied, we must calculate the volume of the gas at the higher temperature and then calculate the density at that temperature. One way to solve this problem is first to solve for the volume at the higher temperature and then calculate both densities and compare them.*

Using the form of Charles' Law shown above, rearrange to solve for V_f:

$$\frac{V_i}{T_i} = \frac{V_f}{T_f} \rightarrow V_f = \frac{T_f}{T_i} V_i$$

Substituting in the values for temperatures (converting to K) and initial volume gives:

$$V_f = \frac{350 + 273}{24 + 273} \, 1.56 \text{ L} = 3.272 \text{ L}$$

Now we can calculate the densities of the gas at the two different temperatures because the mass of the gas has not changed:

$$T = 24\,°C: \quad \rho = \frac{1.21 \times 10^{-2}\,g}{1.56\,L} = 7.74 \times 10^{-3}\,g/L$$

$$T = 350\,°C: \quad \rho = \frac{1.21 \times 10^{-2}\,g}{3.272\,L} = 3.69 \times 10^{-3}\,g/L$$

The density of the nitrogen gas at 350°C is about half the density at 24 °C. From a molecular point of view, this answer makes sense. As the temperature increases, the average kinetic energy of the molecules increases; therefore it takes fewer of these more energetic collisions to exert the same pressure against a surface (such as the wall of a hot-air balloon). As a result, the number of molecules per liter required to exert the same pressure decreases. Because there are fewer molecules per unit volume, the density decreases.

The Combined Gas Law

Another arrangement of the ideal gas law results in an expression that combines both Charles' Law and Boyle's Law:

$$\frac{PV}{T} = nR = \text{constant (for no change in the moles of gas, } n)$$

We can use this expression to calculate changes in three observable quantities: pressure, volume and temperature for a fixed amount of gas:

$$\frac{P_i V_i}{T_i} = \frac{P_f V_f}{T_f}$$

Although this is just another form of the ideal gas law, it can be helpful. It is more effective to examine the problem and rearrange the ideal gas law to fit the problem, rather than memorizing different equations and the conditions to which they apply. Using the molecular point of view gives insight into the appropriate form of the ideal gas law.

Exercise 5.10 A sample of argon occupying 1.53 L at 45 °C and 1.00 atm pressure in a piston and cylinder assembly is compressed and cooled. The final temperature and pressure are 15 °C and 1.35 atm. What is the volume occupied by the gas?

Steps to Solution: From the description of the problem, it is clear that no reaction is occurring that can change the amount of gas in the cylinder. We must consider two other pieces of information. First, the temperature has decreased, so the average kinetic energy of the molecules has also decreased. Therefore the average collision will generate less pressure. However, the final pressure is greater than the initial pressure. In order for the final pressure to be greater than the initial pressure when *the average collision has less energy there must be more collisions occurring. Therefore, in the final state, the number of gas particles (in this case, argon atoms) per unit volume must increase to increase the number of collisions.*

It is useful to draw a picture to help visualize the process that is occurring. In the following sketch, the initial situation is shown on the left and the final situation is shown on the right.

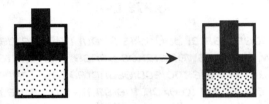

There will be more collisions in the right-hand picture because there is a greater number of gas particles present per unit volume. We will determine the actual volume using the combined gas law since we know that the amount of gas, n, is constant:

$$\frac{P_i V_i}{T_i} = \frac{P_f V_f}{T_f}$$

Solving for the final volume, V_f, we obtain:

$$V_f = V_i \cdot \frac{P_i}{P_f} \cdot \frac{T_f}{T_i}$$

After converting all temperatures to K, and substitution into the equation,

$$V_f = 1.53 \text{ L} \cdot \frac{1.00 \text{ atm}}{1.35 \text{ atm}} \cdot \frac{288 \text{ K}}{318 \text{ K}} = \boxed{0.590 \text{ L}}$$

The answer is reasonable because it is smaller than the initial volume, which we predicted using the molecular perspective.

Determining Molar Mass: Vapor Density Calculations

The ideal gas law states that the pressure and volume of a gas are determined by the number of moles of the gas and its temperature. The mass of the gas molecules is not a factor in the pressure they exert because the average kinetic energy is independent of the mass of the gas molecules(s). However, by measuring the pressure, volume and temperature of a gas, we can calculate the number of moles of that gas. If we also measure the mass of the gas occupying that volume under those conditions, then we can determine the mass of a known number of moles of gas, allowing us to calculate the molar mass, mass/mol:

$$MM = \frac{m}{n}\left(\text{units of }\frac{\text{mass}}{\text{mole}}\right)$$

For an ideal gas, the number of moles is equal to: $n = \dfrac{PV}{RT} = \dfrac{m}{MM}$

Solving for MM gives: $MM = \dfrac{mRT}{PV}$ or $MM = \dfrac{m}{V}\dfrac{RT}{P}$ where vapor density $= \dfrac{m}{V}$

This equation allows us to calculate the molar mass of a gaseous substance that behaves ideally. We need four pieces of data: (1) the mass of the vapor (2) the volume of the vapor (3) the pressure of the vapor and (4) the temperature. The second expression explicitly shows the vapor density, m/V. We see that the vapor density increases with molar mass because in an ideal gas, the number of molecules per volume varies only with pressure and/or temperature.

Different molar masses change the mass of gas molecules contained in a given volume but not the number of gas molecules

Exercise 5.11 One group of compounds that contain only carbon and hydrogen are called the *cycloalkanes* and have the empirical formula CH_2. The vapor from a sample of one of these compounds occupies a volume of 253.2 mL at a temperature of 99.8 °C and a pressure of 754.8 torr. The vapor has a mass of 0.5921g. What is the molecular formula of the compound?

Steps to Solution: *We are asked to determine the molecular formula of a compound. The data supplied are the empirical formula and the mass of the vapor that occupies a stated volume under given temperature and pressure. The strategy we will pursue is to determine the molecular mass from the vapor density data and then find the number of empirical formula units per molecule. The formula for calculating the molar mass is:*

$$MM = \frac{m}{V}\frac{RT}{P}$$

Because the units of R are L • atm/mol • K, the volume, pressure and temperature must be converted to the units of R:

$V = 253.2$ mL $= 0.2532$ L; $P = 754.8$ torr• $\dfrac{1\text{ atm}}{760\text{ torr}} = 0.9931$ atm;

$T = 99.8$ °C $+ 273.15 = 373.0$ K

The units of mass may be left as g because molar mass is usually given in units of g/mol. The molar mass is then found:

$$MM = \frac{0.5921 \text{ g}}{0.2532 \text{ L}}\left(\frac{0.08206 \text{ L·atm/mol·K}(373.0 \text{K})}{0.9931 \text{ atm}}\right) = 72.06 \text{ g/mol}$$

The empirical formula mass is 14.03 g/mol. The number of empirical formula units per molecule, x, is:

$$x = \frac{\text{molar mass}}{\text{empirical formula mass}} = \frac{72.06 \text{g/mol}}{14.03 \text{g/mol}} = 5.137 \approx 5 \text{ empirical formulas/molecule}$$

The 5.137 is rounded down to 5 for two reasons. First, it is not chemically reasonable to have a fractional ratio because that suggests that fractions of atoms will be present in the molecule. Secondly, the molar mass obtained assumes ideal gas behavior and we are dealing with a real molecule, which may not exhibit completely ideal behavior, leading to some deviation from ideal behavior. Therefore, the chemical formula of the compound is:

$$(CH_2)_5 \text{ or } \boxed{C_5H_{10}}$$

Exercise 5.12 The density of air at room temperature and pressure (1 atm) is about 1.16 g/L. Other gases have different vapor densities and may be less dense than air (such as helium) or more dense (propane). Gases with vapor densities greater than air mix slowly with air (because on average, the molecules have smaller velocities) and collect in low-lying areas. In Bhopal, India, a large quantity of methyl isocyanate, CH_3OCN, escaped from a storage tank killing thousands of people. Calculate the vapor density of methyl isocyanate and methane, CH_4, in units of g/L at 298 K and 1.00 atm pressure and compare them to the density of air.

Steps to Solution: The vapor density is the quotient, m/V, which is present in the form of the ideal gas law used to calculate molar mass:

$$MM = \frac{m}{V}\frac{RT}{P}$$

Rearranging the equation to solve for $\frac{m}{V}$ gives:

$$\frac{m}{V} = MM \frac{P}{RT}$$

All the conditions are given in the problem (pressure = 1.00 atm, T = 298 K); all that remains is to calculate the molar masses of the compounds and substitute them into the above equation.

$$CH_3OCN, MM = 57.05 \text{ g/mol}:$$

$$\frac{m}{V} = MM \frac{P}{RT} = 57.05 \text{ g/mol} \frac{1.00 \text{atm}}{0.08206 \text{L} \cdot \text{atm/mol} \cdot \text{K}(298\text{K})}$$

$$\frac{m}{V} = \boxed{2.33 \text{ g/L}}$$

Show yourself that for CH_4 (MM = 16.04 g/mol): $\frac{m}{V} = \boxed{0.656 \text{ g/L}}$

It is clear that methyl isocyanate is more dense and methane is less dense than air (ρ = 1.16 g/L). As a consequence, methyl isocyanate will concentrate in low-lying areas and slowly dissipate (most of the casualties in Bhopal were in low-lying areas), but methane will tend to rise and dissipate more rapidly. Methane will also tend to dissipate more rapidly as its average velocity is greater than that of methyl isocyanate and air.

5.5 Gas Mixtures

QUESTIONS TO ANSWER, SKILLS TO LEARN
1. **How is the pressure of a mixture of gases determined?**
2. **How is the composition of a mixture of gases described quantitatively?**
3. **Calculating partial pressures**

One of the foundations of the ideal gas law is that pressure depends on the average kinetic energy and number of molecules in the system of interest and does not depend on the type of molecules. Therefore the ideal gas law for a system where a mixture of gases is present can be written as:

$$P_{total}V = n_{total}RT \quad \text{or} \quad P_{total} = \frac{n_{total}RT}{V}$$

The total number of mols of gas in the mixture is simply the sum of the mols of the constituent gases. For a sample containing H_2, N_2, and NH_3, the total number of mols would be:

$$n_{total} = n_{H_2} + n_{N_2} + n_{NH_3}$$

Substituting into the above equation for the total pressure gives:

$$P_{total} = \frac{(n_{H_2} + n_{N_2} + n_{NH_3})RT}{V} = \frac{n_{H_2}RT}{V} + \frac{n_{N_2}RT}{V} + \frac{n_{NH_3}RT}{V}$$

Each of the terms is the pressure exerted by that number of moles of gas at P and T:

$$p_{H_2} = \frac{n_{H_2} RT}{V} \qquad p_{N_2} = \frac{n_{N_2} RT}{V} \qquad p_{NH_3} = \frac{n_{NH_3} RT}{V}$$

Because each component is responsible for part of the total pressure in the system, each of these terms is called a **partial pressure**. The first term is called the *partial pressure of hydrogen*, the second term, the *partial pressure of nitrogen*, and the last term is the *partial pressure of ammonia*. The total pressure is the sum of the partial pressures:

$$P_{total} = p_{H_2} + p_{N_2} + p_{NH_3}$$

Dalton's Law of Partial Pressures: The total pressure is equal to the sum of the partial pressures of all the components of the mixture.

For a mixture of "i" components, the total pressure would be:

$$P_{total} = p_1 + p_2 + p_3 + \ldots + p_i$$

where p_1, etc. are the partial pressures of the components.

Exercise 5.13 A 400. L steel tank is filled with a mixture of 20 kg of methane (CH_4), 5 kg of ethane (C_2H_6) and 0.050 g of butanethiol ($C_4H_{10}S$). (This last sulfur containing compound has an intense smell to indicate any leaks of this flammable mixture). What is the total pressure in the tank and the partial pressure of each component if the temperature is 24 °C?

Steps to solution: Probably the most direct way to solve this problem is to calculate the partial pressure of each of the three components and then sum them to obtain the total pressure. The calculation of the partial pressure of methane is illustrated below:

$$p_{CH_4} = \frac{n_{CH_4} RT}{V} = \frac{\frac{20 \times 10^3 g}{16.04 \text{ g/mole}}(0.08206 \text{ L} \cdot \text{atm/mole} \cdot \text{K}) 297 \text{ K}}{400 \text{ L}} = 7.6 \times 10^1 \text{ atm}$$

The partial pressures of ethane and butanethiol are obtained in a similar fashion:

$$p_{C_2H_6} = 1.0 \times 10^1 \text{ atm} \qquad p_{C_4H_{10}S} = 3.4 \times 10^{-5} \text{ atm}$$

The total pressure is the sum of the partial pressures of the components:

$$P_{total} = p_{CH_4} + p_{C_2H_6} + p_{C_4H_{10}S} = 76 + 10 + 3.4 \times 10^{-5} \text{ atm} = \boxed{86 \text{ atm}}$$

Concentration Units for Gas Mixtures

The preceding section illustrates one method of describing the composition of a gas mixture: listing the partial pressures of the components. Alternatively, we might list the moles of the components in the mixture. Another way to describe the concentration of a gas in a mixture might be to use the molarity. In practice, however, the molar concentration of a gas is quite low. There are two other commonly used measures of gas composition, both of which have no units associated with them. They are **parts per million (ppm)** and **mole fraction**. The first unit, ppm, is useful for describing the amount of a component present in small amounts. For substances present in more substantial quantities, the mole fraction is used. The mole fraction, X, is defined as the number of mols of the substance present divided by the total number of mols of substance present in the mixture. For component "i" of a mixture, the mole fraction, X_i, is:

$$X_i = \frac{n_i}{n_{total}}$$

The mole fraction can never be greater than 1.

Related to the mole fraction is the ppm:

$$\text{ppm:} \quad 1 \text{ ppm} = \frac{1}{10^6}$$

When a gas mixture has 1 ppm of methane in it, it means that of 10^6 molecules, only one is methane *or* that one mole of the gas mixture contains 10^{-6} moles of methane. The mole fraction is related to the ppm unit:

$$\text{ppm} = \left(\frac{n_i}{n_{total}}\right) 10^6 \quad or \quad \text{ppm} = (X)\, 10^6$$

The ppm unit is similar to percent, but is more "sensitive" in that the smaller abundance components have larger numbers.

Exercise 5.14 In Exercise 5.13, the partial pressures of the three components of the gas mixture were calculated. Determine the composition of the gas in terms of mole fraction and ppm.

Steps to Solution: There are two methods that may be used to obtain the solution to the problem. The first route, (a), is to find the number of moles of each gas and then the total number of moles of gas. The second, (b), uses the fact that the partial pressure is proportional to the number of moles of gas in the mixture.

(a) methane = 1.25×10^3 mol; ethane = 1.67×10^2 mol; butanethiol = 5.54×10^{-4} mol total moles = 1.41 x 10³ mol

$X \text{(methane)} = \dfrac{1.25 \times 10^3}{1.41 \times 10^3} = 0.88$	ppm(methane) = $X(10^6)$ = 8.8 x 10⁵
$X\text{(ethane)} = \dfrac{1.67 \times 10^2}{1.41 \times 10^3} = 0.12$	ppm(ethane) = $X(10^6)$ = 1.12 x 10⁶
$X\text{(butanethiol)} = \dfrac{5.54 \times 10^{-4}}{1.41 \times 10^3} = 3.9 \times 10^{-7}$	ppm(butanethiol) = $X(10^6)$ = 0.39

(b) In a mixture of gases, the volume of the container and temperature are the same. Because the pressure is directly proportional to the number of moles of a gas and the partial pressure is proportional to the number of moles of that component.

$$P = \frac{nRT}{V} = (\text{constant})n$$

Therefore, the mole fraction should be equal to the partial pressure of that component divided by the total pressure:

$$X = \frac{n_i}{n_{total}} \text{ and } n = \frac{PV}{RT} \text{ gives:}$$

$$X = \frac{\dfrac{p_i V}{RT}}{\dfrac{P_{total} V}{RT}} = \frac{p_i}{P_{total}}$$

The partial pressures and total pressures of the components of this mixture were calculated in the previous problem:

$$X_{methane} = \frac{p_{methane}}{P_{total}} = \frac{76 \text{ atm}}{86 \text{ atm}} = \boxed{8.8 \times 10^{-1}}$$

In a similar way we find:

$$X_{ethane} = \boxed{1.2 \times 10^{-1}} \qquad X_{butanethiol} = \boxed{3.9 \times 10^{-7}}$$

5.6 Gas Stoichiometry

QUESTIONS TO ANSWER, SKILLS TO LEARN
1. What is the pressure at the end of a reaction that consumes or produces gases?
2. Calculation of pressure and or volume changes in reactions with changes in the number of moles of gas

In Chapter 4 we measured the amounts of reactants consumed and products formed in a reaction in terms of mass or volume of solution. We can use the ideal gas law to determine the number of moles or molecules in the gas phase given pressure, volume and temperature, or pressure given the amount of gas under specified temperature and volume. The same strategies employed in Chapter 4 to solve stoichiometry problems may be used with stoichiometry problems involving gases to determine the limiting reagent and the amounts of products formed. We can also calculate the pressure or volume of the gaseous reaction mixture after determining the total number of moles of gas present at the end of the reaction.

Exercise 5.15 Methanol, CH_3OH, is produced by the reaction of methane with oxygen under carefully controlled conditions:

$$2\ CH_{4\ (g)} + O_{2\ (g)} \rightarrow 2\ CH_3OH_{(g)}$$

A 200. L reactor is filled with a mixture of methane to a pressure of 10.5 atm and 3.50 atm oxygen at 298 K. The reactor is then heated to 350 °C.

(a) What is the mass of methanol formed if the reaction proceeds to 100% yield?
(b) What are the partial pressures of each of the substances and the total pressure in the reactor at the end of the reaction if the temperature is 350 °C ?

Steps to Solution: This is a stoichiometry problem with a limiting reagent. The limiting reagent must be determined and the amounts of leftover reactants and the amounts of product formed are to be determined as you did in Chapter 4. The step that makes it different is that the amount of reactants present are expressed in pressure, and must be converted to moles to determine the mass of product formed. The final pressure is determined from the amounts of the products formed and the remaining reactants.

(a) The stoichiometry part of this problem is solved by constructing a table of amounts using the ideal gas law to determine the number of moles of methane and oxygen. The partial pressures are given at 298 K; therefore this temperature is used to solve for the number of moles of the reactants.

$$\text{mol } CH_4 = \frac{P_{CH_4} V}{RT} = \frac{10.5\ \text{atm} \cdot 200.0\ \text{L}}{(0.0821\ \text{L} \cdot \text{atm/mol} \cdot \text{K}) 298 K} = 8.59 \times 10^1\ \text{mol}$$

$$\text{mol } O_2 = \frac{P_{O_2}V}{RT} = \frac{3.50 \text{ atm} \cdot 200.0 \text{ L}}{(0.0821 \text{ L} \cdot \text{atm/mole} \cdot \text{K})298 \text{ K}} = 2.86 \times 10^1 \text{ mol}$$

These amounts are entered into row 3 of the table of amounts. In row 2, the ratios of the mols of the reactants to their stoichiometric coefficients are entered. Because the ratio for oxygen is smaller, oxygen is the limiting reagent. The changes in the amounts of substances in the reaction are entered into row 4.

Quantity/Substance	CH_4	O_2	CH_3OH
$\dfrac{\text{mol reactant}}{\text{stoichiometric coef}}$	$\dfrac{8.59 \times 10^1}{2} =$ 4.29×10^1	$\dfrac{2.86 \times 10^1}{1} =$ $\mathbf{2.86 \times 10^1}$ limiting reagent	----
Starting amount	8.59×10^1 mol	2.86×10^1 mol	0
Change in amount	-5.73×10^1 mol	-2.86×10^1 mol	5.73×10^1 mol
Final amount	2.86×10^1	0	5.73×10^1 mole

change in mol $O_2 = -2.86 \times 10^1$ mol (because it is the limiting reagent)

$$\text{change in mol } CH_4 = -\frac{2 \text{ mol } CH_4}{\text{mol } O_2} \cdot 2.86 \times 10^1 \text{ mol } O_2 = -5.73 \times 10^1 \text{ mol}$$

$$\text{change in mol } CH_3OH = \frac{2 \text{ mol } CH_3OH}{\text{mol } O_2} \cdot 2.86 \times 10^1 \text{ mol } O_2 = 5.73 \times 10^1 \text{ mol}$$

Adding the amounts in rows 3 and 4 give the final amounts of methane, oxygen and methanol. The amount of methanol formed in the reaction is found from the moles and the molar mass of methanol:

$$\text{g } CH_3OH = (5.73 \times 10^1 \text{ mol}) \, 32.04216 \text{ g/mol} = \boxed{1.83 \times 10^3 \text{ g}}$$

(b) *We calculate the final pressure is calculated using Dalton's Law of partial pressures, but two different approaches may be used. We will calculate the pressure of the mixture by calculating the partial pressures of each of the substances knowing the moles, the temperature and the volume of the reactor.*

$$p_{CH_4} = \frac{n_{CH_4}RT}{V} = \frac{(2.86 \times 10^1 \text{ mole}) \, 0.08206 \text{ L} \cdot \text{atm/mole} \cdot 623 \text{ K}}{200 \text{ L}} = 7.32 \text{ atm}$$

$p_{O_2} = 0$ (all oxygen consumed as it is the limiting reagent)

$$P_{CH_3OH} = \frac{n_{CH_3OH}RT}{V} = \frac{(5.73 \times 10^1 \text{mole})0.08206 \text{ L} \cdot \text{atm/mole} \cdot 623 \text{ K}}{200 \text{ L}} = 14.6 \text{ atm}$$

The total pressure is the sum of the three partial pressures:

$$P_T = P_{CH_4} + P_{O_2} + P_{CH_3OH} = 7.32 + 0 + 14.6 \text{ atm} = \boxed{21.9 \text{ atm}}$$

We solved this problem by breaking the large single problem into several smaller problems. It is often easier to divide (the problem into manageable pieces) and conquer (it)!

Exercise 5.16 At an elevation of 10 km above sea level (about 33,000 feet, approximately the cruising altitude of a jet airliner), the atmospheric pressure is 210 torr and the temperature is 230 K.

(a) How many molecules per liter are there at this altitude?
(b) If the composition of air at this altitude is about the same as at sea level, what is the partial pressure of oxygen at 10 km in atm?
(c) How many L of air at this altitude would be required to completely burn 1 kg of jet fuel, $C_{12}H_{26}$?

Steps to Solution: The number of molecules per liter can be found by calculating of the number of mols/L from the ideal gas law, because the pressure and temperature are known.

(a) Solving the gas law for n/V and substituting the values of pressure and temperature at 10 km altitude (210 torr and 230 K) gives:

$$\frac{n}{V} = P \cdot \frac{1}{RT} = \frac{210 \text{ torr}}{760 \text{ torr/atm}} \cdot \frac{1}{0.08206 \text{L} \cdot \text{atm/mole} \cdot \text{K}(230\text{K})} = 1.46 \times 10^{-2} \text{ mol/L}$$

Multiplying the molarity of air molecules by Avogadro's number gives the number of molecules per liter:

$$\frac{\text{number of molecules}}{L} = 1.46 \times 10^{-2} \text{ mol/L} \cdot 6.022 \times 10^{23} \text{ 1/mol} = \boxed{8.82 \times 10^{21} \text{ molecules/L}}$$

(b) This problem asks for the partial pressure of oxygen at this altitude given the total pressure (210 torr) and assuming that the composition of air is similar to that at sea level. From your text, the mole fraction of O_2 in air at sea level is 0.2095. From Dalton's law of partial pressures, the partial pressure can be found given the mole fraction and total pressure:

$$p_{O_2} = X_{O_2} P_{total} = 0.02095(210 \text{ torr}) \cdot \frac{1 \text{ atm}}{760 \text{ torr}} = 5.79 \times 10^{-2} \text{ atm}$$

This answer is small, but it explains why oxygen masks are provided if the airplane loses cabin pressure: there is not much oxygen available at cruising altitude!

(c) *Here, we must find the moles of fuel and then stoichiometry will be used to find the moles of oxygen needed. The partial pressure and temperature will then be used to find the volume of oxygen. First, the balanced chemical equation of reaction is:*

$$2\ C_{12}H_{26\ (l)} + 37\ O_{2\ (g)} \rightarrow 24\ CO_{2\ (g)} + 26\ H_2O_{(g)}$$

The number of moles of fuel is used to find the moles O_2:

$$\text{mols } C_{12}H_{26} = \frac{1000\,g}{170.4\ g/mol} = 5.87\ mol$$

$$\text{mols } O_2 = 5.87\ \text{mols } C_{12}H_{26} \cdot \frac{37\ mol\ O_2}{2\ mol\ C_{12}H_{26}} = 108.6\ mol\ O_2$$

Now we find the moles of air containing 108.6 mols O_2 by rearranging the equation defining mole fraction:

$$X = \frac{\text{mols}}{\text{total mols}} \Rightarrow \text{total mols} = \frac{\text{mols}}{X}$$

$$\text{total mols air} = \frac{108.6\ mol}{0.2095} = 518.4\ mol\ air$$

The volume of this amount of air is found using the ideal gas law for 230 K and 0.276 atm:

$$V_{air} = \frac{nRT}{P} = \frac{518.4\ mol \cdot 0.08206\ L \cdot atm/mol \cdot K(230K)}{210\ torr/760\ torr/atm} = 3.54 \times 10^{4}\ L$$

Test Yourself

1. A sample of N_2 gas is contaminated with one of the noble gases. It is found that the contaminant effuses at 2.69 times the rate of N_2. What is the contaminating noble gas?

2. At 1000 °C and 10.0 torr, the density of a certain element in the gas phase is 2.9×10^{-3} g/L. What is the element?

3. A tank is pressurized with 5.0 atm of N_2 and 10.0 atm of H_2. Ammonia (NH_3) is formed. When the pressure finally remains constant, indicating that the reaction has proceeded as far as it will go, the partial pressure of ammonia is 3.2 atm. What is the total pressure in the tank assuming that neither the temperature nor the volume of the container have changed?

4. How many molecules are there in 1.00 mL in a vacuum of 10^{-10} torr at a temperature of 294 K? What would be the mass of this number of Ar atoms?

Answers

1. He

2. Na

3. 11.8 atm

4. 3. x 10^6 atoms = 2. x 10^{-16} g Ar

Solutions to Chapter Problems

5.1 Molecular speed is related to temperature by the equation, $u = (2\,kE/m)^{1/2}$, and $kE_{avg} = 3RT/2N_A$, so $u_{avg} = (3RT/mN_A)^{1/2}$. Thus, molecular speed increases with the square root of kinetic energy, so molecules will reach the detector sooner at higher temperature, because they have greater speed.

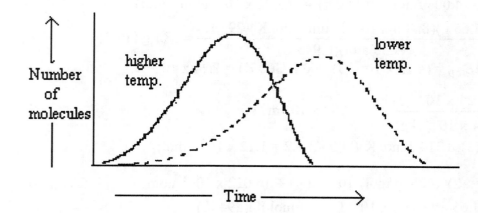

5.3 Molecular speed is related to temperature by the equation, $u = (2\ kE/m)^{1/2}$, and $kE_{avg} = 3RT/2N_A$, so $u_{avg} = (3RT/mN_A)^{1/2}$. Each distribution of molecular speeds will have the same general shape, with the position of the peak depending on $(T/m)^{1/2}$:

5.5 The most probable speed at any temperature is related to most probable kinetic energy by the equation $kE = mu^2/2$, or $u_{mp} = (2\ kE_{mp}/m)^{1/2}$. According to your textbook, the most probable kinetic energy at 300 K is 4.13×10^{-21} J/molecule. Because energy is proportional to temperature, the most probable kinetic energy at any other temperature can be calculated using proportions: $kE_2 = kE_1(T_2/T_1)$. Average kinetic energy, which is not the same as most probable kinetic energy because the distribution of energies is asymmetric, is calculated using $kE_{avg} = 3\ RT/2\ N_A$ per molecule, or $kE_{molar} = 3\ RT/2$.

(a) He: $m = (4.00\ \text{g/mol})(10^{-3}\ \text{kg/g}) \div (6.022 \times 10^{23}\ \text{atoms/mol}) = 6.64 \times 10^{-27}$ kg;

$$MM = \frac{(1.65\ \text{g})(8.206 \times 10^{-2}\ \text{L atm / mol K})(294\ \text{K})}{(1.50\ \text{atm})(0.945\ \text{L})} = 28.1\ \text{g / mol};$$

at 900 K, $kE_{mp} = (4.13 \times 10^{-21}\ \text{J})(900\ \text{K}/300\ \text{K}) = 1.24 \times 10^{-20}$ J.

$$u = \left[\frac{(2)(1.24 \times 10^{-20}\ \text{J})}{(6.64 \times 10^{-27}\ \text{kg})}\right]^{1/2} = 1.93 \times 10^3\ \text{m / s};$$

$kE_{molar} = (3)(8.314\ \text{J/mol K})(900\ \text{K}) \div 2 = 1.12 \times 10^4$ J/mol;

(b) O_2: $m = (2)(16.00\ \text{g/mol})(10^{-3}\ \text{kg/g}) \div (6.022 \times 10^{23}\ \text{atoms/mol}) = 5.31 \times 10^{-26}$ kg;

$$MM = \frac{(1.65\ \text{g})(8.206 \times 10^{-2}\ \text{L atm / mol K})(294\ \text{K})}{(1.50\ \text{atm})(0.945\ \text{L})} = 28.1\ \text{g / mol};$$

at 300 K, $kE_{mp} = 4.13 \times 10^{-21}$ J;

$$u = \left[\frac{(2)(4.13 \times 10^{-21}\ \text{J})}{(5.31 \times 10^{-26}\ \text{kg})}\right]^{1/2} = 3.94 \times 10^2\ \text{m / s};$$

$kE_{molar} = (3)(8.314\ \text{J/mol K})(300\ \text{K}) \div 2 = 3.74 \times 10^2$ J/mol;

(c) SF_6: $MM = [32.066 + 6(18.998)] = 146.0$ g/mol;
$m = (146.0$ g/mol$)(10^{-3}$ kg/g$) \div (6.022 \times 10^{23}$ atoms/mol$) = 2.43 \times 10^{-25}$ kg;

$$MM = \frac{(1.65 \text{ g})(8.206 \times 10^{-2} \text{ L atm / mol K})(294 \text{ K})}{(1.50 \text{ atm})(0.945 \text{ L})} = 28.1 \text{ g/mol};$$

at 900 K, $kE_{mp} = (4.13 \times 10^{-21}$ J$)(900$ K$/300$ K$) = 1.24 \times 10^{-20}$ J.

$$u = \left[\frac{(2)(1.24 \times 10^{-20} \text{ J})}{(2.43 \times 10^{-25} \text{ kg})}\right]^{1/2} = 3.19 \times 10^2 \text{ m/s};$$

$kE_{molar} = (3)(8.314$ J/mol K$)(900$ K$) \div 2 = 1.12 \times 10^4$ J/mol.

5.7 Impulse is proportional to mass and speed (u), and $u = (2 kE/m)^{1/2}$. Further, $kE_{avg} = 3 RT/2 N_A$. Combining these relationships, we find that impulse increases as the square root of the mass and temperature.
(a) At equal temperatures, a xenon atom generates a greater impulse than a neon atom because of its greater mass.
(b) Average speed increases with temperature, so a methane molecule with average speed at 700 K generates a greater impulse than one with average speed at 390 K.
(c) At equal speeds, an F_2 molecule generates a greater impulse than an H_2 molecule because of its greater mass.

5.9 The ideal gas is defined by the conditions that both molecular volumes and intermolecular forces are negligible.
(a) At very high pressure, molecules are very close together, so their volumes are significant compared to the volume of their container; the first condition is not met and the gas is not ideal.
(b) At very low temperature, molecules move very slowly, so the forces between molecules, even though small, are sufficient to influence molecular motion; the second condition is not met and the gas is not ideal.

5.11 (a) If the piston is stationary and there is no friction, the forces and pressures on each side must be equal, so the internal pressure is also 1 atm. This pressure is generated by gas molecules colliding with the face of the piston.
(b) When temperature is doubled, the increase in average molecular speed causes an increase in pressure. The piston will move outward, causing the concentration of gas molecules to decrease. This will lead to a lower frequency of collisions with the wall, reducing the pressure. The piston will stop when the reduced rate of collisions counterbalances the increased impulse per collision and the internal pressure is once again 1 atm.

5.13 A pinhole in the top of the tube would let air leak into the space until the internal pressure matched the external, atmospheric pressure. The barometer would then indicate zero pressure, because the height of the mercury column is determined by the pressure *difference* between inside and outside, and this difference would be zero in the presence of a pinhole.

5.15 The column of liquid must have a height such that the total mass of the column exerts a force equal to the force exerted by the atmosphere. Because water is less dense than mercury, a greater height of water is required. First convert the pressure to torr, which gives the height of a mercury column at this pressure. Then multiply by the ratio of the densities of mercury and water.
$P = (115 \text{ kPa})(760 \text{ torr}/1.01325 \times 10^2 \text{ kPa}) = 862.6 \text{ torr} = 862.6 \text{ mm Hg}$;

$$\text{ht of water} = (862.6 \text{ mm Hg})\left[\frac{(13.59 \text{ g}/\text{mL Hg})}{(1.00 \text{ g}/\text{mL H}_2\text{O})}\right] = 1.17 \times 10^4 \text{ mm};$$

Convert to meters: $(1.17 \times 10^4 \text{ mm})(10^{-3} \text{ m/mm}) = 11.7$ meters.

5.17 The SI unit for pressure is the pascal: $1 \text{ atm} = 1.01325 \times 10^5$ Pa, and the atmosphere-torr conversion factor is 1 atm = 760 torr.
(a) $(455 \text{ torr})(1 \text{ atm}/760 \text{ torr})(1.01325 \times 10^5 \text{ Pa/atm}) = 6.07 \times 10^4$ Pa;
(b) $(2.45 \text{ atm})(1.01325 \times 10^5 \text{ Pa/atm}) = 2.48 \times 10^5$ Pa;
(c) $(0.46 \text{ torr})(1 \text{ atm}/760 \text{ torr})(1.01325 \times 10^5 \text{ Pa/atm}) = 61$ Pa; and
(d) $(1.33 \times 10^{-3} \text{ atm})(1.01325 \times 10^5 \text{ Pa/atm}) = 1.35 \times 10^2$ Pa.

5.19 The question asks for moles, which can be obtained from *P-V-T* data using the ideal gas equation: $n = PV/RT$.
$$n = \frac{PV}{RT} = \frac{(5.00 \text{ atm})(20.0 \text{ L})}{(8.206 \times 10^{-2} \text{ L atm}/\text{mol K})(298 \text{ K})} = 4.09 \text{ mol.};$$

$$V = \frac{nRT}{P} = \frac{(4.09 \text{ mol})(8.206 \times 10^{-2} \text{ L atm}/\text{mol K})(298 \text{ K})}{(1 \text{ atm})} = 1.00 \times 10^2 \text{ L}.$$

5.21 When some variables are held fixed, rearrange the ideal gas equation, $PV = nRT$, to collect fixed values on the right.
(a) n, R, V are constant: $P/T = nR/V$ = constant; $P_i/T_i = P_f/T_f$;
(b) P, R, T are constant: $V/n = RT/P$ = constant; $V_i/n_i = V_f/n_f$;
(c) n, R, T are constant: $PV = nRT$ = constant; $P_iV_i = P_fV_f$;

5.23 When some conditions change but others remain fixed, rearrange the ideal gas equation so the constant terms are grouped on the right. In this problem, n and P are fixed:
$V/T = nR/P$ = constant, so $V_f/T_f = V_i/T_i$.
$V_i = 0.255$ L, $T_i = 25 + 273.15 = 298$ K, $T_f = -10 + 273.15 = 263$ K;
$$V_f = \frac{V_i T_f}{T_i} = \frac{(0.255 \text{ L})(263 \text{ K})}{(298 \text{ K})} = 0.225 \text{ L}$$

5.25 As a gas cools, its molecules move more slowly, so they impart smaller impulses on the walls of their container. This reduces the internal pressure, so the balloon collapses until the increase in gas density inside the balloon brings the internal pressure back up to 1.000 atm.

5.27 The equation is valid only for a gas for which n and T are fixed.
(a) n and T are fixed, so the equation is valid. (b) n changes, so the equation is not valid.
(c) T changes, so the equation is not valid. (d) The equation is not valid for liquids.

5.29 The ideal gas equation, $PV = nRT$, can be used to calculate moles using P-V-T data. Molar mass is related to moles through $n = m/MM$. These equations yield two expressions for n, $n = PV/RT = m/MM$, from which $MM = mRT/PV$.
$T = 25 + 273.15 = 298$ K, $P = (262$ torr$)(1$ atm$/760$ torr$) = 0.345$ atm;

$$MM = \frac{(2.55 \text{ g})(8.206 \times 10^{-2} \text{ L atm / mol K})(298 \text{ K})}{(0.345 \text{ atm})(1.50 \text{ L})} = 121 \text{ g / mol}$$

Knowing that the compound contains only C (12.01 g/mol), F (19.0 g/mol), and Cl (35.45 g/mol), work out the formula by trial and error. In this case, the combination of 1 C, 2 F, and 2 Cl has $MM = 12 + 2(19) + 2(35.45) = 121$ g/mol, matching the experimental value.
The formula is CF_2Cl_2.

5.31 According to Dalton's law of partial pressures, each gas exerts a pressure equal to its mole fraction times the total pressure. Mole fraction can be found from concentration in ppm: $X = (10^{-6}/\text{ppm})(\text{ppm})$.
$p(NO_2) = (0.78 \text{ ppm})(10^{-6}/\text{ppm})(758.4 \text{ torr})(1 \text{ atm}/760 \text{ torr}) = 7.78 \times 10^{-7}$ atm.

5.33 According to Dalton's law of partial pressures, each gas exerts a pressure equal to its mole fraction times the total pressure: $p = XP_{tot}$. Standard atmospheric conditions is $P_{tot} = 1$ atm.

$p(N_2) = (0.7808)(1 \text{ atm})(760 \text{ torr/atm}) = 593.4$ torr;
$p(O_2) = (0.2095)(1 \text{ atm})(760 \text{ torr/atm}) = 159.2$ torr;
$p(Ar) = (9.34 \times 10^{-3})(1 \text{ atm})(760 \text{ torr/atm}) = 7.10$ torr; and
$p(CO_2) = (3.25 \times 10^{-4})(1 \text{ atm})(760 \text{ torr/atm}) = 2.47 \times 10^{-1}$ torr.

5.35 Molecular pictures show the relative numbers of molecules of various substances, which also represents the relative numbers of moles of those substances. The figure contains 10 CO molecules and 20 H_2 molecules.

(a) and (b) There are more H_2 molecules, so the pressure due to H_2 and the mole fraction of H_2 are higher.
(c) $X_{CO} = n_{CO}/n_{tot} = 10/30 = 0.333$.

Chapter 5: The Behavior of Gases

5.37 To calculate partial pressures from a total pressure, mole fractions are required: $p_i = X_i P_{tot}$. Mole fractions can be calculated from the analytical data for the gas sample.

$n(CH_4) = (1.57 \text{ g}) \div (16.04 \text{ g/mol}) = 9.79 \times 10^{-2}$ mol;
$n(C_2H_6) = (0.41 \text{ g}) \div (30.068 \text{ g/mol}) = 1.36 \times 10^{-2}$ mol;
$n(C_3H_8) = (0.020 \text{ g}) \div (44.094 \text{ g/mol}) = 4.54 \times 10^{-4}$ mol;
$n_{tot} = 9.79 \times 10^{-2}$ mol $+ 1.36 \times 10^{-2}$ mol $+ 4.54 \times 10^{-4}$ mol $= 1.120 \times 10^{-1}$ mol;

$X(CH_4) = (9.79 \times 10^{-2} \text{ mol}) \div (1.120 \times 10^{-1} \text{ mol}) = 0.874$;
$X(C_2H_6) = (1.36 \times 10^{-2} \text{ mol}) \div (1.120 \times 10^{-1} \text{ mol}) = 0.121$;
$X(C_3H_8) = (4.54 \times 10^{-4} \text{ mol}) \div (1.120 \times 10^{-1} \text{ mol}) = 4.06 \times 10^{-3}$;

$p(CH_4) = (0.874)(2.35 \text{ atm}) = 2.05$ atm;
$p(C_2H_6) = (0.121)(2.35 \text{ atm}) = 0.284$ atm;
$p(C_3H_8) = (4.06 \times 10^{-3})(2.35 \text{ atm}) = 9.5 \times 10^{-3}$ atm.

5.39 Stoichiometric calculations require moles and a balanced chemical equation. Balance the equation, determine moles of CO_2 produced; and apply the ideal gas equation to calculate V.

Balanced equation: $C_6H_{12}O_6 + 6 O_2 \rightarrow 6 CO_2 + 6 H_2O$;
$MM_{glucose} = 180.16$ g/mol;
$n(\text{glucose}) = m/MM = (4.65 \text{ g}) \div (180.16 \text{ g/mol}) = 2.581 \times 10^{-2}$ mol;
$n(CO_2) = 6\, n(\text{glucose}) = (6)(2.581 \times 10^{-2} \text{ mol}) = 1.549 \times 10^{-1}$ mol;

$$V = \frac{nRT}{P} = \frac{(1.549 \times 10^1 \text{ mol})(8.206 \times 10^{-2} \text{ L atm / mol K})(310 \text{ K})}{(1.00 \text{ atm})} = 3.94 \text{ L}.$$

5.41 Because there are limited amounts of both chemical reactants, this is a limiting reagent problem. Use an amounts table. For chlorine gas, use the ideal gas equation to do pressure-mole conversions. The data must have the same units as those for R.

$n(Na) = (6.90 \text{ g}) \div (22.97 \text{ g/mol}) = 0.300$ mol;
$P(Cl_2) = (1.25 \times 10^3 \text{ torr})(1 \text{ atm}/760 \text{ torr}) = 1.64$ atm;

$$n(Cl_2) = \frac{PV}{RT} = \frac{(1.64 \text{ atm})(3.00 \text{ L})}{(8.206 \times 10^{-2} \text{ L atm / mol K})(300 \text{ K})} = 0.200 \text{ mol};$$

Reaction:	2 Na$_{(s)}$	+	Cl$_{2(g)}$	$\rightarrow$	2 NaCl$_{(s)}$
Initial amount (mol):	0.300		0.200		0
Change in amount (mol):	-0.300		-0.150		+0.300
Final amount (mol):	0		0.050		0.300

$$P = P(Cl_2) = \frac{nRT}{V} = \frac{(5.0 \times 10^{-2} \text{ mol})(8.206 \times 10^{-2} \text{ L atm / mol K})(320 \text{ K})}{(3.00 \text{ L})} = 0.438 \text{ atm}.$$

5.43 Because there are limited amounts of both chemical reactants, this is a limiting reagent problem. Use an amounts table. Use the ideal gas equation to determine the initial amounts of each gas. The reactor initially contains the gases in 1:1 mole ratio, so $p_i = 270$ atm ÷ 2 = 135 atm for each gas.

$$n_i = \frac{p_i V}{RT} = \frac{(135 \text{ atm})(8.75 \times 10^3 \text{ L})}{(8.206 \times 10^{-2} \text{ L atm / mol K})(450 + 273 \text{ K})} = 1.99 \times 10^4 \text{ mol};$$

Reaction:	$N_{2(g)}$	+	$3 H_{2(g)}$	→	$2 NH_{3(g)}$
Initial amount (10^4 mol):	1.99		1.99		0
Change in amount (mol):	-0.66		-1.99		+1.33
Final amount (mol):	1.33		0		1.33

$m(NH_3) = n\, MM = (1.33 \times 10^4 \text{ mol})(17.01 \text{ g/mol}) = 7.26 \times 10^5$ g.

5.45 The yield in a reaction is the ratio of the actual amount produced to the theoretical amount. The calculation in Problem 5.43 gives the theoretical amount:

actual amount = (% yield)(10^{-2}/%)(theoretical amount)
actual amount = (13%)(10^{-2}/%)(2.26×10^5 g) = 2.9×10^4 g.

5.47 Table 5-3 gives the mole fractions of dry air (0% humidity), from which $X(O_2) = 0.2095$. Table 5-4 gives the vapor pressure of water at 20 °C, 17.535 torr; this is the partial pressure at 100% humidity for that temperature. To find the mole fraction of O_2 at 100% humidity, first determine the partial pressure of O_2.

$p(H_2O) = (17.535 \text{ torr})(1 \text{ atm}/760 \text{ torr}) = 2.307 \times 10^{-2}$ atm;
$p(N_2 + O_2) = 1.000 \text{ atm} - 2.307 \times 10^{-2}$ atm = 0.9977 atm;
$p(O_2) = (0.2095)(0.9977 \text{ atm}) = 0.2090$ atm;
$X(O_2) = p(O_2)/P_{total} = (0.2090 \text{ atm}) \div (1.000 \text{ atm}) = 0.2090$;
change = 0.2095 – 0.2090 = 0.0005.

5.49 The dew point is defined as the temperature at which the partial pressure of water vapor in the atmosphere matches the vapor pressure of water. Refer to Table 5-4 for vapor pressures: at 25 °C, $vP = 23.756$ torr.

$p(H_2O) = (78\%)(10^{-2}/\%)(23.756 \text{ torr}) = 18.53$ torr;
This vapor pressure corresponds to a dew point between 20 and 25 °C. Assume a linear variation in vapor pressure between these two temperatures:

$$\text{dew point} = 20.0 \text{ °C} + \left[\frac{(18.53 - 17.535)}{(23.756 - 17.535)}\right] (5 \text{ °C}) = 20.0 + 0.8 = 20.8 \text{ °C}$$

Chapter 5: The Behavior of Gases

5.51 We must calculate the volume of dry air that contains 10 mg of neon. Use moles, mole fractions from Table 5-3, and the ideal gas equation:
$n_{Ne} = m/MM = (10$ mg$)(10^{-3}$ g/mg$) \div (20.180$ g/mol$) = 4.0 \times 10^{-4}$ mol;
From Table 5-3, $X_{Ne} = 1.82 \times 10^{-5} = n_{Ne}/n_{air}$;
Thus, the number of moles of air containing 10 mg of neon is:
$n_{air} = n_{Ne}/X_{Ne} = (5.0 \times 10^{-4}$ mol$) \div (1.82 \times 10^{-5}) = 27.5$ mol;
The volume occupied by this amount of air can be calculated from the ideal gas equation:
$$V = \frac{nRT}{P} = \frac{(27.5 \text{ mol})(8.206 \times 10^2 \text{ L atm/mol K})(298 \text{ K})}{(1.000 \text{ atm})} = 67 \text{ L}.$$

5.53 First calculate moles of CO_2, then use the ideal gas equation to calculate the final pressure.
$n = m/MM = (15.00$ g$)(44.01$ g/mol$) = 0.3408$ mol;
$$P = \frac{nRT}{V} = \frac{(0.3408 \text{ mol})(8.206 \times 10^2 \text{ L atm/mol K})(273 \text{ K})}{(0.750 \text{ L})} = 10.2 \text{ atm}.$$

5.55 Use the ideal gas equation to find molar density and Avogadro's number to convert to molecular density.
$$\frac{n}{V} = \frac{P}{RT} = \frac{(10^{-10} \text{ atm})}{(8.206 \times 10^2 \text{ L atm/mol K})(310 \text{ K})} = 3.93 \times 10^{-12} \text{ mol/L};$$
$\#/V = N_A(n/V) = (6.02 \times 10^{23}/mol)(3.93 \times 10^{-12}$ mol/L$) = 2.37 \times 10^{12}/$L.

5.57 This is an application of the ideal gas equation to changing gas conditions. The only constant quantity is n, so $PV/T = nR =$ constant.
$P_i = 0.969$ atm, $V_i = 0.963$ L, $T_i = 22 + 273.15 = 295$ K;
$P_f = 1.00$ atm, $V_f = ?$, $T_f = 0 + 273.15 = 273$ K;
$$V_f = \left(\frac{P_i V_i}{T_i}\right)\left(\frac{T_f}{P_f}\right) = \frac{(0.969 \text{ atm})(0.963 \text{ L})(273 \text{ K})}{(295 \text{ K})(1.00 \text{ atm})} = 0.864 \text{ L}.$$

5.59 Concentrations in parts per billion can be converted to mole fractions, and the partial pressure can then be calculated using $p_i = X_i P_{total}$. To determine molecular concentration, first use the ideal gas equation to calculate mol/cm^3, then multiply by N_A.
$X(SF_6) = (1$ ppb$)(10^{-9}/$ppb$) = 10^{-9}$;
$p(SF_6) = (10^{-9})(1$ atm$) = 10^{-9}$ atm;
$$\frac{n}{V} = \frac{p}{RT} = \frac{(10^{-9} \text{ atm})(10^{-3} \text{ L/cm}^3)}{(8.206 \times 10^{-2} \text{ L atm/mol K})(21 + 273 \text{ K})} = 2.92 \times 10^{-14} \text{ mol/cm}^3;$$
$\#/V = (6.022 \times 10^{23}$ molec/mol$)(2.92 \times 10^{-14}$ mol/cm$^3) = 2.49 \times 10^{10}$ molecules/cm^3.

5.61 Use the ideal gas equation to determine moles, then do mole-mass-number conversions.
$$n = \frac{PV}{RT} = \frac{(10.0 \text{ atm})(600 \text{ mL})(10^{-3} \text{ L/mL})}{(8.206 \times 10^2 \text{ L atm/mol K})(1000 + 273 \text{ K})} = 5.72 \times 10^{-2} \text{ mol};$$
$m = n\,MM = (5.72 \times 10^{-2}$ mol$)(83.80$ g/mol$) = 4.81$ g;
atoms $= n\,N_A = (5.72 \times 10^{-2}$ mol$)(6.022 \times 10^{23}$ atoms/mol$) = 3.46 \times 10^{22}$ atoms.

Chapter 5: The Behavior of Gases

5.63 Molecular pictures show the relative numbers of molecules of various substances, which also represents the relative numbers of moles of those substances. Chamber A contains 6 atoms, B contains 12 atoms, and C contains 9 atoms, for a total of 27 atoms, and the chambers have equal volumes and are at the same temperature.

(a) Chamber B contains the most atoms, so it exhibits the highest pressure;

(b) The pressures in the chambers are proportional to the number of atoms:
$P_A = P_B(\#_A/\#_B) = (1.0 \text{ atm})(6/12) = 0.50 \text{ atm}$;

(c) The pressure will increase in proportion to the number of atoms:
$P(\text{new}) = P(\text{old})[\#(\text{new})/\#(\text{old})] = (1.00 \text{ atm})(27/6) = 4.5 \text{ atm}$;

(d) When the valves are opened, the number of atoms in each chamber equalize: $27/3 = 9$.
$P(\text{new}) = P(\text{old})[\#(\text{new})/\#(\text{old})] = (0.50 \text{ atm})(9/12) = 0.38 \text{ atm}$.

5.65 Much of the information in this problem is not needed, because molecular speed can be calculated from temperature and molar mass:

$kE_{avg} = 3\ RT/2\ N_A = mu^2/2$, from which $u^2 = 3\ RT/m\ N_A = 3\ RT/MM$.
The mass must be converted to kg/mol for mass units to cancel (1 J = 1 kg m^2/s^2):

$$u_{avg} = \left(\frac{3\ RT}{MM}\right)^{1/2} = \left(\frac{3(8.314\text{ J / mol K})(-25 + 273\text{ K})}{(48.00\text{ g / mol})(10^{-3}\text{ kg / g})}\right)^{1/2} = 359 \text{ m / s}.$$

5.67 This is an empirical formula problem, with P-V-T data added to allow calculation of the molar mass. First, determine the empirical formula from mass percentages. Consider 100 g of the compound.

C: (71.22 g/100 g) ÷ (12.01 g/mol) = 5.930 mol C/100 g;
H: (14.94 g/100 g) ÷ (1.008 g/mol) = 14.82 mol H/100 g;
N: (13.84 g/100 g) ÷ (14.01 g/mol) = 0.9879 mol N/100 g;

Divide each by the smallest among them, 0.9879 mol N/100 g:
C: (5.930 mol C/100 g) ÷ (0.9879 mol N/100 g) = 6.003 mol C/mol N, round to 6 C/N;
H: (14.82 mol H/100 g) ÷ (0.9879 mol N/100 g) = 15.00 mol H/mol N;

Empirical formula: $C_6H_{15}N$, empirical $MM = 101$ g/mol;

$$MM = \frac{mRT}{PV} = \frac{(250 \text{ mg})(10^{-3}\text{ g / mg})(8.206 \times 10^{-2}\text{ L atm / mol K})(150 + 273\text{ K})}{(435 \text{ torr})(1 \text{ atm / 760 torr})(150 \text{ mL})(10^{-3}\text{ L / mL})} = 101 \text{ g / mol};$$

The molar mass is the same as the empirical molar mass, so the molecular formula is the same as the empirical formula.

5.69 The vapor pressure of water is the partial pressure of gaseous water in the atmosphere under 100% humidity conditions. To determine the dew point, first calculate the partial pressure of water vapor at the existing temperature and then estimate the temperature at which this partial pressure equals the vapor pressure of water, assuming a linear variation of vapor pressure with temperature. Use data in Table 5-4: $p(H_2O) = (vP)$(rel. hum.)/(100%).

(a) $p(H_2O) = (42.175 \text{ torr})(80\%) \div (100\%) = 33.74$ torr;

$$\text{dew point} = 30.0 \,°C + \left[\frac{(33.74 - 31.824)}{(42.175 - 31.824)}\right](5\,°C) = 30.0 + 0.9 = 30.9\,°C$$

(b) $p(H_2O) = (12.788 \text{ torr})(50\%) \div (100\%) = 6.394$ torr;

$$\text{dew point} = 0\,°C + \left[\frac{(6.394 - 4.579)}{(6.543 - 4.579)}\right](5\,°C) = 0.0 + 0.6 = 4.6\,°C$$

(c) $p(H_2O) = (23.756 \text{ torr})(30\%) \div (100\%) = 7.1268$ torr;

$$\text{dew point} = 5\,°C + \left[\frac{(7.1268 - 6.543)}{(9.209 - 6.543)}\right](5\,°C) = 5.0 + 1.1 = 6.1\,°C$$

5.71 There are two interconnected parts to this problem. Calculate moles of oxygen using the ideal gas equation, then use stoichiometric analysis for $KClO_3$.

(a) $n = \dfrac{pV}{RT} = \dfrac{(759.2 \text{ torr})(1 \text{ atm} / 760 \text{ torr})(22.96 \text{ mL})(10^{-3} \text{L/mL})}{(8.206 \times 10^{-2} \text{ L atm/mol K})(25 + 273 \text{ K})} = 9.379 \times 10^{-4}$ mol;

(b) The reaction is $2 \, KClO_3 \rightarrow 2 \, KCl + 3 \, O_2$;
$n(KClO_3) = \frac{2}{3}n(O_2) = \frac{2}{3}(9.379 \times 10^{-4} \text{ mol}) = 6.253 \times 10^{-4}$ mol;

(c) $m(KClO_3) = n \, MM = (6.253 \times 10^{-4} \text{ mol})(122.55 \text{ g/mol}) = 7.663 \times 10^{-2}$ g;
$\%(KClO_3) = (100\%)[m(KClO_3)/m_{total}] = (100\%)(7.663 \times 10^{-2} \text{ g}) \div (0.1054 \text{ g}) = 72.70\%$.

5.73 The molar mass of gaseous oxygen can be determined using the ideal gas law: $MM = mRT/PV$.

Pump out a bulb of known volume, weigh it, fill with oxygen at a measured pressure, and weigh again. If gaseous oxygen were monatomic, this experiment would give $MM = 16.00$ g/mol, while the diatomic gas gives $MM = 32.00$ g/mol.

5.75 Use Dalton's law of partial pressures to determine partial pressures, then apply the ideal gas equation to calculate masses.

$X(C_3H_6) = 1.00/5.00 = 0.200$, $p(C_3H_6) = 0.200$ atm; $X(O_2) = 0.800$, $p(O_2) = 0.800$ atm;

$m(C_3H_6) = \dfrac{pV(MM)}{RT} = \dfrac{(0.200 \text{ atm})(2.00 \text{ L})(42.078 \text{ g/mol})}{(8.206 \times 10^{-2} \text{ L atm/mol K})(23 + 273 \text{ K})} = 0.693$ g;

$m(O_2) = \dfrac{pV(MM)}{RT} = \dfrac{(0.800 \text{ atm})(2.00 \text{ L})(32.00 \text{ g/mol})}{(8.206 \times 10^{-2} \text{ L atm/mol K})(23 + 273 \text{ K})} = 2.11$ g.

Chapter 5: The Behavior of Gases

5.77 Density can be calculated from the ideal gas equation and mole-mass conversions:

$$n = \frac{m}{MM} = \frac{PV}{RT}, \text{ from which } \frac{m}{V} = \frac{P\,MM}{RT}$$

$$\frac{m}{V} = \frac{P\,MM}{RT} = \frac{(755 \text{ torr})(1 \text{ atm} / 760 \text{ torr})(146.05 \text{ g/mol})}{(8.206 \times 10^{-2} \text{ L atm/mol K})(27 + 273 \text{ K})} = 5.89 \text{ g/L}.$$

5.79 The hint for this problem gives a mass relationship that must be satisfied:
$m(\text{air}) = m(\text{He}) + 300$ kg, or $m(\text{air}) - m(\text{He}) = 3.00 \times 10^5$ g;
Use the ideal gas equation to link m with V, the volume of the balloon:

$$n = \frac{m}{MM} = \frac{PV}{RT}, \text{ from which } m = \frac{PV\,MM}{RT}.$$

Because all these quantities except for molar mass are the same for air and helium, common terms can be collected:

$$\frac{PV}{RT}\left[MM(\text{air}) - MM(\text{He})\right] = 3.00 \times 10^5 \text{ g}.$$

Rearrange to solve for V: $\quad V = \dfrac{(3.00 \times 10^5 \text{ g})RT}{P\left[MM(\text{air}) - MM(\text{He})\right]}$

The molar mass of air can be found from the mole fractions of N_2 and O_2 (minor components contribute negligibly):
$MM(\text{air}) = (0.7808)(28.01 \text{ g/mol}) + (0.2095)(32.00 \text{ g/mol}) = 28.6$ g/mol;

$$V = \frac{(3.00 \times 10^5 \text{ g})(8.206 \times 10^{-2} \text{ L atm/mol K})(298 \text{ K})}{(1.00 \text{ atm})\left[28.6 \text{ g/mol} - 4.003 \text{ g/mol}\right]} = 2.98 \times 10^5 \text{ L}$$

5.81 Density can be calculated from the ideal gas equation and mole-mass conversions:

$$n = \frac{m}{MM} = \frac{PV}{RT}, \text{ from which } \frac{m}{V} = \frac{P\,MM}{RT}.$$

First, calculate the molar mass of dry air and of moist air. $MM(\text{air}) = \sum_i (X_i)(MM_i)$

$MM(\text{dry air}) = (0.7808)(28.01) + (0.2095)(32.00) + (9.34 \times 10^{-3})(39.948) = 28.96$ g/mol;
For moist air at 298 K, $p(H_2O) = (23.756 \text{ torr})(1 \text{ atm}/760 \text{ torr}) = 3.13 \times 10^{-2}$ atm, and $X(H_2O) = 0.0313$. The total mole fraction of the dry air components is reduced from 1 to $(1 - .0313) = 0.9687$. Each dry air mole fraction must be multiplied by this factor to obtain the mole fractions in moist air.
Thus,
$MM(\text{moist air}) = (3.13 \times 10^{-2})(18.01 \text{ g/mol}) + (0.9687)(28.96 \text{ g/mol}) = 28.62$ g/mol;

$$\frac{m}{V}(\text{dry air}) = \frac{(1 \text{ atm})(28.96 \text{ g/mol})}{(8.206 \times 10^{-2} \text{ L atm/mol K})(298 \text{ K})} = 1.18 \text{ g/L}.$$

$$\frac{m}{V}(\text{moist air}) = \frac{(1 \text{ atm})(28.62 \text{ g/mol})}{(8.206 \times 10^{-2} \text{ L atm/mol K})(298 \text{ K})} = 1.17 \text{ g/L}.$$

5.83 Think about the final conditions for the bottle. It contains air and helium, with $P = p_{air} + p_{He}$. Each partial pressure must be calculated separately.

The air pressure can be calculated as a variation in conditions. Temperature changes, and although the volume of the *bottle* is fixed, the *gas* volume changes because initially there is 100 mL of liquid helium present. Thus, $pV/T = nR = $ constant.
$p_i = 1.00$ atm, $V_i = 2.00$ L $-$ 100 mL $= 1.90$ L, $T_i = 90$ K;
$p_f = ?$, $V_f = 2.00$ L, $T_f = 25 + 273 = 298$ K;

$$p_f(\text{air}) = \left(\frac{p_i V_i}{T_i}\right)\left(\frac{T_f}{V_f}\right) = \frac{(1.00 \text{ atm})(1.90 \text{ L})(298 \text{ K})}{(90 \text{ K})(2.00 \text{ L})} = 3.15 \text{ atm};$$

The pressure due to helium can be calculated by finding the number of moles of liquid helium, then applying the ideal gas law for the final conditions.
$n = m/MM = \rho V/MM = (0.147 \text{ g/mL})(100 \text{ mL}) \div (4.00 \text{ g/mol}) = 3.68$ mol;

$$p(\text{He}) = \frac{nRT}{V} = \frac{(3.68 \text{ mol})(8.206 \times 10^{-2} \text{ L atm / mol K})(298 \text{ K})}{(2.00 \text{ L})} = 44.9 \text{ atm};$$

$P_{total} = 3.15$ atm $+ 44.9$ atm $= 48.1$ atm.

5.85 The concentration in parts per million can be converted to a mole fraction, and the partial pressure can then be calculated using $p_i = X_i P_{total}$. To determine molecular concentration, first use the ideal gas equation to calculate mol/L, then multiply by N_A.
$X(O_3) = (0.50 \text{ ppm})(10^{-6}/\text{ppm}) = 5.0 \times 10^{-7}$;
$p(O_3) = (5.0 \times 10^{-7})(762 \text{ torr}) = 3.8 \times 10^{-4}$ torr;

$$\frac{n}{V} = \frac{p}{RT} = \frac{(3.8 \times 10^{-4} \text{ torr})(1 \text{ atm} / 760 \text{ torr})}{(8.206 \times 10^{-2} \text{ L atm / mol K})(28 + 273 \text{ K})} = 2.0 \times 10^{-8} \text{ mol / L};$$

$\#/V = (6.022 \times 10^{23} \text{ molec/mol})(2.0 \times 10^{-8} \text{ mol/L}) = 1.2 \times 10^{13}$ molecules/L.

5.87 Break this problem into two parts. First, calculate the mass of oxygen in the tank, using the ideal gas equation. Then determine what time this mass of oxygen will support.
$p(O_2) = X(O_2)P_{tot} = (0.2095)(170 \text{ atm}) = 35.6$ atm;

$$n = \frac{pV}{RT} = \frac{(35.6 \text{ atm})(12.5 \text{ L})}{(8.206 \times 10^{-2} \text{ L atm / mol K})(27 + 273 \text{ K})} = 18.1 \text{ mol};$$

$m = n\ MM = (18.1 \text{ mol})(32.00 \text{ g/mol}) = 579$ g;
time $= (579 \text{ g}) \div (14.0 \text{ g/min}) = 41.4$ min;
Allowing 6 min for your return to the surface, you can dive for 35.4 min.

5.89 Your explanation should include the fact that sulfur in coal combines with oxygen from the air to form SO_2, and that when SO_2 reacts with water, sulfuric acid is an end product.
Chemical equations: $S + O_2 \rightarrow SO_2$;
$2\ SO_2 + O_2 \rightarrow 2\ SO_3$;
$SO_3 + H_2O \rightarrow H_2SO_4$.

5.91 (a) The most probable speed is the speed possessed by the largest number of molecules. Reading from the graph, this is 4×10^2 m/s (the peak position cannot be determined to better than one significant figure).

(b) $kE = mu^2/2$; because J = kg m^{2}s^{-2}, molecular mass must be expressed in kg; $m = MM/N_A = (17.03 \text{ g/mol})(10^{-3} \text{ kg/g}) \div (6.022 \times 10^{23} \text{ mol}^{-1}) = 2.83 \times 10^{-26}$ kg; $kE = (1/2)(2.83 \times 10^{-26} \text{ kg})(4 \times 10^2 \text{ m})^2 = 2 \times 10^{-21}$ J.

5.93 (a) This part is a stoichiometry problem. From the balanced equation, every 2 mol of TNT yields $12 + 5 + 3 = 20$ mol of gases. MM(TNT) = 227.14 g/mol.

n(gas) = 10 n(TNT) = 10(1.0 kg)(10^3 g/kg)(227.14 g/mol) = 44.0 mol;

(b) $V = \dfrac{(44 \text{ mol})(8.206 \times 10^{-2} \text{ L atm / mol K})(298 \text{ K})}{(1.00 \text{ atm})} = 1.1 \times 10^3$ L;

(c) X(CO) = (12/20) = 0.60; X(H$_2$) = (5/20) = 0.25; X(N$_2$) = (3/20) = 0.15; p(CO) = (0.60)(1 atm) = 0.60 atm; p(H$_2$) = 0.25 atm; p(N$_2$) = 0.15 atm.

5.95 Rearrange the ideal gas equation to determine how each of the graphs should appear.

(a) $V = nRT/P$, so a graph of V vs. P will appear as a parabola.
(b) $PV = nRT$ = constant, so a graph of PV vs. P is a horizontal straight line.
(c) $P = nRT/V$ = constant/V, so a graph of P vs. $1/V$ is an upward-sloping straight line.

(a)

(b)

(c)
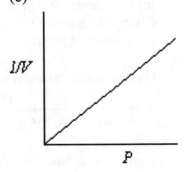

5.97 The chemical reaction in the Haber synthesis is $N_2 + 3\,H_2 \rightarrow 2\,NH_3$. The molecular picture shows 4 mol of N_2 and 12 of H_2, which represents stoichiometric proportions. If the reaction goes to completion, all molecules will react to form 8 mol of NH_3:

5.99 This is an example of an ideal gas thermometer. Because n and V are fixed, $P/T = nR/V =$ constant.

$$T_f = \frac{T_i P_f}{P_i} = \frac{(273\text{ K})(745\text{ torr})}{(345\text{ torr})} = 590\text{ K}; \qquad T_f = 590\text{ K} - 273 = 317\ ^\circ\text{C}.$$

5.101 In addition to a yield problem, this is a limiting reagent problem that involves gases. Solve for the theoretical yield using an amounts table. Water can be ignored, as CH_4 is the product of interest, and pressures can be used as the measures of amounts:

Reaction:	$3\,H_{2(g)}$	+	$CO_{(g)}$	$\rightarrow$	$CH_{4(g)}$	+ $3H_2O$
Initial amount (atm):	20.0		10.0		0	
Change in amount (atm):	-20.0		-6.67		+6.67	
Final amount (atm):	0		3.33		6.67	

Use the ideal gas equation and mole-mass conversion to obtain the theoretical yield in g:

$$n = \frac{PV}{RT} = \frac{(6.67\text{ atm})(100\text{ L})}{(8.206 \times 10^{-2}\text{ L atm / mol K})(600\text{ K})} = 13.5\text{ mol};$$

theoretical amount = $n\,MM$ = (13.5 mol)(16.042 g/mol) = 216.6 g;

% yield = 100%(actual amount/theoretical amount) = (100%)(150 g) ÷ (216.6 g) = 69.3%.

Chapter 6 Atoms and Light

6.1 Characteristics of Atoms

QUESTIONS TO ANSWER, SKILLS TO LEARN
1. What are some important characteristics of atoms?
2. How much space does an atom occupy?

All matter is composed of atoms, uncombined or combined to give chemical substances. By definition, matter has mass and occupies space. With these facts in mind, consider this list of the important characteristics of atoms:

Mass — Atoms are the building blocks of matter and must possess mass in order for matter to possess mass.

Positive nuclei — The protons and neutrons in the nucleus are held close to each other by extremely strong forces called nuclear forces. The atomic number, Z, is the number of protons per atom and thus the positive charge on the nucleus.

Negative electrons — It is rare to encounter a nucleus with no electrons. The positive charge of the nucleus is partially or completely neutralized by electrons. The charge on an electron is equal in size, but opposite in charge to that on a proton.

Volume — Matter occupies space and is made up of atoms. Therefore atoms occupy space and occupy a characteristic volume. Because the nucleus of an atom is so small, *the volume of an atom exists as a result of the amount of space occupied by the electrons associated with the nucleus.* We can estimate the **atomic volume** (the volume occupied by an atom) knowing the element's density and the molar mass for solids.

Other Properties — Atoms of different elements have different properties such as size, mass, chemical reactivity and combining power. The last two properties depend upon the number of protons in the nucleus and the number of electrons of that ion or atom.

Mutual Attraction — Atoms interact with other atoms, usually attracting each other to form molecules or aggregates of atoms. These interactions can be classified as *intra-molecular* (within a molecule) and bond-forming, or *inter-atomic* (between atoms), where chemical bonds are not formed.

Exercise 6.1 The volume occupied by a uranium atom in uranium metal is 2.07×10^{-29} m³. Calculate the density of uranium metal in g/mL.

Steps to Solution: The density is defined as mass per unit volume:

$$\rho = \frac{m}{V}$$

The density of uranium will be equal to the mass of a uranium atom divided by the volume occupied by a uranium atom. In order to obtain the correct units the molar mass must be converted to g/atom and the volume /atom to mL/atom. The molar mass of uranium is 238.0289 g/mol.

$$\rho = \frac{\dfrac{238.0289\,g/mol}{6.022 \times 10^{23}\,1/mol}}{2.07 \times 10^{-29}\,m^3\, 10^3\,L/m^3\, 10^3\,mL/L} = \boxed{19.05\ g/mL}$$

6.2 Characteristics of Light

QUESTIONS TO ANSWER, SKILLS TO LEARN
1. **What is electromagnetic radiation?**
2. **What parameters can be used to characterize a wave?**
3. **Conversion of wavelength to frequency and vice versa**
4. **Calculating the amount of energy carried by light**
5. **What happens when atoms absorb energy?**

Visible light is a small section of a phenomenon called **electromagnetic radiation**. Electromagnetic radiation has no mass but can carry energy called **radiative energy**. Electromagnetic radiation shows wavelike properties in that it is characterized by an oscillation that varies periodically in space and time. One of the properties used to characterize waves is the distance between the corresponding points in the wave, called the *wavelength*, λ. In the illustration below, two different sets of points are used to show the wavelength.

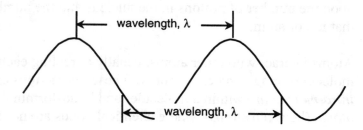

Chapter 6: Atoms and Light

Any two corresponding points could be used to measure the wavelength. A photograph of the ocean (a photo shows waves frozen in time) shows regular spaces between the tops of waves. The distance between the tops of the successive waves is the wavelength.

If you are on a boat sitting still in the water, you have a different perception of the wave. As time passes, the boat is moved up and down by the waves. The waves can be described by the amount of time that passes between successive waves. The waves also can be described by the number of times the boat bobs on the waves in a given amount of time. This latter quantity is the *frequency*, ν, of the waves and has the units of 1/time or, in SI units, s^{-1}. The unit s^{-1} is also called the Hertz (Hz). The frequency, ν, and the wavelength, λ, are related by the speed of the wave, u:

$$\text{velocity of a wave} = \lambda \nu$$

The speed of light (or electromagnetic radiation in general) in a vacuum is sufficiently important to warrant its own symbol, c.

$$\text{speed of light}, c = 2.997925 \times 10^8 \text{ m/s}$$

For electromagnetic radiation, the frequency may be calculated if the wavelength is known and likewise the wavelength if the frequency is known:

$$\nu = \frac{c}{\lambda} \qquad \text{or} \qquad \lambda = \frac{c}{\nu}$$

Although the speed of electromagnetic radiation is a constant, there is a wide range of wavelengths (and therefore frequencies) for electromagnetic radiation, from meters to picometers.

Exercise 6.2 For the following sources, calculate the missing member of the wavelength/frequency pair.
(a) FM radio waves with a frequency of 94.7 MHz
(b) A laser with a wavelength of 1064 nm
(c) An X-ray source, emitting X-rays with λ = 175.4 pm

Steps to Solution: We can solve each of these problems by using one of the two equations obtained from $c = \lambda \nu$. We must convert the units of frequency to s^{-1} and the wavelength to units of meters.

> **COMMON PITFALL: Not Converting Units.**
> **Make sure to convert wavelengths to m and frequency**
> **to s^{-1} *before* use in these equations.**

(a) The frequency of the radiation is given; we are to find the wavelength. The wavelength can be calculated from the frequency using the equation:

$$\lambda = \frac{c}{v}$$

The frequency is 94.7 MHz. The prefix M indicates a factor of 10^6; therefore the frequency is 94.7×10^6 Hz which must be converted to units of frequency, s^{-1}:

$$v = 94.7 \times 10^6 \text{ Hz} \cdot \frac{s^{-1}}{\text{Hz}} = 94.7 \times 10^6 \text{ s}^{-1}$$

Substitution into the expression for the wavelength gives:

$$\lambda = \frac{2.998 \times 10^8 \text{ m/s}}{94.7 \times 10^6 \text{ s}^{-1}} = \boxed{3.17 \text{ m}}$$

(b) *In this problem the wavelength is supplied (1064 nm) and the frequency is to be calculated. The equation for the frequency will be used. The wavelength must be converted from nm to m; the prefix "n" indicates a 10^{-9} multiplication factor. Substitution into the expression gives:*

$$v = \frac{c}{\lambda} = \frac{2.998 \times 10^8 \text{ m/s}}{1064 \times 10^{-9} \text{ m}} = \boxed{2.818 \times 10^{14} \text{ s}^{-1}}$$

(c) *We need to calculate the frequency as in part (b).* $v = \boxed{1.709 \times 10^{18} \text{ s}^{-1}}$

Light is a form of energy, where the amount of energy depends, in part, on the intensity or brightness of the source. All electromagnetic radiation carries energy; because we can only see certain wavelengths, we may not perceive an intense source of radiation outside the visible spectrum. However, the energy of electromagnetic radiation also depends on its frequency. Albert Einstein won the Nobel Prize for explaining that the energy of light depends on both frequency and intensity.

The Photoelectric Effect

The photoelectric effect is a phenomenon observed when light is absorbed by a metal surface. If the frequency of the light is high enough, electrons can be observed leaving the surface of the metal. Electrons do not leave the surface of the metal without light shining on it, so, to be ejected, they require energy that is obtained from the light. A detector measures both the number of the electrons and their kinetic energy. The observations made by a large number of experiments are:

1. Below a frequency characteristic of the metal, v_o, no electrons are emitted, regardless of the intensity of the light.

2. At frequencies greater than v_0, the kinetic energy of the ejected electrons increases in direct proportion to the frequency of the absorbed electromagnetic radiation. There is *no* dependence of the kinetic energy on the intensity of the light.

3. At frequencies greater than v_0, the number of ejected electrons increases with increasing intensity of light.

4. All metals exhibit the same features described in observations 1 through 3, but each metal has a different threshold frequency.

Einstein proposed a relatively simple equation that explained all of these observations. However, the theory underlying the equation required a complete change in perception of the nature of light by the current scientists. The theory assumed that light is a stream of individual packages of energy called **photons** that exhibit properties of particles and waves. The energy of each photon, E_{photon}, was proportional to the frequency of the light, v_{photon}:

$$E_{photon} = h v_{photon} \quad (h = 6.626 \times 10^{-34} \text{ J·s})$$

where the constant h is **Planck's constant**. There are two factors that determine the energy of light: the frequency (the energy per photon) and the intensity (the number of photons).

Exercise 6.3 Calculate the energy of (a) one photon of light of the X-ray in Exercise 6.2(c) and (b) one mole of these photons.

Steps to Solution: The energy of the photon depends on the frequency of the photon. The frequency, v_{photon}, was found in Exercise 6.2(c) to be $1.709 \times 10^{18} \text{ s}^{-1}$. We apply the equation:

$$E_{photon} = h v_{photon}$$

(a) The energy of one photon of the X-ray radiation is found from the above equation:

$$E_{photon} = h v_{photon} = 6.626 \times 10^{-34} \text{ J·s} \, (1.709 \times 10^{18} \text{ s}^{-1}) = \boxed{1.133 \times 10^{-15} \text{ J/photon}}$$

(b) The energy of one mole of these X-ray photons is the product of the energy of one photon and Avogadro's number:

$$E \text{ (one mole)} = (1.133 \times 10^{-15} \text{ J/photon})(6.022 \times 10^{23} \text{ /mol}) = 6.82 \times 10^{8} \text{ J/mol}$$

The photoelectric effect is explained using the law of conservation of energy and the idea that a photon contains energy proportional to its frequency. Upon being absorbed by the metal, all of the photon's energy is transferred to an electron. Because energy is conserved, some of the energy must be used to eject the electron from the metal (overcome the *binding energy*, hv_0) and the rest of the photon's energy is converted to the electron's kinetic energy.

Chapter 6: Atoms and Light

$$E_{photon} = E_{electron}$$

The photon's energy is $h\nu$; the energy absorbed by the electron, $E_{electron}$, is the sum of the energy needed to eject the electron ($h\nu_0$) and the kinetic energy of the electron ($kE_{electron}$):

$$h\nu = h\nu_0 + kE_{electron}$$

The *number of electrons* ejected depends on the number of photons absorbed whose energy is greater than the binding energy. The *kinetic energy of ejected electrons* depends on the amount of energy remaining after the binding energy has been overcome *(how much E_{photon} is greater than the binding energy, $h\nu_0$)* but does not depend at all on the intensity (number of photons).

The energy of electromagnetic radiation depends on two factors:
(1) The energy of each photon as determined by its frequency
(2) The number of photons, or intensity of the radiation.

Exercise 6.4 The photoelectric effect experiment is performed on strontium metal. When light with $\lambda = 389$ nm shines on the metal, electrons are ejected at a velocity of 3.97×10^5 m/s. What is the threshold frequency, ν_0, for strontium metal? What is the wavelength of this frequency of light?

Steps to Solution: The question asks for the threshold frequency, ν_0. The data supplied are the wavelength of incident light (389 nm) and the velocity of the ejected electrons (3.97×10^5 m/s). The solution is obtained by using the photoelectric effect equation. The energy corresponding to the threshold frequency is $h\nu_0$. Calculation of ν_0 allows us to calculate the wavelength. Rearranging the photoelectric effect equation to solve for the threshold energy gives:

$$h\nu_0 = h\nu - kE_{electron}$$

The energy required to eject the photon is the difference between the energy of the absorbed photon and the kinetic energy of the ejected electron. The energy of the absorbed light, $h\nu$, is:

$$h\nu = h\frac{c}{\lambda} = 6.626 \times 10^{-34} \text{ J} \cdot \text{s} \; \frac{3.00 \times 10^8 \text{ m/s}}{389 \times 10^{-9} \text{ m}} = 5.11 \times 10^{-19} \text{ J}$$

The kinetic energy of the electron is found from the equation for kinetic energy:

$$kE = \frac{1}{2} m u^2$$

The mass of an electron is 9.1095×10^{-31} kg and the velocity is supplied in the problem.

$$kE_{electron} = \frac{1}{2}\ 9.1095 \times 10^{-31}\ kg\ (3.97 \times 10^5\ m/s)^2 = 7.16 \times 10^{-20}\ J$$

The threshold energy is:

$$h\nu_o = 5.11 \times 10^{-19} - 7.16 \times 10^{-20}\ J = 4.39 \times 10^{-19}\ J$$

The frequency of light with this energy is found via:

$$\nu = \frac{E}{h} = \frac{4.39 \times 10^{-19}\ J}{6.626 \times 10^{-34}\ J \cdot s} = \boxed{6.63 \times 10^{14}\ 1/s}$$

The wavelength of this light is:

$$\lambda = \frac{c}{\nu} = \frac{3.00 \times 10^8\ m/s}{6.63 \times 10^{14}\ s^{-1}} = 4.52 \times 10^{-7}\ m = \boxed{452\ nm}$$

The above discussion has shown two different ways of dealing with light; it can be treated as a wave, but it also has properties that are usually attributed to a particle. For example, the energy of light depends on the number of photons and the energy of each photon. These are particle-like properties. However, the relationship between wavelength and frequency is a wave-like property. Light has both wave and particle properties, and to understand light one must consider both these properties. Wave properties dominate the interaction of light with large objects; the particle properties of light are more apparent when light interacts with molecules, atoms or electrons.

Light, Atoms and What Happens When They Collide

The photoelectric effect shows the result of the absorption of electromagnetic radiation with electrons bound to a metal surface. Single atoms in the gas phase are different than a metal surface. An atom is a collection of electrons bound by electrical forces to a positively charged nucleus and the electrons will be more stable near the nucleus. However, electrons repel one another so some arrangements will lead to stronger electron-electron repulsions. The most stable arrangement of electrons is called the **ground state** and is one that minimizes the electron-electron repulsions and gives the strongest attraction to the nucleus. Other arrangements that either increase electron-electron repulsions or lower the attraction to the nucleus are less stable and are called **excited states**.

When light is absorbed by atoms that are in the gas phase, the energy of the photon is transferred to the atom. After absorption of a photon, an atom has more energy and is in an excited state. Atoms of a given element have a number of excited states, each with a different energy. The lowest energy state that an atom can occupy is called the **ground state**. The relative energies of these excited states can be shown using an **energy level diagram** as shown below, in which the energy increases along the vertical or y axis and the various states of the atom are shown as horizontal lines. The lowest line represents the energy of the ground state. The horizontal lines above the ground state are the energies of the various excited states. The higher the line, the less stable that excited state is (we say it has a higher energy than the ground state).

Atoms in the ground state may absorb a photon and become excited. That process is shown by the arrow labeled (a) where the ground state atom absorbs energy and moves up in energy to excited state 2. Atoms that are in excited states may lose energy (by emitting a photon) as shown by the downward pointing arrow (b), going from excited state 3 to the ground state.

Atoms of different elements have different energies for their ground and excited states. Therefore, different types of atoms absorb light of different energy (that is, different wavelength) corresponding to the process shown by the arrow labeled (a) in the diagram above. In addition, atoms that are in excited states emit light as shown by the arrow labeled (b) . Since different elements have different energy levels, they emit different wavelengths of electromagnetic radiation and different colors will be seen.

Absorption of energy to change the state of the atom from the ground state to one of the excited states will occur only when the energy difference between the two states (the energy change of the atom, ΔE_{atom}) is equal to the energy of the photon ($h\nu$):

$$\Delta E_{atom} = h\nu$$

6.3 Absorption and Emission Spectra

QUESTIONS TO ANSWER, SKILLS TO LEARN
1. **What is an absorption spectrum?**
2. **What is an emission spectrum?**
3. **Understanding that only certain energies are allowed for atoms (and molecules)**
4. **Calculating the energies of transitions between the different states for hydrogen atoms**

A **spectrum** is a graph of the intensity of light (or the number of photons) as a function of energy or wavelength. An **absorption spectrum** is a record of the number of photons absorbed as a function of wavelength or frequency. If we shine light through a sample of gaseous sodium atoms, the color of the light that emerges from the sample chamber will not be the same as that which entered it. Some of the light will have been absorbed by sodium atoms, but only at specific energies corresponding to the energy between the different states of the sodium atom. The intensity or brightness of the light at those wavelengths will have decreased.

An **emission spectrum** is a graph of the number of photons (intensity of light) plotted on the y-axis versus the wavelength or energy of photons emitted by atoms in excited states going to lower energy states. Fireworks are an example of the emission of light by atoms. Fireworks contain salts of metal ions. When the rocket explodes, the reaction mixture converts the salts to metal atoms and the heat is sufficient to produce the atoms in excited states. The atoms decay to the ground state by emitting a photon of light. An emission spectrum shows peaks at the wavelengths where light is emitted.

One of the striking features of atomic absorption and emission spectra for atoms is that the absorptions are very sharp (they are called lines); the width of a peak may be several nanometers or less under appropriate conditions. This observation suggests that the energy changes in atoms being excited or in excited atoms decaying to the ground state have specific values; therefore the energies of the ground and excited states in atoms must have fixed values. The energies of the ground and excited states for a hydrogen atom, E_n, can be calculated by a simple formula which involves only a constant and an integer, n:

$$E_n = -\frac{2.18 \times 10^{-18} \text{ J}}{n^2} \quad (n = 1, 2, 3, 4, \ldots)$$

The most negative energy (where $n = 1$) is the most stable energy and is the ground state. The energy is zero when the electron and proton are separated and do not interact with each other. The hydrogen atom may have only the energies calculated by this simple formula; it may have not have an energy not found by this equation. When the energy (or other quantity) is restricted to certain values, it is said to be **quantized**. Exercise 6.4 illustrates the construction of an energy level diagram for hydrogen.

Exercise 6.4 Draw the energy level diagram for hydrogen in the ground state ($n = 1$) to $n = 5$.

Steps to Solution: *The energy level diagram is a graph that shows the energy of different states (n values) on the y-axis and uses horizontal lines to indicate the energy of each state. We calculate the energies of the ground state (n = 1) and the excited states (n = 2, 3, 4 and 5) from the equation for the energy of the hydrogen atom.*

$$E_n = -\frac{2.18 \times 10^{-18} \text{ J}}{n^2}$$

$n = 1$: $\quad\quad\quad E_1 = -\dfrac{2.18 \times 10^{-18} \text{ J}}{1^2} = -2.18 \times 10^{-18} \text{ J}$

The energy of the hydrogen atom in the ground state is 2.18×10^{-18} J more stable than a nucleus and electron that do not undergo any interaction. This answer makes sense because positive and negative charges undergo a mutual attraction.

$n = 2$: $\quad E_2 = -5.45 \times 10^{-19}$ J

This answer is somewhat larger (less negative) than that for $n = 1$ which indicates that the atom is less stable when $n = 2$. The result is expected because $n = 2$ corresponds to the first excited state for the hydrogen atom. Using the same method we find the energies for $n = 3, 4$ and 5:

n	E(J)
3	-2.42×10^{-19}
4	-1.36×10^{-19}
5	-8.72×10^{-19}

Plotting these data using a horizontal line to indicate the energy of the state gives the energy level diagram for hydrogen shown above at right.

The energy level diagram for hydrogen is relatively straightforward to generate. If the atom has more than one electron, the energy levels cannot be calculated by a simple equation. Nonetheless, energy levels are quantized in atoms with more than one electron.

The energy change, ΔE, involved when a hydrogen atom changes states from E_1 to E_2 can be calculated from the energy difference of the two states:

$$\Delta E = E_2 - E_1$$

$$\Delta E = -\dfrac{2.18 \times 10^{-18} \text{ J}}{n_2^2} + \dfrac{2.18 \times 10^{-18} \text{ J}}{n_1^2} = 2.18 \times 10^{-18} \text{ J} \left(\dfrac{1}{n_1^2} - \dfrac{1}{n_2^2} \right)$$

The energy change can be converted to the wavelength of light corresponding to that energy change because we know:

$$\Delta E = h\nu = h\dfrac{c}{\lambda}$$

These relationships are illustrated in Exercise 6.5.

Exercise 6.5 What is the energy change when a hydrogen atom changes from the ground state to the state where $n = 4$? What wavelength light does this correspond to and is that photon emitted or absorbed?

Steps to Solution: *We are asked to calculate the energy difference between the ground state ($n = 1$) and the excited state where $n = 4$. We can find this energy by substitution into the equation:*

$$\Delta E = 2.18 \times 10^{-18} \text{ J} \left(\frac{1}{n_1^2} - \frac{1}{n_2^2} \right)$$

where $n_1 = 1$ (initial state) and $n_2 = 4$ (the final state). Thinking molecules (atoms) we can see that this process will require the input of energy because the atom will be in a higher energy state at the end of the process (look at the energy level diagram in Exercise 6.6). Substitution into the equation gives:

$$\Delta E = 2.18 \times 10^{-18} \text{ J} \left(\frac{1}{1^2} - \frac{1}{4^2} \right) = 2.04 \times 10^{-18} \text{ J}$$

The wavelength of the photon that corresponds to this energy is:

$$\lambda = \frac{hc}{\Delta E} = \frac{6.626 \times 10^{-34} \text{ J} \cdot \text{s} (3.00 \times 10^8 \text{ m/s})}{2.04 \times 10^{-18} \text{ J}} = 9.73 \times 10^{-8} \text{ m}$$

$$\lambda = 9.73 \times 10^{-8} \text{ m} \cdot \frac{1 \text{ nm}}{10^{-9} \text{ m}} = \boxed{97.3 \text{ nm}}$$

To state whether the photon is emitted or absorbed, remember that the atom is in a higher energy state at the end of the process; therefore, energy must have been supplied to the atom. To show this transition on the energy level diagram drawn in Exercise 6.6. we would draw an arrow pointing up from the line where $n = 1$ to the $n = 4$ level (an "uphill" process). For energy to be conserved, **a photon of wavelength 97.3 nm must be absorbed to provide the energy to excite the atom from $n = 1$ to $n = 4$.**

Although energy level diagrams for other atoms are more complex and show less regularity than that for the hydrogen atom, the principle used in Exercise 6.4 applies equally to them:

The energy of a photon emitted or absorbed by an atom is equal to the difference in energy of the initial and final states involved in the process.

Conservation of energy and mass also apply in these problems as shown in the next exercise.

Exercise 6.6 The neodymium-YAG laser is a workhorse in industry used for welding, and cutting. The YAG simply serves to keep the neodymium(III) ions in the proper environment so they will be able to work in the laser. The wavelength of light

emitted from the Nd-YAG laser is 1064 nm, but visible light in the range of 400 to 750 nm is used to excited the Nd ions.
(a) How many photons of the 1064 nm light will be required to supply 1.00 J of energy?
(b) What mass of Nd must be present to supply 1 J of energy in a pulse of the laser?

Steps to Solution: (a) *We are asked to find the number of photons of the Nd-YAG laser to give 1 J of energy. Since the wavelength is supplied, we can find the number of photons from the energy required divided by the energy per photon:*

$$\text{number photons} = \frac{\text{energy}}{\text{energy/photon}}$$

We calculate the energy per photon:

$$E/\text{photon} = \frac{6.626 \times 10^{-34} \text{ J} \cdot \text{s} \cdot 3.00 \times 10^8 \text{ m/s}}{1064 \times 10^{-9} \text{ m}} = 1.87 \times 10^{-19} \text{ J/photon}$$

and then find the number of photons:

$$\text{number photons} = \frac{1.00 \text{ J}}{1.87 \times 10^{-19} \text{ J/photon}} = \boxed{5.35 \times 10^{18} \text{ photons}}$$

Alternatively, we can combine the two steps into a single equation:

$$\text{number photons} = \frac{1.00 \text{ J}}{\frac{hc}{\lambda}} = \frac{1.00 \text{ J}}{\frac{6.626 \times 10^{-34} \text{ J} \cdot \text{s} \cdot 3.00 \times 10^8 \text{ m/s}}{1064 \times 10^{-9} \text{ m}}} = \boxed{5.35 \times 10^{18} \text{ photons}}$$

This answer is reasonable: the energy per photon is small so it takes a significant number of photons to supply 1.00 J of energy.

(b) One Nd atom is needed to supply one photon of 1064 nm light. The stoichiometry of this process is $\frac{1 \text{ Nd}}{1 \text{ photon}}$. Therefore the mass of Nd required is calculated using this ratio and the moles of photons from part (a) times the molar mass of Nd:

$$\text{mass Nd} = 5.35 \times 10^{18} \text{ photons} \cdot \frac{1}{6.022 \times 10^{23}/\text{mol}} \cdot \frac{1 \text{ Nd}}{1 \text{ photon}} \cdot 144.24 \text{ g/mol}$$

$$\text{mass Nd} = \boxed{1.28 \times 10^{-3} \text{ g}}$$

The Electromagnetic Spectrum

Visible light accounts for only a small portion of the phenomenon called electromagnetic radiation, with wavelengths from about 400 nm to 750 nm. Electromagnetic radiation has wavelengths from hundreds of meters to 10^{-12} m with a corresponding range of energies per photon. It is no coincidence that Nature evolved our visual senses for energies which correspond to the promotion of electrons to excited states in molecules (the visible region) because there are significant numbers of photons of that wavelength present in daylight. At very short wavelengths, the high energy of the photons involved can penetrate matter and destroy living tissue (gamma and X-rays); under control, this electromagnetic radiation is used for imaging and cancer treatment. The longer wavelengths do not have enough energy to promote electrons to excited states, but can make the molecule vibrate (infrared) or rotate (microwave), to transfer energy into objects, heating them. Microwaves are used in cooking (microwave ovens) because they excite only water molecules. The rotating water molecules transfer their energy to the other molecules of the food, and the result is heated food!

6.4 Sunlight and the Earth

QUESTIONS TO ANSWER, SKILLS TO LEARN
1. **How is the earth's temperature regulated?**
2. **How does the atmosphere's temperature vary with altitude?**
3. **What is the ozone layer?**

The earth is an system where energy flows in and also flows out. If one of these two processes is disrupted, then either more heat will enter than leave causing the earth to heat up, a phenomenon called **global warming**. If more heat leaves than enters, the earth would cool down in a process called **global cooling**. To examine this problem, we must identify sources of energy to the earth and the routes that energy uses to leave the earth.

Most of the earth's energy is from the sun, which emits large amounts of electromagnetic radiation. The wavelengths of this radiation at the surface of the earth are principally in the infrared, visible and ultraviolet regions of the electromagnetic spectrum. The earth loses energy by emission of infrared radiation. Heating of the earth can occur either by reduction of heat lost by the emission process, or by an increase of the heat input from the sun. Cooling of the earth could occur from an increase in the amount of emission infrared radiation or a decrease in the energy output of the sun. The human race has no control over the sun, but we can control what we put into the atmosphere.

Substances that prevent sunlight from reaching the surface of the earth can lead to cooling of the earth. Such an effect can be caused by clouds of dust or mist high in the atmosphere ; the eruption of large volcanoes is known to lead to global cooling. Heating could result from decreasing the amount of heat radiated by the earth into space. An increase in the concentrations of certain gases in the atmosphere leads to capture of emitted radiation; these molecules then emit a significant amount of this radiation back to the earth, reducing the amount of energy being lost by the earth, which may lead to global warming.

6.5 Properties of Electrons

QUESTIONS TO ANSWER, SKILLS TO LEARN
1. **What are properties of electrons?**
2. **Calculating wavelengths of electrons or other objects; matter waves**
3. **How accurately can an electron's position be determined?**
4. **What are the properties of electrons in atoms and molecules?**

In the chapters ahead, you will clearly see that electrons play an important part in the physical and chemical properties of the elements. The following table summarizes certain properties of electrons, some of which have been previously discussed and others which are new and are not obvious from your experiences in the macroscopic world.

Mass and Charge:
Electrons have mass (9.109×10^{-31} kg) and electrical charge (1.602×10^{-19} C).

Magnetic Behavior:
Experiments show that electrons behave like magnets, as a result of a property called **spin**. The magnetic property of electrons always has the same magnitude but may have either a + or - sign.

Wave Properties:
Electrons (and other matter) have wave properties. The de Broglie equation allows us to determine the wavelength, λ, of a particle given its momentum (mass, m, times velocity, u) :

$$\lambda = \frac{h}{mu}$$

Although all matter has wavelike properties, only in particles with small mass such as electrons can wavelike behavior be observed. Wave behavior in electrons was confirmed by the experiments of Davison and Germer, which showed the wave phenomenon, diffraction, in electrons.

Electromagnetic radiation and small particles have both particle and wave like properties. Electromagnetic radiation shows wavelike behavior when interacting with macroscopic objects as observed in phenomena such as diffraction and interference, but shows particle behavior in that the energy of electromagnetic radiation is "grainy"; it comes in quantized units of $h\nu$.

Chapter 6: Atoms and Light

Electrons show particle properties in that they possess mass and an electrical charge. However, electrons also show wave properties that are explained by the de Broglie equation. Therefore, at the microscopic level, electromagnetic radiation and matter show a dual nature: we can observe properties particle-like or wave-like properties, depending on the experiment performed. This dual nature of light and matter should not be considered contradictory, these are simply different properties, both of which are required for a complete description of light and matter.

Exercise 6.7 In this problem we will examine calculations involving the wave-particle dual nature of the microscopic world. The equations relating energy, wavelength and speed for photons and particles are summarized below.

Property	Photon	Particle
Energy	$E = h\nu = \dfrac{hc}{\lambda}$	$E = \dfrac{1}{2} mu^2$
Wavelength	$\lambda = \dfrac{hc}{E}$ or $\lambda = \dfrac{c}{\nu}$	$\lambda = \dfrac{h}{mu}$
Speed	$c = 3.00 \times 10^8$ m/s	$u = \sqrt{\dfrac{2E}{m}}$

(a) Compare the energy of a photon and an electron that both have a wavelength of 0.10 nm, a distance slightly smaller than the distance between atoms in molecules.

(b) Calculate the wavelength of a macroscopic object, a 6.0-ounce hockey puck moving at the speed of 125 km/hour.

Steps to Solution:
(a) To compare the energies, we must first calculate them. First, we will calculate the energy of the photon. The energy of the photon depends only on the wavelength, which must be expressed in the correct units (m).

$$E = \dfrac{hc}{\lambda} = \dfrac{6.626 \times 10^{-34} \text{ J} \cdot \text{s}(3.00 \times 10^8 \text{ m/s})}{0.10 \text{ nm} (10^{-9} \text{ m/nm})} = \boxed{2.0 \times 10^{-15} \text{ J}}$$

We determine the energy of the electron in two steps. First we calculate the velocity from the wavelength. Then the kinetic energy is calculated. The deBroglie equation relates the wavelength to the velocity:

$$\lambda = \dfrac{h}{mu}$$

Rearranging the equation to solve for the velocity and substituting the values for the mass of the electron and the wavelength gives:

$$u = \frac{h}{m\lambda} = \frac{6.626 \times 10^{-34} \text{ J}\cdot\text{s}}{(9.109 \times 10^{-31} \text{ kg})(0.10 \times 10^{-9} \text{ m})} = 7.27 \times 10^6 \text{ m/s}$$

Substitution of this velocity and the electron mass into the expression for kinetic energy gives:

$$E = \frac{1}{2} mu^2 = \frac{1}{2}(9.109 \times 10^{-31} \text{ kg})(7.27 \times 10^6 \text{ m/s})^2 = \boxed{2.4 \times 10^{-17} \text{ J}}$$

We round the answer to 2 significant figures because the wavelength only had 2 significant figures. The photon has more kinetic energy than the electron.

(b) *We calculate the wavelength of the hockey puck using the deBroglie equation. Be careful to use the correct units! We must convert the mass (6.0 oz) from ounces to kg and the velocity (125 km/hr) to km/s.*

$$\lambda = \frac{h}{mv} = \frac{6.626 \times 10^{-34} \text{ J}\cdot\text{s}}{(6.0 \text{ oz} \cdot 28.35 \text{ g/oz} \cdot 10^{-3} \text{ kg/g})(125 \text{ km/hr} \cdot 1 \text{ hr/60 min} \cdot 1 \text{ min/60 s})}$$

$$\lambda = 1.1 \times 10^{-31} \text{ m}$$

This wavelength is smaller than the diameter of an atomic nucleus. To exhibit wavelike behavior, objects must be about the size of the wavelength. Since this wavelength is so small, one cannot perform an experiment to observe wavelike behavior of the hockey puck. In other words, this macroscopic object behaves like a particle and does not display wavelike behavior.

The dual nature of electrons (wave and particle) leads to the next property of electrons, which is better understood by considering the wave nature of electrons.

The exact position of an electron cannot be determined

A macroscopic particle occupies a single spot and has a definite location. The wavelength of a macroscopic object is so small that the wave properties cannot be observed. A wave does not have specific location; it is distributed over a region of space. Consider the difference between an ocean wave and a pelican sitting in the water. The pelican's position may be determined at any time within millimeters; the wave is spread out over meters. Because an electron has a significant wavelength, it does not have a specific location and thus is described as being delocalized.

Werner von Heisenberg demonstrated that one cannot simultaneously measure two related quantities such as position and velocity to high precision. If one measures the position to high precision, then the velocity has a very large uncertainty. Measurement of velocity to high precision leads to low precision in the measurement of the position of the particle.

Chapter 6: Atoms and Light

Heisenberg's Uncertainty Principle: One cannot measure two related quantities such as position and velocity to equally high precision.

Bound electrons are delocalized and are described by waves

The preceding sections have described free electrons as waves. In an atom or molecule, the electron is not free but is bound to one or more nuclei; the wave property of the electron must show this. The electron has three dimensions in which it can move, and its wave form must show its delocalization over three dimensions. Rather than trying to draw the three dimensional wave, we usually show the overall shape of the space over which the electron is delocalized. The three-dimensional waves for the bound electrons are called orbitals. Chapter 7 describes the shapes and sizes of different orbitals.

The energies of bound electrons are quantized

The observation of sharp lines in the emission spectra of atoms was used to argue that the energy levels of electrons in atoms are quantized; that is, only certain energies are allowed. This deduction has been shown to be true for electrons bound to atoms and molecules. Transitions of electrons between energy levels in an atom moves them from one orbital to another orbital of different energy.

Test Yourself

1. What wavelength light would you expect for C^{5+} (a possible component of the solar atmosphere) to emit going from the $n = 2$ to the $n = 1$ state? The energy of such an ion is $E = -2.18 \times 10^{-18} \text{ J} \cdot \frac{Z^2}{n^2}$.

2. High speed electrons are used in electron microscopy; the size of the objects imaged is bigger than the wavelength of the electrons. What is the momentum, mu, of electrons with $\lambda = 100$ pm? (1 Å)

3. How many photons of $\lambda = 45$ mm would need to be absorbed to transfer 6 kJ of energy?

4. The photoelectric effect experiment is performed on gold metal. If the wavelength of the light is 254 nm and the velocity of the ejected electrons is 1.47×10^5 m/s, what is the binding energy for gold?

Answers

1. 3.4 nm

2. 6.626 x 10^{-24} kg•m/s

3. 1.4 x 10^{26} photons

4. 7.72 x10^{-19} J

Solutions to Chapter Problems

6.1 Calculate the molar volume of each metal from its density and molar mass, then convert to atomic volume using Avogadro's number. Then estimate atomic diameter from the atomic volume by assuming that each atom occupies a cubical volume:

(a) $V_{atom} = \dfrac{[MM(g/mol)][10^{-3} kg/g]}{[\rho(kg/m^3)][N_A]}$;

(b) $d \cong (V)^{1/3}$; and (c) Layer thickness = (# of atoms)(atomic diameter).

Ag: $V_{atom} = \dfrac{(107.868 \text{ g/mol})(10^{-3} \text{ kg/g})}{(1.050 \times 10^4 \text{ kg/m}^3)(6.022 \times 10^{23} \text{ atoms/mol})} = 17.06 \times 10^{-30} \text{ m}^3/\text{atom}$;

$d \cong (V)^{1/3} = (17.06 \times 10^{-30} \text{ m}^3)^{1/3} = 2.57 \times 10^{-10}$ m;
Thickness = $(6.5 \times 10^6 \text{ atoms})(2.57 \times 10^{-10} \text{ m}) = 1.67 \times 10^{-3}$ m;

Pb: $V_{atom} = \dfrac{(207.2 \text{ g/mol})(10^{-3} \text{ kg/g})}{(1.134 \times 10^4 \text{ kg/m}^3)(6.022 \times 10^{23} \text{ atoms/mol})} = 30.34 \times 10^{-30} \text{ m}^3/\text{atom}$;

$d \cong (V)^{1/3} = (30.34 \times 10^{-30} \text{ m}^3)^{1/3} = 3.12 \times 10^{-10}$ m;
Thickness = $(6.5 \times 10^6 \text{ atoms})(3.12 \times 10^{-10} \text{ m}) = 2.03 \times 10^{-3}$ m.

6.3 Gas pressure results from transfer of momentum between atoms and container walls, indicating that atoms have momentum and mass; mass spectrometers sort atomic and molecular ions according to their masses; all matter is made up of atoms and has mass.

6.5 At the atomic level, a layer of metal atoms is spheres nestled closely together:

Chapter 6: Atoms and Light

6.7 The energy of one photon of light is described by the equation, $E = h\nu = hc/\lambda$. To obtain energy per mol of photons, multiply by Avogadro's number.

(a) $E_{photon} = \dfrac{(6.626 \times 10^{-34} \text{ Js})(2.998 \times 10^8 \text{ m/s})}{(490.6 \text{ nm})(10^{-9} \text{ m/nm})} = 4.049 \times 10^{-19}$ J;

$E_{molar} = (4.049 \times 10^{-19} \text{ J})(6.022 \times 10^{23}/\text{mol})(10^{-3} \text{ kJ/J}) = 2.438 \times 10^2$ kJ/mol;

(b) $E_{photon} = \dfrac{(6.626 \times 10^{-34} \text{ Js})(2.998 \times 10^8 \text{ m/s})}{(25.5 \text{ nm})(10^{-9} \text{ m/nm})} = 7.79 \times 10^{-18}$ J;

$E_{molar} = (7.79 \times 10^{-18} \text{ J})(6.022 \times 10^{23}/\text{mol})(10^{-3} \text{ kJ/J}) = 4.69 \times 10^3$ kJ/mol;

(c) $E_{photon} = (6.62608 \times 10^{-34} \text{ J s})(2.5437 \times 10^{10} \text{ s}^{-1}) = 1.6855 \times 10^{-23}$ J;

$E_{molar} = (1.6855 \times 10^{-23} \text{ J})(6.0221 \times 10^{23}/\text{mol})(10^{-3} \text{ kJ/J}) = 0.10150$ kJ/mol.

6.9 Convert from molar energy to energy per photon by dividing by Avogadro's number, then use $E = hc/\lambda$ to find $\lambda = hc/E$.
$E_{photon} = (745 \text{ kJ/mol})(10^3 \text{ J/kJ}) \div (6.022 \times 10^{23}/\text{mol}) = 1.237 \times 10^{-18}$ J;
$\lambda = (6.626 \times 10^{-34} \text{ J s})(2.998 \times 10^8 \text{ m/s}) \div (1.237 \times 10^{-18} \text{ J}) = 1.61 \times 10^{-7}$ m or 161 nm.

6.11 (a) $\lambda = c/\nu = (2.998 \times 10^8 \text{ m/s}) \div (1.30 \times 10^{15} \text{ s}^{-1}) = 2.31 \times 10^{-7}$ m or 231 nm;

(b) First determine the energy of the photon, then use $bE = E_{photon} - kE_{electron}$
$E_{photon} = h\nu = (6.626 \times 10^{-34} \text{ J s})(1.30 \times 10^{15} \text{ s}^{-1}) = 8.61 \times 10^{-19}$ J;
$bE = 8.61 \times 10^{-19}$ J $- 5.2 \times 10^{-19}$ J $= 3.4 \times 10^{-19}$ J;

(c) The longest wavelength that will eject electrons corresponds to a photon with energy equal to the binding energy: $\lambda = hc/bE$.
$\lambda = (6.626 \times 10^{-34} \text{ J s})(2.998 \times 10^8 \text{ m/s}) \div (3.4 \times 10^{-19} \text{ J}) = 5.8 \times 10^{-7}$ m or 580 nm.

6.13 Refer to Figure 6-8 for an example of an energy level diagram. The binding energy for cesium (Prob. 6.11) is 3.4×10^{-19} J, and the binding energy for chromium (Prob. 6.12) is 7.21×10^{-19} J. Your energy level diagram should show that chromium has a substantially greater binding energy than cesium:

6.15 (a) Radio waves have wavelengths from 1 to 100 m;
(b) Light with a wavelength of 5.8×10^{-7} m is yellow;
(c) A frequency of 4.5×10^8 Hz lies in the TV, FM region.

6.17 Figure 6-14 does not show an emission wavelength directly to the ground state from the state that emits 404 nm light, but the combination of emissions at 436 nm and 254 nm connects this state to the ground state. Use $E = hc/\lambda$ for each photon:

$$\Delta E \text{ (kJ/mol)} = \frac{hcN_A}{\lambda_1} + \frac{hcN_A}{\lambda_2} = hcN_A\left(\frac{1}{\lambda_1} + \frac{1}{\lambda_2}\right)$$

$hcN_A = (6.626 \times 10^{-34} \text{ J s})(2.998 \times 10^8 \text{ m/s})(10^{-3} \text{ kJ/J})(6.022 \times 10^{23}/\text{mol})$
$= 1.196 \times 10^{-4}$ kJ m/mol

ΔE (kJ/mol) $= (1.196 \times 10^{-4}$ kJ m/mol$)(10^9$ nm/m$)\left(\dfrac{1}{436 \text{ nm}} + \dfrac{1}{254 \text{ nm}}\right) = 745$ kJ/mol

6.19 Energies and wavelengths for transitions in hydrogen atoms can be calculated from the equation for hydrogen atom energy levels:

$$E_n = \frac{-2.18 \times 10^{-18} \text{ J}}{n^2}.$$

$\Delta E_{8-1} = (E_8 - E_1) = \dfrac{-2.18 \times 10^{-18} \text{ J}}{8^2} - \dfrac{-2.18 \times 10^{-18} \text{ J}}{1^2} = 2.146 \times 10^{-18}$ J;

$\lambda = hc/E = (6.626 \times 10^{-34}$ J s$)(2.998 \times 10^8$ m/s$) \div (2.146 \times 10^{-18}$ J$) = 9.28 \times 10^{-8}$ m;

$\Delta E_{9-1} = (E_8 - E_1) = \dfrac{-2.18 \times 10^{-18} \text{ J}}{9^2} - \dfrac{-2.18 \times 10^{-18} \text{ J}}{1^2} = 2.153 \times 10^{-18}$ J;

$\lambda = hc/E = (6.626 \times 10^{-34}$ J s$)(2.998 \times 10^8$ m/s$) \div (2.153 \times 10^{-18}$ J$) = 9.23 \times 10^{-8}$ m;

These photons lie in the ultraviolet region of the spectrum (see Figure 6-4).

6.21 (a) Calculate the energies of the photons using $E = hc/\lambda$:
$E_{488} = (6.626 \times 10^{-34}$ J s$)(2.998 \times 10^8$ m/s$) \div (488$ nm$)(10^{-9}$ m/nm$) = 4.07 \times 10^{-19}$ J;
$E_{514} = (6.626 \times 10^{-34}$ J s$)(2.998 \times 10^8$ m/s$) \div (514$ nm$)(10^{-9}$ m/nm$) = 3.86 \times 10^{-19}$ J;

(b) Each transition leaves the ion in the same state, so the state from which 488 nm emission occurs must be slightly higher in energy than that from which 514 nm emission occurs:

(c) Calculate frequency and wavelength using $\nu = E/h$ and $\lambda = c/\nu$:
$\nu = E/h = (2.76 \times 10^{-18}$ J$) \div (6.626 \times 10^{-34}$ J s$) = 4.17 \times 10^{15}$ s^{-1};
$\lambda = c/\nu = (2.998 \times 10^8$ m/s$) \div (4.17 \times 10^{15}$ s$^{-1}) = 7.19 \times 10^{-8}$ m or 71.9 nm.

6.23 Each picture should show a molecule absorbing a photon and fragmenting appropriately:
N_2:

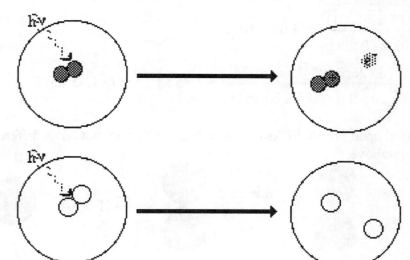

O_2:

6.25 Calculate the energy using $E = hcN_A/\lambda$:

$hcN_A = (6.626 \times 10^{-34}$ J s$)(2.998 \times 10^8$ m/s$)(10^{-3}$ kJ/J$)(6.022 \times 10^{23}$/mol$)$
$= 1.196 \times 10^{-4}$ kJ m/mol
$E = (1.196 \times 10^{-4}$ kJ m/mol$) \div (340$ nm$)(10^{-9}$ m/nm$) = 352$ kJ;

The kinetic energy of the fragments is the difference between the energy of the photon causing the fragmentation and the minimum energy required for fragmentation:

$$kE = E_{photon} - fE = \frac{(6.626 \times 10^{-34} \text{ J s})(2.998 \times 10^8 \text{ m/s})}{(250 \text{ nm})(10^{-9} \text{ m/nm})} - \frac{(352 \text{ kJ/mol})(10^3 \text{ J/kJ})}{(6.022 \times 10^{23}/\text{mol})}$$

$kE = (7.95 \times 10^{-19}$ J$) - (5.85 \times 10^{-19}$ J$) = 2.10 \times 10^{-19}$ J.

6.27 (a) Below 200 nm, O_2, N_2, and O_3 all absorb effectively;
(b) From 200-240 nm, O_2 and O_3 absorb effectively;
(c) From 240-310 nm, only O_3 absorbs effectively;
(d) From 310-700 nm, no gases in the atmosphere absorb effectively (although NO_2, a component of polluted atmospheres, does absorb light in this region);
(e) From 700-2000 nm, CO_2 and H_2O absorb effectively.

6.29 According to Figure 6-16, a temperature of -30 °C = 243 K occurs at four different altitudes:
≈ 7 km, ≈ 35 km, ≈ 65 km, and ≈ 110 km, so the information is insufficient.

6.31 Use the mass of the electron and Avogadro's number to calculate the molar mass:

$m_{molar} = (9.1093897 \times 10^{-31}$ kg$)(10^{-3}$ g/kg$)(6.0221367 \times 10^{23}$ mol$^{-1})$
$= 5.4857990 \times 10^{-4}$ g/mol.

6.33 The wavelength of an electron is determined using $u = \sqrt{2E/m}$, and $\lambda = h/mu$:

$$u = \sqrt{\frac{2(1.15 \times 10^{-19} \text{ J})}{(9.109 \times 10^{-31} \text{ kg})}} = 5.03 \times 10^5 \text{ m/s};$$

$$\lambda = \frac{(6.626 \times 10^{-34} \text{ J s})}{(9.109 \times 10^{-31} \text{ kg})(5.03 \times 10^5 \text{ m/s})} = 1.45 \times 10^{-9} \text{ m or } 1.45 \text{ nm}.$$

6.35 An hourglass-shaped orbital looks the same from each side, but has a different shape when viewed from above:

6.37 To calculate photon properties, use $\lambda \nu = c$ and the equations for photons in Table 6-2:
(a) $\lambda = c/\nu = (2.998 \times 10^8 \text{ m/s}) \div (4.5 \times 10^{13} \text{ s}^{-1}) = 6.6 \times 10^{-6}$ m;
(b) $E = h\nu = (6.626 \times 10^{-34} \text{ J s})(4.5 \times 10^{13} \text{ s}^{-1}) = 3.0 \times 10^{-20}$ J;
(c) t = dist/speed = $(2.86 \times 10^5 \text{ mi})(1.609 \text{ km/mi})(10^3 \text{ m/km}) \div (2.998 \times 10^8 \text{ m/s})$ = 1.54 s.

6.39 In order to construct an energy level diagram, first determine the energies of the various photons, using $E = hc/\lambda = (6.626 \times 10^{-34} \text{ J s})(2.998 \times 10^8 \text{ m/s})(10^9 \text{ nm/m}) \div (\lambda \text{ in nm})$
$E = (1.986 \times 10^{-16} \text{ J nm}) \div (\lambda \text{ in nm})$;

λ, nm:	422.7	272.2	239.9	672.1	504.2
$E, 10^{-19}$ J:	4.699	7.296	8.281	2.955	3.940

E (272.2 nm – 671.8 nm) = (7.296 – 2.955) $\times 10^{-19}$ J = 4.341 $\times 10^{-19}$ J;
E (239.9 nm – 504.2 nm) = (8.281 – 3.940) $\times 10^{-19}$ J = 4.341 $\times 10^{-19}$ J;

Next, construct an energy level diagram showing the relationships among the various levels. Absorptions are upward transitions, emissions are downward transitions:

(a) The state reached by 504.2 nm emission is 4.341×10^{-19} J above the lowest state.
$\lambda = (1.986 \times 10^{-16}$ J nm$) \div \Delta E = 457.5$ nm;
(b) $\Delta E = (7.296 - 4.699) \times 10^{-19}$ J $= 2.597 \times 10^{-19}$ J;
$\lambda = (1.986 \times 10^{-16}$ J nm$) \div \Delta E = 764.7$ nm;
(c) Yes, the 272.2 nm – 671.8 nm sequence and the 239.9 – 504.2 nm sequence both lead to the state at 4.341×10^{-19} J above the ground state.

6.41 Calculate the wavelength using $\lambda = hcN_A/E$:
$hcN_A = 1.196 \times 10^{-4}$ kJ m/mol
$\lambda = (1.196 \times 10^{-4}$ kJ m/mol$) \div (242.7$ kJ/mol$) = 4.929 \times 10^{-7}$ m or 492.9 nm;
This wavelength is in the visible region of the spectrum, which reaches the surface of the Earth. Thus, Cl_2 in the troposphere will be dissociated by sunlight.

6.43 First, calculate the energy of one photon; then use $E_{total} = n\, E_{photon}$ to determine n:
$$E_{photon} = \frac{(6.626 \times 10^{-34}\text{ Js})(2.998 \times 10^{8}\text{ m/s})}{(337.1\text{ nm})(10^{-9}\text{ m/nm})} = 5.893 \times 10^{-19}\text{ J};$$
n = (10 mJ)(10^{-3} J/mJ) ÷ (5.893×10^{-19} J/photon) = 1.7×10^{16} photons.

6.45 Calculate the wavelength using $\lambda = hcN_A/E$:
$hcN_A = 1.196 \times 10^{-4}$ kJ m/mol
$\lambda = (1.196 \times 10^{-4}$ kJ m/mol$) \div (216.4$ kJ/mol$) = 5.529 \times 10^{-7}$ m or 552.9 nm.

6.47 The photoelectric effect is described by an energy balance equation:
$kE_{electron} = E_{photon} - bE$.
(a) $bE = E_{photon} - kE_{electron} = 6.00 \times 10^{-19}$ J $- 2.70 \times 10^{-19}$ J $= 3.30 \times 10^{-19}$ J;
(b) $\lambda = hc/E = (6.626 \times 10^{-34}$ J s$)(2.998 \times 10^{8}$ m/s$) \div (6.00 \times 10^{-19}$ J$) = 3.31 \times 10^{-7}$ m;
(c) For electrons, $\lambda = h/mu$;
$$u = \left(\frac{2E}{m}\right)^{1/2} = \left[\frac{(2)(2.70 \times 10^{-19}\text{ kg m}^2/\text{s}^2)}{(9.109 \times 10^{-31}\text{ kg})}\right]^{1/2} = 7.70 \times 10^{5}\text{ m/s};$$
$$\lambda = \frac{(6.626 \times 10^{-34}\text{ kg m}^2/\text{s})}{(9.109 \times 10^{-31}\text{ kg})(7.70 \times 10^{5}\text{ m/s})} = 9.45 \times 10^{-10}\text{ m}.$$

6.49 For electromagnetic radiation, $E = h\nu$ and $\lambda = c/\nu$;

$\lambda = (2.998 \times 10^{8}$ m/s$) \div (27.3$ MHz$)(10^{6}$ Hz/MHz$)(1$ s^{-1}/Hz$) = 11.0$ m;
$E = (6.626 \times 10^{-34}$ J s$)(27.3$ MHz$)(10^{6}$ Hz/MHz$)(1$ s^{-1}/Hz$) = 1.81 \times 10^{-26}$ J.

6.51 Make use of the figure legends to determine the answers to the questions.

(a and b) At an altitude of 60 km, $T \approx 165$ K and $P \approx 2 \times 10^{-4}$ atm;
(c and d) The nearest temperature maximum is at lower altitude, $T \approx 272$ K at 50 km.

6.53 Calculate the frequencies using $v = c/\lambda$ and the energies using $E = hcN_A/\lambda$.
$hcN_A = 1.196 \times 10^{-4}$ kJ m/mol

λ (nm)	487	514	543	553	578
v (10^{14} s^{-1})	6.16	5.83	5.52	5.42	5.19
E (kJ/mol)	246	233	220	216	207

6.55 See Figure 6-13 for an energy level diagram for hydrogen atoms:

6.57 A wave with half the amplitude has peaks that are half as high; a wave with half the frequency has wavelengths that are twice as long:

original wave   wave with half the amplitude and half the frequency

6.59 Use the equation for electrons from Table 6-2. The neutron mass is 1.6749×10^{-27} kg.

$$u = \frac{h}{m\lambda} = \frac{(6.626 \times 10^{-34} \text{ Js})}{(1.6749 \times 10^{-27} \text{ kg})(75 \text{ pm})(10^{-12} \text{ m/pm})} = 5.27 \times 10^3 \text{ m/s};$$
$$kE = mu^2/2 = (1/2)(1.6749 \times 10^{-27} \text{ kg})(5.27 \times 10^3 \text{ m/s})^2 = 2.3 \times 10^{-20} \text{ J}.$$

6.61 Under normal conditions, absorptions occur between the lowest state and any higher energy state. Emission originates from any higher state, not only the lowest state:

6.63 This problem combines hydrogen atom emission properties with photoelectron emission. Use the equation for hydrogen atom energy levels for the first two calculations and the equation for the photoelectric effect for the last two calculations.

(a) $E_4 = \dfrac{-2.18 \times 10^{-18} \text{ J}}{4^2} = -1.36 \times 10^{-19}$ J;

(b) $E_{photon} = \Delta E_{atom} = E_4 - E_2 = -1.36 \times 10^{-19}$ J $-(-2.18 \times 10^{-18}$ J $\div 2^2) = 4.09 \times 10^{-19}$ J;
$\lambda = hc/E = (6.626 \times 10^{-34}$ J s$)(2.998 \times 10^8$ m/s$) \div (4.09 \times 10^{-19}$ J$) = 4.86 \times 10^{-7}$ m;

(c) $kE = E_{photon} - bE = (4.09 \times 10^{-19}$ J$) - (3.23 \times 10^{-19}$ J$) = 0.84 \times 10^{-19}$ J;

(d) $u = \left(\dfrac{2E}{m}\right)^{1/2} = \left[\dfrac{(2)(0.84 \times 10^{-19} \text{ kg m}^2/\text{s}^2)}{(9.109 \times 10^{-31} \text{ kg})}\right]^{1/2} = 4.3 \times 10^5$ m/s;

$\lambda = \dfrac{h}{mu} = \dfrac{(6.626 \times 10^{-34} \text{ Js})}{(9.109 \times 10^{-31} \text{ kg})(4.3 \times 10^5 \text{ m/s})} = 1.7 \times 10^{-9}$ m.

6.65 The major mechanism by which the Earth loses energy at night is by emission of infrared radiation. Clouds reflect infrared light back toward the Earth's surface, so on cloudy nights much of this infrared light is reabsorbed by the Earth's surface, keeping the Earth warmer than on clear nights, when nearly all the infrared radiation escapes from the atmosphere.

6.67 Consult Figure 6-16 to answer this question. Pressure drops uniformly as altitude increases. Temperature drops with increasing altitude in the troposphere, is nearly constant in the lower stratosphere, and rises to a maximum at the stratosphere/mesosphere boundary. Temperature decreases with increasing altitude through the mesosphere, reaching a minimum at the mesosphere/thermosphere boundary; then it increases with altitude in the thermosphere.

Chapter 6: Atoms and Light

6.69 The wavelengths associated with particles are calculated from $\lambda = h/mu$, and 5.000% of the speed of light is $u = (5.000\%)(10^{-2}/\%)(2.998 \times 10^8 \text{ m/s}) = 1.499 \times 10^7$ m/s.

electron: $\lambda = \dfrac{h}{mu} = \dfrac{(6.626 \times 10^{-34} \text{ Js})}{(9.109 \times 10^{-31} \text{ kg})(1.499 \times 10^7 \text{ m/s})} = 4.853 \times 10^{-11}$ m;

proton: $\lambda = \dfrac{h}{mu} = \dfrac{(6.626 \times 10^{-34} \text{ Js})}{(1.673 \times 10^{-27} \text{ kg})(1.499 \times 10^7 \text{ m/s})} = 2.642 \times 10^{-14}$ m.

6.71 The only information required from Figure 6-17 is the wavelength where O_2 absorbs best.

(a) $\lambda_{max} \approx 190$ nm;
(b) $\nu = c/\lambda = (2.998 \times 10^8 \text{ m/s}) \div (190 \text{ nm})(10^{-9} \text{ m/nm}) \approx 1.6 \times 10^{15}$ s^{-1};
(c) $E = hcN_A/\lambda$.
$hcN_A = 1.196 \times 10^{-4}$ kJ m/mol
$E = (1.196 \times 10^{-4} \text{ kJ m/mol}) \div (190 \text{ nm})(10^{-9} \text{ m/nm}) \approx 6.3 \times 10^2$ kJ/mol.

6.73 The energy level diagram shows the energy differences among the various levels. By examining it, we can see that the smallest energy difference is associated with Photon$_1$ and the largest energy difference with Photon$_2$. The energy of a photon is inversely proportional to its wavelength, so Photon$_1$ has the longest wavelength and Photon$_2$ the shortest wavelength. Thus the assignments are as follows:

Photon$_1$ = 565 nm ($n = 6$ to $n = 4$)
Photon$_2$ = 121 nm ($n = 4$ to $n = 2$)
Photon$_1$ = 152 nm ($n = 2$ to $n = 1$)

Chapter 7 Atomic Structure and Periodicity

7.1 Quantum Numbers

QUESTIONS TO ANSWER, SKILLS TO LEARN
1. What are the four quantum numbers that determine electron energy?
2. What are the restrictions on the values of quantum numbers?

The electrons in atoms have four properties which were described in Chapter 6: (1) energy, which is quantized, (2) the shape of the electron-wave with that energy, (3) its orientation with respect to the x-, y- and z-axes, and (4) their spin. These properties are determined by four numbers which have specific values that are either integers or half integers. These four numbers are the **quantum numbers** and together they describe the energy, orbital shape, orientation of the orbital, and spin state of the electron.

Principal Quantum Number, n

The energy of an electron is important for many reasons. It is quantized, as we deduced during the discussion of emission and absorption spectra. The quantum number for energy is called the **principal quantum number** and is given the symbol, n. The quantum number n may have only certain values; it must be a nonzero positive integer.

$$n, \text{ the } principal \ quantum \ number = 1, 2, 3, \ldots$$

The smaller the value of n, the more stable the electron described by the wave function. For hydrogen, the energy of the electron depends only on the value of n. For atoms with more than one electron, the value of n does not give the exact energy, but all electrons with a given value of n have nearly the same energy compared to electrons with different values of n. A change in the value of n for an electron results in a significant change in energy.

Energy and the sizes of orbitals are also related. The larger n is, the higher (more positive) the energy is. Because the total energy increases with increasing n, the more energetic electron occupies more space, resulting in a larger orbital. In a larger orbital, the electron is on average further away from the nucleus of the atom than in a smaller orbital.

Azimuthal Quantum Number, ℓ

In addition to size and energy, atomic orbitals have shape. The quantum number that describes the shape of the orbital is the **azimuthal quantum number**, ℓ. The values of ℓ are restricted by the value of n:

ℓ can only have integer values less than n: $\ell = 0, 1, 2, \ldots, (n - 1)$

Chapter 7: Atomic Structure and Periodicity

The orbital shapes that correspond to different values of ℓ are referred to by a letter rather than the value of ℓ. The table below gives the letter designations for $\ell = 0$ to 4 and the shapes of the orbitals for $\ell = 0$ to 2.

value of ℓ	0	1	2	3	4
orbital letter	s	p	d	f	g
orbital shape	●	∞	✾ ⚇		

There are two shapes given for the orbital where $\ell = 2$ (the *d* orbital) because there are two different shapes for the *d* orbitals. This will be explained in Section 7.2. For a given value of n, there are n values of ℓ. Orbitals are identified with the value of the principal quantum number and the letter designation for the value of ℓ. The following exercise illustrates this process.

Exercise 7.1 Name the orbitals which have the following values of *n* and *l*.

n, ℓ values	(a)	(b)	(c)
$n =$	5	2	2
$\ell =$	2	0	3

Steps to Solution: To name orbitals, first check to make sure that the combination of quantum numbers is allowed. If allowed, the value of n is listed and then the letter designation follows.

(a) The value of n, 5, is legitimate (a nonzero integer) and the value of ℓ, 2, is also allowed (ℓ is an integer less than n). The orbital letter for $\ell = 2$ is d; the orbital is a 5d (pronounced five-d) orbital.

(b) The values of n (2) and ℓ (0, an integer less than n) are valid. The letter for the orbital where $\ell = 0$ is s. The orbital name is 2s (pronounced two-s).

(c) The value of n (2) is valid (a nonzero positive integer). However, the value of ℓ (3) is not valid because it is greater than n. Therefore this combination does not describe or name an atomic orbital!

Magnetic Quantum Number, m_ℓ

The azimuthal quantum number defines the shape of an orbital. The s orbital is spherical, so the electron wave described by the s orbital is distributed evenly in space around the nucleus (alternatively, the probability of finding an electron in an s orbital at some distance from the nucleus is the same no matter which direction one looks). For orbitals such as the p orbital which give the electron-wave a preferred direction, shape alone is an inadequate description because the probability of finding the electron in a p orbital does depend on the direction we look. The orientation of orbitals is also a quantized quantity designated by a third quantum number, the magnetic quantum number, m_ℓ. The allowed values of the magnetic quantum number, m_ℓ, for the orbitals with azimuthal quantum number ℓ are:

$$m_\ell = -\ell, (-\ell + 1), (-\ell + 2), \ldots -2, -1, 0, 1, 2, \ldots (\ell - 2), (\ell - 1), \ell$$

For a given value of ℓ, there are $2\ell + 1$ values of m_ℓ.

Spin Quantum Number, m_s

The electron has a property called spin, discussed in Chapter 6, Section 6.5. The spin of the electron is a quantized property and is described by the **spin quantum number**, given the symbol m_s, which has only two values: $+\frac{1}{2}$ and $-\frac{1}{2}$.

$$m_s = \pm \frac{1}{2}$$

One analogy to an orbital is a room in a large hotel with rooms that contain two beds each. The n value is analogous to the floor the room is on (the higher the floor, the higher the energy); the ℓ value is the wing of the building in which the room is located; the m_ℓ value is the room number in the wing; and the m_s value determines which of the two beds in the room is yours. All four quantum numbers are necessary for a complete description of the electron where a complete description of an electron means knowledge of energy and most probable volume of space in which the electron may be found.

Exercise 7.2 Write all the valid combinations of quantum numbers for the 3p orbitals.

Steps to Solution: The description 3p tells us that $n = 3$ and $\ell = 1$. The only quantum numbers that can change (and still describe a 3p orbital) are m_ℓ and m_s. The quantum number m_ℓ may have integer values from ℓ to $-\ell$; for $\ell = 1$, m_ℓ can have three different values, 1, 0, -1. There are two possible values of m_s for each value of m_ℓ: $\frac{1}{2}$ and $-\frac{1}{2}$. Tabulating these possible values for m_ℓ and m_s gives us the *table that follows. There are six allowed combinations of quantum numbers for 3p orbitals.*

n	3	3	3	3	3	3
l	1	1	1	1	1	1
m_l	1	1	0	0	-1	-1
m_s	1/2	-1/2	1/2	-1/2	1/2	-1/2

7.2 Shapes of Atomic Orbitals

QUESTIONS TO ANSWER, SKILLS TO LEARN
1. **What information can we obtain from the different ways of depicting orbitals?**
2. **What are the shapes of the orbitals with different values of l?**
3. **Drawing *s*, *p* and *d* orbitals**
4. **How does the size of an orbital depend on n?**

The shapes and sizes of atomic orbitals are important in the interactions of atoms and molecules in chemical reactions. An orbital is the three-dimensional wave function, Ψ, of a bound electron with certain values of n, l, m_l and m_s. However, it is difficult to perceive all the important aspects of size and shape from any single two-dimensional illustration. Three different representations are used in the text: (1) plots of Ψ^2 vs. r; (2) "pictures" of electron density and (3) pictures of electron contour surfaces.

Plots of the wave function, Ψ, vs. distance from the nucleus provide a direct numerical representation, but it is the square of the wave function, Ψ^2, that is has physical significance. The square of the wave function at a point, x, y, z, is proportional to finding the electron at that point. Plots of a related function vs. r for the 1*s*, 2*s* and 2*p* orbitals are shown below. These graphs show how the probability of finding the electron varies with distance from the nucleus. The high point in these graphs for a particular orbital can be considered the size of the orbital; the maximum for the 2*s* orbital is at about 2.9Å. However, such graphs do not show the shape of the orbital. Pictures of orbitals show the shape of orbitals; however, other information (size, etc.) is not as readily apparent.

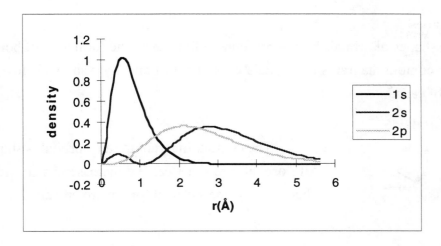

Another commonly used type of orbital pictures is the **orbital density picture** in which dots are used to represent the probability of finding an electron. The greater the number of dots, the greater the probability of finding an electron in that region of space (the larger the **electron density**). These pictures show a "slice" of the orbital through the nucleus and accurately portray the shape of the orbital and the electron density distribution, but require considerable care in their drawing.

The last type of orbital picture commonly used is the **electron contour drawing** in which an outline of the orbital is drawn that contains most (usually 90%) of the electron density; that is, the probability of finding the electron within that volume is 90%. The contour drawing gives a good picture of the orbital but provides little detail. We will make frequent use of electron contour pictures of orbitals in the section, "Orbital Shapes."

Orbital Sizes

The size of an orbital has many important consequences that we will discuss in later chapters. The principal quantum number has especially important effects. The more significant trends in orbital sizes are listed below.

In an atom, as n increases, the orbital size increases.

>As n increases, the energy of the electron increases. A more energetic electron occupies more space; therefore the orbital describing the electron must be larger.

Orbitals with the same value of n but different values of ℓ, m_ℓ, and m_s have similar sizes.

>The quantum number n determines energy and size; the other quantum numbers have only a small effect on orbital size.

Orbitals become smaller as the nuclear charge increases.

>An electron is attracted to the nucleus by electrical forces; as the size of the positive charge on the nucleus increases, so does the force on the electron. The increased force will keep the electron closer to the nucleus, so the orbital for that electron will be smaller.

Orbital Shapes

The shapes of orbitals have a profound influence on the formation of bonds between atoms. The contour diagrams for orbitals where $\ell = 0$, 1 and 2 are shown below with a brief discussion of each.

$\ell = 0$: *s* orbitals

The contour diagram for an *s* orbital is shown to the left. All *s* orbitals are spherical and increase in size with increasing n. There is only one *s* orbital for any given value of n.

$\ell = 1$: *p* orbitals

When $\ell = 1$, there are three allowed values for m_ℓ: 1, 0, -1. These are the three *p* orbitals. The *p* orbitals are nonspherical; the orbitals place most of the electron density in two "lobes" along one of the x-, y- or z-axes on either side of the nucleus. The axis along which the electron density lies is the subscript for the orbital.

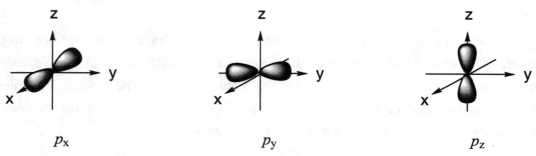

As n increases, the overall shapes of the orbitals remain the same.

$\ell = 2$: *d* orbitals

For $\ell = 2$, there are five allowed values for m_ℓ (2, 1, 0, -1, -2) so there are five *d* orbitals, the $d_{xy}, d_{xz}, d_{yz}, d_{z^2}, d_{x^2-y^2}$ orbitals. Of these orbitals, two lie along the x-, y- or z-axes and three lie between the axes. The latter three are shown below.

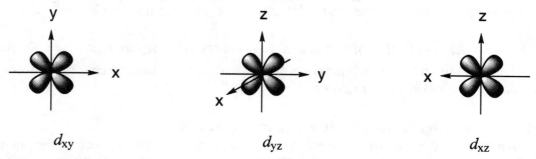

The orbitals are named for the planes in which they lie; the d_{xy} orbital lies in the xy plane, etc. The other two *d* orbitals place electron density directly on the x-, y- and z-axes. One, the

$d_{x^2-y^2}$ orbital, looks like the previous three, but the lobes lie along the x- and y-axes. The other orbital, the d_{z^2} orbital, looks quite different from the other four d orbitals. Most of the electron density is directed along the z-axis with a small "doughnut" of electron density in the xy plane.

$d_{x^2-y^2}$ d_{z^2}

Exercise 7.3 Atomic orbitals on different atoms can interact with each other, with different results. For now, consider two atoms, A and X. Draw contour pictures of the interactions described below. (In each of these cases, have the same signs of the wave function pointing toward each other.)
(a) The 2s orbital on A with the 2p orbital on X, with the preferred axis of the p orbital (one of the lobes) pointing at the s orbital
(b) The 2s orbital on A with the 2p orbital on X, with the preferred axis of the p orbital perpendicular to the s orbital
(c) The 2p orbital on A and the 2p orbital on X where the preferred axes are parallel (a side-by-side interaction).

Steps to Solution: The first step in solving this problem is to remember the shapes of the orbitals required. **It is imperative to memorize the shapes of the s, p and d orbitals.** The next step is to draw the orbitals in the requested orientation. In each example it will be assumed that you have committed the shapes of the orbitals to memory and can draw reasonable representations of them.

(a) Here the 2p orbital on X is pointing at the 2s orbital on A. Think small! (i) Draw the nuclei and then (ii) the orbitals. It may help to connect the two nuclei with a line.

(b) Here the 2p orbital's preferred axis is perpendicular to the line joining the nuclei. Again, first draw the nuclei and then the orbitals:

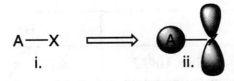

(c) Here the preferred axes of the 2p orbitals are parallel to each other in this part. If the z-axis contains the line connecting A and X, then the p_x orbitals of A and X would be in this orientation. The p_y orbitals on A and X would also satisfy this condition. The diagram below shows how these orbitals would look.

A—X   ⇒
i.   ii.

The p_x orbital on A and the p_y orbital on X would not fulfill the conditions stated in the question because in that case, the preferred axes would be perpendicular.

7.3 Orbital Energies

QUESTIONS TO ANSWER, SKILLS TO LEARN
1. **How does nuclear charge affect the size of an orbital?**
2. **What effect do additional electrons have on orbital size?**
3. **Predicting the effects of screening**

Effect of Nuclear Charge

Electrons are bound to the nucleus by electrical forces, which depend on the size of the charges. Therefore, if the nuclear charge increases, we expect that the force binding the electron to the nucleus will be stronger and the electron will be held closer to the nucleus. The electron will be more stable as the size of the nuclear charge increases. One way to test this idea is to measure the amount of energy required to remove an electron from the atom, illustrated below in a chemical equation.

$$A_{(g)} \rightarrow A^+_{(g)} + e^-_{(g)} \qquad \Delta E = IE_A$$

The energy required for this process is called the **ionization energy**, IE. The values for the ionization energy for the $1s$ electron in the following chemical species are tabulated below. In these species, there is only the nucleus and one electron, so there are no other factors except nuclear charge affecting the size of the ionization energy.

Chemical species	H	He$^+$	Li^{2+}
Ionization energy	2.178 x 10^{-18} J	8.7159 x 10^{-18} J	1.9613 x 10^{-17} J

The strong dependence of ionization energy and nuclear charge is apparent: in tripling the nuclear charge in the lithium cation ($Z = 3$), the ionization energy has increased by a factor of 9. In systems with more than one electron, the dependence is not as simple, but generally:

Orbitals with given n and ℓ grow more stable with increasing nuclear charge.

Screening: Partial Cancellation of Nuclear Charge

In atoms with more than one electron, other factors than the nuclear charge affect the energy of electrons. One factor is the change in the electrical force experienced by the electron in a $2p$ orbital when the $1s$ orbital is filled. Compare the relative sizes of these orbitals:

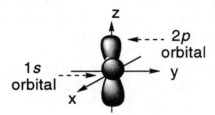

Because the $1s$ electrons have a high probability of being between the $2p$ electron and the nucleus, the $2p$ electron experiences a charge that is less than the nuclear charge Z: it is the nuclear charge, Z, minus some amount due to the presence of the $1s$ electrons, s. This reduced nuclear charge is called the **effective nuclear charge**. The reduction of the nuclear charge by the presence of electrons in the atom is called **screening**. To envision screening, consider the nuclear charge experienced by an electron as being the sum of the nuclear charge and the amount of electron density contained in a sphere whose radius is related to the most probable radius of the electron being screened. This concept is illustrated in the figure below.

To an electron, the nucleus plus electron density is effectively the same as that of a nucleus with a smaller charge.

Note that s will generally not be an integer because screening by electrons is usually not 100% effective. The amount of screening depends on the following factors:
- *The n value of the electron being considered and that of the occupied orbitals*

Electrons with smaller n values than the electron being considered are, on average, much closer to the nucleus and are quite efficient at screening. Those with the same n value are inefficient at screening compared to electrons of smaller n value.

- *The shape of the orbital (value of l) of the electron being considered and those of the occupied orbitals*

    For electrons with the same n, the orbitals of lower l are more stable than orbitals of higher l values.

- *Electrons of the same n and l quantum numbers are inefficient at screening.*

Exercise 7.4 Carbon atoms contain two electrons each in the 1s, 2s and 2p orbitals. Consider an electron in the 2p orbital. List the orbitals in order of increasing effectiveness at screening the 2p electron.

Steps to Solution: To answer this question, we must consider the factors that determine the effectiveness of screening. The electron in question is the 2p electron; to be effective at screening, the orbital must be small (most of its electron density is between the nucleus and the 2p electron). The 1s orbital will be most effective at screening because it has the lowest value of n, and will be the smallest orbital. The 2s and 2p orbitals remain, which have the same value of n. The 2s orbital has a smaller value of l and therefore has more electron density near the nucleus than the 2p orbital. Therefore the 2s orbital will be more effective at screening than the 2p orbital. Now to list the orbitals in **increasing** effectiveness of screening, remembering that the 2p orbital is the least effective:

$$\boxed{\text{(least effective) 2p} < \text{2s} < \text{1s (most effective)}}$$

7.4 The Periodic Table

QUESTIONS TO ANSWER, SKILLS TO LEARN
1. **What are the Pauli Exclusion and Aufbau Prinicples**
2. **What are "valence" electrons?**
3. **Identifying the number of valence electrons and the group number**

The four quantum numbers lead to a complete description of an electron (description means knowledge of energy and most probable volume of space in which the electron may be found). Because n may have any value through infinity, there are an infinite number of sets of quantum numbers. In an atom with more than one electron each electron can be represented by a set of these quantum numbers. (Use of the word *atom* here has a specific meaning; the atom in question is in the gas phase and is unaffected by any other gas phase atoms that might be present. No bonding or interaction with other atoms is present in atoms being discussed for the balance of this chapter.) Only one of these sets of quantum numbers for each electron represents the most stable (lowest energy) arrangement of electrons. There are two guiding principles to help determine which values of the quantum numbers (orbitals) represent the most stable arrangement of electrons in an atom: the **Pauli exclusion principle** and the **Aufbau principle**.

The Pauli Exclusion Principle limits the number of electrons in an orbital.

The Pauli exclusion principle: *No two electrons may have the same four quantum numbers.*

There are no exceptions to this principle and as a consequence, only two electrons may occupy an atomic orbital. The Aufbau (building-up) Principle determines the order in which the electrons are "added" to the orbitals in determining the ground state configuration.

The Aufbau principle: *Each successive electrons is placed in the most stable orbital whose quantum numbers have not already been assigned to another electron.*

What is now the most commonly used version of the periodic table was originally developed through the observed variations in elements' physical and chemical properties. Elements with similar chemical properties are arranged vertically in columns, with the elements with lighter molar mass at the top of the column. However, the number of electrons increases in each row until all the orbitals have been filled and a new value of n must be used to accommodate the electrons. The boxes containing the elements' symbols are placed so that the electron configuration of the non-closed shell (highest-energy) electrons is the same in each column. The exception is helium, which is at the top of the noble gas column, even though it has the closed shell $1s^2$ configuration rather than the ns^2np^6 configurations of the other noble gases. Sodium, lithium, potassium and the other alkali metals, for example, all share the ns^1 electron configuration above the closed shell configuration. With the brief background in quantum mechanics provided in the preceding sections we can deduce another basis for the periodic table: changes in n and the number of **valence electrons** (electrons that are in the highest energy levels of the stable atom).

The chemical properties of the elements were responsible for their original placement in columns; the reason for the similar properties is the number of valence electrons. These electrons are highest in energy and the furthest away from the nucleus and are the electrons which participate in the formation of bonds with other atoms. In contrast, the electrons in the closed shell have lower values of n and are closer to the nucleus of the atom and lower in energy than the valence electrons. These electrons are called the **core electrons**.

Elements with nearly degenerate ns and $(n-1)d$ or $(n-2)f$ orbitals present a dilemma because of the similar energies of the orbitals. Chemical reactivity of the transition metals and the actinides and lanthanides supports the following rule:

Valence electrons are all those of highest principal quantum number occupied in the ground state plus those in partially filled d and f orbitals

Exercise 7.11 How many valence electrons do each the following elements have: (a) B (b) Ca (c) Co (d) Sb?

Steps to Solution: The fastest way to determine the number of valence electrons is to consult the periodic table to find the number of electrons beyond the closed shell. We then note whether the d or f shell is partially filled; if so, we must add the number of electrons in the partially filled shell to the number of electrons in the orbitals of highest n.

(a) B has the electron configuration of $[He]2s^2 2p^1$. There are 3 electrons with $n = 2$; therefore B has $\boxed{3 \text{ valence electrons}}$.

(b) Ca has the electron configuration $[Ar]4s^2$. There are 2 electrons with $n = 4$, which is the highest principal quantum number. Therefore Ca has $\boxed{2 \text{ valence electrons}}$.

(c) Co has the electron configuration $[Ar]4s^2 3d^7$. Not only does Co have 2 electrons in the 4s orbital, but there are 7 electrons in the partially filled 3d orbital. The number of valence electrons is the sum of these two. Therefore Co has 2 + 7 = \X(9 valence electrons).

(d) Confirm for yourself that Sb has $\boxed{5 \text{ valence electrons}}$.

7.5 Electron Configurations

QUESTIONS TO ANSWER, SKILLS TO LEARN
1. **What is an electron configuration?**
2. **Writing the ground state electron configurations for atoms and ions**
3. **Identifying excited states and the number of unpaired electrons in atoms and ions**

The electron configuration tells us what orbitals are occupied by electrons in an atom. We are most often interested in the lowest energy arrangement of electrons, the **ground state**

Chapter 7: Atomic Structure and Periodicity

electron configuration. We can find the ground state electron configuration by applying the Pauli exclusion principle and the Aufbau principle. The lowest energy orbitals are filled first, making sure that only two electrons go into each orbital. The following diagram is helpful in determining the order of filling orbitals (their relative energy). We find the order of filling the orbitals by following the arrows proceeding from left to right, that is, 1s, 2s, 2p, 3s, etc. The following exercise will make use of this diagram.

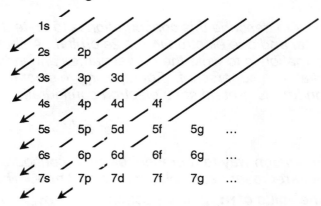

Exercise 7.6 Write the quantum numbers for the electrons in the ground state electron configuration of Li.

Steps to Solution: Lithium has three electrons. According to the Aufbau principle, the lowest energy atomic orbitals will be filled first. The most stable orbital will be the 1s orbital ($n = 1$, $\ell = 0$), followed by the 2s ($n = 2$, $\ell = 0$) and 2p ($n = 2$, $\ell = 1$) orbitals. Only two electrons may be placed in the 1s orbital, where the only different quantum number is m_s:

$n =$	$\ell =$	$m_\ell =$	$m_s =$	orbital name
1	0	0	1/2	1s
1	0	0	-1/2	1s

No more electrons may be placed in the 1s orbital without violating the Pauli exclusion principle, so the last electron will be placed in the 2s orbital, the next orbital on the diagram shown above (lower in energy than the 2p orbital).

$n =$	$\ell =$	$m_\ell =$	$m_s =$	orbital name
2	0	0	1/2	2s

These three sets of quantum numbers describe the lowest-energy arrangement of electrons in the lithium atom:

Chapter 7: Atomic Structure and Periodicity

$n =$	$\ell =$	$m_\ell =$	$m_s =$
1	0	0	1/2
1	0	0	-1/2
2	0	0	1/2

Another way to describe this configuration is to state that there are two electrons in the 1s orbital and one electron in the 2s orbital. A shorthand way to convey this same information is to write the orbital names and indicate the number of electrons in the orbital as a superscript. For Li, using this notation would give the following expression for the ground state electron configuration:

$$1s^2\ 2s^1$$

The third common way to represent electron configurations is to use the energy level diagram and represent each electron as an arrow. The direction of the arrow denotes the value of m_s of the electron: up for $m_s = 1/2$ and down for $m_s = -1/2$. Using this notation gives:

The previous exercise illustrated the three common ways to show electron configurations. The most compact way is the second one, where just the orbital names and numbers of electrons are used. The first and third methods convey information that will be important in determining the electron configurations of elements with partially filled orbital sets where $\ell > 0$.

For fixed values of n and ℓ, when $\ell > 0$, there are several possible values for m_ℓ, all of which have the same energy. When orbitals have the same energy, they are said to be **degenerate**. In order to determine the lowest-energy electron configuration where there are partially filled degenerate orbitals, we must apply a third principle, **Hund's rule**, which ensures that the configuration selected will have the minimum of electron-electron repulsions.

Hund's rule states: *In the ground state configuration, the number of unpaired electrons will be the maximum consistent with the Aufbau principle and the unpaired electrons will have the same spin orientation (same m_s).*

Exercise 7.7 applies Hund's rule.

Exercise 7.7 Determine the ground state electron configuration of N.

Steps to Solution: First we must find the number of electrons and then allocate them to orbitals consistent with the Pauli and Aufbau principles and Hund's rule. Nitrogen has seven electrons. The order of filling the orbitals is obtained as above in Exercise 7.6: 1s, 2s, then 2p. Two electrons go to each of the 1s and 2s orbitals, leaving 7-2(2) = 3 electrons. These electrons go to the 2p orbitals in the ground state. Hund's rule states that the number of unpaired electrons in degenerate orbitals must be the maximum consistant with the Aufbau principle and that the unpaired electrons must have the same value of m_s. The systematic way to achieve this result is to add the electrons in the degenerate orbitals one at a time with the same spin until all the degenerate orbitals are half-filled. The remaining electrons then complete the filling of the orbitals. This is shown in the energy level diagram below.

There are three unpaired electrons in a nitrogen atom in the ground state. The shorthand electron configuration is $1s^2 2s^2 2p^3$. There are other ways to arrange the three electrons in the 2p orbitals, but they are higher in energy than the one shown above.

In the atoms we have examined to this point, the shorthand electron configuration has been compact enough that the complete listing of orbitals has been convenient. However, in elements with more electrons, this notation may become cumbersome. The inert gases (Group 18) have special properties, which are a consequence of having filled the *ns* and *np* orbitals. For example, the configuration for He is $1s^2$; the $1s^2$ configuration is often written as [He]. The electron configuration for N could then be written as [He]$2s^2 2p^3$. The closed shell configurations are especially useful with elements of high atomic number. The following table lists the filled shell configurations for the inert gases.

Noble Gas	Number Of Electrons	Configuration =	Closed Shell
He	2	$1s^2$	[He]
Ne	10	$[He]2s^2 2p^6$	[Ne]
Ar	18	$[Ne]3s^2 3p^6$	[Ar]
Kr	36	$[Ar]4s^2 3d^{10} 4p^6$	[Kr]
Xe	54	$[Kr]5s^2 4d^{10} 5p^6$	[Xe]
Rn	86	$[Xe]6s^2 5d^{10} 4f^{14} 6p^6$	[Rn]

Exercise 7.8 Write the electron configurations and determine the number of unpaired electrons for the ground states of atoms of (a) Mg (b) Se and (c) Cs atoms. Consult the above table to aid in determining closed shells.

Steps to Solution: One can use the mnemonic diagram shown above for all the electrons or count the number of electrons that the atom has in excess of the nearest noble gas. The latter strategy will be used in this problem.

(a) Mg has 12 electrons. Mg (Z = 12) has two more electrons than Ne (10 electrons), which has a closed shell configuration. Therefore two electrons remain to be assigned orbitals. Using the orbital filling diagram shows that the next orbital to be filled after the 2p orbital is the 3s orbital. The two electrons can occupy the 3s orbital; one with m_s = 1/2 and the other with m_s = - 1/2. The electron configuration is $\boxed{[Ne]3s^2}$. Because each orbital with a given set of n, ℓ, and m_ℓ has two electrons, there are no unpaired electrons.

(b) Se (Z = 34) has 34 electrons. Looking at the periodic table, we see that Se is in the fourth row, following the closed shell corresponding to Ar, which has 18 electrons, the last of which fill the 3p orbital. The closed shell is [Ar]; there are 34 - 18 = 16 electrons to assign to orbitals. We now add electrons to orbitals following the Aufbau principle and total up the electrons as they are added as shown below:

Configuration:	[Ar]	$4s^2$	$3d^{10}$	$4p^4$
Total electrons:	18	20	30	34

The electronic configuration of Se is: $\boxed{[Ar]4s^2 3d^{10} 4p^4}$

The 4p orbitals are filled using Hund's rule, as shown above. The number of unpaired electrons is 2.

Chapter 7: Atomic Structure and Periodicity

(c) *Confirm to yourself using the methods in parts (a) and (b) that the electronic configuration of Cs is [Xe] 6s^1 and that there is 1 unpaired electron in a Cs atom.*

Exceptions: Near Degenerate Orbitals

The orbital filling diagram gives a fairly accurate order of orbital filling for most of the elements that we will encounter in this course. A few elements, especially transition metals, do not strictly follow this order. In these elements the energies of the two orbitals that change filling order or violate the "rules" are very similar so that electron-electron repulsions change the normal order. The most notable violations are Cr and Mo [$ns^1(n-1)\,d^5$] and Cu, Ag, Au [$ns^1\,(n-1)\,d^{10}$].

Ionic Configurations.

Much as monatomic ions are made from addition of electrons to form anions and cations are made by the removal of electrons, the electron configurations of ions are determined by adding electrons to or removing them from the electron configurations of the atoms, *with the exception of cations of the transition metals*. In writing the electron configuration of a cation, there is one less electron per positive charge than in the atom. For anions, there is one more electron per unit of negative charge. The element that has the same number of valence electrons as the ion is **isoelectronic** with the ion. With the exception of transition metal cations, the electron configuration will be the same as that element's configuration.

In transition metals, the energy of the *s* and *d* orbitals is very similar, which is responsible for the exceptions to the Aufbau order discussed above. Upon forming a cation, the *ns* orbitals are higher in energy than the (*n* - 1)*d* orbitals. The result of this effect is that the non-closed shell electrons will occupy the *d* orbitals and the configuration of the atom will not be the same as that of the isoelectronic atom.

In all transition metal cations, the electrons after the closed shell occupy d orbitals.

Exercise 7.9 Write the electron configurations for (a) Mg^{2+}, (b) Ru^{2+}, (c) O^{2-}.

Steps to Solution: To write the electron configurations of these ions we must first find the number of electrons in the cation or anion and then fill the orbitals in order of increasing energy. We may use the shortcut of recognizing closed shell configurations for nontransition metal ions.

(a) *There are 10 electrons in an Mg^{2+} ion. In Exercise 7.10 (a) the electron configuration for Mg was found to be 1s^2 2s^2 2p^6 3s^2 or [Ne] 3s^2. Therefore we must remove two electrons to obtain the configuration for Mg^{2+}. Removing the two highest energy electrons gives the configuration [Ne]. One obtains the same answer by finding the isoelectronic element, Ne, and assigning its electronic configuration to Mg^{2+} because Mg is not a transition metal. The configuration is* [Ne] .

(b) There are 42 electrons in the Ru^{2+} cation (44 - 2 = 42). Of the 42 electrons, 36 will be in the closed shell configuration, [Kr], leaving six electrons (42 - 36 = 6). The next orbitals in the filling order specified by the diagram are 5s, then 4d. However, because Ru^{2+} is a transition metal cation, the 4d orbitals are lower in energy than the 5s orbitals and therefore the electrons will fill the 4d orbitals, giving a [Kr] $4d^6$ configuration. Using Hund's rule we find that there are 4 unpaired electrons (the d orbitals are singly occupied until the electrons must pair): [Kr] $4d^6$ or

[Kr] ↑↓ ↑ ↑ ↑ ↑ 4d

(c) Use the methods illustrated above to show that the electron configuration of O^{2-} is [Ne] .

Magnetic Behavior of Atoms

In Chapter 6, we explained that electrons have spin and can act as magnets. This phenomenon can be used to test the accuracy of electron configurations because electrons paired in orbitals cancel each other's magnetic behavior.

Atoms, ions or molecules that have no unpaired electrons are **diamagnetic** *and are repelled by a magnetic field.*

Atoms, ions or molecules that have unpaired electrons are **paramagnetic** *and are attracted into magnetic fields.*

Exercise 7.10 Which of the atoms or ions in Exercises 7.8 and 7.9 are paramagnetic? (Consult those electron configurations.)

Steps to Solution: To determine if an atom or ion is paramagnetic, the number of unpaired electrons must be determined. If the number of unpaired electrons is zero, the atom or ion will be diamagnetic; otherwise it will be paramagnetic. In Exercise 7.8, the atoms are Mg, Se and Cs. Mg has no unpaired electrons, Se has two unpaired electrons and Cs has one unpaired electron. Therefore, Se and Cs are paramagnetic.

In Exercise 7.9 Mg^{2+} and O^{2-} have no unpaired electrons and are diamagnetic, not paramagnetic. However Ru^{2+} has four unpaired electrons and is paramagnetic.

Of the atoms and ions in these exercises, Se, Cs and Ru^{2+} are paramagnetic.

Remember that the properties we are discussing are the properties of the *gas phase atoms or ions* of that element, and not those of atoms of the element that have formed bonds to themselves (in a pure element) or to other atoms (in a chemical compound). For example, Cs atoms are paramagnetic, but in Cs metal, the Cs atoms have formed bonds to one another. Because of this interaction, the electrons have paired, so Cs metal is diamagnetic.

7.6 Periodicity of Atomic Properties

QUESTIONS TO ANSWER, SKILLS TO LEARN
1. **Understanding how periodic properties change with Z and n**
2. **What are atomic radii, ionization energies and electron affinity?**
3. **What are the trends in atomic radii, ionization energies and electron affinity as one moves across and/or down the periodic table?**
4. **Identifying trends in the periodic properties for isoelectronic species**

The physical properties of an atom that show a marked dependence on the electron configuration are the atomic radius, ionization energies and electron affinity. All of these reflect the size of orbitals or the energy of orbitals (stability of an electron in that orbital) in an atom. The three factors listed below affect these properties

Principal quantum number	As n increases, atomic orbitals become larger and less stable (electrons in that orbital have a weaker attraction to the nucleus).
Nuclear charge	As Z increases, atomic orbitals become smaller and more stable (electrons in that orbital have a stronger attraction to the nucleus).
Screening	Screening has the effect of reducing the nuclear charge experienced by an electron in a selected orbital. Efficient screening leads to significant reduction in the positive charge exerted by the nucleus on the electron.

As we move down a column in the periodic table, both Z and n increase. For example, in comparing F and Cl, note that the valence electronic configurations are $2s^2 2p^5$ and $3s^2 3p^5$, respectively. The nuclear charge in Cl has increased by 8 over that in F. However, F has 2 core electrons compared to 10 in Cl. The core electrons are very efficient at screening, so the additional 8 core electrons reduce the nuclear charge experienced by the valence electrons by very nearly the amount that the nuclear charge has increased. As one proceeds down a column, orbitals get larger and less stable.

Moving from left to right across a row in the periodic table, n remains constant, but Z increases. Because the valence electrons are poor at screening, the valence electrons feel a stronger attraction to the nucleus and therefore the orbitals become smaller and more stable. The general trends for orbitals' energies and size are summarized below.

Chapter 7: Atomic Structure and Periodicity

Moving down a group (increasing n, same number of valence electrons):
Orbitals become larger and less stable.

Moving left to right in a row (n is constant, increasing number of valence electrons):
Orbitals become smaller and more stable.

Atomic Radii

Atomic size is determined by the volume occupied by the electrons, which in turn is determined by the sizes of the orbitals. A convenient measure of atomic size is the atomic radius which is defined as one half the distance between the nuclei in a substance containing identical atoms. For example, the distance between the two Cl nuclei in Cl_2 is 198 pm; the atomic radius of Cl is 99 pm.

The atomic radius changes as predicted by the trends in orbital size:

Moving down a group, atoms become larger.

Moving left to right in a row, atoms become smaller.

These general trends are followed throughout the periodic table with only slight variations. Examination of the radii of the transition metal atoms in a row (or the lanthanides or actinides) shows that their radii change by only small amounts. In the transition metals, the electron configurations are $ns^x (n-1)d^y$, where x = 1 or 2 and y is the number of *d* electrons. In these elements, the size of the atom is determined by the size of the *ns* orbital. However the $(n-1)d$ orbitals are quite effective at screening the *ns* orbitals. An increase in the atomic number that would shrink the *ns* orbital is almost completely compensated by the efficient screening by the $(n-1)d$ electron. Therefore the changes in atomic radius in the transition metals are not as dramatic as in the main group elements.

Exercise 7.12 For each of the following pairs of atoms, predict which atom is larger:
(a) Ga or As, (b) Re or Pt, (c) Sr or Ba.

Steps to Solution: We can answer these questions by application of the rules regarding periodic trends. We may also analyze the electron configurations to understand the trends in orbital size in terms of effective nuclear charge. We will use both approaches in developing our answers below.

(a) Ga and As are in the same row of the periodic table. As lies to the right of Ga, so we expect As to be smaller than Ga. Therefore, Ga atoms are larger than As atoms.

The electron configurations of Ga and As are [Ar] $4s^2 3d^{10} 4p^1$ and [Ar] $4s^2 3d^{10} 4p^3$ respectively. The nuclear charge of As (33) is larger than that of Ga (31), so the *nucleus exerts a stronger force on the electron cloud in As. The two additional electrons of As will not be effective at screening the nucleus because they have the same l value as the orbitals that determine the size of the atom. The orbitals that are responsible for most of the screening are those in the [Ar] core and, to a lesser extent, the 4s² electrons, so the additional 4p electrons in As will provide little screening, and the electron cloud in As will be more tightly bound. Therefore, Ga should be the larger atom.*

(b) *Re and Pt are in the same row of the periodic table and have incompletely filled d orbitals (Re:[Xe] 6s² 4f¹⁴ 5d⁵ and Pt: [Xe] 6s² 4f¹⁴ 5d⁸). Periodic trends suggest that Pt might be smaller because it lies to the left of Re.*

A more careful analysis is based on the nature of the orbitals occupied by the three additional electrons on Pt and the orbitals that determine the sizes of the atoms. The size of an atom depends on the size of the orbital of highest n; in Pt, this is the 6s orbital. The 3 additional electrons in Pt occupy the 5d orbital and are quite efficient at screening the 6s orbitals. This screening offsets the increase in nuclear charge; therefore it is expected that Pt and Re will have nearly the same size.

(c) Use both types of analysis to show that we expect Ba to be larger than Sr.

Ionization Energy

The **ionization energy** (*IE*) is the amount of energy required to remove an electron from an atom, ion or molecule:

$$A_{(g)} \rightarrow A^+_{(g)} + e^-_{(g)} \qquad \Delta E = IE$$

The energy required to remove the first electron is that of the highest-energy (least stable) orbital in the atom and is called the *first ionization energy*. In previous sections we have seen that the highest-energy orbital becomes more stable from left to right across a row because the effective charge acting on the electron increases. We have also seen that the energy of the valence orbital increases (becomes less stable) proceeding down a column, due to increasing n.

The first ionization energy:
- *increases from left to right across a row.*
- *decreases from top to bottom down a column.*

The reason for this trend is the same as for the trends observed in atomic radius: they reflect the increasing stability from left to right and decreasing stability going down a column of atomic orbitals. The *d* and *f* block elements show behavior similar to that observed in the atomic radius; the increase in Z proceeding from left to right is offset by the efficient screening of the electrons added to the *d* or *f* orbitals, leading to relatively small changes in ionization energy.

Removal of additional electrons from ions requires more energy because the electron must be removed from a more highly positively charged species. The energies of the subsequent ionization steps to remove the second and third electrons are called the *second ionization energy* and *third ionization energy*, respectively.

$$A^+_{(g)} \rightarrow A^{2+}_{(g)} + e^-_{(g)} \qquad \Delta E = IE(2)$$

$$A^{2+}_{(g)} \rightarrow A^{3+}_{(g)} + e^-_{(g)} \qquad \Delta E = IE(3)$$

The first four ionization energies for Al are listed and plotted below:

Step	E(kJ/mole)
IE(1)	580
IE(2)	1815
IE(3)	2740
IE(4)	11600

Note that the first three ionization energies increase steadily, due to the increasing charge on the ion and electrical force acting on the electrons. These three ionization energies correspond to removal of the valence electrons in the $3p$ and $3s$ orbitals. The very large increase in the fourth ionization energy is due to the fact that the electron being removed now is a core electron ($2p$ orbital) for Al which was quite stable before ionization occurred. Removal of a core electron requires large amounts of energy.

Subsequent ionization energies increase steadily until all the valence electrons have been removed. Further ionization energies are much higher, indicating that a stable core electron is being removed.

Electron Affinity

We need to measure how stable an anion is compared to the atom or anion of lower negative charge. The energy required to remove an electron from an anion is called the **electron affinity**, *EA*:

$$A^-_{(g)} \rightarrow A_{(g)} + e^-_{(g)} \qquad \Delta E = EA$$

In writing the electron configuration of anions, we learned that the electrons occupy the most stable empty orbitals of the atom. The more stable these orbitals are, the greater the electron affinity will be. Compare the electron affinities for Cl (348.7 kJ/mole) and Br (324.5 kJ/mole). The empty orbital in Br is a $4p$ orbital; that in Cl is a $3p$ orbital. Because the empty orbital's n value is higher for Br, this orbital is less stable and the electron affinity is expected to be smaller.

The electron affinity:
- *increases from left to right across a row*
- *decreases from top to bottom down a column*

The transition metals (*d*-block) and the actinides and lanthanides (*f*-block) show relatively small changes in the electron affinity on moving from left to right. This is because effective screening by the *d* or *f* electrons largely cancels the increase in *Z* across the row.

Exceptions to the Trends

There are some exceptions to the trends outlined above if we carefully examine the data. The exceptions usually occur when the electron configuration of the reactant or product atom or ion:

- contains a single electron in an orbital of different ℓ or
- has all the orbitals of a given ℓ half filled.

In the second row elements, B and O fulfill these requirements and have ionization energy lower than we would expect based on effective nuclear charge. A plot of the first ionization energies for the second row elements is shown below.

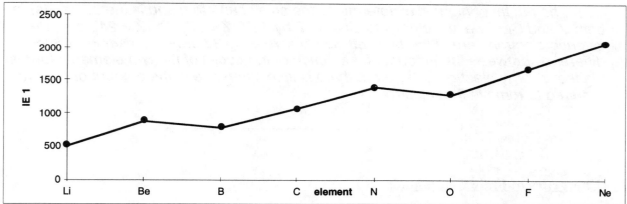

Periodic Properties in Isoelectronic Series

Isoelectronic series are atoms and ions with the same number of electrons but different net charges. An example of an isoelectronic series is:

$$Ca^{2+}, K^+, Ar, Cl^-, S^{2-}$$

where each of the species listed has 18 electrons with the electron configuration [Ar]. In these cases n and ℓ are the same for the valence electrons (highest-energy electrons) and the periodic properties will be affected only by the nuclear charge. As the nuclear charge increases, the highest energy orbital will become more stable. The highest nuclear charge coincides with the highest positive charge in an isoelectronic series.

Exercise 7.13 Use only the periodic table and the periodic properties to answer the following questions:

(a) Arrange the isoelectronic series, Ca^{2+}, K^+, Ar, Cl^-, S^{2-}, in order of decreasing radius.
(b) Why are the ionization energies of V and Cr nearly the same?

Steps to Solution: The answer is obtained by considering the effects of Z and n on the periodic property of interest.

(a) In this question we are asked to determine the relative sizes of the ions or atoms and list them in order from the largest to the smallest. Size depends on both Z and n. In an isoelectronic series, the electron configurations are the same, so only Z changes. In an isoelectronic series, Z increases with increasing positive charge; therefore the largest ion will that with the most negative charge, S^{2-}. The smallest ion is that with the greatest positive charge, Ca^{2+}. The order is:

$$S^{2-} > Cl^- > Ar > K^+ > Ca^{2+}$$

(b) First, we determine the electronic configurations for V ([Ar] $4s^2\,3d^3$) and Cr ([Ar] $4s^1\,3d^5$). The orbital of highest n determines the size of an atom and is the same (4s) in both V and Cr. The two elements differ in Z by 1 [V (Z = 23), Cr (Z = 24)], but the valence electrons are in the fairly efficiently screening 3d orbitals, leading to little difference between the amount of attraction of the nuclei of the two elements for the highest energy electrons; therefore there is little difference in the amount of energy required to remove that electron.

7.7 Energetics of Ionic Compounds

QUESTIONS TO ANSWER, SKILLS TO LEARN
1. **How do electron configurations help predict the stability of anions and cations?**
2. **Predicting the monatomic ions formed for metals and nonmetals**
3. **Predicting the formulas of ionic compounds based on position in the periodic table**

The same electrical forces that lead to the attraction of electrons to nuclei can also be used to describe the interaction of ions of opposite charge, which form a stable ionic compound where the sum of the positive and negative charges is zero. For example, a potassium ion and a chloride ion separated by a small distance, d, certainly experience an attractive force:

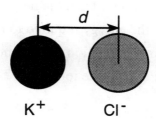

K^+ Cl^-

In fact, energy is released when potassium metal reacts with chlorine gas:

$$2K(s) + Cl_{2(g)} \rightarrow 2KCl_{(s)} \quad \Delta E_{experimental} = -870 \text{ kJ} (\Delta E = -435 \text{ kJ/mole KCl})$$

Because energy is conserved, the release of energy must be caused by the lower energy (stronger bonding) in the product compared to the reactant. How can we estimate the energy released in this process?

We can estimate the energy of attraction between the K^+ and the Cl^- ions by assuming all the attraction is due to electrical forces. We can calculate the energy of attraction due to the electrical force using Coulomb's Law:

$$E_{Coulomb} = \frac{1.389 \times 10^5 \text{ kJ} \cdot \text{pm/mol} (q^+)(q^-)}{d}$$

q^+ = charge on cation in units of electronic charge; q^- = charge on anion in units of electronic charge; d = distance between ions in pm

The distance between the ions in KCl is 319 pm, the charge on the cation is +1 and the charge on the anion is -1. Substitution into the equation for the Coulomb energy gives:

$$E_{Coulomb} = \frac{1.389 \times 10^5 \text{ kJ} \cdot \text{pm/mol} (1)(-1)}{319 \text{pm}} = -435 \text{ kJ/mole}$$

This amount is the energy released in the following process, not the reaction of Cl_2 and K metal!

$$K^+_{(g)} + Cl^-_{(g)} \rightarrow KCl_{(g)}$$

We must form K^+ ions and Cl^- ions from K and Cl. The energy required to form a cation may be estimated by the ionization energy (or energies for cations with charge more than +1). To form K^+,

$$K_{(g)} \rightarrow K^+_{(g)} + e^-_{(g)} \quad\quad IE = 418.9 \text{ kJ/mole}$$

The energy required to form an anion is found from the negative of the electron affinity Remember that the electron affinity is the energy required to remove an electron from an ion: because the process is reversed in forming an anion, energy numerically equal to the *EA* is released. The energy change in forming chloride ions is:

$$Cl_{(g)} + e^-_{(g)} \rightarrow Cl^-_{(g)} \qquad \Delta E = -EA = -348.8 \text{ kJ/mole}$$

The table below summarizes the energy changes for these three processes: (1) interaction of ions, (2) formation of K^+, and (3) formation of Cl^-. The last line sums up the chemical processes and the energy changes. We treat the arrow ($\rightarrow$) like an equals sign (=). The experimental energy change for the formation of solid $KCl_{(s)}$ from potassium metal and chlorine gas is -435 kJ/mole KCl, considerably larger than the value that has been calculated.

Process	Chemical Equation	Energy Change
attraction of K^+ and Cl^-	$K^+_{(g)} + Cl^-_{(g)} = KCl_{(g)}$	-435. kJ
ionization of $K_{(g)}$ to give $K^+_{(g)}$	$K_{(g)} = K^+_{(g)} + e^-_{(g)}$	418.9 kJ
formation of $Cl^-_{(g)}$ from $Cl_{(g)}$	$Cl_{(g)} + e^-_{(g)} = Cl^-_{(g)}$	-348.8 kJ
sum of above three steps	$K_{(g)} + Cl_{(g)} \rightarrow KCl_{(g)}$	-365. kJ/mole KCl

The difference between the two values is the result of ignoring some other processes. The molecular pictures below compare the reaction used in the above calculations with the reaction for the formation of $KCl_{(s)}$ from potassium metal and chlorine gas.

Chapter 7: Atomic Structure and Periodicity

The reason for the difference between the energy change calculated for the first reaction and that found for the second reaction lies mainly in three terms.

(1) In one reaction, the potassium is in the form of gaseous atoms whereas in the second reaction it is a metal where forces are holding the metal atoms together.

(2) Chlorine molecules (Cl_2) are different than Cl atoms; to convert the molecules to atoms requires energy to break the bond holding the atoms together in the molecule. To separate the atoms from each other in their chemically combined forms requires energy.

(3) In $KCl_{(s)}$, each ion is surrounded by several ions of opposite charge rather than being attracted to only one ion of opposite charge as in $KCl_{(g)}$. This difference leads to the release of more energy than that calculated by the simple Coulombic term for two ions of opposite charge.

Despite these glaring deficiencies in the model, the energy calculated is in error by only about 16%. This very simple description of the energy released in formation of an ionic compound reveals two important features:

(1) Most of the energy released is a result of the attraction of cations to anions.

(2) The main energy requirement is the ionization of the cation.

Exercise 7.14. Use an analysis like that shown above to estimate the energy released on forming KCl_2 where K has a +2 charge.

Steps to Solution: There are three processes used in the above model. The table below lists them. In the electrical attraction term, the positive charge on the cation is now 2, not 1, so this term will be:

$$E_{Coulomb} = \frac{1.389 \times 10^5 \text{ kJ} \cdot \text{pm/mole } (2)(-1)}{319 \text{ pm}} = -870 \text{ kJ}$$

The K^{2+} ion is attracted to two Cl^- ions, so the total energy of interaction of the ions will be:
2(-870 kJ) = -1740 kJ.

The energy released by forming 2 Cl^- ions from 2 Cl atoms will be double that released by forming 1 Cl^- ion.

The energy required for the formation of the K^{2+} ion will be the sum of the first and second ionization energies. The second ionization energy is quite large because it corresponds to the removal of a core electron.

Process	Chemical Equation	Energy Change
attraction of K^{2+} and 2 Cl^-	$K^{2+}_{(g)} + 2\,Cl^-_{(g)} = KCl_{2\,(g)}$	$\Delta E = -1740$ kJ/mole
ionization of $K_{(g)}$ to give $K^{2+}_{(g)}$	$K_{(g)} = K^+_{(g)} + e^-_{(g)}$ $K^+_{(g)} = K^{2+}_{(g)} + e^-_{(g)}$	$IE(1) = $ 418.9 kJ/mole $IE(2) = $ **3051.4 kJ/mole !!**
formation of 2 $Cl^-_{(g)}$ from 2 $Cl_{(g)}$	$2\big(Cl_{(g)} + e^-_{(g)} = Cl^-_{(g)}\big)$	$\Delta E = -697.6$ kJ/mole
sum of above three steps	$K_{(g)} + 2Cl_{(g)} \rightarrow KCl_{2\,(g)}$	$\Delta E_{total} = 1031.$ kJ/mole KCl_2

As a result, even though the energy of the attraction of the ions is 4 times that in KCl, **the large amount of energy required to remove the second electron from K leads to a positive energy change:** *energy would need to be supplied to form KCl_2. The high energy input required suggests that KCl_2 will not be found.*

Cation Stability and Anion Stability

Examining the trends in successive ionization energies we see that a very large increase in ionization energy when the electron removed is a core electron. Cations will be stable only when the number of electrons removed does not exceed the number of valence electrons. Exercise 7.14 showed that even the extra energy released from the electrical attractions in forming KCl_2 would not compensate for the energy required to remove the second electron from K. Anions also have predictable electron configurations. Electron configurations of stable anions have closed shell configurations like those of the noble gases. The monatomic anions of the halogens all have -1 charges. Anions with higher charges (S^{2-}, N^{3-}, O^{2-}) also have the closed shell configuration.

Cations will not have charges that require removing core electrons.

Anions will have charges that give them the closed shell electron configuration of a noble gas.

Chapter 7: Atomic Structure and Periodicity

Test Yourself

1. Write the ground state electron configuration and state the number of unpaired electrons for the following atoms and ions: (a) Cl; (b) Se; (c) Zn^{2+}; (d) I^-

2. Name the orbital in the legitimate set(s) of four quantum numbers ($\{n, \ell, m_\ell, m_s\}$) for an electron in an atom and explain why the other(s) cannot exist: (a) {4, 2, -1, +1/2}; (b) {5, 0, -1, -1/2}; (c) {4, 4, -1, -1/2}; (d) {4, 2, 3, +1/2}.

3. Which atom or ion in each pair has the smaller radius? (a) Li^+ or Na^+; (b) Cl or Cl^-; (c) Cl or S; (d) N^{3-} or O^{2-}; (e) Mg^{2+} or Al^{3+}.

4. Explain why MgN would not be expected to be stable.

5. For each of the following pairs of atoms indicate which has the higher first ionization energy: (a) S or P; (b) Rn or At; (c) Al or Mg; (d) Sr or Rb.

Answers

1. (a) [Ne] $3s^2 3p^5$; 1 unpaired electron; (b) [Ar] $4s^2 3d^{10} 4p^4$; 2 unpaired electrons; (c) [Ar] $3d^{10}$; no unpaired electrons; (d) [Xe]; no unpaired electrons.

2. (a) a 4d orbital; (b) the 5s orbital; (c) cannot exist, ℓ must be smaller than n; (d) cannot exist, m_ℓ cannot be greater than ℓ.

3. (a) Li^+; (b) Cl; (c) Cl; (d) O^{2-}; (e) Al^{3+}.

4. Because Mg forms stable cations only with a +2 charge and N forms a stable anion with a -3 charge. A compound with the stoichiometry MgN would require that either Mg have a +3 charge or N have a -2 charge, neither of which are expected to be stable.

5. (a) S; (b) Rn; (c) Al; (d) Sr.

Solutions to Chapter Problems

7.1 The designation 6p specifies that $n = 6$ and $\ell = 1$; the other two quantum numbers m_ℓ and m_s, can take any of their possible values. Remember that m_ℓ must be an integer with value between $+\ell$ and $-\ell$, while m_s is either +1/2 or –1/2. There are 6 valid sets:

n	ℓ	m_ℓ	m_s
6	1	+1	+1/2
6	1	+1	–1/2
6	1	0	+1/2
6	1	0	–1/2
6	1	–1	+1/2
6	1	–1	–1/2

7.3 Remember that ℓ is restricted to positive integers less than n, m_ℓ is restricted to integers between $-\ell$ and $+\ell$, and $m_s = +1/2$ or $-1/2$. For $n = 3$, ℓ can be 0 with $m_\ell = 0$, ℓ can be 1 with $m_\ell = +1, 0,$ or -1; and 1, or 2, and ℓ can be 2 with $m_\ell = +2, +1, 0, -1,$ or -2.

7.5 Remember that n must be a positive integer, ℓ is restricted to zero and positive integers less than n, m_ℓ is restricted to integers between $-\ell$ and $+\ell$, and $m_s = +1/2$ or $-1/2$.
(a) non-existent: m_s must be +1/2 or –1/2; (b) actual; (c) non-existent: ℓ must be less than n; and (d) actual.

7.7 The designation 3d specifies that $n = 3$ and $\ell = 2$; the largest possible value for m_ℓ is $+\ell$, in this case +2; and "spin up" means $m_s = +1/2$. Thus, the values are 3, 2, 2, +1/2.

7.9 The designation $n = 2, \ell = 1$ refers to a 2p orbital. This orbital has no nodes, so its electron density rises and then falls. Figure 7-10 shows a contour diagram. For this orbital, an electron density drawing looks very similar to a contour drawing.

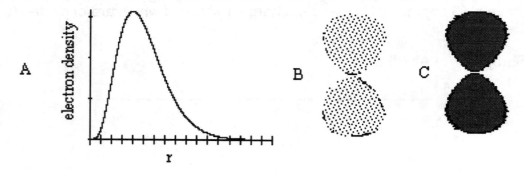

7.11 Orbital B has the shape characteristic of a p orbital, while orbitals B and C have shapes characteristic of d orbitals. Because orbital A appears smaller than the 3s orbital, it has $n < 3$ and is a 2p orbital; because orbitals B and C have similar sizes as the 3s orbital, they have $n = 3$ and are 3d orbitals. A = 2p, $n = 2, \ell = 1$; B, C = 3d, $n = 3, \ell = 2$.

7.13 Electron density plots accurately show how the wave function varies along one axis, but they fail to show whether or not an orbital has spherical symmetry, and they do not directly show electron densities. Thus, they fail to give a good picture of how an orbital appears in three dimensions.

7.15 A contour drawing of an *s* orbital is a spherical surface. The orbitals should grow larger as n increases:

7.17 Orbital stability in multi-electron atoms depends on the value of n (stability decreases with increasing value), Z (stability increases with increasing value), l (stability decreases with increasing value), and amount of screening (stability increases as other electrons are removed).

(a) He 1*s* is more stable than He 2*s* because of its lower value of n.
(b) Kr 5*p* is more stable than Kr 6*p* because of its lower value of n.
(c) He$^+$ 2*s* is more stable than He 2*s* because it is less screened by 1*s* electrons.
(d) Ar 4*s* is more stable than Ar 4*p* because of its lower value of l.

7.19 A hydrogen atom contains just one electron, so there is no screening effect. In the absence of screening, orbital energy depends only on n and Z, so all $n = 3$ orbitals have identical energy. In a helium atom, on the other hand, an electron in an $n = 3$ orbital is screened from the nucleus by the second electron, and the amount of screening decreases as l increases.

7.21 (a) The ionization energy (I.E.) of the He 2*p* orbital (0.585 x 10^{-18} J) is not much larger than that of the H 2*p* orbital (0.545 x 10^{-18} J). The similar values indicate nearly equal effective nuclear charges due to nearly complete screening; in the absence of screening, the He 2*p* orbital would have four times the ionization energy as the H 2*p* orbital.
(b) The I.E. of the He$^+$ 2*p* orbital (2.18 x 10^{-18} J) is four times larger than that of the H 2*p* orbital (0.545 x 10^{-18} J). A four-fold increase when Z doubles indicates a Z^2 dependence.

7.23

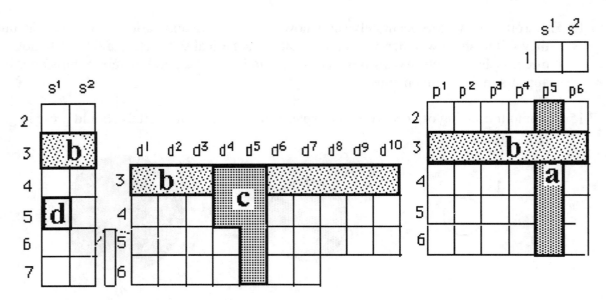

7.25 The element below lead in the periodic table would have atomic number 114 (count from Mt, element 109: 3 elements needed to complete the *d* block, 2 more to reach column 14 in the *p* block). The block that is filling would match that of lead, but with one higher **n**-value: 7*p*.

7.27 Element 111 would fall directly below gold in Group 11, in row 7 (**n** = 6 for the *d* block).

7.29 The column location of an element is the indicator of how many valence electrons it has. Remember that *s* electrons as well as those in the filling block count as valence electrons. O, fourth column of *p* block, 4 + 2 = 6 valence electrons; V, third column of *d* block, 3 + 2 = 5 valence electrons; Rb, 1 valence electron; Sn, second column of *p* block, 2 + 2 = 4 valence electrons; and Cd, end of *d* block, 2 valence electrons.

7.31 Use the Aufbau principle to determine the orbital occupancies and quantum numbers for the electrons in a configuration. When there are equivalent configurations, unpair as many electrons as possible in accordance with Hund's rule.

Be, 4 electrons:

n	ℓ	m_ℓ	m_s
1	0	0	+1/2
1	0	0	−1/2
2	0	0	+1/2
2	0	0	−1/2

O, 8 electrons:

n	ℓ	m_ℓ	m_s
1	0	0	+1/2
1	0	0	−1/2
2	0	0	+1/2
2	0	0	−1/2
2	1	1	+1/2
2	1	0	+1/2
2	1	−1	+1/2
2	1	1	−1/2

Ne, 10 electrons:

n	ℓ	m_ℓ	m_s
1	0	0	+1/2
1	0	0	−1/2
2	0	0	+1/2
2	0	0	−1/2
2	1	1	+1/2
2	1	1	−1/2
2	1	0	+1/2
2	1	0	−1/2
2	1	−1	+1/2
2	1	−1	−1/2

P, 15 electrons: first 10 electrons are the same as for Ne; remaining 5 electrons are:

n	ℓ	m_ℓ	m_s
3	0	0	+1/2
3	0	0	−1/2
3	1	1	+1/2
3	1	0	+1/2
3	1	−1	+1/2

7.33 Atoms with unpaired electrons in their configurations are paramagnetic. Of the atoms in Problem 7.31, O and P have unpaired electrons. Here are the orbital occupancy diagrams for the least stable occupied orbitals:

7.35 (a) This is Pauli-forbidden: s orbitals can hold no more than two electrons.
(b) This is Pauli-forbidden: s orbitals can hold no more than two electrons.
(c) This is an excited-state configuration: the $1s$ orbital is not full.
(d) This is an excited-state configuration: the $2s$ orbital is not full.
(e) This is the ground-state configuration.
(f) This configuration uses a non-existent orbital: there is no $1p$ orbital.
(g) This configuration uses a non-existent orbital: there is no $2d$ orbital.

7.37 Mo has ground-state configuration [Kr] $5s^1\ 4d^5$; that of Tc is [Kr] $5s^2\ 4d^5$:

Mo: $5s$ ↑ $4d$ ↑ ↑ ↑ ↑ ↑

Tc: $5s$ ↑↓ $4d$ ↑ ↑ ↑ ↑ ↑

7.39 (a) When orbitals are near-degenerate, exceptions to the normal filling order exist. Normally, the $4s$ orbital fills immediately after $3p$, starting with element 19. Exceptions that indicate near-degeneracy of $3d$ and $4s$ are Cr, $4s^1\ 3d^5$; and Cu, $4s^1\ 3d^{10}$.

(b) The first two elements in column 6, Cr and Mo, have $s^1\ d^5$ configurations, while the second two elements in this column, W and Sg, have $s^2\ d^4$ configurations.

(c) The $6d$ and $5f$ orbitals fill in the second f-block, the actinides. In this block, four elements have both these orbitals partly filled, indicating near-degeneracy: Pa, U, Np, and Cm.

7.41 The ground state for N is $1s^2\ 2s^2\ 2p^3$. Excited states with no electron having $n > 2$ can be formed by moving electrons out of the $1s$ and/or $2s$ orbital and placing them in the $2p$ orbital.

There are various ways to do this:
$1s^1\ 2s^2\ 2p^4$; $2s^2\ 2p^5$; $1s^1\ 2s^1\ 2p^5$; $1s^2\ 2s^1\ 2p^4$; $1s^2\ 2p^5$; $2s^1\ 2p^6$; and $1s^1\ 2p^6$.

7.43 Ionization energy (*IE*) decreases with increasing n and increases with increasing Z. Cs, which has the largest n-value, has the smallest *IE*; K, which has next larger n-value, has the next smallest *IE*; because of its larger Z value, Ar has larger *IE* than Cl: Ar > Cl > K > Cs.

7.45 The value of IE_2 is almost ten times that of IE_1, indicating that the second electron removed is a core electron rather than a valence electron. The elements in column 1 contain only one valence electron, so this is a column 1 element. An electron affinity around + 50 kJ/mol matches this assignment (see Figure 7-26). (The element is Cs).

7.47 The valence configurations are N, $2s^2 2p^3$; Mg, $3s^2$; and Zn, $4s^2 3d^{10}$.

N: $2p$ ↑ ↑ ↑ —add electron→ $2p$ ↑↓ ↑ ↑

The added electron adds to an already-occupied orbital. Electron-electron repulsion makes this a disfavored process.

Mg: $3s$ ↑↓ —add electron→ $3s$ ↑↓ $3p$ ↑ _ _

The added electron adds to the next higher orbital, which is significantly higher in energy.

Zn: $4s$ ↑↓ $3d$ ↑↓ ↑↓ ↑↓ ↑↓ ↑↓ —add electron→ $4s$ ↑↓ $3d$ ↑↓ ↑↓ ↑↓ ↑↓ ↑↓ $4p$ ↑ _ _

The added electron adds to the next higher orbital, which is significantly higher in energy.

7.49 Br⁻ has 36 electrons. Isoelectronic ions with less than 4 units are P^{3-}, S^{2-}, Cl^-, K^+, Ca^{2+}, and Sc^{3+}. For isoelectronic ions, size decreases with increasing Z:

$Sc^{3+} < Ca^{2+} < K^+ < Cl^- < S^{2-} < P^{3-}$.

7.51 Stable atomic anions form for those elements that are one electron short of filled p orbitals; these elements have high electron affinities: F, Cl, Br, I, (At).

7.53 The sequence of reactions for doing this calculation is:

$K \rightarrow K^+ + e^-$, $\Delta E = IE_1 = 418.8$ kJ/mol;
$I + e^- \rightarrow I^-$, $\Delta E = EA = -295.3$ kJ/mol;
$K^+ + I^- \rightarrow KI$, $\Delta E = E_{coulomb}$; (use Equation 7-1).

$$E_{coulomb} = \frac{(\text{Const.})(q^+)(q^-)}{d} = \frac{(1.389 \times 10^5 \text{kJ pm/mol})(+1)(-1)}{(133 \text{ pm} + 220 \text{ pm})} = -393 \text{ kJ/mol};$$

$\Delta E_{total} = (418.8 \text{ kJ/mol}) - (295.3 \text{ kJ/mol}) - (393 \text{ kJ/mol}) = -270 \text{ kJ/mol}$.

7.55 Stable anions form from elements in columns 16 (dianions) and 17 (monoanions). Stable cations form from various metals.

Ca, Cu, Cs, and Cr, all metals, may be found in ionic compounds as cations.
Cl may be found in ionic compounds as a monoanion.
C is not found in ionic compounds.

7.57 Metals are characterized by ease of forming cations and ability to share valence electrons in their solid metallic form. These features arise from the presence of loosely bound valence electrons. With increasing value of n, the valence electrons are less tightly bound to the atom; within a row, valence electrons are less tightly bound as Z decreases. Thus, the lower left region of the p block is where valence electrons are least tightly bound and the elements most metallic in character.

7.59 Non-metals are found in the upper right portion of the periodic table, metalloids along a diagonal running through the p block, and all other elements are metals. In Column 16, Te is classified as a metalloid. The elements above it, O, S, and Se, are non-metals, and the element below it, Po, is a metal.

7.61 A $4p$ electron must have $n = 4$, $l = 1$, but m_l has three possible values and m_s has two possible values. There are six possible sets of values:

n	l	m_l	m_s
4	1	1	1/2
4	1	1	−1/2
4	1	0	1/2
4	1	0	−1/2
4	1	−1	1/2
4	1	−1	−1/2

7.63 Use the filling rules to determine the correct configuration, being aware of exceptions.
C (6 electrons): $1s^2\, 2s^2\, 2p^2$;
Cr (24 electrons, exception): $1s^2\, 2s^2\, 2p^6\, 3s^2\, 3p^6\, 4s^1\, 3d^5$;
Sb (51 electrons): $1s^2\, 2s^2\, 2p^6\, 3s^2\, 3p^6\, 4s^2\, 3d^{10}\, 4p^6\, 5s^2\, 4d^{10}\, 5p^3$;
Br (35 electrons): $1s^2\, 2s^2\, 2p^6\, 3s^2\, 3p^6\, 4s^2\, 3d^{10}\, 4p^5$;
Xe (54 electrons): $1s^2\, 2s^2\, 2p^6\, 3s^2\, 3p^6\, 4s^2\, 3d^{10}\, 4p^6\, 5s^2\, 4d^{10}\, 5p^6$.

7.65 Most metallic elements (alkali metals, alkaline earths, transition metals, lanthanides and actinides) have vacant valence p orbitals, so the only metals with occupied valence p orbitals are found in the lower left portion of the p block. In row 6, Po, with valence configuration $s^2 p^4$, is classified as metallic.

7.67 Total electron spin depends on the number of unpaired electrons. In partially filled shells, each unpaired electron contributes 1/2 to the total spin.
P: [Ne] $3s^2\, 3p^3$; each $3p$ orbital is half filled, so $S = 3(1/2) = 3/2$;
Br⁻: [Kr], all orbitals are filled, so $S = 0$;
Cu⁺: [Ar] $3d^{10}$, all orbitals are filled, so $S = 0$;
Gd: [Xe] $6s^2\, 5d^1\, 4f^7$, each $4f$ orbital is half filled, so $S = (7+1)(1/2) = 4$;
Sr: [Kr] $5s^2$, all orbitals are filled, so $S = 0$.

7.69 Consult Figures 7-6, 7-7, and 7-16 for examples of this kind of drawing.

The shaded portions of the 3s and 3p indicate the region that is ineffectively screened by the 2p orbital. A 3s electron has greater probability of being in this region.

7.71 Use the periodic table to determine the correct configurations.

$Z = 9$ is F: [He] $2s^2 2p^5$; $Z = 20$ is Ca: [Ar] $4s^2$; $Z = 26$ is Fe: [Ar] $4s^2 3d^6$; $Z = 33$ is As: [Ar] $4s^2 3d^{10} 4p^3$; and $Z = 92$ is U: [Rn] $7s^2 6d^1 5f^3$ (exception).

7.73 Ionization energy decreases with increasing value of n and, for the same value of n, increases with increasing Z (increasing Coulombic attraction). Na is the only species in this set with an $n = 3$ valence electron, so it has the smallest I.E.. Na^+ has the largest Z, so it has the largest IE. O has its last two electrons paired, reducing its IE below that of N despite its higher Z because of increased electron-electron repulsion.

Na < O < N < Ne < Na^+.

7.75 Size increases with increasing n value and, for the same value of n, decreases with increasing Z (increasing Coulombic attraction). Cl, Cl^-, and K^+ all have $n = 3$ for the valence electrons, while Br^- has $n = 4$, so Br^- is the largest of this set. Among the others, the extra electron in Cl^- makes it larger than Cl, and K^+ has higher Z than Cl.

Br^- > Cl^- > Cl > K^+.

7.77 First identify the orbitals from their shapes and relative sizes, given that A has $n = 2$. A, being spherical, is 2s; the cloverleaf shape of B identifies it as 3d; C has the hourglass shape of a p orbital and is the same size as A, so it is 2p.

(a) Any orbital can hold up to two electrons;
(b) $n = 2$, $\ell = 1$, $m_\ell = +1$, $m_s = +1/2$; $n = 2$, $\ell = 1$, $m_\ell = +1$, $m_s = -1/2$;
$n = 2$, $\ell = 1$, $m_\ell = 0$, $m_s = +1/2$; (there are three additional possibilities);
(c) A is more stable than C, and B ($n = 3$) is least stable;
(d) There are four other 3d orbitals, all of equal stability as orbital B;
(e) Beryllium has a filled 2s orbital but empty 2p orbitals;
(f) Boron and carbon have partially filled 2p orbitals and at least one empty 2p orbital;
(g) The fluoride and oxide anions have all three 2p orbitals filled.

7.79 This problem provides an exercise in graph reading.
  (a) The greatest drop within a row is from element 80 (Hg) to 81 (Tl);
  (b) Cs, element 55, has the lowest value shown in the figure;
  (c) The value changes least from $Z = 56$ to 71, $Z = 39$ to 46, and $Z = 20$ to 29;
  (d) Elements with values between 925 and 1050 kJ/mol are 16 (S), 33 (As) and 34 (Se).

7.81 Remember that in the transition metal cations, the ns orbital is less stable than $(n-1)d$:
  Cu^+ is $4s^0\, 3d^{10}$ Mn^{2+} is $4s^0\, 3d^5$ Au^{3+} is $6s^0\, 5d^8$

7.83 Francium is an alkali metal (column 1) whose properties should closely resemble those of cesium: low IE, highly reactive, forms +1 cation, soft metal, melts at a low temperature.

7.85 The filling pattern after $7p$ should mirror the pattern after $6p$: $8s$, then $5g$ (or perhaps $6f$).

7.87 Atoms with odd numbers of electrons must have magnetism, because there will always be at least one electron whose magnetism does not cancel. Atoms with even numbers of electrons may or may not have magnetism, depending on their configurations:
Be (4 electrons, $1s^2\, 2s^2$) and Kr (36 electrons, filled shells) have no net magnetism;
K (19 electrons) and I (53 electrons) must have net magnetism.

7.89 The permitted values for m_l are determined by the value of l.
  For a d orbital ($l = 2$), $m_l = -2, 1, 0, -1,$ and -2 are permitted;
  For an f electron ($l = 3$), $m_l = 3, 2, 1, 0, -1, -2,$ and -3 are permitted.

7.91 In an excited state configuration, one electron is shifted to a less stable orbital. Start by constructing the ground state configuration, then move an electron from the most stable occupied to the least stable unoccupied orbital:

  Be: [He] $2s^2$; the next orbital is $2p$: [He] $2s^1\, 2p^1$;
  O^{2-}: [He] $2s^2\, 2p^6$; the next orbital is $3s$: [He] $2s^2\, 2p^5\, 3s^1$;
  Br^-: [Ar] $4s^2\, 3d^{10}\, 4p^6$; the next orbital is $5s$: [Ar] $4s^2\, 3d^{10}\, 4p^5\, 3s^1$;
  Ca^{2+}: [Ar]; the next orbital is $4s$: [Ne] $3s^2\, 3p^5\, 4s^1$;
  Sb^{3+}: [Kr] $4d^{10}\, 5s^2$; the next orbital is $5p$: [Kr] $4d^{10}\, 5s^1\, 5p^1$.

7.93 Examine the configurations of the atoms and ions to determine the reason for the ionization energy differences:

  Li: $1s^2\, 2s^1$ Be: $1s^2\, 2s^2$;
  Li^+: $1s^2$ Be^+: $1s^2\, 2s^1$;

The first ionization energies involve the $2s$ orbital for both atoms, so Be, with larger Z, has a larger IE. The second electron removed from Li is a core electron, so the IE is much greater than the second IE for Be.

7.95 Calculate the energy of the photon that causes the transition. By conservation of energy, the ionization energy of the excited state is lower than that of the ground state by this amount:

$$E_{photons} = \frac{(6.626 \times 10^{-34}\text{ Js})(2.998 \times 10^8 \text{ m/s})}{(589 \text{ nm})(10^{-9} \text{ m/nm})}(6.022 \times 10^{23} \text{ mol}^{-1}) = 203.1 \text{ kJ/mol};$$

$IE_{\text{excited state}} = 496.5$ kJ/mol $- 203.1$ kJ/mol $= 293.4$ kJ/mol.

7.97 Use the relationships between quantum numbers to answer the first parts of this question.

(a) The number of m_l values is $(2l + 1) = 9$;
(b) The possible values range from $+l$ to $-l$: 4, 3, 2, 1, 0, –1, –2, –3, and –4;
(c) n must be greater than l, so $n = 5$;
(d) In analogy with the near-degeneracy of 5f, 6d, and 7s orbitals, we expect near-degeneracy of 5g, 6f, 7d, and 8s orbitals;
(e) Because of the likelihood of near-degeneracy, it is difficult to know when the 5g will begin to fill, but it is reasonable to assume that 5g will begin filling right after 8s, just as 5f begins to fill right after 7s. The 6d orbitals are full at Z = 112, 7p orbitals are full at Z = 118, and 8s is full at Z = 120. Thus Z = 121 is the first element that is likely to have a 5g electron.

7.99 The value of l can never be as great as the value of n.
Thus, 4g ($n = 4, l = 4$) and 2d ($n = 2, l = 2$) are non-existent orbitals.

7.101 The 1s orbital is spherical and significantly smaller than the $2p_z$ orbital:

7.103

Quantum number	Associated Property	Restrictions
n	electron energy	any positive integer
l	orbital shape	positive integer less than n
m_l	orbital orientation	any integer from $+l$ to $-l$
m_s	spin orientation	either $+1/2$ or $-1/2$

Chapter 7: Atomic Structure and Periodicity

7.105 The periodic table would be expanded by 50% in Morspin compared to our universe because of the extra possible spin value. All other laws would remain the same, so orbital shapes and periodic trends would be the same as in our universe.

(a) Elements 18 and 30 would be in the same column of the periodic table. Element 18, with a smaller valence n value, would have the larger ionization energy.

(b) Elements 15 and 17^{2+} would have the same number of electrons, so the one with smaller Z, element 15, would have the larger radius.

(c) Elements 47 and 48 both would fall in the p block. Element 47 would have p^2 configuration while 48 would have p^3 configuration. An electron added to 48 would be added to an already-occupied orbital, which would generate significant electron-electron repulsion, so 48 would have a smaller electron affinity than 47.

7.107 The periodic table would be expanded by 50% in Morspin compared to our universe because of the extra possible spin value. All other laws would remain the same, so orbital shapes and periodic trends would be the same as in our universe.

(a) The noble gases are at the ends of the rows: the first three would be 3, 15, and 27.

(b) Element 4 would have one valence 2s electron: $n = 2$, $l = 0$, $m_l = 0$, $m_s = +1/2$;

Element 7 would have three valence 2s electrons and one valence 2p electron:

n	l	m_l	m_s
2	0	0	+1/2
2	0	0	−1/2
2	0	0	0
2	1	1	+1/2

Element 32 would have three valence 4s and two valence 3d electrons:

n	l	m_l	m_s
4	0	0	+1/2
4	0	0	−1/2
4	0	0	0
3	2	2	+1/2
3	2	1	+1/2

Chapter 8 Fundamentals of Chemical Bonding

8.1 Overview of Bonding

QUESTIONS TO ANSWER, SKILLS TO LEARN
1. **What keeps atoms together as molecules?**
2. **Working with bond lengths and bond energies**
3. **What factors mainly affect the overlap of orbitals?**

Let's contrast NaF and H_2 to examine the differences in bonding of ionic compounds versus *covalently* bonded compounds.

NaF:

Sodium fluoride is composed of a positively charged sodium ion and a negatively charged fluoride ion, and the electrons of each ion are tightly bound. Sodium fluoride in water exists as sodium ions and fluoride ions, indicating the stability of these ions.

H_2: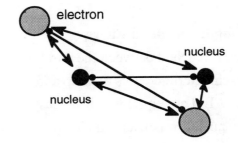

Hydrogen is not composed of ions; it dissolves only to a small extent in water, and then as H_2 molecules, not as ions. How are the two hydrogen atoms attracted to each other?

Electrical forces bind the two H atoms together, but differently than in ionic substances. Each hydrogen atom is composed of a nucleus and one valence electron. The sketch shows the forces that are present in a two nucleus/two electron molecule. There are four attractive forces acting between each nucleus and the two electrons; these are shown by the double-headed arrows. In addition, there are two repulsive forces shown by the lines: one acting between the two nuclei and the other acting between the two electrons. These forces are shown by the lines ending with heavy dots (•). The hydrogen molecule is stable because the sum of the attractive forces overcomes the sum of the repulsive forces, resulting in stabilization. Both nuclei interact with both electrons, so the electrons are "shared" between the nuclei. This "sharing" of electrons is the identifying feature of the **covalent bond**. This term describes atoms bound to one another by attraction to shared electrons.

Recall from Chapters 6 and 7 that electrons are best described by wave functions. We can obtain a qualitatively accurate picture of the wave functions describing electrons in molecules by considering the atomic orbitals of the valence electrons. In hydrogen atoms, the electrons are in $1s$ orbitals. One wave function describing electrons in the hydrogen molecule can be obtained by adding the wave functions. Important features of the new wave function obtained from adding the atomic orbitals are:

1. The new wave function is large between the two nuclei, placing electron density in the region where it interacts favorably (attractive interactions) with both nuclei.

2. The electron density is delocalized over both nuclei (the electrons are "shared").

3. The amount of overlap depends on the types of orbitals involved in the interaction.

4. The amount of overlap depends on the distance between the atoms.

The new wave function describes the electron in the molecule and replaces the wave functions used to describe the electron in the atom.

Measurable Quantities of Chemical Bonds: Bond Length and Bond Energy

In the formation of any chemical bond, a balance is reached between the attractive forces and the repulsive forces. Moving the nuclei closer to each other increases overlap and increases the attractive forces between the nuclei and the electrons. But at very close distances, decreasing the distance between the nuclei increases the nucleus-nucleus and electron-electron repulsions more than the attraction between the nuclei and the electrons increases. This change in the energy of the two atoms (or molecule) is shown in the figure on the next page. The energy is plotted as the vertical axis on the graph and is the change in energy of the molecule compared to the energy of the separated atoms. At the top of the plot are figures schematically showing the amount of overlap between *s* orbitals interacting in bond formation. The lighter shading indicates the valence orbitals, and the darker shading shows the core of the atom.

At infinite separation, the atoms do not interact and the energy of interaction is defined as zero. When the two atoms are far apart, there is very little overlap of wave functions. The energy of the interaction of the two atoms is very small but is negative (stabilizing). As the distance between the atoms decreases, the stabilization increases and energy will be released (the energy becomes more negative). When the energy of the bonding interaction is lowest, the distance between the atoms is called the **bond length** or **bond distance**. The energy at the bond length reflects the stability of the bond and is equal to the energy required to separate the atoms in the bond. This energy is called the **bond energy, *BE***. At distances shorter than the bond length, the energy increases rapidly because of the repulsive forces acting between the nuclei.

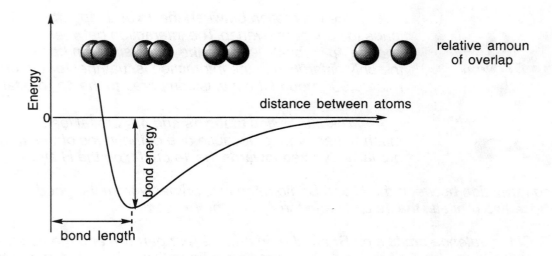

For hydrogen, H_2, the bond energy corresponds to the energy change for the following reaction:

$$H_{2\,(g)} \rightarrow 2\,H_{(g)} \qquad \Delta E = BE$$

Exercise 8.1 Draw pictures showing the interaction of:
(a) the valence orbitals of lithium atoms in the formation of Li_2
(b) the valence orbitals of H and Br in the formation of HBr
(c) the valence orbitals of Cl and Br in the formation of BrCl.

Steps to Solution: To solve this problem we need to decide which orbitals are the valence orbitals (as described in Chapter 7) and then determine how they would interact/overlap to form the new orbital. The orbitals with the best overlap will correspond to the "best" combination of orbitals. The final answer will be a drawing of the contour surface of the atomic orbitals showing the overlap.

(a) In lithium atoms, the electronic configuration is [He] $2s^1$. The valence orbitals are the highest energy partially filled orbitals, in this case the 2s orbitals. The 2s orbitals are spherical and have no directional dependence; therefore they would overlap as shown shown below, where the core orbitals and nucleus are represented by the solid black circles.

(b) In this molecule, the valence orbitals are the H 1s orbital and the Br 4p orbitals. The three possible interactions between these orbitals are shown below. We assume that the line that joins the H and Br nuclei is the z-axis. The first sketch shows the interaction between the 1s and $4p_x$ orbitals. The second drawing shows the interaction between the 1s and $4p_z$ orbitals.

Chapter 8: Fundamentals of Chemical Bonding

Bond formation between the H and Br atoms results primarily from the good overlap of these two orbitals, the 1s on H and the $4p_z$ on the Br atom.

(c) In BrCl the valence orbitals on Br are the 4p orbitals [per part (b)] and the valence orbitals for Cl are the 3p orbitals. The electron configurations show that one of the p orbitals is half filled for both Cl and Br. We assume that the line joining the two atoms is the z-axis. The different possible interactions between p_x orbitals are: p_z with p_x; p_z with p_y; p_z with p_z; p_x with p_x; p_x with p_y. Other combinations of orbitals will be similar to those already listed, except that one of the orbitals may be rotated around the z-axis by 90°. The combinations are illustrated below in order of increasing overlap.

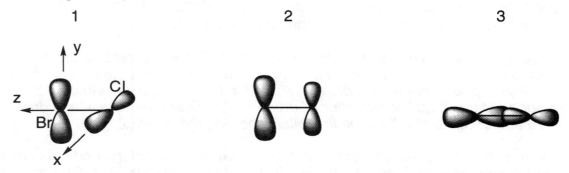

When perpendicular orbitals like the p_x and p_y orbitals are combined, the overlap is very poor because the orbitals do not point at the nucleus of the other atom in the molecule and also point in different directions. Note that in drawing 1, the p orbitals on Cl and Br have somewhat different sizes as we expect from the different values of *n* for the orbitals. When the "p" orbitals are parallel (for example, the $4p_y$ and $3p_y$), in drawing 2, the overlap is better than that shown above because the orbitals point along the same direction. Interaction of the $4p_z$ and $3p_z$ orbitals of Br and Cl leads to the best overlap, shown in 3. The orbitals point at each other so the maximum amount of overlap occurs between the lobes of the "p" orbitals. Bonding in the BrCl molecule result sprimarily from overlap of the $4p_z$ and $3p_z$ orbitals on the Br and Cl atoms, respectively

Exercise 8.2 Draw a diagram of energy vs. distance similar to that above for the interaction of the H and Br atoms in 8.1 (b). Include sketches of the relative amounts of overlap of the orbitals. The bond energy of HBr is 363 kJ/mole and the H-Br bond length is 140.8 pm.

Steps to Solution: Several features should be present in this diagram. The energy at very long distance is zero. As the distance between the two nuclei decreases, overlap between the orbitals increases and the energy decreases. The energy reaches a minimum at the bond length; the energy at the bond length is the negative of the bond energy. The energy increases very rapidly as the distance grows smaller than the bond length, because valence orbitals start to overlap significantly with the core orbitals and and the core electrons repel each other. For the sake of clarity, only the H 1s and Br $4p_z$ orbitals are shown.

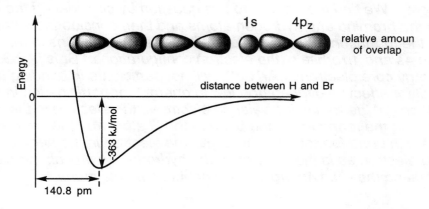

Describing Bonding

In the previous exercises, the molecules' bonds were formed by the overlap of two half-filled orbitals to produce an orbital containing two electrons with paired spins. However, in molecules with more atoms and more half-filled orbitals, the situation becomes more complicated. Orbitals (wave functions) for electrons in molecules can be described as *localized* (extending over two atoms at most) or *delocalized* (extending over more than two atoms). These two descriptions both have some common features.

- Each electron is described by a wave function, which describes the shape of the orbital.

- No more than two electrons can occupy a single orbital.

- The Aufbau principle is used to "fill" the orbitals: the most stable orbitals are filled first.

- The valence orbitals are the most important orbitals to consider in describing the bonding in a molecule.

In the following pages we will describe bonding with the localized model, where there are three types of electrons:

- **Core electrons** (which we will ignore) are described by purely atomic orbitals.

- **Bonding electrons** are valence electrons between the two nuclei that lead to bonding and occupy orbitals formed from the overlap of atomic orbitals on the two atoms.

- **Nonbonding electrons** are valence electrons that are not involved in bonding and can be described by purely atomic orbitals.

Exercise 8. 3 Exercise 8.2 provided a description of the bonding in HBr. Describe the orbitals in Br in the molecule HBr in terms of core, bonding and nonbonding electrons.

Steps to Solution: We find the answer to this question by considering the orbitals occupied in the bromine and hydrogen atoms and then considering what changes occurred in the formation of the molecule. The valence electrons in the bromine atom are the 4s and 4p orbitals (the electron configuration of Br is [Ar] $3d^{10}\ 4s^2\ 4p^5$). The low-energy core electrons ([Ar] $3d^{10}$) will not participate in bonding because they are too small for effective overlap (too low in energy) for participation in effective bond formation. Of the remaining valence electrons, the electron in the half-filled 4p orbital is used in the formation of the bond to the H atom and the other electrons do not participate in bond formation. Therefore, the electron in the half-filled 4p orbital is a bonding electron as is the electron in the hydrogen 1s orbital; the electrons in the 4s orbital and the filled 4p orbitals are nonbonding electrons.

8.2 Unequal Electron Sharing: Electronegativity

QUESTIONS TO ANSWER, SKILLS TO LEARN
1. **Understanding the effects of different effective nuclear charges on covalent bonds**
2. **What is a polar bond?**
3. **Understanding how electronegativity affects the distribution of shared electrons**

Sharing of electron density leads to the binding of two nuclei in a covalent bond. In the localized picture that we developed in the above sections, the electrons are described by orbitals made from the valence orbitals of the atoms linked by the covalent bond. The electrons in the bond are now subject to the effects of both nuclei in the bond. The figure below shows a bond between atoms A and X and an electron, e, shown as the larger circle:

If A and X are the same atom, for example hydrogen, the forces acting on the electron will be symmetrical; the effective nuclear charge on X will be the same as that on A. When X is different from A, then the attractive force towards A will not be the same as the attractive force towards X; the forces are *asymmetric*. The atom with the larger effective nuclear charge in the bond will attract electrons more strongly; the higher stability is achieved when the electron is nearer the atom with the greater effective nuclear charge. The electron will not be equally shared between the nuclei. The electron density in this situation is shown in the figure above at right.

The figure below expands on this point. If the total charge (nucleus + electron density) is added up for each atom where each atom "owns" the electron density to half the bond length, the unequal sharing of electron density leads to the formation of a partial positive charge on A and a partial negative charge on X (partial means less than the charge on one electron). Sometimes the charge resulting from this unequal sharing of electron density is represented by writing partial charges on the atoms concerned; $X^{\delta-}$ — $A^{\delta+}$ ($0 < \delta < 1$).

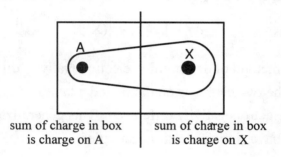

sum of charge in box is charge on A | sum of charge in box is charge on X

When there is unequal sharing of electron density in a bond, that bond is described as being **polarized** or as a **polar bond**. The covalent bond in A-X would be described as a *polar covalent bond*. The degree to which a bond is polarized (the size of the fractional charges on A and X) depends on the ability of the atoms to attract electron density to themselves. This ability is an important quantity called **electronegativity**, a quantity represented by the symbol, χ (Greek letter chi).

Electronegativity, χ: *the ability of an atom to attract electron density to itself in chemical bonds*

Electronegativity should not be confused with electron affinity. Electronegativity is a measure of the ability of an *atom in a molecule* to attract electron density from another atom in a bond. Electron affinity measures the ability of a *gas phase atom* (or ion) to bind a nonbound, free electron. The most widely used values for electronegativity were developed by Linus Pauling and are called the Pauling electronegativities. Figure 8-8 in the text tabulates electronegativities in the context of the periodic table. The most apparent trends are:

Electronegativity:
- *increases* as we proceed from *left to right across a row*.
- *decreases* as we proceed *down a column*.

Chapter 8: Fundamentals of Chemical Bonding

Electronegativities range from 0.8 for Cs to 4.0 for F, as we would expect from the trends described above. Metals tend to have low electronegativities (less than 2.0) and nonmetals have higher electronegativities (greater than 2.2). The metalloids have intermediate electronegativities, from 1.8 for Si to 2.0 for As.

The electronegativity difference, $\Delta\chi$, is a measure of a bond's polarity and may be calculated from electronegativity values.

For a bond containing the atoms A and X, $\Delta\chi$ is the absolute value of the difference of the electronegativities of the atoms in the bond:

$$\Delta\chi = |\chi(A) - \chi(X)|$$

The more common method of determining electronegativity difference is qualitative; one compares the positions of the two elements on the periodic table. In general, the further apart the two atoms, the greater the electronegativity difference. This generalization breaks down for the *f*-block elements which all have similar electronegativities.

The further apart two elements are in the periodic table, the greater the electronegativity difference.

Exercise 8.4 For each of the following bonds, calculate the electronegativity difference and predict which atom would be positively charged: (a) N-O (b) Al-C (c) Cl-Cs.

Steps to Solution: We solve this problem by consulting a table of the electronegativities of the elements.(text, Figure 8-8). After finding the electronegativities of both elements we find the difference and take the absolute value so that the answer is a positive number. We expect the positive charge to be on the atom of lower electronegativity.

(a) *The electronegativities of N and O are 3.0 and 3.5, respectively. Remember that electronegativities have no units.*

$$\Delta\chi = |3.0 - 3.5| = 0.5$$

Because N has the lower electronegativity, we expect it to have the positive charge.

(b) *The electronegativities of Al and C are 1.5 and 2.5 respectively.*

$$\Delta\chi = |1.5 - 2.5| = 1.0$$

Because Al has lower electronegativity, we expect it to have the positive charge.

(c) *The electronegativities of Cl and Cs are 3.0 and 0.7, respectively.*

Chapter 8: Fundamentals of Chemical Bonding

$$\Delta\chi = |3.0 - 0.7| = 2.3$$

Because Cs has the lower electronegativity, we expect it to have the positive charge.

The size of the electronegativity difference helps you place the bond's polarity in a range from completely ionic to nonpolar covalent bond.

$\Delta\chi = 0$	$0 < \Delta\chi < 1.7$	$\Delta\chi > 2.0$
nonpolar bond	polar covalent bond	ionic bond

For the cases where $1.6 < \Delta\chi < 2.0$ the classification of bond type depends on the identities of the elements involved. If one of the elements is a metal, the bond is considered ionic. If both elements are nonmetals, then the bond is considered polar but covalent.

Exercise 8.5 Classify the bonds in Exercise 8.4 as polar covalent, nonpolar covalent, or ionic.

Steps to Solution: To determine the bond type, we compare the electronegativity difference in the bond with the ranges shown above. We can then classify the bond type as ionic or polar covalent.

(a) In the N-O bond, $\Delta\chi = 0.5$. Because this value is greater than 0 but less than 1.7, the bond is polar covalent.
(b) For the Al-C bond, $\Delta\chi = 1.0$. Because this value is greater than 0 but less than 1.7, the bond is polar covalent.
(c) For the Cl-Cs bond, $\Delta\chi = 2.3$. This value is greater than 2.0 ; therefore the Cl-Cs bond is ionic.

Exercise 8.6 Without using electronegativity values, rank the following bonds from most polar to least polar:

Cl-O, Mg-O, O-O, C-O

Steps to Solution: To answer this question we determine how close the elements lie to each other in the periodic table. The closer the elements, the smaller the electronegativity difference and the lower the polarity of the bond. The common element in each bond is oxygen. The elements furthest from O in the periodic table will form the most polar bonds, and those closer will form less polar bonds. Mg lies furthest away from O; therefore the most polar bond is the Mg-O bond and the least polar (actually nonpolar) is the O-O bond. The remaining elements are Cl and C; Cl is diagonally adjoining O, and C is separated from O by one element. Cl is closer to O; therefore the less polar bond between Cl-O and C-O is the Cl-O bond. The order of bond polarity starting with the most polar is :

(most polar) Mg-O > C-O > Cl-O > O-O (least polar)

Chapter 8: Fundamentals of Chemical Bonding

8.3 Lewis Structures

QUESTIONS TO ANSWER, SKILLS TO LEARN
1. **Drawing Lewis structures**
2. **What are formal charges?**
3. **Using multiple bonds to reduce formal charge**
4. **What if there are several similar Lewis structures for the same species?**

Lewis structures are drawings of molecules that show the following features of molecules:
- the connectivity of the atoms: which atoms are connected to others in a molecule
- the number of bonds to each atom
- the number of nonbonding valence electrons on each atom.

The information provided by a Lewis structure is very useful in (1) formulating a description of the bonding in a molecule in terms of the orbitals used in bonding and (2) deducing the shape of a molecule. The following guidelines are used in the construction of Lewis structures:

1. Only the valence electrons appear.
2. A line joining two atoms represents a bond (two electrons shared between two atoms).
3. Dots placed next to, above, or below an atom represent nonbonding electrons.

Drawing Lewis Structures

Following the steps outlined below will make it easier to draw Lewis structures.

1. In an ionic compound, draw the Lewis structure for the cation and anion separately.

2. Count the valence electrons. In a cation, subtract the appropriate number of valence electrons to obtain the correct charge; in an anion, add the appropriate number of electrons.

3. Assemble the bonding framework: find the different ways in which atoms may be connected to each other. There are two types of atoms in molecules or polyatomic ions: those bonded to only one atom (**outer atoms**) and those bonded to more than one atom (**inner atoms**). The following guidelines help eliminate many unreasonable frameworks:

    (a) In this text, H and F atoms will always be outer atoms because they can form only one localized bond.

    (b) Outer atoms are usually those with higher electronegativities, because these atoms are less likely to "share" electron density.

    (c) The order of listing of atoms in chemical formulas usually reflects the connectivity.

    (d) Quite often, the number of bonds to an atom is given by the difference between 8 and the number of valence electrons (8 - VE).

Chapter 8: Fundamentals of Chemical Bonding

Although these guidelines limit the number of molecular skeletons we obtain, often several are obtained that meet the criteria in (a), (b) and (c). You must retain all of these skeletons in the next steps until other criteria (step 7) make some skeletons less acceptable.

4. Atoms except H tend to have a total of eight valence electrons around them in a molecule. *Outer atoms* have this requirement fulfilled first, with the nonbonding electrons remaining after allotment of electrons to bonds.

5. *Inner atoms* are allotted nonbonding electrons only after each outer atom has received three pairs of nonbonding electrons. The structures obtained at this point will be referred to as *provisional Lewis structures*.

6. We now calculate the formal charges (FC) on all the atoms in the molecule (or polyatomic ion) in preparation for step 7. The formal charge is defined as:

    FC = (#valence e⁻ s in neutral atom) - (#valence e⁻ s of atom in Lewis structure)

    We find the number of valence electrons of the atom in the Lewis structure by adding the number of bonds and the number of nonbonding (unshared) electrons. To check if you have calculated the formal charges correctly, remember that the sum of the formal charges in a molecule is zero and the sum of the formal charges in a polyatomic ion is equal to the charge on the ion.

7. Of the Lewis structures obtained using the preceding steps, the one with the smallest formal charges on the atoms usually represents the "best" distribution of electrons. Nonbonding pairs of electrons may be shifted to bonding positions to decrease the formal charges. This may lead to formation of double (4 electrons shared) or even triple (6 electrons shared) bonds in order to obtain the Lewis structure with the smallest formal charges.

The following exercises develop and expand on these steps and guidelines.

Exercise 8.7 Draw the Lewis structures for each of these two industrial chemicals: (a) CF_2Cl_2 (Freon) and (b) $CaCO_3$, calcium carbonate.

Steps to Solution: *The steps listed above will be applied to draw the desired Lewis structures.*

(a) CF_2Cl_2

Step 1 *If the substance is ionic, the ions must be treated separately. Freon is not ionic; it does not contain either a Group 1 or 2 metal, nor does it contain one of the common polyatomic ions. Therefore CF_2Cl_2 is a covalent molecule.*

Step 2 *The next step is to count the number of valence electrons. We first determine the number of valence electrons (VE) per element in the compound, multiply by the number of atoms of that element, and then sum them. The table below shows this explicitly.*

Chapter 8: Fundamentals of Chemical Bonding

Element	# of atoms	# of VEs/atom	Total VEs
C	1	4	4
Cl	2	7	14
F	2	7	14
	total valence electrons =		32

Step 3 *Assemble the framework. The more electronegative elements will be the outer atoms. The least electronegative element is C, so it will be an inner atom. One possibility is to have C as the center atom and the four other atoms bonded to it, as shown in (i) below. Another possibility is that the C and Cl atoms are inner atoms, as shown in (ii). In the remaining steps we will retain both of these structures to illustrate how one of them will be eliminated. We will not consider other structures such as F-Cl-Cl-C-F. Substances with symmetrical formulas usually have symmetrical structures.*

(i)

```
      F
      |
  Cl—C—F
      |
      Cl
```

(ii)

```
  F—Cl—C—Cl—F
```

Electron count: 32 - 8 = 24   32 - 8 = 24

In both structures four bonds have been formed that require 4(2) = 8 electrons. There are now 32 - 8 = 24 valence electrons left to add to complete the molecule.

Step 4 *Each outer atom is now given three pairs of electrons. In (i), this requires 4(6) = 24 electrons, using up all the valence electrons. In (ii), only 2(6) = 12 electrons are used in adding electrons to the outer atoms. There are 12 VEs remaining for (ii).*

(i)

```
       ..
      :F:
       |
  ..   |   ..
 :Cl—C—F:
  ..   |   ..
      :Cl:
       ..
```

(ii)

```
   ..           ..
  :F—Cl—C—Cl—F:
   ..           ..
```

Electron count: 24 - 24 = 0   24 - 12 = 12

Step 5 *Assign the remaining electrons to inner atoms. The target number of nonbonding electron pairs and bonding pairs is 4 (8 electrons around each inner atom). For (i), this step is not necessary. In (ii), each Cl atom has 2 bonding pairs, and 2 nonbonding pairs are required for each Cl atom, a total of 2 Cl (4e/Cl) = 8. A total of 12 - 8 = 4 electrons remain. The C atom has only 2 bonds; adding the remaining 4 electrons to C gives the target 4 pairs (8 electrons). These are the provisional Lewis structures.*

Chapter 8: Fundamentals of Chemical Bonding

Electron count: 0   12 - 12 = 0

Step 6 Calculate the formal charges on all the atoms. The formal charge is the difference between the number of valence electrons in an atom in the molecule and the number of valence electrons in the free atom.

In step 2, we used the number of valence electrons in the free atoms. We find the number of electrons around the atom in the molecule by adding the number of unshared (nonbonding) electrons and the number of bonds to that atom (each atom effectively "owns" one of the two electrons in the shared pair forming the bond between the two).

(i)

FC = 7-(6+1) = 0
FC = 7-(6+1) = 0
← number of bonds
FC = 4-(0+4) = 0
← number of unshared electror

(ii)

FC = 4-(4+2) = -2
FC = 7-(6+1) = 0
FC = 7-(4+2) = 1

Step 7 *Structure (i) has formal charges of 0 on each atom; structure (ii) has nonzero formal charges on three of the atoms (C and the two Cl atoms). These results indicate that (i) is a better Lewis structure than (ii). There is no need to shift electrons from nonbonding positions to form double bonds to reduce formal charges; therefore structure (i) is the Lewis structure for Freon.*

(b) $CaCO_3$

1. This compound is ionic because it contains a Group 2 metal and a polyatomic anion, CO_3^{2-}. Therefore we must deal separately with the Ca^{2+} ion and the CO_3^{2-} ions.

2. We count the valence electrons on each ion beginning with the Ca^{2+} ion; the electron configuration of Ca is [Ar] $4s^2$. The calcium ion, Ca^{2+}, has lost two electrons and thus has the [Ar] electronic configuration; it has no valence electrons. The electron count for the carbonate anion is shown below.

Element	# of atoms	# of VE s/atom	total VE s
C	1	4	4
O	3	6	18
Total for CO_3			22
To complete anionic charge			2
Total for CO_3^{2-}			24

279

3. Assemble the framework of the carbonate ion. The least electronegative element is C, so it will certainly be an inner atom. The most likely structure is the one where the three O atoms are bonded to a central C atom; two other possibilities are shown.

$$\text{O} \quad \text{O} \quad \text{O}$$
$$\text{O-C-O} \quad \text{O-C-O-O} \quad \text{C-O-O-O}$$

The two structures to the right incorporate O as an inner atom, which is not favorable, leaving the left-most structure. There are three bonds between C and O in the left-most structure requiring 3(2) = 6 electrons. There are now 24 - 6 = 18 valence electrons left to add to the atoms of this ion.

4. We give each outer atom three pairs of electrons.

This requires 3(6) = 18 electrons, consuming the rest of the valence electrons present in the ion.

5. There are no electrons remaining to assign to inner atoms.

6. The formal charges on the C and O atoms are calculated below (note that because the O atoms have an identical number of bonds and nonbonding electrons, the formal charge is calculated for only one of them). Note that C has no nonbonding electrons and three bonds in the ion, leading to a formal charge of +1.

FC = 4-(0+3) = 1

FC = 6-(6+1) = -1

7. Because all the O atoms and the C atom show formal charges, converting a nonbonding or lone pair to a bonding pair will lower the formal charge. Remember that the sum of the formal charges on the atoms must equal the charge on the ion, so we cannot eliminate all formal charges in an ion. The only nonbonding pairs are on the O atoms; the shift can be depicted as shown below.

FC = 6-(4+2) = 0

FC = 4-(0+4) = 0

FC = 6-(6+1) = -1

*The structure on the right is a good Lewis structure for the carbonate ion because the number of atoms with formal charge has been reduced. However, we can draw two other Lewis structures for the carbonate ion that differ only in the placement of the double bond; these are **resonance** structures. The structure above and the other two resonance structures are shown below.*

The double bond does not switch positions; experimentally all three C-O bonds are the same (the C-O bond lengths and all the O-C-O angles), suggesting that a better picture would spread the electrons of one of the bonds in the double bond around the three C-O bonds so that they are the same. This is shown in the next picture where the dashed lines indicate that the C-O bonds are not double bonds with four electrons being shared.

*The reason three different Lewis structures are needed to represent the carbonate ion is that Lewis structures localize two electrons per bond shown by the line; in Chapter 9 we will show that bonds need not be localized between only two atoms. Therefore the need for resonance structures **indicates a limitation of Lewis structures, not a "flip-flopping" of bonds in the molecule or ion.***

8.4 Tetrahedral Molecules: Carbon

QUESTIONS TO ANSWER, SKILLS TO LEARN
1. **What Lewis structures don't show or Most molecules aren't flat!**
2. **Drawing 3D structures**
3. **What are hybrid orbitals?**
4. **What are alkanes?**

Lewis structures are a picture of a molecule or polyatomic ion on a flat surface which show the distribution of valence electrons. The bonds in molecules are the result of electron pairs occupying space between nuclei; however, the forces acting between the electrons are repulsive. For maximum stability, we expect the electron pairs in a molecule to be as far apart as possible *in three dimensions*. Only a few molecules have all of their atoms lying in a common plane. In this section, we will describe the common ways of showing the 3D aspects of molecules in drawings and also describe hybrid orbitals, which point in the right directions to form the strongest bonds to the outer atoms.

The ammonium ion will be used for illustration. In the table below are four representations of the ammonium ion.

(a) Chemical Formula	(b) Lewis Structure	(c) "3D" Tetrahedral Structure	(d) Shorthand 3D Structure

The ammonium ion is known to adopt a tetrahedral structure where all four N-H bond lengths are the same and all four H-N-H bond angles are 109.48°. The tetrahedral structure of the ammonium ion is shown in (c) where the tetrahedron is drawn by placing the N atom in the center of a cube and the H atoms at alternating corners of the cube. *An important feature of this structural type is that the four bonded atoms (and the bonding pairs of electrons) are as far away from each other as possible.* The structure shown in (d) is a more concise way of showing the 3D character of the ammonium ion. Imagine that the 3D structure in (c) has been rotated so that the N atom and the two upper H atoms are in the plane of the paper. One of the lower H atoms will be in front of the plane of the paper and the other will lie behind it. In (d), the normal lines indicate bonds that lie in the plane of the paper. The thick, wedge shaped line indicates a bond that extends out of the plane towards the reader and the dashed line indicates a bond that extends behind the plane of the paper away from the reader.

Bonding in a Tetrahedral Structure: Hybrid Orbitals

The atomic orbitals we have previously used in the discussion of bonding (ns and np) do not have the right geometry to give a tetrahedral geometry (bond angles of 109.5°). The np orbitals are oriented at angles of 90° with respect to each other and the spherical ns orbitals have no preferred direction. We can solve this problem by recognizing that atomic orbitals are mathematical wave functions that can be combined following certain rules to give new wave functions that have different shapes and orientations but are still legitimate wave functions. The three p orbitals and the s orbital can mix to give **hybrid orbitals** which point towards the outer atoms in a tetrahedral structure.

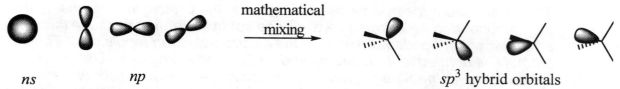

ns   np sp^3 hybrid orbitals

These four orbitals are called *sp³* (pronounced s-p-three) **hybrid orbitals**, because they are generated from the *s* and three *p* orbitals. The hybrid orbitals point at the bonds formed in tetrahedral molecules. Each hybrid orbital can overlap with an orbital from an outer atom to form the bond between the inner and outer atom or hold a nonbonding pair of electrons.

The hybrid orbital on the inner atom can overlap with any orbital that is properly oriented. Because outer atoms form a bond to only one atom, hybridization is not required to describe the orbitals they use in forming bonds. The overlap of an sp³ orbital with *s* and *p* orbitals is illustrated below in (a) and (b) in the formation of the A-X bond.

Hybrid orbitals may also overlap with other hybrid orbitals as shown in (c). This situation is necessary in the formation of molecules where there are several inner atoms. The following exercise illustrates the use of hybrid orbitals in describing the bonding in several molecules.

Outer atoms do not require hybrid orbitals to form bonds to the inner atoms.

Exercise 8.8 In terms of the localized model, describe the bonding in:
(a) methylamine, CH_3NH_2
(b) 1,2-dichloroethane, $ClCH_2CH_2Cl$.

Steps to Solution: To describe the bonding in terms of the localized bonding model, we must first determine the Lewis structure. Then we will choose the hybridization of the inner atom(s) and the appropriate orbitals on the outer atoms to give the best overlap.

(a) *The Lewis structure for methylamine is shown below (remember that formulas are often written with bonded groups of atoms listed together).*

The carbon atom methylamine is analogous to the C atom in methane: there are four atoms bonded to it. Therefore we will use sp³ hybridization because this arrangement of bonds will minimize the electron-electron repulsions. The other inner atom, N, is bonded to three atoms and has a nonbonding pair. Using sp³ hybridization at the nitrogen atom will minimize the electron-electron repulsions between the bonding pairs themselves and between the bonding pairs and the unshared pair. The outer atoms are all H atoms and have the 1s orbital available to overlap with the hybrid orbitals. A picture of the orbitals and their overlap is shown below along with a "shorthand" 3D structure. Only the large lobes of the sp³ hybrid orbitals are shown.

The bonding between the atoms can be described as resulting from the overlap of the four sp³ orbitals on the C atom with 1s orbitals on three H atoms and with one sp³ orbital from the N atom. Three of the four sp³ orbitals on the N atom overlap with 1s orbitals on the two H atoms and with one sp³ orbital on the C atom. The nonbonding pair of electrons occupies the remaining sp³ orbital on the N atom.

(b) The Lewis structure for 1,2-dichloroethane is shown below.

The inner atoms are both C atoms with four bonding pairs forming four bonds to four atoms. Therefore sp³ hybrid orbitals are necessary on the C atoms because this hybridization minimizes electron-electron repulsion between the bonding pairs. The H atoms have only 1s orbitals available for overlap with the C atom's sp³ hybrids; therefore the C-H bonds will be formed from overlap between the C atoms' sp³ hybrids with the 1s orbitals on the H atoms. The C-C bond will be formed from the overlap of sp³ hybrid orbitals on the two C atoms. The C-Cl bonds will be formed from the overlap of the C atoms' sp³ orbital with an orbital on the Cl atom. The valence orbitals on Cl are 3s and 3p orbitals. The 3p orbitals will be used because (1) they are larger than the 3s orbitals and (2) they overlap better with the sp³ orbital on the C atom due to the preferred direction of the p orbitals. For clarity, the illustration below does not show the lone pairs that remain on the Cl atoms; the nonbonding electrons are in unhybridized orbitals on the Cl atoms.

Chapter 8: Fundamentals of Chemical Bonding

Isomerism: Same Chemical Formula; Different Arrangements of Atoms

There are many compounds that have the same chemical formulas but have distinctly different chemical properties. The different properties are a result of different structures of the the compounds even though they have the same formula. Compounds with the same formula but different structures are called **isomers**. For example, there are two possible different ways of arranging the atoms in $C_2H_4Cl_2$: one where both chlorine atoms are bonded to one C atom and the other where one Cl atom is bonded to each C atom.

These two molecules are isomers of the formula $C_2H_4Cl_2$. The two molecules are **structural isomers**: they have the same chemical formula, but different structures or arrangements of atoms.

Structural isomers **have the same chemical formula but different structures.**

However, we must be careful not to confuse the structure or connectivity of a molecule with the way it is drawn. The two illustrations below look different but represent the same molecule, which in the figure at right has been rotated so that it "faces" the other way. Structural isomerism is common in the class of compounds that contain just C and H with the general formula C_nH_{2n+2}, the *alkanes*, where the number of carbons, n, is greater than 3.

Exercise 8.9 Draw the structural isomers for pentane, C_5H_{12}.

Steps to Solution: Structural isomers will differ in the arrangements of atoms and <u>cannot</u> be interconverted into one another by rotation of the molecule or of groups around bonds. Because there is only one type of outer atom (H), the variations in the structures must arise from different arrangements of the inner atoms: the five C atoms. One way to determine the different arrangements of the C atom framework is to start with a straight chain and then increase the number of "branches" from the chain while decreasing the length of the chain.

There are three different carbon skeletons and therefore three structural isomers. Hydrogen atoms are added to the skeletons so that each C atom has four bonds.

8.5 Other Tetrahedral Molecules

QUESTIONS TO ANSWER, SKILLS TO LEARN
1. **What is the steric number of an atom?**
2. **Determining the shapes of molecules that have nonbonding pairs and bonds**
3. **Naming the shapes of molecules**

A tetrahedral arrangement of four electron pairs around inner atoms is quite common and is found most often in compounds of C, N, O and Si. The sp^3 hybridization is used in all of these molecules to accommodate the electron pairs and minimize electron-electron repulsions. Compare water, ammonia, and methane. A common factor about the bonding in these three molecules is that the number of atoms bonded to the inner atom (*coordination number*) plus the number of lone pairs is 4. This sum is called the **steric number**. The four electron pairs, bonding and nonbonding, in these molecules are accommodated in the four sp^3 orbitals. The steric number helps us choose the appropriate hybridization of the inner atom from its Lewis structure.

The steric number is the sum of the coordination number and the number of nonbonding electron pairs.

Once we have decided the orbital geometry, we can name the shape of the molecule. We deduce the geometries of the three compounds by remembering that the sp^3 orbitals point towards the vertices of a tetrahedron with an angle of 109.5° between the center lines of the orbitals. When naming the shape of a molecule or deciding the geometry around a particular inner atom, we consider only the positions of the atoms bonded to the particular inner atom under scrutiny. Although nonbonded electrons influence the chemical reactivity and the positions of the other atoms, they are *not* included in the naming of structures.

Chapter 8: Fundamentals of Chemical Bonding

For steric number 4, sp³ hybridization (tetrahedral orbital geometry) describes the bonding.

Lewis Structure of Molecule	Coordination Number	Number Nonbonding Electron Pairs	Steric Number	Structure (lone pairs shown as shaded lobes)	Name
H–O–H (with 2 lone pairs on O)	2	2	4	(bent structure with lone pairs shown)	bent
H–N–H with H (one lone pair)	3	1	4	(pyramidal structure with lone pair)	trigonal pyramidal
H–C–H with H, H	4	0	4	(see Section 8.4)	tetrahedral

> **COMMON PITFALL: Confusing Orbital Geometry With Molecular Geometry!**
> Name the shape that the atoms around the inner atom form–
> ignore the nonbonding pairs when naming shapes!

Exercise 8.10 Describe the geometry of the inner atoms of (a) methyl amine and (b) 1,2-dichloroethane. See Exercise 8.8 for the Lewis structures.

Steps to Solution: Determining the structure of a molecule or the geometry about an inner atom requires several steps. First, we draw the Lewis structure for the molecule. Second, we determine the steric number of the inner atom. Then we use the orientation of the hybrid orbitals to determine the positions of the neighboring atoms relative to the inner atom.

(a) The Lewis structure for methylamine is shown in Exercise 8.8. There are two inner atoms, N and C. Both atoms have a steric number of 4 and therefore will use sp³ hybrid orbitals in forming bonds or holding lone pairs. Because the C atom has no lone pairs, the geometry around the C atom will be tetrahedral with the 3 H atoms and the N atom bonded to it. The N atom has one lone pair and three bonds. The geometry around the N atom will be trigonal pyramidal, where the N atom is the peak of the pyramid and the two H atoms and the C atom form the base of the pyramid.

(b) Use the Lewis structure of 1,2-dichloroethane in Exercise 8.8 to show that the geometry around both C atoms is expected to be tetrahedral.

8.6 Other Molecular Shapes

QUESTIONS TO ANSWER, SKILLS TO LEARN
1. **What hybrid orbitals are used for inner atoms with steric numbers other than 4?**
2. **Determining geometries of inner atoms with steric numbers other than 4**

There are many molecules where the inner atoms have steric numbers different from 4. The more common steric numbers, the hybrid orbitals used to describe the bonding in terms of localized bonding, and geometries for different combinations of bonds and lone pairs are described below.

Steric Number 3

Aluminum trichloride, $AlCl_3$, is found in the gas phase and has a steric number of 3. The Lewis structure is shown below on the left. Note that there are three bonds to Al and no lone pairs.

The structure that maximizes the distance between the pairs of electrons is a trigonal planar structure, where all four atoms lie in a plane and the Cl-Al-Cl bond angles are 120°. The orbitals that have the proper orientation for this geometry are the three *sp²* hybrid orbitals (pronounced "s-p-two"), formed from two *p* orbitals and an *s* orbital. The bonding between the Al and Cl atoms is shown in the right-hand figure where the Cl atoms use the $3p$ orbitals to overlap with the three *sp²* orbitals on the Al atom.

The molecule $SnCl_2$ has a steric number of 3 (two bonds and a lone pair). The two bonds might be be colinear, leading to a linear molecule (both Sn-Cl bonds in line) or the two bonds can meet at an angle, leading to a bent molecule. Because three orbitals are required to describe the electrons in the two bonds and the lone pair, we use *sp²* hybrid orbitals which meet at an angle. The lone pair occupies space and repels the bonding pairs. The $SnCl_2$ molecule is described as having a **bent** geometry.

Chapter 8: Fundamentals of Chemical Bonding

Steric Number 2

The Lewis structure of beryllium chloride, $BeCl_2$, is shown below to left. The steric number of $BeCl_2$ is 2. The structure that maximizes the distance between the electron pairs for steric number 2 places the electron pairs on opposite sides of the inner atom, leading to a bond angle, defined by the atoms, Cl-Be-Cl, of 180°. The orbitals that are used to form bonds to the outer atoms are hybrid orbitals, generated from mixing s and p orbitals, called **sp hybrid orbitals** (pronounced "s-p"), shown to the right of the Lewis structure. In this case the sp hybrid orbitals overlap with $3p$ orbitals on the Cl atoms.

Steric Number 5: Trigonal bipyramidal Orbital Geometry

Phosphorus pentafluoride, PF_5, has the Lewis structure shown below at left. The steric number of 5 results from the five P-F bonds that are present in the molecule.

The hybrid orbitals that most effectively reduce electron-electron repulsions are the five **dsp^3 hybrid orbitals** shown in the center figure: they are obtained by mixing s, d and p orbitals. For clarity, we do not show the p orbitals used by the F atoms to form the bonds to the P atom.

The five positions occupied by the fluorine atoms in PF_5 are not the same. In the right-hand sketch, two types of positions are labeled: **eq** (short for equatorial) and **ax** (short for axial: the two bonds form a single line, like an axle). These two types of position differ in the bond angles to their nearest neighbors. The axial positions have only the equatorial sites as nearest neighbors, and the F_{ax}-P-F_{eq} bond angles are all 90°. The equatorial sites have both axial and equatorial sites as neighbors; the F_{eq}-P-F_{eq} bond angles are 120°. The equatorial sites have more room for larger groups bound to the inner atom.

Lone pairs are larger than bonding pairs;
in steric number 5, lone pairs always occupy larger equatorial sites.

Steric Number 5

Lone Pairs	1	2	3
Example	SeF_4	IF_3	ICl_2^-
Sketch	F_{ax}—Se—F_{ax}, F_{eq}, F_{eq}	F_{ax}, F_{eq}—I, F_{ax}	Cl—I—Cl
Structure Name (Angles Between Atoms)	Seesaw (F_{ax} - Se - F_{eq} < 90°) (F_{eq} - Se - F_{eq} < 120°)	T-shape (F_{ax} - I - F_{eq} < 90°)	Linear (Cl - I - Cl = 180°)

Steric Number 6: Octahedral Orbital Geometry

The phosphorus hexafluoride anion, PF_6^-, is an example of steric number 6 and has the Lewis structure shown below at left. The structure that minimizes electron-electron repulsions for steric number 6 is an *octahedral* arrangement. Such an arrangement places the electron groups along the plus and minus x, y and z axes of the Cartesian coordinate system. The hybrid orbitals that point in those directions are the d^2sp^3 **hybrid orbitals**; they are formed from a mixture of *d*, *s*, and *p* orbitals. The right-hand figure shows the orientation of these orbitals in hexafluorophosphate; the 2*p* orbitals of the fluorine atoms (not shown) would overlap with the d^2sp^3 hybrids in bond formation. In octahedral geometry, the bond angles between neighboring atoms are all 90°.

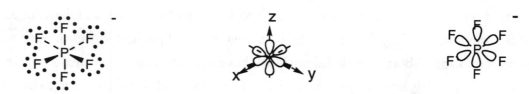

In compounds with inner atoms of steric number 6 where lone pairs replace bonds, the most common examples are where there are one or two lone pairs. These are illustrated in the following table.

Steric Number 6

Lone pairs	1	2
Example	TeF_5^-	BrF_4^-
Sketch	(F,F,F,F,F around Te with lone pair)$^-$	(F,F,F,F around Br with two lone pairs)$^-$
Structure name (bond angles)	square pyramid F - Xe - F < 90°	square planar F - Br - F = 90°

Exercise 8.11 Expand the line structure of thiophane (C_4H_8S, shown below) and describe the bonding at the inner atoms in terms of the hybrid orbitals.

Steps to Solution: First, we must expand the line structure so that the total number of atoms bonded to the inner atoms is evident. This is done below.

line structure → structural formula

There are three types of inner atoms: the S atom; the C atoms bonded to S, C and H; and the C atoms bonded only to other C atoms and H atoms. The S atom has a steric number of 4 (two lone pairs and two bonds); therefore it forms bonds using two sp^3 hybrid orbitals, and the two lone pairs reside in the other two hybrid orbitals. The C atoms nearest the S atom have steric number 4 (bonds to S, one C atom and two H atoms), and therefore use sp^3 hybrid orbitals to form bonds to the S, H and C atoms. The other two C atoms also have steric number 4 (bonds to two H atoms and two C atoms) and also use sp^3 hybrid orbitals.

The S and C atoms bond through overlap of the sp^3 orbitals on the S atom and the sp^3 orbitals on the C atoms. The C atoms form bonds to the H atoms by overlap of sp^3 hybrid orbitals with the 1s orbitals on the H atoms. The C-C bonds are formed by overlap of sp^3 hybrid orbitals of the two C atoms.

8.7 Confirmation of Molecular Shape

QUESTIONS TO ANSWER, SKILLS TO LEARN
1. What features are used to compare molecular structures?
2. Lone pairs effects on changing bond angles from the ideal values
3. What is a polar molecule?
4. Predicting the presence of a permanent dipole moment

Bond Angles

In Section 8.1 we described bond length as the distance between the nuclei of two atoms joined by a chemical bond. In discussing the shape and geometry of molecules, the notion of bond angle is useful. The bond angle is the angle defined by the two lines connecting three bonded nuclei. Earlier we predicted approximate bond angles by considering the geometry of the orbitals containing the bonding and nonbonding electrons. This provides a first approximation for the determination of bond angles. The angles between the hybrid orbitals for the various steric numbers are listed below.

Steric number	2	3	4	5	6
Bond angle(s)	180°	120°	109.48°	two types: eq - eq 120° ax - eq 90°	adjacent bonds 90°

However, equating the bond angles with the angles between the hybrid orbitals assumes that the bonding and lone pairs are the same size. This is not true. Recall that lone pairs "prefer" the less crowded positions because lone pairs are larger and they are attracted to only one nucleus. Therefore we expect that as bonding pairs are replaced by lone pairs for a given steric number, the remaining bonds will be forced closer to each other to relieve electron repulsions and the bond angles will decrease from the values expected from the pure hybrid orbitals. This trend is seen for the molecules CH_4 (no lone pairs, 109.5°) , NH_3 (one lone pair, 107°) and H_2O (two lone pairs, 104.5°).

When lone pairs are present, bond angles will be smaller than the angles between the ideal hybrid orbitals.

Molecules and Charge Separation

Bonds between different elements lead to the formation of *polar* bonds, as described in Section 8.2. If there are only two atoms in the molecule, the polar bond results in a molecule with a positive end and a negative end: a molecule with a **permanent dipole moment**, also called a **polar molecule**. Knowing if a molecule is polar helps in predicting some of its physical properties, such as what solvent the substance may dissolve in, or whether its melting point will be high or low. When there is more than one bond in a molecule, predicting whether or not the molecule has a dipole moment is more complicated than simply determining if the bonds are polar. One must also consider the directions in which the bond dipoles point. Compare linear carbon dioxide, CO_2, and bent sulfur dioxide, SO_2, two triatomic molecules. Because the atoms in the bonds have different electronegativities, both molecules have polar bonds. The Lewis structures of these molecules are shown below; the arrows correspond to the polarity of the bond.

$\delta+ \longrightarrow \delta-$ $\ddot{\text{O}}=\text{C}=\ddot{\text{O}}$	$\delta+$ $\ddot{\text{O}}=\text{S}=\ddot{\text{O}}$ $\delta-$; $\downarrow$
no net dipole moment; bond dipoles cancel	dipoles do not cancel; resulting dipole moment

In CO_2, the bond dipoles are equal in size (the C=O bonds are identical) but the dipoles point in opposite directions. The two bond dipoles cancel out; therefore CO_2 does not have a permanent dipole moment. In SO_2 the bond dipoles do not point in opposite directions and do not completely cancel out. As a result, SO_2 has a permanent dipole moment with the positive charge at the sulfur end of the molecule and the negative charge at the oxygen end of the molecule.

Requirements for a polar molecule:
- polar bonds
- asymmetric structure so bond dipoles do not cancel out.

Any molecule (1) whose steric number is the same as its coordination number and (2) all of whose outer atoms are the same will not have a permanent dipole moment because all the bond dipoles cancel each other out. Replacing an outer atom with a lone pair or a different atom will often lead to the asymmetry required for a permanent dipole moment. However, we should always determine the structure of the molecule to ensure that the bond dipoles do or do not cancel out.

Exercise 8.12 Determine whether each of the following molecules possesses a permanent dipole moment: (a) SeF_6 (b) IF_3 (c) BrF_4^- (d) thiophane, with the line structure shown in Exercise 8.10.

Steps to Solution: The two criteria for a permanent dipole moment are (1) the presence of polar bonds and (2) asymmetry so that the bond dipoles do not cancel. All three species listed above have polar bonds; therefore the structures must be examined to determine how symmetric they are.

(a) The SeF_6 molecule has an octahedral structure with no lone pairs. Therefore, although it has polar bonds, each bond dipole is canceled out by one directly opposite to it and as a result, SeF_6 has no permanent dipole moment.

(b) The structure of IF_3 has steric number 5, with two lone pairs. The structure of the molecule will be based on the trigonal bipyramid. The position of the lone pairs is critical; if they were to occupy the axial positions, the trigonal planar arrangement would result in no net dipole moment. However, the T-shaped structure (obtained because the lone pairs occupy the roomier equatorial sites) does not permit the bond dipoles to cancel and as a result, IF_3 will have a dipole moment where the negative end of the molecule is directed towards the equatorial fluorine atom.

(c) BrF_4^- has steric number 6 and two lone pairs so it adopts the square planar structure shown above. Although the bonds are polar, each bond dipole is opposite one of the same size. Therefore the bond dipoles cancel, and the BrF_4^- ion has no permanent dipole moment.

(d) The molecule SC_4H_8 has polar bonds (S-C bonds) and the molecule has low enough symmetry so that the bond dipoles do not cancel out (the C-S-C angle $\neq 180°$). Therefore thiophane has a permanent dipole moment.

Test Yourself

1. (a) Draw the Lewis structure of phosgene, $COCl_2$ (b) determine the geometry about the central atom and (c) name the hybrid orbitals used by the central atom.

2. Draw the isomers of $C_2H_3Cl_3$ and determine which isomer has the greater dipole moment.

3. Draw combinations of a d_{xy} orbital and a p_x orbital that lead to good overlap.

4. (a) Draw the Lewis structure of BrF_5, and predict if it has a dipole moment. (b) What is the geometry about the central atom and hybrid orbitals used by the central atom?

Answers

1. (a) [Lewis structure: O double-bonded to C, with two Cl atoms single-bonded to C; O and Cl atoms have lone pairs shown]

   (b) trigonal planar (c) sp^2

2. [Two structures of C₂H₂Cl₄ isomers shown: 1,1,1,2-tetrachloroethane (CCl₃–CH₂Cl type) on left; 1,1,2,2-tetrachloroethane on right]

   greater dipole moment [under right structure]

3. [Diagram showing σ bond (end-to-end orbital overlap) and π bond (side-by-side orbital overlap)]

   σ bond   π bond

4. (a) [Structure of BrF₅: square pyramidal with Br center, 5 F atoms] (lone pairs or F omitted)

   Yes, BrF₅ has dipole moment.

   (b) Square pyramidal geometry, d^2sp^3 hybrid orbitals at Br

---

Solutions to Chapter Problems

8.1 Determine configurations from the position of an element in the periodic table. The electrons with the highest principal quantum number will be involved in bond formation. Be has the configuration $1s^2\, 2s^2$. Its two $n = 2$ electrons will be involved in bond formation.

8.3 Atoms with a single valence electron form bonds by overlap of their valence orbitals. The configuration of Na is [Ne] $3s^1$. Two sodium atoms form a bond by overlap of their $3s$ orbitals to produce an orbital with high electron density between the nuclei:

Chapter 8: Fundamentals of Chemical Bonding

8.5 The orbitals that participate in bond formation are the valence orbitals; that is, the occupied orbitals of largest principal quantum number and any partially-filled orbitals of lower principal quantum number. Consult the configuration of the atom to determine what these orbitals are:
Al, [Ne] $3s^2\ 3p^1$: the $3s$ and $3p$ orbitals participate in bonding;
As, [Ar] $3d^{10}\ 4s^2\ 4p^3$: the $4s$ and $4p$ orbitals participate in bonding;
F, [Ne] $3s^2\ 3p^5$: the $3s$ and $3p$ orbitals participate in bonding;
Sn, [Kr] $4d^{10}\ 5s^2\ 5p^2$: the $5s$ and $5p$ orbitals participate in bonding.

8.7 One valence $4p$ orbital of each bromine atom is directed along the axis between the two atoms, and these two orbitals can overlap to form a bond:

$4p$ orbital is shaded

8.9 The periodic trends predict that electronegativity increases across a row (because of increasing Z which binds electrons more tightly) and decreases down a column (because of increasing n, which means less tightly bound electrons). Consult Figure 8-7 in your textbook to find exceptions to these trends:
Exceptions across the rows: Cr (1.6) and Mn (1.5); Cu (1.9) and Zn (1.6);
Pd (2.2), Ag (1.9), and Cd (1.7); Au (2.4) and Hg (1.9);
Exceptions down the columns: Cr (1.6) and Mo (1.8); Mn (1.5) and Tc (1.9).

8.11 Ionic substances have Δ EN > 2.0, non-polar substances have Δ EN = 0, and polar substances have intermediate values:
Non-polar: F_2; ionic: NaF and CaO; polar: HF and NaH.

8.13 Electronegativities describe the tendency of each element to attract bonding electrons from another. In any pair, the element with higher electronegativity attracts bonding electrons more strongly:
(a) N; (b) S; (c) I; and (d) S.

8.15 Bond polarity increases with the difference in electronegativity of the bond-forming elements. In these compounds, H is the least electronegative element, so bond polarity increases with the electronegativity of the other element. The EN order is P < S < N < O, so the order of bond polarity is $PH_3 < H_2S < NH_3 < H_2O$.

8.17 Ionic substances contain a metal or ammonium cation and/or one of the polyatomic anions. All other substances are covalent. Ionic: KBr and NH_4Br; Covalent: HBr, CBr_4, and Br_2.

8.19 To determine the number of valence electrons in a species, count the valence electrons of each atom and, for ions, adjust for ionic charge:
HBr, 1 + 7 = 8 valence electrons; KBr (ionic): K^+, 1 – 1 = 0; Br^-, 7 + 1 = 8;
NH_4Br (ionic): NH_4^+, 5 + 4(1) – 1 = 8; Br^-, 8; CBr_4, 4 + 4(7) = 32; Br_2, 2(7) = 14.

Chapter 8: Fundamentals of Chemical Bonding

8.21 Ionic substances contain a metal or ammonium cation and/or one of the polyatomic anions. All other substances are covalent. To determine the number of valence electrons in a species, count the valence electrons of each atom and, for ions, adjust for ionic charge: NH_3 is covalent with $5 + 3(1) = 8$ valence electrons; NH_4NO_3 (ionic); NH_4^+, $5 + 4(1) - 1 = 8$, NO_3^-, $5 + 3(6) + 1 = 24$ valence electrons; HNO_3 is covalent with $1 + 5 + 3(6) = 24$ valence electrons.

8.23 Identify polyatomic ions from memory or from the chemical formula: LiOH, OH^-; KH_2PO_4, $H_2PO_4^-$; $NaBF_4$, BF_4^-; $LiIO_4$, IO_4^-; NaCN, CN^-; and $(NH_4)_2CrO_4$, NH_4^+ and CrO_4^-.

8.25 To determine the number of valence electrons in a polyatomic ion, count the valence electrons of each atom and adjust for ionic charge: OH^- has $6 + 1 + 1 = 8$; $H_2PO_4^-$ has $2(1) + 5 + 4(6) + 1 = 32$; BF_4^- has $3 + 4(7) + 1 = 32$; IO_4^- has $7 + 4(6) + 1 = 32$; CN^- has $4 + 5 + 1 = 10$; NH_4^+ has $5 + 4(1) - 1 = 8$; and CrO_4^- has $6 + 4(6) + 2 = 32$ valence electrons.

8.27 A bonding framework links atoms in the species with pairs of electrons, represented by lines:

[Structures shown: $H_2PO_4^-$, BF_4^-, IO_4^-, CN^-, OH^-, NH_4^+, CrO_4^{2-}]

8.29 To form a provisional Lewis structure, add three pairs of electrons around each outer atom except hydrogen, add electron pairs to inner atoms until all valence electrons are placed. For outer and/or inner atoms of different elements, place electrons first around the more electronegative atom. Formal charge is FC = (valence electrons) − (assigned electrons).

[Structure of OH^-] 8 valence electrons, all placed.

[Structure of $H_2PO_4^-$ provisional] 24 electrons placed, 8 left. → [Final structure] all 32 valence electrons placed. $FC_P = 5 - 4 = +1$.

297

8.31 Adjust provisional Lewis structures by shifting electron pairs to form multiple bonds. For second row atoms with less than an octet of electrons, shift pairs to make multiple bonds until the atom possesses an octet; for atoms beyond the second row, shift pairs to make multiple bonds until the formal charge on the atom is reduced to zero.

The provisional Lewis structures of OH^-, BF_4^-, and NH_4^+ in Problem 8.29 are satisfactory.

$FC_{Cr} = 6 - 4 = +2$.

plus several additional equivalent structures

8.33 A description of bonding and geometry always starts with a Lewis structure and a count of bonds and non-bonding pairs around inner atoms.

The steric number of the carbon atom is 4. The bonding about it can be described using sp^3 hybrid orbitals overlapping with $1s$ orbitals of H and $3p$ orbitals of Cl.

The geometry about the carbon atom is tetrahedral. The bonds can be depicted as follows:

C–H bond

C–Cl bond

8.35 Construct the bonding framework from the formula, knowing that the more electronegative atoms (Cl) will be in outer positions. In 1,2-dichloroethane, each carbon atom is bonded to the other, and there is one Cl atom bonded to each carbon atom. There are $2(7) + 2(4) + 4(1) = 26$ valence electrons. The bonding framework contains 7 bonds, and there are three lone pairs on each Cl atom, accounting for all the valence electrons. Thus, the bonding framework is the correct Lewis structure of cyclohexane. Each carbon atom has SN = 4 and tetrahedral geometry. Your ball-and-stick model should reflect this.

8.37 To draw structural isomers of an alkane, start with the "straight chain" compound, then rearrange the bonding arrangement to make other isomers. It is convenient to work with only the carbon skeleton:

"Straight chain:" C — C — C — C — C — C

Putting the branched carbon in position 4 gives the same isomer as putting it in position 2

8.39 Determine the Lewis structure following the standard procedures.

1.) There are $2(4) + 6(1) + 6 = 20$ valence electrons.
2.) The chemical formula indicates that the bonding framework includes an –OH group attached to one of the carbon atoms. The bonding framework contains 8 bonds = 16 electrons, leaving 4 valence electrons.
3.) All the outer atoms are hydrogen, so no electrons can be placed on outer atoms.
4.) Place the remaining two pairs on the oxygen atom. The resulting structure has an octet around each row 2 atom, so it is the correct Lewis structure.
Each inner atom has SN = 4, so the appropriate hybridization is sp^3 and the geometry about each inner atom is tetrahedral.

Lewis Structure:

```
    H       H
     \     /
  H — C — C — Ö:
     /     \  \
    H       H   H
```

methane molecule with left hydrogen replaced with methyl, top hydrogen replaced with –OH

8.41 Determine the Lewis structure following the standard procedures.

1.) There are $4(4) + 4 + 12(1) = 32$ valence electrons.
2.) The chemical formula indicates that the bonding framework includes four –CH$_3$ groups attached to the silicon atom. The bonding framework contains 16 bonds = 32 electrons. All electrons are placed, so this structure is the correct Lewis structure. Each inner atom has
SN = 4, sp^3 hybridization, and the geometry about each inner atom is tetrahedral.

8.43 Determine the Lewis structure following the standard procedures.

1.) There are $8 + 7(4) + 6 = 42$ valence electrons.
2.) Xe is the only inner atom, so the bonding framework has Xe surrounded by the other atoms. Five bonds require 10 electrons, leaving 32 valence electrons to place.
3.) Place three pairs around each of outer atom for a total of 30, leaving 2 valence electrons.
4.) Place the remaining pair on the xenon atom. $FC_{Xe} = 8 - 5 - 2 = +1$.
5.) Make one double bond to Xe by shifting an electron pair from the least electronegative outer atom, oxygen:

Provisional: Complete:

Xe has SN = 6, octahedral electron pair geometry, and the appropriate hybridization is d^2sp^3. One position is occupied by a lone pair, giving a square pyramid with bond angles near 90°.

8.45 Determine the Lewis structures following the standard procedures.

Compound:	GeF$_4$	SeF$_4$	XeF$_4$
Valence Electrons:	$4 + 7(4) = 32$	$6 + 7(4) = 34$	$8 + 7(4) = 36$
Bonding Electrons:	8	8	8
Electrons on F:	$4(6) = 24$	$4(6) = 24$	$4(6) = 24$
Electrons on inner atom:	0	2	4
FC on inner atom:	0	0	0
SN of inner atom:	4	5	6
Molecular geometry:	tetrahedron	see saw	square plane
Hybridization:	sp^3	dsp^3	d^2sp^3

8.47 Bond angles can be predicted from the steric number of the inner atom, which determines the preferred electron pair geometry. The presence of lone pairs on the inner atom reduces bond angles slightly due to increased electron-electron repulsion for lone pairs.

GeF_4, SN = 4, tetrahedron, 109°; SeF_4, SN = 5, trigonal bipyramid with lone pair, < 120° and < 90°; XeF_4, SN = 6, octahedron with two lone pairs in axial positions, 90° (the lone pairs are above and below the square and do not distort the angles).

8.49 Determine Lewis structures using the standard procedures. Only asymmetric molecules have dipole moments. Use the Lewis structures to determine steric numbers and ascertain whether or not the molecules are asymmetric.

SiF_4 is tetrahedral and symmetric; it has no dipole moment; H_2S, like H_2O, is bent and has a dipole moment; XeF_2 has SN =5, with its 3 lone pairs in equatorial positions, so the molecule is linear without a dipole moment; $GaCl_3$ has SN = 3, is trigonal planar and symmetric, and has no dipole moment; NF_3, like NH_3, is pyramidal and has a dipole moment.

8.51 Determine Lewis structures following the standard procedures. Draw ball-and-stick models that illustrate the molecular geometries determined by the steric numbers of the inner atoms.
(a) Cl_2O has 2(7) + 6 = 20 valence electrons. Its Lewis structure has two bonding pairs, three lone pairs on each chlorine atom, and two lone pairs on the inner oxygen atom. SN_O = 4, leading to tetrahedral electron pair geometry and bent molecular shape:

(b) C_6H_6 has 6(4) + 6(1) = 30 valence electrons. After constructing the bonding framework, six electrons remain to be placed. Place these on alternate carbon atoms; then complete octets around the other carbon atoms by shifting lone pairs to make double bonds. There are two equivalent resonance structures:

$SN_C = 3$, leading to trigonal planar geometry about each carbon atom and a flat, hexagonal molecular shape.

(c) C_2H_4O has $2(4) + 4(1) + 6 = 18$ electrons. When the bonding framework is complete, four electrons remain. These are placed on the oxygen atom. The triangular ring constrains this molecule to have 60° bond angles, even though the steric numbers of the atoms are four.

(d) NO_2 has $5 + 2(6) = 17$ electrons, making it one of the few stable species with an odd number of valence electrons.

Two pairs are used in the bonding framework and three pairs are placed on each oxygen atom, leaving a single electron which must be placed on the inner nitrogen atom.

Shift one electron pair into a double bond, so the nitrogen atom has seven electrons (one short of an octet) associated with it. There are two equivalent structures. $SN_N = 3$, so the molecule has trigonal planar electron geometry and bent molecular shape:

8.53 All the species in this problem have four halogen atoms bonded to an inner atom. All have 32 electrons except for SeF$_4$ (34 valence electrons) and XeF$_4$ (36 valence electrons). All the 32-electron species have the same Lewis structure and the same geometry.

The 32-electron species have SN = 4 and are tetrahedral; SeF$_4$ is a seesaw and XeF$_4$ is a square plane.

8.55 Determine the number of different isomers by constructing the different possible geometric possibilities. Two X atoms can be placed at opposite ends of one Cartesian axis; all structures with this arrangement are equivalent, because the molecule can be rotated about an axis to superimpose these positions. The three X atoms can all be placed at right angles to one another, giving a different isomer. Thus, there are two isomers:

8.57 A description of bonding in a molecule starts with its Lewis structure, from which steric number, geometry, and hybridization can be deduced. AlCl$_3$ has 3 + 3(7) = 24 valence electrons. The bonding framework requires three pairs, and the remaining 18 electrons are placed around the three chlorine atoms. FC$_{Al}$ = 3 – 3 = 0, so this is the correct structure. The structure has SN$_{Al}$ = 3, indicating that the molecule has trigonal planar structure with Al-Cl bonds that can be represented using sp^2 hybrids on Al and $3p$ orbitals on Cl.

304

8.59 The shape of a molecule indicates the steric number of its inner atom and the number of atoms bonded to the inner atom. Consult Table 8-6 of your textbook for details. Tellurium, in column 16 of the periodic table, has six valence electrons, and each fluorine contributes seven valence electrons. Species with an even number of fluorine atoms are neutral, but those with an odd number must have one unit of charge (+1 or –1) in order to contain an even number of electrons.

(a) bent indicates TeF_2; (b) T-shape indicates SN = 5 on Te, three F atoms, and two lone pairs on Te, for a total of 4 + 3(8) = 28 valence electrons. TeF_3 would have 6 + 3(7) = 27 valence electrons, so the formula of this species must be TeF_3^-; (c) square pyramid indicates SN = 6 on Te, five F atoms, and one lone pair on Te, for a total of 2 + 5(8) = 42 valence electrons. TeF_5 would have 6 + 5(7) = 41 valence electrons, so the formula of this species must be TeF_5^-; (d) trigonal bipyramid indicates SN = 5 on Te and five F atoms, for a total of 5(8) = 40 valence electrons. TeF_5 would have 6 + 5(7) = 41 valence electrons, so the formula of this species must be TeF_5^+; (e) octahedron indicates SN = 6 and six F atoms, formula TeF_6; seesaw indicates SN = 5 on Te, four F atoms, and one lone pair on Te, formula TeF_4.

8.61 Determine Lewis structures following the standard procedure.
(a) Bromate, BrO_3^-, contains 7 + 3(6) + 1 = 26 valence electrons. Three pairs are required for the bonding framework, nine pairs go on the oxygen atoms, and the remaining pair is placed on the bromine atom:

$FC_{Br} = 7 - 5 = +2$
Shift 2 electron pairs to make 2 double bonds

There are 2 additional equivalent structures

(b) Nitrite, NO_2^-, contains 5 + 2(6) + 1 = 18 valence electrons. Two pairs are required for the bonding framework, six pairs go on the O atoms, and the remaining pair is placed on N. Shift one electron pair to make a double bond and complete the octet on nitrogen. There are two equivalent structures:

2 equivalent structures

(c) Phosphate, PO_4^{3-}, contains 5 + 4(6) + 3 = 32 valence electrons. Four pairs are required for the bonding framework and 12 pairs go on the oxygen atoms, giving a provisional structure with $FC_P = 5 - 4 = +1$. Shift one electron pair to make a double bond and reduce the formal charge to zero. There are four equivalent structures.

Chapter 8: Fundamentals of Chemical Bonding

[Lewis structure diagrams showing PO₄³⁻ resonance structures with "Shift one electron pair" arrow, and note: "There are two additional equivalent structures"]

(d) Hydrogencarbonate, HCO₃⁻, contains $1 + 4 + 3(6) + 1 = 24$ valence electrons. Four pairs are required for the bonding framework, three pairs go on each outer oxygen atom, and the remaining two pairs go on the inner oxygen atom. Shift one electron pair from an outer O atom to form a double bond to C and complete its octet. There are two equivalent structures:

[Two resonance Lewis structures of HCO₃⁻]

8.63 Determine the Lewis structure following the standard procedure.

H₃CNH₂ has $5(1) + 4 + 5 = 14$ valence electrons. Six pairs are required to complete the bonding framework and the remaining pair goes on the nitrogen atom, giving a structure in which both carbon and nitrogen have complete octets.

SN = 4 for both C and N, so the bond angles about carbon are 109° and those about nitrogen are < 109° because of the extra repulsion created by the lone pair on the nitrogen atom.

8.65 Determine Lewis structures following the standard procedure. Shift electron pairs to form multiple bonds until each second row element has eight electrons associated with it.

8.67 Determine molecular geometries and hybridizations from steric numbers, which can be found using the Lewis structures.

(a) $SN_O = 4$ with two lone pairs, bent geometry, sp^3 hybridization;
(b) $SN_C = 2$, linear geometry, sp hybridization;
(c) $SN_C = 2$, linear geometry, sp hybridization;
(d) $SN_C = 3$, trigonal planar geometry, sp^2 hybridization;
(e) $SN_S = 4$ with two lone pairs, bent geometry, sp^3 hybridization;
(f) for the methyl carbon, $SN = 4$, tetrahedral geometry, sp^3 hybridization; for the other carbon, $SN = 2$, linear geometry, sp hybridization; and
(g) $SN_N = 4$ with one lone pair, pyramidal geometry, sp^3 hybridization.

8.69 A description of bonding and geometry starts with determination of the Lewis structure. $(CH_3)_3C^+$ has $4(4) + 9(1) - 1 = 24$ electrons. All the electrons are required to complete the bonding framework, so there are no lone pairs to shift and the provisional structure is correct even though the central carbon atom has only six electrons associated with it (as this suggests, carbocations are highly reactive species).

The central carbon has $SN = 3$, trigonal planar geometry, and its bonding orbitals can be described using sp^2 hybrid orbitals. The methyl carbons have $SN = 4$, tetrahedral geometry, and their bonding orbitals can be described using sp^3 hybrid orbitals.

8.71 A bond angle of 92.2° indicates that the bonding can be well described using p orbitals rather than hybrid orbitals, whereas an angle of 104.5° indicates significant distortion from pure p orbital bonding. Space-filling models of the two molecules shows that the smaller oxygen atom cannot accommodate two H atoms at right angles, but the larger S atom can.

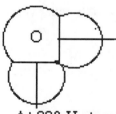

At 90°, H atoms would interpenetrate

At 104.5°, H atoms do not interpenetrate

H atoms do not interpenetrate at 90°

8.73 Determine Lewis structures using the standard procedures. Both ions have the same number of valence electrons, 5 + 4 + 6 + 1 = 16. Two pairs are required to complete the bonding framework and the remaining six pairs are placed on the two outer atoms. Then shift electron pairs to make two additional bonds to the central atom, completing its octet. The cyanate ion has two near-equivalent structures, each of which has one atom with formal charge −1:

$$[\ddot{\underset{..}{N}}^{-1}=C=\ddot{\underset{..}{O}}]^{-} \longleftrightarrow [:N\equiv C-\ddot{\underset{..}{O}}^{-1}:]^{-}$$

The isocyanate ion also has two near-equivalent structures, but the formal charge on the nitrogen atom cannot be reduced to zero:

$$[:\overset{-1}{C}\equiv \overset{+1}{N}-\ddot{\underset{..}{O}}^{-1}:]^{-} \longleftrightarrow [:\overset{-2}{\underset{..}{C}}=\overset{+1}{N}=\ddot{\underset{..}{O}}]^{-}$$

8.75 The silicate structures of various minerals are described in Section 8.5. The appropriate matches are as follows:

quartz and silica; mica and silicate sheets; zircon and orthosilicate; asbestos and silicate fibers; and jade and metasilicate.

8.77 The electronegativity of hydrogen (2.1) is the smallest among the elements involved in these chemical bonds, so the ranking of bond polarity is from smallest to largest electronegativity of the bonding partner. Remember that EN decreases down a column and increases across a row:

Si-H < C-H < N-H < O-H < F-H.

8.79 The Lewis structures show octets around the inner atom and $FC_X = 0$ for XF_3, making these compounds stable. Formula XF_5 also has $FC_X = 0$ but has five electron pairs associated with the inner atom. This is possible for P, a 3rd row element with d orbitals available for bonding. It is not possible for N, a 2nd row element which lacks valence d orbitals.

X = N or P

308

8.81 There are no inner atoms in the diatomic hydroxide anion, so the bonding in this species is described using a $2p$ orbital from the oxygen atom. In the other species, the oxygen atom is inner, has SN = 4, and its bonding can be described using sp^3 hybrid orbitals:

8.83 (a) Square planar XY_2Z_2 molecules may have like atoms adjacent to or opposite each other:

(b) When like atoms are opposite each other, their bond polarities oppose each other and cancel, so this isomer has no dipole moment. When like atoms occupy adjacent positions, the bond polarities do not entirely cancel and the molecule has a net dipole moment.

8.85 Electron affinity is the strength with which an atom binds a free electron. Electronegativity describes the ability of an atom to attract electrons that are bound to other atoms.

8.87 A compound XY_7 is possible if X has d orbitals available for bonding, but it is sterically crowded because of the large number of Y atoms around the central X atom. For this compound to exist, Y must be as small as possible and X must be as large as possible. Thus, Y is the halogen with the smallest size, fluorine; and X is iodine, the halogen with the largest size (apart from astatine, which is not involved). The compound is IF_7.

8.89 Silica has the chemical formula SiO_2, but it contains a network of bonds (see Fig. 8-16) in which each silicon atom bonds to four oxygen atom and each oxygen atom bonds to two silicon atoms. Each oxygen atom also has two lone pairs, so all atoms have SN = 4 and the network is tetrahedral about each silicon atom and bent about each oxygen atom. Silicate minerals contain SiO_4^{4-} ions, each of which has four terminal oxygen atoms bonded to a central silicon atom, which displays tetrahedral geometry. Metasilicates also contain silicon atoms bonded to four oxygen atoms and displaying tetrahedral geometry, but there are two types of oxygen atoms in metasilicates: Half the oxygen atoms form bonds to two silicon atoms and display bent geometry, while the other half are bonded to only one silicon atom.

8.91 Molecules have dipole moments only if they are unsymmetric. A linear N–O–N structure would not have a dipole moment, because the N–O dipole would exactly cancel the O–N dipole. Thus N_2O must have the structure N–N–O.

8.93 All positions around a tetrahedral center are equivalent, so there is only one isomer of CH_2Br_2. The two structures show different views of the same compound.

Chapter 9 Chemical Bonding: Multiple Bonds

9.1 Multiple Bonds in Carbon Compounds

QUESTIONS TO ANSWER, SKILLS TO LEARN
1. What is a multiple bond?
2. What is the difference between π (pi) and σ (sigma) bonds?
3. Using localized orbitals to describe double and triple bonds.

A multiple bond exists when the Lewis structure of a molecule or polyatomic ion has more than one electron pair between two nuclei. Compare the provisional and final Lewis structures for propylene, C_3H_6, shown below.

Provisional Lewis structure   Final Lewis Structure

In the final structure there are four electrons between the two C atoms, leading to a *double bond* (there are twice as many electrons between the nuclei as in a single bond). The steric number of each C atom in the double bond is 3 because each C atom is bonded to three other atoms and no unshared electrons are present in the Lewis structure. Steric number 3 indicates sp^2 hybridization; the four valence electrons of the C atoms joined by the double bond are distributed among the four valence orbitals in C: the 3 sp^2 hybrid orbitals and the $2p$ orbital not used in hybridization. The C-H bonds and the C-C bond can be described as arising from overlap of the hybrid orbitals with the H $1s$ orbitals and the hybrid orbitals on the other C atoms as shown belown at left. The shaded p orbitals form a bond where the electron density is not along the line joining the nuclei.

To the right is shown the electron density resulting from overlap of the orbitals in the illustration at left. Five of these orbitals have maximum electron density on the line joining the two nuclei being bonded. The sixth orbital, formed from the "side-by-side" overlap of the p orbitals not used in hybridization, is different; the maximum electron density lies *above and*

below the line joining the nuclei, not along the line. These different types of bonds are called σ (sigma) and π (pi) bonds respectively.

Sigma (σ) bonds are formed from the overlap of orbitals pointing *directly along the bond axis* and have the highest electron density on the line joining the nuclei in the bond.

Pi (π) bonds are formed from the *side-by-side overlap of orbitals* and have the *highest electron density above and below* the line joining the nuclei.

COMMON PITFALL: Miscounting π Bonds.
You should remember that a π bond has electron density above and below the line joining the nucleus; a common mistake is to consider each of the π lobes as a bond. The figure above shows one π bond.

The C-C double bond is formed from the combination of the σ bond made by overlap of the *sp²* hybrid orbitals and the π bond made from overlap of the leftover 2*p* orbitals. This is true for any double bond.

A *double* bond consists of a σ bond *plus* a π bond.

Exercise 9.1 Acetaldehyde, CH$_3$-(C=O)H, contains a double bond. Describe the bonding around both C atoms, naming the hybridization (if needed) of the orbitals. Draw a sketch showing the orbitals.

Steps to Solution: The first step in describing the bonding is to draw a Lewis structure. Shown below are the provisional and final Lewis structures for acetaldehyde.

provisional Lewis structure   final Lewis structure

The 2 C atoms are the inner atoms in the molecule. The right-C atom has steric number 4 and coordination number 4 (it is bonded to 3 H atoms and the other C atom) and therefore uses sp³ hybrid orbitals to form bonds. The left C atom has *steric number 3 and coordination number 3 (it is bonded to a C atom, an O atom and an H atom) and needs to use sp^2 hybrids to form bonds.*

The bonds to the H atoms on the right C atom result from the overlap of the sp^3 hybrid orbitals on the C atom with the 1s orbitals on the H atoms. The C-C bond is formed by the overlap of the sp^3 orbital of one C atom with the sp^2 orbital on the other C atom. The C-H bond on the left C atom results from the overlap of the 1s orbital of the H atom with the sp^2 hybrid orbital on the C atom. The sketch below at left shows the orbitals in the σ bonds. The C-O double bond results from a σ bond formed from the overlap of the sp^2 orbital on the C atom with a 2p orbital on the O atom plus the π bond which is formed from the overlap of the unhybridized 2p orbital on the C atom with a 2p orbital on the O atom. At right is an illustration of the orbitals in the π bond.

σ bonding   π bonding

Triple Bonds. Acetonitrile, CH_3CN, is used as a solvent in many reactions in research and industry. The Lewis structure of acetonitrile requires a *triple bond*, three pairs of electrons between the C and N atoms. The central C atom has a steric number and coordination number

Provisional Lewis structure   Final Lewis structure

of 2; therefore, *sp* hybrid orbitals are used to form the σ bonds to the outer C and N atoms. The σ bonding framework is shown below at left. The other two bonds in the triple bond are π bonds, because double and triple bonds are formed from one σ bond and one or two π bonds, respectively. The σ and π components of the triple bond are shown below.

σ bonding   π bonds: The heavy lines emphasize the orbital lobes that interact.

The σ bond between the C and N atoms results from overlap of a *sp* hybrid on C with the $2p_z$ orbital on the N atom. The C-C bond results from overlap of the *sp* hybrid on the central C atom with the sp^3 hybrid orbital on the CH_3 carbon. The C-H bonds are formed as described in the acetaldehyde example above. The two π bonds result from overlap of the $2p_x$ and $2p_y$ orbitals on N with the unhybridized $2p_x$ and $2p_y$ orbitals on C.

Exercise 9.2 The line structure of a steroid, norethindrone, is shown below. Convert the line structure to a full structural formula and determine the hybridization at each C atom.

Steps to Solution: We convert the line structure to the full structural formula by adding the appropriate number of hydrogens to the C atoms at the vertices to give the structure on the left. We find the hybridization of the C atoms by determining the steric number at each C atom. Because there are no lone pairs on the C atoms, the steric number in this case will be the same as the coordination number. For steric number 2, sp hybridization is used; for steric number 3, sp² hybridization; and for steric number 4, sp³ hybridization. Most of the carbons are sp³ hybridized in this molecule: only the sp² and sp hybridized C atoms are labeled.

9.2 Bond Lengths and Energies

QUESTIONS TO ANSWER, SKILLS TO LEARN
1. What is an "average" bond length?
2. Understanding the relation between orbital type and bond length
3. Understanding the effect of atomic size on overlap

In this section we will revisit two parameters used to characterize chemical bonds: bond length and bond energy. We will describe the factors influencing bond length and bond energy, as well as methods to estimate the energy released or consumed in a chemical process.

Bond Length

The bond length is the distance between the nuclei of two atoms joined by a chemical bond. In the ethane molecule there are several C-H distances, two of which are shown below. The term, *bond length*, refers only to the distance between directly bonded atoms.

The bond length is affected by the following factors:

- The smaller the principal quantum number of the valence orbitals of the atoms in the bond, the shorter the bond.

  Rationale: Because covalent bonds are formed from the overlap of orbitals on the two atoms joined by the bond, the bond distance is affected by the sizes of those orbitals. Smaller orbitals will place the bonded atoms closer to each other.

- The higher the bond multiplicity, the shorter the bond.

  Rationale: The greater number of electrons localized between nuclei in multiple leads to stronger attraction of the nuclei toward the electrons and a shorter bond.

- The higher the effective nuclear charges (Z_{eff}) of the bonded atoms, the shorter the bond.

  Rationale: The higher effective nuclear charge will tend to make the valence orbitals smaller. Smaller orbitals lead to shorter bonds.

- The larger the electronegativity difference ($\Delta\chi$) of the bonded atoms, the shorter the bond.

  Rationale: The forces that bind atoms together in covalent bonds are electrostatic in nature. Charge separation resulting from the different electronegativities of atoms in a bond leads to attraction of the unlike electrical charges, strengthening the bond.

Exercise 9.3 Explain, in terms of the four factors listed above, the differences in bond lengths in the following pairs of bonds:

(a) The Si-Si bond is longer than the P-P bond.
(b) The Si-Cl bond is shorter than the Si-Br bond.
(c) The C-C bond in ethylene is longer than the C-C bond in acetylene.
(d) The P=O bond is longer than the O=O bond.

Steps to Solution: To rationalize the difference in bond lengths in terms of the four factors listed above, we must determine which of the factors are relevant and then determine which is the most significant.

(a) <u>Si-Si vs. P-P.</u> Both of these bonds are single bonds, nonpolar, and between elements of the 3rd row, so factors 1, 2 and 4 are not applicable. However, P is further to the right of the periodic table than Si and thus has a higher effective nuclear charge. The smaller orbitals on the P atoms would be expected to result in the P-P bond being shorter than the Si-Si bond.

(b) <u>Si-Cl vs. Si-Br.</u> These bonds are both single bonds and the effective nuclear charges on both atoms are expected to be very nearly the same (Cl and Br are both in Group 17). These facts eliminate factors 2 and 3. The differences that lead to the bond length difference are that the valence orbitals of Cl have principal quantum number 3, whereas in Br, $n = 4$. Also, the electronegativity of Cl is somewhat higher than that of Br. The major factor is expected to be the difference in n, because Br and Cl are quite close to each other in the periodic table and have similar electronegativities (Cl, 3.0; Br, 2.8).

(c) <u>C-C bond in ethylene vs. acetylene.</u> The significant difference between these two bonds is that the C-C bond in ethylene is a double bond and the C-C bond in acetylene is a triple bond. None of the other factors are relevant to this problem (same n, χ, and effective nuclear charge). Because the higher multiplicity bond will be shorter, the triple C-C bond in acetylene, HC≡CH, is expected to be shorter than the C-C double bond in ethylene, $H_2C=CH_2$.

(d) <u>P=O vs. O=O.</u> These two bonds have the same bond multiplicity, but the other three factors, n, Z_{eff}, and $\Delta\chi$, are different. The electronegativity difference between P and O is fairly small ($\Delta\chi = |2.1 - 3.5| = 1.4$). The difference in effective nuclear charge is expected to be small because P and O are in neighboring columns. More important is the fact that the valence orbitals in P have $n = 3$ and in O, $n = 2$. Therefore, the main reason for the shorter O=O bond is that the valence orbitals in O have a smaller principal quantum number.

Bond Energy

Another number used to characterize chemical bonds is bond energy. Bond energy is a measure of the strength and stability of a bond; however, it is defined as the amount of energy required to break the bond into fragments in which the electrons in the bond are evenly divided between the fragments.

The *bond energy* is the amount of energy required to break one mole of chemical bonds.

Chapter 9: Chemical Bonding: Multiple Bonds

In diatomic molecules, bond energies are quantities that may be determined exactly because there is only one bond to break in the molecule:

$$HF_{(g)} \rightarrow H_{(g)} + F_{(g)} \qquad \Delta E = BE = 565 \text{ kJ/mole}$$

The values of bond energies are always positive because breaking a chemical bond requires energy. The formation of a bond will release energy.

$$H_{(g)} + F_{(g)} \rightarrow HF_{(g)} \qquad \Delta E = -BE = -565 \text{ kJ/mole}$$

In more complex molecules, there may be several bonds that involve the same atoms, but are different because of different neighboring atoms. The following molecule has two types of carbon atoms: one bonded to an OH group, 2 H atoms and a C atom and one bonded to an O atom, 2 H atoms and a C atom. There will be 2 types of C-H bonds and 2 types of C-O bonds, which will have different bond energies. There is only one type of O-H bond in this molecule.

Reaction	Bond Broken
$O(CH_2C'H_2OH)_2 \rightarrow O(CH_2C'H_2OH)(CH_2C'H_2O) + H$	O-H (1)
$O(CH_2C'H_2OH)_2 \rightarrow O(CH_2C'H_2OH)(CH_2C'H_2) + OH$	C-O (2)
$O(CH_2C'H_2OH)_2 \rightarrow O(CH_2C'H_2OH) + CH_2C'H_2OH$	C-O (3)
$O(CH_2C'H_2OH)_2 \rightarrow O(CH_2C'H_2OH)(CHC'H_2OH) + H$	C'-H (4)
$O(CH_2C'H_2OH)_2 \rightarrow O(CH_2C'H_2OH)(CH_2C'HOH) + H$	C-H (5)

$C_4H_{10}O_3$ or $O(CH_2CH_2OH)_2$

However, the two C-O bond energies in this molecule (and C-O single bond energies in other molecules) are similar. The two C-H bond energies are also similar. Tables of bond energies have two sets of values: one set are exact values from measurements on diatomic molecules and the others are average values which are determined from the values for a large number of measurements of those bonds in different molecules. The average values are useful for estimating the energy changes in chemical reactions when other data are unavailable.

Factors Affecting Bond Energy

Bond energies and bond lengths are related; shorter bonds tend to have higher bond energies. The factors that affect bond lengths will also affect bond energies. The relative importance of the factors is different than for bond lengths.

- The higher the bond multiplicity, the stronger the bond and the larger the bond energy.

Rationale: The greater number of electrons in multiple leads to stronger attraction of the nuclei toward the electrons and the greater the bond energy.

- The greater the electronegativity difference, $\Delta\chi$, the stronger the bond.

    Rationale: The forces that bind atoms together in covalent bonds are electrostatic in nature. Polar bonds between atoms have a fractional negative charge on the more electronegative atom and a corresponding positive charge on the less electronegative atom. The electrical attraction between the unlike charges strengthens the bond.

- The better the orbital overlap, the stronger the bond.

    Rationale: Good overlap results from having similar sized atomic orbitals which result in molecular orbitals that concentrate electron density between the nuclei.

Reaction Energy

The average values of bond energies are useful in estimating the energy change in a chemical reaction, because chemical reactions result in the formation or breaking of chemical bonds. These rules and definitions are imperative to remember:

1. Breaking chemical bonds **always requires** energy.

2. Formation of chemical bonds **always releases** energy.

3. *Exothermic* processes release energy ($\Delta E < 0$), and *endothermic* processes absorb energy ($\Delta E > 0$).

We can estimate the energy change in a chemical reaction by estimating the energy required to break the bonds and the energy released by the formation of bonds in the reaction. The energy change, ΔE, is the difference between the energy released from formation of the new bonds in the products and the energy required to break the bonds in the reactants:

$$\Delta E = - \text{(sum of } BE\text{s of bonds formed)} + \text{(sum of } BE\text{s of bonds broken)}$$

(the negative sign before the first term makes that sum negative)

To use this relation, we must be able to draw structures of the molecules involved so we can identify the bonds broken and formed. Molecular pictures are useful for this purpose.

Exercise 9.4 The solvent dichloromethane, CH_2Cl_2, is made from the reaction of methane, CH_4, with chlorine gas, Cl_2. The other product of the reaction is hydrogen chloride, HCl. Draw a molecular picture representing the balanced chemical reaction for this process and then determine the energy change for the reaction of one mole of methane to give one mole of dichloromethane.

Steps to Solution: The first step is to write the balanced chemical equation of the reaction. Then we determine the structures of the molecules and draw the molecular *picture. From the molecular picture, we see the bonds that are broken and formed. The energy change in the reaction can be determined from the bonds broken and formed using the bond energies listed in the table in your text.*

The balanced chemical reaction is written using the fact that the reactants (CH_4 and Cl_2) and products (CH_2Cl_2 and HCl) are identified as such in the problem. The balanced chemical equation is given below.

$$CH_4 + 2\ Cl_2 \rightarrow CH_2Cl_2 + 2\ HCl$$

The structures of the molecules are determined using the methods described in Chapter 8. The molecular picture for the reaction is shown below.

At this point, we need to determine the type and number of bonds broken and formed.

Bonds Broken			Bonds Formed		
Type	BE/bond	Number	Type	BE/bond	Number
C-H	415	2	C-Cl	325	2
Cl-Cl	239.7	2	H-Cl	428	2

The energy change is determined from the difference of the energy required to break the bonds in the reactants and the energy released upon formation of the new bonds in the products.

$$\Delta E = 2\ mol\ C\text{-}H\ (415\ kJ/mol\ C\text{-}H) + 2\ mol\ Cl\text{-}Cl\ (239.7\ kJ/mol\ Cl\text{-}Cl)$$

$$-\ (2\ mole\ C\text{-}Cl(325\ kJ/mole\ C\text{-}Cl) + 2\ mole\ H\text{-}Cl\ (428\ kJ/mole\ H\text{-}Cl))$$

$$\Delta E = \boxed{-197\ kJ/mole}$$

The reaction is exothermic, releasing an estimated 197 kJ for each mol of dichloromethane produced. The reaction is exothermic mainly due to the large difference between the C-Cl and Cl-Cl bond energies; the larger bond energy in the products leads to net release of energy.

Notice that only the bonds that are broken or formed are used in this calculation even though there are other bonds present; for example, in Exercise 9.4, the 2 C-H bonds are not considered in the calculation. If included in the calculation, there will be no difference in the answer because the energy required to break those bonds will be "returned" when those bonds are formed in the products.

Exercise 9.5 Nylon is an important component in many products. One step in the production of nylon is the reaction of cyclohexane with oxygen to give cyclohexanone and water, shown below in the unbalanced reaction in line formula notation.

⬡ + O=O → ⬡=O + H_2O

(a) Write the balanced chemical reaction and estimate the energy change for this process.
(b) Identify which bond energies are mainly responsible for the reaction being endothermic or exothermic.

Steps to Solution: The solution to this problem is similar to that in Exercise 9.4 with the main difference in the initial stage of the problem: the line structures must be converted to structural formulas before the equation is balanced. Then we can identify the bonds that are broken and formed in the reaction. In part (b), we need to find which bond energies lead to the difference in energy.

(a) The reaction above is converted to full structural formulas below; the C atom labels have been omitted for clarity. The reaction is balanced as written.

[structural formula of cyclohexane] + O=O → [structural formula of cyclohexanone] + H_2O

Now we can determine the type and number of bonds broken and formed.

Bonds Broken

Type	BE/bond	Number
C-H	415	2
O=O	493.6	1

Bonds Formed

Type	BE/bond	Number
C=O	750	1
H-O	460	2

The energy change is the difference of the energy required to break the bonds in the reactants and the energy released upon formation of the new bonds in the products.

ΔE = 2 mole C-H (415 kJ/mole C-H) + 1 mole O=O (493.6 kJ/mole O=O)

− (1 mole C=O(750 kJ/ mole C=O) + 2 mole H-O (460 kJ/mole H-O))

ΔE = $\boxed{-346 \text{ kJ/mole}}$

(b) *The bonds that are most important in determining the energy change are the C=O and O=O bonds, with bond energies of 750 kJ and 494 kJ, respectively. The C-H and O-H bond energies have much more similar values of 415 and 460 kJ. Therefore the fomation of the strong C=O bond is responsible for most of the energy released in the reaction.*

9.3 Second Row Diatomic Molecules

QUESTIONS TO ANSWER, SKILLS TO LEARN
1. What is constructive and destructive interference?
2. New ways of mixing orbitals besides hybridization; generation of bonding and antibonding orbitals
3. Electron configuration and bond order in diatomics

Qualitative Energies and Orbital Mixing in Heteronuclear Diatomic Molecules

In the localized picture of bonding described above, new orbitals are formed from the overlap of hybrid orbitals on the inner atoms and unhybridized orbitals on the terminal or outer atoms. This model of bonding has some shortcomings; one is the occasional necessity to write resonance structures where different arrangements of electrons can be written. Another problem with the localized bonding model is that it may not predict whether a molecule is paramagnetic (attracted to a magnet because of unpaired electrons) or diamagnetic (all electron spins paired). Oxygen, O_2, is a good example. The Lewis structure of oxygen is:

$$\ddot{\underset{..}{O}}=\ddot{\underset{..}{O}}$$

where there is an O-O double bond and no unpaired electrons. Experiments show that there are two unpaired electrons in O_2. A different bonding theory that successfully describes the bonding in O_2 and does not require the use of resonance structures is described below.

Use of the localized bonding model for O_2 would lead to the formation of the double bond through the formation of a σ bond and a π bond. The σ bond would result from the overlap of the $2p$ orbitals that point at the oxygen atoms. This orbital is called the σ_p orbital because it is made from the overlap of two $2p$ orbitals.

$$\infty\infty \Rightarrow \infty\infty \quad (\sigma_p)$$

The unshaded portion of the new orbital puts electron density between the two nuclei, leading to a bonding interaction. The new orbital, called a *bonding orbital*, results from a phenomenon of waves called *constructive interference*: when two waves interact, a new wave is formed that is the sum of the two component waves. If the "crests" align, the new wave has higher amplitude than either of the old waves, as shown below at left.

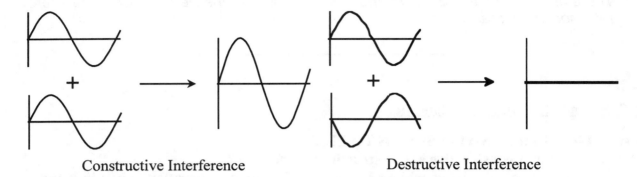

Constructive Interference   Destructive Interference

Another combination of these waves results in a different wave. In the new combination, one of the waves has been "inverted". In this case, the two waves combine to cancel one another out. This is called *destructive interference*, shown above at right.

Wavefunctions of orbitals also show constructive and destructive interference. The combination of the two $2p$ orbitals of oxygen shown earlier where electron density is greater between the two nuclei is an example of constructive interference. The combination of $2p$ orbitals leading to destructive interference is shown below:

$$\infty\text{○}\infty\text{○} \Longrightarrow \text{○}\infty\text{○}\infty\text{○} \quad (\sigma^*_p)$$

This new orbital leads to a lower amount of electron density between the two nuclei than either of the $2p$ orbitals uncombined. Because of the low electron density between the nuclei (it is actually zero at one point) the electrons do not attract the nuclei to each other as strongly as in constructive interference. This new orbital is an *antibonding molecular orbital* or just an *antibonding orbital*.

Because both of these orbitals, σ_p and σ^*_p, have maximum electron density on a line drawn through the two nuclei, they are called σ bonds. These two combinations are the only allowed combinations for the $2p$ wavefunctions lying along the bond axis. These two new bond types and a rule about the number of orbitals that can be obtained from atomic orbitals are:

Constructive **interference of wave functions** → *bonding* **orbital**

Destructive **interference of wave functions** → *antibonding* **orbital**

The combinations of x atomic orbitals will result in exactly x orbitals (bonding and antibonding) in the molecule.

Exercise 9.6 Show the bonding and antibonding orbitals obtained from:
(a) The 2s orbitals in Li_2
(b) The $2p_x$ orbitals in F_2 (the z-axis contains the F-F bond).

Steps to Solution: To answer this question, we must draw the orbitals on the atoms of the molecule and show the two combinations obtained by constructive and destructive interference.

(a) In Li_2 the two 2s orbitals can either have the same sign, leading to constructive interference, or different signs, leading to destructive interference. These two combinations are shown below.

Constructive : σ_s bonding orbital   Destructive : σ_s^* antibonding orbital

(b) The $2p_x$ orbitals in F_2 can form two combinations through (i) addition of like-oriented wave functions (constructive interference) or (ii) adding the two wave functions so destructive interference will occur:

Constructive Interference: π_p bonding orbital   Destructive Interference: π_p^* antibonding orbital

We have seen above the σ and σ^* orbitals that result from the combinations of the two s orbitals and the two p_z orbitals. The π and π^* orbitals result from the combinations of two p orbitals that have parallel axes but are perpendicular to the σ bond. In diatomic molecules of the second row elements, there are two sets of π and π^* orbitals which differ only in orientation which are generated from the p_x and p_y orbitals. In summary:

Two s orbitals give: σ_s and σ_s^* orbitals

Six p orbitals give: σ_p and σ_p^* orbitals and π_x, π_y and π_x^* and π_y^* orbitals

In general, the bonding orbitals will be more stable than the parent atomic orbitals because formation of bonds leads to the release of energy. The antibonding orbitals will be less stable than the parent atomic orbitals, as shown in the diagram to the right.

The combinations for the p orbitals in diatomic molecules are more complex. The six p orbitals form 6 new orbitals in the molecule. The most stable is the σ_p because overlap is most effective along the bond axis. The π bonding orbitals formed from the p_x and p_y orbitals will be of equal energy and more stable than the parent atomic orbitals, but not nearly as stable as the σ bonding orbital because of the poorer overlap of the "side-by-side" orientation. The π^* orbitals will be higher in energy than the π orbitals, and the highly destabilizing σ^* orbitals will have the highest energy of the orbitals formed.

Energy Levels for MO's from p orbitals

In diatomic molecules, both s and p orbitals are participating in the formation of bonds. The s orbitals are more stable than the p orbitals, so orbitals obtained from ns orbitals are expected to be more stable than those obtained from np orbitals. The energy level diagram that reflects this greater stability of the s orbitals is shown below. This energy level diagram is qualitative in nature but will give the correct answers for nearly all of the diatomic molecules.

MO Diagram for O_2

The electron configuration of a diatomic molecule is obtained by filling the molecular orbitals with the valence electrons of the molecule, in keeping with the Aufbau and Pauli principles. For oxygen there are 12 valence electrons, resulting in this electron configuration:

$$(\sigma_s)^2 (\sigma_s^*)^2 (\sigma_p)^2 (\pi_x)^2 (\pi_y)^2 (\pi_x^*)^1 (\pi_y^*)^1$$

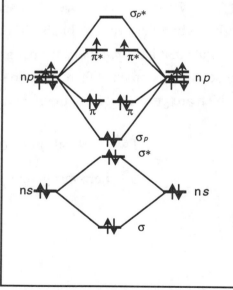

Note that the sum of the superscripts is the same as the number of valence electrons. To determine the number of bonds in the O_2 molecule using this bonding scheme is not as straight-forward as counting the bonding pairs between the nuclei in the Lewis structure. However, examination of the diagram at right shows that population of an antibonding orbital destabilizes the molecule by about the same amount of energy that population of a bonding orbital of the same type stabilizes the molecule. Because a bond has been previously defined as two nuclei sharing a pair of electrons, the **bond order** is defined as the *net number of bonding pairs of electrons*, calculated by the following formula:

$$\text{Bond order} = \frac{\text{number of bonding electrons} - \text{number of antibonding electrons}}{2}$$

In O_2 there are 8 bonding electrons and 4 antibonding electrons, so the bond order is:

$$\text{Bond order} = \frac{8-4}{2} = 2$$

It is comforting that the Lewis structure and this bonding theory give the same result for the bond order. This analysis also predicts that the O_2 molecule has 2 unpaired electrons in the ground state, a fact verified by experiment, but not predicted by the localized bonding theory description.

Determining electron configuration and bond order in diatomic second row molecules and molecular ions.

1. Determine the number of valence electrons.

2. Add the electrons to the molecular orbitals in accordance with the Pauli and Aufbau principles. The order of the orbitals from lowest to highest energy is:

$$(\sigma_s) < (\sigma_\sigma^*) < (\sigma_p) < (\pi_x), (\pi_y) < (\pi_x^*), (\pi_y^*) < (\sigma_p^*)$$

3. Calculate the bond order by the above formula.

Exercise 9.6 Predict the bond order and number of unpaired electrons in (a) O_2^+ (b) O_2^- (c) NO^+ (d) F_2^+.

Steps to Solution: To determine the bond order in the given molecular ions listed we need to apply the steps listed above. We can then determine the number of unpaired electrons by counting the electrons in singly occupied orbitals.

(a) The number of valence electrons in O_2^+ is 11. The electron configuration for this molecular ion is developed below. The orbitals for the diatomic second row molecules are shown, and the number of electrons in each orbital is tallied in the row beneath the orbital label. Electrons in antibonding orbitals are designated by italics.

(σ_s) (σ_σ^*) (σ_p) (π_x) (π_y) (π_x)
(2+   *2*+   2+   2+   2+   1) = 11 electrons

Chapter 9: Chemical Bonding: Multiple Bonds

The electron configuration lists the orbital names and the number of electrons in that orbital, as shown below:

$$(\sigma_s)^2 (\sigma_\sigma*)^2 (\sigma_p)^2 (\pi_x)^2 (\pi_y)^2 (\pi_x*)^1$$

$$\text{Bond order} = \frac{2+2+2+2-(2+1)}{2} = \boxed{2.5}$$

There is one unpaired electron in a $\pi*$ orbital.

(b) The number of valence electrons in O_2^- is 13. The orbitals for the diatomic second row molecules are shown and the number of electrons in each orbital is tallied in the row beneath the orbital label. Electrons in antibonding orbitals are designated by italics.

$(\sigma_s)\quad (\sigma_s*)\quad (\sigma_p)\quad (\pi_x)\quad (\pi_y)\quad (\pi_x*)\quad (\pi_y*)$
$(2+\quad 2+\quad 2+\quad 2+\quad 2+\quad 2+\quad 1\;)\qquad = 13$ electrons

The electron configuration is;

$$(\sigma_s)^2 (\sigma_\sigma*)^2 (\sigma_p)^2 (\pi_x)^2 (\pi_y)^2 (\pi_x*)^2 (\pi_y*)^1$$

$$\text{Bond order} = \frac{2+2+2+2-(2+2+1)}{2} = \boxed{1.5}$$

There is one unpaired electron in a $\pi*$ orbital.

(c) Use the methods in parts (a) and (b) to show the electron configuration and bond order are:

$$(\sigma_s)^2 (\sigma_\sigma*)^2 (\sigma_p)^2 (\pi_x)^2 (\pi_y)^2 \qquad \text{Bond order} = \boxed{3.0}$$

There are no unpaired electrons.

(d) The number of valence electrons in F_2^+ is 13, the same as in O_2^-. The electron configuration and bond order are the same as those of the isoelectronic O_2^- ion.

$$(\sigma_s)^2 (\sigma_\sigma*)^2 (\sigma_p)^2 (\pi_x)^2 (\pi_y)^2 (\pi_x*)^2 (\pi_y*)^1 \qquad \text{Bond order} = \boxed{1.5}$$

There is one unpaired electron.

Exercise 9.7 Sketch the highest energy occupied orbital in: (a) part (c) of Exercise 9.6. and (b) part (d) of Exercise 9.6.

Steps to Solution: To answer this question, we must first identify the highest-energy occupied orbital. Then we may remember the shapes of the orbitals or use of the *ideas of constructive/destructive interference to draw the orbitals or (less preferable) refer to the figures in the text. Then draw the orbital. (This takes practice!)*

(a) *The highest-energy occupied orbitals in NO⁺ (part [c]) are the π_x and π_y orbitals. Because these two orbitals have the same energy (they differ only in their orientation to the x and y axes), a drawing of either one will answer the question. The π orbitals place electron density between the nuclei but above and below the bond axis.*

(b) *The highest-energy occupied orbital of F_2^+ (part [d]) is a π^* orbital. Again, the two π^* orbitals are degenerate, so either π_x^* or π_y^* are acceptable answers. The π^* orbitals do not place electron density between the atoms being bonded, but on the side opposite the atom being bonded and above and below the line that contains the nuclei.*

9.4 Delocalized π Orbitals

QUESTIONS TO ANSWER, SKILLS TO LEARN
1. **What does *delocalized* mean?**
2. **Making orbitals from combinations of wave functions from more than two atoms**
3. **Understanding the consequences of delocalized bonding in some molecules**
4. **Describing bonding in terms of localized and delocalized bonding**

Delocalized bonding is a term used to describe bonding where electron density in an orbital is distributed over more than two atoms. In the molecules that will be discussed in this text, delocalization is observed only in π bonds.

Delocalized bonding: Electron density in a π bond distributed over more than two atoms

The allyl cation, $C_3H_5^+$, will be used to illustrate a method of describing bonding in molecules and molecular ions in terms of localized and delocalized bonding.

1. **Determine the Lewis structure and important resonance structures.**

   The allyl cation has 16 valence electrons and the following Lewis and resonance structures:

Chapter 9: Chemical Bonding: Multiple Bonds

[Resonance structures of the allyl cation shown at top of page]

2. Use localized bonding and appropriate hybridization to describe the σ bond framework.

In this ion (a) each C atom has a steric number of 3 in both resonance structures and (b) the two structures vary only in the position of the π bond; that is, no changes have occurred in the σ bonding framework. The σ bonding framework may be described by assigning sp^2 hybridization to the three C atoms (consistent with steric number 3) and forming C-C and C-H σ bonds, as shown in the illustration below.

3. For molecules with multiple bonds, construct the π bonding system with the unhybridized p orbitals. The presence of resonance structures indicates delocalized π bonds.

The presence of two resonance structures suggests that π bonding should involve all three C atoms. The following figures show combinations of (a) the unhybridized $2p$ orbitals on the C atoms that form new orbitals that involve all three C atoms and (b) the new orbitals that result from these combinations.

π (bonding)   π*(antibonding) $π_n$ (nonbonding or lone pair)

Note that three molecular orbitals are obtained from three atomic orbitals.

4. Each σ bond will require 2 valence electrons. Next place a pair of valence electrons in each atomic orbital that has not been used in either hybridization or delocalized orbital formation. (Achieve an 8 electron count on outer atoms and fill valence atomic orbitals.)

The σ bonding requires 14 electrons in the 7 σ bonds. The outer H atoms require no more electrons. All of the other orbitals have been used either in hybrid orbital formation or in the formation of the π bond; therefore there are 2 valence electrons to be placed in the π orbitals formed.

5. **Nonbonding orbitals obtained by this method show where lone pairs (if present) are delocalized.**

In some molecules or polyatomic ions, atoms may have different numbers of different resonance structures (or lone pairs may "change positions" in different resonance structures) as in the carbonate ion). These lone pairs are delocalized over several atoms which are shown in the π_n molecular orbital.

9.5 π Bonding in Polyatomic Anions

QUESTIONS TO ANSWER, SKILLS TO LEARN
1. **What indicates that delocalized bonds are present in polyatomic anions?**
2. **Identifying the types of orbitals used in forming π bonds in polyatomic anions.**

Polyatomic anions that contain oxygen, such as sulfate, nitrate and phosphate, demonstarate the importance of delocalized bonding. Also, different orbitals are utilized for bonding in the second and third row compounds. The bonding in NO_3^- can be described using the sequence of steps described above in Section 9.4.

Exercise 9.8 Describe the bonding in the polyatomic anion, nitrate.

Steps to Solution: *The bonding can be described using the five step procedure used above in Section 9.4 and Exercise 9.7.*

1. *Determine the Lewis structure and important resonance structures. These are shown below.*

2. Use localized bonding and appropriate hybridization to describe the σ bond framework.
In the nitrate ion, the inner N atom has a steric number of 3 so that sp² hybridization is indicated for that atom. The σ bond framework can then be constructed from overlap of p orbitals on the O atoms with the sp² hybrids on the N atom as shown at right. The ion is expected to have a trigonal planar structure.

3. For molecules with multiple bonds, construct the π bonding system with the unhybridized p orbitals. The presence of resonance structures indicates delocalized π bonds. Because there are three resonance structures, delocalized π bonding is expected. The unhybridized 2p orbital on the N atom can combine with the three 2p orbitals on the oxygen atoms that point above and below the plane of the molecule to give four new orbitals. Two of these combinations and the shape of the resulting π orbitals are shown below.

	Combination of Atomic Orbitals	π Molecular Orbital
Constructive interference places electron density between atoms if this orbital is occupied. This combination is a π bonding orbital that will be of low energy.		
Destructive interference results in little electron density between the atoms when this orbital is occupied. This is a π* orbital that is of higher		

energy than the π bonding and the nonbonding π_n orbitals. The other two orbitals obtained from the four atomic orbitals will be nonbonding (lone pair) π_n orbitals that delocalize the lone pairs onto the oxygen atoms. There will be one π bonding, two nonbonding π_n and one antibonding π* orbital. The bonding orbital places electron density between the N and O nuclei; this orbital delocalizes one pair of electrons (the π that "migrates") over the four nuclei. The non-bonding orbitals hold the two lone pairs that are not localized on the oxygens; the two lone pairs are delocalized over the three oxygen atoms.

9.6 Band Theory of Solids

QUESTIONS TO ANSWER, SKILLS TO LEARN
1. What are the properties of metals that suggest delocalized bonds?
2. Understanding the nature of the orbitals obtained by considering many atoms
3. Learning that bands represent many orbitals separated by very small energies
4. What are the differences between an electrical conductor, an insulator and a metalloid?

The bonding in metals is different than the bonding in molecules. Think small: a piece of metal is a large number of metal atoms that must be held together by some form of covalent bonding (ionic bonding is not possible because the electronegativity difference is zero). The hardness of metal varies tremendously, from mercury, which is a liquid at room temperature, to tantalum and tungsten which are very hard and melt at very high temperatures. However, there are some properties that are common to all metals and are indicative of delocalized bonding:

- Electrical current is readily conducted through metals by the motion of electrons.

- Metals are easily deformed: hammered thin (malleable) or drawn into wires (ductile). The bonds between the metal atoms do not point in a specific direction.

- All clean metal surfaces have metallic luster (are shiny).

One way to view the bonding in a piece of metal is that the valence atomic orbitals overlap and new orbitals are fomed that extend over the entire piece of metal. The number of orbitals in the metal will be equal to the number of atomic orbitals; each of these new orbitals has a specific energy. The interaction of three *s* orbitals is shown to the right; this leads to three energy levels. The interaction of a very large number

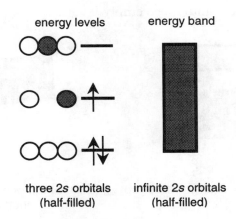

of *s* orbitals leads to so many different orbitals that the energy difference between the sucessive orbitals becomes very small. The energy difference becomes so small that, effectively, there is a continuous range of energies between some minimum and maximum values. This range of energies called an **energy band**. This theory of bonding is called **band theory**; it applies to metals, nonmetals and metalloids.

Important features of energy bands:

- **There is a continuous range of allowed energies for electrons within a band.**

- **Electrons in a pure substance may only have energies contained within a band.**

- **The different orbitals (*s*, *p*, *d* and *f*) all form bands which may overlap.**

Electricity is conducted by metals because it requires very little energy to promote (or excite) an electron in the band into a higher energy state. This is illustrated below where the energy of the band is lower nearer the positive charge and more electrons occupy states at the low-energy end. This is possible only with partially filled bands.

Solids can conduct electricity if they have a partially filled band such as that for lithium or if filled bands overlap with empty bands. The figures at right illustrate these two situations. In both of these scenarios, there are empty energy levels with only a very small amount of energy required for electrons to be "excited" into those empty levels.

Insulator (large band gap)   Tilting of the energy of a band of an insulator upon applying voltage   Semiconductor (small band gap)

If a band is filled (corresponding to a filled valence shell) there is no shifting of electron density towards one end of the solid (shown above at center); therefore no current flows. This is characteristic of an electrical insulator. The figure at left shows the band structure for an insulator. A filled band is separated from an empty band by a large enough amount of energy so that electrons may not be promoted by thermal energy.

The energy between the filled band and the empty band is called the **band gap**. Large band gaps are characteristic of insulators. A large band gap ensures that very few electrons have sufficient energy to be able to jump the gap into the empty band where they are mobile.

If the band gap is small, different behavior is observed. The substances are poor conductors of electricity, but much better than insulators. In addition, the conductivity of these substances increases with increasing temperature, in contrast to a conductor. Due to the modest conductivity that these substances possess, they are called **semiconductors**. The band structures of an insulator and a semiconductor are shown above.

Heating a semiconductor increases the number of electrons that have enough energy to jump from the filled band to the empty band where they are mobile, therefore increasing the conductivity. Alternatively, absorption of a photon can also provide the energy to promote an electron into these levels.

Electrical Conductivity	Band Gap	Examples
very low: insulator	large (>300 kJ/mole)	quartz (SiO_2), diamond
modest: semiconductor	small (60-120 kJ/mole)	Si, Ge, GaAs
high: conductor	none (0)	Cu, Zn, Au, Ag, Al

Doping Semiconductors

Pure semiconductors have relatively low conductivity, but the controlled addition of impurities (**doping**) can increase the conductivity. For pure Si, addition of small amounts of phosphorus (1 atom of phosphorus per billion Si atoms) which has one *more* valence electron than Si, increases the conductivity. Curiously enough, the addition of similar quantities of B (which has one *less* valence electron than Si) also increases the conductivity. When phosphorus is added, there are more valence electrons than in pure Si; therefore this is called an *n-type* semiconductor ("n" for the excess of negative charges). The addition of B leads to a deficit of valence electrons; the B doped material is called a *p-type* semiconductor.

Exercise 9.9 Semiconductors are used in solar energy conversion in devices called photovoltaic cells. For high efficiency, the cell should be able to utilize as much light as possible. What band gap (in kJ/mole) would be needed to utilize light of 550 nm (about the middle of the visible spectrum) and shorter wavelengths?

Steps to Solution: The band gap would correspond to the energy of the lowest-energy photon that will excite an electron. This energy is the energy of a 550 nm photon, as we were told in the problem. The energy of the photon is found from the relation between energy and wavelength:

$$E = \frac{hc}{\lambda} = \frac{6.626 \times 10^{-34} \text{J} \cdot \text{s}(3.00 \times 10^8 \text{m/s})}{550 \times 10^{-9} \text{m}} = \boxed{3.61 \times 10^{-19} \text{ J}}$$

The energy per mol of photons is:

$$E = 3.61 \times 10^{-19} \text{ J } (6.022 \times 10^{23} \text{ 1/mol}) = \boxed{217 \text{ kJ/mol}}$$

The band-gap must be 217 kJ/mol to utilize all wavelengths less than 550 nm.

Exercise 9.10 Draw the band diagram that is appropriate for a ZnS semiconducor that has been doped with copper.

Steps to Solution: The type of semiconductor (p or n) must be determined. Copper has one less valence electrons than Zn, so the resulting solid will have less valence electrons than the undoped semiconductor. Therefore the band diagram will be similar to that for p-type semiconductors and is shown at right.

Test Yourself

1. In which chemical species would you expect the strongest bond: C_2^+; N_2^+; O_2^+? Write the electronic configuration and calculate the bond order for that species.

2. Estimate ΔE for the following reaction:

$$C_2H_{4(g)} + Br_{2\ (g)} \rightarrow C_2H_4Br_{2\ (g)}$$

(BE_{C-Br} = 276 kJ/mole, BE_{Br-Br} = 193 kJ/mole)

3. Draw the bonding and antibonding MO's for NO_2^-.

4. Describe the bonding in formaldehyde, CH_2O, in terms of σ and π bonding.

5. What kind of bonds are obtained by the overlap of the d_{xy} and the p orbitals shown below?

Answers

1. O_2^+; $(\sigma_s)^2(\sigma_s^*)^2(\sigma_p)^2(\pi_x)^2(\pi_y)^2(\pi_x^*)^1$; bond order = 2.5

2. $\Delta E = -359$ kJ

3.

      antibonding MO   bonding MO

4. The Lewis structure of formaldehyde is shown below

The C-H bonds are σ bonds made from overlap of the sp^2 hybrid orbitals on C with the 1s orbitals on the H atoms. The C-O double bond is formed from a σ bond and a π bond. The σ bond is formed from overlap of the sp^2 orbital on the C atom with a p orbital on the O atom. The π bond is formed from overlap of the unhybridized p orbital on the C atom with a p orbital on the O atom.

5. A σ bond has electron density directly on the line joining the two nuclei; a π bond puts electron density above and below the line joining the nuclei. The assignment of σ bonding and π bonding combinations is shown below.

      σ bond   π bond

Chapter 9: Chemical Bonding: Multiple Bonds

Solutions to Chapter Problems

Problems 9.1, 3, 5 all deal with organic molecules whose line structures are provided. To determine the bonding pattern from a line structure, first convert the line structure into a molecular structure by adding C and H atoms. Molecular structures contain enough information to deduce steric numbers of C, N, and O inner atoms without determining the complete Lewis structure. An inner C, N, or O atom with no multiple bonds has SN = 4 and bonding that can be described using sp^3 hybrid orbitals. One with one π bond has SN = 3 and bonding that can be described using sp^2 hybrid orbitals and one p orbital. One with two π bonds has SN = 2 and bonding that can be described using sp hybrid orbitals and two p orbitals.

9.1 Acetone has three inner C atoms. Two have only single bonds (one C–C and three C–H); these have SN = 4. The atom bonded to O has one π bond and SN = 3. The σ bonds are:

LIST OF σ BONDS:	Bond	Number	Orbitals
	C–H	6	sp^3 - $1s$
	C–C	2	sp^3 - sp^2
	C–O	1	sp^2 - $2p$

Between C and O there is also a π bond formed by side-by-side overlap of $2p$ orbitals on each atom. Here are sketches of the bonding orbitals:

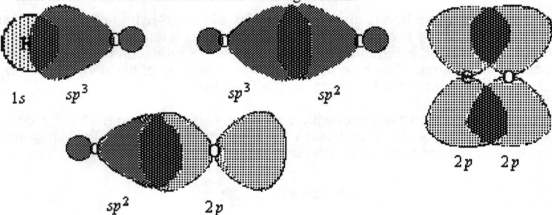

9.3 The line structure of neocembrene in Box 9-1 shows that it contains only inner carbon atoms and outer hydrogen atoms. Some of the inner atoms have all single bonds, SN = 4, sp^3 hybridization, and tetrahedral geometry; others have one double bond, SN = 3, sp^2 hybridization, and trigonal planar geometry. The SN values for each of the 20 carbon atoms are shown in the line structure below.

9.5 From the line structures of the compounds, we determine the steric number of each carbon atom. These are shown on the line structures below:

1,3-pentadiene   1-pentyne   cyclopentene

C atoms designated "4" have only single bonds, SN = 4, sp^3 hybridization, and tetrahedral geometry. Those with a double bond, designated "3," have SN = 3, sp^2 hybridization, and trigonal planar geometry. Two C atoms, designated "2," have two double bonds, sp hybridization, and linear geometry. The σ bonds are described by overlap of hybrid orbitals, with each C–H bond described as a hybrid overlapping with a hydrogen $1s$ orbital. The π bonds are described by side-by-side overlap of $2p$ orbitals. In 1–pentyne, the two C atoms designated "2" have three bonds: a σ bond formed from sp hybrids and two π bonds, at right angles to each other, as in acetylene. Here are sketches of the different types of bonds:

9.7 The determinants of bond length, in order of importance, are principal quantum number of the valence orbitals, bond order, and bond polarity. Among these bonds, H–N is shortest because it involves an n = 1 orbital and Cl–N is longest because it involves an n = 3 orbital. Of the three bonds involving n = 2 orbitals, the N–N single bond is longest and the N≡N triple bond is shortest: H–N < N≡N < C=O < N–N < Cl–N

9.9 Bond strengths, like bond lengths, depend on principal quantum number, bond order, and bond polarity, but the correlation is not as strong as for bond lengths, so tabulated values must be consulted. Here are the values from Table 9.2 of your textbook, listed in increasing order:

Bond	C–C	<	H–N	<	C=C	<	C=O	<	N≡N
Energy (kJ/mol)	345		390		615		750		945

Strong bonding of the n = 1 orbital makes H–N > C–C; Multiple bonding makes C=C > H–N; Bond polarity makes C=O > C=C; and Multiple bonding makes N≡N > C=O.

9.11 To estimate the energy change in a reaction, determine Lewis structures to obtain types of bonds, list the number of bonds of each type in reactants and products, and subtract the sum of average bond energies for products from the sum of average bond energies for reactants, using values from Table 9.2 of your textbook.

The reaction is $N_2 + 2\,O_2 \rightarrow N_2O_4$

Here are the Lewis structures (resonance forms not shown):

Reactants		
Bond	No.	Energy
N≡N	1	945
O=O	2	495

Products		
Bond	No.	Energy
N–N	1	165
N–O	2	200
N=O	2	607

$\Delta E_{rxn} \cong [1(945) + 2(495)] - [1(165) + 2(200) + 2(605)] = 1935 - 1775 \cong 160$ kJ

9.13 To estimate the energy change in a reaction, determine Lewis structures to obtain types of bonds, list the number of bonds of each type in reactants and products, and subtract the sum of average bond energies for products from the sum of average bond energies for reactants, using values from Table 9.2 of your textbook.

$2\,H_2 + O_2 \rightarrow 2\,H_2O$

Reactants		
Bond	No.	Energy
H–H	2	435
O=O	1	495

Products		
Bond	No.	Energy
O–H	4	460

$\Delta E_{rxn} \cong [2(435) + 1(495)] - [4(460)] = 1365 - 1840 = -475$ kJ

$3\,H_2 + N_2 \rightarrow 2\,NH_3$

Reactants		
Bond	No.	Energy
H–H	3	435
N≡N	1	945

Products		
Bond	No.	Energy
N–H	6	390

$\Delta E_{rxn} \cong [3(435) + 1(945)] - [6(390)] = 2250 - 2340 = -90$ kJ

(continued on next page)

9.13 (continued)

$$3\,H_2 + CO \rightarrow CH_4 + H_2O$$

Reactants			Products		
Bond	No.	Energy	Bond	No.	Energy
H–H	3	435	C–H	4	415
C≡O	1	1070	O–H	2	460

$\Delta E_{rxn} \cong [3(435) + 1(1070)] - [4(415) + 2(460)] = 2375 - 2580 = -205$ kJ

9.15 Orbital energy ladders for diatomic molecules have the form shown in Figure 9-17 of your textbook. Count valence electrons, then place them in available orbitals using the Pauli and Aufbau principles. Stability is determined by the number of bonding and antibonding electrons. Na_2 has two valence electrons, one from each atom:

The configuration has two bonding and no antibonding electrons, indicating a stable molecule with a single bond.

9.17 Stability of a diatomic molecule is determined by the number of bonding and antibonding electrons. To compare species, determine how many valence electrons each species possesses, then place them in the available orbitals following the Pauli and Aufbau principles. Here are the configurations for nine, ten, and eleven valence electrons:

(a) CO has 10 valence electrons. To make CO^-, an electron is added to a π^* orbital, which destabilizes the ion, so CO has a stronger bond.
(b) N_2 has 10 valence electrons. To make N_2^+, an electron is removed from a bonding orbital, which destabilizes the ion, so N_2 has a stronger bond.
(c) CN^- has 10 valence electrons. To make CN, an electron is removed from a bonding orbital, which destabilizes the species, so CN^- has a stronger bond.

9.19 Determine how many valence electrons the molecule possesses, then place them in the available orbitals following the Pauli and Aufbau principles. Stability is determined by the number of bonding and antibonding electrons.

ClO⁻ has 14 valence electrons, six from O, seven from Cl, plus one due to its charge.

There are eight bonding electrons and six antibonding electrons, for a net of two bonding electrons and a single bond for this ionic species.

9.21 Orbital sketches for diatomic molecular orbitals are constructed by overlapping the appropriate atomic orbitals. See the figures in Section 9.3 of your textbook. The configuration of CO (10 valence electrons) is $(\sigma_s)^2 (\sigma_s^*)^2 (\pi_x)^2 (\pi_y)^2 (\sigma_p)^2$, indicating that there are four bonding orbitals to depict. The two π orbitals appear identical except for their orientations relative to each other, so three pictures are needed:

9.23 Determine the Lewis structure using the usual procedures. The molecule has $3(4) + 2(6) = 24$ valence electrons:

:Ö—C—C̈—C—Ö: → :Ö=C=C=C=Ö:

Each carbon atom has SN = 2, so the molecule is linear and the σ bond network can be described using sp hybrids. There are two sp-sp bonds between carbon atoms and two sp-$2p$ bonds between carbon and oxygen atoms:

If the molecular axis is the z-axis, a p_x and a p_y orbital on each atom overlaps side-by-side, giving two sets of delocalized π orbitals, each extending over all five atoms:

9.25 Determine the Lewis structure using the usual procedures. The molecule has $2(6) + 4 = 16$ valence electrons:

$$\ddot{\underset{..}{S}} = C = \ddot{\underset{..}{S}}$$

The bonding pattern for CS_2 is like that for CO_2, except that the S atoms have $n = 3$ valence orbitals rather than $n = 2$ orbitals. The inner atom can be described using sp hybrids which overlap with $3p$ orbitals from the S atoms to form two σ bonds. There are two delocalized π systems at right angles to each other, each made up of a $2p$ orbital on the C atoms overlapping side-by-side with a $3p$ orbital on each S atom. As in CO_2, eight electrons occupy the delocalized π orbitals, four in bonding and four in non-bonding orbitals.

9.27 To determine the Lewis structure, first expand the line structure into a complete structural formula, then add lone pairs and look for resonance structures. The structure of acrolein is straightforward:

Each carbon atom has SN = 3, has trigonal planar geometry, and can be described using sp^2 hybridization. This leaves one p orbital on each atom, plus one from the oxygen atom, to form a delocalized π network extending over four atoms, as in butadiene.

9.29 Conjugation stabilizes a molecule whenever there are more than two adjacent atoms involved in π bond formation, thereby leading to delocalized π orbitals. Of the molecules whose structures appear in Box 9-1 of your textbook, neocembrene and limonene have only isolated π bonds, hence no stabilization by conjugation. All five π bonds in vitamin A and all 11 π bonds in carotene are parts of networks of delocalized bonding, so these two compounds are extensively stabilized by conjugation.

9.31 We can determine hybridizations directly from line structures: Each carbon atom with no double bonds, and each inner oxygen atom, has SN = 4 and sp^3 hybrids can describe the bonding. Carbon and oxygen with one double bond have SN = 3 and sp^2 hybridization.
(a) Hybridization is as shown on the line drawing ($3 = sp^3$, $2 = sp^2$).
(b - c) The Lewis structure shows 2 π bonds, on adjacent atoms, so as in butadiene there are four electrons in the extended π system
(d) The four atoms screened in the line drawing lie in the same plane.

9.33 Describe bonding and geometry starting with a Lewis structure and a count of bonds and non-bonding pairs around inner atoms. The provisional structure of each compound has a positive formal charge on the third-row sulfur atom, which is reduced to zero by making double bonds to each oxygen atom:

The sulfur atom in each compound has SN = 3, so sp^2 hybrids overlapping with oxygen $2p$ orbitals describe the σ bonds: two in SO_2 and three in SO_3. SO_2 is bent, like O_3, and SO_3 is trigonal planar, like NO_3^-. The Lewis structures indicate the presence of two π bonds in SO_2 and three π bonds in SO_3. All the π orbitals extend over the entire molecule. Because sulfur is a third-row element, its $3d$ orbitals contribute to the extended π bonding orbitals, which form from side-by-side overlap of oxygen $2p$ orbitals with sulfur $3p$ and $3d$ orbitals.

9.35 A description of bonding and geometry always starts with a Lewis structure and a count of bonds and non-bonding pairs around inner atoms. The provisional structure has three pairs of electrons associated with each carbon atom. Make one double bond to each C atom to complete its octet, and each C atom bonds to either O atom, giving four resonance structures:

Each carbon atom has SN = 3, so sp^2 hybrids overlapping with oxygen $2p$ orbitals describe the σ bonds. The geometry around each carbon atom is trigonal planar. All six atoms contribute p orbitals to the π system, which extends over the entire structure.

9.37 Doped semiconductors are p-type if the dopant has fewer valence electrons than the bulk element and n-type if the dopant has more valence electrons than the bulk element:
(a and c) GaP and CdSe are stoichiometric compounds, hence are undoped.
(b) InSb (Groups 13 and 15) doped with Te (Group 16) is n-type.

9.39 The bonding of any metal can be described by constructing bands from the valence orbitals and then distributing the valence electrons into these bands. Iron and potassium are both fourth-row elements with similar band structures, but whereas each potassium atom contributes just one valence electron to the valence (bonding) band of the solid, each iron atom contributes eight electrons. Thus iron has much greater bonding, making it harder and giving it a higher melting point than potassium.

9.41 To find exceptions to normal trends, consult the appropriate listings in tabulated values. Table 9-1 lists bond lengths by n-value. For $n = 2$ it lists four X–X bonds, C–C, N–N, O–O, and F–F; and three X–Y bonds, C–N, C–O, and C–F.
(a) C–N is shorter but weaker than C–C; (b) C–N is also longer but stronger than N–N.

9.43 Semiconductor type is determined by the number of valence electrons of the host element and its dopant. When the dopant contains more valence electrons than the host, the semiconductor is n-type; when it contains fewer valence electrons, the semiconductor is p-type.
(a) Ge (Group 14) doped with Sb (Group 15) is n-type; (b) As (Group 15) doped with Si (Group 14) is p-type; and (c) Si (Group 14) doped with In (Group 13) is p-type.

9.45 To predict stability from bond energies, tabulate the number of bonds of each type and add up their contributions. Start with Lewis structures in order to count bonds of each type:

Ethanol: 5 C–H bonds, 1 C–O bond, 1 O–H bond, 1 C–C bond

Diethyl ether: 6 C–H bonds, 2 C–O bonds

We can ignore the 5 C–H bonds and 1 C–O bond that the two compounds have in common, leaving O–H and C–C in ethanol vs. C–H and C–O in diethyl ether:
Ethanol: 460 (O–H) + 345 (C–C) = 805 kJ/mol
Diethyl ether: 415 (C–H) + 360 (C–O) = 775 kJ/mol.
Based on average bond energies, ethanol is more stable by 30 kJ/mol.

9.47 The bonding in second-row diatomic molecules can be described using the orbital energy level diagrams shown in Figure 9-17 of your textbook. NF has 12 valence electrons and the orbital energy level diagram for $Z \geq 7$:

The highest occupied orbital is π^*, so removing an electron decreases the amount of antibonding and increases stability. The lowest orbital with a vacancy is also π^*, so an added electron goes into an antibonding orbital and decreases stability.

Here is a sketch of the orbital involved in these changes:

9.49 If the carbonate anion had single bonds, it would show bond lengths around 143 pm, and if it had double bonds, it would show bond lengths around 122 nm. The actual bond lengths are 129 nm, in between these two values. This indicates that the bonds for carbonate are intermediate between single and double bonds. Intermediate bond orders are characteristic of bonds involved in extended π bonding.

9.51 Follow the standard procedure to obtain the Lewis structure of a molecule, then use the Lewis structure to determine hybridization, geometry, and identify the extended π system.
(a) This molecule has 4 N (5 e⁻) + 2 C (4 e⁻) + 2 O (6 e⁻) + 2 H (1 e⁻) = 44 valence electrons, 22 of which are used for the 11 bonds in the framework. Add 6 electrons around each outer O atom, then place the remaining 5 pairs on inner N atoms, which are more electronegative than C atoms, giving a provisional structure in which the three second-row atoms marked with *3* have only three pairs of electrons. Complete their octets by shifting electron pairs to form double bonds in the final Lewis structure:

Provisional structure   Final structure

(b) The end N atoms have SN = 4, sp^3 hybridization, and tetrahedral geometry. The C atoms and the center N atoms have SN = 3, sp^2 hybridization, and trigonal planar geometry.
(c) The four atoms with SN = 3 plus the two outer O atoms form a 6-atom plane with an extended π system. There are three extended π orbitals containing 6 electrons.

9.53 The standard procedure gives a Lewis structure from which hybridization, geometry, and π bonding are identified. Ketene has 2 C (4 e⁻) + 1 O (6 e⁻) + 2 H (1 e⁻) = 16 valence electrons, 8 of which are used for the 4 bonds in the framework. Add 6 electrons around the outer O atom, then place the remaining pairs on one C atom. Complete the octet around the other C atom by shifting electron pairs to form 2 double bonds in the final Lewis structure:

The steric numbers of the two C atoms differ, as indicated on the Lewis structure. All atoms lie in a common plane, with a σ bond framework described as shown in the sketch.

The C atom with SN = 3 has trigonal planar geometry and 120° angles. The C atom with SN = 2 has linear geometry and 180° angles. There is a two-center π bond in the plane of the molecule and an extended π bond perpendicular to the molecular plane:

9.55 A calculation using average bond energies requires a list of all bonds in the reactants and products. To obtain an estimate for reaction energy per mole of material, use the equation, $\Delta E \cong \Sigma B.E._{reactants} - \Sigma B.E._{products}$:

Ethane: H_3C-CH_3 +   7/2 O_2 →   2 CO_2 +   3 H_2O
Bonds: 1 C–C, 6 C–H   7/2 (1 O=O)   2 (2 C=O)   3 (2 O–H)
Energies (kJ/mol):   345   415   495   800   460
$\Delta E \cong [1(345) + 6(415) + 7/2(495)] - [4(800) + 6(460)] = 4570 - 5960 = -1390$ kJ/mol

ΔE(per gram) = $\dfrac{\Delta E}{MM} = \dfrac{-1390 \text{ kJ/mol}}{30.08 \text{ g/mol}} = -46.2$ kJ/g

Ethylene: $H_2C=CH_2$ +   3 O_2 →   2 CO_2 +   2 H_2O
Bonds: 1 C=C, 4 C–H   3 (1 O=O)   2 (2 C=O)   2 (2 O–H)
Energies (kJ/mol):   345   415   495   800   460
$\Delta E \cong [1(615) + 4(415) + 3(495)] - [4(800) + 4(460)] = 3760 - 5040 = -1280$ kJ/mol

ΔE(per gram) = $\dfrac{\Delta E}{MM} = \dfrac{-1280 \text{ kJ/mol}}{28.06 \text{ g/mol}} = -45.6$ kJ/g

Acetylene: HC≡CH +   5/2 O_2 →   2 CO_2 + H_2O
Bonds: 1 C≡C, 2 C–H   5/2 (1 O=O)   2 (2 C=O)   1 (2 O–H)
Energies (kJ/mol):   345   415   495   800   460
$\Delta E \cong [1(835) + 2(415) + 5/2(495)] - [4(800) + 2(460)] = 2900 - 4120 = -1220$ kJ/mol

ΔE(per gram) = $\dfrac{\Delta E}{MM} = \dfrac{-1220 \text{ kJ/mol}}{26.04 \text{ g/mol}} = -46.9$ kJ/g

These estimates indicate that acetylene releases slightly more energy per unit mass than ethane or ethylene, but experimental values show that ethane releases the most. Remember that average bond energy calculations provide estimates, not exact values.

9.57 A description of bonding in a molecule starts with its Lewis structure, from which steric number, geometry, and hybridization can be deduced.

HNCO has $1 + 5 + 4 + 6 = 16$ valence electrons. The bonding framework requires three pairs, 6 electrons are placed around the O atom, and the remaining 4 electrons are placed on the more electronegative inner atom, N. This leaves C with only 2 pairs, so shift one electron pair each from N and O to make two double bonds and complete the octet:

N has SN = 3, bent geometry, and uses sp^2 hybrid orbitals to form two σ bonds.
C has SN = 2, linear geometry and uses sp hybrids to form two σ bonds.
There are two π systems at right angles to each other, between C and O in the molecular plane and extending over N, C, and O perpendicular to the molecular plane:

OCS has $6 + 4 + 6 = 16$ valence electrons. The bonding framework requires two pairs and
6 electrons are placed O and S. This leaves C with only 2 pairs, so shift one electron pair each from S and O to make two double bonds and complete the octet:

C has SN = 2, linear geometry and uses sp hybrids to form two σ bonds.
There are two π systems at right angles to each other, each extending over O, C, and S:

9.59 To estimate the energy change in a reaction, determine Lewis structures to obtain types of bonds, list the number of bonds of each type in reactants and products, and subtract the sum of average bond energies for products from the sum of average bond energies for reactants, using values from Table 9.2 of your textbook.

$$H-Cl + H_2C=CH_2 \rightarrow H_3C-CH_2Cl$$

Reactants			Products		
Bond	No.	Energy	Bond	No.	Energy
H–Cl	1	430	C–Cl	1	325
C–H	4	415	C–H	5	415
C=C	1	615	C–C	1	345

$\Delta E_{rxn} \cong [1(430) + 4(415) + 1(615)] - [1(325) + 5(415) + 1(345)] = 2705 - 2745 = -40$ kJ

$$Cl_2 + H_2C=CH_2 \rightarrow ClH_2C-CH_2Cl$$

Reactants			Products		
Bond	No.	Energy	Bond	No.	Energy
Cl–Cl	1	240	C–Cl	2	325
C–H	4	415	C–H	4	415
C=C	1	615	C–C	1	345

$\Delta E_{rxn} \cong [1(240) + 4(415) + 1(615)] - [2(325) + 4(415) + 1(345)] = 2515 - 2655 = -140$ kJ

9.61 We can determine hybridizations directly from line structures (remember that C–H bonds and lone pairs are not explicitly shown in line structures):: Each C or N atom with no double bonds, and each inner O atom, has SN = 4 and sp^3 hybrids can describe the bonding. C, N, and O with one double bond have SN = 3 and sp^2 hybridization.
(a) The Lewis structure show three π bonds, all isolated.
(b) Atoms A, B, and D have SN = 3, so sp^3 hybrids are appropriate. Atom C has SN = 2, so it is appropriately described using sp^2 hybrids.
(c) Atoms E and F have SN = 4, tetrahedral geometry, and 109.5° bond angles.
Atom G has SN = 3, trigonal planar geometry, and 120° bond angles.
(d) Any atom with SN = 3 is the center of a four-atom plane. The three such atoms in saxitonin are circled in the redrawn structure:

9.63 The determinants of bond length, in order of relative importance, are principal quantum number of the valence orbitals, bond order, and bond polarity.
The bonds found in these molecules are H–O, H–C, C=O, and C≡N. Of these, H–O and H–C are shorter than the others because they form from an $n = 1$ orbital; H–O is shorter than H–C because it is a more polar bond. C=O has bond order 2 and is longer than C≡N because bond order is a stronger determinant of length than bond polarity.

9.65 The appearance of a band gap diagram depends on the nature of the semiconductor.

Silicon (Group 14) doped with antimony (Group 15) is an *n*-type semiconductor, with the dopant providing an excess of electrons. Thus its band gap diagram shows a full valence band and a small population of electrons in the conduction band:

9.67 A description of bonding always begins with a Lewis structure. Trigonal planar O_4 is isoelectronic with NO_3^- and CO_3^{2-} and has the same Lewis structure:

Average bond energies allow us to estimate the energy change associated with the reaction,

$2 O_2 \rightarrow O_4$: $\Delta E_{rxn} \cong [2(495)] - [2(145) + 1(495)] = 990 - 785 = + 205$ kJ

The resonance structures signal an extended π system distributed over all four O atoms that would stabilize the molecule relative to this calculation, but resonance energies are on the order of 50 kJ/mol, not enough to counteract the large positive energy change given by the bond energy calculation. Another approach to stability is to look at formal charges in the Lewis structure: the inner O atom has FC = +2, an unfavorable situation for this highly electronegative atom (notice that in the stable nitrate and carbonate anions, the formal charge on the inner atom is smaller - +1 on N and 0 on C). Thus both energy estimates and formal charge considerations predict that O_4 is not stable.

9.69 A description of bonding and geometry starts with determination of the Lewis structure. ClO_2 has $7 + 2(6) = 19$ electrons. Thus it is unusual in having an odd number of electrons.

The bond framework requires two pairs, six additional electrons are placed on each O atom, leaving three lone electrons on the Cl atom. $FC_{Cl} = 7 - 2 - 3 = +2$, so shift two pairs to make double bonds:

The Cl atom has SN = 4 (remember that a lone electron requires an orbital, just as an electron pair does), so its geometry is tetrahedral, the σ bonds can be described using sp^3 hybrids from Cl, and the molecule has a bond angle near 109°. There is an extended π system formed from *d* orbitals on Cl overlapping side-by-side with $2p$ orbitals from the two O atoms.

9.71 To determine which orbitals atoms use in a molecule, we must determine steric numbers from either the line structure or the Lewis structure. Any inner C, N, or O atom with no double bonds has SN = 4; with one double bond, an inner C or N atom has SN = 3. Remember that outer atoms (including H) always use atomic orbitals.

[Structures showing: ketone with O←atomic, sp² on carbonyl C, sp³ on methyl C's; alkene with sp² and sp³ labels; benzene ring labeled "All C's use sp^2"; hydrazine H₂N–NH₂ labeled "Both N's use sp^3"; methylamine H₃C–NH₂ labeled "C and N use sp^3"]

9.73 Before calculating estimated reaction energies, we must determine what the net reaction is and how many bonds are involved. When a Y–Y chain is converted into a Y–O–Y chain, each Y–Y bond is replaced by two Y–O bonds:

Reaction:	–Si–Si– +	1/2 O_2 →	–Si–O–Si–
Bonds:	1 Si–Si	1/2 O=O	2 Si–O
Energy (kJ/mol):	225	495	450

$\Delta E_{rxn} \cong [1(225) + 1/2(495)] - [2(450)] = 470 - 900 = -430$ kJ

Reaction:	–C–C– +	1/2 O_2 →	–C–O–C–
Bonds:	1 C–C	1/2 O=O	2 C–O
Energy (kJ/mol):	345	495	360

$\Delta E_{rxn} \cong [1(345) + 1/2(495)] - [2(360)] = 590 - 720 = -130$ kJ

As the results imply, Si–O chains are considerably more stable than C–O chains.

9.75 The reaction caused by light is $RC-Cl \xrightarrow{h\nu} RC + Cl$. To determine what frequency of light is needed for this reaction, we need to know the energy of the bond (Table 9-2) and the equation relating frequency to photon energy, $E = h\nu$. Bond energy must be converted from a molar quantity to the energy per bond:

$E_{per\ bond}$ = (325 kJ/mol) ÷ (6.022 x 10^{23}/mol) = 5.40 x 10^{-22} kJ;

$\nu = \dfrac{(5.40 \times 10^{-22} \text{kJ})(10^3 \text{J/kJ})}{(6.63 \times 10^{-34} \text{J s})} = 8.14 \times 10^{14}$ s^{-1}.

Chapter 9: Chemical Bonding: Multiple Bonds

9.77 To generate the configuration of an excited state, move an electron from an occupied orbital to an unoccupied orbital. The most stable excited state results from moving an electron from the least stable occupied orbital to the most stable unoccupied orbital:

The excited state has one more antibonding electron and one less bonding electron than the ground state, so the excited state has a weaker N–N bond than the ground state.

9.79 When a molecule forms, atomic orbitals interact to form bonding and antibonding molecular orbitals. Bonding molecular orbitals are more stable than the atomic orbitals from which they form, but antibonding orbitals are less stable than the atomic orbitals from which they form. Ionization energies measure the amount of stability of orbitals. N_2 has a bonding molecular orbital as its least stable occupied orbital. Removing an electron from N_2 requires more energy than removing an electron from N. O_2 has an antibonding orbital as its least stable occupied orbital. Removing an electron from O_2 requires less energy than removing an electron from O. The diagram shows the energy relationships in schematic fashion:

9.81 There is only one isomer of 1,2-dichloroethane because the C–C bond is a σ bond only, and the molecule can freely rotate without disrupting the amount of overlap. When rotation occurs about the C=C double bond in 1,2-dichloroethylene, the side-by-side overlap of the π bond is disrupted:

As the molecule rotates about its C=C bond, the double bond breaks, leaving only a single bond. Thus we can estimate the energy required to convert the isomers from the difference between a C=C and a C–C bond: 615 – 345 = 270 kJ/mol.

9.83 Any time a set of p orbitals interact to form molecular orbitals, the number of molecular orbitals equals the number of interacting orbitals. In C_{60}, 60 $2p$ orbitals interact to form the π network, generating 60 molecular orbitals, 30 of which are π (bonding) and 30 of which are π* (antibonding). The 30 bonding orbitals are full, giving a total of 60 π electrons in extended bonding orbitals.

9.85 The threshold energy is the minimum energy required to remove an electron from a metal surface. This corresponds to the stability of the electrons at the top of the filled band, as shown in the diagram.

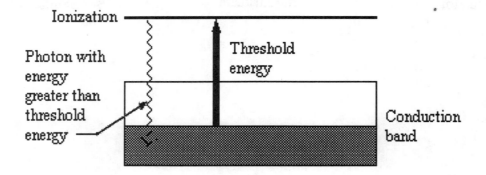

A photon with energy greater than the threshold energy can eject an electron from the top of the conduction band, giving it a large amount of kinetic energy. Such a photon can also eject an electron from deeper within the conduction band, giving it a smaller amount of kinetic energy. Thus photons with energy greater than the threshold energy produce electrons with a range of kinetic energies.

9.87 To draw orbital overlap pictures, start with sketches of atomic orbitals and then allow them to interact. Remember that σ bonds always overlap end-on, giving axially-symmetric interactions, while π bonds always involve side-by-side overlap.

9.89 Silicon carbide has the same type of bonding as diamond, except that half the atoms in the tetrahedral network are Si rather than C. Each carbon atom bonds to four silicon atoms, and each silicon atom bonds to four carbon atoms. Both atoms have SN = 4, tetrahedral geometry, and can be described using sp^3 hybrids that give a network of σ bonds. Recall that silicon has an intermediate band gap and diamond has a large band gap. SiC would be expected to have a band gap intermediate between these two materials:

Chapter 10 Effects of Intermolecular Forces

10.1 The Nature of Intermolecular Forces

QUESTIONS TO ANSWER, SKILLS TO LEARN
1. **Intermolecular vs. intramolecular forces**
2. **A molecular view of physical properties**
3. **Strength of intermolecular forces vs. thermal energy available determines phase**
4. **Properties of liquids**

Chapters 8 and 9 focused on the nature of the forces that bind atoms together to form molecules or ionic compounds. These forces are generally described as *intramolecular forces* because forces act between atoms to form molecules or between ions to form an ionic substance. At temperatures greater than absolute zero, we expect that molecules have some kinetic energy; in the absence of any forces holding the molecules together, the molecules should all be moving independently, colliding with each other and the walls of the container like an ideal gas. The diatomic molecular halogens offer a useful perspective on the difference between forces that bind atoms together to form molecules and the much weaker forces that bind *molecules* to each other in the phases where molecules are in contact with each other: liquids, solids and solutions. These latter forces are *intermolecular forces*.

***Intramolecular* forces bind atoms together in molecules and polyatomic ions.**

***Intermolecular* forces attract molecules together.**

The volumes occupied by one mole each of fluorine, chlorine, bromine and iodine at 298 K, their states and the atomic radii of each of the atoms are listed in the following table. The molar volume of the two gases are the same and much greater than the liquid, which in turn has a slightly greater molar volume than the solid.

Compound	F_2 (gas)	Cl_2 (gas)	Br_2 (liquid)	I_2 (solid)
Atomic radius (pm)	71	99	114	133
Molar volume (L, 298 K, 1 atm)	24.5	24.5	0.05152	0.05148

The two gases have the same molar volume even though the atomic radius of chlorine is 28 pm greater than that of fluorine. Because the molar volumes of the two gases are the same, the volume of the molecules must be insignificant compared to the empty space between the molecules of the gas. The themal energy at room temperature overcomes the forces acting between the molecules in these two gases, so a gas can be pictured as molecules in rapid motion, colliding with each other and the container, but not sticking to either.

The molar volume of bromine, on the other hand, is much smaller than fluorine and chlorine, but still greater than the molar volume of the larger molecule iodine. This observation suggests that there is some empty space between the bromine molecules (they are not packed together as tightly as possible). The forces that bind the Br_2 molecules to each other are comparable to their kinetic energy but they are not strong enough to keep the Br_2 molecules fixed in position relative to one another; the molecules can break the forces acting between one pair, and move on to another molecule(s) and move freely through the liquid. The molecules in contact with any one molecule change, but on average the number of molecules around a given Br_2 molecule is the same. Because of this constant motion, the bromine molecules have some empty space around them.

Iodine is a solid; the molecules in this solid are held in place and are not mobile because the intermolecular forces attracting one molecule to the others are stronger than their kinetic energy. The molar volume of iodine is about the same as that of bromine even though iodine has a significantly larger atomic radius. This is because the iodine molecules are arranged in a manner that maximizes the attractive forces between them. This specific packing and lack of motion in the solid state leads to more molecules per unit volume in the solid than in the liquid. The phase of a substance may be changed by changing the kinetic energy of the molecules (that is, by changing the temperature). For example, bromine becomes a solid at −7.2°C.

In summary:
- In gases, the energy of stabilization of molecules by intermolecular forces is much smaller than the thermal kinetic energy.

- In liquids, the energy of stabilization of molecules by intermolecular forces is not much different from the thermal kinetic energy.

- In solids, the energy of stabilization of molecules by intermolecular forces is significantly larger than the thermal kinetic energy.

- Phase changes to the system (solid to liquid, liquid to gas) may be induced by the addition or removal of thermal energy.

Exercise 10.1 A block of frozen benzene is placed in a pan and slowly heated until it turns to liquid and then finally boils away. Describe the events that occur at the molecular level during this process.

Steps to Solution: *To answer this question, we must remember the qualities of the solid, liquid and gas phases and the effects of adding energy to molecules. First, the molecules in the solid are in contact with each other and their relative positions are fixed; molecules do not move past one another. As energy is added to the solid, the molecules obtain more and more thermal energy until the thermal energy is about the same as the energy of stabilization because of the intermolecular forces. At this point molecules can break the intermolecular force that hold them in place in the solid and move past one another. However, the forces still keep the molecules stuck to each other. The molecules may flow past one another so that the liquid now can take the shape of the container. After considerably more energy has been added to the molecules in the liquid form, more and more molecules have sufficient energy to break free of the forces that bind them in the liquid phase and enter the gas phase. In the gas phase, the molecules travel freely, affected only by collisions with other molecules and the walls of the container.*

The Distance Dependence of the Energy of Intermolecular Forces

Intermolecular forces depend on the distance between the molecules involved. A simplified model of intermolecular forces gives plots of the energy of stabilization vs. distance that are quite similar qualitatively to those for bond formation. However, the energies involved are much smaller (J, not kJ) and the distances are larger. The general features of this diagram that you should remember are that at large distances, the forces are almost negligible; as the atoms or molecules approach one another, attractive forces act, causing energy to decrease and eventually reach a minimum. At that point, the electron clouds of the atoms or molecules involved start to interact with repulsive forces from bringing the two electron clouds near each other.

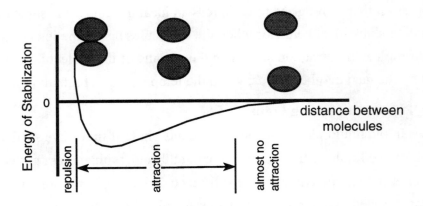

The phase of a substance depends on the balance between the energy of stabilization caused by intermolecular forces and the thermal energy of the molecules involved.

Non-Ideal Behavior of Gases: Real Gases

An ideal gas (a) undergoes no attractive interactions with the other gas molecules and (b) its molecules occupy a volume so small that it can be ignored compared to the volume the gas occupies. The ideal gas law can be rearranged to give a ratio called the compressibility.

Chapter 10: Effects of Intermolecular Forces

$$PV = nRT \qquad \text{compressibility} = \frac{PV}{nRT}$$

For an ideal gas, the compressibility is equal to one under all conditions; however, the fact that real molecules have attractive intermolecular forces and a finite size results in compressibilities other than one under different conditions of temperature and pressure.

The compressibility is less than one when the pressure is less than that expected from the ideal gas law: the pressure decreases from the ideal value because attractive intermolecular forces subtract from the energy that molecules would normally have in collisions. A compressibility less than one is expected of real gases under the following conditions:

- Low temperature (less kinetic energy to be overcome by the intermolecular forces)

- High molecular density (the distance between molecules is relatively small; the attractive forces between molecules reduce the force of the collisions of the molecules with the walls of the container)

The compressibility is greater than one when the pressure is high and the actual size of the molecules begins to occupy a significant volume of the gas, and repulsive forces between molecules become important. Under these conditions, the volume of the gas requires more pressure to compress it than is required to compress an ideal gas under the same conditions.

Gases tend to show ideal behavior when the molecules have kinetic energy greater than the energy of intermolecular attraction and when the volume of the molecules is small compared to that occupied by the gas; in other words, when the temperature is high and the pressure is low.

Normal Boiling Point and Melting Point

Molecules in the liquid phase escape into the gas phase if they have sufficient energy. At a given temperature, the liquid will be in dynamic equilibrium with enough molecules in the gas phase to exert a pressure characteristic of the substance called the vapor pressure. Boiling is the phenomenon characterized by bubbles of the vapor forming in the liquid phase.

> The temperature at which the vapor pressure of a liquid
> is one atmosphere is the **normal boiling point**.

The highly ordered solid phase can be converted to the liquid phase by the addition of sufficient energy to break some of the intermolecular forces so the molecules (or ions) of the solid can move about, resulting in the less ordered phase called the liquid phase.

> The temperature at which a solid converts to a liquid with 1 atm pressure
> exerted on the phases is the **normal melting point**.

Chapter 10: Effects of Intermolecular Forces

The normal boiling point and normal melting point depend on the strength of the intermolecular forces holding the molecules together.

__Higher normal boiling and melting points indicate stronger intermolecular forces than compounds with lower normal boiling and melting points.__

Properties of Liquids

A liquid is comprised of molecules that are in contact because of attractive intermolecular forces, but are able to move past one another, by breaking the attractive forces to one group of molecules while developing such forces to another group of molecules. The strength of the intermolecular forces is related to the following physical properties:

Surface tension: the property of a liquid that allows one to "float" a steel needle on water.

Viscosity: the resistance of a liquid to flow (slow-flowing liquids have high viscosity).

Cohesion: the attraction of molecules in a liquid to the other molecules of the liquid.

Adhesion: the attraction of molecules in a liquid to the container walls or other surfaces.

Surface tension is a consequence of the very different environments of a molecule at the surface of a liquid and a molecule that is in the interior region of the liquid. The diagram on the next page shows the different environments. Molecules in the interior of the liquid are attracted to molecules on all sides of them and therefore are not strongly pulled in any particular direction. Those at the surface experience attraction only to the molecules that are below them; as a result, the surface molecules are pulled down, making the surface a more tightly packed region of the liquid. Therefore energy is required to increase the surface area of a liquid, and liquids tend to take a shape that minimizes surface area. For a needle to sink into water, the water must touch every part of the needle, increasing the liquid's surface area. The needle will float because gravity alone does not supply enough energy to change the surface area of the liquid water to coat the surface of the needle. A gentle push will send the needle below the surface of the water, at which point the needle sinks.

Chapter 10: Effects of Intermolecular Forces

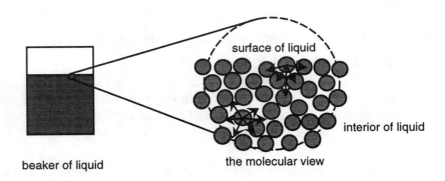

Viscosity is the resistance to flow of a liquid. The greater the viscosity, the more slowly the liquid will flow (high-viscosity liquids are "syrupy"). For a liquid to flow, molecules must move past one another and the walls of the container. When stronger intermolecular forces bind molecules together, the molecules do not move past one another easily. Therefore liquids in which the molecules have strong intermolecular forces tend to have high viscosities.

Adhesion and *cohesion* describe the behavior of a liquid when it is in contact with a solid such as the container or an object like a tube or rod is immersed in it. *Cohesion* is the attraction of molecules to one another. *Adhesion* is the attraction of molecules in a liquid to a substance in the solid phase such as the container.

The two limiting cases are the one where the cohesive forces are stronger than the adhesive forces and the one where the adhesive forces are greater than the cohesive forces. These two cases are described below.

When cohesion is stronger than adhesion, the attraction of molecules in the liquid to each other is stronger than their attraction to the surface. The most stable situation occurs when the amount of surface contacted by the liquid is at a minimum, because more stabilization energy is obtained by molecule-molecule interaction than by molecule-solid interaction. In a narrow enough container, the liquid will minimize the amount of solid it touches by forming a convex meniscus as shown at right.

When adhesion is stronger than cohesion, the molecules of the liquid are strongly attracted to the surface of the solid. The attractive forces overcome gravitational force by climbing up the solid to touch more of the surface. This results in a concave meniscus (shown at right) that is commonly observed when water wets clean glass.

Exercise 10.2 For each of the following pairs of substances, predict that which would have the greater viscosity as a liquid at the same temperature;
(a) water or methanol (CH_3OH)   (b) ozone (O_3) or oxygen (O_2)

Chapter 10: Effects of Intermolecular Forces

Steps to Solution: To identify the substance with the greater viscosity, we must determine the relative strength of the intermolecular forces by examining the structures of the two substances. Because viscosity increases with increasing intermolecular forces, the substance with the stronger intermolecular forces will have the higher viscosity.

(a) Water and methanol are both polar molecules capable of making strong hydrogen bonds. The hydrogen bonding interactions will dominate the intermolecular forces because similar dipole moments result from the similar bent geometry at the O atom. Each water molecule may make as many as four H-bonds to neighboring molecules (two H-bond donors (O-H bonds) and two pairs of nonbonding electrons), whereas methanol, with only one polar O-H bond and two pairs of nonbonding electrons can form only three H-bonds. Because water can undergo more H-bonding interactions, it will have the stronger intermolecular forces and be more viscous.

(b) Ozone is a bent molecule that possesses a dipole moment, and oxygen is a linear diatomic molecule with no dipole moment. In addition, ozone will have stronger dispersion forces because it has more electrons. Therefore liquid ozone will have stronger intermolecular forces and have greater viscosity than liquid oxygen.

Exercise 10.3 On a freshly waxed car, water forms spherical beads, whereas on a car that badly needs waxing, water coats the surface evenly.
(a) Explain the difference between the surfaces of the two cars.
(b) What type of molecules are on the surface of the freshly waxed car?

Steps to Solution: This problem addresses the behavior of a liquid on a surface; therefore we must consider the difference between the adhesive and cohesive forces acting in the system.

(a) On the freshly waxed car, the water beads up because the cohesive forces in water are much stronger than the adhesive forces between the car's surface and water. Therefore water forms the shape that will minimize the amount of surface of the car it will wet: a sphere. On the weathered car's surface, the water wets (coats) the surface evenly. This observation shows that the adhesive forces between water and the car's surface are strong enough to compete with the cohesive forces in water so that water will coat the surface. This suggests that the surface must be able to interact with hydrogen bonding molecules or polar molecules.

(b) Because the adhesive forces must be quite weak, the surface must be incapable of hydrogen bonding or interacting with polar molecules. The molecules must have low polarity and contain no bonds between H and the highly electronegative atoms N, O or F.

10.2 Types of Intermolecular Forces

QUESTIONS TO ANSWER, SKILLS TO LEARN
1. **What are dispersion forces?**
2. **What is the difference between a "compact" and an "extended" structure**
3. **What physical factors vary as structure changes?**
4. **Deducing the presence of molecular dipoles**
5. **What is a hydrogen bond?**
6. **Recognizing the factors necessary to form a significantly strong hydrogen bonding interaction**
7. **In what chemical systems does hydrogen bonding make a difference?**

Dispersion Forces

Dispersion forces are intermolecular forces possessed by any chemical species that has electrons. Dispersion forces occur because the motions of electrons in nearby molecules are correlated. Correlated electron motion means that the electrons in molecules are repelled by the electrons in nearby molecules, as illustrated in the following sketches.

| A spherical electron distribution. No net attraction between molecules is expected. | Distortion of electron cloud in one molecule results in the charge buildup indicated by the + and - signs, resulting in attractive forces. | The electron cloud shifts rapidly; the electron cloud distortion shifts so attractive forces still exist between the molecules. |

The strength of dispersion forces depends on how tightly the electrons are bound; the less tightly the electrons are bound, more likely that they will be distributed asymmetrically. A less tightly bound electron cloud is more easily distorted by the asymmetric distribution of electrons in nearby molecules.

In molecules that have similar molar masses and bonding but different shapes, the relative strength of dispersion forces depends on how much of the molecule is in "contact" with other molecules. The line structures and boiling points of two isomers of C_4H_{10} are shown below.

n-butane (bp = -0.5 °C)   2-methylpropane (bp = -12 °C)

In *n*-butane, more of the molecule is exposed to other molecules because there are no branches in the structure; therefore we refer to *n*-butane as having an extended structure.

Structures with few or no branches are called extended.

The structure of 2-methylpropane is branched; it is not a linear string of C atoms bonded to each other. There is less of the molecule exposed to the other molecules in the liquid to interact with; this type of structure is called compact.

Highly branched structures are called compact.

Compact structures cannot undergo as many effective dispersion force interactions so, for molecules of similar molar mass and polarity, compact molecules will have lower boiling points than extended one. The following structural features affect polarizability :

Higher polarizability (stronger dispersion forces):	Lower polarizability (weaker dispersion forces):
• valence electrons in orbitals of higher n • delocalized orbitals • larger, more extended molecule	• valence electrons in orbitals of lower n • localized orbitals • smaller, more compact molecule

Dipolar Forces

Dipolar forces are forces that act between molecules that possess a permanent dipole moment; that is, because of polar bonds and low symmetry, one portion of the molecule has a small net positive charge and another has a small net negative charge. The negative end of the molecule will tend to be attracted to the positive end of another molecule leading to stabilization of the liquid (or solid) phase due to these interactions as illustrated below for 2 HCl molecules:

Dipolar forces are inherently stronger than dispersion forces. However, dispersion forces can become significant as molecular size increases. For example, the nonpolar hydrocarbons that contain more than 18 C atoms (paraffin wax) have higher melting points than the polar molecule, water (solid wax is melted in hot [liquid] water). It is difficult to predict the overall combined effects of dispersion forces and dipolar forces in molecules of significantly different molar mass on physical properties such as the melting or boiling point that depend on intermolecular forces. However, in molecules with roughly the same shapes and molar masses, the presence of a dipole moment in one of the molecules indicates that intermolecular forces will be stronger in the molecule with the dipole moment.

Exercise 10.4 For the three molecules whose structures and boiling points are shown below, explain the trend in boiling points in terms of the strength and types of intermolecular forces acting between the molecules in the pure liquids.

structure formula, name normal boiling point	(CH$_3$-O-CH$_3$) C_2H_6O, methyl ether -24.8 °C	(CH$_3$-CH$_2$-CH$_3$) C_3H_8, propane -42.1 °C	(CH$_3$-S-CH$_3$) C_2H_6S, methylsulfide 38 °C

Steps to Solution: We must first identify the types of intermolecular forces active in each molecule and then compare them using the experimental data (boiling points) to help assign their strengths. It is helpful to list the compounds in order of either increasing or decreasing boiling point. In this example the compounds will be listed in order of increasing boiling point.

compound	propane	methyl ether	methyl sulfide
intermolecular forces present	**dispersion** forces no polar bonds; therefore no dipolar forces	**dispersion** forces molecule has a permanent dipole moment; therefore **dipolar** forces are present	**dispersion** forces molecule has a permanent dipole moment; therefore **dipolar** forces are present

The molar masses and shapes of methyl ether and propane are nearly the same, so the dispersion forces are nearly the same. The higher boiling point of methyl ether compared to propane results from dipolar forces because methyl ether has a permanent dipole moment and propane does not.

The higher boiling point of methyl sulfide compared to methyl ether is less straightforward. They both have dipolar and dispersion forces, yet methyl sulfide has a significantly higher boiling point. The dipole moments of the two molecules are not likely to be dramatically different; the geometries at the S and O atoms are expected to be similar but the electronegativities of O and S are different: 3.5 and 2.5, respectively (the electronegativity of S is very close to that of C!). The lower electronegativity of S suggests that the methyl sulfide should be less polar. However, the valence electrons of S are in the 3p orbitals and those of O are in the smaller, more tightly bound 2p orbitals. Therefore we expect that the dispersion forces will be stronger in methyl sulfide than in methyl ether, and the higher boiling point results from the stronger dispersion forces. (The dipole moment of methyl sulfide (1.50 D) is somewhat greater than that of methyl ether (1.30 D), although this is not apparent from the understanding expected at this level. The increase in boiling point is caused by increased dispersion forces and dipolar forces.)

Hydrogen Bonding

A hydrogen bond is formed from the interaction of a highly polar X-H bond with the lone pair of an electronegative atom.

$$\overset{\delta-}{X}-\overset{\delta+}{H} \cdots Y$$

Hydrogen bonding may occur between different chemical species. The hydrogen-bond interaction is much weaker than a "full-fledged" chemical bond, but is often stronger than dipolar and dispersion forces. For strong hydrogen bonding to occur, the following conditions must be present:

- A polar X-H bond; X must be N, O or F to donate the H for the H bond (H-bond donor).

- Nonbonding electrons on a highly electronegative atom that interact with the polar X-H bond to form the H-bond (H-bond acceptor).

- The H-bond is not restricted to one type of molecule. The H-bond acceptor and the H-bond donor can be located on two molecules that may or may not be the same. In some very large molecules, the donor and acceptor can be located on different parts of the same molecule.

The following exercises illustrate the use of these guides for recognizing some the various types of H-bonds that occur.

Exercise 10.5 In which of the following pure substances will hydrogen bonding be an important intermolecular force? (a) dichloromethane, CH_2Cl_2 (b) CH_3CH_2OH (c) methylamine (d) trimethylamine, $N(CH_3)_3$

Steps to Solution: We must first determine the Lewis structure and see if the structures meet the criteria listed above, namely the presence of a highly polar bond to H and a nonbonding electron pair on an electronegative atom.

(a) *The structure of CH_2Cl_2 (dichloromethane) is shown below.*
Dichloromethane contains nonbonding electrons on electronegative atoms (on the chlorines) but has no polar bonds to H (only C-H bonds). H-bonding will not be an important intermolecular force in pure dichloromethane.

(b) *The structure of CH_3CH_2OH (ethanol) is shown below.*

Ethanol contains nonpolar C-H bonds, but also the very polar O-H bond that can serve as an H-bond donor. In addition, nonbonding electrons are present on the O atom that may serve as H-bond acceptors. The presence of both an H-bond acceptor and an H-bond donor suggests that hydrogen bonding will be an important intermolecular force in ethanol.

(c) The structure of methylamine (CH$_3$NH$_2$) is shown below.

First, note that the CH$_3$ group has no polar bonds to H. Methylamine contains polar N-H bonds that can act as H bond donors and nonbonding electrons on the N atom that may serve as H-bond acceptors. The presence of both an H-bond acceptor and an H-bond donor suggests that hydrogen bonding will be an important intermolecular force in methylamine.

(d) The structure of trimethylamine (N(CH$_3$)$_3$) is shown below:

Trimethylamine contains no polar N-H bonds for H bond donors but does have nonbonding electrons on the N atom that may serve as H-bond acceptors. The lack of an H-bond donor tells us that hydrogen bonding will not be an important intermolecular force in pure trimethylamine.

Exercise 10.6 Proteins are formed from many (hundreds to thousands) repeating units of the type -[-NH-CHR-(C=O)-]- as shown in the structure below (it is missing lone pairs but shows many of the features central to the structure of a protein).

The R group need not be the same in each repeating unit, and in proteins, adjacent R groups will rarely be the same.
(a) Will proteins form hydrogen bond with water?
(b) Can hydrogen bonds occur in an *intramolecular* fashion, from one part of a protein molecule to another part of the same molecule?

Steps to Solution: We are asked to determine if hydrogen bonding can occur in a protein molecule that has the structural elements shown in the structure. We need to complete the structure by adding lone pairs; one lone pair on each N atom and two lone pairs on each O atom are required (this Lewis structure has formal charges of 0 on the atoms). Because the R group is not specified, the analysis will focus on the part of the molecule that is shown.

(a) Water contains both H-bond donors and acceptors; so do proteins (the N-H bonds are H bond donors and the lone pairs on the O atom and the N atom are H-bond acceptors). Proteins would be expected to form H-bonds with water.

(b) *Proteins have H-bond donors and H-bond acceptors listed above in part (a). Proteins could have hydrogen bonding as an intramolecular phenomenon because they can be such large molecules. In fact, H-bonding in proteins is important in determining the structure that the long chain of repeating groups eventually forms.*

In summary, the important intermolecular forces are dispersion forces, dipolar forces and hydrogen bonds. Theses forces give energies of stabilization that are small compared to the covalent bonds (hundreds of kJ/mole); intermolecular forces have energies of stabilization that range from less than 1 kJ/mole to tens of kJ/mole.

10.3 Forces in Solids

QUESTIONS TO ANSWER, SKILLS TO LEARN
1. **How do we classify solids based on the intermolecular forces holding them together?**
2. **What is the difference between a crystalline and an amorphous solid?**

One way in which solids are classified is by the types of forces holding together the constituent parts (atoms, molecules or ions) to form the solid.

Type of Force	Type of Solid and Examples
Electrical attraction between cations and anions	**Ionic solids** such as NaCl and CsI
Delocalized metallic bonding	**Metals** such as Fe, Na, U
Covalent bonds	**Covalent solids** such as quartz (SiO_2), diamond and graphite (forms of C)
Intermolecular forces	**Molecular solids** that are composed of molecules held together by dispersion, dipolar and/or H-bonding interactions

The last class of solids, molecular solids, has relatively weak forces holding it together (compared to ionic attraction, covalent bonds, and most metallic bonding), so molecular solids tend be soft and have low melting points.

Predicting Physical Properties of Crystalline Solids

A number of physical properties of crystalline solids are affected by the type of forces holding together the constituent particles (atoms, ions or molecules). The melting point and hardness of a solid are expected to increase with increasing strength of the forces binding the particles.

Molecules that form crystals are held together in the solid state by dispersion, dipolar or hydrogen bonding forces. These forces are usually relatively weak (kJ to 10's of kJ/mole) and thus, crystals formed from molecules (**molecular crystals**) are soft and have low melting points.

Bonding in ionic solids was described in Chapter 7; they have strong forces holding the ions together in a lattice and thus are hard substances with high melting points. However, the attractive forces are converted to repulsive forces when part of the crystal is moved (see below); the crystal shatters, rather than flattening out as a metal would. Ionic solids are brittle.

Different metals have different lattice strengths. Metals can be soft with low melting points (mercury is a liquid at room temperature) or very hard with high melting points (tungsten melts at 3410 °C). Metals conduct electricity, a property that requires the presence of charged particles that can easily move through the solid. Of the solids that have been discussed to this point, only metals with their delocalized electrons readily conduct electricity.

The fourth general type of forces holding solids together are covalent bonds such as those found in graphite, diamond and quartz. These substances have high melting points and, with the exception of graphite, high hardness. These compounds are called **covalent solids**. In order to move the fixed atoms of a covalent solid, strong covalent bonds must be broken (hundreds of kJ/mole of bonds), and the attractive forces are not readily converted to repulsive forces as in ionic compounds. Covalent solids have high melting points and are very hard and are used where hardness and durability are important such as abrasives (sandpaper) and coatings on cutting tools.

10.4 Order in Solids

QUESTIONS TO ANSWER, SKILLS TO LEARN
1. **What is a unit cell and what are common unit cells?**
2. **Using unit cells to determine macroscopic properties**

In addition to these categories, two other classes are used that describe the organization and structure of any type of solid at the microscopic level.

- **Crystalline solids** are characterized by high regularity because the solid and many of its properties can be described by a small arrangement of molecules/ions/atoms that repeats in three dimensions.
- **Amorphous solids** do not possess the same degree of order and regularity as a crystalline solid. They have no small repeating unit that describes the solid and its properties.

Close-Packed Crystals

Some ways of packing atoms or molecules in a regular, repeating form are more efficient than others (there is less empty space between the objects being packed). The most efficient arrangements of packing have a minimum of empty space and are called **close packed** or sometimes closest packed. For spherical objects of the same size (atoms such as metals, nearly spherical molecules and the noble gases), the most efficient arrangement in two dimensions has six spheres packed around each other sphere, as shown in the following diagram. The hexagon is drawn to emphasize the six spheres around the chosen sphere. This first layer will be called the A layer.

To minimize empty space, the next layer of spheres must be placed in the depressions formed between any three spheres in the first layer. One such location is indicated on the figure by the triangle. Filling all the available depressions gives a second layer of spheres, the B layer. The third layer has two possible locations in the depressions formed by the B layer: the spheres may be located directly above the A layer (below left), or form a layer in which the spheres do not lie directly above either the A spheres or the B spheres (below right). To designate the different position of these spheres relative to the A and B layers, this is called the C layer.

Chapter 10: Effects of Intermolecular Forces

Side view of layers	A B A (ABA layers)	C B A (CBA layers)
Order of layers	ABABAB…	ABCABCABC…
Structure name	hexagonal close packed	cubic close packed (or face-centered cubic)

The most common close packed crystalline solids have the atoms packed in repeating sequences of AB or ABC layers. These close-packed structures are called hexagonal close packed and cubic close packed, respectively. Side views are shown below. See your text for perspective views of these packings.

For stable ionic compounds, the positive charges and negative charges must alternate to minimize repulsive attractions. Two common structures are found in crystalline ionic substances: the NaCl (sodium chloride or rock salt) structure and the CsCl structure. In the rock salt structure, each ion is surrounded by an octahedral arrangement of six ions of the other charge. In the CsCl structure, each ion is surrounded by a cube of eight ions of the opposite charge. These structures are illustrated in the text.

Unit Cells

A crystal consists of atoms/ions/molecules in the solid state in which there is a distinct repeating arrangement, the simplest of which is called the **unit cell**. The unit cell contains sufficient information to reconstruct the entire crystal and determine the physical properties of actual crystals. Exercise 10.7 illustrates some of the characteristics of the unit cell.

Exercise 10.7. The diagrams that follow show two different crystal structures adopted by several metals. In the simple cubic crystal structure at left, the unit cell is outlined in heavy lines that intersect to form the corners at the center of the spheres.

(a) simple cubic

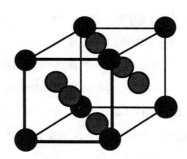

(b) face-centered cubic

Chapter 10: Effects of Intermolecular Forces

The face-centered cubic unit cell to the right has been shaded to add clarity; the balls of in the corners are black: those in the faces are gray. If the corners of the unit cells are at the center of the corner spheres, how many atoms are in one unit cell of (a) a simple cubic and (b) a face-centered cubic unit cells?

Steps to Solution: To solve this problem, we must determine the positions of all atoms that fall in the unit cell and then determine what fraction of those atoms falls inside the walls that form the shape of the unit cell. One way to determine the fraction of the sphere in the unit cell is to determine how many unit cells share that sphere. If the sphere is shared equally among a number of unit cells, n, then the fraction of the sphere in the unit cell is $\frac{1}{n}$. Then we sum the fractions of atoms in the unit cell.

(a) The only types of positions in the unit cell are corners: the eight corner spheres. Each of those spheres are shared among eight unit cells. Note the central sphere to the right of the unit cell; it is shared between four unit cells that are shown in the sketch and 4 more in the space where atoms are not shown. The sphere is shared between eight unit cells; therefore only $\frac{1}{8}$ of each sphere lies inside a given unit cell. Another way to picture the division of the sphere is to "cut" the sphere along each side of the unit cell it lies on. The corner cells lie at the intersection of three walls, and the three slices cut the sphere into eighths (try it with an orange). The total number of atoms in a simple cubic unit cell is:

$$\text{Number of spheres/unit cell} = 8\left(\frac{1}{8}\right) = \boxed{1}$$

(b) There are two types of positions in the face-centered cubic unit cell; the eight corner positions and the six spheres located in the centers of the faces. The eight atoms in the corner positions have the same geometry as those in part (a) of this exercise and contribute $8\left(\frac{1}{8}\right) = 1$ sphere. The spheres on the faces have different geometry; they are shared between two unit cells, and therefore $\frac{1}{2}$ of each sphere on the face of the unit cell lies inside of it. The total number of spheres in a face centered unit cell is:

$$\text{Number of spheres/unit cell} = 8\left(\frac{1}{8}\right) + 6\left(\frac{1}{2}\right) = \boxed{4}$$

10.5 The Nature of Solutions

QUESTIONS TO ANSWER, SKILLS TO LEARN
1. **What are the different types of solutions?**
2. **What is the nature of a solution and the molecules that constitute the solution?**
3. **Predicting the solubility of substances in a solvent**

Solutions: General properties

Solutions are **homogeneous** in their properties: they consist of a single phase (gas, liquid or solid) containing two or more chemical substances. A solution is characterized by the components of the solution and the **concentrations** of the components. The component in highest concentration is usually called the **solvent**, and those in lower concentration are called the **solutes**. Concentrations are measured in various units; two common concentration units are **molarity** and **mass percent**. The concentration at which no more solute will dissolve is called the **solubility**; if the concentration of a solute is at its solubility, the solution is said to be **saturated** in that solute. The solubility of a substance depends on the solvent and the temperature. With few exceptions, the solubility increases with increasing temperature.

The Solubility Rule: Like Dissolves Like

We can picture the dissolving of a solute in a solvent as occurring in three steps:

1. Breaking the intermolecular forces holding the solvent molecules together,
2. Breaking the intermolecular forces holding the solute molecules together, and
3. Forming new intermolecular attractions between the solute and solvent molecules.

The first two steps require energy, and the third step will release energy. For a substance to have a significant solubility, the energy change should be at worst modestly endothermic; that is, the energy released from the solute-solvent interactions should be similar to that required to disrupt the solvent-solvent and solute-solute interactions. To fulfill this condition, the dominant intermolecular forces active in the pure solvent should be the dominant intermolecular forces active in the solute; like forces are active in the solute and the solvent. Specifically:

Dominant force active in solvent	Dominant force active in soluble solutes
Dispersion	Dispersion
Dipolar	Dipolar, ionic (electrical)
H-bonding	H-bonding, dipolar

Chapter 10: Effects of Intermolecular Forces

The high energy cost of breaking the bonds in covalent solids and metals makes metals and covalent solids insoluble in liquid solvents unless they undergo chemical transformation by reaction with the solvent. Often, several types of forces are active in both the solvent and solute and the solvent may be able to dissolve a wide variety of solutes; we may be surprised by unexpected solubility (or insolubility) of a substance. The following guideline is very useful in choosing a solvent to dissolve a particular substance. When applying this rule, remember that ionic substances are considered to be quite polar.

Like dissolves like:
polar solvents for *polar solutes*;
nonpolar solvents for *nonpolar solutes*

Exercise 10.8 A solute and a pair of solvents are provided in each part of this exercise. Choose the solvent in which you expect the solute will be more soluble and provide a brief explanation of your choice.
(a) Dextrose, $C_6H_{12}O_6$ (see figure, below; solvents = water or benzene
(b) Butanol, $CH_3CH_2CH_2CH_2O$; solvents = water or acetone
(c) Sulfur hexafluoride; solvents = benzene or acetone
(d) Potassium nitrate; solvents = acetone or water

Steps to Solution: *To determine which of the solvents will be more effective at dissolving a given solute, we must compare the intermolecular forces present in the pure solute and the pure solvent and the forces that would act between solvent and solute. If the forces are similar or stronger between the solute and the solvent than between the pure substances, the substances will be more soluble than if the forces are quite different or weaker.*

(a) *Dextrose has the structural features shown below; there are 5 O-H bonds present in dextrose to participate in H-bonding in the pure solute. Therefore we expect that the predominant intermolecular force acting in dextrose is hydrogen bonding. In benzene, C_6H_6, the predominant intermolecular forces are dispersion forces, whereas in water, H-bonding is most important (along with dipolar forces). Therefore we expect dextrose to be more soluble in the solvent with more similar forces, water.*

(b) *Butanol has several types of intermolecular forces present: hydrogen bonding from the OH group, dipolar forces and dispersion forces arising from the CH_2-CH_2CH_3 "tail". Butanol is not very polar due to this hydrocarbon portion of the molecule. Water has hydrogen bonding dominating its intermolecular forces and is quite polar. Acetone (H_3C-[C=O]-CH_3) has dispersion forces and dipolar forces present in the pure solvent; acetone is less polar than water.*

Water interacts most strongly with butanol through hydrogen bonding, and will not be able to replace the dispersion forces that will be lost when butanol dissolves. Acetone can undergo significant dispersion forces (the two CH_3 groups) and dipolar

371

forces. In addition, the oxygen atom of acetone has lone pairs and may act as a H-bond acceptor to the O-H bonds of butanol. Because the intermolecular forces between acetone and butanol are more similar than those between water and butanol, we expects butanol to be more soluble in acetone.

(c) Sulfur hexafluoride is a nonpolar molecule; the forces acting between SF_6 molecules are dispersion forces. The intermolecular forces in benzene are dispersion forces; those in acetone are dipolar forces and, to a smaller extent, dispersion forces. To dissolve SF_6 in acetone would require breaking dipolar forces in acetone; these interactions would be replaced by weaker dispersion forces in the solution. The forces acting between benzene and SF_6 would be dispersion forces, the same sort of forces that would be broken on dissolving SF_6 in benzene. SF_6 is expected to be more soluble in benzene than acetone because similar intermolecular forces exist between the solvent and the solute.

(d) Use the type of reasoning illustrated above to show that KNO_3 is expected to be more soluble in water than in acetone.

10.6 Dual-Nature Molecules: Surfactants and Biological Membranes

QUESTIONS TO ANSWER, SKILLS TO LEARN
1. What is a dual-nature molecule?
2. Recognizing hydrophobic and hydrophilic portions of molecules
3. What molecular pieces are necessary in a dual-nature molecule?
4. What are the uses and sources of dual-nature molecules?

Water is probably the most abundant chemical substance on earth and is important in biochemical and industrial processes; for this reason, substances and molecular structures are classified by their ability to interact with water. Those molecules that can interact strongly with water through hydrogen bonding or dipolar forces are called hydrophilic. Those molecules or structures that cannot interact strongly with water (their structures allow dispersion forces as the dominant intermolecular force) are called hydrophobic.

Hydro*philic*: water *loving*   **Hydro*phobic*: water *fearing***

Exercise 10.9 Classify the following molecules or fragments as hydrophobic or hydrophilic. A line from an atom indicates where a bond would be formed to another fragment.
(a) the $-CH_2SO_3^-$ group (b) CH_3CH_2OH (c) the $-CH_2(CH_2)_{11}CH_3$ group (d) the $-NH_3^+$ group

Steps to Solution: The way to recognize hydrophobic and hydrophilic molecules or molecular sections is to determine what intermolecular forces would be active between water and the structure in question. If only dispersion forces are possible, then the structure is hydrophobic. The presence of strong hydrogen bonding acceptors or donors, dipolar or ionic groups indicates that the structure is hydrophilic.

(a) The $-CH_2SO_3^-$ group is ionic and also has O atoms that are good H-bond acceptors. Therefore this group is expected to be hydrophilic because it is able to strongly interact with water.

(b) The ethanol molecule, CH_3CH_2OH, has a short hydrocarbon portion (the CH_3CH_2- group) but also has the polar, hydrogen bonding -OH group. The H-bond accepting and donating properties of this group make the ethanol molecule hydrophilic and completely soluble with water in all proportions (miscible).

(c) The $-CH_2(CH_2)_{11}CH_3$ group will only be able to interact with water through dispersion forces because there are no polar groups or polar bonds to H. Therefore it is expected to be hydrophobic.

(d) The $-NH_3^+$ group is ionic and also contains three N-H bonds that may act as H-bond donors. Therefore the $-NH_3^+$ group is hydrophilic.

In a **dual-nature molecule**, a hydrophobic structure is bonded to a hydrophilic structure such as $CH_3(CH_2)_{11}CH_2-CH_2 SO_3^-$.

Chemical formula: $\quad CH_3(CH_2)_{11}CH_2-CH_2— SO_3^-$

hydrophobic portion or *tail*   hydrophilic portion or *head*

Schematic representation:

In aqueous solution, the most stable configuration for these dual-nature molecules is to have the hydrophilic heads interacting with water and the hydrophobic tails in a configuration that minimizes their unfavorable interaction with water and maximizes interaction with nonpolar structures like themselves. Three configurations satisfy this like-dissolves-like requirement:

1. **Monolayers**: In a monolayer, the dual-nature molecules move to a surface of the aqueous phase. The heads of the molecules are in contact with the aqueous phase, and the tails point away from the aqueous phase where they may interact with each other.

2. **Micelles**: In a micelle, the dual-nature molecules form a spherical array where the tails point toward the center of the sphere and the heads form the outer surface of the sphere in contact with the water.

3. **Bilayers** or **vesicles**: A bilayer is like two monolayers that have been placed together with the tails in contact, not unlike a hook and loop (Velcro) fastener. The outer surface of the bilayer that contacts the aqueous phase is made up of the hydrophilic heads.

Dual-nature molecules are manufactured as soaps, detergents, and other products that fall into a general category called *surfactants* (short for *surf*ace *act*ive age*nts*). They are also found in life forms as part of cell membranes; they are called **phospholipids,** where the *lipid* portion is the hydrophobic part of the molecule and the *phospho-* portion is the hydrophilic part.

Exercise 10.10 Water beads when it is put on a shiny, newly waxed car, but soapy water forms a film that coats the surface. Explain why soapy water does not bead up on the waxed surface.

Steps to Solution: *We are asked to describe the different behaviors of water and a soap solution. We must first identify the differences between the solutions (solute$_{(s)}$, etc.) and see how these differences relate to the behavior observed.*

In responding to Exercise 10.6, we deduced that the waxed surface of a car is hydrophobic: this is why water beads (forms small spheres) when applied to such a surface. Soapy water contains soap, a surfactant (dual-nature molecules). The tail (hydrophobic end) of the surfactant molecules has strong adhesive forces to the hydrophobic waxed surface. The molecular picture on the next page shows the result of the adhesive forces.

The structure looks a lot like a monolayer: the tails are pointing away from the aqueous phase but the tails are in contact with a surface. The hydrophilic heads of the surfactant molecules are pointing away from the waxed surface, and the aqueous phase adheres much better to these hydrophilic groups than to the hydrophobic waxed surface. Therefore the aqueous phase covers the waxed surface evenly because the dual-nature molecules in soap have temporarily changed the character of the surface from hydrophobic to hydrophilic. The soap molecules can be washed away with water so that the hydrophobic nature of the surface can be restored.

Chapter 10: Effects of Intermolecular Forces

10.7 Properties of Aqueous Solutions

QUESTIONS TO ANSWER, SKILLS TO LEARN
1. **Visualizing a solution at the molecular level as a mix of solvent and solute** *particles*
2. **How does the presence of the solute change some properties of the solution compared to the pure solvent?**
3. **What is the relationship between substance type and the number of particles it forms when dissolved?**
4. **Calculations using solution concentrations and colligative properties**

Not surprisingly, the addition of solute to a substance leads to the formation of a solution whose physical properties are different than those of the pure solvent. For example, the addition of sugar to water leads to a change in density of the solution. A solution with an equal concentration of a different solute would probably have a different density. However, there are some properties of solutions that, surprisingly, don't depend on the type of solute, but only on the number of solute molecules or ions present in solution. We will discuss the temperatures for the phase transitions between liquid and solid (the *melting point*) and between liquid and vapor (the *boiling point*), and a new property, the *osmotic pressure*.

The Melting Point; Without and With a Solute

At the melting point, a dynamic equilibrium is occurring. Consider benzene, which has a normal melting point of 5.4 °C. If we place a piece of frozen benzene into liquid benzene at a temperature somewhat higher than the melting point, two processes will occur. Some molecules will leave the crystal of benzene and enter the liquid phase:

$$C_6H_6\ (s) \rightleftarrows\ C_6H_{6(l)} \text{ (solid benzene to liquid benzene)}$$

Simultaneously, molecules of benzene in the liquid phase will collide with the crystal and become part of the crystal:

$$C_6H_6\ (l) \rightleftarrows\ C_6H_{6(s)} \text{ (liquid benzene to solid benzene)}$$

At temperatures higher than the melting point, the conversion of solid benzene to liquid benzene is faster than the conversion of liquid to solid. The melting point is significant because only at that temperature is no net change observed: the number of molecules of benzene being converted from solid to liquid is the same as the number of molecules being converted from liquid to solid; *the system is at equilibrium and if no energy is added or removed from the system, the amounts of liquid and solid benzene will not change.* The melting point does depend on pressure; we use a standard pressure of 1 atm for what is called the **normal melting point**.

The transition from solid to liquid is called *melting* or *fusion*; the transition from liquid to solid is called *freezing*. The temperatures will be the same for both processes; the different names signify only the initial and final phases of the process. The freezing point of a substance of solution is abbreviated as T_f.

Addition of a small amount of solute to any liquid will change the equilibrium between solid and liquid. Consider a solution of naphthalene, $C_{10}H_8$, in benzene.

napthalene, $C_{10}H_8$:

Because the solution is a mixture of solvent and solute molecules, the molecules in the liquid phase that collide with the crystal are both solvent and solute molecules. Only collisions by solvent molecules lead to crystal formation; therefore the rate of solid formation is slowed, but the rate of liquid formation (melting) from the solid remains the same. The solvent will be in equilibrium with the solid (freeze) only at lower temperatures than the melting point of the pure solvent. The identity of the solute molecule is not important; adding the same number of *any* solute molecules will lead to lowering of the freezing point. The amount by which the freezing point is lowered is called the *freezing point depression*, ΔT_f.

$$\Delta T_f = T_f(\text{pure solvent}) - T_f(\text{solution})$$

The more concentrated the solute, the fewer collisions are made by solvent molecules with the crystal of frozen solvent and the greater the freezing point depression. The freezing point depression is found to depend on the molal concentration of solute, c_m:

$$\Delta T_f = K_f\, c_m \qquad K_f = \text{freezing point depression constant (°C•kg/mol)}$$

The freezing point depression constant has units of °C•kg/mol and depends only on the solvent.

The Boiling Point: Without and With a Solute

The concept of liquid-vapor equilibrium was discussed in earlier chapters. Using benzene as an example, the processes that occur at equilibrium are:

$$C_6H_{6\,(l)} \rightleftarrows C_6H_{6\,(g)} \qquad\qquad C_6H_{6\,(g)} \rightleftarrows C_6H_{6\,(l)}$$

(liquid benzene to gaseous benzene)   (gaseous benzene to liquid benzene)

Addition of a nonvolatile solute (one whose vapor pressure is much lower (less than 1%) than that of the solvent) leads to an increase in the temperature (average kinetic energy) required to obtain boiling; this increase in temperature is called the *boiling point elevation*, ΔT_b.

Chapter 10: Effects of Intermolecular Forces

$$\Delta T_b = T_b(\text{solution}) - T_b(\text{pure solvent})$$

The boiling point elevation depends on the molal concentration of solute as well:

$$\Delta T_b = K_b \, c_m \qquad K_b = \text{boiling point elevation constant } (°C\cdot kg/mol)$$

The boiling point elevation constant has units of °C•kg/mol and depends only on the solvent.

Ionic Solutions

The freezing point depression and boiling point elevation of a solution depend only on the number of solute particles in solution, not on their identity. If we prepare a mixture of substances dissolved in a solvent, the freezing point depression and boiling point elevation depend only on the total molality of all the solutes present. We must recognize this fact when dissolving ionic substances such as $CaCl_2$:

$$CaCl_2 \rightarrow Ca^{2+} + 2\,Cl^-$$

Dissolving *one* mole of $CaCl_2$ leads to the formation of *three* moles of solute particles. One way to deal with this problem is to correct for the number of ions produced per mole of ionic solute with a factor, i. For $CaCl_2$, the formation of three moles of ions from one mole of substance indicates that $i = 3$. For ionic substances, the formulas for freezing point depression and boiling point elevation are modified by this factor, i.

$$\Delta T_f = i\, K_f\, c_m \qquad \text{and} \qquad \Delta T_b = i\, K_b\, c_m$$

Exercise 10.11 An 0.457 g sample of a pure substance with empirical formula C_4H_3N is dissolved in 11.21 g C_6H_6 (K_f = 5.12 °C•kg/mol).
(a) Calculate the freezing point depression for the empirical formulas.
(b) The actual freezing point depression is 1.48 °C. Use this information and that in part (a) to determine the molecular formula.

Steps to Solution: To answer this question we must first use the empirical formula mass to calculate the freezing point depression and then use this value to find the molecular formula from the empirical formula. Part (a) involves calculating the mole fraction based on the empirical formula weight. We can solve part (b) in either of two ways. We can use the equation for freezing point depression to find the MM of the unknown; an alternative is to use trial and error and add another empirical formula mass at a time to the molar mass used in calculating the mole fraction of solute.

(a) The freezing point depression is calculated from the formula:

$$\Delta T_f = K_f\, c_m$$

The empirical formula mass of C_4H_3N is 65.07 g/mole; the molality of the solute is:

$$c_m = \frac{\text{mols solute}}{\text{kg solvent}} = \frac{\frac{g}{MM}}{\text{kg solvent}} = \frac{\frac{0.457 g}{65.07 g/mol}}{0.01121 \text{ kg}} = 0.626 \text{ mol/kg}$$

Substitution of this trial value for X_{solute} into the ΔT_f expression gives:

$$\Delta T_f = c_m \cdot K_f = 0.626 \text{ mol/kg} \cdot 5.12 \text{ °C} \cdot \text{kg/mol} = 3.20 \text{ °C}$$

(b) Because the observed freezing point depression and the calculated value are not the same (about 100% difference), the molecular formula and the empirical formula are not the same. The next possible molecular formula would have two C_4H_3N units which would result in a molecular formula of $C_8H_6N_2$ with molar mass 130.1 g/mole. The freezing point depression calculated using this molar mass is:

$$c_m = \frac{\frac{g}{MM}}{\text{kg solvent}} = \frac{\frac{0.457 g}{131.0 \text{ g/mol}}}{0.01121 \text{ kg}} = 0.313 \text{ mol/kg}$$

This value is a little larger than half the mole fraction calculated in part (a). The freezing point depression found using the mole fraction based on the higher molar mass is:

$$\Delta T_f = c_m \cdot K_f = 0.313 \text{ mol/kg} \cdot 5.12 \text{ °C} \cdot \text{kg/mol} = 1.60 \text{ °C}$$

This value is nearly the same as the observed freezing point depression (1.45 °C) and so the molecular formula is probably $C_8H_6N_2$. It is important to note that the freezing point depression decreases with increasing molar mass.

Exercise 10.12 What is the boiling point elevation of a solution of 12.0 g $Ce(NO_3)_3 \cdot 6\ H_2O$ in 93.0 g H_2O?

Steps to Solution: This compound is ionic because it is a compound containing the polyatomic ion, nitrate. Therefore the modified version of the boiling point elevation equation must be used,

$$\Delta T_b = i\ K_b\ c_m$$

What is the value of i to be used in this equation? First, think at the molecular level and consider what occurs when this substance is dissolved in water.

$$Ce(NO_3)_3 \cdot 6\ H_2O_{(s)} \rightarrow Ce^{3+}{}_{(aq)} + 3\ NO_3^-{}_{(aq)} + 6\ H_2O_{(l)}$$

Note that four ions and six molecules of water are produced by dissolving the salt. The 6 molecules of water become part of the solvent water; therefore the "i" factor is four since four solute particles are produced. The mole fraction of the solvent, water, must include the water released by dissolving the hydrated salt.

$$\text{mole } Ce(NO_3)_3 \cdot 6\ H_2O = \frac{12.0 \text{ g}}{434.23 \text{ g/mole}} = 2.76 \times 10^{-2} \text{ mole}$$

g H₂O = mass solvent + mass waters of hydration

g H₂O = 93.0 g + 18.01 g/mole H₂O(2.76 × 10⁻² mole Ce(III)•(6 mole H₂O/mole Ce(III))

g H₂O = 96.0 g

$$c_m = \frac{\frac{g}{MM}}{\text{kg solvent}} = \frac{2.76 \times 10^{-2} \text{mol}}{0.0960 \text{ kg}} = 0.288 \text{ mol/kg}$$

$\Delta T_b = i\, K_b\, c_m = 4\,(0.510\ °C\cdot kg/mol)0.288\ mol/kg = 0.58\ °C$

Osmotic Pressure

Membranes were briefly described earlier; they are sheets of material through which only molecules can pass. If a membrane allows one type of molecule to pass through but does not allow others, it is called a **semipermeable** membrane. When water moves through a semipermeable membrane, the process is called **osmosis**. Three different situations are described in the drawing below.

(a) In this case, identical solutions are present on either side of the membrane. The rate of water transport in both directions across the membrane is the same, so there is no change in the levels of the solutions.

(b) In this case, a solution with a higher concentration of solute is present on the left-hand side, as indicated by the different shading. More water moves from left to right than from right to left through osmosis. The system is not at equilibrium because the volume of solution on the right-hand side will increase over time.

(c) Again, a solution with a higher concentration of solute is present on the left-hand side, as indicated by the different shading. However, the levels of the solutions are not the same: the solution at right is higher, and exerts more pressure on the membrane. The higher pressure leads to more collisions of water with the membrane so that the rate of water transfer in both directions is the same (but greater than that in case [a]).

The height in part (c) can be converted to pressure. This pressure is the pressure difference between the two solutions required to maintain dynamic equilibrium and is called the **osmotic pressure**, Π. The relationship between osmotic pressure and solution concentration is given by:

$$\Pi = MRT$$

The concentration of *all* solutes, M, is expressed in moles/L; T is the temperature in Kelvin; and with $R = 0.08206$ L·atm/mole·K, the osmotic pressure, Π, is obtained in units of atm.

> **COMMON PITFALL: Forgetting the "i" Factor.**
> Don't forget the "i" factor when dealing with ionic salts in osmotic pressure problems.

Exercise 10.13. Consider a sample of lysozyme, an enzyme found in tears whose function is to break down bacterial cell walls to kill the bacteria and protect the eye from infection. The molar mass of this enzyme is rather small for proteins, about 1.4×10^3 g/mole. A solution is made that contains 0.100 g of lysozyme in 150 g of water ($\rho = 1.00$ g/mL) at 25 °C. Calculate (a) the boiling point elevation, (b) the freezing point depression and (c) the osmotic pressure of this solution.

Steps to Solution: To solve this problem we must calculate the concentration of the solute and then apply the appropriate formulas for ΔT_b, ΔT_f and Π.

(a) The boiling point elevation is found using :

$$\Delta T_b = K_b c_m$$

The boiling point elevation constant for water is 0.512 °C·kg/mol; we must calculate the molality of lysozyme.

$$c_m = \frac{\frac{0.100 \text{ g}}{64,000 \text{ g/mol}}}{0.100 \text{ kg}} = 1.56 \times 10^{-5} \text{ mol/kg}$$

Substituting this value and that of K_b into the formula gives:

$$\Delta T_b = K_b c_m = 0.512 \text{ °C·kg/mol}(1.56 \times 10^{-5} \text{ mol/kg}) = 8.0 \times 10^{-6} \text{ °C}$$

(b) *The freezing point depression is found using by the formula below, where the freezing point depression constant K_f = 18.60 °C•kg/mol. The molality was calculated in part (a) of this exercise.*

$$\Delta T_f = K_f c_m = 18.60 \text{ °C•kg/mol } (1.56 \times 10^{-5} \text{ mol/kg}) = 2.9 \times 10^{-4} \text{ °C}.$$

It is clear that ΔT_b and ΔT_f are quite small and could not be measured with accuracy and high precision.

(c) *The osmotic pressure is calculated from the equation for the osmotic pressure.*

$$\Pi = MRT$$

(Assume that the volume of the solution is the same as the volume of the solvent.)

$$\Pi = \frac{0.100 \text{ g}/1.4 \times 10^3 \text{ g/mole}}{0.150 \text{ L}} (0.0821 \text{ L•atm/mole•K})(298 \text{ K}) = 1.2 \times 10^{-2} \text{ atm}$$

$$\Pi = 1.2 \times 10^{-2} \text{ atm } (760 \text{ mm Hg/atm}) = \boxed{8.9 \text{ mm Hg}}$$

Determining Molar Masses With Osmotic Pressures

The molar mass, *MM*, is related to the osmotic pressure through the equation:

$$MM = \frac{mRT}{\Pi V}$$

where *m* is the mass of the sample, *V* is the volume of the solution, and *R* and *T* have the usual meanings. From Exercise 10.14, it should be clear that only osmotic pressure will be useful in determination of the molar masses of molecules with large molar masses like proteins. The other properties (freezing point depression and boiling point elevation) give such small values they cannot be reliably measured.

Test Yourself

1. List the following three compounds in order of increasing boiling point:

    i   ii   iii

2. Predict which of the pairs of substances will have the greater viscosity.
   (a) CH_3CH_2OH, CH_3CH_2Cl; (b) $CH_3(CH_2)_{12}OH$, $CH_3(CH_2)_{12}CH_3$;
   (c) $CH_3(CH_2)_3C(CH_3)_3$, $CH_3(CH_2)_6CH_3$

3. A sample of adrenaline has the following elemental composition: 59.0 % C; 26.2 % O; 7.10 % H; 7.65 % N. When 0.64 g of adrenaline is dissolved in 36.0 g benzene (K_f = 62.7 °C), the freezing point is lowered by 0.50 °C. Determine the empirical formula, molar mass and molecular formula of adrenaline.

4. In which of the two solvents, methanol (CH_3OH) or benzene (C_6H_6) will the following solutes be more soluble? (a) napthalene ($C_{10}H_8$) (b) KNO_3 (c) SF_6 (d) O_2.

5. Calculate the mole fractions of the cations in following aqueous solutions: (a) 0.10 m NaCl (b) 0.20 m Na_2CO_3 (c) 10.0 % KNO_3.

Answers

1. ii, iii, i

2. (a) CH_3CH_2OH (b) $CH_3(CH_2)_{12}OH$ (c) $CH_3(CH_2)_6CH_3$

3. $C_9H_{13}O_3N$; 1.8 x 10² g/mole; $C_9H_{13}O_3N$

4. (a) C_6H_6 (b) CH_3OH (c) C_6H_6 (d) C_6H_6

5. (a) X_{Na^+} = 1.7 x 10⁻³ (b) X_{Na^+} = 7.1 x 10⁻³ (c) X_{K^+} = 1.9 x 10⁻²

Solutions to Chapter Problems

10.1 The boiling point of a substance measures the strength of intermolecular forces, so the graph for Cl_2 should show larger intermolecular attractions than for F_2 but smaller than for Br_2:

10.3 For any given substance, intermolecular attractions are constant. Their relative importance depends on two features: How close together molecules are, on average; and how much kinetic energy of motion molecules have, on average.
(a) Molecules are farther apart, making intermolecular attractions less significant.
(b) Molecules are closer together, so intermolecular attractions are more significant.
(c) Molecular kinetic energies are smaller, so intermolecular attractions are more significant.

10.5 Pictures of atomic arrangements should show a high degree of order for a solid, close packing but some disorder for a liquid, and large distances between atoms for a gas:

10.7 To determine deviation from ideal behavior, calculate the pressure using the ideal gas equation and compare the calculated result with the experimental value:

$$P_{ideal} = \frac{nRT}{V} = \frac{(1.00 \text{ mol})(8.206 \times 10^{-2} \text{ L atm/mol K})(40.0 + 273.2 \text{ K})}{(1.20 \text{ L})} = 21.4 \text{ atm}$$

$$\% \text{ deviation} = (100\%)\left(\frac{P_{actual} - P_{ideal}}{P_{ideal}}\right) = (100\%)\left(\frac{19.7 \text{ atm} - 21.4 \text{ atm}}{21.4 \text{ atm}}\right) = -7.9\%$$

10.9 To calculate pressure using the van der Waals equation, we must know n, V, T, a, and b:

$$P = \left(\frac{nRT}{V-nb}\right) - \left(\frac{n^2 a}{V^2}\right)$$

Table 10-1 in your textbook lists values for Cl_2: $a = 6.493$ L^2 atm/mol^2, $b = 0.0562$ L/mol.

$$n_{Cl_2} = \frac{m}{MM} = \frac{(1.25 \text{ kg})(10^3 \text{ g/kg})}{(70.906 \text{ g/mol})} = 17.63 \text{ mol}$$

$(V - nb) = 15.0 \text{ L} - (17.63 \text{ mol})(0.0562 \text{ L/mol}) = 14.01 \text{ L}$

$$P = \frac{(17.63 \text{ mol})(8.206 \times 10^{-2} \text{ L atm/mol K})(295 \text{ K})}{(14.01 \text{ L})} - \frac{(17.63 \text{ mol})^2 (6.493 \text{ L}^2 \text{ atm/mol}^2)}{(15.0 \text{ L})^2}$$

$P = 30.46$ atm $- 8.97$ atm $= 21.5$ atm

$$P_{ideal} = \frac{nRT}{V} = \frac{(17.63 \text{ mol})(8.206 \times 10^{-2} \text{ L atm/mol K})(295 \text{ K})}{(15.0 \text{ L})} = 28.5 \text{ atm}$$

% deviation $= (100\%)\left(\dfrac{P_{actual} - P_{ideal}}{P_{ideal}}\right) = (100\%)\left(\dfrac{21.5 \text{ atm} - 28.5 \text{ atm}}{28.5 \text{ atm}}\right) = -25\%$

10.11 The viscosities of a set of similar substances increases with increasing chain length, because of increased dispersion forces (stronger intermolecular attractions make it harder for molecules to slide by one another) and increased "tangling" (longer chains are more entangled). Thus the order of increasing viscosity is pentane (C_5) < gasoline (C_8) < fuel oil (C_{12}).

10.13 Ease of liquefaction depends on the magnitude of intermolecular attractions: The larger the attractions, the easier the substance is to liquefy. All these molecules are symmetric (tetrahedral), so the ranking depends entirely on dispersion forces. Ease of liquefaction will increase with orbital size: CH_4 (hardest to liquefy) < CF_4 < CCl_4 (easiest to liquefy).

10.15 Polarizability increases with the size of the orbitals. The molecule with larger orbitals (CCl_4) should be shown with greater distortion than the molecules with smaller orbitals (CH_4). These molecules are close to spherical in the absence of polarization:

Chapter 10: Effects of Intermolecular Forces

10.17 Boiling point depends on the magnitude of intermolecular attractions: The larger the attractions, the higher the boiling point. The size of intermolecular attractions depends on amount of dispersion forces (size of molecule and size of orbitals), presence of polar bonds, and hydrogen bonding. Among these substances, ethanol forms hydrogen bonds, so it has the highest boiling point. Propane, being smaller than *n*-pentane, has the lowest boiling point: Propane (lowest bp) < *n*-pentane < ethanol (highest bp).

10.19 Hydrogen bonding capability requires the presence of an electronegative atom with lone pairs (O, F, or N) and an H-*X* bond to a highly electronegative atom (*X* = O, N, or F). (b and d) hydrogen bonding; (a and c) no hydrogen bonding.

10.21 Hydrogen bonding capability requires the presence of an electronegative atom with lone pairs (O, F, or N) and an H–*X* bond to a highly electronegative atom (*X* = O, N, or F).
(a) Ammonia has one N atom with a lone pair that can form a hydrogen bond to any H atom of another ammonia molecule.

(b) Two types of hydrogen bonds are possible, between an N atom of NH$_3$ and an H atom of H$_2$O, and between an O atom of H$_2$O and an H atom of NH$_3$.

10.23 The position of elements in the periodic table, the chemical formula, and a knowledge of polyatomic ions all help in identifying types of solids:
Sn is a metal, so is a metallic solid; S$_8$ has a specific molecular formula, so is a molecular solid; Se is not a metal, so is a network solid; SiO$_2$ is not ionic, so is a network solid; and Na$_2$SO$_4$ contains Na$^+$ cations and SO$_4^{2-}$ polyatomic anions, so is an ionic solid.

10.25 Consult your textbook for information about the differences between different types of solids.
(a) The bonding in metals comes from extended networks of delocalized electrons, while the bonding in network solids includes many individual covalent bonds.
(b) Metals conduct electricity, are malleable and ductile, and are shiny in appearance. Network solids are non-conductors or semiconductors, are brittle, and often have dull appearances.

10.27 Consult your textbook for information about the differences among different types of solids. (a) The ability to conduct electricity is characteristic of metals.
(b) Ionic solids do not conduct electricity but dissolve in water to give conduction solutions.
(c) Molecular solids have discrete neutral particles so are low-melting and give non-conducting liquids.

10.29 The density of a solid depends not only on the nature of the elements it contains but also on the tightness of bonding. A structure with more open bonding pattern has a lower density than one with a more compact bonding pattern. Graphite uses sp^2 hybrids and has an extended π bonding network, making it planar and leaving a relatively open structure between successive planes. Diamond uses sp^3 hybrids and has a compact network of bonds:

Diamond: all distances relatively short

Graphite: large distance between successive layers

10.31 To determine a chemical formula from a unit cell, count the number of atoms of each element contained within the unit cell. Atoms on a face count 1/2 (each is part of two unit cells), those on an edge count 1/4 (each is part of four unit cells), and those at a corner count 1/8 (each is part of eight unit cells).
There is 1 Y atom in the center of the cell and 2 Ba atoms within the cell at either end.
Cu: 1/4 (8 edge atoms) + 1/8 (8 corner atoms) = 3 atoms;
O: 1/2 (8 face atoms) + 1/4 (12 edge atoms) = 7 atoms.
Chemical formula is $YBaCu_3O_7$.

10.33 The problem states that carborundum is diamond-like. Diamond is composed entirely of tetrahedral carbon atoms (SN = 4), and the empirical formula of carborundum, SiC, indicates that alternate C atoms should be replaced with Si atoms. Each C atom bonds to four Si atoms and each Si atom bonds to 4 C atoms:

10.35 The immiscibility of liquids can be explained using the concept that "like dissolves like." Molecules of salad oil are almost entirely hydrocarbons, which have dispersion-type intermolecular forces but low polarity and no hydrogen-bonding capability. A solution of acetic acid in water has high polarity and a large hydrogen-bonding capability. Thus these two liquids are not at all like each other and do not mix. In terms of balance of forces, hydrogen bonds would have to be broken for the two liquids to mix, making mixing energetically unfavorable.

Chapter 10: Effects of Intermolecular Forces

10.37 The types of solids that dissolve in a solvent can be predicted using the concept that "like dissolves like." Liquid ammonia is similar to liquid water: It is a polar molecule with a strong hydrogen-bonding capability. Thus liquid ammonia should dissolve the same sorts of substances as water, ionic salts and polar organic materials such as alcohols.

10.39 To determine how many hydrogen bonds to water a molecule can form, count its polar H–X bonds and its electronegative atoms with lone pairs of electrons. Glycerol has three of each, so it can form six hydrogen bonds to water molecules:

10.41 In a chromatography column, components that are attracted to the support material move more slowly than those that are not attracted, so the least polar substance moves fastest through a polar column. The areas under chromatographic peaks are a measure of relative amounts.

(a) Heptane is least polar and comes off first, followed by methylpentyl ether. The most polar compound, 1-hexanol, comes off last.
(b) The last peak is largest, so 1-hexanol is present in largest amount; The middle peak is smallest, so methylpentyl ether is present in smallest amount.

10.43 Surfactants must have a highly polar (or ionic) end and a non-polar end. Of these three substances, propionic acid has too short a non-polar end and would be the worst. Lauryl alcohol and sodium lauryl sulfate have similar non-polar portions, but the ionic portion of sodium lauryl sulfate makes it the best surfactant of these three substances:

10.45 Surfactants have a polar (or ionic) end and a non-polar end. The polar end of the surfactant molecule associates with a polar substance, while the non-polar end of a surfactant molecule associates with a non-polar substance. A bilayer membrane between two aqueous layers will have its polar "heads" associated with the aqueous layers and its non-polar "tails" associated with one another, as the drawing shows:

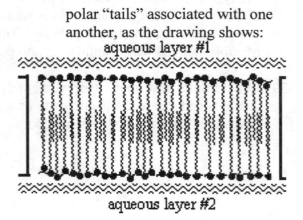

10.47 Freezing points are calculated using Equation 10-2 from your textbook: $\Delta T_f = K_f c_m$. For water, K_f is 1.858 °C kg/mol. First, the molality of the solution must be calculated:
c_m = (mol solute)/(kg solvent)
Consider 100 g of a solution that is 12% by mass ethanol. It contains 12 g ethanol and 88 g of solvent (water). Thus

mol solute = $\dfrac{12 \text{ g}}{46 \text{ g/mol}}$ = 0.26 mol   and   kg solvent = (88 g)(10^{-3} kg/g) = 8.8 x 10^{-2} kg.

c_m = (0.26 mol) ÷ (8.8 x 10^{-2} kg) = 2.95 mol/kg
$\Delta T_f = K_f c_m$ = (1.858 °C kg/mol)(2.95 mol/kg) = 5.5 °C.
This is the depression in freezing point, so the new freezing point is 0.0 − 5.5 = − 5.5 °C.

10.49 Boiling points can be calculated for solutions of non-volatile solutes, using the boiling point elevation constant. However, ethanol is volatile, so the equation does not apply and the boiling point of the solution cannot be calculated from the information provided.

10.51 Molar masses can be calculated from osmotic pressures using Equation 10-5 in your textbook.
First convert Π into atm: Π = (64.8 torr)(1 atm/760 torr) = 8.53 x10^{-2} atm.

(a) $MM = \dfrac{mRT}{\Pi V} = \dfrac{(1.00 \text{ g})(8.206 \times 10^{-2} \text{ L atm/mol K})(25 + 273 \text{ K})}{(8.53 \times 10^{-2} \text{ atm})(1.00 \text{ L})}$ = 287 g/mol

(b) The non-polar tail contains 11 C atoms and 2 H atoms for every C except the first, which has 3 H atoms, giving 23 H atoms altogether. It contributes 11(12.01) + 23(1.008) = 155 g/mol, leaving for the polar head group, MM = 287 − 155 = 132 g/mol.

10.53 Gas samples are conveniently analyzed using gas chromatography. A gas sample mixed with a carrier gas passes through a column containing a stationary phase that attracts different gases to different extents. The more strongly attracted a particular component is to the stationary phase, the slower it passes through the column. Thus the column separates a gas mixture into a series of "peaks" that emerge from the column at different times. Each component can be identified by comparing its time of passage through the column with those of standard samples of known substances.

10.55 Molar masses can be calculated from osmotic pressures using Equation 10-5 in your textbook.

$$MM = \frac{mRT}{\Pi V} = \frac{(1.00 \text{ g})(8.206 \times 10^{-2} \text{ L atm / mol K})(25 + 273 \text{ K})}{(1.36 \text{ atm})(1.00 \times 10^2 \text{ mL})(10^{-3} \text{ L / mL})} = 180 \text{ g/mol}$$

10.57 The solubilities of organic substances in water depends on the presence of polar groups and groups that are capable of forming hydrogen bonds with water. Alcohols contain OH groups that enhance solubility. Among these compounds, benzene (C_6H_6) is non-polar and has the lowest solubility, while erythritol has four OH groups, making it highly soluble:

benzene (lowest solubility) < pentanol < erythritol (highest solubility).

10.59 Hydrophilicity is conveyed by the presence of ionic, polar, or hydrogen-bonding groups. DDT contains none of these, so it is highly hydrophobic. Thus it is stored in fatty tissues (which are non-polar) rather than being excreted from the body. Consequently, the effects of DDT are cumulative, especially for organisms high up on the "food chain" such as predator birds.

10.61 Differential solubilities can be explained using the concept that "like dissolves like." Here, iodine is a non-polar solid which will differentially dissolve in a non-polar liquid. Water is highly polar and CCl_4 is non-polar, so these two liquids are immiscible and I_2 preferentially dissolves in CCl_4.

10.63 All substances are subject to dispersion forces. In addition, look for polarity and hydrogen bonding ability:
(a) NH_3 is polar and forms hydrogen bonds in addition to its dispersion interactions;
(b) $CHCl_3$ is a polar molecule in addition to having dispersion interactions;
(c) CCl_4 is symmetric (tetrahedral) so has only dispersion interactions; and
(d) CO_2 is symmetric (linear) so has only dispersion interactions.

10.65 All substances are subject to dispersion forces. In addition, look for polarity and hydrogen bonding ability:
(a) CH_3OH can hydrogen bond, so it boils at a higher temperature than CH_3OCH_3;
(b) SiO_2 is a network solid with covalent bonds that must break for it to boil, so it has a higher boiling point than SO_2;
(c) HF can hydrogen bond, so it boils at a higher temperature than HCl; and
(d) I_2 has larger orbitals, giving it higher polarizability, larger dispersion forces, and a higher boiling point than Br_2.

10.67 When two substances with otherwise similar structures have different boiling points, look for differences in polarity and/or hydrogen bonding behavior. The Lewis structures of the two isomers, shown with bond polarities drawn in, indicate why the isomers have different boiling points:

In the *trans* isomer, the two polar bonds oppose each other and cancel, making this a non-polar compound. In the *cis* isomer, the two polar bonds add, making this a polar compound. Thus *cis* 1,2-dichloroethylene has the higher boiling point.

10.69 The notion of "like dissolves like" can be extended to the wetting behavior of one substance for another. Solder is a metal that will adhere well to another metal only if unlike substances have been removed from the surface. Non-polar substances such as oils and greases as well as network solids such as metal oxides must be removed by flux.

10.71 Calculate osmotic pressure using Equation 10-4 in your textbook: $\Pi = MRT$.
Here, M is the total molarity of the solution. Consider 100 g of brackish water; it contains 0.5 g NaCl and has a volume of 1.00×10^2 mL = 0.100 L.

$$n_{NaCl} = \frac{m}{MM} = \frac{0.5 \text{ g}}{58.44 \text{ g/mol}} = 8.6 \times 10^{-3} \text{ mol, and } n_{total} = 2\, n_{NaCl} = 1.7 \times 10^{-2} \text{ mol}$$

$$\Pi = \frac{(1.7 \times 10^{-2} \text{mol})(8.206 \times 10^{-2} \text{L atm/mol K})(298 \text{ K})}{(0.100 \text{ L})} = 4 \text{ atm}$$

This result has only one significant figure, because the concentration (0.5 %) is stated to only one significant figure.

10.73 Elements that form metallic solids are found on the left side of the periodic table, those that form molecular solids are found on the far right side, and network solids are restricted to carbon and the semi-metals in the diagonal region of the *p* block:

Metals: Na (Group 1), Fe (Group 8), Al (Group 13), etc.
Molecular: Xe (Group 18), I_2 (Group 17), P_4 (Group 15), etc.
Network: C (Group 14), Se (Group 16), As (Group 15), etc.

10.75 If water were linear rather than bent, the two polar O–H bonds would oppose each other and the molecule would be non-polar like CO_2. Non-polar water molecules could not associate with ions through ion-dipole interactions, so linear water would not be a good solvent of salts.

10.77 The magnitudes of dispersion forces depend on how extended the valence electrons are. In a molecule, valence electrons are spread over a larger volume because they are shared between atoms. This feature is illustrated by sketches of the electron clouds of H₂ and He:

10.79 Pentane boils at a higher temperature than dimethylpropane, indicating that dispersion forces are greater for the linear chain than the branched chain. Both compounds have the same number of atoms and electrons, so the effect must be due to extension of the electron clouds. Line structures show that the branched compound is more compact than the linear compound, accounting for the difference:

10.81 Sea water has a higher osmotic pressure than fresh water, so the effect of sea water ingested into the body is to leach water from the insides of cells into the fluids that are contaminated with sea water. Water molecules cross the cell membranes, but proteins and ions do not. This causes the cells to collapse.

10.83 To determine a chemical formula from a unit cell, count the number of atoms of each element contained within the unit cell. Atoms on a face count 1/2 (each is part of two unit cells), those on an edge count 1/4 (each is part of four unit cells), and those at a corner count 1/8 (each is part of eight unit cells).
There is one Ti atom within the cell.
Ca: 1/8 (8 edge atoms) = 1
O: 1/2 (6 face atoms) = 3.
The formula is $CaTiO_3$.

10.85 The solubility properties of water and gasoline can be predicted from the concept of "like dissolves like." Water is highly polar and hydrogen-bonding, while gasoline is a mixture of non-polar hydrocarbons. Thus water will not dissolve gasoline. Gasoline, being less dense, floats on water, so if water is used on a gasoline fire the burning gasoline will be transported to other places by the water. Rather than extinguishing a gasoline fire, water actually causes it to spread!

10.87 Deviations from ideal gas behavior can be predicted based on the strengths of intermolecular interactions. The larger the intermolecular interactions, the greater will be the deviations from ideality. Both pairs of substances are non-polar, so dispersion interactions dominate their intermolecular forces. The substance with the higher n value will have larger dispersion forces and deviate more from ideality:

(a) Cl_2
(b) SnH_4

10.89 Boiling points are determined by the strengths of intermolecular interactions, which in turn are determined by dispersion forces, polarity, and hydrogen bonding interactions. CH_3F and CH_3OH have similar dispersion forces and polarities, but whereas CH_3OH forms hydrogen bonds, CH_3F does not, because CH_3F has no polar H–X bond. Thus CH_3OH has a higher boiling point.

10.91 Among otherwise similar substances, the magnitudes of dispersion forces determine phase behavior. The more extended the molecule, the greater its polarizability and dispersion forces. Thus, benzene, the smallest of the set, is a liquid; and anthracene, the largest, is the less volatile solid. Naphthalene is the relatively volatile solid.

Chapter 11 Macromolecules

11.1 Polymerization of Alkenes

QUESTIONS TO ANSWER, SKILLS TO LEARN
1. What is a monomer?
2. Drawing pictures of polymers (or pieces of them)
3. What is crosslinking?

Some polymers are formed from compounds with carbon-carbon multiple bonds. The formation of these polymers is shown below as a line drawing.

monomer   polymer

The fragment in brackets repeats.

An important feature of this type of polymer formation reaction (or *polymerization*) is that many small molecules are linked to form a large molecule; *there is no product other than the polymer molecules*. The bonds that link the monomers in the final polymer are σ (sigma) bonds that are formed from electrons that were formerly used for π bonding in the monomers. In the reactants, the hybridization at the C atoms is sp^2; in the products, the C atoms are sp^3 hybridized. The rearrangement of electrons to form the polymer is not spontaneous; some chemical species must make these π-bonding electrons reactive to forming σ bonds. The species responsible is called the *initiator*. One type of initiator is a substance with unpaired electrons called *free radicals*. In the following reaction, the free radical initiator is designated as •I and the radicals that result from radical reactions are labeled with a heavy dot, •:

Initiation

The reaction of the free radical with the alkene CH_2CHR gives a new free radical that can react with other alkenes.

Propagation

The result of this reaction is another free radical, which may add another alkene to make the molecule longer. This cycle of addition of alkene molecules can proceed as long as the free radical's reactivity is present. This cycle of addition of alkenes is stopped by reaction with another free radical called a *terminator*, T•.

Termination

Many chemical species may act as T•; examples include other growing polymer chains, initiator fragments and H atoms abstracted from other molecules. The critical property of a terminator is that it kills the free radical reactivity of the chain that leads to propagation.

Exercise 11.1 Draw the structure of a polymer obtained from the free radical polymerization of styrene (line structure is shown below). The resulting polymer is called polystyrene, one of the more common polymers.

styrene =

Steps to Solution: Free radical polymerization of styrene will take place at the alkene double bond, not in the benzene ring structure. The polymer is made of repeating styrene units with an initiator and a terminator group. Because the I and T groups are such a small part of the polymer (there are often hundreds of monomer groups in a polymer chain), we ignore them in drawing the structure of the polymer.

In forming the polymer, the double bond of the alkene is converted to a single bond to form the σ bonds that link the monomers together. We can start by drawing several monomer units near each other, as is done below for three monomers. The relative orientation of the monomers is not important.

The next step is to remove the double bonds and join the monomers with single bonds. The dashed lines in the above diagram show where the new bonds will be formed. Remember that bonds will extend from this part of the polymer to the rest of the polymer. These bonds are shown as lines that end in "mid-air". The use of brackets emphasizes that this structure is just part of a longer chain.

Copolymers

The extremely high reactivity of radicals leads to a different type of polymer when two different alkenes (monomers) are present in the reaction mixture. If two alkenes such as styrene and propene ($CH_2=CH[CH_3]$) are present in a polymerization reaction, a polymer that contains the two different types of monomers will be formed. The order in which the monomers are added to the chain will often be random; in free radical polymerization, the free radicals involved in chain propagation are so reactive that they usually react with the first alkene they collide with. Copolymers often have properties that are in between the properties of polymers made of either monomer. Many materials, such as tire rubber and plastic (PET) soda bottles, are copolymers.

Tying Polymer Chains Together: Crosslinking

At the molecular level, the polymers we have discussed to this point are long molecules somewhat like chains where each link in the chain can be likened to a monomer unit. We can remove a single chain from a pile of chains without much difficulty. However, if we link several chains to one another, it is much more difficult to remove a chain. The more links between one chain and another the greater the rigidity; at the extreme, the pile of chains resembles one three-dimensional molecule, not unlike a covalent solid. Vulcanization of rubber is a process that introduces crosslinks into polymer molecules, making the final product more rigid and durable.

Exercise 11.2 The compound 1,5-hexadiene (below) has two double bonds that can participate in adding to polymers.

1,5-Hexadiene:

(a) Draw a segment of a copolymer of 1,5-hexadiene and ethylene in which only one double bond of 1,5-hexadiene is involved in the polymerization process. The ratio of ethylene to 1,5-hexadiene is 4:1.
(b) Draw the structure of a polymer section where 1,5-hexadiene acts as a crosslinking agent.

Steps to Solution To draw the structure of a polymer molecule, it is useful to line up the *pieces (monomers) and then draw in the new bonds.*

(a) *In a copolymer, two different monomers are incorporated into one polymer chain. We know that the ratio of ethylene to 1,5-hexadiene is 4 to 1; therefore the section of polymer chain to be drawn should contain four ethylene units and one 1,5-hexadiene unit. One such arrangement is shown below (others are possible and entirely correct). The bonds that will be present in the polymer are indicated by dashed lines in the figure to left. To complete the sketch, we convert the double bonds to single bonds and draws the new bonds that link the monomers as shown to right.*

line up monomers   add bonds for structure of polymer segment

(b) *1,5-Hexadiene may act as crosslinking molecules in polymers because there are two double bonds that can be added to growing polymer chains. In part (a) only one of the double bonds was added to a polymer. The other end may also be added into a polymer. Filling in the bonds as done in part (a) shows the crosslinked section of the polymer, shown below at right.*

line up monomers   add bonds for structure of polymer segment

---

11.2 Functional Groups

QUESTIONS TO ANSWER, SKILLS TO LEARN
1. **What is a functional group?**
2. **What are linkage groups?**
3. **Learning the reactions between functional groups which link smaller molecules together**

A **functional group** is an atom or a group of atoms that has a characteristic reactivity, especially when bonded to a hydrocarbon fragment. The reactivity may be caused by the polarity of the bond between the functional group and a C atom, or by the presence of a multiple bond or by a combination of these factors. For example, the C=C double bond in alkenes leads to their reactivity, including that of polymerization. Double bonds undergo similar reactions even though they might have different alkyl or other groups bonded to the alkene. This characteristic reactivity is caused by a structural feature that allows us to predict the reactivity.

Some reactions between functional groups lead to the joining of smaller molecules to give larger molecules. Some common functional groups are listed in the following table. The unlabeled lines in the structure are bonds to C atoms.

Structure of group	Name	Compound Names	Reacts with:
—O—H	hydroxyl	alcohols: *react by loss of proton or by loss of OH group*	acids, amines, aldehydes, alcohols
—S—H	sulfhydryl	thiols	thiols
—N(H)—H	amino	amines: *react by accepting proton (Lewis base) or by loss of proton to OH-*	acids, alcohols
—C(=O)H	aldehyde	aldehydes: *contain the polar $C^{\delta+}=O^{\delta-}$ group*	alcohols, amines
—C(=O)—O—H	carboxylic acid	carboxylic acids or acids: *react by proton loss or OH- loss*	alcohols, amines

The presence of a functional group will lead to reactivity characteristic of that group. An important class of reactions in polymerization chemistry is the *condensation reaction*.

In a condensation reaction, two molecules are joined together by covalent bond(s) formed by reaction of functional groups accompanied by the formation of a smaller molecule.

A condensation reaction is illustrated by the reaction of acetic acid ($CH_3[C=O]OH$) and ethanol (CH_3CH_2OH) to give the ester, ethyl acetate, and the byproduct, water. The box encloses

[Reaction: acetic acid + ethanol → ethyl acetate + H₂O, with O-H from acid and H-O from alcohol boxed]

the parts of the functional groups eliminated to form water. Condensation reactions where the smaller product molecules are not water are common, but these reactions involve functional groups that not included in the short list provided above.

Linkage groups formed by reactions between functional groups	Name	Compound name/ General type of polymer	Formed from the reaction of:
[ester linkage structure]	ester	esters/*polyesters*	acids and alcohols
[amide linkage structure]	amide	amides/*polyamides and proteins (polypeptides)*	acids and amines
[ether linkage structure]	ether	ethers/ *polyethers and cellulose, starch*	alcohols

Exercise 11.3 Draw the products of the condensation reactions of (a) dimethylamine (HN(CH₃)₂) and formic acid (H-C=O)-OH) (b) 1-butanol condensed with itself.

Steps to Solution: It is helpful to draw the structures of the molecules involved in the reaction and then identify which parts of the functional groups will be eliminated. We then draw the bond linking the remaining fragments to give the product of the reaction. In determining the product, it is helpful to draw the molecules so that the reactive functional groups are "pointed" at each other.

(a) *Formic acid and dimethylamine have the structures shown below. Note that these structures are suggested by the way the molecular formulas are written.*

[Structures of formic acid and dimethylamine]

H- (C= O) - OH   H - N -(CH₃)₂

The parts of the functional groups that will produce water are outlined. The new bond (shown by a dashed line in the reactants) will form between the C atom where the OH was bonded and the N atom where the H atom was bonded.

The product, an amide, is shown in the box.

(b) The reaction of the condensation of 1-butanol with itself is the reaction of two alcohols. The structure of two molecules of 1-butanol are shown below, oriented so that the two functional groups are near each other. Again, the bond to be formed is shown by a dashed line.

The product,, di-butyl ether, is shown in the box.

Exercise 11.4 Draw the structures of two molecules that would form in the molecule shown below in a condensation reaction.

Steps to Solution The way to solve this problem is to look for one of the three linking groups shown above: ester, amide or ether. After determining the linking group, we may reconstruct the reactants by adding the appropriate fragments of water (H or OH) to the atoms involved in the linking bond.

On examination of this molecule, it is clear that it has three functional groups: an amine, a carboxylic acid, and an amide. Of these three groups, only the amide is a linking group, so the C-N bond is the linking bond (shown below as the thicker bond). Because amides are formed from the reaction of an acid and an amine, the OH of water must be attached to the C atom and the H to the N atom to give an acid and an amine.

Chapter 11: Macromolecules

11.3 Condensation Polymers

QUESTIONS TO ANSWER, SKILLS TO LEARN
1. **What structural features are needed to form polymers by condensation reactions?**
2. **What are the general classes of condensation polymers?**

Exercise 11.4 illustrated the deduction of the starting materials for a product obtained through a condensation reaction. The amide in this exercise has similarities to the amide in Exercise 11.3 (a) but also has several significant differences.

A (from 11.4)   vs.   **B** (from 11.3(a))

The main difference is the number of functional groups *excluding* the amide link. In **A**, the molecule has two functional groups that can react to form further links, the amine and the acid; however **B** has no additional linking groups. This difference means that **A** can react with another molecule of acid (at the amine end) or another amine molecule (at the acid end), whereas **B** will be unreactive toward carboxylic acids or amines under conditions in which **A** would readily react with them. In other words, **A** can continue forming links via condensation reactions whereas **B** can no longer undergo condensation reactions. **A** is still reactive toward condensation reactions because it has reactive functional groups left over from the monomers used to form it. Each monomer contained two functional groups, only one of which was used to form the link, leaving a reactive group. This structural feature of the monomer is necessary to form polymers by condensation reactions.

> *If formation of a polymer by condensation reactions is to occur,*
> *the monomers must contain two (or more) functional groups.*

As a final note, the functional groups in the monomers used to form **A** were different, but in many cases the functional groups in the monomers are the same.

General Types of Condensation Polymers

Polyamides

The linking group in polyamides is the amide group. Polyamides are of two types, as shown below. In one type there is one amide link and a single "spacer" group (a fragment of a molecule that is often a hydrocarbon; in **A**, the spacer groups were -CH(R)- groups) separating the amide links per repeat unit. Remember that R is an abbreviation for any hydrocarbon group or an H atom.

In Nylon 6, the spacer unit is the linear five-carbon alkane fragment, $-(CH_2)_5-$.

In the other type of polyamide, there are two types of spacer groups and the repeat unit is derived from two monomers.

In one important nylon, the spacer between the N atoms is a six-carbon fragment, $-(CH_2)_6-$ and the spacer between the C atoms is a ten-carbon fragment, $-(CH_2)_{10}-$. The difference between the two types of polyamide is found in the types of monomers used to form them.

The first type needs a monomer that contains both the acid and amine functional groups. The diagram that follows shows the general type of monomer and the groups that will be eliminated in condensation reactions. The two functional groups in this monomer molecule are different. This is the type of monomer used in the formation of compound **A**.

The second type of polyamide requires two different monomers, each of which contains two functional groups that are the same; that is, one monomer bears two acid groups and the other bears two amine groups. These are shown below, where the different size boxes indicates the spacer groups, which need not be the same in both monomers.

Polyesters

The linking group in polyesters is the ester group. It is possible to make two types of polyesters as found in polyamides, but in practice only the type that is built from two different monomers is of significance because of the difficulty of synthesizing a monomer that contains both an alcohol and an acid group. The general structure is shown below left.

general polyester

commercial polyester, Dacron

The spacer groups in the highest production polyester are the $-(CH_2CH_2)-$ group between the oxygens and a benzene ring bonded at opposite positions between the C atoms, above at right. The monomers used to manufacture polyesters are dialcohols (below at left) and diacids (below at right).

Polymers from Formaldehyde ($H_2C=O$) and Urea ($O=C(NH_2)_2$)

Formaldehyde and urea are among the top 50 industrial chemicals because of their use in polymer manufacture. The reactions leading to the formation of polymers from these substances are condensation reactions. The initial reaction leading to polymers based on formaldehyde is the reaction of formaldehyde with some X-H bond:

$$X\text{-}H + H_2C=O \rightarrow X\text{-}CH_2\text{-}OH$$

The hydroxyl group resulting from this process can then participate in condensation reactions by reacting with hydrogen atoms in other molecules, shown here as Y-H:

$$Y\text{-}H + X\text{-}CH_2\text{-}OH \rightarrow X\text{-}CH_2\text{-}Y + H_2O$$

Chapter 11: Macromolecules

11.4 Polymer Properties

QUESTIONS TO ANSWER, SKILLS TO LEARN
1. **How does the structure of a polymer affect its physical properties?**
2. **Using intermolecular forces to predict polymer properties**

Predicting the physical properties of the material that results in a polymerization reaction is a full-time job for the chemists and engineers involved in the design of new materials for construction, medicine and industrial products. There are three general classes of polymers:

- *Plastics*: polymers that form rigid solids and may be molded into shapes.

- *Fibers*: polymers in the form of long, thin strands that are flexible and can be woven into cloth or spun into threads or ropes.

- *Elastomers*: polymers that can be stretched and return to the original shape and size.

The general properties of these different types of polymers can be understood by considering their structures and the nature of the forces holding the molecules of the polymer together.

Plastics

Two classes of plastics are known, *thermoplastics* and *thermosets*. Thermoplastics have few crosslinks between the polymer molecules and when heated, the chains can slide past one another. This results in the material becoming more fluid as it is heated (it can be melted), and the soft material can be molded into desired shapes.

Thermoset plastics have a high degree of crosslinking and have a network of covalent bonds between polymer chains. To move one chain away from others requires breaking covalent bonds, which requires high levels of energy. Therefore thermoset polymers do not soften or melt upon heating, but decompose when heated to high enough temperatures.

Fibers

Fibers are made from thermoplastic polymers, usually polyamides, polyesters or polyacrylonitrile, which are converted to a fluid state (by melting or dissolving in a solvent that is easily removed). The fluid is forced through small holes to form thin strands that solidify either by cooling or evaporation of the solvent. Changes occur at the observable level (the polymer is now in the form of fine threads) and at the molecular level (forcing the molecules through the small opening leads to more parallel alignment of the polymer chains). The latter change can be envisioned as pouring spaghetti noodles through a funnel: the noodles emerge from the spout mostly parallel to one another. The alignment of chains along the length of the fiber leads to higher strength of the fiber compared to the plastic because of more effective intermolecular interactions.

Elastomers

Elastomers are solids made from polymers that change shape and return to the original size, like a rubber band or ball. To accomplish this, the polymer chains must be able to move past one another without much energy input and also have a structure that can stretch without breaking covalent bonds. One structure that will permit this sort of reversible deformation is a helix or coil that can be elongated and then rebound like a stretched coil spring. A molecular coil can be formed by a polymer chain that has carbon atoms with both trigonal planar (from sp^2 hybridization) and tetrahedral (from sp^3 hybridization) geometries. In addition, the structure should not readily form crystals because the greater intermolecular forces acting in a crystal will make it more difficult for the molecules to move past one another. In all elastomers, there are both carbon-carbon single and double bonds present, leading to the twisted, coil-like structures that are sufficiently irregular so that these solids are amorphous.

The flexibility of an elastomer depends upon how far the polymer chains can move when a given force is applied. The motion of polymer chains is inhibited by increasing the number of crosslinks between them. The rubber used in an automobile's bumper is hard and inflexible; it has a high degree of crosslinking (only a few monomer units between crosslinks). In contrast, a rubber band with few crosslinks is quite stretchy.

Exercise 11.5 Nylon rope, a polyamide fiber product, has a tendency to stretch when it becomes wet (during a rainstorm, for example). Explain this phenomenon.

Steps to Solution: To explain changes in the properties of a polymer, we must examine its structural features. Nylon is a polyamide with many hydrogen bonding groups present (C=O bonds, N-H bonds). The molecules are held together by hydrogen bonds, but in the presence of water, the nylon molecules may H-bond to something other than themselves. The presence of water disrupts the H-bonds between the nylon molecules and the rope changes shape (lengthens).

Exercise 11.6 Butadiene, $CH_2=CH-CH=CH_2$ (line drawing below), forms polymers that are used in elastomers. Write the structure of polybutadiene and explain why this polymer would be more elastic than polyethylene.

Steps to Solution: We need to examine the structures (think molecules) to determine the properties. The structure of polyethylene is in your text and is similar to that shown on the first page of this chapter (let R = H). Butadiene polymerizes to form a *molecule with double bonds. A section of a poly-butadiene molecule is shown below. A section that contains one butadiene monomer is boxed. Note the different orientations about the double bonds.*

The polybutadiene molecule can form twisted structures that deform more easily than the linear chains formed in polyethylene, which has no double bonds present. The more readily deformable polymer would be more elastic, much as it is easier to strech a coil spring compared to a straight wire.

11.5 Carbohydrates

QUESTIONS TO ANSWER, SKILLS TO LEARN
1. **What are carbohydrates and saccharides?**
2. **Drawing monosaccharide structures**
3. **Drawing oligosaccharides and the glycoside linkages**
4. **What are the important polysaccharides?**

A large class of biological macromolecules is derived from linking together sugar molecules; the entire class of molecules from monomers to polymers are called **carbohydrates**. Carbohydrates' empirical chemical formulas are the same: they correspond to a 1:1 combination of carbon and water, $C_x(H_2O)_x$. Carbohydrates are often called *saccharides* after the Latin for sugar. There are several types of carbohydrates:

- *Monosaccharides*: carbohydrates with formula $C_n(H_2O)_n$; n = 3 to 6. These are the building blocks of other carbohydrates.

- *Oligosaccharides*: carbohydrates formed by condensation reactions containing two to ten monosaccharides.

- *Polysaccharides*: carbohydrates formed by condensation reactions containing many monosaccharides.

Monosaccharides

Monosaccharides are classified by the number of C atoms they contain; the number of C atoms is denoted by a prefix like those used in chemical nomenclature (3 = tri, 4 = tetr, 5 = pent-) followed by -ose. Within a given number of carbon atoms there is a large number of isomers only some of which are naturally occurring. Below are shown two isomers of the hexose, glucose which, like other saccharides, forms ring structures that incorporate an O atom. Features common to other cyclic saccharides' structures include:

- An O atom in the ring structure
- A CH$_2$OH group as a side group
- H and OH groups bonded to each C atom

α-glucose

A six membered ring structure is formed by some monosaccharides

β-glucose

Side view of the ring structure of the six membered ring.

Note the "chair" configuration formed by the ring.

In six-membered ring structures, the more stable form adopted by the ring is described as a chair; the ends form the back and footrest of the chair as seen in the partial picture of β-glucose above at right. Several bonds to H and OH groups are shown in heavy lines to emphasize the difference in their orientation compared to the other bonds. Those in heavy lines are perpendicular to the "seat" of the chair; the others are nearly parallel to the plane of the seat.

Exercise 11.7 Drawing monosaccharides and interpreting the drawings to see structural differences.
(a) Which of the following three pictures is the pentose, arabinose?

i.   ii.   iii.

(b) Draw a picture of β-glucose where the O atom in the ring is located as shown below.

Chapter 11: Macromolecules

Steps to Solution: To identify or draw sugars, it is important to start at a reference point, usually the ring oxygen and then move around the ring (usually toward the C-substituted ring C atom), noting the orientation of the OH groups with respect to that group. Usually the orientation of the CH_2OH group is considered "up."

(a) *In arabinose (see your text), there are three OH groups, in order down, up, up. In molecule i, the order is the same:*

In ii, the order is down, down, up; therefore this saccharide is not arabinose. In iii, the order is up, down, up; therefore this saccharide is not arabinose. Only molecule i is arabinose.

(b) *In this case the molecule is to be drawn from the given orientation of the ring. In the drawing above, the CH_2OH group is to the left of the ring O.*

After placing the reference group, and adding the OH groups, we obtain:

Polysaccharides

Alcohols undergo condensation reactions to give ethers; saccharides have many hydroxy groups that may react with other saccharides to join the units via an ether linkage:

When saccharides are joined in this way, the ether link is called a **glycosidic bond**. The number of saccharide units linked in an oligosaccharide is noted by the standard prefix: two linked saccharides form a **disaccharide**, three, a **trisaccharide**, and so on. Natural disaccharides are always formed with one end of the glycosidic bond at the position next to the O atom contained in the ring. This is illustrated above where the reacting OH in the six-membered ring is on the C atom adjacent to the O atom.

Linking a large number of monosaccharides leads to the formation of polymers with saccharide repeating units called polysaccharides. Because there are a variety of hydroxy groups available for the glycosidic bonds, a large number of types of polysaccharides might be formed from a given monosaccharide. The major polysaccharides are formed from two monomer units: α-glucose and β-glucose.

Monomer	Polysaccharide	Characteristic Structural Features
α-glucose	amylose	A molecule with end-to-end glycosidic bonds, about 200 glucose units per molecule. The α-structure leads to kinking of the polymer strand.
α-glucose	amylopectin	End-to-end glycosidic bonds but with branching caused by multiple glycosidic links at some glucose units
α-glucose	glycogen	Like amylopectin, but with a higher degree of branching
β-glucose	cellulose	A linear polymer where the β-linkage leads to a linear molecule containing up to 3000 monomer units.

11.6 Nucleic Acids

QUESTIONS TO ANSWER, SKILLS TO LEARN
1. **What are the building blocks of nucleic acids?**
2. **What is the difference between RNA and DNA?**
3. **Learning the complentary base pairs of nucleotides**

Nucleic acids are condensation polymers that are constructed of several monomers called **nucleotides**, which have the general structure shown below at left:

Chapter 11: Macromolecules

General nucleotide structure (pentose (ribose or deoxyribose), phosphate group, nitrogen containing organic base)

Pentoses found in nucleotides: ribose, deoxyribose

Nucleotides are formed from the condensation reaction of phosphoric acid with a **nucleoside**, the pentose-base pair. The nitrogen containing bases have the ring structures and abbrevations shown below.

Guanine, G	Adenine, A	Thymine, T	Uracil, U	Cytosine, C
(structure)	(structure)	(structure)	(structure)	(structure)
Found in :	Found in :	Found in :	Found in :	Found in :
DNA	DNA	DNA		DNA
RNA	RNA		RNA	RNA

The polymers formed from nucleotides are **DNA** (deoxyribonuleic acid, molar mass about 10^{12} g/mole) and **RNA** (ribonucleic acid, molar mass 20,000-40,000 g/mole). DNA contains the genetic information necessary for reproduction, and RNA transmits the information from the DNA to the site of biochemical synthesis. The main change in the monomer units is derived from the nitrogen bases. The H atom in outline type (H) is the H atom that reacts in the condensation reaction with the sugar portion of the nucleic acid.

nucleoside + phosphoric acid → nucleotide and H_2O

Exercise 11.8 Draw the structures of (a) the nucleoside and (b) the nucleotide formed from the base cytosine.

Steps to Solution

(a) Nucleosides are the condensation reaction product of the sugar unit and the

409

nitrogen-containing base. The condensation reaction occurs between the H atom indicated in the list of nitrogen bases and the OH group bonded to the C atom adjacent to the O atom in the sugar ring to give the C-N bond. It is helpful to draw the two molecules involved in the reaction oriented so that the eliminated portions are adjacent to each other.

[base + ribose → nucleoside + H_2O]

The product of the reaction is the nucleoside, cytosine.

(b) The nucleotide is formed from the condensation reaction of the nucleoside with phosphoric acid. The phosphoric acid reacts with the OH group on the C atom that is not part of the ring structure. Again, it is helpful to to draw the molecules oriented so that the fragments to be eliminated (and the atoms between which the new bond will be formed) are near each other.

The product of this reaction is the nucleotide, cytosine monophosphate.

Polymerization of Nucleotides; Primary Structure of Nucleic Acids

Nucleic acids are polymers of the nucleotides formed from the sugar, the base and the phosphate group. Condensation occurs between the phosphate of one nucleotide and the OH group of the sugar of another nucleotide. The backbone of the polymer is constructed of sugar-phosphate units; the variety in the polymer is derived from the different bases that are bound to the sugars. The **primary structure** of the polymer is the *order of the bases from the end of the polymer which bears the phosphate to the end that bears the sugar*. Rather than using the names of the bases, the one letter abbreviations are used.

or

```
      OH    base₁   OH    base₂         OH   base₃       OH   base₄
      |      |      |      |            |     |          |     |
HO—P·O—sugar—P—O—sugar——O—P—O·sugar————P—O-sugar—OH
      ‖             ‖                   ‖                ‖
      O             O                   O                O
```

Exercise 11.9 Draw the four nucleotide nucleic acid with primary structure TTGC. Use the shorthand notation shown above.

Steps to Solution: The primary structure is a list of the bases that are bonded to the sugar fragments, starting from the phosphate end to the sugar end of the molecule. The molecule will contain four phosphate-sugar-base units:

```
      OH    base₁   OH    base₂         OH   base₃       OH   base₄
      |      |      |      |            |     |          |     |
HO—P—O—sugar——P—O—sugar——O—P—O—sugar————P—O—sugar-OH
      ‖             ‖                   ‖                ‖
      O             O                   O                O
```

The names of the bases are then substituted for base₁, base₂, and so on:

```
      OH   thymine  OH   thymine        OH  guanine      OH   cytosine
      |      |      |      |            |     |          |     |
HO—P—O—sugar——P—O—sugar——O—P—O—sugar————P—O—sugar-OH
      ‖             ‖                   ‖                ‖
      O             O                   O                O
```

Secondary Structure of DNA; the Double Helix and Complementary Bases

Secondary structure is the structure that the long polymer chain takes. The most common secondary structure of DNA formed is a double helix which two chains are bound to one another by hydrogen bonding from a base on one chain to a base on the other chain. The strongest hydrogen bonding interactions are between select pairs of the bases: A is matched with T and G is matched with C. These base pairs are called *complementary base pairs*. Therefore, given the primary structure of one segment of DNA, we can deduce the primary structure of the matching segment in the double helix using this complementary relationship.

Exercise 11.10 If the primary structure of a segment of DNA is TTGCATTGC, what is its complementary structure?

Steps to Solution: The complementary structure is found by remembering that T and A are complements and G and C are complements. Therefore, to obtain the complementary structure, we substutute A for T (or T for A) and G for C (or C for G).

```
                T    T    G    C    A    T    A    C    G
Substituting:   ↓    ↓    ↓    ↓    ↓    ↓    ↓    ↓    ↓
                A    A    C    G    T    A    T    G    C
```

The complementary sequence is: AACGTATGC .

The Structure of RNA

RNA and DNA are similar, but have several important differences.
1. RNA is a smaller molecule, with molar masses between 20,000 and 40,000 g/mole.
2. The sugar unit is ribose, not deoxyribose.
3. In RNA, the complement of adenine is uracil, U. Thymine is not found in RNA.
4. RNA's secondary structure is single stranded.

Despite these differences, remember that both RNA and DNA are formed from the condensation reactions of nucleotides, and each polymer strand has a backbone structure of sugar and phosphate units to which are bound the various nitrogen bases.

11.7 Proteins

QUESTIONS TO ANSWER, SKILLS TO LEARN
1. **What are amino acids?**
2. **Putting together amino acids to form polyamides or polypeptides**
3. **What is meant by primary, secondary and tertiary structure in proteins?**

Proteins are polyamides and are among the most important biological macromolecules. They are structural tissue for the body and serve to accelerate the rate of nearly all necessary biological chemical processes. The general structure of a protein is shown below (left) and at right is the structure of one of the monomer units. Unlike the synthetic polyamides we

polypeptide   amino acid (monomer)

discussed, the "monomer" units of a protein are not identical; they contain an amino functional group and an acid functional group and thus are called **amino acids**. Proteins are formed from the condensation reaction of amino acids. There are twenty R groups found in nature and they are listed below; they lead to a wide variety of structures of proteins. These properties stem largely from the hydrophilic and hydrophobic nature of the R groups.

Hydrophobic R Groups

-H	-CH$_2$CH(CH$_3$)$_2$	-CH(CH$_3$)$_2$	-CH$_3$
Glycine, Gly	Leucine, Leu	Valine, Val	Alanine, Ala
-CH$_2$CH$_2$SCH$_3$	-CH$_2$SH	CH(CH$_3$)(-CH$_2$CH$_3$)	-CH$_2$-C$_6$H$_5$ (benzyl)
Methionine, Met	Cysteine, Cys	Isoleucine, Ile	Phenylalanine, Phe

Hydrophilic R Groups

(indole-CH$_2$-)	-CH$_2$OH	-CH$_2$-CO$_2$H	-CH$_2$CH$_2$CO$_2$H
Tryptophan, Trp	Serine, Ser	Aspartic acid, Asp	Glutamic acid, Glu
-(CH$_2$)$_3$NH(C=NH)NH$_2$	(pyrrolidine-CO$_2$H)	-CH$_2$-(C=O)-NH$_2$	-(CH$_2$)$_4$NH$_2$
Argenine, Arg	Proline, Pro	Asparagine, Asn	Lysine, Lys
-CH(OH)(CH$_3$)	-(CH$_2$)$_2$-(C=O)-NH$_2$	-CH$_2$-C$_6$H$_4$-OH	-CH$_2$-(imidazole)
Threonine, Thr	Glutamine, Gln	Tyrosine, Tyr	Histidine, His

    The R groups have a significant influence on the properties and shape of the resulting macromolecule through their hydrophobic or hydrophilic character. Nine of the naturally occuring amino acids have nonpolar groups (hydrophobic) and the remaining 11 have polar R groups, many of which are capable of strong hydrogen bonding. As an exercise, identify the amino acid side groups as hydrophobic or hydrophilic without referring to the text.

    Amino acids link together through the formation of amide bonds, called peptide bonds.

Primary Structure of Polypeptides

Amino acids linked together form *peptides*; two amino acids form a *dipeptide*, three a *tripeptide*, and so on. Many amino acids linked together form a **polypeptide**. All proteins are polypeptides that contain between 50 and 1800 amino acid units. The order of the amino acids in a polypeptide is very important and is called its **primary structure**. Each amino acid has two functional groups, so reaction with another amino acid may occur at either end.

The condensation of alanine with glycine may occur at either end of alanine to give either of two distinct products, as shown below. The boxes outline the groups involved in the condensation reaction.

Ala-Gly

Gly-Ala

The accepted way of listing the amino acids in a polypeptide is to start with the amino acid that bears the unreacted amine group and list the amino acids in order, ending with the amino acid that contains the unreacted acid group (amine-left, acid-right).

Exercise 11.11 Write out the primary structures for the different tripeptides we could obtain from the three amino acids Glu, Tyr and His.

Steps to Solution: *The primary structures of polypeptides are written by listing the amino acids. Remember that the first amino acid is the one with the unreacted amine group. It is easier to write structures where the same amino acid is first and then vary the others.*

Glu-Tyr-His   Tyr-Glu-His   His-Tyr-Glu

Glu-His-Tyr   Tyr-His-Glu   His-Glu-Tyr

Exercise 11.12 Draw the line structure of the tripeptide Glu-Tyr-Ala

Steps to Solution: To draw the line structure of the tripeptide from the primary structure, we first draw the skeleton of the tripeptide chain:

To complete the line structure, we add the appropriate R groups on the tripeptide chain: for Glu, R = $CH_2CH_2CO_2H$; for Tyr, R = $CH_2(C_6H_4OH)$ and for Ala, R = CH_3. Therefore the line structure is:

Secondary Structure in Polypeptides

The primary structure of a polypeptide is important, but it only describes part of the overall structure of a protein. The primary structure (sequence of amino acids) is analogous to naming the colors of beads on a string; however, the string of beads may be looped into a variety of shapes. Similarly, a polypeptide contains a number of different amino acids that have groups capable of hydrogen bonding (the N-H and the C=O groups) which lead to different arrangements of the chain of amino acids called **secondary structure**. Two common types of secondary structure are the α-**helix** and the **pleated sheet**.

Tertiary Structure in Polypeptides

In proteins, the structure is not limited to α-helixes or pleated sheets, but is more complex. There are regions of helical structure, pleated sheet structure and other less common secondary structures formed by the peptide chain of the protein. The final structure of the protein is determined by the nature of the R groups on the peptide. The protein will be most stable when the structure maximizes H-bonding, and polar interactions with the water it is immersed in and minimizes the interactions of the nonpolar R groups with water. To achieve this situation, the protein will be folded and twisted so that the polar and H-bonding groups will be facing the aqueous surroundings (or polar groups in other parts of the protein) and nonpolar groups will directed to the inner part of the protein where little water is found.

The tertiary structure of proteins is related to the function of the protein; the shape of proteins leads to two general classes of proteins, *globular proteins* and *fibrous proteins*. Fibrous proteins lend themselves to the construction of fibers; they are long helical molecules. Globular proteins have more complex structures. They serve a variety of biochemical purposes, including transport and storage of molecules required by cells, mediating biochemical reactions required for metabolism and detroying infections. The shape of the protein determined by the tertiary structure is very important in determining its role; modification of the protein by changing a few amino acids in its primary structure may render it inactive.

Test Yourself

1. Draw a the structure of the polymer made from cyclopentene, whose structure is shown below.

2. What type of polymer (elastomer, fiber, plastic) would be best used in each of the following items?
   (a) a rope (b) a bicycle innter tube (c) a circuit board for a computer that gets warm.

3. What are the primary structures of the tripeptides that can be made from the following amino acids: Pro, Pro, Trp?

4. What is the complementary structure for the following DNA base sequence: TGGCTTAAT?

Answers

1.

2. (a) fiber (b) elastomer (c) thermoset plastic

3. Pro -Pro - Trp; Pro - Trp - Pro; Trp - Pro -Pro

4. ACCGAATTA

Solutions to Chapter Problems

11.1 Polymers made from substituted ethylenes form by breaking the C=C double bond to form two new single C–C bonds to other monomers. In a copolymer, each monomer occurs randomly:

11.3 To deconstruct a polymer into its monomers, break C–C single bonds and form double bonds:
(a)

(b)

11.5 Polybutadiene forms from butadiene by a combination of double bond breakage and migration. The resulting polymer differs from polyethylene in having one C=C double bond in each repeat unit, whereas polyethylene is entirely CH_2 units connected by C–C single bonds:

electron shifts   bond formation

11.7 Identify functional groups from memory and convert structures to line drawings following the standard rules:

(a) ester, phenyl

(b) amine

(c) aldehyde

(d) alkene, bromide

(e) phenyl, alcohol

11.9 To draw examples of compound types, first identify the chemical formulas of its functional groups. Remember that carbon always forms four bonds. There are many possible correct answers to this problem. We provide just one example for each.

(a) an amine contains N bonded to C. The simplest amines contain $-NH_2$ units:

(b) An ester contains $C-CO_2-C$. There must be five additional C atoms, arranged in any fashion:

(c) The functional group in an aldehyde is O=C–H, MM = 29 g/mol. To have a molar mass greater than 80, an aldehyde must contain at least four additional C atoms:

(d) The ether linkage is –O–, and phenyl is the benzene ring:

418

Chapter 11: Macromolecules

11.11 In a condensation reaction, two monomers react to form a larger unit, eliminating a small molecule such as H_2O in the process.

(a) The N–C linkage in the center of this molecule forms in a condensation reaction between a carboxylic acid and an amine:

[Structure: secondary amine + benzoic acid → N-substituted benzamide + H_2O]

(b) The ester linkage in this molecule forms in a condensation reaction between a carboxylic acid and an alcohol:

[Structure: methanol + cyclopentylacetic acid → methyl cyclopentylacetate + H_2O]

(c) The –O– linkage in this molecule forms in a condensation reaction between two alcohols:

[Structure: two alcohols → ether + H_2O]

11.13 All amino acids can condense with each other in two ways. In addition, the extra carboxylic acid of glutamic acid could undergo condensation, so there are three possible products:

[Three structures shown: glutamic acid–isoleucine dipeptides via different linkages]

11.15 To determine the monomers from which a condensation polymer has formed, decompose the polymer into its monomeric parts and add components of a small molecule, for example –H and –OH. Here, the structure shows a single component whose repeat unit is $(CH_2)_{10}$, and the linkage is an amide, which breaks down into an amine and a carboxylic acid:

11.17 Construct polyethylene oxide by breaking a C–O bond and linking the fragments:

11.19 Cross-linking leads to increased rigidity, because chemical bonds between polymer chains restrict the ability of polymer chains to slide past one another. Thus relatively rigid tires have more extensive cross-linking than flexible surgeon's gloves.

11.21 The categories of polymers and their characteristic properties are as follows: plastics, which exist as blocks or sheets; fibers, which can be drawn into long threads; and elastomers, which can be stretched without breaking. (a) Balloons must stretch, so they are made of elastomers; (b) rope is made of fibers; (c) camera cases are rigid, so they are made of plastics.

11.23 Dioctyl phthalate is an example of a liquid plasticizer, which reduces the amount of cross-linking in a polymer as well as adding a fluid component; both these changes result in improved flexibility of the polymer.

11.25 See Figure 11-22 for an example of the "chair" shape of the 6-membered cyclohexane ring. For emphasis, we show the C atoms and C–H bonds at the CHOH positions:

11.27 To draw the structures of sugars that are related to glucose, start with the glucose structure (Figure 11-22) and make modifications as needed:

α–glucose   α–talose   β–talose

11.29 As described in your textbook, polymers of β-glucose coil upon themselves. Glycogen is this type of polymer. Polymers of α-glucose, of which cellulose is an example, form planar sheets. In the drawings, the dark solid lines indicate the direction of the continuing chain:

11.31 The complementary strands of DNA form from hydrogen bond linkages between specific nucleic acids. A pairs with T and G pairs with C. Thus the complementary sequence of A-A-T-G-C-A-C-T-G   is   T-T-A-C-G-T-G-A-C.

11.33 The structure of DNA consists of a phosphate-sugar-nucleic acid trio bonded to others through its phosphate groups. Apart from the structure of the nucleic acid, each unit is identical. On the left is shown the backbone linkage, with A and T attached and * indicating the point of attachment. The two other nucleic acids are shown to the right, with * indicating where they link to the backbone:

11.35 The strands of DNA fit together as a result of the hydrogen-bonding interactions illustrated in Figure 11-36. The complementary sequence for A-T-C is T-A-G. The backbones of complementary strands run in opposite directions. For clarity in showing the hydrogen bonding interactions, we represent the backbone with a solid line:

11.37 Consult Figure 11-40 for the structures of the different amino acids, all of which have the same amino acid backbone.

(a) Tyr (b) Phe (c) Glu (d) Met

11.39 Hydrophilic side chains are characterized by the presence of N or O atoms that generate polar bonds and hydrogen bonding capability, or an S–H bond that is polar. Among the structures shown in Figure 11.37, Tyr (O–H bond) and Glu (CO_2H group) are hydrophilic, while Met and Phe are hydrophobic.

11.41 To identify an amino acid, examine the side chain attached to the carbon atom between the N atom and the C=O bond in the amino acid backbone. (a) the "side chain" is –H, making this glycine, the simplest amino acid, with a hydrophobic side chain; (b) the side chain is $-CH_2OH$, so this is serine, with a hydrophilic side chain; (c) the side chain is $-CH_2SH$, so this is cysteine, with a hydrophilic side chain.

11.43 When two amino acids condense, the amino end of either molecule can link to the carboxylic acid end of the other. A water molecule is eliminated, creating a C–N bond. Three amino acids can combine in six different ways, A–B, B–A, A–C, C–A, B–C, and C–B:

11.45 When amino acids condense to form polypeptides, each added amino acid reacts to eliminate a water molecule. The chemical formula is alanine is $C_3H_7NO_2$, and the net reaction to form poly-Ala is: $n\ C_3H_7NO_2 \rightarrow n\ H_2O + [C_3H_5NO]_n$;
Thus the empirical formula of this polymer is C_3H_5NO, $MM = 71$ g/mol;

Divide the molar mass of the polypeptide by the molar mass of the empirical formula unit to obtain the number of amino acid units: $(1.20 \times 10^3$ g/mol$) \div (71$ g/mol$) = 17$ amino acids.

11.47 There are four possible choices for each base in a DNA strand, so the number of ways to connect 12 bases is $4^{12} = 16{,}777{,}216$ (this is an exact number, because 4 is an exact number).

11.49 (a) The monomers from which hair spray is made are substituted ethylenes, which polymerize in a free radical process in which each ethylene has an equal probability of adding to the growing chain. Thus this polymer will have a random arrangement;
(b) The backbone of a polyethylene is a carbon chain with the substituents connected to every other carbon atom. Here is a line structure of six monomer units:

(c) The backbone of this polymer is hydrophobic, but its side chains contain highly polar C=O groups that interact with one another and with polar and H–bonding groups on hair.

11.51 A nucleotide is a combination of 1 base, 1 sugar, and 1 phosphate group. A duplex is a pair of nucleotides bound together by hydrogen-bonding interactions. A duplex consists of either A and T or G and C. Here is C–G:

Chapter 11: Macromolecules

11.53 To draw the structure of a condensation product, start with the structures of the two molecules undergoing condensation and then connect them together, eliminating a small molecule such as water:

11.55 Polystyrene has the polyethylene backbone of carbon atoms, with a benzene ring attached to every second carbon atom. A divinylbenzene monomer reacts with the growing chain by means of one of its C=C double bonds, leaving the second C=C double bond available to cross-link by becoming incorporated into another polystyrene chain. Thus the benzene rings form "bridges" between polystyrene chains:

11.57 The condensation reaction between an alcohol and an amine releases a water molecule and leaves a C–N–C linkage:

Chapter 11: Macromolecules

11.59 All sugar molecules have multiple –OH functional groups, each of which can participate in hydrogen bond formation with water molecules. α-glucose has 5 –OH groups; in addition, its ring O atom can participate in formation of hydrogen bonds.

11.61 The bases of the mRNA molecule must be complementary to the bases of the template DNA on which they are modeled, meaning A generates U, C generates G, G generates C, and T generates A (remember that in RNA, U appears rather than T). Thus the sequence generated by this strand is the same as that of Strand B except for U appearing in place of T: 1 = cytosine, 2 = uracil, 3 = adenine, and 4 = guanine.

11.63 All polyethylene has the same empirical formula, $(CH_2)_n$, but whereas high-density polyethylene is all straight chains that nest together readily, low-density polyethylene has many side chains that cannot nest easily. Thus low-density polyethylene has more open space, accounting for the lower number of CH_2 groups per unit volume. See Figure 11-17 for a visual representation.

11.65 This problem is a "simple" stoichiometric problem that addresses the composition of a polymer. Use the percent by mass composition of copper to determine the mass of copper in 1 mol of the enzyme; convert to moles to obtain the molar ratio of copper to enzyme, which also gives the number of copper atoms per molecule:
(0.40%)(10^{-2}/%)(64,000 g) = 256 g of Cu in 1 mol of enzyme;
(256 g Cu) ÷ (63.546 g/mol Cu) = 4.0 mol Cu/mol enzyme.
There are 4 copper atoms in each molecule of this enzyme.

11.67 The names of the two types of proteins indicate their major large-scale features. Globular proteins are globe-shaped. They tend to form balls, whose structures often are stabilized by cross-linking. The interiors of these balls are generally hydrophobic, while the exteriors are hydrophilic. Globular proteins act as enzymes because they can take on highly specific shapes that allow them to bind to various materials. They also act as transport agents because their compact shapes and hydrophilic exteriors allow them to move relatively easily through the aqueous fluids of the body. Fibrous proteins have the shapes of fibers, long and narrow. Fibrous proteins serve as structural components, for example of muscles, because they can be quite strong. The fibrous shape is stabilized by α-helical structures in which hydrophobic amino acids point inward and hydrophilic amino acids point outward.

11.69 The three steps of free radical polymerization are initiation, which generates a free radical; propagation, in which monomer units add to the free radical end of the polymer chain; and termination, in which two free radical chains link together. The functional group on an ethylene monomer does not participate in any of these processes:

Initiation:

Propagation:

Termination:

11.71 A cyclic product forms when one end of a monomer molecule can condense with the other end of the same molecule. The monomer illustrated in this problem contains five carbon atoms in addition to its terminal functional groups, so it can easily form a ring containing six carbon atoms and one oxygen atom:

11.73 Polyethylene is a carbon backbone (with two C–H bonds to each C atom). The butadiene-styrene polymer, while it also contains an all-carbon backbone, has benzene rings attached at intervals of approximately 6 carbon atoms, as well as a double bond in the backbone at intervals of approximately 6 carbon atoms. The extra bulk of the benzene rings and the geometry changes where the double bonds exist (trigonal planar as opposed to tetrahedral) make it impossible for the carbon chains to stack as closely together in butadiene-styrene as in polyethylene. With less close stacking, the dispersion forces holding the chains together are smaller, so butadiene-styrene is not as rigid as polyethylene.

polyethylene:

butadiene-styrene:

11.75 In globular proteins, amino acids with polar *R* groups will tend to be on the outside, in contact with the aqueous environment. Those without polar *R* groups will tend to be on the inside, in contact with other hydrophobic groups. (a) Arg is polar, not found inside; (b) Val is non-polar, found inside; (c) Met is non-polar, found inside; (d) Thr is polar, not found inside; and (e) Asp is polar, not found inside.

11.77 Trigonal planar geometry is best described by sp^2 hybrid orbitals. The simple Lewis structure of the peptide linkage indicates SN = 4, which normally requires sp^3 hybrid orbitals. The trigonal planar geometry indicates that the lone pair of electrons on the N atom is delocalized in a π bonding network, which we illustrate by the Lewis structure of a gly-gly dimer. There is a resonance structure which places a π bond between C and N. Studies of bond lengths and strengths indicate that the peptide linkage is stronger than a single bond; this along with the trigonal planar geometry about the N atom confirms this bonding arrangement:

11.79 RNA and DNA share many characteristics but also have some significant differences, as indicated in the completed table:

	RNA	DNA
Sugar	ribose	deoxyribose
One-ring bases	U and C	T and C
Two-ring bases	G and A	G and A

Chapter 12 Chemical Energetics

12.1 Thermodynamic Definitions

QUESTIONS TO ANSWER, SKILLS TO LEARN
1. **What is the vocabulary used to describe energy changes?**
2. **Dividing the universe into the system and the surroundings**
3. **Understanding the difference between "paths" and "states"**

Energy is conserved: it flows from place to place, but is not created or destroyed. Thermodynamics is the study of energy transfer and the effects of energy changes. In the study of energy transfer, it is useful to divide the universe into separate pieces from which energy flows to and from. The following definitions are essential:

Universe: In the universe, there is no change in the amount of energy.

System: The system is the part of the universe that is to be studied.

Surroundings: The surroundings are the rest of the universe that is not the system.

In other words, the system plus the surroundings is the universe. There are several types of systems, which are described by the following terms:

Open system: can exchange both matter and energy with the surroundings.

Closed system: can exchange energy but not matter with the surroundings.

Isolated system: exchanges neither matter nor energy with the surroundings.

Exercise 12.1 For each of the following systems, determine whether it is open, closed or isolated.
(a) An ice cube and water in a perfect, stoppered thermos (perfect implies that the thermos allows no energy transfer to or from the interior of the thermos).
(b) The same amount of ice and water in an Erlenmeyer flask.
(c) Water in a sealed glass tube.

Steps to Solution: To determine the type of system we must remember the fundamental definitions listed above. Then we scrutinize each system to see if matter or energy (thermal, radiant, etc.) can be transferred in or out of the system, allowing us to classify the system.

(a) *The ice cube and water form the system and are contained in a stoppered, perfectly insulated thermos. Because the flask is stoppered, no matter can leave or enter the flask. Also, because the thermos is perfect, no energy may leave or enter the flask. Therefore this system fits the definition of an* isolated *system.*

(b) *If the ice and water are in an Erlenmeyer flask whose top is open, water may escape or enter. The flask may be heated or cooled and the contents will have energy transferred to them or away from them. Because energy and matter may be exchanged with the surroundings, the system is an open system.*

(c) *In a sealed glass tube, matter may not enter or leave. However, the contents may be cooled or warmed and light of many frequencies may shine on it as well. Because energy may be transferred to the system but matter cannot leave or enter, it is a closed system.*

In part (a) of Exercise 12.1, the system of ice and water in a thermos is described as isolated. Systems are chosen to make it easier for the observer to calculate important quantities that describe the system, such as the temperature after the ice has melted. To do so, we sometimes can mentally separate components in a mixture, which amounts to a different definition of the system and the surroundings. In the ice and water system, the thermos isolates these components from the rest of the universe; the system could be defined as the ice and the water as the surroundings. The energy transfer would occur from the water to the ice and also from the water to the water produced by melting the ice.

To describe a system, we need measurable quantities that will allow someone else to construct the system in the same condition that the observer noted. Another word that is used for conditions is **state,** and the quantities describe the **state of the system**. The measurable quantities that describe the state of the system are called **state variables**. State variables may be related through equations called **equations of state** such as the ideal gas law for systems consisting of ideal gases. When a process occurs such that the values of one or more of the state variables change, that process is said to have produced a **change of state**.

Changes in state occur through two types of processes:

Physical processes: No chemical bonds are broken or formed so the molecular identities are the same at the beginning and end of the process. This is a *physical change of state*.

Chemical processes: Bonds are broken and/or formed in the process so the chemical identities of the substances change during the process. This is a *chemical change of state*. To completely describe a chemical change of state we must know the amounts of reactants and products present at the beginning and the end of the process.

The melting of ice in water described above is a physical process; some water is undergoing a phase change from solid to liquid, and the temperature of the water is changing. However, the substances present in the initial and final states of the system have the same molecular identity. The combustion of hydrogen in air is a chemical change of state; the substances present in the initial state of the system (hydrogen, oxygen and other gases) do not have the same identity or amounts of the substances present in the final state of the system. These definitions are essentially the same as those given in Chapter 1 for physical and chemical changes.

Changes of state, either physical or chemical, can occur in many ways. State variables depend only on the initial and final states of the system; other variables depend on how the change is effected. The variables that depend on how the change takes place (or the *path* of the change) are called **path functions.** The variables that do not depend on path are called **state functions**. The difference between path and state functions is explored in Exercise 12.2.

Exercise 12.2 Samples of gas in two identical cylinders with movable pistons each occupy a volume of 1.0 L at 1.0 atm pressure at 325 K. Cylinder A undergoes the following sequence of operations: the gas is heated to 650 K and the piston locked in place while the gas is cooled back to 325K. Cylinder B undergoes this sequence of operations: a strong student pulls the piston back very slowly until the volume is doubled but there is no temperature change. Then the piston is locked in place. Analyze the changes and name some of the state functions involved in the two changes.

Steps to Solution: It is important to remember the defining characteristic of a state function: it depends only on the initial and final states of the system and does not depend on how the change was brought about. Sometimes it is helpful to draw the system before and after the process occurred to visualize the differences between the initial and final states. If possible we should also summarize data that describe quantitatively the state of the system. This is done below.

Summary of changes:

Cylinder A	Cylinder B
Before process:	Before process:
$P = 1.0$ atm $V = 1.0$ L $T = 325$ K	$P = 1.0$ atm $V = 1.0$ L $T = 325$ K
After process:	After process:
$P = 0.5$ atm $V = 2.0$ L $T = 325$ K	$P = 0.5$ atm $V = 2.0$ L $T = 325$ K

The sketch below shows the change schematically.

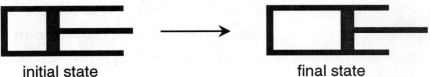

initial state   final state

In both cases, the volume is increased to 2.0 L and the pressure decreased to 0.5 atm while the temperature is the same at the initial and final states. The state functions in this exercise are temperature, pressure and volume, because their values do not depend on how the change was brought about (the path) in cylinder A or B.

12.2 Energy

QUESTIONS TO ANSWER, SKILLS TO LEARN
1. **Predicting energy flow to and from systems: endothermic and exothermic processes**
2. **Predicting temperature changes from the amount of energy transferred**
3. **Calculating work that may be done knowing the amount of energy transferred**
4. **What is the First Law of Thermodynamics?**

Endothermic and Exothermic Processes

In thermodynamics, the universe is divided into two pieces, the system and the surroundings. In open and closed systems, energy may be transferred from the system to the surroundings *or* from the surroundings to the system. In the diagrams that follow, the direction of energy flow is indicated by the arrow.

If a process releases energy, the energy may be transferred to the surroundings. Because energy goes *out* of the system, the process is called **exothermic**. For a chemical reaction, *exothermic* means that the *sum of bond energies of the bonds being formed is greater than the sum of the bond energies of the bonds being broken.* If you hold a flask in which an exothermic reaction takes place, the flask will feel warm because energy is being transferred from the flask and its contents to your hand. If a process requires energy, the energy may be absorbed from the surroundings. Because energy goes *into* the system, the process is called **endothermic**. For a chemical reaction, *endothermic* means that the *sum of the bond energy(ies) of the bonds being formed is less than the sum of the bond energy(ies) of the bonds being broken.* If you hold a flask in which an endothermic reaction takes place, the flask will feel cool because energy is being transferred from your hand to the flask.

Exercise 12.3 Glauber's salt, $Na_2SO_4 \cdot 10\ H_2O$, melts at 32 °C.

(a) Is this process exothermic or endothermic?
(b) As molten Glauber's salt is cooled, it will solidify and form crystals. Is this process exothermic or endothermic? (*Hint:* you should consider the $Na_2SO_4 \cdot 10\ H_2O$ to be the system.)

Steps to Solution: To answer this question, we must determine whether energy is required for the process to occur or if energy is released by the process. The system is the $Na_2SO_4 \cdot 10\ H_2O$; if energy is required, the process is endothermic and if energy is released, the process is exothermic.

(a) *When any solid melts, the forces holding together the particles that make up the solid are being broken, a process that requires energy.*

$$Na_2SO_4 \cdot 10\ H_2O_{(s)} \rightarrow Na_2SO_4 \cdot 10\ H_2O_{(l)} \text{ (requires energy)}$$

Because energy is required to overcome the attractive forces, this process requires that energy is absorbed by the system, the $Na_2SO_4 \cdot 10\ H_2O$, and therefore the process is endothermic.

(b) *In a solid, the intermolecular forces between the particles in the solid are stronger than those in the liquid.*

$$Na_2SO_4 \cdot 10\ H_2O_{(l)} \rightarrow Na_2SO_4 \cdot 10\ H_2O_{(s)} \text{ (releases energy)}$$

Therefore energy will be released as the weaker forces in the liquid are replaced by the stronger interactions in the solid. Because energy is released from the system to the surroundings, the process is exothermic.

The preceding exercise illustrates an important principle in the energy changes that occur in chemical processes, including phase changes and chemical reactions:

A process or reaction that is exothermic in one direction is endothermic in the other direction.

Energy and Heat

Energy manifests itself in many forms; one is thermal energy, the energy of motion at the molecular level, which is indicated by the temperature of the material.

Thermal energy is heat!
Heat is given the symbol, q

Transfer of heat proceeds from warmer objects to cooler objects. The transfer of thermal energy results in temperature changes in both objects (unless a phase change or chemical reaction occurs in one of them). The temperature change in a substance involved in a heat transfer may be calculated by the following formula:

$$q = nC_p \Delta T \quad \text{gives} \quad \Delta T = T_{\text{final}} - T_{\text{initial}} = \frac{q}{nC_p}$$

where:

    q = heat transferred in Joules (J)

    n = mols of substance absorbing or releasing heat (mol)

    C_p = molar heat capacity, an intensive property of the substance (J/mol•K)

The molar heat capacity, C_p, is the amount of energy required to raise the temperature of one mol of that substance by one Kelvin (or 1 °C because the size of a degree is the same in both the Celsius and Kelvin scales).

Exercise 12.4 Al_2O_3 is the raw material for aluminum manufacture and must be heated considerably before it can be converted to aluminum. Calculate the amount of heat (in kJ) required to heat 200 kg of Al_2O_3 from 25 °C to 1000 °C and compare that to the amount of heat required to heat 200 kg of Al metal from 25 °C to 550 °C.

Steps to Solution: *This problem requires the amount of heat required to change the temperature of two substances by a given amount. The data supplied are mass and identity of the substances and initial and final temperatures. The equation*

$$q = nC_p\Delta T$$

can be used after the molar heat capacities are found. The values obtained from Table 12-1 in the text are given below.

Substance	Al	Al_2O_3
C_p (J/mol • K)	24.35	79.04

The heat required to warm the Al_2O_3 is calculated using:

$$q = nC_p\Delta T$$

$$\Delta T = T_f - T_i = 1000\ °C - 25\ °C = 975\ °C$$

$$n = \frac{200 \times 10^3\ g\ Al_2O_3}{101.96\ g/mole} = 1.96 \times 10^3\ mol\ Al_2O_3$$

Therefore,

$$q = nC_p\Delta T = 1.96 \times 10^3\ mol\ Al_2O_3 (79.04\ J/mol•K)(975\ K) = 1.51 \times 10^8\ J$$

$$q = \boxed{1.51 \times 10^5\ kJ}$$

The energy required to heat the aluminum metal to 550 °C is found similarly:

Chapter 12: Chemical Energetics

$$\Delta T = T_f - T_i = 550\ °C - 25\ °C = 525\ °C$$

$$n = \frac{200 \times 10^3\ g\ Al}{26.98\ g/mole} = 7.41 \times 10^3\ mol\ Al$$

$q = nC_p\Delta T = 17.41 \times 10^3\ mol\ Al(24.35\ J/mol \cdot K)(525\ K) = 9.48 \times 10^7\ J = \boxed{9.48 \times 10^4\ kJ}$

The energy requirements just to heat the aluminum oxide in preparation for refining into aluminum are much greater than the energy required to heat Al metal to the melting point.

Work and Energy

Another form of energy is called **work**, w, where energy is used to overcome a force to move an object. Lifting an object to a shelf requires overcoming the force of gravity. Work is calculated through the relation:

$$w = F \cdot d$$

where: F is the force that is acted against in Newton (1 N = 1 kg·m/s²) and
d is the displacement (distance) the force is acted against in meters.

Exercise 12.5 A CRC Handbook of Chemistry and Physics has a mass of about 2.83 kg. How much work must be done against gravity to lift the book from the floor to the shelf 2.0 m off the floor? (Note: the force due to gravity is $F_g = m(9.8\ m/s^2)$ where m is the mass of the object.)

Steps to Solution: In this problem the work to move the book requires acting against gravity. We must solve for the force that gravity exerts on the book and then calculate the work using this equation:

$$w = F \cdot d$$

We solve for the force gravity exerts on the book using the supplied equation:

$$F_g = m(9.8\ m/s^2) = 2.83\ kg(9.8\ m/s^2) = 27.7\ kg \cdot m/s^2 = 28\ N$$

We calculate the work knowing that the book is to be moved 2.0 m against gravity:

$$w = F \cdot d = 27.7\ kg \cdot m/s^2\ (2.0\ m) = 55.5\ kg \cdot m^2/s^2 = \boxed{5.6 \times 10^1\ J}$$

To lift the book from the floor to the high shelf requires 56 J of energy to be expended in work against gravity.

Expansion (PV) Work

One type of work that is often encountered in chemistry is the work done against pressure when a system increases or decreases its volume: this kind of work is called **PV work**. In such cases the work may be calculated in terms of a constant external pressure, P_{ext}, and the change in volume, ΔV:

$$w = -P_{ext}\Delta V \qquad \Delta V = V_{final} - V_{initial}$$

The negative sign in the equation for work results from the fact that to expand ($\Delta V > 0$), the system must do work on the surroundings (push back against the external pressure) by transferring the kinetic energy of the molecules of the system to the walls of the container (or the molecules of the surroundings). For energy to be conserved, the system must lose that amount of work on the surroundings. Conversely, if a gas is compressed, $\Delta V < 0$. In this case the surroundings are transferring energy to the molecules in the system by increasing the number of collisions of the system molecules with the surroundings. The next two exercises examine these two cases.

Exercise 12.6 A sample of 14.5 g of sodium bicarbonate is added to 150 mL of 6.0 M HCl in an open flask and the following reaction occurs at 298 K and 1.04 atm pressure:

$$HCO_3^-{}_{(aq)} + H_3O^+{}_{(aq)} \rightarrow CO_2{}_{(g)} + 2\,H_2O\,{}_{(l)}$$

Assume that the reaction goes to completion and that the amount of CO_2 that dissolves in the solution can be neglected. How much work is done by the gas evolved in the reaction?

Steps to Solution: This is a multiple skill problem. We must first determine the limiting reagent in the reaction and then the amount of gas that is produced by the reaction. After those quantities have been found we can determine the work done by the CO_2 expanding into the atmosphere, by determining the volume of CO_2 evolved by the reaction and using $w = -P\Delta V$ (even though the gas is escaping into the air). The work done by the reaction is the same whether the gas just escapes into the air or is used to moved a piston forward; in the first case the work is lost, and in the second case it is harnessed.

Limiting reagent?

$$\text{mol } H_3O^+ = M(V) = 6.0\,M\,(0.150\,L) = 0.90 \text{ mol HCl}$$

$$\text{mol } HCO_3^- = \frac{g(NaHCO_3)}{MM(NaHCO_3)} = \frac{14.5\,g}{84.01\,g/mole} = 0.173 \text{ mol } NaHCO_3$$

The limiting reagent is $\boxed{NaHCO_3}$.

Now we need to calculate the amount of work. We obtain the volume of the gas using the ideal gas law and the data supplied in the problem.

$$V = \frac{nRT}{P}$$

n = mol $NaHCO_3$ = n_{CO_2}; P = 1.04 atm (the external pressure); T = 298 K (given)

$$V_{CO_2} = \frac{0.173 \text{ mole}(0.0821 \text{ L} \cdot \text{atm/mole} \cdot \text{K})298 \text{ K}}{1.04 \text{ atm}} = 4.06 \text{ L}$$

No other gases are produced or consumed in the reaction, so this value represents the volume change for the reaction. Because the work depends on the external pressure, the work done is found by substituting the external pressure and the volume of CO_2 evolved into the expression for expansion work:

$$w = -P_{ext} \Delta V = -1.05 \text{ atm}(4.06 \text{ L}) = -4.26 \text{ L} \cdot \text{atm}$$

However, we need to convert the units from L•atm to J. We must convert atm to Pa (N/m^2) and L to m^3:

$$w = -4.26 \text{ L} \cdot \text{atm} \cdot \frac{101325 \text{ N/m}^2}{1 \text{ atm}} \cdot \frac{1 \text{ m}^3}{1000 \text{ L}} = -432 \text{ N} \cdot \text{m} = \boxed{-432 \text{ J}}$$

The evolution of gas does 432 J of work on the surroundings.

Converting L•atm to J: $1 \text{ L} \cdot \text{atm} = \frac{101325 \text{ N/m}^2}{1 \text{ atm}} \cdot \frac{1 \text{ m}^3}{1000 \text{ L}} = 101.325 \text{ J}$

The First Law of Thermodynamics

In a system energy may be transferred to or from the system by work performed by or on the system. Heat may also be transferred from the surroundings to the system (endothermic) or from the system to the surroundings (exothermic). Because energy is conserved, the energy change of the system, ΔE, is the sum of the heat and work:

First Law of Thermodynamics: The energy change in a system, ΔE, is equal to the sum of heat transferrred to the surroundings, q, and the work done on the surroundings, w.

$$\Delta E = q + w$$

The lack of subscripts indicates that the work and heat used in this equation are the work and heat with respect to the system. The signs of the work done and heat transfer are very important; the sign convention is summarized below.

Work	Sign	Rationale
Work done by system on surroundings	$w < 0$	Energy is required to perform work, so any work done by the system reduces the energy of the system.
Work done by surroundings on system	$w > 0$	Energy is required to perform work. Therefore any work done on the system adds energy to the system.

Heat	Sign	Rationale
Heat transferred to surroundings from system	$q < 0$	Heating the surroundings requires energy that comes from the system, reducing the energy of the system.
Heat added to system from surroundings	$q > 0$	The addition of heat to the system leads to an increase in the energy of the system.

The energy change, ΔE, is a state function. Consider the reaction of hydrogen and oxygen:

$$2\,H_{2\,(g)} + O_{2\,(g)} \rightarrow 2\,H_2O_{(g)} \quad \Delta E \approx -484 \text{ kJ}$$

This reaction may be performed in a number of ways, each of which releases different amounts of heat and work. In the space shuttle, hydrogen and oxygen are used in the main rocket motors to launch the shuttle, a process that releases large amounts of heat as well as doing work to lift the spacecraft. However, hydrogen and oxygen also react in a device called a fuel cell that converts the substances to water with much less heat produced but with the production of work (as electricity).

$$2\,H_{2\,(g)} + O_{2\,(g)} \rightarrow 2\,H_2O_{(g)} \quad \Delta E \approx -484 \text{ kJ}$$

+ Energy

The bond energies between the reactants and products are the same in both cases; therefore the energy change per mol of hydrogen burned, ΔE, is the same in both devices. However, in the case of the fuel cell, more energy is obtained as work and less wasted as heat.

Exercise 12.7 Predict the sign of the thermodynamic quantities in the given system for each of the following processes:
(a) The work, w, if the system is an automobile tire hitting a bump.
(b) The heat, q, if the water in the ocean absorbs solar radiation.
(c) The work, w, if the system is a burrito heated until it is *steaming* hot.
(d) The heat, q, if the system is an automobile brake assembly being used to slow an automobile.

Steps to Solution: To answer these questions, we must determine the direction of energy transfer with regard to the system: in the process described, is energy being transferred to the system or is energy being transferred to the surroundings? If energy is being transferred to the system, ΔE will be positive and so will the sign of the relevant quantity, q or w.

(a) When an automobile tire hits a bump, the tire's shape deforms to the shape of the bump, resulting in a reduction of the volume of air inside the tire. Because energy is needed to compress gas, work is being done on the system (the tire) by the surroundings to increase the energy of the system. Therefore w is positive. Another method of analysis is based on this rationale: because ΔV of the system is negative (the volume of the tire hitting the bump is smaller than the volume before hitting the bump), the work ($w = - P_{ext}\Delta V$) is positive.

(b) When the ocean water absorbs solar radiation, it becomes warmer. The increase in temperature indicates that energy is being transferred to the water; therefore q is positive.

(c) The key to determining the sign of the work done by this system is the word steaming. Steam indicates water vapor (a gas) is being given off by the burrito. This means the water vapor is doing work to push back the atmosphere, thus transferring energy to the surroundings. Therefore the sign of the work in this process is negative because the process lowers the energy of the burrito.

(d) *In this example, the surroundings (the automobile) is losing kinetic energy (it is slowing down) and this energy is being transferred to the brake assembly. The energy of the system is increasing by this process; therefore the sign of q is positive.*

Exercise 12.8 How many kJ of energy will be converted to heat when a 1.4×10^3 kg automobile stops from a speed of 67 km/hr?

Steps to Solution: An automobile has kinetic energy while moving, but none when stopped (u = 0). Because $kE = \frac{1}{2} mu^2$, the kinetic energy of the moving car is all converted to heat, with some heat being transferred to the road, some to the tires and a large amount to the brake assembly. The critical step in this problem is to use the correct units. The kinetic energy is given by:

$$kE = \frac{1}{2} mu^2$$

The mass and velocity are supplied in the problem. For the energy to be obtained in J, the mass must be in kg and the velocity in m/s.

$$m = 1.4 \times 10^3 \text{ kg}$$

$$u = 67 \text{ km/hr} \cdot \frac{1000 \text{ m}}{\text{km}} \cdot \frac{1 \text{ hr}}{3600 \text{ s}} = 1.8 \times 10^1 \text{ m/s}$$

$$kE = \frac{1}{2} mu^2 = \frac{1}{2} \cdot 1.4 \times 10^3 \text{ kg} \cdot (1.8 \times 10^1 \text{ m/s})^2 = 2.4 \times 10^5 \text{ kg} \cdot \text{m}^2/\text{s}^2$$

$$kE = 2.4 \times 10^5 \text{ J} = 2.4 \times 10^2 \text{ kJ}$$

12.3 Heat Measurements: Calorimetry

QUESTIONS TO ANSWER, SKILLS TO LEARN
1. **How is the energy in heat transfers accurately measured?**
2. **Using calorimeter data to calculate heat transfer**
3. **Understanding the difference between constant volume and constant pressure calorimeters**

The energy change, ΔE, is measured in two parts, work and heat. The instruments used to measure the heat produced in processes (such as phase changes and chemical reactions) are called **calorimeters**. The fundamental features of a calorimeter are illustrated in the diagram that follows. An ideal calorimeter would exchange no energy with the rest of the universe so all of the energy of a process would be transferred to or obtained from the calorimeter. The heat transfer can be measured by the change in temperature, ΔT, of the "surroundings" whose heat capacity, C_{cal}, is known:

$$q = C_{cal}\Delta T$$

C_{cal} is the heat capacity of the entire calorimeter. Often the sample is immersed in water and the temperature change measured is that of the water and the calorimeter (the container, thermometer and insulation). In many cases the heat transferred to the water is much greater than that transferred to the calorimeter, so the heat transferred to the calorimeter is neglected. For accurate work, the heat capacity of the entire calorimeter system is determined using electrical heating or the reaction of a standard substance that generates a known amount of heat.

Exercise 12.9 The combustion of an organic compound releases 35.6 kJ of energy and causes the temperature of a calorimeter to increase by 4.76 °C. What is the heat capacity of the calorimeter?

Steps to Solution: The heat capacity of the calorimeter is the amount of energy required to change the temperature of the calorimeter 1 K (or equivalently, 1 °C). Because the reaction is exothermic, the temperature of the calorimeter increases. The amount of energy supplied (35.6 kJ) and the temperature change (4.76 °C) are given in the problem; therefore we can calculate the heat capacity of the calorimeter from the equation obtained by rearranging the heat transfer equation:

$$q = C_{cal}\Delta T \quad \text{is rearranged to give} \quad C_{cal} = \frac{q}{\Delta T}$$

Substituting numerical values for q and ΔT;

$$C_{cal} = \frac{35.6 \times 10^3 \text{ J}}{4.76 \text{ K}} = \boxed{7.48 \times 10^3 \text{ J/K} = 7.48 \text{ kJ/K}}$$

Calorimeters

There are two common types of calorimeters: those in which the pressure of the system remains constant throughout the course of the experiment (*constant pressure calorimeters*) and those in which the volume of the system remains constant throughout the course of the experiment (*constant volume calorimeters*). Both devices are used to measure heat evolved (or consumed) under different conditions.

A constant pressure calorimeter keeps the pressure constant by being "leaky" to gases; the device is not sealed shut. Therefore gases evolved during a process may escape; conversely, if gas is consumed (or cooled) during a process, air from the surrounding atmosphere can enter the calorimeter, thus keeping the pressure in the calorimeter constant.

A constant volume calorimeter is designed so that the process takes place in a sealed, strong container whose volume does not change. The container then transfers heat to the rest of the calorimeter system. The two different types of calorimeters measure related but different quantities. They both measure heat transferred, but under different conditions. The heat transferred in a constant volume calorimeter is the energy change, ΔE, whereas that transferred in a constant pressure calorimeter is the topic of the next section.

Exercise 12.10 A constant pressure "coffee-cup" calorimeter is used to measure the heat of reaction between 0.10 L of 1.0 M NaOH and 0.10 L of 1.0 M HCl. Both solutions initially are at 24.0 °C; after the solutions are mixed, the final temperature is 30.4 °C. If the heat capacity of the resulting solution is assumed to be the same as water (C_p = 75.291 J/mol•K), how much heat was evolved from the reaction? Assume the density of all solutions is 1.0 g/mL.

Steps to Solution: *In any calorimeter experiment, the heat transfer is found using the equation:*

$$q = C_{cal}\Delta T$$

In a coffee cup calorimeter, we neglect the heat capacity of the calorimeter and use only the heat capacity of the aqueous solution contained in it. We find the heat capacity of the solution by converting the volume of the solution to mass and then using the mass of the solution to calculate the heat capacity:

$$V_{solution} = V_{NaOH} + V_{HCl} = 0.10 + 0.10 \text{ L} = 0.20 \text{ L} = 2.0 \times 10^2 \text{ mL}$$

$$m_{solution} = V(\rho) = 200 \text{ mL} \left(1.0 \frac{g}{mL}\right) = 2.0 \times 10^2 \text{ g}$$

$$C_{cal} \approx n_{solution}(75.291 \text{ J/mol} \cdot \text{K}) = \frac{2.0 \times 10^2 \text{ g}}{18.0 \text{ g/mole}} (75.291 \text{ J/mole} \cdot \text{K}) = 8.36 \times 10^2 \text{ J/K}$$

The temperature increases from 24.0 to 30.4 °C:

$$\Delta T = T_f - T_i = 30.4 - 24.0 = 6.4 \text{ °C} = 6.4 \text{ K}$$

Therefore the heat transferred from the reacting substances to the aqueous solution is found:

Chapter 12: Chemical Energetics

$$q = C_{cal}\Delta T = 8.36 \times 10^2 \text{ J/K}(6.4 \text{ K}) = 5.35 \times 10^3 \text{ J} = 5.4 \text{ kJ}$$

Energy Changes and Calorimetry

The amount of energy transferred by a chemical reaction depends on the reaction and the amounts of reactants consumed; energy is an extensive property. The energy transferred by a specific chemical reaction is often stated in units of energy per mol, usually kJ/mol. Energy changes are reported in units of kJ, being calculated from the product of standard energy change and the number of moles of substance that reacted.

Exercise 12.11 The reaction that took place in Exercise 12.10 is an acid-base reaction with the net equation of reaction being:

$$H_3O^+_{(aq)} + OH^-_{(aq)} \rightarrow 2 H_2O_{(l)}$$

What is the energy change per mol of $H_2O_{(l)}$ formed?

Steps to Solution: The energy change for the reaction between 0.10 L of 1.0 M HCl with the same volume and molarity of NaOH was found to be 5.4 kJ. The number of moles of each reactant is the same; therefore the number of moles of water formed is the same as the number of moles of either reagent:

$$\text{mol } H_3O^+ = 0.10 \text{ L}(1.0 \text{ M}) = 0.10 \text{ mol } H_3O^+$$

$$\text{mol } OH^- = 0.10 \text{ L}(1.0 \text{ M}) = 0.10 \text{ mol } OH^-$$

$$\text{mol } H_2O_{(l)} \text{ formed} = \text{mol } H_3O^+ = 0.10 \text{ mol } H_2O_{(l)}$$

To calculate the energy change per mole water formed, we divide the energy change by the number of moles of water formed:

$$\Delta E(\text{molar}) = \frac{\Delta E}{n} = \frac{5.4 \text{ kJ}}{0.10 \text{ mole}} = \boxed{5.4 \times 10^1 \text{ kJ/mole}}$$

12.4 Enthalpy

QUESTIONS TO ANSWER, SKILLS TO LEARN
1. **What is the difference between energy, ΔE, and enthalpy, ΔH?**
2. **What is an enthalpy of formation?**
3. **Calculating enthalpy changes for chemical reactions**

Most processes that occur naturally occur under conditions of constant pressure. Constant pressure calorimetry measures the heat transfer under conditions in which combustion usually occurs, thus the need for a quantity that describes the heat content of fuels (chemical or biochemical). This quantity is called the **enthalpy**, H. Enthalpy is a thermodynamic quantity that has been defined for convenience and is equal to the energy plus the product of pressure and volume:

$$H = E + PV$$

The enthalpy is not directly measured; instead the enthalpy change, ΔH, is the quantity that is observed.

$$\Delta H = \Delta E + \Delta(PV)$$

The enthalpy change is equal to the heat transferred under conditions of constant pressure:

$$\Delta H = q_p$$

The difference between the enthalpy and energy changes is the $\Delta(PV)$ term. However, for solids and liquids involved in chemical reactions, the volume change with temperature is small; therefore

$$\Delta(PV) \approx 0 \quad \text{so} \quad \Delta H \approx \Delta E \text{ for reactions involving } \textit{only} \text{ solids and liquids}$$

For gases, volume changes can be significant and the $\Delta(PV)$ term may be computed by using the ideal gas law

$$\Delta(PV) = \Delta(nRT)$$

where the Δ can apply to the temperature *and/or* the number of moles of gases present in the system. In the problems encountered in this course only the number of moles of gas in the system will change, so:

$$\Delta(PV) = \Delta n(RT) \quad \text{where} \quad \Delta n = n_{\text{gaseous products}} - n_{\text{gaseous reactants}}$$

The larger the number of moles of gas produced or consumed, the more significant this term is.

Exercise 12.12 The compound nitroglycerine decomposes into several gases, as shown in the following reaction.

$$2\ NO_3CH_2CHNO_3CH_2NO_{3\ (l)} \rightarrow 3\ N_{2\ (g)} + 6\ CO_{2\ (g)} + 5\ H_2O_{(g)} + \frac{1}{2}\ O_{2\ (g)}$$

Calculate the difference between the enthalpy change, ΔH, and the energy change, ΔE, for this reaction in kJ.

Steps to Solution: The difference between the enthalpy and energy changes in a chemical reaction is the $\Delta(PV)$ term. For the given reaction, the only gases are those present in the products:

$$n_{reactants} = 0$$

$$n_{products} = 3\ (mol\ N_2) + 6\ (mol\ CO_2) + 5\ (mol\ H_2O) + \frac{1}{2}\ (mole\ O_2) = 14.5$$

$$\Delta n = n_{products} - n_{reactants} = 14.5 - 0 = 14.5$$

At 298 K, the quantity $\Delta(PV)$ is found:

$$\Delta(PV) = \Delta nRT = 14.5\ mol(8.314\ J/mol \cdot K)298\ K = 3.59 \times 10^5\ J = \boxed{35.9\ kJ}$$

This number may seem significant, but the enthalpy change for this reaction is quite large: $\Delta H = 1501$ kJ/mol nitroglycerine. The term above is for 2 mols of nitroglycerine; therefore the difference, δ, between the enthalpy and energy changes is about 1%:

$$\delta = \frac{35.9\ kJ}{2(1501\ kJ)} = 1.2 \times 10^{-2} = 1.2\ \%$$

This reaction is an extreme case where the difference between enthalpy and energy changes will be quite large because there is a large difference in the number of mols of gaseous reactants and products. The energy change, ΔE, and enthalpy change, ΔH, are often used interchangeably because the difference between the two quantities is often quite small.

Heats (Enthalpies) of Formation

Heats of formation (also called enthalpies of formation) allow the calculation of enthalpy changes in chemical reactions. The standard enthalpy of formation of a substance is symbolized as ΔH_f^o and is the enthalpy change for the formation of a substance from the elements in their most stable forms at 298 K and 1 atm pressure. In addition, the formation reaction produces only one mole of the desired product and *no other products*.

> **COMMON PITFALL: Incorrect identification of formation reactions**

Exercise 12.13 Which of the following chemical equations of reaction are formation reactions?
(a) $H_2\,(g) + S\,(s) + 2\,O_2\,(g) \rightarrow H_2SO_4\,(l)$
(b) $NO\,(g) + \frac{1}{2}\,O_2\,(g) \rightarrow NO_2\,(g)$
(c) $CH_4\,(g) + O\,(g) \rightarrow CH_3OH\,(g)$
(d) $2\,H_2\,(g) + C\,(s,\,graphite) + \frac{1}{2}\,O_2\,(g) \rightarrow CH_3OH\,(g)$

Steps to Solution: *A formation reaction has certain defining characteristics: there must be only one product that is present in its most stable form at 298 K and 1 atm pressure; the coefficient in the balanced equation of reaction is 1; and the reactants must be elements in their most stable states at 298 K and 1 atm pressure. We must examine the above equations of reactions to see if they meet these criteria.*
(a) *This equation is that for a formation reaction: a single product in its most stable state at 298 K and 1 atm pressure is formed from elements in their most stable forms.*
(b) *This equation is not that of a formation reaction because one of the reactants (NO) is not an elemental substance.*
(c) *This equation is not a formation reaction for two reasons: first, one of the reactants is not an elemental substance (CH_4) and second, the other reactant (O) is an elemental substance, but not the most stable form of oxygen, which is O_2.*
(d) *This equation is that for a formation reaction: a single product is formed in its most stable state at 298 K and 1 atm pressure from elements in their most stable forms.*

There are two statements that make formation reactions particularly useful in predicting the energy changes in chemical reactions:

The enthalpy change in a formation reaction is the heat of formation (the enthalpy of formation) of that substance.

The enthalpy of formation of an element in its standard state is 0.

Enthalpy is a state function because it is derived from state functions: the energy change, pressure, and volume. Energy changes in chemical reactions depend on the different bond energies in the products and reactants. The bond energies in a molecule are independent of the starting material$_{(s)}$. For example, methanol may be manufactured from methane and oxygen or by heating wood under low-oxygen conditions. There is no difference between molecules of methanol derived from these different sources; the bond energies and strengths in methanol are independent of the source of the molecules. Because enthalpy is a state function, enthalpies of chemical reactions may be estimated quite accurately using enthalpies of formation rather than calorimetry. Consider the reaction in Exercise 12.13(b):

$$NO_{(g)} + \frac{1}{2} O_{2\,(g)} \rightarrow NO_{2\,(g)}$$

This reaction can be obtained by adding the formation reactions of NO and NO_2. However, the reactants are not being formed, they are being destroyed (a more dramatic way of saying changed to products); the enthalpy change of the reverse of the formation reaction is the negative of the enthalpy of formation, $-\Delta H_f^o$(reactants). The product is being formed; the enthalpy change is ΔH_f^o (products).

[Diagram showing: N≡O + 1/2 O=O → O-N=O with arrows down to box containing 1/2 N≡N + 1/2 O=O + 1/2 O=O, and from the product box 1/2 N≡N + O=O; $-\Delta H_f^o(NO) = -90.4$ kJ/mol; $1/2\, \Delta H_f^o (O_2) = 0$ kJ/mol; $\Delta H_f^o(NO_2) = 33.85$ kJ/mol]

The enthalpy change in this (and any) reaction is the sum of the enthalpies of formation of the products minus the sum of the enthalpies of formation of the reactants:

$$\Delta H^o{}_{reaction} = \sum_{all\ products}^{i} \Delta H_f^o(i) - \sum_{all\ reactants}^{j} \Delta H_f^o(j)$$

$$\Delta H^o{}_{reaction} = 1\ mol\ NO_2(33.85\ kJ/mol) - (1\ mol\ NO(90.4\ kJ/mol) + \frac{1}{2}\ mol\ O_2(0\ kJ/mol))$$

$$\Delta H_{reaction} = -56.6\ kJ$$

Therefore with a table of enthalpies of formation we may calculate enthalpy changes of many reactions, even at temperatures that are not standard conditions, because enthalpies of formation change little with moderate changes in temperature. In addition, we can use the measurement of an enthalpy of reaction to determine the enthalpy of formation of a substance provided we know the enthalpies of formation of the other substances.

Exercise 12.14 Calculate (a) the enthalpy change for the reaction in Exercise 12.13(c) (b) the enthalpy of formation of nitroglycerine using the data in Exercise 12.12.

Chapter 12: Chemical Energetics

Steps to Solution: To solve these problems, we must use enthalpies of formation. In both cases we should set up the problem to solve for the enthalpy change of the reaction in terms of the standard enthalpies of formation of the products and reactants and their stoichiometric coefficients (using the variable names-no numbers yet). Then we rearrange the equation to solve for the desired quantity. At this point, we substitute numerical values for the enthalpies of formation and crunch the numbers.

(a) The equation of reaction is:

$$CH_{4\ (g)} + O_{(g)} \rightarrow CH_3OH_{(g)}$$

The enthalpy change is found using the general relation:

$$\Delta H°_{reaction} = \sum_{all\ products}^{i} \Delta H_f°(i) - \sum_{all\ reactants}^{j} \Delta H_f°(j)$$

$$\Delta H° = \Delta H_f°(CH_3OH_{(g)}) - [\Delta H_f°(CH_{4\ (g)}) + \Delta H_f°(O_{(g)})]$$

The coefficient of each of these terms is 1. The desired quantity is the enthalpy change for the reaction. We find the values of the enthalpies of formation of the reactants and products in tables (either in the text or in sources such as the CRC Handbook of Chemistry and Physics).

Compound	$\Delta H°_f$ (kJ/mol)	Number of mols, n	$n \cdot \Delta H°_f$ (kJ)
$CH_3OH_{(g)}$	-201.5	1	-201.5
$CH_{4\ (g)}$	-74.4	1	-74.4
$O_{(g)}$	249.	1	249.

$$\Delta H° = -201.5 - (-74.4 + 249)\ kJ = \boxed{-376.1\ kJ}$$

(b) The equation for the decomposition of nitroglycerine is given below.

$$2\ NO_3CH_2CHNO_3CH_2NO_{3\ (l)} \rightarrow 3\ N_{2\ (g)} + 6\ CO_{2\ (g)} + 5\ H_2O_{(g)} + \frac{1}{2} O_{2\ (g)}$$

$$\Delta H° = -3002\ kJ$$

Writing the enthalpy change in terms of the enthalpies of formation of the reactants and products, we obtain the following equation:

$$\Delta H° = 3 \cdot \Delta H°_f(N_{2(g)}) + 6 \cdot \Delta H°_f(CO_{2(g)}) + 5 \cdot \Delta H°_f(H_2O_{(g)}) + \frac{1}{2} \Delta H°_f(O_{2(g)}) -$$

$$(2 \cdot \Delta H°_f(NO_3CH_2CHNO_3CH_2NO_{3\ (l)})$$

We solve this equation for the enthalpy of formation of nitroglycerin to give:

$$\Delta H°_f (NO_3CH_2CHNO_3CH_2NO_{3\,(l)}) = \frac{3 \cdot \Delta H°_f(N_{2(g)}) + 6 \cdot \Delta H°_f(CO_{2(g)}) + 5 \cdot \Delta H°_f(H_2O_{(g)}) + \frac{1}{2}\Delta H°_f(O_{2(g)}) - \Delta H°}{2 \text{ mole}}$$

Substituting values for the enthalpies of formation and reaction into the above equation gives:

$$\Delta H°_f (NO_3CH_2CHNO_3CH_2NO_{3(l)}) = \frac{3 \cdot 0 + 6 \cdot (-393.51) + 5 \cdot (-241.83) + \frac{1}{2}(0) - (-3002) \text{ kJ}}{2 \text{ mole}}$$

$$\Delta H°_f (NO_3CH_2CHNO_3CH_2NO_{3(l)}) = \boxed{171 \text{ kJ/mole}}$$

12.5 Enthalpy and Intermolecular Forces

QUESTIONS TO ANSWER, SKILLS TO LEARN
1. **Calculation of the energy and enthalpy changes in processes involving phase changes**

Heat of Solution

To dissolve *any* solid in a liquid requires that the intermolecular forces binding the particles (atoms, ions or molecules) in the solid must be broken to obtain separated atoms, ions or molecules, a process that requires energy. The solvent then interacts with the atoms, ions, or molecules providing energy of stabilization, a process that releases energy.

The heat of solution, ΔH_{soln}, is the sum of the energy required to separate the constituent particles of a solid, ΔH_{solute}, and the energy of stabilization of these particles by the solvent, $\Delta H_{solute\text{-}solvent}$

$$\Delta H_{soln} = \Delta H_{solute} + \Delta H_{solute\text{-}solvent}$$

Dissolving compounds is exothermic if the solute-solvent interaction is stronger than the solute-solute interaction and endothermic if the converse is true. Common calculations involve the heats of solution of ionic compounds.

Exercise 12.15 Preparation of sodium hydroxide solutions is common because NaOH is an inexpensive base. However, some care must be taken because ΔH_{soln}(NaOH) = -44.5 kJ/mol. Estimate the highest temperature if a 6 M solution of NaOH is prepared when both starting materials, water and NaOH, are at 24 °C. Assume that the heat capacity and density of the solution will be the same as the heat capacity and density of water.

Steps to Solution: The heat of solution is exothermic; therefore we expect the temperature to increase. The highest temperature reached would be that if no heat were transferred to the surroundings; therefore we will treat the problem as though all the energy released from dissolving the NaOH will be used to heat the resulting solution. The following diagram illustrates our approach:

Because energy is conserved, the sum of the energy released by the solution process, q_{soln}, and the energy absorbed by the solution, q_{warm}, must be 0 if no energy escapes to the outside world (our assumption).

$$0 = q_{soln} + q_{warm}$$

The energy released, q_{soln}, is found from the number of moles of solute, n, and the heat of solution, ΔH_{soln}:

$$q_{soln} = n\Delta H_{soln}$$

The temperature change, ΔT, is calculated from heat transferred to the solution, q_{warm}:

$$q_{warm} = n_{solvent} C \Delta T$$

Substitution for q_{soln} and q_{warm} into $0 = q_{soln} + q_{warm}$ gives:

$$0 = n\Delta H_{soln} + n_{solvent} C \Delta T \quad \text{or} \quad n_{solvent} C \Delta T = - n\Delta H_{soln}$$

Solving for ΔT gives:

$$\Delta T = - \frac{n\Delta H_{soln}}{n_{solvent} C}$$

The heat of solution and heat capacity of water are known, but what values of n and $n_{solvent}$ should we use? The problem states only that a 6 M solution is to be prepared. For simplicity, we will perform the calculation for 1 L of solution. To obtain 1 L of 6 M NaOH requires 6 mols of NaOH giving 1 L of solution. The mols of solvent can be estimated as the number of mols of water in 1 L. Therefore:

$$n = 6 \text{ mol NaOH}$$

$$n_{solvent} = \frac{1000 \text{ g}}{18.0 \text{ g/mole}} = 55.5 \text{ mol H}_2\text{O}$$

$$\Delta T = -\frac{6 \text{ mole} \cdot 44.5 \times 10^3 \text{ J/mole}}{55.5 \text{ mole} \cdot 75.291 \text{ J/mole} \cdot \text{K}} = 64 \text{ K}$$

Note that the units of energy for the heat of solution have been changed from kJ to J so that they are the same as the units in the heat capacity. The final temperature will be the initial temperature plus the temperature change, ΔT:

$$T_f = T_i + \Delta T = 24 \text{ °C} + 64 \text{ °C} = \boxed{88 \text{ °C}}$$

The highest temperature that we would expect to obtain in the preparation of a 6 M NaOH solution starting with the components at 24 °C is 88 °C, certainly hot enough to be uncomfortable. Preparation of more concentrated solutions leads to higher temperatures.

Energies of Phase Changes

Phase changes involve changes in the magnitude of intermolecular forces between the constituents of the substance, as discussed in Chapter 10. The names of the processes for common phase changes are summarized in the following schematic figure.

All processes that break intermolecular attractions require energy.

All processes in which intermolecular attractions are formed release energy.

The energy requirements for these various processes depend on the strength of the active intermolecular forces and the initial and final phase. The energy required for a phase change is usually expressed as the enthalpy change for changing the phase of one mole of the substance. The following table summarizes the phase changes and the enthalpy changes.

Chapter 12: Chemical Energetics

Initial Phase	Final Phase	Name Given to Phase Transition	Enthalpy Change (kJ/mol)
solid	liquid	fusion or melting	ΔH_{fus} (molar heat of fusion)
liquid	solid	solidification	$-\Delta H_{fus}$
liquid	gas	vaporization	ΔH_{vap} (molar heat of vaporization)
gas	liquid	condensation	$-\Delta H_{vap}$
solid	gas	sublimation	ΔH_{subl} (molar heat of sublimation)
gas	solid	deposition	$-\Delta H_{subl}$

Invariably, the molar heat of vaporization is greater than the heat of fusion for any given compound because in vaporization all the intermolecular forces are ruptured, whereas in fusion, forces are active in the liquid phase, but because of molecular motion are not at their strongest.

Exercise 12.16 On humid days, we often observe condensation of water vapor on containers of cold drinks. If a bottle of 400 g of water containing 200 g of ice sits outside on a warm day, how many mL of water vapor will need to condense to melt the ice?

Steps to Solution: This question involves the energetics of phase transitions. We must determine how much energy is required for the endothermic process of melting the ice and then find how much water vapor must condense (an exothermic process) to supply this amount of energy. A sketch helps us visualize this process. The exothermic process of condensation (vapor to liquid) will provide the energy needed for fusion (solid to liquid).

The molar heat of fusion of water is 6.01 kJ/mol and the molar heat of vaporization is 40.79 kJ/mol. The amount of ice to be melted is 200. g; the energy required to melt n mols of a substance is:

$$q = n\Delta H_{fus}$$

The amount of ice to be melted is:

$$n = \frac{200.\text{ g}}{18.01 \text{g/mole}} = 11.1 \text{ mol}$$

so the heat required is:

$$q = 11.1 \text{ mol}(6.01 \text{ kJ/mol}) = 66.7 \text{ kJ}$$

Melting 200 g of ice at 0 °C to give 200 g of liquid water at 0 °C requires 66.7 kJ of energy. Condensation of water releases energy because liquid water has strong hydrogen bonding to stabilize the liquid phase:

$$H_2O_{(g)} \rightarrow H_2O_{(l)} \qquad \Delta H = -\Delta H_{vap} = -40.79 \text{ kJ/mol}$$

The heat released from condensing n moles of water vapor is:

$$q = -n\Delta H_{vap}$$

The number of moles of water vapor required to supply 66.7 kJ (q = -66.7 kJ) through this process is:

$$n = \frac{66.7 \text{ kJ}}{40.79 \text{ kJ/mole}} = 1.64 \text{ mol}$$

The mass of water that will condense is:

$$g\ H_2O = 1.64 \text{ mol } (18.01 \text{ g/mol}) = 29.5 \text{ g}$$

The density of water is about 1.00 g/mL; the volume of water that must condense is $\boxed{29.5 \text{ mL}}$

In general, the energy requirements involved in phase changes are much larger than the heat involved in warming or cooling the same mass of substance.

Test Yourself

1. A piece of cauliflower was combusted in a calorimeter with a heat capacity of 5.24 kJ/°C. The temperature increased from 24.35 °C to 25.07 °C. What was the enthalpy change for the combustion of the cauliflower?

2. Assume that coal is essentially all carbon and has a density of 1.5 g/mL. How much heat is produced if a piece of coal 7cm x 5 cm x 6 cm undergoes complete combustion by the following reaction:

$$C_{(s)} + O_{2\ (g)} \rightarrow CO_{2\ (g)} \qquad \Delta H = -394 \text{ kJ}$$

What mass of water could be heated from 25 °C to 100 °C with this amount of heat?

3. The oxidation of sulfur dioxide to sulfur trioxide is important in the manufacture of sulfuric acid:

$$2\ SO_{2\ (g)} + O_{2\ (g)} \rightarrow 2\ SO_{3\ (g)}$$

Find the enthalpy change of this reaction given the following:

$$S_{(s)} + O_{2(g)} \rightarrow SO_{2(g)} \quad \Delta H° = -296.83 \text{ kJ}$$

$$2 S_{(s)} + 3 O_{2(g)} \rightarrow 2 SO_{3(g)} \quad \Delta H° = -791.44 \text{ kJ}$$

4. Write the chemical equations of reaction whose enthalpy changes are the standard enthalpies of formation for the following substances: (a) $AlCl_{3(g)}$ (b) $KClO_{3(s)}$ (c) alanine, $H_2NCH(CH_3)COOH_{(s)}$.

Answers

1. -3.8 kJ

2. $q = -1. \times 10^5$ kJ; mass of water heated from 25 °C to 100 °C = 3.3×10^4 g

3. $\Delta H° = -197.78$ kJ

4. (a) $Al_{(s)} + \frac{3}{2} Cl_{2(g)} \rightarrow AlCl_{3(g)}$

   (b) $K_{(s)} + \frac{1}{2} Cl_{2(g)} + \frac{3}{2} O_{2(g)} \rightarrow KClO_{3(s)}$

   (c) $\frac{1}{2} N_{2(g)} + 3 C_{(s)} + \frac{7}{2} H_{2(g)} + O_{2(g)} \rightarrow H_2NCH(CH_3)COOH_{(s)}$

Solutions to Chapter Problems

12.1 A system is isolated if it exchanges neither energy nor matter with its surroundings, closed if it exchanges energy but not matter, and open if it exchanges both:
(a) open; (b) (almost) isolated; (c) open; and (d) closed.

12.3 State variables are properties that are independent of the past history of the system.
(a) not a state variable; (b) state variable; (c) state variable (mass is conserved); and (d) not a state variable.

12.5 A system is any part of the universe that can be defined with specified boundaries. Among possible systems are the automobile and driver, the driver, the automobile, the cooling system, the engine, a cylinder, and the battery.

12.7 Calculate the temperature change resulting from an energy input using Equation 12-1 and C_p values from Table 12.1 of your textbook: $q = nC_p\Delta T$, which can be rearranged to $\Delta T = q/nC_p$

(a) $n = m/MM = (10.0 \text{ g}) \div (26.98 \text{ g/mol}) = 0.3706$ mol;

$$\Delta T = \frac{(25.0 \text{ J})}{(0.3706 \text{ mol})(24.35 \text{ J/mol K})} = 2.77 \text{ K} = 2.77 \text{ °C}.$$

$T_f = T_i + \Delta T = 15.0 + 2.77 = 17.8$ °C.

(b) $n = m/MM = (25.0 \text{ g}) \div (26.98 \text{ g/mol}) = 0.9266$ mol;

$$\Delta T = \frac{(25.0 \text{ J})}{(0.9266 \text{ mol})(24.35 \text{ J/mol K})} = 1.108 \text{ K}. \quad T_f = T_i + \Delta T = 295 + 1.108 = 296 \text{ K}.$$

(c) $n = m/MM = (25.0 \text{ g}) \div (107.9 \text{ g/mol}) = 0.2317$ mol;

$$\Delta T = \frac{(25.0 \text{ J})}{(0.2317 \text{ mol})(25.351 \text{ J/mol K})} = 4.256 \text{ K}. \quad T_f = T_i + \Delta T = 295 + 4.256 = 299 \text{ K}.$$

(d) $n = m/MM = (25.0 \text{ g}) \div (18.02 \text{ g/mol}) = 1.387$ mol;

$$\Delta T = \frac{(25.0 \text{ J})}{(1.387 \text{ mol})(75.291 \text{ J/mol K})} = 0.2394 \text{ K} = 0.2394 \text{ °C}.$$

$T_f = T_i + \Delta T = 22.0 + 0.2394 = 22.2$ °C.

12.9 Calculate an energy change accompanying a temperature change using Equation 12-1 and C_p values from Table 12.1 of your textbook: $q = nC_p\Delta T$.

$\Delta T = 200.0 - 23.0 = 177.0$ °C = 177.0 K;

$$n_{\text{iron}} = \frac{m}{MM} = \frac{(1.35 \text{ kg})(10^3 \text{ g/kg})}{(55.85 \text{ g/mol})} = 24.17 \text{ mol}$$

$q_{\text{iron}} = (24.17 \text{ mol})(25.10 \text{ J/mol K})(177.0 \text{ K}) = 1.07 \times 10^5$ J;

To find the amount of air in the oven, use the ideal gas equation.

$$n_{\text{air}} = \frac{PV}{RT} = \frac{(1.00 \text{ atm})(0.150 \text{ m}^3)(10^3 \text{ L/m}^3)}{(0.08206 \text{ L atm/mol K})(273.1 + 23.0 \text{ K})} = 6.09 \text{ mol}$$

$n_{\text{nitrogen}} = (78\%)(10^{-2}/\%)(6.09) = 4.75$ mol; $n_{\text{oxygen}} = (22\%)(10^{-2}/\%)(6.09) = 1.34$ mol;

$q_{\text{air}} = (4.75 \text{ mol})(29.125 \text{ J/mol K})(177.0 \text{ K}) + (1.34 \text{ mol})(29.355 \text{ J/mol K})(177.0 \text{ K})$;

$q_{\text{air}} = (2.449 \times 10^4 \text{ J}) + (0.696 \times 10^4 \text{ J}) = 3.14 \times 10^4$ J.

12.11 When two objects at different temperatures are placed in contact, energy flows from the warmer to the cooler object until the two objects are at the same temperature. Let that temperature be T. The energy lost by the warmer object equals the energy gained by the cooler object: $q_{\text{cool}} = -q_{\text{warm}}$, and $q = nC_p\Delta T$.

Water: $n = (37.5 \text{ g}) \div (18.02 \text{ g/mol}) = 2.081$ mol;
$q = (2.081 \text{ mol})(75.351 \text{ J/mol K})(T - 20.5 \text{ °C})$.
Coin: $n = (27.4 \text{ g}) \div (107.9 \text{ g/mol}) = 0.2539$ mol;
$q = (0.2539 \text{ mol})(25.351 \text{ J/mol K})(T - 100.0 \text{ °C})$.

Substitute and solve for T:
$(2.081 \text{ mol})(75.351 \text{ J/mol K})(T - 20.5 \text{ °C}) = -(0.2539 \text{ mol})(25.351 \text{ J/mol K})(T - 100.0 \text{ °C})$.
$156.8\,T - 3214.5 = -6.437\,T + 643.7$;
$163.2\,T = 3858.2$, and $T = 23.6$ °C.

Chapter 12: Chemical Energetics

12.13 Calculate the energy change for a temperature change using Equation 12-1 and C_p values from Table 12.1 of your textbook: $q = nC_p\Delta T$, and $\Delta T = 87.6 - 21.5 = 66.1$ °C = 66.1 K;

$$n_{water} = \frac{m}{MM} = \frac{(475 \text{ mL})(1.00 \text{ g/mL})}{(18.02 \text{ g/mol})} = 26.36 \text{ mol}$$

$q_{water} = (26.36 \text{ mol})(75.291 \text{ J/mol K})(66.1 \text{ K}) = 1.31 \times 10^5$ J = 131 kJ.

12.15 Expansion work can be calculated using Equation 12-3, $w_{sys} = -P_{ext}\Delta V_{sys}$. The external pressure opposing inflation of a balloon is atmospheric pressure, 1.00 atm:
$w_{sys} = -P_{ext}\Delta V_{sys} = -(1.00 \text{ atm})(2.5 \text{ L}) = -2.5$ L atm.

Convert to joules:
$w_{sys} = (-2.5 \text{ L atm})(101.325 \text{ J/L atm}) = -2.5 \times 10^2$ J.

12.17 To work this problem, use data in Box 12-1. First determine the energy difference(ΔE) between chicken and beef:
Chicken: (250 g)(6.0 kJ/g) = 1500 kJ; beef: (250 g)(16 kJ/g) = 4000 kJ; ΔE = 2500 kJ;

According to data in Box 12-1, a 55-kg person walking 6 km/hr consumes 1090 kJ/hr:
t = (2500 kJ) ÷ (1090 kJ/hr) = 2.3 hr. Distance = (2.3 hr)(6 km/hr) = 14 km.

12.19 The heat capacity of a calorimeter can be found from energy and temperature data using Equation 12-5, $q_{calorimeter} = C_{cal} \Delta T$.
Here, $q_{calorimeter} = -q_{glucose} = -(1.7500 \text{ g})(-15.57 \text{ kJ/g}) = 27.25$ kJ
$\Delta T = 23.34 - 21.45 = 1.89$ °C
$C_{cal} = q_{calorimeter} / \Delta T = (27.25 \text{ kJ}) \div (1.89 \text{ °C}) = 14.4$ kJ/°C.

12.21 (a) In a combustion reaction, the products are CO_2 and H_2O:
$$C_7H_6O_2 + O_2 \rightarrow CO_2 + H_2O \text{ (unbalanced)}$$
Give CO_2 a coefficient of 7 to balance C, H_2O a coefficient of 3 to balance H, O_2 a coefficient of 15/2 to balance O, then multiply by 2 to clear fractions:
$$2\ C_7H_6O_2 + 15\ O_2 \rightarrow 14\ CO_2 + 6\ H_2O$$

(b) Find energy per mole from energy per gram using the molar mass (122.1 g/mol):
ΔE (kJ/mol) = [ΔE (kJ/g)][MM (g/mol)] = (−35.61 kJ) (122.1 g/mol) = − 4.349 × 10^3 kJ/mol

(c) 15 mol O_2 is consumed for each 2 mol of benzoic acid, so the energy released per mol of O_2 is (−4.349 × 10^3 kJ/mol)(2/15) = −5.80 × 10^2 kJ/mol.

12.23 Calculate the energy released during combustion from the temperature increase and the total heat capacity of the calorimeter. Then convert to molar energy using molar mass.
$q = C\Delta T$ = (7.85 kJ/K)(302.04 K − 297.65 K) = 34.46 kJ
$\Delta E = -q = -34.5$ kJ
ΔE (kJ/mol) = [ΔE (kJ/g)][MM (g/mol)] = (−34.46 kJ/g) (166.2 g/mol) = −5.73 × 10^3 kJ/mol

12.25 Standard enthalpy changes are calculated from Equation 12-10 using standard enthalpies of formation, which can be found in Appendix E of your textbook:

$$\Delta H^\circ_{rxn} = \sum coeff_p \, \Delta H^\circ_f(\text{products}) - \sum coeff_r \, \Delta H^\circ_f(\text{reactants})$$

(a) $\Delta H^\circ = 2(-393.5) + 2(-285.8) - 1(52.26) - 3(0) = 1411$ kJ;
(b) $\Delta H^\circ = 1(0) + 3(0) - 2(-46.1) = 92.2$ kJ;
(c) $\Delta H^\circ = 1(-2984.0) + 5(0) - 5(-277.4) - 4(0) = -1597$ kJ;
(d) $\Delta H^\circ = 1(-910.94) + 4(-92.307) - 1(-687.0) - 2(-285.830) = -21.5$ kJ.

12.27 Reaction energies are related to reaction enthalpies (see Prob. 12.25) via Equation 12-9:

$$\Delta H_{rxn} \cong \Delta E_{rxn} + \Delta(nRT)_{gases} \quad \text{so} \quad \Delta E_{rxn} \cong \Delta H_{rxn} - \Delta(nRT)_{(gases)}$$

(a) $\Delta n_{gases} = 2 - (3 + 1) = -2$ mol;
$\Delta E_{rxn} \cong 1411$ kJ $- (-2$ mol$)(8.314$ J/mol K$)(298$ K$)(10^{-3}$ kJ/J$) = 1411 + 4.96 = 1416$ kJ;
(b) $\Delta n_{gases} = 3 + 1 - 2 = 2$ mol;
$\Delta E_{rxn} \cong 92.2$ kJ $- (2$ mol$)(8.314$ J/mol K$)(298$ K$)(10^{-3}$ kJ/J$) = 92.2 - 4.96 = 87.2$ kJ;
(c) $\Delta n_{gases} = 0$ mol;
$\Delta E_{rxn} \cong -1597$ kJ;
(d) $\Delta n_{gases} = 4$ mol;
$\Delta E_{rxn} \cong -21.5$ kJ $- (4$ mol$)(8.314$ J/mol K$)(298$ K$)(10^{-3}$ kJ/J$) = -21.5 - 9.9 = -31.4$ kJ.

12.29 A formation reaction has elements in their standard states as reactants and 1 mol of a single product:
(a) $3 K_{(s)} + P_{(s)} + 2 O_{2(g)} \rightarrow K_3PO_{4(s)}$;
(b) $2 C_{(graphite)} + 2 H_{2(g)} + O_{2(g)} \rightarrow CH_3CO_2H_{(l)}$;
(c) $3 C_{(graphite)} + 9/2 H_{2(g)} + 1/2 N_{2(g)} \rightarrow (CH_3)_3N_{(g)}$;
(d) $2 Al_{(s)} + 3/2 O_{2(g)} \rightarrow Al_2O_{3(s)}$.

12.31 Standard enthalpy changes are calculated from Equation 12-10 using standard enthalpies of formation, which can be found in Appendix E of your textbook:

$$\Delta H^\circ_{rxn} = \sum coeff_p \, \Delta H^\circ_f(\text{products}) - \sum coeff_r \, \Delta H^\circ_f(\text{reactants})$$

(a) $\Delta H^\circ_{rxn} = 4(90.25$ kJ/mol$) + 6(-285.8$ kJ/mol$) - 4(-46.1$ kJ/mol$) - 5(0) = -1169.4$ kJ;

(b) $\Delta H^\circ_{rxn} = 2(0) + 6(-285.8$ kJ/mol$) - 4(-46.1$ kJ/mol$) - 3(0) = -1530.4$ kJ.

12.33 Calculate the energy released during the dissolving process from the temperature increase and the total heat capacity of the calorimeter. Then convert to molar energy using molar mass.
Assume that the heat capacity of the calorimeter is the heat capacity of its water contents:

$$C = \frac{(110.0 \text{ g})(75.291 \text{ J / mol K})}{(18.02 \text{ g / mol})} = 459.6 \text{ J/K} = 459.6 \text{ J/°C};$$

$q = C\Delta T = (459.6 \text{ J/°C})(29.7 - 22.0)(°C) = 3.54 \times 10^3$ J;
$\Delta H = -q/m = (-3.54 \times 10^3 \text{ J}) \div (4.75 \text{ g}) = -745$ J/g
ΔH (J/mol) $= [\Delta H$ (J/g)$][MM$ (g/mol)$] = (-745$ J/g$)(111.0$ g/mol$) = -8.27 \times 10^4$ J/mol.

12.35 (a) The positive value for ΔH signals that reaction to form solid $MgSO_4$ is endothermic. Dissolving, the reverse reaction, is exothermic, releasing energy which is absorbed by the water.
(b) Use the molar enthalpy and number of moles dissolving to calculate the energy released:
$n = m/MM = (2.55 \text{ g}) \div (120.371 \text{ g/mol}) = 2.118 \times 10^{-2}$ mol;
$q_{water} = -q_{salt} = -n \Delta H = - (2.118 \times 10^{-2} \text{ mol})(-91.3 \text{ kJ/mol}) = 1.93$ kJ.
(c) Use Equation 12-1, $q = nC_p\Delta T$, to determine the temperature change:
$\Delta T = q/nC_p$; $n = m/MM = (5.00 \times 10^2 \text{ mL})(1.00 \text{ g/mL}) \div (18.02 \text{ g/mol}) = 27.747$ mol;
$\Delta T = \dfrac{(1.93 \text{ kJ})(10^3 \text{ J/kJ})}{(27.747 \text{ mol})(75.291 \text{ J/mol K})} = 0.925$ K

12.37 The formation of a solid from ions in solution involves two opposing energy effects. Energy is released because of the coulombic attraction between cations and anions, but energy is absorbed to overcome the ion-dipole attractions between ions and water molecules. Solid formation is endothermic when the sum of all ion-dipole attractions in solution is greater than the ion-ion interactions in the solid.

12.39 The heats of phase changes indicate the strengths of intermolecular forces; the larger the intermolecular forces, the larger the heat of the phase change.
(a) Methane has a lower heat of vaporization than ethane because, being a smaller molecule, it has smaller dispersion forces.
(b) Ethanol has a significantly higher heat of vaporization than diethyl ether because of strong hydrogen bonding.
(c) Argon (18 electrons) has a higher heat of fusion than methane (10 electrons) because it has a higher polarizability due to its larger number of electrons.

12.41 Because E is a state function, the energy released (ΔE) when one gram of gasoline burns is the same regardless of the conditions. Work (w) is done when an automobile accelerates but no work is done when an automobile idles. Thus $q_{idle} = \Delta E$ and $q_{accel} = \Delta E - w_{accel}$, so more heat must be removed under idling conditions (this is part of the reason why automobiles tend to overheat in traffic jams).

12.43 To work a problem involving heat transfers, it is useful to set up a block diagram illustrating the process. In this problem, a copper block transfers energy to ice:

Cu block, m = 12.7 g, T = 200.0 °C → Cu block, m = 12.7 g, T = 0.0 °C
            ↓ q
         ice → water

Thus, $q_{ice} = -q_{Cu}$, $q_{Cu} = n_{Cu} C_p \Delta T$, and $q_{ice} = n_{ice}\Delta H_{fus}$, so $n_{ice}\Delta H_f = - n_{Cu} C_p \Delta T$
Here are the data: $n_{Cu} = m/MM = (12.7 \text{ g}) \div (63.546 \text{ g/mol}) = 0.1999$ mol;
$C_p = 24.435$ J/mol K, $\Delta T = (0.0 \text{ °C} - 200.0 \text{ °C}) = -200$ °C = -200 K; $\Delta H_f = 6.01$ kJ/mol;
Substitute and solve for the amount of ice that melts:
$n_{ice} = \dfrac{-n_{Cu} C_p \Delta T}{\Delta H_f} = \dfrac{-(0.1999 \text{ mol})(24.435 \text{ J/mol K})(\pm 200 \text{ K})}{(6.01 \text{ kJ/mol})(10^3 \text{ J/kJ})} = 0.1625$ mol
Finally, convert to mass: $m = n\,MM = (0.1625 \text{ mol})(18.02 \text{ g/mol}) = 2.93$ g of ice melts.

Chapter 12: Chemical Energetics

12.45 (a) In a combustion reaction, a substance reacts with molecular oxygen to form CO_2 and H_2O:
$$C_6H_{12}O_6 + O_2 \rightarrow CO_2 + H_2O$$
Give CO_2 a coefficient of 6 and H_2O a coefficient of 6 to balance C and H; then give O_2 a coefficient of 6 to balance O:
$$C_6H_{12}O_6 + 6\,O_2 \rightarrow 6\,CO_2 + 6\,H_2O$$
(b) To determine the molar heat of combustion, multiply the heat released (negative, indicating an exothermic reaction) in burning one gram by the molar mass:
$\Delta H_{molar} = (-15.7 \text{ kJ/g})(180.15 \text{ g/mol}) = -2.83 \times 10^3$ kJ/mol;
(c) Use the molar heat of combustion along with Equation 12-10 and data from Appendix E to determine the heat of formation of glucose:
$$\Delta H^o_{rxn} = \sum \text{coeff}_p\, \Delta H^o_f(\text{products}) - \sum \text{coeff}_r\, \Delta H^o_f(\text{reactants})$$
-2.83×10^3 kJ/mol $= 6(-285.8$ kJ/mol$) + 6(-393.5$ kJ/mol$) - 6(0) - \Delta H^o_f$(glucose)

ΔH^o_f(glucose) $= (2.83 \times 10^3 - 1.71 \times 10^3 - 2.36 \times 10^3)$ kJ/mol $= -1.24 \times 10^3$ kJ/mol

12.47 Living organisms require both matter and energy to carry out their life processes. An isolated system receives neither matter nor energy from its surroundings, so a living organism that is isolated will quickly die.

12.49 (a) The energy required in a heating process is calculated from Equation 12-1: $q = nC_p\Delta T$
$n = m/MM = (155 \text{ m}^3)(10^2 \text{cm/m})^3(1.00 \text{ g/cm}^3) \div (18.02 \text{ g/mol}) = 8.60 \times 10^6$ mol;
$C_p = 75.291$ J/mol K, $\Delta T = (31 - 21) = 10\,°C = 10$ K;
$q = (8.60 \times 10^6 \text{ mol})(75.291 \text{ J/mol K})(10 \text{ K}) = 6.5 \times 10^9$ J
(b) If heating is 80% efficient, the heat required is $(6.5 \times 10^9 \text{ J}) \div (0.80) = 8.1 \times 10^9$ J;
Given that the heat of combustion of methane is -803 kJ/mol, the amount of methane required is $(8.1 \times 10^9 \text{ J})(10^{-3} \text{ kJ/J}) \div (803 \text{ kJ/mol}) = 1.0 \times 10^4$ mol;
Multiply by MM to convert to grams: $(1.0 \times 10^4 \text{ mol})(16.04 \text{ g/mol}) = 1.6 \times 10^5$ g methane.

12.51 In this process, neither p nor V is constant, so we cannot calculate q directly. Instead, we can calculate w and ΔE. An ideal gas has no intermolecular forces, so its energy does not change when it expands or contracts. Thus for $\Delta T = 0$, $\Delta E = 0$.
Calculate w using $w = -P_{ext}\Delta V$: $w = -(4.00 \text{ atm})(20.0 \text{ L} - 30.0 \text{ L}) = 40.0$ L atm;
Convert to joules, recalling that 1 L atm = 101.325 J:
$w = (40.0 \text{ L atm})(101.325 \text{ J/L atm}) = 4.05 \times 10^3$ J
$q = \Delta E - w = 0 - 4.05 \times 10^3$ J $= -4.05 \times 10^3$ J.

12.53 Standard enthalpy changes are calculated from Equation 12-10 using standard enthalpies of formation, which can be found in Appendix E of your textbook:
$$\Delta H^o_{rxn} = \sum \text{coeff}_p\, \Delta H^o_f(\text{products}) - \sum \text{coeff}_r\, \Delta H^o_f(\text{reactants})$$
(a) $\Delta H^o = 2(-454.51) - 2(-296.83) - 1(0) = -315.36$ kJ;
(b) $\Delta H^o = 1(9.16) - 2(33.18) = -57.20$ kJ;
(c) $\Delta H^o = 1(-1675.7) + 2(0) - 1(-824.2) - 2(0) = -851.5$ kJ.

12.55 An enthalpy of formation can be calculated from the heat of a reaction provided all other enthalpies of formation are known. For this reaction, formation enthalpies of three of the four reagents appear in Appendix E of your textbook:

$$\Delta H^o_{rxn} = \sum \text{coeff}_p \, \Delta H^o_f \, (\text{products}) - \sum \text{coeff}_r \, \Delta H^o_f \, (\text{reactants})$$

$-3677 \text{ kJ} = 1(-2984.0) + 3(-296.8) - \Delta H^o_f \, (P_4S_3) - 8(0)$

$\Delta H^o_f \, (P_4S_3) = 3677 - 2984.0 - 890.4 = -197 \text{ kJ}$

12.57 To work a problem involving heat transfers, it is useful to set up a block diagram illustrating the process. In this problem, the unknown metal transfers energy to water:

Thus, $q_{metal} = -q_{water} = (nC_p\Delta T)_{water}$
$n = m/MM = (80.0 \text{ g}) \div (18.02 \text{ g/mol}) = 4.44 \text{ mol}$; $C_p = 75.291$ J/mol °C,
and $\Delta T = (28.4 - 24.8) = 3.6$ °C;
$q_{metal} = -(4.44 \text{ mol})(75.291 \text{ J/mol °C})(3.6 \text{ °C}) = -1203$ J.
Without knowing the identity of the metal, we cannot determine n_{metal}, but heat capacity values are provided in units of J/g K, so we can use $q = mC\Delta T$:
$\Delta T = 28.4 - 100.0 = -71.6$ °C $= -71.6$ K;

$$C = \frac{q}{m\Delta T} = \frac{(-1203 \text{ J})}{(44.0 \text{ g})(-71.6 \text{ K})} = 0.382 \text{ J/g K}$$

This matches the heat capacity of Cu, so the metal is copper.

12.59 The process shown is condensation of a gas to form a solid, the reverse of sublimation. Sublimation always is endothermic, as energy must be provided to overcome intermolecular forces of attraction; thus the depicted process is exothermic.
(a) ΔH_{sys} is negative (exothermic process); (b) ΔE_{surr} is positive (surroundings must absorb the energy released in the process); (c) $\Delta E_{universe} = 0$ (total energy always is conserved).

12.61 The enthalpy of a reaction can be calculated from Equation 12-10 and standard heats of formation (see Appendix E):

$$\Delta H^o_{rxn} = \sum \text{coeff}_p \, \Delta H^o_f \, (\text{products}) - \sum \text{coeff}_r \, \Delta H^o_f \, (\text{reactants})$$

For this "reaction," the calculation is simple:

$\Delta H^o_{rxn} = 1(-285.8) + 1(-241.8) = 44.0$ kJ

Compare this value with $\Delta H_{vap} = 40.79$ kJ/mol. The values differ because ΔH_{vap} refers to vaporization of liquid water at the normal boiling point, 373 K, while ΔH^o_{rxn} refers to vaporization of liquid water under standard thermodynamic conditions, 298 K.

12.63 The enthalpy of a reaction can be calculated from Equation 12-10 and standard heats of formation (see Appendix E):

$$\Delta H^\circ_{rxn} = \sum coeff_p \, \Delta H^\circ_f(products) - \sum coeff_r \, \Delta H^\circ_f(reactants)$$

(a) The question asks for ΔH when 1 mol of hydrazine burns, so the appropriate reaction is:

$$N_2H_4 + 1/2 \, N_2O_4 \rightarrow 3/2 \, N_2 + 2 \, H_2O_{(g)}$$

$\Delta H^\circ_{rxn} = 3/2(0) + 2(-241.8) - 1(50.6) - 1/2(9.16) = -538.8$ kJ/mol

(b) When hydrazine burns in oxygen, the reaction is:

$$N_2H_4 + O_2 \rightarrow N_2 + 2 \, H_2O_{(g)}$$

$\Delta H^\circ_{rxn} = 1(0) + 2(-241.8) - 1(50.6) - 1(0) = -534.2$ kJ/mol

The reaction with dinitrogen tetroxide is slightly more exothermic because that compound is slightly less stable than the elements from which it forms. Dinitrogen tetroxide is used in preference to molecular oxygen because the reaction occurs on contact, while oxygen would require an ignition device.

12.65 The difference between ΔH_{rxn} and ΔE_{rxn} can be estimated using Equation 12-9;

$$\Delta H_{rxn} \cong \Delta E_{rxn} + \Delta(nRT)_{gases} \quad so \quad \Delta H_{rxn} - \Delta E_{rxn} \cong \Delta(nRT)_{(gases)}$$

(a) $\Delta n_{gases} = 0$, so $\Delta H_{rxn} - \Delta E_{rxn} \cong 0$;
(b) $\Delta n_{gases} = 1 + 1 - 0 = 2$; $\Delta H_{rxn} - \Delta E_{rxn} \cong$ (2 mol)(8.314 J/mol K)(298 K) = 4.96×10^3 J;
(c) The reaction is $C_4H_9OH_{(l)} + 6 \, O_{2(g)} \rightarrow 4 \, CO_{2(g)} + 5 \, H_2O_{(l)}$; $\Delta n_{gases} = 4 - 6 = -2$;
$\Delta H_{rxn} - \Delta E_{rxn} \cong$ $-(2 \text{ mol})(8.314 \text{ J/mol K})(298 \text{ K}) = -4.96 \times 10^3$ J.

12.67 The standard elemental state of carbon, graphite, is the most stable state of the element under standard conditions. Therefore, the heat of formation of buckminsterfullerene must be positive: $60 \, C_{(graphite)} \rightarrow C_{60}$, $\Delta H^\circ > 0$.

The combustion reactions of graphite and buckminsterfullerene are:
$C_{(graphite)} + O_{2(g)} \rightarrow CO_{2(g)}$ and $C_{60} + 60 \, O_{2(g)} \rightarrow 60 \, CO_{2(g)}$;
For graphite, $\Delta H_{combustion} = 1(-393.5 \text{ kJ/mol}) = -393.5$ kJ/mol;
For buckminsterfullerene, $\Delta H_{combustion} = 60(-393.5 \text{ kJ/mol}) - \Delta H^\circ$. This is much more negative than $1(-393.5 \text{ kJ/mol})$.
On a per gram basis, the values are:

Graphite: $\dfrac{(-393.5 \text{ kJ/mol})}{(12.01 \text{ g/mol})} = -132.76$ kJ/g;

Buckminsterfullerene:

$$\dfrac{(60)(-393.5 \text{ kJ/mol}) \pm \Delta H^\circ}{(60)(12.01 \text{ g/mol})} = -132.76 \text{ kJ/g} \pm \dfrac{\Delta H^\circ}{(60)(12.01 \text{ g/mol})};$$

Because ΔH° is positive, the value for buckminsterfullerene is slightly more negative than for graphite.

12.69 (a) This is a cooling process, for which $q = n\, C_p\, \Delta T$:
$n = m/MM = (2.50\text{ g}) \div (18.02\text{ g/mol}) = 0.1387$ mol;
$\Delta T = (100.0 - 37.5)\,°C = 62.5\,°C = 62.5$ K;
$q = (0.1387\text{ mol})(75.291\text{ J/mol K})(62.5\text{ K}) = 653$ J.
(b) This is a phase change, for which $q = n\, \Delta H_{phase}$, followed by the same cooling process as in part (a):
$q = (0.1387\text{ mol})(40.79\text{ kJ/mol})(10^3\text{ J/kJ}) + 653\text{ J} = 5658\text{ J} + 653\text{ J} = 6.31 \times 10^3$ J.
Steam at 100.0 °C releases nearly ten times as much heat as boiling water.

12.71 (a) Total energy is conserved in any process;
(b) The energy of the universe is a constant;
(c) $\Delta E = q + w$.

12.73 Data in Problem 12.45 indicate that glucose combustion releases 15.7 kJ/g of energy. First determine how much glucose would be required at 100% efficiency, then correct for inefficiency and sugar content:

Mass required = $\dfrac{(220\text{ kJ})}{(15.7\text{ kJ / g glucose})} \dfrac{(100\%)}{(30\%)} \dfrac{(100\text{ g cereal})}{(35\text{ g glucose})}$ = 133 g cereal

12.75 The cooling process takes place at constant pressure, so ΔH can be calculated using $\Delta H = nC_p\Delta T$. To determine ΔE, use $H = E + PV$, from which $\Delta E = \Delta H - \Delta(PV)$:
$n = m/MM = (1.25\text{ g}) \div (132.93\text{ g/mol}) = 9.40 \times 10^{-3}$ mol; $\Delta T = (20.0 - 50.0) = -30$ K;
$\Delta H = (9.40 \times 10^{-3}\text{ mol})(80.7\text{ J/mol K})(-30\text{ K}) = -22.8$ J;
$\Delta(PV) = P\,\Delta V = (1.00\text{ atm})(248 - 274\text{ mL})(10^{-3}\text{ L/mL})(101.325\text{ J/L atm}) = -2.63$ J;
$\Delta E = -22.8\text{ J} - (-2.63\text{ J}) = -20.2$ J.

12.77 To work a problem involving heat transfers, it is useful to set up a block diagram illustrating the process. In this problem, a silver spoon transfers energy to coffee (water):

Thus, $q_{water} = -q_{Ag}$, $q_{Ag} = (nC_p\Delta T)_{Ag}$, and $q_{water} = (nC_p\Delta T)_{water}$
For Ag, $n = m/MM = (99\text{ g}) \div (107.9\text{ g/mol}) = 0.918$ mol; $C_p = 25.351$ J/mol K, and $\Delta T = (x - 280)$ K;
For water, $n = m/MM = (205\text{ mL})(1.00\text{ g/mL}) \div (18.02\text{ g/mol}) = 11.38$ mol;
$C_p = 75.291$ J/mol K, and $\Delta T = (x - 350)$ K;
Substitute and solve for x:
$(0.918\text{ mol})(25.351\text{ J/mol K})(x - 280)\text{ K} = -(11.38\text{ mol})(75.291\text{ J/mol K})(x - 350)$ K;
$23.26 x - 6513 = -856.8 x + 299880$, $\qquad$ $880.1 x = 306390$ $\qquad$ $x = 348$ K.
To determine whether an Al spoon would be more or less effective, compare nC_p for the two spoons. Al: $= (99\text{ g})(24.35\text{ J/mol K}) \div (26.982\text{ g/mol}) = 89.34$ J/K;
$\qquad$ Ag: $= (99\text{ g})(25.351\text{ J/mol K}) \div (107.9\text{ g/mol}) = 23.26$ J/K;
An Al spoon requires more energy per degree temperature increase, so an Al spoon would be more effective in cooling the coffee.

Chapter 12: Chemical Energetics

12.79 First compute the quantity of gasoline consumed in traveling 1 km, then determine the energy content of that quantity of gasoline:

$$\text{Mass of gasoline} = \frac{(1\,\text{km})}{(6.0\,\text{km/L})}(0.68\,\text{g/mL})(10^3\,\text{mL/L}) = 113\,\text{g}$$

Energy consumed = $(113\,\text{g})(48\,\text{kJ/g}) = 5.4 \times 10^3$ kJ.

12.81 The conversion of water at 25.0 °C into steam at 100.0 °C takes place in two steps, heating to 100.0 °C and vaporization at that temperature. The total heat required is the sum of the heats for the two steps: $q = nC_p\Delta T + n\Delta H_{vap}$:

$n = m/MM = (2.50\,\text{kg})(10^3\,\text{g/kg}) \div (18.02\,\text{g/mol}) = 138.7$ mol;
$q = (138.7\,\text{mol})[(75.291\,\text{J/mol K})(75\,\text{K})(10^{-3}\,\text{kJ/J}) + 40.79\,\text{kJ/mol}] = 6.44 \times 10^3$ kJ;
The heat of combustion of methane could be calculated from Equation 12-10 using data from Appendix E; the value is given in Problem 12.49: $\Delta H_{combustion} = -803$ kJ/mol.

$$\text{Mass required} = \frac{(6.44 \times 10^3\,\text{kJ})}{(803\,\text{kJ/mol})}(16.043\,\text{g/mol}) = 129\,\text{g}.$$

12.83 A molar heat capacity can be calculated from a temperature change using $q = nC_p\Delta T$:
$\Delta T = 299.6\,\text{K} - 280.0\,\text{K} = 19.6\,\text{K}$; $n = (52.5\,\text{g}) \div (207.2\,\text{g/mol}) = 0.2534$ mol;

$$C_p = \frac{q}{n\Delta T} = \frac{(100.0\,\text{J})}{(0.2534\,\text{mol})(19.6\,\text{K})} = 20.1\,\text{J/mol K}.$$

12.85 The process shows an expansion at constant temperature. As a gas expands, it does work on its surroundings, but if temperature remains constant, the energy of the gas does not change:
(a) w_{sys} is negative (system does work on surroundings);
(b) $\Delta E_{surr} = 0$ because $\Delta E_{sys} = 0$ and total energy is conserved;
(c) q_{sys} is positive because w_{sys} is negative and $\Delta E_{sys} = 0$.

12.87 First calculate the heat required to warm the bear. Then calculate the heat of combustion of arachadonic acid. Finally, determine the amount of arachadonic acid required to do the job:
$q = mC_p\Delta T = (500\,\text{kg})(10^3\,\text{g/kg})(4.186\,\text{J/g °C})(25\,°\text{C} - 5\,°\text{C}) = 4.186 \times 10^7$ J;
$\Delta H_{comb} = 20(-393.5\,\text{kJ/mol}) + 16(-285.8\,\text{kJ/mol}) - 1(-636\,\text{kJ/mol}) - 27(0) = -1.18 \times 10^4$ kJ

$$\text{Mass required} = \frac{(4.186 \times 10^7\,\text{J})(304.45\,\text{g/mol})}{(1.18 \times 10^4\,\text{kJ/mol})(10^3\,\text{J/kJ})} = 1.08 \times 10^3\,\text{g}.$$

Chapter 13 Spontaneity of Chemical Processes

13.1 Spontaneity

QUESTIONS TO ANSWER, SKILLS TO LEARN
1. **What is "spontaneity"?**
2. **Qualitative estimation of relative disorder at the molecular level**
3. **What is the second law of thermodynamics?**

A process is called spontaneous when it can proceed under the existing conditions without any further input from the surroundings. The process may be chemical (a reaction) or physical (a phase change, for example) and has a preferred direction. In addition, once the process has apparently stopped, it will not proceed back to the initial state (concentrations of reactants, etc.) if the conditions (temperature, pressure, etc.) remain the same. For example, the figure below shows two tanks, one of which is empty, that are connected by a valve and tube. When the valve is opened, the liquid flows from one tank to the other until the two levels are the same. This movement is the spontaneous process and corresponds to the direction of the arrow. After the two levels are equal, the liquid will not flow from one tank to the other (resulting in unequal levels) unless work is put into the system to pump the liquid from one tank to the other. Therefore the process moving from right to left is *not* spontaneous.

Chemical processes have definite directions of spontaneity as seen in Exercise 13.1.

Exercise 13.1 Identify each of the following processes as spontaneous or nonspontaneous.
 (a) Salt dissolving in water.
 (b) Sour milk becoming sweet and fresh.
 (c) Water evaporating from sea water to give solid salt.
 (d) A cut tree decomposing.

Steps to Solution: We must rely on experience and sometimes help from outside sources to determine the spontaneity of processes. In the next section some further criteria will be supplied to determine the spontaneity of reactions.
(a) *This is a spontaneous process as it proceeds on its own.*

(b) *This is a nonspontaneous process; in fact, it is the reverse of a process known to be spontaneous.*
(c) *Even though this is the reverse of the process described in (a), it also is spontaneous, as one can confirm by experience. Many processes can proceed in both directions with the input of sufficient energy.*
(d) *This is a spontaneous process.*

Energy changes: Do they determine spontaneity?

Chemical and physical processes nearly always result in an energy change. The energy change alone (exothermic vs. endothermic) is not sufficient to predict whether a process will be spontaneous or not. Consider the formation of solutions of $CaCl_2$ and NH_4NO_3. Dissolving calcium chloride is exothermic and dissolving ammonium nitrate is endothermic, yet both are spontaneous processes. Therefore the enthalpy (or energy change) is not the sole predictor of spontaneity.

Molecular Disorder

Much of the following discussion will be concerned with the notion of order and disorder in molecular systems. Although order and disorder are opposites, there are degrees of disorder. Consider a system that consists of a checkerboard and four checkers.

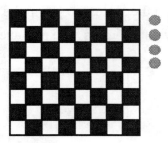

There are many ways to place the four checkers on the board when only one checker is allowed per square. Consider the following three arrangements:

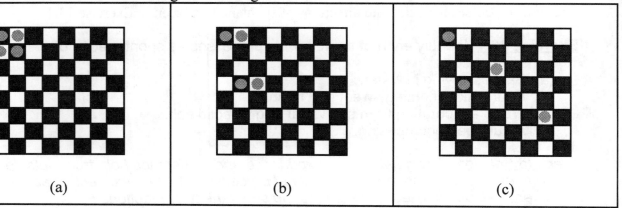

(a)　　　　　　　　　　(b)　　　　　　　　　　(c)

The arrangement in (a) has the highest degree of order of the three. What do we mean by the term *order*? Notice that the four occupied squares all share at least a corner if not an edge. Also, the four occupied squares form a corner of the checkerboard. There are relatively few arrangements (four: one at each corner) that have these characteristics. The arrangement in (b) is somewhat less ordered than that in (a); notice that the checkers are in pairs and are separated by several spaces. There are more arrangements of checkers that are disordered like (b) than ordered like in (a), where we find two pairs of checkers. Arrangement (c) has even less order; the pieces are spread out over much of the board and there is no discernible pattern to their placement. There are many ways that one may arrange the four pieces so they do not touch each other. Of the three arrangements, that shown in (c) would be considered the *least ordered* (or sometimes we say the *most disordered* or the *most random*).

These figures could be considered analogous to solids (a), liquids (b) and gases (c). In (a), there is a regular pattern filling the squares, much like the regular packing seen in crystalline solids. In (b) there is a less regular pattern, although there are some adjacent spaces filled, much as there is contact between the particles (atoms, ions or molecules) present in a liquid. Finally, in (c), the markers are pretty much randomly scattered around the board, just as gas molecules tend to occupy random positions in a gas.

Consider the idea of order in a molecular system. Two identical vessels linked by a valve contain neon and helium, respectively, at the same pressure and temperature. When the valve

He   Ne   mixture of Ne and He

is opened, the two gases will mix until the concentrations of He and Ne are the same in each vessel. The energy change in this process is quite small since the noble gases He and Ne have very weak (and similar) intermolecular forces. In addition, the two gases will never separate spontaneously; that is, the vessels will always contain a mixture unless we undertake to separate the two components. Why is this process spontaneous? The answer lies in the fact that the mixture of gases is more random or disordered than the two pure gases. This example and other experience show:

Processes proceed toward more disordered states.

The formation of a crystal of sodium acetate trihydrate is an example of an exothermic

Chapter 13: Spontaneity of Chemical Processes

$$Na^+_{(aq)} + CH_3CO_2^-_{(aq)} \rightarrow Na(CH_3CO_2)\cdot 3\,H_2O_{(s)} \quad \Delta H < 0$$

process that proceeds with formation of a ordered crystalline structure. This event appears to violate the above statement of disorder being favored. However, both the system *and* the surroundings must be considered to use the above principle. If we consider the system to consist of the sodium acetate crystal, then the crystal shows an increase in order. However, energy is lost as heat to the surroundings (aqueous solution, the air around the container, etc.). This thermal energy increases the motion present in the surroundings by increasing the average kinetic energy, leading to changes in the surroundings. For example, a heated gas or liquid occupies a greater volume; therefore the same number of molecules occupy a greater volume, which is more disordered (using the checkerboard analogy, one has a larger board on which to put the pieces). Therefore the surroundings undergo an increase in their disorder. The total change in disorder (that in the system *and* the surroundings) in any spontaneous process will always be positive (disorder will increase).

Second Law of Thermodynamics: Total disorder increases in all spontaneous processes.

In the next section, disorder will be quantified.

Exercise 13.2 Liquids composed of different molecules have different degrees of order. Compare the liquid phases of H_2S and H_2O and suggest which is more "ordered." (Hint: what intermolecular forces are different in the two liquids?)

Steps to Solution: The degree of order will depend on how the molecules are oriented with respect to each other; order implies some preferred orientation of the molecules in the liquids. Use the hint and compare the types of intermolecular forces. In H_2S, the predominant intermolecular force will be dipole-dipole attractions because H-bonding is not strong in H-S bonds. In water, the predominant intermolecular force will be H-bonding. The energy of stabilization of H-bonding will be greater than the dipole forces in H_2S. Also, H-bonding leads to more specific orientations of the molecules for H-bonding to occur.

Therefore we expect that the H-bonded liquid to be more ordered than the liquid that has no H-bonding interactions.

Chapter 13: Spontaneity of Chemical Processes

13.2 Entropy: The Measure of Disorder

QUESTIONS TO ANSWER, SKILLS TO LEARN
1. **What are the two ways of calculating entropy?**
2. **How can entropy changes be used to predict the spontaneity of a process?**

The quantitative measure of disorder is *entropy*, which is symbolized by the letter S. The second law of thermodynamics can be restated in terms of the entropy change:

In every spontaneous process, the *total* entropy of the *universe* increases.

There are two definitions of entropy, one based on statistics and the other on heat transfer. The first definition is given by the Boltzmann equation:

$$S = k \ln W$$

The constant k is Boltzmann's constant ($k = 1.3807 \times 10^{23}$ J/K), and W is the number of ways the microscopic parts of the system can be arranged such that the macroscopic properties of the system are the same. For example, in the checkerboard system described above, there are relatively few ways of placing the four pieces next to each other [situations like (a)] compared to the number of ways they may be arranged in separated spaces [like (c)]. Therefore W is smaller for arrangements like (a) than for those like (c). In the checkerboard system, there are four pieces and 64 positions, which leads to a relatively small number of possibilities compared to real molecular systems with 10^{20} or more molecules and many possible positions. This definition will be little used here because W can be very difficult to determine.

The other definition that will be used calculates the entropy change, ΔS, corresponding to the heat, q_T, transferred at constant temperature, T:

$$\Delta S = \frac{q_T}{T} \qquad \text{units}: \frac{\text{energy}}{\text{temperature}} = \frac{\text{J}}{\text{K}}$$

COMMON PITFALL: Using wrong temperatures
Using temperatures in °C instead of K when calculating ΔS results in large errors.

This equation may be used when heat is transferred at constant or nearly constant temperature. Situations where heat is transferred at nearly constant temperature are:

- Phase changes such as fusion, vaporization, etc.

Chapter 13: Spontaneity of Chemical Processes

- Chemical processes occurring in situations where the temperature is regulated by removing heat or adding heat. An example is the digestion of food.

- Systems where the surroundings are sufficiently large so that any heat transfer between them will result in an insignificant temperature change, such as adding an ice cube to a swimming pool.

Exercise 13.3 Calculate the entropy changes for the fusion of 1 mole of hydrogen sulfide (mp = -85.6 °C, ΔH_{fus} = 2.39 kJ/mole) and for the fusion (melting) of 1 mole of water.

Steps to Solution: Fusion is a phase change that occurs at constant temperature; therefore we may use the following equation to calculate the entropy change:

$$\Delta S = \frac{q_T}{T} \text{ (units are J/K)}$$

For a phase change at constant pressure, the heat is given by the enthalpy change, which must be converted to units of J. The temperature is the normal melting point, which must be converted to Kelvin.

For H_2S, the entropy change for fusion under standard conditions is:

$$\Delta S = \frac{2.39 \times 10^3 \text{ J}}{(-85.6 + 273.15) \text{ K}} = 12.8 \text{ J/K}$$

For water the entropy change for fusion under standard conditions is:

$$\Delta S = \frac{6.01 \times 10^3 \text{ J}}{(0 + 273.15) \text{ K}} = 22.0 \text{ J/K}$$

Notice that the entropy change for the fusion of water is greater than that for H_2S.

In the exercise above, the entropy change we solved for was the entropy change of a system composed of one mole of substance. Determining the entropy change of the universe requires calculating the entropy change of the surroundings as well. The entropy change of the surroundings depends upon the temperature of the surroundings and the amount of heat transferred to the surroundings. In the next exercise the entropy change of the universe will be calculated for water melting at three different temperatures.

Exercise 13.4 Calculate the entropy change of the universe for 1 mole of solid water at 0 °C melting at -5 °C, 0 °C and +5 °C.

Steps to Solution: The entropy change of the universe, ΔS_{univ}, is the sum of the entropy change of the system, ΔS_{sys}, and the entropy change of the surroundings, ΔS_{surr}:

$$\Delta S_{univ} = \Delta S_{sys} + \Delta S_{surr}$$

Chapter 13: Spontaneity of Chemical Processes

ΔS_{sys} *(the entropy change for solid water converting to liquid water) was calculated in Exercise 13.2 and is 22.0 J/K. The entropy change in the surroundings is calculated using the same general formula:*

$$\Delta S = \frac{q_T}{T}$$

The heat in this case is the heat exchanged with the surroundings, which are at the supplied temperatures. The critical part of this step is determining the sign of q. A simple rule to remember is:

The sign of q_{surr} is the opposite of q_{sys}

If heat is released from the system ($q_{sys} < 0$), then heat must be absorbed by the surroundings, increasing the energy of the surroundings ($q_{surr} > 0$). In this problem, heat must be transferred to the system (the ice) to melt it, so $q_{sys} > 0$. The surroundings must lose heat; therefore $q_{surr} < 0$.

The amount of heat that must be supplied by the surroundings is the heat of fusion of the solid water; because the heat is being transferred to the system, $q < 0$.

$$q_{surr} = -n\Delta H_{fus} = -1 \text{ mole } (6.01 \text{ kJ/mole}) = -6.01 \text{ kJ}$$

The amount of water being melted is the same in all three cases, so q_{surr} will be the same in all three cases, but the temperature will be different.

$T = 5\ °C$:
$$\Delta S_{univ} = \Delta S_{sys} + \Delta S_{surr}$$

$$\Delta S_{surr} = \frac{q_{surr}}{T} = \frac{-6.01 \times 10^3 \text{ J}}{(5 + 273.15) \text{ K}} = -21.6 \text{ J/K}$$

$$\Delta S_{univ} = 22.0 + (-21.6) \text{ J/K} = \boxed{0.4 \text{ J/K}} \qquad \Delta S_{univ} > 0$$

$T = 0\ °C$:
$$\Delta S_{univ} = \Delta S_{sys} + \Delta S_{surr}$$

$$\Delta S_{surr} = \frac{q_{surr}}{T} = \frac{-6.01 \times 10^3 \text{ J}}{(0 + 273.15) \text{ K}} = -22.0 \text{ J/K}$$

$$\Delta S_{univ} = 22.0 + (-22.0) \text{ J/K} = \boxed{0.0 \text{ J/K}}$$

$T = -5\ °C$: Use the above method to show that $\boxed{\Delta S_{univ} = -0.4 \text{ J/K}} \qquad \Delta S_{univ} < 0$

Notice that only when the temperature is greater than the melting point of ice is the entropy change of the universe is greater than zero, indicating that the melting of ice is spontaneous at those conditions of pressure and temperature. When the temperature is less than the melting point of ice, the entropy change of the universe is less than zero, indicating that melting ice at -5 °C is not spontaneous, a fact known from common experience. It is known that liquid water will freeze (solidify) at -5 °C, the reverse of the process for which the entropy change was calculated.

If the entropy change of the universe is less than zero ($\Delta S_{univ} < 0$), the process described is not spontaneous, but the reverse process is spontaneous.

At 0 ° C , the entropy change of the universe is zero, suggesting that melting ice is neither spontaneous nor nonspontaneous at 0 °C. Therefore at 0 °C ice and water will coexist in equilibrium.

If the entropy change of the universe is zero ($\Delta S_{univ} = 0$), the process described is at equilibrium.

The processes may be physical processes such as the phase change described above or may also be chemical changes.

13.3 Entropies of Substances

QUESTIONS TO ANSWER, SKILLS TO LEARN
1. **How much disorder is there in substances (what is their standard entropy)?**
2. **What is the effect of temperature on entropy?**
3. **How is entropy affected by changes in molecular complexity and concentration?**
4. **Calculating entropy changes using standard entropies**

In the discussion of randomness using a checkerboard system, we made an analogy between certain arrangements of the pieces and phases of matter. The most random and separated arrangements of the checkers were analogous to gases, and the most ordered arrangements were analogous to crystalline solids. In molecular systems, molecules have different degrees of order depending on the phase; in addition, the thermal motion of the particles of the substance (molecules, etc.) changes with temperature. The more motion the particles exhibit (kinetic energy), the greater the disorder in the system. The entropy of a particular substance depends on:

- **Phase** (solid, liquid or gas)
- **Temperature** (entropy increases with increasing temperature)

Chapter 13: Spontaneity of Chemical Processes

To determine the entropy present in a substance at a given temperature, the entropy present at absolute zero must be known. The third law of thermodynamics defines the entropy of a pure, perfect crystal (one with no missing or misplaced particles) at absolute zero equal to zero.

Third Law of Thermodynamics: S (pure, perfect crystal at $T = 0$ K) $= 0$

The entropy of substances is usually tabulated at standard conditions of temperature and pressure and given the symbol $S°$, where the $°$ superscript indicates the standard conditions. Unlike standard enthalpies of formation, these values are not determined relative to other compounds (elements in their standard states). They are absolute values; no standard entropy has a value of zero. The relative entropies of substances may be predicted qualitatively by use of the following guidelines, two of which were referred to above.

Guidelines for prediction of relative entropies of substances.

- The entropy of a substance depends on its phase. $S°_{(g)} > S°_{(l)} > S°_{(s)}$
- The entropy of substances of similar structure increases with increasing molar mass.
- The entropy of substances increases with increasing molecular size.
- The entropy of a substance increases with increasing temperature.

Exercise 13.5 Predict which of the following pairs of substances (or groups of substances) will have the greater standard entropy and briefly list the reason(s).
  (a) One mole of CH_4 $_{(g)}$ or one mole of $C_{(graphite)}$ and two moles of H_2 $_{(g)}$
  (b) One mole of $TiCl_4$ $_{(l)}$ or one mole of $SiCl_4$ $_{(l)}$
  (c) One mole of NH_3 $_{(g)}$ or one mole of N_2H_4 $_{(g)}$
  (d) One mole of Au $_{(s)}$ at 195 K or one mole of Au $_{(s)}$ at 298K
  (e) One mole of H_2O $_{(l)}$ or one mole of H_2O $_{(g)}$

Steps to Solution: The difference in entropies between the choices should be based on the guidelines supplied above. We can also consult tables and compare the numerical values of the standard entropies.
(a) The comparison is between the same number of atoms in different chemical forms. Methane is a larger molecule but is only one mole of gas. The other choice, two moles of hydrogen and one mole of graphite, contain two moles of gas and one mole of a crystalline solid. Despite the low entropy expected for a crystalline solid, the second choice contains more moles of gas, and it is expected that one mole of C and two moles of H_2 will have a higher entropy than one mole of methane. The numerical values obtained using standard entropies are given below and support this reasoning.

Substance	S° (J/K • mole)	Moles Substance	Entropy (J/K)
$CH_{4(g)}$	186.16	1	186.16
total			**186.16**
$C_{(graphite)}$	5.740	1	5.740
$2\ H_{2(g)}$	130.574	2	261.15
total			**266.89**

(b) The difference between these two substances is their molar mass; they have similar formulas and structures and are in the same phase. Titanium tetrachloride has a higher molar mass than silicon tetrachloride and therefore is expected to have the greater standard entropy of the pair. The numerical values obtained using standard entropies are given below and support this reasoning.

$$S° (TiCl_{4(l)}) = 252.34\ J/K \cdot mole\ >\ S° (SiCl_{4(l)}) = 239.74\ J/K \cdot mole$$

(c) In this pair of substances the compounds are in the same phase, have the same number of moles of gas, and are composed of the same elements. However, the molecular formulas are different, and the compound with the greater molar mass, N_2H_4, is expected to have the higher entropy. The experimental values for the entropies listed below are in accord with this reasoning.

$$S° (N_2H_{4\ (g)}) = 238.3\ J/K \cdot mole\ >\ S° (NH_{3\ (g)}) = 192.5\ J/K \cdot mole$$

(d) In this case the substance and its phase are the same in both choices, However, the temperature is different. Because increasing temperature corresponds to increasing motion at the molecular/ atomic scale, higher temperature will result in greater disorder in the crystal. Therefore we expect gold at higher temperature to have the greater entropy, and this expectation is borne out by the experimental data given below.

$$S° (Au_{(s)}\ (298\ K)) = 47.40\ J/K \cdot mole\ >\ S° (Au_{(s)}\ (195\ K)) = 36.8\ J/K \cdot mole$$

(e) In this case we are comparing the same substance at the same temperature, but in different phases. The less ordered phase is expected to have the greater entropy. The gas phase is less ordered than the liquid phase and therefore is expected to have the greater entropy. This expectation is borne out by the experimental data given below.

$$S° (H_2O_{\ (g)}) = 188.72\ J/K \cdot mole\ >\ S° (H_2O_{\ (l)}) = 69.94\ J/K \cdot mole$$

Chapter 13: Spontaneity of Chemical Processes

Entropy and Concentration

Because the less dense and more molecularly dispersed phases have higher entropies, entropy depends on concentration or partial pressure, depending on the phase of the substance. The equation for calculating the entropy at nonstandard concentrations or pressures, S, is:

$$S = S° - R \cdot \ln c$$

($S°$ = standard entropy and c = concentration in M or partial pressure in atm)

Note that for concentrations and molarities less than one (more dilute than standard conditions), $R \ln c$ is greater than zero, making $S > S°$. Conversely, for $c > 1$ (more concentrated than standard conditions), $R \ln c$ is less than zero, making the entropy less than the standard entropy. This is in agreement with the qualitative notion that more disperse (dilute) systems are more disordered or random (have higher entropy) than concentrated solutions.

Exercise 13.6 In adjusting the acidity of swimming pool water, concentrated aqueous HCl is added. What is the entropy change for the chloride ions when 0.5 L of 12.0 M HCl is added to a swimming pool of 9.6×10^4 L (about 25,000 gallons)? ($S°\ Cl^-_{(aq)}$ = 56.5 J/mole • K)

Steps to Solution: The above equation can be used to calculate the molar entropies of chloride at the different concentrations. The concentration is supplied in this problem; we must calculate the final concentration. Then the molar entropies must be converted to entropies by multiplying by the number of moles of Cl^- involved. Finally, the entropy change is calculated by subtracting the entropy of the initial state (6 M) from the entropy of the final state (Cl^- diluted in the pool).

Before addition to the pool, the entropy of the chloride ions is $S_{initial}$:

$S_{initial} = S_{6M} = S° - R \ln c = $ 56.5 J/mole•K $- $ 8.314 J/mole • K(ln(12.0)) = 36. J/mole • K

After addition, the total volume is 9.6×10^4 L (the 0.5 L added is not significant in calculating the total volume). The number of moles of Cl^- added are:

mole Cl^- = 0.5 L (12.0 mole/L) = 6 mole

The final concentration of the Cl^- is:

$$M_{Cl^-} = \frac{6.\ \text{mole}}{9.6 \times 10^4 L} = 6 \times 10^{-5}\ M$$

The molar entropy of Cl^- at this concentration is calculated using the above equation:

$S_{final} = S° - R \ln c = $ 56.5 J/mole • K $-$ 8.314 J/mole • K(ln (6×10^{-5})) = 81. J/mole • K

Chapter 13: Spontaneity of Chemical Processes

The entropy change is now determined. Because six moles of Cl⁻ undergo the process, the molar entropies calculated below must be multiplied by the number of moles of Cl⁻ that undergo dilution:

$$\Delta S = S_{final} - S_{initial} = 6 \text{ mole}(81 \text{ J/mole} \cdot K) - 6 \text{ mole}(35 \text{ J/mole} \cdot K) = \boxed{\Delta S = 271 \text{ J/K}}$$

This answer makes sense because one expects the entropy of the system (chloride ions in this case) to increase when concentration decreases.

Calculating Standard Reaction Entropy Changes

Entropy changes in chemical reactions are calculated by much the same method as the enthalpy change because entropy is a state function:

$$\Delta S = \Sigma \text{ entropies of products} - \Sigma \text{ entropies of reactants}$$

You must remember to use the appropriate stoichiometric coefficients to insure that the final entropy change has units of J/K *not* units of J/mole • K.

Exercise 13.7 What is the entropy change for the reaction of graphite and hydrogen to give methane?

Steps to Solution: As in all chemical processes, the first step is to write a balanced chemical equation of reaction. Then we must find the standard entropies of the substances involved from tables in the text. We then calculate the entropy change by finding the difference of the entropies of the products and the reactants.

The balanced chemical reaction is:

$$C_{(graphite)} + 2 H_{2(g)} \rightarrow CH_{4(g)}$$

The standard entropies for the products and reactants can be found in Exercise 13.4(a). Note the stoichiometric coefficient of 2 for the entropy of hydrogen.

$$\Delta S = S°(CH_{4(g)}) - \left(S°(C_{(graphite)}) + 2S°(H_{2(g)})\right)$$

$$\Delta S = 1 \text{ mole CH}_4 \cdot 186.16 \text{ J/mole} \cdot K - (1 \text{ mole C}_{(graphite)} \cdot 5/740 \text{ J/mole} \cdot K + 2 \text{ mole H}_2(g) \cdot 130.574 \text{ J/mole} \cdot K)$$

$$\boxed{\Delta S = -80.73 \text{ J/K}}$$

The entropy change for this process is negative, showing that the entropy of the system decreases as reactants are converted to products.

13.4 Spontaneity and Free Energy

QUESTIONS TO ANSWER, SKILLS TO LEARN
1. **How can spontaneity be predicted?**
2. **What is "free energy"?**
3. **Calculating free energy changes under standard and nonstandard conditions**
4. **Predicting conditions that will be conducive to spontaneous reactions knowing only the signs of enthalpy and entropy changes**

Predicting spontaneous processes or the conditions under which a process will be spontaneous is very important in the manufacture of chemicals and in the understanding of chemical processes in biological systems. The criterion for a spontaneous process is that the entropy of the universe must increase. However, a more useful criterion focuses on the system, and the most commonly used criterion is a defined state function, Gibb's Free Energy, G, (or free energy) which can be used to predict whether a process will be spontaneous if the pressure and temperature are the same at the inital and final states of the system. The free energy change, ΔG, is:

$$\Delta G = \Delta H - T\Delta S$$

ΔH is the enthalpy change for the system and ΔS is the entropy change for the system.

The quantity ΔG under conditions of constant pressure and temperature is proportional to the entropy change of the universe:

$$\Delta G = -T\Delta S_{univ}$$

Therefore we can determine the fate of a process by examining the sign of ΔG:

Sign of ΔG	Fate of Process
< 0	Spontaneous: The entropy of the universe increases in the process.
> 0	Not spontaneous: The entropy of the universe decreases for process as described. The reverse process would be spontaneous.
= 0	The process is at equilibrium. It will not spontaneously proceed in *either* direction.

The free energy change of a reaction is a measure of the spontaneity of that reaction.

Chapter 13: Spontaneity of Chemical Processes

The more negative the free energy change, the more spontaneous the reaction.

The free energy change in chemical reactions can be calculated either by using standard molar free energies of formation, $\Delta G_f°$, or the standard enthalpy and entropy changes.

The standard free energy of formation of a substance, $\Delta G_f°$, is the free energy change when one mole of that substance is formed from the elements in their standard states.

The free energy change of a reaction under standard conditions is obtained in a fashion totally analogous to that used to find enthalpy and entropy changes for reactions:

$$\Delta G° = \text{sum } \Delta G_f°(\text{products}) - \text{sum } \Delta G_f° (\text{reactants})$$

$$\Delta G° = \Sigma \Delta G_f°(\text{products}) - \Sigma \Delta G_f°(\text{reactants})$$

Alternatively, if we know $\Delta H°$ and $\Delta S°$, then we can substitute into the equation for $\Delta G°$ and calculate the standard free energy change:

$$\Delta G° = \Delta H° - T\Delta S°$$

The second expression is more flexible in that we may use it estimate free energy changes at temperatures other than 298 K, whereas the use of free energies of formation to find $\Delta G°$ is only valid at 298 K.

Exercise 13.8 Calculate the free energy for the oxychlorination of ethylene to vinyl chloride, CH_2CHCl, under standard conditions. The equation of reaction is:

$$CH_2CH_{2\ (g)} + HCl_{\ (g)} + \frac{1}{2} O_{2\ (g)} \rightarrow CH_2CHCl_{\ (g)} + H_2O_{\ (g)}$$

Steps to Solution: The free energy change will be calculated by both methods mentioned above.

Using standard free energies of formation: this method is faster if you only need ΔG at 298K. Thermodynamic data for the reactants and products are summarized in the following table.

$$\Delta G° = \Sigma \Delta G_f°(\text{products}) - \Sigma \Delta G_f°(\text{reactants})$$

$$\Delta G° = \Delta G_f°(H_2O_{(g)}) + \Delta G_f°(CH_2CHCl_{(g)}) - \{\Delta G_f°(C_2H_{4(g)}) + \Delta G_f°(HCl_{(g)}) + 1/2\ \Delta G_f°(O_{2(g)})\}$$

Chapter 13: Spontaneity of Chemical Processes

Substance	$\Delta G°_f$ (kJ/mole)	$\Delta H°_f$ (kJ/mole)	$S°$ (J/mole • K)
$CH_2CH_{2\,(g)}$	81.3	36.36	122.17
$HCl_{(g)}$	-95.3	-92.31	186.80
$O_{2\,(g)}$	0	0	205.03
$CH_2CHCl_{(g)}$	51.9	35.56	263.88
$H_2O_{(g)}$	-241.8	-228.59	188.71

$\Delta G° = $ 1 mole H_2O • -241.8 kJ/mole + 1 mole CH_2CHCl • 51.9kJ/mole -{ 1 mole $C_2H_{4(g)}$• 81.3 kJ/mole + 1 mole HCl• -95.30 kJ/mole + 1/2 mole O_2 • 0 kJ/mole}

$$\boxed{\Delta G° = -162.7 \text{ kJ}}$$

Using standard enthalpy and entropy change is necessary for estimating ΔG at temperatures other than 298 K.

The standard enthalpy change for this reaction is calculated as before and is found to be -150.3 kJ. The entropy change using the standard entropies is 41.1 J/K. The standard free energy change is calculated using these quantities where the temperature is 298 K (standard temperature):

$$\Delta G° = \Delta H° - T\Delta S° = -150.3 \text{ kJ} - 298K(41.1 \text{ J/K})\frac{1 \text{ kJ}}{1000 \text{ J}} = \boxed{-162.7 \text{ kJ}}$$

COMMON PITFALL: Entropy conversion
Not converting J/K to kJ/K when calculating $\Delta G°$ leads to large and surprising errors

---

The fairly large negative free energy change indicates a quite spontaneous reaction for the formation of vinyl chloride under standard conditions.

Effect of Nonstandard Conditions on Free Energy

<u>Nonstandard Temperatures</u>

At nonstandard temperatures, the free energy change, $\Delta G°_T$, can be written:

$$\Delta G°_T = \Delta H°_T - T\Delta S°_T$$

However, the enthalpy and entropy *changes* (ΔH and ΔS) in a reaction do not change very much with temperature. The free energy change at standard concentrations and/or pressures can be reasonably accurately estimated using the enthalpy and entropy changes at standard temperatures:

Chapter 13: Spontaneity of Chemical Processes

$$\Delta G°_T = \Delta H° - T\Delta S°$$

The remaining source of temperature dependence is the $T\Delta S°$ term. This term is significant when the product $T\Delta S$ is on the order of kJ. The following table summarizes the how the standard free energy change is affected by temperature for the different possibilities of the signs of the enthalpy and entropy changes.

$\Delta H°$	$\Delta S°$	$\Delta G°$ and Spontaneity
> 0	< 0	$\Delta G°$ always > 0. The process is never spontaneous as written.
> 0	> 0	$\Delta G° > 0$ at low temperature and the process is not spontaneous. At high enough temperature $T\Delta S° > \Delta H°$ and the process will be spontaneous.
< 0	< 0	At low temperatures, $\Delta G° < 0$ and the process is spontaneous. At high enough temperatures, $T\Delta S°$ may be greater than $\Delta H°$ and then the process will not be spontaneous.
< 0	> 0	$\Delta G°$ will always be less than zero and the process will become more spontaneous as temperature increases.

Exercise 13.9 One of the more important industrial chemicals is hydrogen. One process for hydrogen production is called "steam reforming", in which hydrocarbons react with water to give hydrogen and CO. The equation of reaction for reforming methane is written below.

$$CH_{4\ (g)} + H_2O_{\ (g)} \rightarrow CO_{\ (g)} + 3\ H_{2\ (g)}$$

(a) Calculate the free energy change for this reaction at standard conditions.
(b) Estimate the temperature at which the process becomes spontaneous.

Steps to Solution: The first step is to determine the enthalpy and entropy changes for this reaction at standard temperature and pressure. For part (b) we need to find the temperature at which the reaction first becomes spontaneous, or when $\Delta G°$ is no longer greater than zero. Usually this means finding the temperature at which the standard free energy change is zero.

(a) The standard enthalpy change is calculated by methods already described above using the values of enthalpies of formation and standard entropies listed in the table that follows.

Chapter 13: Spontaneity of Chemical Processes

Substance	$CH_{4(g)}$	$H_2O_{(g)}$	$CO_{(g)}$	$H_{2(g)}$
$\Delta H°_f$ (kJ/mole)	-75	-241.8	-110.5	0
$S°$ (J/mole·K)	186	188.7	197.9	131

$$\Delta H° = \Delta H°_f (CO_{(g)}) + 3 \cdot \Delta H°_f (H_{2(g)}) - \{\Delta H_f° (H_2O_{(g)}) + \Delta H_f° (CH_{4(g)})\}$$

$$\Delta H° = -110.5 + 3 \text{ mole } H_2 \cdot 0 - (-241.8 + -75) \text{ kJ} = \boxed{206 \text{ kJ}}$$

$$\Delta S° = S° (CO_{(g)}) + 3 \cdot S° (H_{2(g)}) - \{\Delta S_f° (H_2O_{(g)}) + \Delta S_f° (CH_{4(g)})\}$$

$\Delta S = 1 \text{ mole CO} \cdot 197.9 \text{ J/K} \cdot \text{mole CO} + 3 \text{ mole } H_2 \cdot 131 \text{ J/K} \cdot \text{mole } H_2 - $

$(1 \text{ mole } H_2O \cdot 188.7 \text{ J/K} \cdot \text{mole } H_2O + 1 \text{ mole } CH_4 \cdot 186 \text{ J/K} \cdot \text{mole } CH_4)$

$$\boxed{\Delta S = 216 \text{ J/K}}$$

$$\Delta G° = \Delta H° - T\Delta S° = 206 \text{ kJ} - 298K(216 \text{ J/K})\left(\frac{1 \text{kJ}}{1000 \text{ J}}\right) = \boxed{142 \text{ kJ}}$$

(b) Because $\Delta G° > 0$, the process is not spontaneous. However, $\Delta S > 0$, so as temperature increases, the $T\Delta S$ term will become larger than $\Delta H°$ and because of the negative sign, will make $\Delta G° < 0$. When $\Delta G° = 0$, the reactants and products at standard concentrations will coexist; the free energy change for conversion of the substances from one side of the arrow to the other is zero in either direction. When the temperature is greater than this temperature, the reaction will become spontaneous. Therefore, to determine the temperature at which the process will become spontaneous, we will use the equation for the standard free energy change to solve for the temperature at which $\Delta G° = 0$.

$$\Delta G° = 0 = \Delta H° - T\Delta S°$$

$$\Delta H° = T\Delta S° \text{ or } T = \frac{\Delta H°}{\Delta S°} = \frac{206 \text{ kJ}}{216 \times 10^{-3} \text{ kJ/K}} = \boxed{9.54 \times 10^2 \text{ K}}$$

Because of the approximations that ΔS and ΔH do not change significantly with temperature, we should not use all the significant figures. The reaction will become spontaneous when the temperature is greater than about 1000 K.

Concentration Dependence of Free Energy

Free energy changes also depend on the concentration or pressures of the reactants or products. For concentrations or pressures not equal to 1M or 1 atm, the free energy change is written ΔG without the superscript, °. For the general reaction:

$$aA + bB \rightarrow cC + dD$$

$$\Delta G = \Delta G° + RT \ln \frac{[C]^c[D]^d}{[A]^a[B]^b} \quad \text{for solutions and} \quad \Delta G = \Delta G° + RT \ln \frac{p_C^c \, p_D^d}{p_A^a \, p_B^b} \quad \text{for gases}$$

The ratio of concentrations or pressures is called the ***concentration quotient*** (sometimes the reaction quotient) and is encountered often enough to merit its own symbol, Q.

$$Q = \frac{[C]^c[D]^d}{[A]^a[B]^b} \quad \text{for solutions and, for gases,} \quad Q = \frac{p_C^c \, p_D^d}{p_A^a \, p_B^b}$$

In words, the concentration quotient is the multiplicative product of the products of the reaction raised to the power of their stoichiometric coefficients divided by the multiplicative product of the reactants of the reaction raised to the power of their stoichiometric coefficients. In short, products over reactants, each raised to the power of its stoichiometric coefficient. The effects of relative reactant and product concentrations on Q and ΔG are summarized in the following table.

Conditions	Size of Q	Effect on ΔG
Product concentrations greater than reactant concentrations	$Q > 1$	$\Delta G > \Delta G°$, making reaction less spontaneous.
Reactant concentrations greater than product concentrations	$Q < 1$	$\Delta G < \Delta G°$, making reaction more spontaneous.

Manipulation of reactant and product concentrations (or pressures) may affect the size and even sign of the free energy change, leading to a spontaneous reaction ($\Delta G < 0$) when the standard free energy change indicates a nonspontaneous reaction ($\Delta G° > 0$).

Exercise 13.10 This exercise examines the dependence of free energy change on the concentrations of reactants and products.
(a) What is the free energy change, ΔG, for reforming methane at 298 K when the partial pressure of the products is 5.0 x 10⁻⁵ atm and the partial pressure of the reactants is 1.0 atm?

(b) What is the free energy change, ΔG, for reforming methane at 298 K when the partial pressure of the products is 1.0 atm and the partial pressure of the reactants is 5.0×10^{-5} atm?

(c) If the reactants are supplied at equal pressures ($p_{CH_4} = p_{H_2O}$), what pressures of methane and water would be required to spontaneously produce hydrogen and CO at a pressure of 1 atm when the temperature is 600 K?

Steps to Solution: *This multipart question addresses the concentration dependence and the temperature dependence of free energy changes. In parts (a) and (b), we examine the effect of changing Q on ΔG; in part (c) we must calculate the pressures required to induce spontaneity at a higher temperature (make $\Delta G = 0$).*

(a) *Here we must find the value of ΔG at standard temperature when the partial pressures of the gaseous substances are not at standard conditions. The equation to be used for systems where all the reactants and products are gases is:*

$$\Delta G = \Delta G° + RT \ln Q = \Delta G° + RT \ln \frac{p_C^c \, p_D^d}{p_A^a \, p_B^b}$$

For the reaction: $\quad CH_{4\,(g)} + H_2O_{(g)} \rightarrow CO_{(g)} + 3\,H_{2\,(g)}$

the concentration quotient, Q, is:
$$Q = \frac{p_{CO} \, p_{H_2}^3}{p_{CH_4} \, p_{H_2O}}$$

Substituting the numerical values for the partial pressures gives:

$$Q = \frac{5.0 \times 10^{-5}(5.0 \times 10^{-5})^3}{1 \cdot 1} = 6 \times 10^{-18}$$

$$\Delta G = 142 \text{ kJ} + 8.314 \text{ J/mole} \cdot K \frac{1 \text{ kJ}}{1000 \text{ J}} (298) \ln(6 \times 10^{-18}) = \boxed{44 \text{ kJ}}$$

The decrease in pressure (concentration) of the products has made the free energy change less positive, so the reaction is more spontaneous than it was under standard conditions.

(b) *This part of the problem is similar to part (a) except that the concentration of the products (1 atm) is greater than the concentration of the reactants (5×10^{-5} atm). Use the method outlined in part (a) to find:*

$$\Delta G = \boxed{191 \text{ kJ}}$$

The more positive free energy change indicates that the reaction is less spontaneous (less likely to convert reactants to products). Notice that the difference between the free energy change and the standard free energy change is not the same in parts (a) and (b), because the pressure of one product is raised to the third power.

(c) *In this part, the desired answer is the minimum partial pressures of the reactant gases to produce the products at 1 atm as a spontaneous process. To find this, we must first calculate the standard free energy change. Then, using the expression for free energy change at nonstandard conditions, we can determine the pressures of reactants required for $\Delta G = 0$. We choose $\Delta G = 0$, because the reactants and products will be at equilbrium at those pressures; any increase in the pressure of any reactant will lead to $\Delta G < 0$.*

First, we determine $\Delta G°$ at 600K ($\Delta H°$ and $\Delta S°$ were found in Exercise 13.8).

$$\Delta G°(600\ K) = 206\ kJ - 600\ K(216\ J/K)\frac{1\ kJ}{1000\ J} = 77\ kJ$$

Now for calculation of the pressures of reactants when $\Delta G = 0$:

$$\Delta G = \Delta G° + RT \ln \frac{p_{CO}p_{H_2}^3}{p_{CH_4}p_{H_2O}} = 0 \quad \text{so} \quad RT \ln \frac{p_{CO}p_{H_2}^3}{p_{CH_4}p_{H_2O}} = -\Delta G°(600\ K)$$

$$\ln \frac{p_{CO}p_{H_2}^3}{p_{CH_4}p_{H_2O}} = \frac{-\Delta G°(600 K)}{RT} = \frac{-77.0 \times 10^3 J}{8.314\ J/mole \cdot K(600\ K)} = -15.44$$

Taking the antilog (using natural logarithms) of both sides removes the logarithmic factor and allows us to deal with the pressures:

$$\frac{p_{CO}p_{H_2}^3}{p_{CH_4}p_{H_2O}} = \exp(-15.44) = 1.98 \times 10^{-7}$$

The pressures of methane and water are equal and are replaced by the less bulky variable, x.

$$p_{CH_4} = p_{H_2O} = x$$

Substituting into the equation and solving for x gives:

$$\frac{p_{CO}p_{H_2}^3}{x \cdot x} = 1.98 \times 10^{-7} \quad \text{gives} \quad x = \sqrt{\frac{p_{CO}p_{H_2}^3}{1.98 \times 10^{-7}}}$$

The pressures of CO and H_2 are given in the problem; these products are to be produced at 1 atm. Substitution of the values of pressures gives:

$$x = \sqrt{\frac{1 \cdot 1^3}{1.98 \times 10^{-7}}} = \boxed{2 \times 10^3 \text{ atm}} \text{ (only one significant figure)}$$

Note that the pressure of the reactants must be high to form the products at 1 atm because the standard free energy change is positive (77 kJ). However, when the pressure of the reactants is increased, significant amounts of product may be formed despite the positive standard free energy change.

13.5 Some Applications of Thermodynamics

QUESTIONS TO ANSWER, SKILLS TO LEARN
1. Using free energy changes to investigate phase equilibria
2. What are phase diagrams and how are they used?

Free energy changes depend on the enthalpy change and the entropy change. The more condensed phases are favored by enthalpy (there are more and stronger intermolecular forces acting as the substances are more ordered) and the less ordered phases are favored by entropy. In separating mixtures, it is useful to have information about the conditions of pressure and temperature under which particular phases exist and the conditions under which different phases may be interconverted (liquid to gas, for example). Such information can be found in a *phase diagram*.

A **phase diagram** is a two-dimensional plot with pressure as the y axis and temperature as the x axis. Lines divide the area into regions corresponding to solid, liquid and gas phases. In the sample phase diagram drawn below, general conditions of temperature and pressure are included with the phase labels to help you remember the phases are found in these areas. The lines indicate the pressure and temperature at which the two phases on either side of the line can coexist at equilibrium. The processes and regions are labeled in the following figure:

Remember that solids are stable at low temperatures (low thermal energy) and higher pressure. Gases are stable when the number of molecules per unit volume is low and the thermal energy is high (higher temperatures and/or lower pressures). Liquids are present when the thermal energy is high enough to disrupt the order of the solid, but the pressure is insufficient to completely disrupt the interactions between molecules. Therefore the solid phase tends to be the leftmost phase, and the gas phase will be found to the far right.

To draw a qualitative phase diagram, we need several data: the normal boiling and freezing points and the temperature and pressure of the triple point. The normal boiling and freezing points have already been described, but the triple point has not been discussed. The **triple point** is the point of temperature and pressure at which all three phases coexist at equilibrium. On the phase diagram, the three curves intersect at the triple point. For example, if we place water under a vacuum, it will start to bubble off water vapor and cool. When the temperature is 0.01°C, solid will appear in the presence of liquid and the vapor pressure will be 0.00604 atm. All three phases are now in equilibrium with each other. If the water temperature cools further (moving left on the phase diagram), the liquid will freeze.

To draw a phase diagram of a compound with a triple point of less than 1 atm, we draw the pressure and temperature axes and locate the positions of the freezing, boiling and triple points on the chart. We then draw smooth curves from the triple point through the melting point, from the triple point through the boiling point and a last curve from near the origin (0 K, 0 atm) to the triple point. These steps are illustrated in Exercise 13.10.

Exercise 13.11 Oxygen is a component of the atmosphere. Draw the phase diagram for O_2 given the normal melting point (-218 °C), the normal boiling point (-183 °C) and the triple point (-219 °C and 1.10 mm Hg).

Steps to Solution: The two main steps in drawing a phase diagram are to locate the points characteristic of the substance (melting, boiling, and triple) and then to draw the smooth curves connecting the appropriate points. The temperatures and pressures should all be converted to K and atm, respectively, prior to location of the points on the graph as done below.

mp: $T(K) = -218 + 273.15 = 55$ K bp: $T(K) = -183 + 273.15 = 90$ K

tp: $T(K) = -219 + 273.15 = 54$ K

$$P(atm) = \frac{1.1 \text{ mm Hg}}{760 \text{ mm Hg/atm}} = 1.4 \times 10^{-3} \text{ atm}$$

Drawing the axes and locating the points gives the figure below. Note that the pressure scale is not "true" to scale because the difference in pressures between the normal boiling point and the triple point cannot be accurately represented in this small a space. However, the temperature scale is linear. Addition of the smooth curves connecting the origin and the triple point, the triple point and the melting point and the triple point and the normal boiling point and addition of labels for the liquid, solid and gas phases gives the figure at right, the completed qualitative phase diagram. For greater accuracy, the curves could be drawn from equations or experimental data.

Exercise 13.12 A phase diagram may also be used to determine which phase will be predominant under specific conditions of temperature and/or pressure. Use the phase diagram of oxygen to:
(a) Determine the phases oxygen passes through starting at 0.001 atm and 100 K, cooling and increasing the pressure steadily to 25 K at 1 atm.
(b) Estimate the temperature at which liquid oxygen will boil under 0.5 atm external pressure and compare that to the normal boiling point.
(c) Estimate the temperature at which liquid oxygen will freeze under 0.5 atm external pressure and compare that to the normal freezing point.

Steps to Solution: Part (a) requires use of the phase diagram to obtain information regarding phases under different conditions of temperature and pressure and parts (b) and (c) require use of the phase diagram to estimate temperatures at which O_2 undergoes phase changes under a nonstandard pressure.

(a) Probably the easiest way to solve this problem is to draw an arrow from the initial temperature and pressure to the final temperature and pressure on the phase diagram. The phases through which the substance passes are deduced by following the arrow. The initial conditions are in the gas phase region of the phase diagram. Cooling and increasing the pressure lead to passage through the liquid phase, and the final conditions correspond to the solid phase. The phases are: gas; liquid; then solid.

(b) In this part of the problem (and part (c)), the temperature at which a phase change occurs under a given pressure is estimated using the phase diagram. To obtain this information, remember that phase changes occur at the temperature-pressure combinations that occur at the line separating the two phases. One way to solve this problem is to draw a line at the desired pressure and find where it intersects the curve separating the liquid and gas phases. Then we find the temperature at this intersection, about 80 K. This temperature is much lower than the normal boiling point of oxygen, which is 90 K. This can be attributed to the large entropy change that occurs upon vaporization.

(c) Using a similar strategy to that in part(b), you should find that the melting point of O_2 is nearly the same at 0.5 atm as the normal melting point (perhaps a bit lower). The melting point is much less sensitive to pressure than the boiling point.

13.6 Bioenergetics

QUESTIONS TO ANSWER, SKILLS TO LEARN
1. Understanding the principle of coupled chemical processes
2. What are some of the principal chemical processes in biochemical energy production?

Coupled Reactions

Chemical processes do occur in systems, especially in biochemical systems, which have positive free energies, such as the synthesis of sugars by plants from CO_2 and water:

$$n\, CO_{2\,(g)} + n\, H_2O \rightarrow n\, O_{2\,(g)} + [CH_2O]_n$$

The reason this synthesis of carbohydrates occurs is that this chemical process does not occur by itself; other reactions occur that are linked to the process and have negative free energy changes. Because the free energy change is a state function, the total free energy change is the sum of the free energy changes of the two reactions. Consider the simple reaction of dissolving lead chloride:

$$PbCl_{2\,(s)} \rightarrow Pb^{2+}_{\,(aq)} + 2\, Cl^-_{\,(aq)} \qquad \Delta G° = 7.41 \text{ kJ}$$

The positive free energy change indicates that this reaction is not very spontaneous. However, the precipitation of silver chloride is fairly spontaneous, as judged by its free energy change:

$$Ag^+_{(aq)} + Cl^-_{(aq)} \rightarrow AgCl_{(s)} \qquad \Delta G° = -13.3 \text{ kJ}$$

The addition of silver ion to a solution containing lead chloride will result in the formation of silver chloride and aqueous lead ion. The balanced chemical equation of reaction can be considered as being obtained by adding the separate reactions above:

	$PbCl_{2\,(s)} \rightarrow Pb^{2+}_{(aq)} + 2\,Cl^-_{(aq)}$	$\Delta G° =$	7.41 kJ
+	$2\,(Ag^+_{(aq)} + Cl^-_{(aq)} \rightarrow AgCl_{(s)})$	$\Delta G° = 2(-13.3 \text{ kJ}) =$	-26.6 kJ
	$2\,Ag^+_{(aq)} + PbCl_{2\,(s)} \rightarrow Pb^{2+}_{(aq)} + 2\,AgCl_{(s)}$	$\Delta G =$	-19.2 kJ

Notice that the negative free energy for the precipitation of AgCl more than compensates for the positive free energy for dissolving $PbCl_2$. These reactions are **coupled**: the negative free energy of the formation of AgCl makes possible the otherwise nonspontaneous reaction, dissolving $PbCl_2$. The key features of coupled reactions are:

¥ The two reactions involve a common substance appearing as a reactant in one reaction and as a product in the other reaction. In the lead chloride reaction listed above, the common substance is the chloride ion.

¥ One of the reactions has a nonspontaneous free energy change and the other reaction has a spontaneous free energy change; the free energy of the coupled process is negative.

One analogy to a coupled reaction is a seesaw with a weight on one end that we want to lift; a nonspontaneous process. To lift this weight, we place a heavier weight on the other end of the seesaw. The spontaneous process of the heavy weight moving down drives the nonspontaneous process of lifting the smaller weight up:

Coupled reactions are very important in bioenergetics, because many biochemical processes form large ordered molecules and other processes which, uncoupled, would not be spontaneous.

Metabolism: Energy From Fats and Saccharides

Most life processes require energy which is supplied by the reaction of carbohydrates or fats with oxygen. The process of converting theses substances to their products in a controlled fashion through a series of chemical reactions is called **metabolism**. The combustion reactions of carbohydrates and fats are illustrative of aerobic metabolism and show the greater energy content of fats per C atom.

Glucose: $\qquad C_6H_{12}O_6 + 6\ O_2 \rightarrow 6\ CO_2 + 6\ H_2O$

$$\Delta G° = \frac{-2870\ \text{kJ/mole}\ C_6H_{12}O_6}{6\ \text{mole C/mole}\ C_6H_{12}O_6} = -478\ \text{kJ/mole C}$$

Palmitic acid: $\qquad C_{15}H_{31}CO_2H + 23\ O_2 \rightarrow 16\ CO_2 + 16\ H_2O$

$$\Delta G° = \frac{-9790\ \text{kJ/mole}\ C_{15}H_{31}CO_2H}{15\ \text{mole C/mole}\ C_{15}H_{31}CO_2H} = -611\ \text{kJ/mole C}$$

These "combustion" reactions occur through a complex series of steps that varies from organism to organism and depends even on the nature of the activity engaged in. The energy released in these reactions is utilized to drive biochemical processes and also stored by forming compounds which can release energy on demand, the most common of which is adenosine triphosphate, ATP.

Exercise 13.13 The following questions deal with bioenergetics:
(a) Write the thermochemical equation of reaction for the energy releasing decomposition of ATP to ADP.
(b) Estimate how many moles of ATP would be required to ride a bicycle at 35 km/hour for 1 hour, an activity that would require about 1500 kJ of energy.
(c) Glucose is metabolized at a rate of about 38% efficiency. How many grams of glucose would supply the energy to do the bicycle ride in part (b).

Steps to Solution: Answering this question requires knowing the structures of ATP and ADP and the free energy change. These may be found in the text. Given this information, we can use thermochemical methods and stoichiometry to solve the other two parts.

(a) $\qquad\qquad\qquad\qquad$ ATP $\rightarrow$ ADP + H$_2$O $\qquad\qquad\qquad\qquad \Delta G° = -30.6$ kJ

(b) *The number of moles of ATP that would be required to perform this activity can be estimated by using the standard free energy change for the ATP to ADP transformation.*

$$\text{mole ATP} = \frac{1500 \text{ kJ}}{30.6 \text{ kJ/mole ATP}} = \boxed{49 \text{ mole ATP}}$$

Confirm to yourself that this is about 24 kg of ATP. For practical reasons, your body cannot store all its energy needs as ATP, but must "recycle" the ADP produced by metabolic processes.

(c) *The problem asks for the grams of glucose that will be required for this exercise given that glucose metabolism is only about 38% efficient. The free energy change for the aerobic "burning" of glucose is:*

$$C_6H_{12}O_6 + 6\ O_2 \rightarrow 6\ CO_2 + 6\ H_2O \qquad\qquad \Delta G° = -2780 \text{ kJ/mole glucose}$$

Only 38% of this energy is utilized, so the usable energy available from glucose "burning" is:

usable energy = (0.38)(2780 kJ/.mole) = 1056 usable kJ/mole glucose

Therefore the cyclist will require:

$$\text{moles glucose} = \frac{1500 \text{ kJ}}{1056 \text{ usable kJ/mole glucose}} = 1.4 \text{ mole glucose}$$

g glucose = 1.4 mole glucose(180.158 g/mole glucose) = $\boxed{2.6 \times 10^2 \text{ g glucose}}$

Note that this is about a half a pound! However, the exercise is pretty vigorous, so this figure is reasonable.

Test Yourself

1. State whether the entropy of each of the following systems increases or decreases in the stated process without consulting any tables:
   (a) $2\ S_{(s)} + 3\ O_{2\ (g)} \rightarrow 2\ SO_{3\ (g)}$
   (b) $CO_{2\ (g)} \rightarrow CO_{2\ (s)}$
   (c) heating water from 5 °C to 75 °C

2. Calculate the standard reaction entropy, $\Delta S°$, for each of the following reactions:
   (a) $4\ KClO_{3\ (s)} \rightarrow 3\ KClO_{4\ (s)} + KCl_{(s)}$
   (b) $CaO_{(s)} + CO_{2\ (g)} \rightarrow CaCO_{3\ (s)}$

3. An electric heater releases 1200 J of energy every second. How much does the entropy of the surroundings increase when the heater runs for 10 minutes in a room that is at 25 °C?

4. Determine whether iron(III) oxide can be reduced by carbon at 1000 K by either of the following two reactions. The following free energies of formation are valid at

1000 K: $\Delta G°_f$ ($CO_{(g)}$) = -200 kJ/mole; $\Delta G°_f$ ($CO_{2(g)}$) = -396 kJ/mole; $\Delta G°_f$ ($Fe_2O_{3(s)}$) = -562 kJ/mole; assume $\Delta G°_f$ of the elements = 0
(a) $Fe_2O_{3(s)} + 3\ C_{(s)} \rightarrow 2\ Fe_{(s)} + 3\ CO_{(g)}$
(b) $2\ Fe_2O_{3(s)} + 3\ C_{(s)} \rightarrow 2\ Fe_{(s)} + 3\ CO_{2\ (g)}$

5. Calculate the equilbrium constant at 298 K for the following two reactions:
(a) $4\ KClO_{3\ (s)} \rightarrow 3\ KClO_{4\ (s)} + KCl_{(s)}$
(b) $CaO_{(s)} + CO_{2\ (g)} \rightarrow CaCO_{3\ (s)}$

6. A gas sample is heated in a cylinder, absorbing 850 kJ of energy. The piston, pushed by the gases, does 535 kJ of work on the surroundings. What is ΔE?

Answers

1. (a) decrease (b) decrease (c) increase

2. (a) -36.8 J/K   (b) -160.6 J/K

3. ΔS_{surr} = 2.4 x 10^3 J/K

4. Reduction can occur in both cases because $\Delta G°$ is negative. (a) $\Delta G°$ = -38 kJ (b) $\Delta G°$ = -64 kJ

5. (a) 2.1 x 10^{23} (b) 7.1 x10^{22}

6. ΔE = 315 kJ

Solutions to Chapter Problems

13.1 (a) A sand castle represents an ordered structure constructed by a person. Waves destroy that structure, returning the sand grains to a disordered arrangement.
(b) Two separate liquids represent an ordered arrangement (all molecules of one kind in one container). Upon mixing, the molecules in the liquids are distributed randomly throughout the container, an increase in disorder.
(c) Sticks in a bundle are ordered (all aligned in the same direction). When dropped, the sticks lose their alignment and become more disordered.
(d) Water in a puddle is relatively ordered, as it is confined to a small volume. When the water evaporates, the molecules spread over a much larger volume and become more disordered.

13.3 The molecules in a drop of ink are relatively ordered, because they occupy one particular part of the total volume. As they move randomly about, they spread throughout the volume of the container and become more disordered.

13.5 (a) The air molecules in a tire are relatively ordered, because they are confined to a specific, small volume. A puncture allows gas molecules to escape from the tire and fill a much larger volume, becoming less ordered in the process.
(b) Like air in a tire, the fragrant molecules in a perfume bottle are relatively ordered because they are confined to a specific, small volume. When the bottle is open, molecules escape from the confined volume into a much larger space of the room, becoming more disordered in the process.

13.7 Calculate W by determining how many ways each symbol can be placed. The first X can be placed in any of the nine compartments, the first O in any of the eight empty compartments, so $W = (9)(8) = 72$.

13.9 The entropy change accompanying a constant-temperature process is $\Delta S = q/T$.
(a) Melting involves absorption of heat:
$$q_{ice} = n\Delta H_{fus} = \frac{(13.8 \text{ g})}{(18.02 \text{ g/mol})}(6.01 \text{ kJ/mol})(10^3 \text{ J/kJ}) = 4.60 \times 10^3 \text{ J};$$
$\Delta S_{ice} = (4.60 \times 10^3 \text{ J}) \div (273.15 \text{ K}) = 16.8 \text{ J/K}$.
(b) The heat absorbed by the ice cube must be supplied by the pool water: $q_{pool} = -q_{ice}$;
$\Delta S_{pool} = (-4.60 \times 10^3 \text{ J}) \div (27.5 + 273.15 \text{ K}) = -15.3 \text{ J/K}$.
(c) $\Delta S_{total} = \Delta S_{ice} + \Delta S_{pool} = (16.8 \text{ J/K}) + (-15.3 \text{ J/K}) = 1.5 \text{ J/K}$.

13.11 (a) The system (water) absorbs heat to boil, so q_{sys} is positive and ΔS_{sys} is positive. The surroundings (stove) release heat to the system, so q_{surr} is negative and ΔS_{surr} is negative.
(b) The system (ice cubes) absorbs heat to melt, so q_{sys} is positive and ΔS_{sys} is positive. The surroundings (table top) release heat to the system, so q_{surr} is negative and ΔS_{surr} is negative.
(c) The system (coffee) absorbs heat, so q_{sys} is positive and ΔS_{sys} is positive. The surroundings (microwave oven) release heat to the system, so q_{surr} is negative and ΔS_{surr} is negative.

13.13 The entropy change accompanying a constant-temperature process is $\Delta S = q/T$.
Condensation involves release of heat:
$$q_{steam} = -n\Delta H_{vap} = -\frac{(15.5 \text{ g})}{(18.02 \text{ g/mol})}(40.79 \text{ kJ/mol})(10^3 \text{ J/kJ}) = -35.09 \times 10^3 \text{ J};$$
$\Delta S_{steam} = (-35.09 \times 10^3 \text{ J}) \div (373.15 \text{ K}) = -94.0 \text{ J/K}$.
Knowing that ΔS_{total} must be positive, we can say without doing further calculations that $\Delta S_{surr} > 94.0$ J/K.

13.15 The entropy change accompanying fusion is $\Delta S = n\Delta H_{fus}/T_{fus}$.
(a) $\Delta S_{molar} = \dfrac{(1 \text{ mol})(1.3 \text{ kJ/mol})(10^3 \text{ J/kJ})}{(83 \text{ K})} = 16$ J/mol K;
(b) $\Delta S_{molar} = \dfrac{(1 \text{ mol})(0.84 \text{ kJ/mol})(10^3 \text{ J/kJ})}{(90 \text{ K})} = 9.3$ J/mol K;
(c) $\Delta S_{molar} = \dfrac{(1 \text{ mol})(7.61 \text{ kJ/mol})(10^3 \text{ J/kJ})}{(156 \text{ K})} = 48.8$ J/mol K;
(d) $\Delta S_{molar} = \dfrac{(1 \text{ mol})(23.4 \text{ kJ/mol})(10^3 \text{ J/kJ})}{(234 \text{ K})} = 1.00 \times 10^2$ J/mol K;

13.17 (a) Both are ionic solutions, but $MgCl_2$ produces three moles of ions per mole of substance, while NaCl produces two moles of ions per mole of substance, so $MgCl_{2(aq)}$ has the larger molar entropy;
(b) Both are ionic solids, but HgS has a higher molar mass than HgO, so HgS has the larger molar entropy;
(c) Both are diatomic molecules, but $Br_{2(l)}$ is liquid while $I_{2(s)}$ is solid, so $Br_{2(l)}$ has the larger molar entropy.

13.19 All these substances are small gaseous molecules. Methane is a highly symmetric molecule, so it has fewer distinguishable orientations in space. Thus it has lower entropy than the other two despite having five atoms per molecule. Ozone has more entropy than O_2 because it has three atoms per molecule while O_2 has only two.

13.21 Absolute entropies are tabulated in Appendix E of your textbook. To obtain the entropy per mole of atoms, divide each value by the number of atoms in one molecule:

He: $S^o = 126.15 \div 1 = 126.15$ J/mol K; H_2: $S^o = 130.684 \div 2 = 65.34$ J/mol K;
CH_4: $S^o = 186.26 \div 5 = 37.25$ J/mol K; C_3H_6: $S^o = 226.9 \div 9 = 25.21$ J/mol K.
Entropy per mole of atoms decreases as the number of atoms in a molecule increase, because tying together atoms into a molecule increases the amount of order among those atoms.

13.23 Standard entropy changes can be calculated using tabulated values of absolute entropies (Appendix E of your textbook) and Equation 13-5:
$$\Delta S^o_{rxn} = \Sigma coeff_p\, S^o_{(products)} - \Sigma coeff_r\, S^o_{(reactants)}$$

(a) $\Delta S^o = 2(192.45) - 1(191.61) - 2(130.684) = -198.76$ J/K;
(b) $\Delta S^o = 2(238.93) - 3(205.138) = -137.55$ J/K;
(c) $\Delta S^o = 1(64.81) + 2(37.99) - 1(68.6) - 2(29.86) = 12.5$ J/K;
(d) $\Delta S^o = 2(213.74) + 2(69.91) - 1(219.56) - 3(205.14) = -267.68$ J/K.

13.25 Reactions have negative entropy changes when products are more ordered than reactants.

(a) There is a relatively large negative value because of the reduction in number of moles of gaseous substances: 4 mol of gaseous reactants are converted to 2 mol of gaseous products;
(b) There is a relatively large negative value because of the reduction in number of moles of gaseous substances: 3 mol of gaseous reactants are converted to 2 mol of gaseous products;
(c) This reaction has a near-zero entropy change because all reagents are relatively ordered solid substances;
(d) There is a relatively large negative value because of the reduction in number of moles of gaseous substances: 4 mol of gaseous reactants are converted to 2 mol of gaseous products.

Chapter 13: Spontaneity of Chemical Processes

13.27 Entropies of substances at pressures different from 1.00 atm are calculated using Equation 13-4: $S_{(c \neq 1)} = S^o - R \ln c$. Take into account amounts different from 1 mol by multiplying by the number of moles.
(a) $S = (2.50 \text{ mol})[154.843 \text{ J/mol K} - (8.314 \text{ J/mol K}) ln(0.25 \text{ atm})] = 416 \text{ J/K}$;
(b) $S = (0.75 \text{ mol})[238.93 \text{ J/mol K} - (8.314 \text{ J/mol K}) ln(2.75 \text{ atm})] = 1.7 \times 10^2 \text{ J/K}$;
(c) Calculate the entropy of each component separately, using $n_i = X_i n_{tot}$ and $p_i = X_i p_{tot}$:
$S_{N_2} = (0.78)(0.45 \text{ mol})\{191.61 \text{ J/mol K} - (8.314 \text{ J/mol K}) ln[(0.78)(1.00 \text{ atm})]\} = 68 \text{ J/K}$;
$S_{O_2} = (0.22)(0.45 \text{ mol})\{205.138 \text{ J/mol K} - (8.314 \text{ J/mol K}) ln[(0.22)(1.00 \text{ atm})]\} = 20 \text{ J/K}$;
$S = S_{N_2} + S_{O_2} = 88 \text{ J/K}$.

13.29 False statements can be made true in various ways; we choose the simplest change that corrects each statement:
(a) $\Delta G_{system} < 0$ for any spontaneous process at constant T and P.
(b) $\Delta G = \Delta H - \Delta(TS)$.
(c) The free energy of a system decreases in any process at constant T and P.

13.31 Standard free energy changes are calculated from Equation 13-8 using standard free energies of formation, which can be found in Appendix E of your textbook:

$$\Delta G^o_{rxn} = \Sigma \text{coeff}_p \Delta G^o_f \text{ (products)} - \Sigma \text{coeff}_r \Delta G^o_f \text{ (reactants)}$$

(a) $\Delta G^o_{rxn} = 2(-16.45 \text{ kJ/mol}) - 1(0) - 3(0) = -32.90 \text{ kJ}$;
(b) $\Delta G^o_{rxn} = 2(163.2 \text{ kJ/mol}) - 3(0) = 326.4 \text{ kJ}$.
(c) $\Delta G^o_{rxn} = 1(0) + 2(-211.7 \text{ kJ/mol}) - 1(-217.33 \text{ kJ/mol}) - 2(0) = -206.1 \text{ kJ}$;
(d) $\Delta G^o_{rxn} = 2(-394.359 \text{ kJ/mol}) + 2(-237.129 \text{ kJ/mol}) - 1(68.15 \text{ kJ/mol}) - 3(0) = -1331.13 \text{ kJ}$.

13.33 Use Equation 13-10 to estimate the standard free energy change at a temperature different from 298 K: $\Delta G^o_{rxn, T} \cong \Delta H^o_{rxn, 298} - T \Delta S^o_{rxn, 298}$. Standard entropy changes are calculated in Problem 13-23, but standard enthalpy changes need to be calculated from standard enthalpies of formation:
(a) $\Delta H^o_{rxn} = 2(-46.11 \text{ kJ/mol}) - 1(0) - 3(0) = -92.22 \text{ kJ}$;
$\Delta G^o_{rxn, T} \cong -92.22 \text{ kJ} - (425 \text{ K})(-198.76 \text{ J/K})(10^{-3} \text{ kJ/J}) = -7.7 \text{ kJ}$.
(b) $\Delta H^o_{rxn} = 2(142.7 \text{ kJ/mol}) - 3(0) = 285.4 \text{ kJ}$;
$\Delta G^o_{rxn, T} \cong 285.4 \text{ kJ} - (425 \text{ K})(-137.55 \text{ J/K})(10^{-3} \text{ kJ/J}) = 343.9 \text{ kJ}$;
(c) $\Delta H^o_{rxn} = 1(0) + 2(-239.7 \text{ kJ/mol}) - 1(-277.4 \text{ kJ/mol}) - 2(0) = -202.0 \text{ kJ}$;
$\Delta G^o_{rxn, T} \cong -202.0 \text{ kJ} - (425 \text{ K})(-12.5 \text{ J/K})(10^{-3} \text{ kJ/J}) = -207.3 \text{ kJ}$;
(d) $\Delta H^o_{rxn} = 2(-393.509 \text{ kJ/mol}) + 2(-285.83 \text{ kJ/mol}) - 1(52.26 \text{ kJ/mol}) - 3(0) = -1410.94 \text{ kJ}$;
$\Delta G^o_{rxn, T} \cong -1410.94 \text{ kJ} - (425 \text{ K})(-267.68 \text{ J/K})(10^{-3} \text{ kJ/J}) = -1297 \text{ kJ}$.

13.35 Use Equations 12-10, 13-5, and 13-10 to calculate standard thermodynamic values:

$$\Delta H^o_{rxn} = \sum coeff_p \Delta H^o_f(products) - \sum coeff_r \Delta H^o_f(reactants)$$
$$\Delta S^o_{rxn} = \sum coeff_p S^o(products) - \sum coeff_r S^o(reactants)$$
$$\Delta G^o_{rxn} = \sum coeff_p \Delta G^o_f(products) - \sum coeff_r \Delta G^o_f(reactants)$$

ΔH^o_{rxn} = 1(–365.56 kJ/mol) + 1(–285.83 kJ/mol) – 2(–46.11 kJ/mol) – 2(0) = –559.17 kJ;
ΔS^o_{rxn} = 1(151.08) + 1(69.91) – 2(192.45) – 2(205.14) = –574.19 J/K;
ΔG^o_{rxn} = 1(–183.87 kJ/mol) + 1(–237.13 kJ/mol) – 2(–16.45 kJ/mol) – 2(0) = –388.10 kJ.

13.37 A phase diagram is a plot of P vs. T, and phase boundary lines meet at the triple point and cross the horizontal line at P = 1 atm at the normal freezing and boiling points:

13.39 Phase diagrams provide "road maps" allowing us to determine what phase changes occur as temperature and pressure vary in particular ways.
(a) At T = 400 K, P = 1.00 atm, Br_2 is a gas. As it cools under a pressure of 1.00 atm, it liquefies at 331.9 K and solidifies at 265.9 K.
(b) At T = 265.8 K, P = 1.00 x 10^{-3} atm, Br_2 is a gas. Compressing it at this temperature causes it to liquefy at about P = 6 x 10^{-2} atm and solidify at about 0.5 atm.
(c) P = 2.00 x 10^{-2} atm is below the triple point of Br_2. Heating a solid at this pressure from 250 to 400 K causes sublimation to the vapor at about 265 K.

13.41 A pair of weights and a pulley are a particularly simple example of a coupled system. The lifting of the 1-kg weight is "non-spontaneous" in that it cannot occur without outside intervention. The falling of the 2-kg weight is spontaneous. By linking the two together by a rope over a pulley, the spontaneous process is coupled to the non-spontaneous, and the 1-kg weight is driven uphill by the falling 2-kg weight.

13.43 The reactions that are coupled in this example are the following:
ATP + H_2O → ADP + H_3PO_4 $\Delta G^o = -30.6$ kJ
$C_6H_{12}O_6$ (fructose) + $C_6H_{12}O_6$ (glucose) → $C_{12}H_{22}O_{11}$ + H_2O $\Delta G^o = 23.0$ kJ;
Coupled reaction: ATP + $C_6H_{12}O_6$ (fructose) + $C_6H_{12}O_6$ (glucose) → ADP + $C_{12}H_{22}O_{11}$
$\Delta G^o = 23.0$ kJ – 30.6 kJ = –7.6 kJ.

13.45 (a) Energy is always conserved, so $\Delta E_{universe} = 0$;
(b) The teaspoon warms, so $\Delta E_{teaspoon} > 0$;
(c) This is a spontaneous process, so $\Delta S_{universe} > 0$;
(d) The teaspoon warms, so $\Delta S_{teaspoon} > 0$; and (e) The teaspoon warms, so $q_{teaspoon} > 0$.

13.47 Entropy depends on amount, phase, temperature, and concentration. The order is:
(0.5 mol, l, 298 K) < (1 mol, l, 298 K) < (1 mol, l, 373 K) < (1 mol, g, 1 atm, 373 K)
< (1 mol, g, 0.1 atm, 373 K).

13.49 Spontaneity is determined by free energy, heat flow by enthalpy, and change in order by entropy: Use Equations 12-10, 13-5, and 13-10 to calculate standard thermodynamic values:

$$\Delta H^o_{rxn} = \Sigma \text{coeff}_p \Delta H^o_f(\text{products}) - \Sigma \text{coeff}_r \Delta H^o_f(\text{reactants})$$
$$\Delta S^o_{reaction} = \Sigma \text{coeff}_p S^o(\text{products}) - \Sigma \text{coeff}_r S^o(\text{reactants})$$
$$\Delta G^o_{rxn} = \Sigma \text{coeff}_p \Delta G^o_f(\text{products}) - \Sigma \text{coeff}_r \Delta G^o_f(\text{reactants})$$

(a) $\Delta G^o_{rxn} = 2(0) + 3(-237.129) - 1(-1582.3) - 3(0) = 870.9$ kJ. The reaction is not spontaneous.

(b) $\Delta H^o_{rxn} = 2(0) + 3(-285.830) - 1(-1675.7) - 3(0) = 818.2$ kJ. The reaction absorbs heat.

(c) $\Delta S^o_{reaction} = 2(28.33) + 3(69.91) - 1(50.92) - 3(130.684) = -176.58$ J/K. Entropy decreases, so the products are more ordered than the reactants.

13.51 Use Equation 13-5 to calculate the entropy change of a reaction:

$$\Delta S^o_{reaction} = \Sigma \text{coeff}_p S^o(\text{products}) - \Sigma \text{coeff}_r S^o(\text{reactants})$$

(a) $\Delta S^o_{reaction} = 2(162) - 2(101) - 1(205.138) = -83$ J/K;
(b) $\Delta S^o_{reaction} = 2(87.40) + 6(223.066) - 4(142.3) - 3(205.138) = 328.6$ J/K;
(c) $\Delta S^o_{reaction} = 6(210.761) + 6(69.91) - 3(121.21) - 4(238.93) = -478.37$ J/K

13.53 Use Equation 13-10 to calculate the standard free energy change of a reaction:

$$\Delta G^\circ_{rxn} = \sum \text{coeff}_p \Delta G^\circ_f(\text{products}) - \sum \text{coeff}_r \Delta G^\circ_f(\text{reactants})$$

(a) $\Delta G^\circ_{rxn} = 2(-3) - 2(17) - 1(0) = -40$ kJ;

(b) $\Delta G^\circ_{rxn} = 2(-742.2) + 6(0) - 4(-334.00) - 3(0) = -148.4$ kJ;

(c) $\Delta G^\circ_{rxn} = 6(86.55) + 6(-237.129) - 3(149.34) - 4(163.2) = -2004.3$ kJ.

13.55 To determine ΔG° at non-standard temperature, calculate ΔS° and ΔH° using Equations 13-5 and 12-10, then use Equation 13-10. The standard entropy changes are calculated in Problem 13.51.

$$\Delta S^\circ_{reaction} = \sum \text{coeff}_p S^\circ_{(products)} - \sum \text{coeff}_r S^\circ_{(reactants)}$$

$$\Delta H^\circ_{rxn} = \sum \text{coeff}_p \Delta H^\circ_f(\text{products}) - \sum \text{coeff}_r \Delta H^\circ_f(\text{reactants})$$

$$\Delta G^\circ_{rxn,T} \cong \Delta H^\circ_{rxn,298} - T \Delta S^\circ_{rxn,298}$$

(a) $\Delta H^\circ_{rxn} = 2(-104) - 2(-67) - 1(0) = -74$ kJ;

$\Delta G^\circ_{rxn,350} = (-74 \text{ kJ}) - (350 \text{ K})(-83 \text{ J/K})(10^{-3} \text{ kJ/J}) = -45$ kJ;

(b) $\Delta H^\circ_{rxn} = 2(-824.2) + 6(0) - 4(-399.49) - 3(0) = -50.44$ kJ;

$\Delta G^\circ_{rxn,350} = (-50.44 \text{ kJ}) - (350 \text{ K})(328.6 \text{ J/K})(10^{-3} \text{ kJ/J}) = -165$ kJ;

(c) $\Delta H^\circ_{rxn} = 6(90.25) + 6(-285.83) - 3(50.63) - 4(142.7) = -1896$ kJ.

$\Delta G^\circ_{rxn,350} = (-1896 \text{ kJ}) - (350 \text{ K})(-478.37 \text{ J/K})(10^{-3} \text{ kJ/J}) = -1729$ kJ;

13.57 The figure shows a chemical reaction in which a set of diatomic molecules fragments into atoms with no change in volume or temperature.
(a) There is no volume change, so $w_{sys} = 0$; (b) Energy must be supplied to break the chemical bonds, so $q_{sys} > 0$; and (c) Because $q_{sys} > 0$, $q_{surr} < 0$, so $\Delta S_{surr} < 0$.

13.59 A reaction is thermodynamically feasible if $\Delta G < 0$.
(a) $\Delta G^\circ = 2(-74.72) + 1(86.55) - 3(51.31) - 1(-237.129) = 20.31$ kJ; not thermodynamically feasible under standard conditions.

(b) $\Delta G = \Delta G^\circ + RT \ln Q$, $\quad Q = \left[\dfrac{(p_{HNO_3})^2(p_{NO})}{(p_{NO_2})^3}\right]$;

$\Delta G = 20.31 \text{ kJ} + RT \ln \left[\dfrac{(10^{-6})^2(10^{-6})}{(1)^3}\right] = 20.31 \text{ kJ} - 102.69 \text{ kJ} = -82.38$ kJ;

thermodynamically feasible under these conditions.

13.61 Entropy changes for constant-temperature processes can be calculated using Equation 13-1,

$\Delta S = \dfrac{q_T}{T}$, and q for water freezing can be found using $q = -n\Delta H_{fus}$:

$n_{water} = m/MM = (155\text{ g}) \div (18.02\text{ g/mol}) = 8.60\text{ mol}$; $T_{water} = (0.0 + 273.15) = 273.2\text{ K}$;
$q_{water} = -(8.60\text{ mol})(6.01\text{ kJ/mol})(10^3\text{ J/kJ}) = -51.7 \times 10^3\text{ J}$; $T_{surr} = (-20.0 + 273.15) = 253.2\text{ K}$;

(a) $\Delta S_{water} = (-51.7 \times 10^3\text{ J}) \div (273.2\text{ K}) = -189\text{ J/K}$;
(b) $\Delta S_{surr} = (51.1 \times 10^3\text{ J}) \div (253.2\text{ K}) = 204\text{ J/K}$; $\Delta S_{universe} = -189 + 204 = 15\text{ J/K}$;
(c) Cooling the ice to $-20\ °C$ is an irreversible change (because the temperatures of ice and freezer are not the same except at the end of the process), for which the $\Delta S_{universe} > 0$. Thus the cooling must generate an additional entropy increase for the universe.

13.63 Spontaneity is determined by free energy, heat flow by enthalpy, and change in order by entropy: Use Equations 12-10, 13-5, and 13-10 to calculate standard thermodynamic values:

$$\Delta H^o_{rxn} = \sum \text{coeff}_p\, \Delta H^o_f(\text{products}) - \sum \text{coeff}_r\, \Delta H^o_f(\text{reactants})$$

$$\Delta S^o_{rxn} = \sum \text{coeff}_p S^o(\text{products}) - \sum \text{coeff}_r S^o(\text{reactants})$$

$$\Delta G^o_{rxn} = \sum \text{coeff}_p\, \Delta G^o_f(\text{products}) - \sum \text{coeff}_r\, \Delta G^o_f(\text{reactants})$$

(a) $\Delta G^o_{rxn} = 1(-261.905) + 1(-131.228) - 1(-384.138) = -8.995\text{ kJ}$. The reaction is spontaneous.

(b) $\Delta H^o_{rxn} = 1(-240.12) + 1(-92.307) - 1(-411.153) = 78.73\text{ kJ}$. The reaction absorbs heat.

(c) $\Delta S^o_{rxn} = 1(59.0) + 1(56.5) - 1(72.13) = 43.4\text{ J/K}$. Entropy increases, so the amount of order decreases during this reaction.

13.65 In general, we expect the standard entropies of gaseous substances to increase with the number of atoms per molecule; thus the lower entropy of methane (5 atoms) than ammonia (4 atoms) is surprising. Methane, however, is highly symmetric, so it has fewer different possible rotational orientations in space. This leads to less rotational disorder and accounts for the difference.

13.67 (a) When dry ice at its sublimation temperature is placed in water at its freezing temperature, energy flows as heat from molecules of H_2O to molecules of CO_2, allowing CO_2 molecules to sublime into the gas phase and causing H_2O molecules to form solid ice. The CO_2 system undergoes its phase transition at 195 K, while the H_2O system undergoes its phase transition at 273.15 K.
(b) Entropy changes for constant-temperature processes are calculated using Equation 13-1,

$\Delta S = \dfrac{q_T}{T}$, and q for dry ice subliming can be found using $q = n\Delta H_{subl}$:

$n_{dry\ ice} = m/MM = (12.5\text{ g}) \div (44.01\text{ g/mol}) = 0.284\text{ mol}$;
$q_{dry\ ice} = (0.284\text{ mol})(25.2\text{ kJ/mol})(10^3\text{ J/kJ}) = 7.16 \times 10^3\text{ J}$;
$\Delta S_{dry\ ice} = (7.16 \times 10^3\text{ kJ}) \div (195\text{ K}) = 36.7\text{ J/K}$;
$\Delta S_{water} = (-7.16 \times 10^3\text{ kJ}) \div (273.15\text{ K}) = -26.2\text{ J/K}$; $\Delta S_{universe} = 36.7 - 26.2 = 10.5\text{ J/K}$;

13.69 Use bond energy, intermolecular forces, and order-disorder to predict signs for ΔH and ΔS:
(a) Bonds form, so energy is released and ΔH is negative. The number of independent particles decreases, so order increases and ΔS is negative.
(b) When a solid melts, energy must be added; q is positive, ΔH is positive, and ΔS is positive.
(c) Combustion reactions release energy, so we expect ΔH to be negative. The number of molecules of gas remain the same during this reaction, so ΔS is expected to be small, but it is not possible to predict whether it is positive or negative without doing actual calculations.

13.71 The standard free energies of formation of oxides and sulfides are generally highly negative, so reactions of elements with elemental oxygen or sulfur are highly spontaneous. Thus most elements have long since reacted to form oxides and/or sulfides. For nitrogen, however, the standard free energies of formation of compounds are generally positive, so molecular nitrogen does not spontaneously react to form oxides or sulfides.

13.73 The attractive forces in CaO are substantially larger than the attractive forces in KCl, so the individual Ca^{2+} and O^{2-} ions cannot as easily vibrate in the crystal. Thus the amount of vibrational disorder is substantially less in CaO than in KCl.

13.75 (a) Water falling from high to low altitude is a spontaneous process that drives the turbine and "organizes" electrons into an electric current, a non-spontaneous process.
(b) Gasoline burns in the engine in a spontaneous process that drives the pump and pushes water uphill, a non-spontaneous process.

13.77 Use Equations 12-10 and 13-10 to calculate standard thermodynamic values:

$$\Delta H^\circ_{rxn} = \sum coeff_p \, \Delta H^\circ_f(products) - \sum coeff_r \, \Delta H^\circ_f(reactants)$$

$$\Delta G^\circ_{rxn} = \sum coeff_p \, \Delta G^\circ_f(products) - \sum coeff_r \, \Delta G^\circ_f(reactants)$$

CCl_4: $\Delta H^\circ_{rxn} = 1(-393.509) + 4(102.5) - 1(-135.44) - 5(0) = 151.9$ kJ;

$\Delta G^\circ_{rxn} = 1(-394.359) + 4(120.5) - 1(-65.21) - 5(0) = 152.9$ kJ;

CS_2: $\Delta H^\circ_{rxn} = 1(-393.509) + 2(-296.830) - 1(89.70) - 3(0) = -1076.87$ kJ;

$\Delta G^\circ_{rxn} = 1(-394.359) + 2(-300.194) - 1(65.27) - 3(0) = -1060.02$ kJ.

The calculations indicate that combustion of CS_2 is highly spontaneous, while that of CCl_4 is not. Thus CCl_4 is not a fire hazard, but plants using CS_2 must be extremely cautious about fire hazards, as this solvent has a similar flammability as gasoline.

13.79 Make use of thermodynamic definitions to identify the state functions:
(a) $q_V = \Delta E$; (b) $q_p = \Delta H$; (c) $q_T = T\Delta S$.

13.81 (a) Whenever a substance cools, ΔS is negative.
(b) Whenever a gas is compressed (at constant T), ΔS is negative.
(c) Whenever two substances mix, disorder increases and ΔS is positive.

13.83 Ammonia is a toxic gas, while urea and ammonium nitrate both are relatively non-toxic solids. Thus the transport and application of ammonia entail significant risks to humans. Even in aqueous solution, ammonia is highly irritating, as anyone knows who has used strong ammonia as a cleanser.

13.85 A phase diagram provides a "road map" that allows us to determine the temperatures and pressures at which phase changes occur.
(a) A constant temperature process is a vertical line on the phase diagram. $T = 70$ K is in between the triple point (0.124 atm, 63 K) and normal boiling point (1 atm, 77 K) of nitrogen. At 70 K, 1 atm, nitrogen is liquid. When the pressure falls below about 0.5 atm, the liquid boils, and at 70 K, 0.1 atm, the substance is gaseous.
(b) A constant temperature process is a vertical line on the phase diagram. $T = 298$ K is within the gaseous region at all pressures, so compression from 1.00 atm to 50.0 atm leaves the substance gaseous.
(c) A constant pressure process is a horizontal line on the phase diagram, and 1 atm represents "normal" pressure, shown on the figure as a dotted line. When the temperature of the gas reaches 77 K, the gas liquefies, and when the temperature reaches 63 K, the liquid solidifies. At 50 K, N_2 is solid.

13.87 Opening the refrigerator door will result in overall heating rather than cooling, because in order to remove heat from its contents, the refrigerator must transfer additional heat to the surroundings. Coolers such as air conditioners reduce the internal temperature at the cost of increasing the external temperature by transferring more heat to the surroundings than is removed from the system.

13.89 According to your textbook, the complete oxidation of 1 mol of palmitic acid produces 130 ATP molecules. The reactions are:
$$ADP + H_3PO_4 \rightarrow ATP + H_2O \quad \Delta G° = +30.6 \text{ kJ}$$
Palmitic acid: $C_{15}H_{31}CO_2H + 23\ O_2 \rightarrow 16\ CO_2 + 16\ H_2O \quad \Delta G° = -9790 \text{ kJ}$

Thus the amount of energy stored is (130)(30.6 kJ) = 3978 kJ and the efficiency is:
Efficiency = (100 %)(Energy stored)/(Energy released) = (100 %)(3978 kJ) ÷ (9790 kJ)
Efficiency = 40.6 %.

13.91 Both of these processes are spontaneous. Thus, assuming that each of these processes occurs at constant T and P (perhaps more reasonable for the firefly than for lightning), both must have
$\Delta G < 0$.

13.93 The heat flow described in this problem is from body + water to the atmosphere. Here is a flow diagram of the process:

The key feature is that heat is transferred from system at 37.5 °C to surroundings at 23.5 °C.

$q = n \Delta H_{vap}$ = (1.0 g)(40.79 kJ/mol)(10^3 J/kJ) ÷ (18.02 g/mol) = 2.3 x 10^3 J.
ΔS_{sys} = (– 2.3 x 10^3 J) ÷ (37.5 + 273.15 K) = – 7.4 J/K;
ΔS_{surr} = (2.3 x 10^3 J) ÷ (23.5 + 273.15 K) = 7.7 J/K;
ΔS_{total} = 7.7 – 7.4 = 0.3 J/K.

Chapter 14 Mechanisms of Chemical Reactions

14.1 What Is a Mechanism?

QUESTIONS TO ANSWER, SKILLS TO LEARN
1. **What is the difference between a balanced equation of reaction and a mechanism?**
2. **What are elementary processes?**
3. **Learning to recognize chemically valid mechanisms**
4. **What are intermediates in mechanisms?**
5. **What is the "rate-determining step" in a mechanism?**

A balanced equation of a chemical reaction can be compared to two pictures: one of a person before a haircut and another of the person after the haircut with a bag containing the hair that was cut off. The pictures identify the appearance and amounts of the reactants (person with uncut hair) and products (person with cut hair and trimmings; remember, mass is conserved). What is missing, however, is the process by which the reactants are converted to the products. The process of the haircut involves many different steps, perhaps trimming the same area several times, which are not evident from examination of the before (reactants) and after (products) pictures. The same is true of a chemical reaction.

Consider a chemical reaction like the conversion of palmitic acid and oxygen to CO_2 and water:

$$C_{15}H_{31}CO_2H + 23\ O_2 \rightarrow 16\ CO_2 + 16\ H_2O$$

There are 24 molecules of reactants and 32 molecules of products. What is involved in the process of the conversion? Common sense dictates that atoms that are bonded to one another must at some point come close enough to form the bond.

For chemical bonds to be formed, the atoms involved in the bond must come in contact with one another. This event usually occurs through a collision.

In the case of palmitic acid, the fact that the reaction takes place in a series of steps was mentioned in Chapter 13 as part of the gradual release of energy available from the molecule in the process of metabolism. However, another reason that the process must take place in a series of steps is that 1 molecule of palmitic acid and 23 molecules of O_2 are very unlikely to collide in the same place in space at the same time in the right orientation to form the products. In fact, the reaction proceeds through simpler reactions involving breaking down the palmitic acid into smaller fragments and then the eventual conversion of the fragments to CO_2 and H_2O. The sequence of chemical reactions describing the conversion is called the **mechanism** of the reaction.

The series of steps that is proposed to account for the net conversion of products to reactants is called the *mechanism* of the reaction.

The formation of isopropyl alcohol (2-propanol) from water and propene in the presence of sulfuric acid is a two-step mechanism.

$$H_2C = CHCH_3 + H_2O \xrightarrow{H_2SO_4} CH_3CH(OH)CH_3$$

The first step in the mechanism is the attack of sulfuric acid on the propene molecule to give isopropyl sulfate:

$$H_2C = CHCH_3 + H_2SO_4 \rightarrow CH_3CH(OSO_2(OH))CH_3$$

The second step is the reaction of isopropyl sulfate with water to give the product, isopropyl alcohol:

$$CH_3CH(OSO_2(OH))CH_3 + H_2O \rightarrow CH_3CH(OH)CH_3 + H_2SO_4$$

Each of the reactions in a mechanism is called an *elementary reaction*. The reaction for the formation of isopropyl alcohol in the presence of sulfuric acid has two elementary reactions. In general, most reaction mechanisms have two or more elementary reactions; very few reactions have one elementary reaction.

Elementary Reactions (and: How many molecules are involved?)

The first elementary reaction results from the collision between two molecules, the sulfuric acid and propene molecules. In general, an elementary reaction that involves the collision of two molecules is a **bimolecular reaction**. Note that both of the elementary reactions in the mechanism for 2-propanol synthesis are bimolecular reactions.

Chapter 14: Mechanisms of Chemical Reactions

Elementary reactions that involve just one molecule are called **unimolecular reactions**. Unimolecular reactions are usually isomerization reactions, such as the rearrangement of cyclopropane to propene:

$$\text{cyclopropane} \longrightarrow \text{propene}$$

or reactions in which two or more molecules result from a single molecule such as the conversion of $C_{10}H_{12}$ to cyclopentadiene, an important industrial chemical:

$$C_{10}H_{12} \longrightarrow 2 \text{ cyclopentadiene}$$

The last type of elementary reaction is one in which three molecules must collide, a **termolecular reaction**. Termolecular processes are rare; collisions involving three molecules are much less common than those involving two molecules. In addition, most of the orientations of the moecules in termolecular collisions are unfavorable, making termolecular processes usually unimportant in the mechanisms we will encounter in the following sections. Unimolecular and bimolecular reactions dominate reaction mechanisms.

An important requirement of a reaction mechanism is that the sum of all the elementary reactions in a mechanism results in the net balanced chemical equation of reaction. The mechanism for the formation of isopropyl alcohol illustrates this principle:

$$H_2C = CHCH_3 + H_2SO_4 \rightarrow CH_3CH(OSO_2(OH))CH_3 \quad \text{Step 1}$$

$$\underline{CH_3CH(OSO_2(OH))CH_3 + H_2O \rightarrow CH_3CH(OH)CH_3 + H_2SO_4 \quad \text{Step 2}}$$

$$H_2C = CHCH_3 + H_2O \rightarrow CH_3CH(OH)CH_3 \quad \text{net reaction}$$

The sum of all the elementary reactions in a valid mechanism is the balanced chemical equation of the reaction.

Notice that one chemical species, $CH_3CH(OSO_2(OH))CH_3$, is found in both elementary reactions (as a product in step 1 and as a reactant in step 2), but does not appear in the products. Because this compound is neither a reactant nor a product but is formed and consumed in the course of the reaction, it is called an **intermediate**.

***Intermediates* are formed in one step of a reaction mechanism and consumed in another step of the mechanism. They do not appear in the net, balanced equation of reaction.**

Chapter 14: Mechanisms of Chemical Reactions

Exercise 14.1. The following reaction is observed to proceed in aqueous solution:

$$I^-_{(aq)} + OCl^-_{(aq)} \rightarrow Cl^-_{(aq)} + OI^-_{(aq)}$$

(a) Is the following sequence of reactions a valid *possible* mechanism for this reaction?
(b) If so, identify the chemical species that are intermediates.

$$OCl^-_{(aq)} + H_2O \rightarrow HOCl_{(aq)} + OH^-_{(aq)}$$

$$I^-_{(aq)} + HOCl_{(aq)} \rightarrow HOI_{(aq)} + Cl^-_{(aq)}$$

$$HOI_{(aq)} + OH^-_{(aq)} \rightarrow OI^-_{(aq)} + H_2O$$

Steps to Solution: A possible mechanism has one criterion that the individual steps must satisfy before the mechanism is subjected to experimental verification: the sum of all of the steps must result in the balanced chemical equation for the reaction. Note that this sequence of reactions satisfies this condition. The chemical species shown in **bold** appear twice in the sequence, once as a reactant and once as a product and thus "cancel out".

(a)

$$OCl^-_{(aq)} + H_2O \rightarrow \mathbf{HOCl}_{(aq)} + \mathbf{OH^-}_{(aq)}$$

$$I^-_{(aq)} + \mathbf{HOCl}_{(aq)} \rightarrow \mathbf{HOI}_{(aq)} + Cl^-_{(aq)}$$

$$\mathbf{HOI}_{(aq)} + \mathbf{OH^-}_{(aq)} \rightarrow OI^-_{(aq)} + H_2O$$

$$I^-_{(aq)} + OCl^-_{(aq)} \rightarrow Cl^-_{(aq)} + OI^-_{(aq)}$$

This mechanism meets the stoichiometric test of validity. A proposed reaction mechanism whose steps add up to give the balanced chemical equation of reaction must undergo experimental verification through means we will discuss later.

(b) To identify chemical species that are intermediates, we must examine the mechanism and find the compounds that are neither products nor reactants. The species in **bold face type** in the above mechanism are active in the reaction mechanism without being either products or reactants. However, water is present in relatively large quantities in aqueous solutions compared to other species, and is not considered an intermediate because it is not produced through the mechanism. The intermediates generated from starting materials are HOCl, HOI and OH⁻.

Chapter 14: Mechanisms of Chemical Reactions

How fast can a reaction go? (or, How fast are each of its steps/elementary reactions?)

The speed of a chemical reaction is measured by how quickly molecules of reactants are converted to molecules of products. In a reaction that has several steps in the mechanism, such as that in Exercise 14.1, each step has an inherently different speed. The first and last reactions in Exercise 14.1 are known to be very fast, and the second reaction is slower. The formation of Cl^- depends on the speed of the second reaction, and OI^- will not be formed until HOI is formed in the second reaction. Once HOI is formed, it is rapidly converted to OI^- through the fast reaction in the last step. Therefore the speed of the second step determines the overall speed of the reaction, so the second step is called the **rate-determining step** of the reaction.

> The *rate-determining step* of a reaction is the rate of
> the slowest elementary reaction in the mechanism of the reaction.

A reasonable analogy to a reaction mechanism is the process of sending someone a message. We can write a four-step mechanism for this process:

1. Composing, writing, and putting the letter in an addressed envelope
2. Placing the envelope into a mailbox
3. Pickup and delivery of the envelope to the addressee
4. Opening and reading the letter

This "mechanism" usually will have step 3 as the rate-determining step; this often takes a day or two, whereas the other steps may not take as long. To speed up the process, people now use overnight couriers, fax machines, or even electronic mail (E-mail) to speed step 3. However, this may not always succeed; if the addressee is on vacation, the final step of reading the mail may not occur for several days, thus changing the rate-determining step for the process to step 4. Similarly, the rate-determining step may not not always be the same in chemical reactions.

14.2 Rates of Chemical Reactions

QUESTIONS TO ANSWER, SKILLS TO LEARN
1. **Measuring chemical change as a function of time**
2. **How to express reaction rates in terms of concentrations of reactants or products**

In general, chemical change is measured by the number of molecules of reactant(s) that are converted to products. The speed of a chemical reaction is a measurement of the number of molecules converted per unit time:

Chapter 14: Mechanisms of Chemical Reactions

$$\text{rate} = \frac{\text{change in number of molecules}}{\text{change in time}} = \frac{\Delta \text{ molecules}}{\Delta \text{ time}}$$

The number of molecules is expressed in units of concentration, usually moles/L. The units of time are typically expressed in seconds. Consider the idea of reaction rate with the reaction for the formation of 2-propanol:

$$H_2C = CHCH_3 + H_2O \rightarrow CH_3CH(OH)CH_3$$

If the reaction mixture initially consists of propene and water of unequal concentrations and no 2-propanol is present, then we obtain the points on the graph below at time = 0.

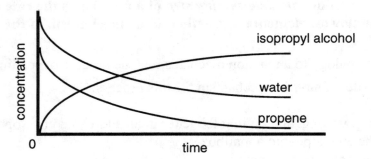

However, as the reaction proceeds, propene and water are converted to 2-propanol (isopropyl alcohol). The volume of the system is kept constant, so as the number of molecules decreases, the concentration decreases as well. The concentration of 2-propanol, however, will increase, because it is being produced. Because the stoichiometric relationship between both reactants and the product is 1:1, an increase in the concentration of isopropanol will result in an equal-sized *decrease* in the concentrations of propene and water. The rate of the reaction is calculated as:

$$\text{rate} = \frac{c(i)_2 - c(i)_1}{t_2 - t_1} = \frac{\Delta c(i)}{\Delta t}$$

$c(i)$ = concentration of a reactant, i, at times t_2 and t_1 where $t_2 > t_1$

If we examine the plot of concentration of propene as a function of time, the following observations can be drawn.

1. For equal increments of Δt, the change in concentration grows smaller as the reaction proceeds. That is, at early times the concentration changes rapidly (the curve is steep), but at later times the curve is almost a horizontal line, indicating nearly constant concentration. This means that the reaction rate changes and the rate changes because the number of reactant molecules has decreased, so the number of collisions that might lead to product formation has also decreased.

2. The size of the time increment, Δt, is important. The rate of the reaction is most accurately represented by the smallest time increments possible. For large increments of Δt, the ratio of $\dfrac{\Delta c}{\Delta t}$ will not be an accurate approximation of the slope of c vs. t.

3. Because of the 1:1 stoichiometry between all the reactants and products in this reaction, the sizes of the concentration changes are equal. In cases where the stoichiometric ratios are not 1:1, the concentration changes would be unequal and $\dfrac{\Delta c}{\Delta t}$ would not be the same for the different reactants.

4. The concentration changes for the reactants are *negative* (decreasing concentration) and those for the product are *positive* (increasing concentration).

Exercise 14.2 The following concentration vs. time data were collected for the reaction:

$$A + 2B \rightarrow C$$

time (s)	[A] (M)	[B] (M)	[C] ((M)
0.00	1.50×10^{-2}	1.00×10^{-1}	0.00
6.00×10^{1}	1.37×10^{-2}	9.74×10^{-2}	1.29×10^{-3}
1.80×10^{2}	1.15×10^{-2}	9.29×10^{-2}	3.55×10^{-3}
4.20×10^{2}	7.99×10^{-3}	8.60×10^{-2}	7.01×10^{-3}
4.80×10^{2}	7.30×10^{-3}	8.46×10^{-2}	7.70×10^{-3}
6.00×10^{2}	6.10×10^{-3}	8.22×10^{-2}	8.90×10^{-3}
7.20×10^{2}	5.09×10^{-3}	8.02×10^{-2}	9.91×10^{-3}
9.00×10^{2}	3.89×10^{-3}	7.78×10^{-2}	1.11×10^{-2}
9.60×10^{2}	3.55×10^{-3}	7.71×10^{-2}	1.14×10^{-2}

Calculate $\dfrac{\Delta c}{\Delta t}$ for A, B and C for the following time differences: 0 and 60 s; 900 and 960 s.

Steps to Solution: *The concentration change over a time period is to be calculated for all three compounds at different times. This exercise will demonstrate the slowing of the concentration change with time and the effect of stoichiometry. The answers are calculated using the formula:*

$$\frac{\Delta c(i)}{\Delta t} = \frac{c(i)_2 - c(i)_1}{t_2 - t_1}$$

Remember that $t_2 > t_1$ and that the $c(i)$'s are the concentrations of the desired compound at t_1 and t_2.

$t_1 = 0$ and $t_2 = 60$ s:

$$\frac{\Delta[A]}{\Delta t} = \frac{1.37 \times 10^{-2} - 1.50 \times 10^{-2}}{60 - 0} = -2.15 \times 10^{-5} \text{ M/s}$$

$$\frac{\Delta[B]}{\Delta t} = \frac{9.74 \times 10^{-2} - 1.00 \times 10^{-1}}{60 - 0} = -4.30 \times 10^{-5} \text{ M/s}$$

$$\frac{\Delta[C]}{\Delta t} = \frac{1.29 \times 10^{-3} - 0}{60 - 0} = 2.15 \times 10^{-5} \text{ M/s}$$

$t_1 = 900$ and $t_2 = 960$ s: Using similar methods you should obtain:

$\frac{\Delta[A]}{\Delta t} = -5.58 \times 10^{-6}$ M/s	$\frac{\Delta[B]}{\Delta t} = -1.12 \times 10^{-5}$ M/s	$\frac{\Delta[C]}{\Delta t} = 5.58 \times 10^{-6}$ M/s

Notice that the reaction speed, $\frac{\Delta c}{\Delta t}$, decreases at greater time, indicating that the rate of chemical change is slowing. Note also the different signs for the rates for A and B vs. C as well as the different sizes of the rates of change of the concentrations of A and B.

Relative Rates: Getting the same value for the rate

Exercise 14.2 shows that the rates of change of concentration of the different reagents in a reaction differ depending on their stoichiometric coefficients and role as reactant or product. However, the rate of a reaction should have the same value no matter what reagent you use to define the rate. In addition, the rate of a process is expected to always be a positive number. To satisfy these requirements, chemists use *relative rates* that use the stoichiometric factor to obtain numerically identical rates from these rates of concentration changes. Rates of disappearance of reactant are multiplied by -1 so that a positive rate is obtained. When chemists speak of the rate of a reaction, they are referring to the relative rate: the rate that is the same no matter what reagent you use and is always positive.

As a general rule, for the reaction where A, B, C and D are substances and a, b, c and d are the stoichiometric coefficients,

$$a A + b B \rightarrow c C + d D$$

the rate of the reaction is given by:

$$\text{rate} = -\frac{1}{a}\frac{\Delta[A]}{\Delta t} = -\frac{1}{b}\frac{\Delta[B]}{\Delta t} = \frac{1}{c}\frac{\Delta[C]}{\Delta t} = \frac{1}{d}\frac{\Delta[D]}{\Delta t}$$

Chapter 14: Mechanisms of Chemical Reactions

Exercise 14.3 Convert the rate of change of concentrations of A, B and C calculated in Exercise 14.2 for t_1 = 420 s and t_2 = 480 s to relative rates.

Steps to Solution: To convert the rate of change of concentration to relative rate, we must first divide $\frac{\Delta c}{\Delta t}$ by the stoichiometric coefficient. If $\frac{\Delta c}{\Delta t}$ is negative (a reactant is being consumed), we must multiply that result by -1.

The concentration change rates are listed below.

$\frac{\Delta[A]}{\Delta t}$ = -1.15 x 10⁻⁵ M/s	$\frac{\Delta[B]}{\Delta t}$ = -2.29 x 10⁻⁵ M/s	$\frac{\Delta[C]}{\Delta t}$ = 1.15 x 10⁻⁵ M/s

The equation of the reaction is: A + 2B → C

A: A is a reactant with a stoichiometric factor of 1. The rate of the reaction is the negative of the rate of change of the concentration of A.

$$\text{rate} = -\frac{1}{1} \cdot \frac{\Delta[A]}{\Delta t} = 1.15 \times 10^{-5} \text{ M/s}$$

B: B is a reactant with a stoichiometric factor of 2. The rate of the reaction is $\frac{1}{2}$ times the negative of the rate of change of the concentration of B.

$$\text{rate} = -\frac{1}{2} \cdot \frac{\Delta[B]}{\Delta t} = -\left(\frac{1}{2}\right)(-2.29 \times 10^{-5} \text{ M/s}) = 1.15 \times 10^{-5} \text{ M/s}$$

C: C is a product with a stoichiometric factor of 1. The rate of the reaction is the rate of change of the concentration of C.

$$\text{rate} = \frac{1}{1} \cdot \frac{\Delta[C]}{\Delta t} = 1.15 \times 10^{-5} \text{ M/s}$$

Note that all the relative rates are the same in magnitude, units and sign.

14.3 Concentration and Reaction Rates

QUESTIONS TO ANSWER, SKILLS TO LEARN
1. **How do changing concentrations of reactants affect reaction rates?**
2. **What is a rate law?**
3. **How does a rate law relate to a reaction mechanism?**
4. **What is the rate constant and what are the units of the rate constant?**

Chapter 14: Mechanisms of Chemical Reactions

Concentration plays an important role in the rate or speed of a chemical reaction; nearly always, increasing the concentration of a reactant will increase the rate of a reaction. Reactions occur between two molecules when they collide. In elementary reactions that involve more than one molecule, the increase in rate is easy to understand: the higher the concentration of reactants, the more collisions occur between reactant molecules and the greater the likelihood of a reaction occurring. In unimolecular elementary processes, the rate increases because there are more reactant molecules present that may react. Other reasons for increased rate will be explained in the discussion of the temperature dependence of reaction rates.

Exercise 14.4 Consider the following three molecular pictures that represent the relative numbers of the two reactants involved in one step of the depletion of stratospheric ozone by chlorine atoms.

(a)   (b)   (c)

The equation for the elementary reaction and a molecular picture of the reaction process are shown below.

$$Cl\bullet + O_3 \rightarrow ClO + O_2$$

reactants   collision   products

If the three samples represented by (a), (b) and (c) are at the same temperature, what are the rates of reaction of (b) and (c) compared to that of (a)?

Steps to Solution: The rate of the reaction depends upon the number of collisions between the Cl• and ozone molecules. The problems states that the temperature is the same in all three samples so the average kinetic energy in all three samples is the same. Therefore the relative rates of reaction will depend upon the relative number of collisions between reactant atoms present in the mixture. The number of collisions will depend upon the number of atoms of each reactant per unit volume (concentration). The following table summarizes the numbers of each reactant present in each molecular picture.

512

Picture	a	b	c
# Cl atoms	2	1	2
# ozone molecules	4	4	8

In (b), the number of ozone molecules is the same as in (a), but the number of Cl atoms in (b) is half that in (a). Therefore the number of collisions between the Cl atoms and ozone will be half that in (a), and we expect that the reaction rate in (b) will be half that in (a).

In (c), both the number of Cl atoms and the number of ozone molecules are twice the numbers of those chemical species in (a). Doubling the number of ozone molecules alone will double the number of Cl atom-ozone collisions. Doubling the ozone concentration alone would double the number of ozone-chlorine atom collisions. Doubling the concentrations of both chlorine and ozone will increase the number of collisions between Cl atoms and ozone molecules by four times compared to (a). Therefore the reaction rate in (c) would be expected to be four times greater than that in (a).

Rate Laws

The quantitative relation between rate and concentration is called the **rate law**. A rate law can be derived from a proposed mechanism, but the true rate law is obtained from meticulous experimental work. Experimental methods of determining rate laws are discussed in Section 14.4. We now examine (a) the general form of the rate law, (b) the concentration-independent portion of the rate law (the **rate constant**) and (c) how a rate law may be obtained given a simple mechanism. Chemists study the kinetics of reaction to determine rate laws and use the rate laws and further experiments to determine the mechanism of reactions.

The general form of the rate law is:

$$\text{rate} = k\,[A]^x[B]^y[C]^z \ldots$$

where A, B, and C are some of the chemical species present in the reaction system and x, y and z are exponents that may be whole numbers or fractions. The k is the **rate constant**, a temperature dependent proportionality constant particular to that reaction.

The chemical species in the rate law may or (more likely) may not represent all of the reactants present in the balanced chemical equation of reaction and may also include other compounds that are not reactants (such hydrogen ion concentration). The exponents x, y and z are called the **order** of the reaction with respect to that particular reactant. The sum of all the exponents is the **overall order of the reaction**. The order of a reactant reflects sensitivity of reaction rate to that particular reactant.

Chapter 14: Mechanisms of Chemical Reactions

> **COMMON PITFALL: Confusing stoichiometry and order.**
> The stoichiometric coefficient and the reaction order of a reagent are usually *not the same*.

Exercise 14.5 The rate law for the reaction of NO with O_2 to give NO_2 is shown below.

$$\text{rate} = k[NO]^2[O_2]$$

(a) What is the overall order of the reaction?
(b) What is the order of the reaction with respect to NO and O_2?
(c) If all other conditions are kept constant, what will be the effect on the rate if the concentration of NO is doubled?
(d) If all other conditions are kept constant, what will be the effect on the rate if the concentration of O_2 is doubled?

Steps to Solution: This problem requires interpretation of the rate law.
(a) The overall order of the reaction is the sum of all the exponents in the rate law. The exponent of NO is 2 and the exponent of O_2 is 1, therefore the overall order = 2 + 1 = 3. The reaction is third order overall.

(b) The order of the reaction with respect to NO is its exponent, 2; therefore the reaction is second order in NO. The exponent for O_2 is 1; therefore the reaction is first order in O_2.

(c) The rate is determined by the rate law. In this problem, we compare the rates of two reactions where one parameter has changed; that is most easily done by taking the ratio of the two rates. In this case, the only thing that changes is the concentration of NO, where the [NO] has increased by a factor of 2. The data are summarized below at left and the ratio of the two rates is calculated at right.

Experiment	[NO]	[O_2]
1	$[NO]_1$	$[O_2]_1$
2	$[NO]_2 = 2[NO]_1$	$[O_2]_2 = [O_2]_1$

The subscripts indicate the different experiments. Because the concentration of [NO] in the second experiment is double that in the first experiment or

$$[NO]_2 = 2[NO]_1$$

$$\frac{\text{rate (2)}}{\text{rate (1)}} = \frac{k([NO]_2)^2[O_2]_2}{k([NO]_1)^2[O_2]_1}$$

Substituting the values so that all concentrations are expressed in terms of the subscript concentrations:

$$\frac{\text{rate (2)}}{\text{rate (1)}} = \frac{k(2[NO]_1)^2[O_2]_2}{k([NO]_1)^2[O_2]_1}$$

$$\frac{\text{rate (2)}}{\text{rate (1)}} = (2)^2 = 4$$

$$\text{rate (2)} = 4 \cdot \text{rate(1)}$$

Increasing the concentration of NO by a factor of 2 increases the rate by a factor of 4!

(d) *Using the type of reasoning in part (c) shows that increasing the concentration of O_2 by a factor of 2 increase the rate by a factor of 2, a consequence of the reaction being first order in O_2.*

One important message from the results of this example is that by performing experiments where we change the concentration of a reactant and measure the rate of a reaction, we can determine the order of the reaction with respect to one reactant. Rates change much more rapidly when the reactant has a larger order. A second message is that it is only by experiments that the true rate law is found. From the rate law and more experiments are deduced the mechanism of the reaction.

14.4 Experimental Kinetics

QUESTIONS TO ANSWER, SKILLS TO LEARN
1. **Experimental evidence gives the real rate law!**
2. **How does concentration change with time in first- and second-order reactions?**
3. **How can we measure the order of just one reactant in a rate law?**

All of the previous discussion focuses on molecular events and how they relate to the speed/rate of a reaction. Kinetics would be relatively simple if we could view the events at the microscopic level and then determine rate laws. However, we generally cannot observe the events at the microscopic level to determine the mechanism. A chemist studying kinetics determines the rate law through experiments observing how concentration changes with time in the course of several reactions, and then proposes mechanisms that are consistent with that rate law. Further tests of the mechanisms are made to find which is the most consistent with experimental data. In this section we review the theory and general methods used to determine rate laws.

The Integrated Rate Laws; Reactant Order by Graphical Methods

The rate laws express the rate as a function of concentration of the reagents. Using calculus, these rate laws can be modified to give a function in concentration and time, the **integrated rate law**. These functions are linear in time so that a plot of the function of concentration vs. time will give a linear plot for a reaction that is of the correct order. The following table summarizes the results for reactions that are first order or second order in a substance, A.

Chapter 14: Mechanisms of Chemical Reactions

Order of Reaction in A	Rate Law	Integrated Rate Law	Linear Plot WHEN	Rate Constant, k
1st	rate = $k[A]$	$\ln[A] = \ln[A]_0 - kt$	$\ln[A]$ plotted vs. t	negative of slope of *linear* plot
2nd	rate = $k[A]^2$	$\dfrac{1}{[A]} = \dfrac{1}{[A]_0} + kt$	$\dfrac{1}{[A]}$ plotted vs. t	slope of *linear* plot

Exercise 14.6 Data for the gas phase decomposition of trioxane, $C_3H_6O_3$, to formaldehyde, CH_2O, at 519 K are listed below. Do the data support a mechanism that is first order or second order in trioxane? Determine the value of the rate constant in the rate law supported by the data and write the rate law for the decomposition of trioxane to formaldehyde.

$C_3H_6O_3\,(g) \rightarrow 3\,CH_2O\,(g)$

t(s)	$[C_3H_6O_3]$ (M)
0	1.50
1200	1.04
2400	0.72
3600	0.50
4800	0.35
6000	0.24
7200	0.17

Steps to Solution: One way to determine whether a reaction is first order or second order in a single reagent is by graphical methods. The plot of a $\ln[A]$ vs t is linear for a first order process, but nonlinear for other order reactions. The plot of $\dfrac{1}{[A]}$ vs t will be linear for a second order reaction, and nonlinear for other orders. Both sets of data will be plotted below.

First Order Analysis: *A plot of the logarithm (or natural logarithm) of concentration of the reagent vs. t is linear for a first order reaction. The data and resulting plot are shown below.*

516

t(s)	ln[$C_3H_6O_3$]
0	0.40
1200	0.039
2400	-0.33
3600	-0.69
4800	-1.06
6000	-1.42
7200	-1.79

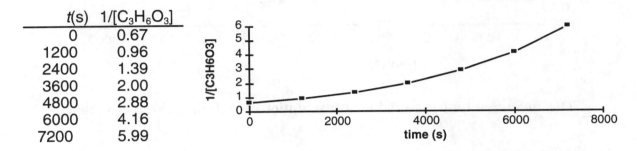

This plot looks quite linear. For comparison purposes, the second-order analysis follows. Here the reciprocal of concentration, 1/[$C_3H_6O_3$], is plotted vs time.

t(s)	1/[$C_3H_6O_3$]
0	0.67
1200	0.96
2400	1.39
3600	2.00
4800	2.88
6000	4.16
7200	5.99

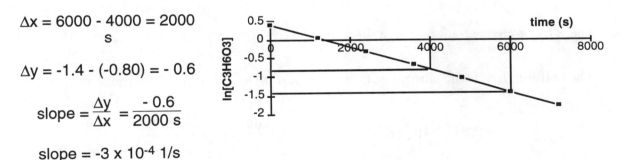

The first order plot is linear, indicating a reaction first order in trioxane. Therefore the rate constant will determined from the slope of the first order plot. The slope of the plot is calculated by choosing two convenient points and finding Δx and Δy. The points are indicated in the plot below

Δx = 6000 - 4000 = 2000 s

Δy = -1.4 - (-0.80) = - 0.6

slope = $\frac{\Delta y}{\Delta x}$ = $\frac{-0.6}{2000 \text{ s}}$

slope = -3 x 10^{-4} 1/s

k = - slope = $\boxed{3 \times 10^{-4} \text{ s}^{-1}}$

*This is the **first-order rate constant** for the reaction. With more precisely ruled graph paper or with linear regression results from a scientific calculator, higher precision could be obtained.*

The rate law from these data is:

$$\text{rate} = -\frac{\Delta[C_3H_6O_3]}{\Delta t} = (3.0 \times 0^{-4} \text{1/s})[C_3H_6O_3]$$

The half life: Another order-determining criterion

Careful examination of a graph of concentration vs. time for the data in Exercise 14.6. shows a trend: approximately every 2400 s, the concentration of trioxane is halved.

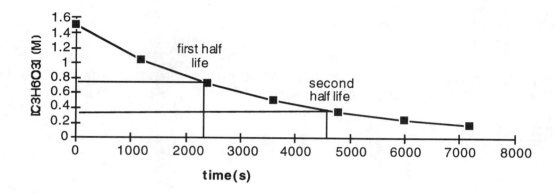

The amount of time required for consumption of half the substance is called the *half life*, $t_{1/2}$.

In the first-order decomposition of trioxane, the half life is constant; however, *only in first-order reactions is the half life constant*. The amount or concentration of substance A remaining in a first-order process is given by the following equation:

$$\ln\frac{[A]_0}{[A]} = kt,$$

where [A] = concentration of A at time t and $[A]_0$ = concentration of A at time = 0.

The half life, $t_{1/2}$, is the amount of time required for $[A] = \frac{1}{2}[A]_0$. Substitution gives:

$$\ln 2 = 0.693 = k(t_{1/2}): \qquad t_{1/2} = \frac{0.693}{k} \text{ and } k = \frac{0.693}{t_{1/2}}$$

Knowing the half life, we can calculate the first-order rate constant and vice versa.

Exercise 14.7 The first-order rate constant for the decomposition of trioxane is known to be 3.05 x 10⁻⁴ 1/s at 519 K. What is the half life of trioxane at 519 K in seconds and hours?

Steps to Solution: *The equation just obtained relates the rate constant and half life. Because the rate constant is known, we solve for the half life using*:

$$t_{1/2} = \frac{0.693}{k}$$

Chapter 14: Mechanisms of Chemical Reactions

After we obtain the half life in seconds, we must convert the units to hours.

Substituting the given value of the rate constant into the equation gives:

$$t_{1/2} = \frac{0.693}{3.05 \times 10^{-4} \text{ 1/s}} = \boxed{2.27 \times 10^3 \text{ s}}$$

This value is quite close to the value of about 2400 s estimated by examination of the data and the graph. Conversion of the time units to hours gives:

$$t_{1/2} = \frac{2.27 \times 10^3 \text{ s}}{3600 \text{ s/hr}} = \boxed{0.631 \text{ hr}}$$

Exercise 14.8 The radioactive isotope strontium-90 is a byproduct of nuclear fission reactors such as those used in the production of electricity. Because of the high levels of radioactivity associated with this isotope, it must be stored until the radioactivity reaches a sufficiently low level. Strontium 90 decays by first-order kinetics (*all* radioactive decay obeys first-order kinetics) with a half life of 29 years. How many years must pass for 99.9% of the strontium-90 in a sample to decay?

Steps to Solution: *In this problem the desired quantity is the time, t, required for the amount of strontium-90 to reach some fraction of its original value. The equation that relates concentration and time is:*

$$\ln\frac{[A]_o}{[A]} = kt \implies t = \frac{1}{k}\left(\ln\frac{[A]_o}{[A]}\right)$$

First the rate constant for the process is calculated in a method similar to that used in Exercise 14.6. Then the concentration terms must be determined before we solve for the time required.

The rate constant is found by the equation relating half life and rate constant:

$$k = \frac{0.693}{t_{1/2}} = \frac{0.693}{29 \text{ yr}} = 2.4 \times 10^{-2} \text{ 1/yr}$$

Next the value of the ratio $\frac{[A]_o}{[A]}$ must be found. The usual error that occurs in these sorts of problems is in the assignment of [A]; remember that [A] will always be smaller than $[A]_o$.

COMMON PITFALL: Assigning the value of [A] in terms of [A]$_o$. If A decomposes, [A] will always be smaller than [A]$_o$

*The problem asks for the amount of time required for 99.9% of the strontium 90 to decay; we must remember that [A] describes the amount of A **remaining**. Therefore [A] will be given by:*

519

Chapter 14: Mechanisms of Chemical Reactions

$[A] = [A]_o$ - (fraction substance consumed)$[A]_o$ = (1 - fraction consumed)$[A]_o$

The fraction of strontium-90 consumed is: $\quad 100(99.9\%) = 0.999$

$[Sr\text{-}90] = [Sr\text{-}90]_o$ - (fraction substance consumed)$[Sr\text{-}90]_o$ = (1 - 0.999)$[Sr\text{-}90]_o$

$$[Sr\text{-}90] = 1.\times 10^{-3}\,[Sr\text{-}90]_o$$

Now we substitute the values into the equation. Note that no numerical values of $[Sr\text{-}90]$ are required because we are concerned with only the time needed for a certain fraction of Sr-90 decay.

$$t = \frac{1}{k}\left(\ln\frac{[A]_o}{[A]}\right) = \frac{1}{2.4\times 10^{-2}\,\text{yr}^{-1}}\left(\ln\frac{[A]_o}{1.\times 10^{-3}\,[A]_o}\right) = \frac{1}{2.4\times 10^{-2}\,\text{yr}^{-1}}\left(\ln\frac{1}{1.\times 10^{-3}}\right)$$

$$t = \boxed{2.9\times 10^2\text{ years}}$$

Isolation Experiments: Finding Orders of Reactants in Systems With Several Reactants

Most chemical systems involve more than one reacting species and therefore include more than one concentration term in the rate law. One way to determine rate laws in such systems is to use a technique called **isolation experiments**. The idea in an isolation experiment is to manipulate the concentrations of the different reactants in the system so that one is much smaller than the others. So, as the reaction proceeds, only one concentration changes significantly. The concentrations of the other reagents are essentially constant and are implicitly included in the rate constant obtained from that experiment. This method is illustrated in the following exercise.

Exercise 14. 9. The above two sets of data for the reaction between NO and O_2 to give NO_2 are run so that the rate dependence is isolated; only the concentration of NO changes significantly over the course of the reaction. The difference between Experiment 1 and Experiment 2 is the O_2 concentration.

$$2\,NO_{(g)} + O_{2\,(g)} \rightarrow 2\,NO_{2\,(g)}$$

What is the rate law for this reaction?

Steps to Solution: *Comparing the concentrations of O_2 and NO in the tables, it is clear that O_2 is in great excess (its concentration is about 2000 times that of NO in Experiment 1) and so will remain essentially constant throughout the reaction. Within significant figures, this actually occurs (what is the change in O_2 concentration if all the NO reacts?). The form of the rate law is expected to be:*

$$\text{rate} = k[NO]^x[O_2]^y \text{ (x and y are the orders of the reaction)}$$

Chapter 14: Mechanisms of Chemical Reactions

Because [O_2] is a constant, the rate law can be written as:

$$\text{rate} = k[O_2]^y[NO]^x = k'[NO]^x \quad \text{where } k' = k[O_2]^y$$

Now the problem is to deduce the order of the reaction with respect to NO using the data from Experiment 1 by graphical methods as in previous problems. The transformed data are in the second table.

Experiment 1

t (s)	[NO]	[O_2]	
0	4.09 × 10⁻¹⁰	8.17 × 10⁻⁷	M
0.06	3.06 × 10⁻¹⁰	8.17 × 10⁻⁷	M
0.12	2.46 × 10⁻¹⁰	8.17 × 10⁻⁷	M
0.18	2.07 × 10⁻¹⁰	8.17 × 10⁻⁷	M
0.24	1.78 × 10⁻¹⁰	8.17 × 10⁻⁷	M
0.3	1.57 × 10⁻¹⁰	8.17 × 10⁻⁷	M
0.36	1.41 × 10⁻¹⁰	8.17 × 10⁻⁷	M
0.42	1.27 × 10⁻¹⁰	8.17 × 10⁻⁷	M
0.48	1.16 × 10⁻¹⁰	8.17 × 10⁻⁷	M
0.54	1.07 × 10⁻¹⁰	8.17 × 10⁻⁷	M
0.6	9.92 × 10⁻¹¹	8.17 × 10⁻⁷	M

Experiment 2

t (s)	[NO]	[O_2]	
0	4.09 × 10⁻¹⁰	1.63 × 10⁻⁶	M
0.06	2.30 × 10⁻¹⁰	1.63 × 10⁻⁶	M
0.12	1.66 × 10⁻¹⁰	1.63 × 10⁻⁶	M
0.18	1.31 × 10⁻¹⁰	1.63 × 10⁻⁶	M
0.24	1.09 × 10⁻¹⁰	1.63 × 10⁻⁶	M
0.3	9.33 × 10⁻¹¹	1.63 × 10⁻⁶	M
0.36	8.17 × 10⁻¹¹	1.63 × 10⁻⁶	M
0.42	7.27 × 10⁻¹¹	1.63 × 10⁻⁶	M
0.48	6.56 × 10⁻¹¹	1.63 × 10⁻⁶	M
0.54	5.98 × 10⁻¹¹	1.63 × 10⁻⁶	M
0.6	5.49 × 10⁻¹¹	1.63 × 10⁻⁶	M

Calculations for Experiment 1

t (s)	ln[NO]	1/[NO]
0	-21.617957	2.45 × 10⁹
0.06	-21.907749	3.27 × 10⁹
0.12	-22.124908	4.06 × 10⁹
0.18	-22.299774	4.84 × 10⁹
0.24	-22.446627	5.60 × 10⁹
0.3	-22.573444	6.36 × 10⁹
0.36	-22.685167	7.11 × 10⁹
0.42	-22.785087	7.86 × 10⁹
0.48	-22.875509	8.60 × 10⁹
0.54	-22.958117	9.35 × 10⁹
0.6	-23.034176	1.01 × 10¹⁰

Calculations for Experiment 2

t (s)	ln[NO]	1/[NO]
0	-21.62	2.45 × 10⁹
0.06	-22.19	4.35 × 10⁹
0.12	-22.52	6.02 × 10⁹
0.18	-22.75	7.62 × 10⁹
0.24	-22.94	9.18 × 10⁹
0.3	-23.09	1.07 × 10¹⁰
0.36	-23.23	1.22 × 10¹⁰
0.42	-23.34	1.37 × 10¹⁰
0.48	-23.45	1.52 × 10¹⁰
0.54	-23.54	1.67 × 10¹⁰
0.6	-23.63	1.82 × 10¹⁰

The plots of the data for experiment 1 are shown below.

Because the $\frac{1}{C}$ vs time plot is linear, the reaction is second order in [NO]. The order of the reaction in NO has been determined; now we must find the order of the reaction in O_2 and the value of the rate constant to complete the rate law. The slope of the second-order plot is called the **observed rate constant**, and is the product of the rate constant for the reaction and the oxygen concentration.

$$\text{slope} = \frac{\Delta y}{\Delta x} = k_{obs}(1) = k[O_2]^{y_1}$$

slope = $1.27 \times 10^{10} \ \frac{1}{s \cdot M}$ (the subscript 1 indicates this calculation is for Experiment 1)

The slope can be determined using the linear regression feature on many calculators.

The order of the reaction in O_2 and the rate constant for the rate law can be determined by finding the observed rate constant in the second experiment where the initial concentration of O_2 is about twice that of the first experiment. Analysis of the second-order plot gives the slope:

Chapter 14: Mechanisms of Chemical Reactions

$$\text{slope} = k_{obs}(2) = k[O_2]^{y_2} = 2.6 \times 10^{10} \ \frac{1}{s \cdot M}$$

(The slope is determined from the plot of 1/[NO] vs. t for experiment 2)

Taking the ratio of the observed rate constants in the two experiments:

$$\frac{k_{obs(1)}}{k_{obs(2)}} = \frac{k[O_2]^{y_1}}{k[O_2]^{y_2}} = \frac{1.27 \times 10^{10}}{2.60 \times 10^{10}} = 0.50$$

$$0.50 = \frac{k[O_2]^{y_1}}{k[O_2]^{y_2}} = \left(\frac{8.17 \times 10^{-7} \ M}{1.63 \times 10^{-6} \ M}\right)^y = (0.5)^y$$

$$y = 1 \text{ by inspection}$$

The reaction is first order in O_2. Now we must determine the value of the rate constant to obtain the complete rate law. Given the definition of the observed rate constant, we can solve for the rate constant:

$$k_{obs} = k[O_2] \quad \text{therefore} \quad k = \frac{k_{obs}}{[O_2]} = \frac{1.27 \times 10^{10} \ M^{-1} \cdot s}{8.17 \times 10^{-7} \ M} = 1.6 \times 10^{16} \ \frac{1}{M^2 \cdot s}$$

The rate law is:

$$\boxed{\text{rate} = 1.6 \times 10^{16} \ \frac{1}{M^2 \cdot s} \ [NO]^2[O_2]}$$

The complete rate law contains the value of the rate constant and the orders of the reactants; the reaction is first order in O_2, second order in NO, and third order overall.

14.5 L:inking Mechanisms and Rate Laws

QUESTIONS TO ANSWER, SKILLS TO LEARN
1. **Deriving rate laws from mechanisms**

A mechanism is a proposed sequence of bimolecular and/or unimolecular events that describes how a chemical process might occur. A rate law is the experimentally determined equation that gives the rate of the reaction as a function of concentration of the reactants. This section reviews how the rate law can be obtained from the rates of elementary reactions and some assumptions about the relative speeds of the different steps. First, three characteristics of a reaction mechanism are:

1. It is a sequence of one or more unimolecular and/or bimolecular elementary processes.

2. The sum of this series of elementary processes gives the balanced net chemical equation of reaction.

3. ***For a proposed mechanism to even be considered a possible description of a reaction, the rate law obtained from the mechanism must agree with the experimental rate law.***

In addition, the rate of each step in a mechanism depends on the concentrations of the reactants in that step and the size of the rate constant. Next, the rate of the overall reaction is determined by the slowest step in the mechanism, called the rate-determining step. Finally, the rate law is expressed in concentrations of reactants, products, and reagents whose concentrations can be directly observed and controlled; in other words, concentrations of intermediates *are not* used in rate laws.

The difficulty of deriving a rate law from a mechanism depends on whether or not the first step is rate determining. If the first elementary reaction of the mechanism is rate determining, the rate of the reaction is the rate of that step, expressed in terms of concentrations of the reactant(s).

If the rate-determining elementary reaction occurs later in the mechanism, the rate of the reaction is the rate of that later step. However, often the rate of that elementary process is expressed in concentrations of a reactant and/or one (or more) intermediates. Therefore these two situations will be treated separately.

First Step Rate Determining

If the first step in the reaction mechanism is rate determining, the rest of the steps in the mechanism proceed rapidly enough so they do not slow the reaction rate below that of the rate-determining step. The rate of that step is proportional to the concentrations of the reactants in that elementary reaction.

**If the first step in a mechanism is rate determining,
the rate of the reaction is the rate of that elementary process.**

Exercise 14.10 The reaction of nitrogen dioxide and fluorine is:

$$2\ NO_2 + F_2 \rightarrow 2\ NO_2F$$

One proposed mechanism has two steps; the first step is rate determining:

Chapter 14: Mechanisms of Chemical Reactions

Step	Elementary Reaction	Relative Rate	Rate of Elementary Process
1	$NO_2 + F_2 \rightarrow NO_2F + F\bullet$	slow	$k_1[NO_2][F_2]$
2	$F\bullet + NO_2 \rightarrow NO_2F$	fast (very reactive fluorine atom)	$k_2[NO_2][F\bullet]$

The experimentally determined rate law is:

$$\text{rate} = k_{obs}[NO_2][F_2]$$

Is the mechanism consistent with the rate law?

Steps to Solution: The mechanism is consistent with the experimental rate law if the rate law derived from the proposed mechanism is the same as the experimental rate law. The proposed mechanism already fulfills the other criteria for a possible mechanism: it is a series of bi- or unimolecular processes that adds up to the net reaction. Because the first elementary reaction is rate determining, the overall rate of the reaction is the rate of that slow step. The rate of step 1 is given in the summary table:

$$\text{rate} = k_1[NO_2][F_2] = \text{reaction rate}$$

This rate law is nearly the same as that determined by experiment:

$$\text{rate} = k_{obs}[NO_2][F_2]$$

Therefore the mechanism is consistent with the experimental rate law. If the mechanism is the correct mechanism, then the value of the observed rate constant is the rate constant of the first elementary process.

$$\frac{\text{rate} = k_1[NO_2][F_2]}{\text{rate} = k_{obs}[NO_2][F_2]} \Rightarrow k_1 = k_{obs}$$

Later Step Rate Determining

In many reactions, the first step is not rate determining and the rate law is not first or second order and may often contain fractional orders or negative orders (the concentration of the substance is in the denominator). In the mechanisms of these reactions, often the first step or steps involve the formation (and sometimes consumption) of intermediate species. The intermediate then reacts in the rate-determining step, leading to a rate law for the reaction that contains the concentration of that intermediate. The rate law must be expressed in terms of concentrations of chemical species that are "real" and externally adjustable; reactants or products or sometimes solvents, but *not proposed intermediates*. The task in these situations is to express the concentration of the intermediate in terms of reactants or products.

The concentration of the intermediate can be often related to concentrations of reactants (and sometimes products) by using the notion that during the course of a reaction, the concentration of the intermediate will be small but nearly constant. Therefore the rate of formation of the intermediate is equal to the rate of the recombination to give starting materials. Because of this *equality of rates*, the rate of the reaction forming the intermediate is equal to the rate of the process consuming the intermediate, so that the concentration of the intermediate can be expressed mathematically in terms of rate constants and concentrations of reactants and perhaps even products. This method is illustrated in the next two exercises.

Exercise 14.11 The decomposition of ozone to oxygen has an unusual rate law:

$$2\ O_{3\ (g)} \rightarrow 3\ O_{2\ (g)} \qquad \text{rate} = k\frac{[O_3]^2}{[O_2]}$$

A proposed mechanism has three steps.

Step	Elementary Reaction	Rate Law	Relative Speed
1	$O_3 \rightarrow O_2 + O$	rate = $k_1[O_3]$	fast
2 (actually the reverse of step 1)	$O_2 + O \rightarrow O_3$ or $O_3 \leftarrow O_2 + O$	rate = $k_{-1}[O][O_2]$ (The rate constant is k_{-1} to emphasize this reaction is the reverse of step 1.)	fast
3	$O + O_3 \rightarrow 2\ O_2$	rate = $k_2[O][O_3]$	slow (rate determining)

Is the mechanism consistent with the experimental rate law?

Steps to Solution: We must determine whether the mechanism is (a) valid in the stoichiometric sense and (b) is the derived rate law consistent with the experimental rate law. The first criterion is satisfied. To derive the rate law, we must identify the slow step and then use the equality of rates to express the rate of this step in terms of reactants and products.

The slow step is step 3; therefore the rate of the reaction is the rate of step 3:

$$\text{rate} = k_2[O][O_3]$$

This rate law has two problems: first it does not have the same concentration dependence as the experimental rate law and second, it contains the concentration of an intermediate. The rate law of the second step does contain [O]. If the concentration of [O] becomes constant during the course of the reaction, then the equality of rates of step 1 and step 2 can be used. Sometimes steps 1 and 2 are combined into the single expression for the sake of compactness and to further emphasize that step 2 is the reverse of step 1:

Chapter 14: Mechanisms of Chemical Reactions

$$O_3 \rightleftarrows O_2 + O$$

If the concentration of the intermediate is constant, then the rates of the fast steps 1 and 2 must be equal:

$$\text{rate (1)} = \text{rate (2)}$$

$$k_1[O_3] = k_{-1}[O][O_2]$$

Solving for [O], we obtain:

$$[O] = \frac{k_1[O_3]}{k_{-1}[O_2]}$$

Substituion of this quantity into the expression for the rate of the rate-determining step gives:

$$\text{rate} = k_2[O][O_3] = k_2[O_3] \cdot \frac{k_1[O_3]}{k_{-1}[O_2]} = \frac{k_1 k_2}{k_{-1}} \frac{[O_3]^2}{[O_2]} = k_{obs} \frac{[O_3]^2}{[O_2]}$$

This rate law has the same concentration dependence as the experimental rate law and supports the proposed mechanism. The proposed mechanism also meets the other criteria for being a potentially valid mechanism: it consists of elementary reactions that are at most bimolecular, and the sum of the elementary processes is the net balanced equation of reaction.

step1 and its reverse:	$O_3 \rightleftarrows O_2 + O$
step 3:	$O + O_3 \rightarrow 2\,O_2$
net:	$2\,O_3 \rightarrow 3\,O_2$

Exercise 14.12 The reaction of NO with O_2 to give oxygen is known to follow a third order rate law (see Exercise 14.9). Two possible mechanisms are shown below.

Mechanism 1	Mechanism 2	
$2\,NO_{(g)} + O_{2\,(g)} \rightarrow 2\,NO_{2\,(g)}$ (slow)	$2\,NO_{(g)} \rightleftarrows (NO)_{2\,(g)}$	(fast)
	$(NO)_2 + O_2 \rightarrow 2\,NO_2$	(slow)

Which of these two mechanisms is a more acceptable mechanism, based on the criteria given above?

Steps to Solution: To decide which of the mechanisms is more acceptable, we must check to see if the sum of the elementary reactions gives the balanced equation of reaction (both mechanisms fit this criterion) and then see if the rate laws obtained from the mechanisms agree with that obtained by experiment. Then we can check for the molecularity criterion. The rate laws for the two mechanisms are derived below.

Mechanism 1. *This is a one-step mechanism, so the rate of the reaction is the rate of this elementary process:*

$$\text{rate} = k[NO]^2[O_2]$$

This rate law agrees with the experimental rate law. However, this is a termolecular process.

Mechanism 2. *This mechanism has a later slow, rate-determining step; the rate of the reaction is the rate of that second elementary reaction:*

$$\text{rate} = k[(NO)_2][O_2]$$

The molecule, $(NO)_2$, is an intermediate. The concentration of this intermediate can be related to other concentrations by assuming that the rates of the fast processes in the first step will be equal. The forward rate constant will be k_1, and the reverse rate constant will be k_{-1}

rate forward step 1 (2 NO → $(NO)_2$ = rate backwards step 1 ($(NO)_2$ → 2 NO)

$$k_1[NO]^2 = k_{-1}[(NO)_2]$$

$$[(NO)_2] = \frac{k_1[NO]^2}{k_{-1}}$$

Substituting into the rate law for step 2 gives:

$$\text{rate} = k \cdot \frac{k_1[NO]^2}{k_{-1}}[O_2] = \frac{kk_1}{k_{-1}}[NO]^2[O_2]$$

This rate law is also consistent with the experimental rate law. However, because this mechanism requires only bimolecular collisions (elementary reactions), it is a more likely mechanism than mechanism 1.

14.6 Reaction Rates and Temperature

QUESTIONS TO ANSWER, SKILLS TO LEARN
1. **How does changing temperature affect reaction rates?**
2. **What happens to molecules when chemical changes occur and what are the energy costs? (Or, what is the *activation energy*?)**
3. **Why and how much does the rate constant change with temperature?**
4. **Calculating rate constants at different temperatures.**

Increasing the temperature of a system in general will increase the rate of a reaction even though the concentrations of the reactants remain unchanged. Because temperature is a measure of the average kinetic energy of molecules in the system, the rate increase is associated with the increased average kinetic energy. The following discussion addresses the source of energy requirements for reactions to proceed.

In a chemical reaction, bonds are formed and/or broken in the conversion of reactants to products. The slow, rate-determining elementary reaction often involves bond breaking that requires energy. One way to illustrate the energy requirements for reaction to occur is to plot the energy of the chemical species involved vs. the reaction coordinate that indicates the amount of rearrangement of reactants into products. The reaction coordinate can involve one internuclear distance (as in Exercise 14.13) or many distances and angles in more complex reactions.

Exercise 14.13 In the dissociation of dinitrogentetroxide, the N-N bond breaks resulting in the formation of two NO_2 molecules:

$$N_2O_4 \rightarrow 2\ NO_2$$

Draw molecular pictures superimposed on a diagram of energy vs. reaction coordinate that illustrate this process.

Steps to Solution: *The structure of N_2O_4 has an N-N single bond that stretches. The overlap of the orbitals decreases as the bond lengthens, leading to a destabilization of the molecule (an increase in energy) until the bond finally breaks. This process is illustrated below.*

The reaction coordinate in this unimolecular reaction corresponds to the N-N distance.

In Exercise 14.10, the mechanism for the reaction of NO_2 and F_2 to give NO_2F involved an initial slow bimolecular reaction in which the F-F bond was broken and an N-F bond was formed.

$$NO_2 + F_2 \rightarrow NO_2F + F\bullet$$

The Lewis representation of the reaction suggests that only certain collisions will be effective at initiating reaction:

$$O_2N \cdot + F-F \longrightarrow O_2N-F + F\cdot$$

In this elementary reaction, more is occuring than in the unimolecular process in Exercise 14.13 because in this reaction one bond is being formed (N-F) and one is being broken (F-F). The reaction coordinate for this reaction represents both of these processes.

Exercise 14.14 The following diagram illustrates the energy changes in the progress of the formation of F• and NO$_2$F in the reaction of NO$_2$ and F$_2$. Give a brief explanation of the shape of the curve.

Steps to Solution: *The explanation lies in remembering that higher energy indicates lower stability. The initial increase in energy is caused by two energy-requiring processes. The more obvious process is breaking the F-F bond. The other is overcoming electron-electron repulsions (rearrangement of the electron clouds) so that the two molecules can come close enough to start forming the N-F bond. After a certain point, the formation of the N-F bond (a process that is always exothermic) starts to stabilize the system and the F-F bond breaks, leading to the products of this step.*

The diagrams shown in Exercises 14.13 and 14.14 are given the name **activation energy diagrams** or sometimes just energy diagrams. The diagrams above were qualitative, but they can be made quantitative, too. Several important quantities should be noted. First, the energy required to reach the high point of the diagram from the reactants is called the **activation energy**.

Activation energy: The minimum energy required to rearrange the reactants before they can form products

The size of the activation energy depends on the amount of molecular rearrangement (bond breaking and forming) that occurs in reaching the transition state.

The figure below is an activation energy diagram for an exothermic reaction. The high point in the diagram represents a chemical species called the **activated complex** or **transition state**.

The activated complex /transition state is formed only by collisions between molecules that have energy greater than the activation energy and are aligned correctly for bond formation.

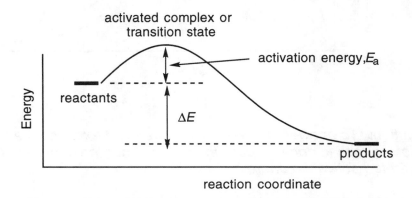

The activated complex must not to be confused with an intermediate; an intermediate has a significant lifetime, such as the (NO)$_2$ species proposed in the mechanism for the reaction of NO with oxygen. An activated complex would be the "molecule" formed by the collision of (NO)$_2$ with O$_2$ that will persist only for *very short periods of time (*10^{-13} s). This species could separate to reform starting materials or products.

Activation Energy and the Rate Constant

Because temperature alone can increase the rate of reactions with no change in concentration, the rate increase must be the result of a change in the magnitude of the rate constant. The underlying reason is that molecular energies are distributed asymmetrically about the average kinetic energy that may be predicted mathematically from the Boltzmann distribution of energies in a sample of molecules.

A sketch of the Boltzmann distributions at two different temperatures is shown below with two activation energies indicated by vertical lines at E_a and E_a'. The total number of molecules is proportional to the area under the entire curve. The number of molecules with energy greater than the activation energy is proportional the area under the curve to the right of the energy corresponding to the activation energy. As temperature increases, more molecules have higher energies and therefore more molecules will have energy greater than the activation energy.

Exercise 14.15 Use the Boltzmann distributions sketched below to answer the following questions;
(a) At temperature T_1, will more molecules have energy greater than E_a or E_a'?
(b) Will a reaction with activation energy E_a have a faster rate at T_1 or T_2?

Steps to Solution: To answer these questions we must examine the Boltzmann distributions supplied and compare the areas under the curve(s) at energies greater than the stated activation energy.

(a) The difference between the numbers of molecules with energies greater than the two activation energies marked is the area under the t_1 curve between E_a and E_a'. Therefore more molecules have energy greater than E_a' than E_a at T_1. In general, at any temperature, more molecules will have sufficient energy to react for smaller activation energies than larger activation energies.

(b) The important factor here is the areas under the curves for T_1 and T_2 at energies greater than E_a. The area under the T_2 curve for $E > E_a$ is greater than that under the t_1 curve; therefore more molecules will have energy greater than the activation energy at T_2 and will be able to react. In general, the fraction of molecules with energy greater than the activation energy increases with increasing temperature; therefore the rate of a reaction will increase with increasing temperature.

The Arrhenius Equation

The relationship between the activation energy, E_a, and the rate constant, k, is given by the Arrhenius equation:

$$k = A \exp\left(-\frac{E_a}{RT}\right) \quad \text{or} \quad \ln(k) = \ln(A) - \frac{E_a}{RT}$$

A = geometric (orientation) factor; E_a = activation energy
R = 8.314 J/mol·K; T = temperature in Kelvin

Chapter 14: Mechanisms of Chemical Reactions

The factor A can be considered constant for small temperature changes ($\Delta T \leq 50$ K), so this expression can be used to obtain information about rate constants at different temperatures given the activation energy or to determine the activation energy with rate constants measured at different temperatures.

Exercise 14.16 The following rate constants were obtained at the stated temperatures for the first-order reaction:

$$A \rightarrow B$$

T (K)	300	310	320	330	340
k (s^{-1})	8.93 x 10^{-8}	1.94 x 10^{-7}	4.01 x 10^{-7}	7.95 x 10^{-7}	1.51 x 10^{-6}

Find the activation energy (in kJ/mole) for this reaction.

Steps to Solution: The expression obtained by taking the natural logarithm of the Arrhenius equation gives $\ln(k)$ as a linear function of $\frac{1}{T}$, so plotting $\ln(k)$ vs $\frac{1}{T}$ should give a straight line with a slope of $-\frac{E_a}{R}$. Multiplying the slope of the line by R will give the activation energy in J/mole. The following table gives the data for 1/T and $\ln(k)$.

1/T	3.33 x 10^{-3}	3.23 x 10^{-3}	3.13 x 10^{-3}	3.03 x 10^{-3}	2.94 x 10^{-3}
$\ln(k)$	-16.232	-15.456	-14.728	-14.045	-13.402

The plot of the data is linear and the slope is calculated.

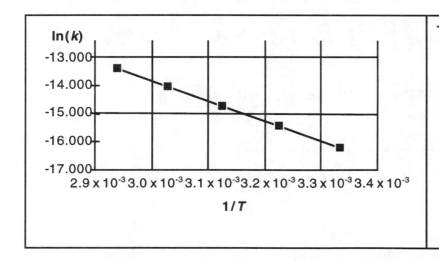

The slope of the line through the points is found to be:

$$m = \frac{\Delta y}{\Delta x} = 7.22 \times 10^3 \text{ K}$$

The slope is $\frac{E_a}{R}$; therefore:

$E_a = m(R) = 7.22 \times 10^3 \cdot \text{K}(8.314 \text{ J/mole} \cdot \text{K}) = 6.00 \times 10^3$ J/mole = $\boxed{60.0 \text{ kJ/mole}}$

Chapter 14: Mechanisms of Chemical Reactions

The same reaction proceeding through the same mechanism with activation energy E_a at different temperatures, t_1 and T_2, will have different rate constants, k_1 and k_2. Thie activation energy equation can be rearranged to solve for the activation energy given rate constants at two different temperatures.

$$\ln\left(\frac{k_2}{k_1}\right) = \frac{E_a}{R}\left(\frac{1}{T_1} - \frac{1}{T_2}\right) \quad \text{or} \quad E_a = R\frac{\ln\left(\frac{k_2}{k_1}\right)}{\left(\frac{1}{T_1} - \frac{1}{T_2}\right)}$$

Exercise 14.17 The reaction of ozone with oxygen atoms to give oxygen has an activation energy of 17.1 kJ/mole with a rate constant at 298 K, $k = 4.8 \times 10^6$ $M^{-1} \cdot s$. Calculate the rate constant for this reaction at 315 K.

Steps to Solution: This problem requires the use of the equation that relates rate constants, temperatures and activation energy:

$$\ln\left(\frac{k_2}{k_1}\right) = \frac{E_a}{R}\left(\frac{1}{T_1} - \frac{1}{T_2}\right)$$

The values at 298 K will be the values with the 1 subscript. Also, the quantity on the right will be calculated first.

$$\ln\left(\frac{k_2}{4.8 \times 10^6 \, M^{-1} \cdot s}\right) = \frac{17.1 \times 10^3 \, J/mole}{8.314 \, J/mole \cdot K}\left(\frac{1}{298} - \frac{1}{315}\right) = 0.372$$

Taking the natural antilog of both sides gives:

$$\exp\left(\ln\left(\frac{k_2}{4.8 \times 10^6 \, M^{-1} \cdot s}\right)\right) = \frac{k_2}{4.8 \times 10^6 \, M^{-1} \cdot s} = \exp(0.372) = 1.45$$

$$k_2 = 1.45 \, (4.8 \times 10^6 \, M^{-1} \cdot s) = \boxed{7.0 \times 10^6 \, M^{-1} \cdot s}$$

14.7 Catalysis

QUESTIONS TO ANSWER, SKILLS TO LEARN
  1. **What is a catalyst and how does one work?**
  2. **What is the difference between a homogeneous and a heterogeneous catalyst?**
  3. **What are some important industrial and biochemical catalysts?**

Chapter 14: Mechanisms of Chemical Reactions

The rate of reactions may be increased in several ways: one way is to increase the concentration of the reactants, and another way is to increase the temperature. Both of these tactics are used, but yet another way to increase the rate of a reaction is to change the mechanism of the reaction by introduction of a substance so that the activation energy is lower. One example of this technique was used in the conversion of SO_2 to SO_3 by the addition of NO:

$$2\,NO + O_2 \rightarrow 2\,NO_2$$

$$2 \cdot (SO_2 + NO_2 \rightarrow SO_3 + NO)$$

$$2\,SO_2 + O_2 \rightarrow 2\,SO_3 \text{ or } 2\,SO_2 + O_2 \xrightarrow{NO} 2\,SO_3$$

Note that neither NO nor NO_2 appears in the balanced equation of reaction, but they are active in the mechanism. However, the rate of the reaction is much faster in the presence of NO. In this reaction NO acts as a **catalyst** because it increases the rate of the reaction without entering into the reaction stoichiometry.

A **catalyst**: (a) increases the rate of a reaction
(b) is not present in the balanced chemical equation of reaction
(c) is often indicated over the arrow in the equation of reaction
(d) is often included in the rate law

Catalysts speed up a reaction by providing a mechanism with a lower activation energy than the uncatalyzed mechanism. The catalyst may achieve lower activation energies by weakening or breaking bonds in one of the reactants or by helping to bring the reactants together in the right geometry. The difference between a catalyzed reaction and an uncatalyzed reaction is shown in the following activation energy diagram. The catalyzed reaction has a lower activation energy so that a larger fraction of molecules has energy greater than the activation energy at any given temperature. Therefore more molecules are able to undergo reaction and the reaction proceeds more rapidly. In both the catalyzed and uncatalyzed path the energy change, ΔE, is the same; only the activation energy is different. E_a is a path function, whereas ΔE is a state function. Specific ways in which different types of catalysts achieve lowering of the activation energy are described below.

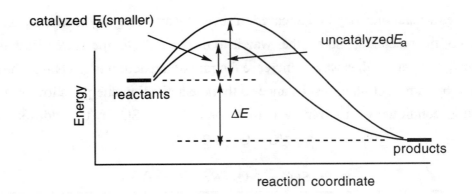

Types of Catalysts

One description of catalysts relates to their phase: are they in the same phase as the reactants or not? **Homogeneous** catalysts are in the same phase as the reactants; **heterogeneous** catalysts are in a different phase than the reactants. The catalysis of the reaction of sulfur dioxide with oxygen can be accomplished with homogeneous or heterogeneous catalysts. The example above where nitrogen monoxide is used is an example of a homogeneous catalyst: NO is in the same phase as the reactants and products.

In the process currently used to manufacture sulfuric acid, a heterogeneous catalyst, vanadium(V)oxide (V_2O_5), is used.

$$2\ SO_2 + O_2 \xrightarrow{V_2O_5(s)} 2\ SO_3$$

The mechanism for the homogeneous catalytic process using NO is well understood, whereas the mechanisms for heterogeneously catalyzed reactions are less well characterized because of their complexity and the technical difficulties encountered in studying reactions running under moderately high temperatures and pressures.

Four processes generally occur at heterogeneous catalysts. These will be illustrated using the example of the heterogeneously catalyzed reaction of nitrogen and hydrogen to give ammonia. In the production of ammonia from nitrogen and hydrogen, an iron catalyst is used with various other metals added to give higher rate enhancement.

$$2\ N_2 + 3\ H_2 \xrightarrow{Fe(s)} 2\ NH_3$$

The reaction is typically run at temperatures of 450 °C and pressures of N_2 and H_2 of 100 to 300 atm. The four steps are:

1. **Adsorption**: One or more of the starting materials bind to the surface, and the bonds in these reactants are weakened or broken. In the production of ammonia, both reactants are adsorbed onto the surface of the catalyst to give atoms bonded to the catalyst. This state is described by the *ads* state designation.

$$N_{2\ (g)} \xrightarrow{Fe(s)} 2\ N_{(ads)}$$

(nitrogen gas forms nitrogen atoms on the Fe surface)

$$H_{2\ (g)} \xrightarrow{Fe(s)} 2\ H_{(ads)}$$

2. **Migration**: The atoms or fragments bound to the surface of the catalyst can move about on the surface of the catalyst until they encounter another reactive adsorbed atom or fragment.

3. **Reaction**: When the adsorbed atoms or fragments encounter one another on the surface, they may react. Two examples of such reactions are given below.

$$N_{(ads)} + H_{(ads)} \rightarrow NH_{(ads)}$$

$$NH_{2\ (ads)} + H_{(ads)} \rightarrow NH_{3\ (ads)}$$

4. **Desorption**: The products escape from the surface of the catalyst.

$$NH_{3\ (ads)} \rightarrow NH_{3\ (g)}$$

Enzymes

A category of catalyst that does not clearly fit into the above two categories is **enzymes**, biologically produced macromolecules (proteins) that assist a wide variety of reactions.

An enzyme is a protein that catalyzes a specific biochemical reaction.

Enzymes catalyze reactions by inducing bond weakening and/or bringing together reactants in a favorable configuration, but they are highly specific with regard to the type of reaction that they catalyze as well as the type and shape of the reactant that is to be transformed. One enzyme, nitrogenase, has a molar mass of 60,000 g/mole, and converts atmospheric N_2 and hydrogen from water to ammonia under atmospheric pressure and room temperature (compare these conditions to those for the commmercial catalyst described above). The key steps in enzymatic catalysis are listed below with a brief explanation. E is used to designate the enzyme, R the reactant, and P the product.

Chapter 14: Mechanisms of Chemical Reactions

E + R
↑↓
(binding)
↑↓
E•R
↑↓
(distortion)
↑↓
{E•R}$_D$
↑↓
(reaction)
↑↓
E•P
↑↓
(release)
↑↓
E + P

The enzyme, E, and the reactant, R, can bind to one another. The reactant binds to a specific position on the enzyme called the **active site** that has a shape that accommodates the reactant molecule. This intermediate is called the enzyme-substrate (substrate is another word for reactant) complex, E•R. The enzyme may distort the reactant molecule, making it more susceptible to reaction. This distorted complex is called {E•R}$_D$ in the reaction scheme to the left. This complex may revert to the undistorted E•R complex or the reactant may be converted to product molecule or molecules bound to the enzyme, E•P. The final step in the reaction process is release of the products, P, to regenerate the enzyme, E, that may then, once again bind reactant and start the catalytic cycle over again. The cyclic nature of this catalysis is illustrated below.

The shape-selective nature of the binding of substrate to enzymes' active sites has led to the name "lock and key" mechanism, because the shape of the substrate is critical to binding and the subsequent steps

Exercise 14.18 For the NO catalyzed reaction of SO_2 with O_2, write the rate laws that one would obtain if the second step is rate determining.

Steps to Solution: The elementary process allows us to write the rate law for that particular step. The second step is the reaction of NO_2 with SO_2:

$$SO_2 + NO_2 \rightarrow SO_3 + NO$$

The rate of this step would be the rate of the reaction if it is the rate-determining step:

$$\text{rate} = k[SO_2][NO_2]$$

Test Yourself

1. Write the overall equation of reaction for the following mechanism and identify the reaction intermediates.
   (1) $Cl_2 \rightarrow 2\, Cl\bullet$
   (2) $Cl\bullet + CO \rightarrow COCl$

(3) $COCl + Cl_2 \rightarrow COCl_2 + Cl\bullet$
(4) $2\ Cl\bullet \rightarrow Cl_2$

2. The rate constant of the reaction,

$$O_{(g)} + N_{2\,(g)} \rightarrow NO_{(g)} + N_{(g)}$$

is 9.7×10^{10} M$^{-1}\bullet$s at 800 K and has an activation energy of 315 kJ/mole. What is the value of the rate constant at 700 K?

3. The data in the following table were collected for the decomposition reaction of C_2H_6 at
   973 K:

$$H_3C\text{-}CH_{3\,(g)} \rightarrow 2\ CH_{3\,(g)}$$

t (s)	0	1000	20000	30000	40000
[H$_3$C-CH$_3$] (mmole/L)	1.59	0.92	0.53	0.31	0.18

What is the order of the reaction in H_3C-CH_3, and what is the rate constant (remember the units!)?

4. What is the reaction rate for the decomposition of N_2O_5 when 4.0 g of N_2O_5 is in a 2.0 L container at 65 °C where $k = 5.2 \times 10^{-3}$ 1/s?

5. The rate law for the reaction

$$2\ H_{2\,(g)} + 2\ NO_{(g)} \rightarrow N_{2\,(g)} + 2\ H_2O_{(g)}$$

is rate = $k[H_2][NO]^2$. Which of the following mechanisms can be ruled out because the derived rate law is not consistant with the observed rate law?

Mechanism 1
$H_2 + NO \rightarrow N + H_2O$ slow
$N + NO \rightarrow N_2 + O$ fast
$O + H_2 \rightarrow H_2O$ fast

Mechanism 2
$H_2 + 2\ NO \rightarrow N_2O + H_2O$ slow
$N_2O + H_2 \rightarrow N_2 + H_2O$ fast

Mechanism 3
$2\ NO \rightleftarrows N_2O_2$ fast equilibrium
$N_2O_2 + H_2 \rightarrow N_2O + H_2O$ slow
$N_2O + H_2 \rightarrow N_2 + H_2O$ fast

Answers

1. $CO + Cl_2 \rightarrow COCl_2$. Intermediates are: $Cl\bullet$ and $COCl$.

2. $k(700\ K) = 2.9 \times 10^9\ M^{-1}\bullet s$

3. First order in $H_3C\text{-}CH_3$; k = negative of slope of plot of $\ln[H_3C\text{-}CH_3]$ vs. t; $5.5 \times 10^{-4}\ s^{-1}$

4. Rate = 9.6×10^{-5} mole $N_2O_5/L\bullet s$

5. Mechanism 1 is not consistent since its rate law would be second order, first order in H_2 and first order in NO.

Solutions to Chapter Problems

14.1 In any sequence of steps, the slowest one will be rate-determining. (a) Pouring the coffee from the urn into the cup; (b) Entering the items on the cash register (if the market has a good laser scanner, paying and receiving change may be rate-determining); (c) Preparing for the jump and passing through the door.

14.3 Every elementary reaction must depict actual molecular processes. (a) $I_2 \rightarrow I + I$; (b) $H_2 + I_2 \rightarrow H_2I_2$; (c) $H_2 + I_2 \rightarrow H + HI_2$.

14.5 A molecular picture of an elementary reaction shows the reactants, the products, and (if necessary) the intermediate collision complex.
(a)

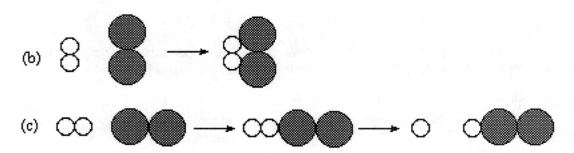

14.7 A satisfactory mechanism must consist entirely of reasonable elementary steps that sum to give the correct overall stoichiometry of the reaction.
(a) $I_2 \rightarrow I + I$
$I + H_2 \rightarrow HI + H$
$H + I \rightarrow HI$
(b) $H_2 + I_2 \rightarrow H_2I_2$
$H_2I_2 \rightarrow HI + HI$
(c) $H_2 + I_2 \rightarrow H + HI_2$
$H + HI_2 \rightarrow HI + HI$

Chapter 14: Mechanisms of Chemical Reactions

14.9 The rate of a reaction can be expressed using the general expression, Rate $= -\dfrac{1}{a}\dfrac{\Delta[A]}{\Delta t}$.

(a) Rate $= -\dfrac{\Delta[Cl_2]}{\Delta t}$

(b) $-\dfrac{\Delta[Cl_2]}{\Delta t} = \dfrac{1}{2}\dfrac{\Delta[NOCl]}{\Delta t}$

(c) Use the result of (b) to calculate that NOCl appears at a rate of 94 M s^{-1}

14.11 In the reaction of NO and Cl_2, two NO molecules react for every Cl_2 that reacts, producing two NOCl molecules.

(a)

(b)

(c)

14.13 (a) To calculate the average rate of production, determine how many moles form during the time interval and divide by the time:

$n = PV/RT = \dfrac{(0.15 \text{ atm})(5.0 \text{ L})}{(0.08206 \text{ L atm / mol K})(550 + 273.15 \text{ K})} = 1.11 \times 10^{-2}$ mol;

$\dfrac{\Delta[CO_2]}{\Delta t} = \dfrac{(1.11 \times 10^{-2} \text{ mol})}{(5 \text{ min})} = 2.2 \times 10^{-3}$ mol / min

(b) The reaction is $CaCO_3 \rightarrow CaO + CO_2$, so the amount of $CaCO_3$ decomposing is the same as the amount of CO_2 produced, 1.1×10^{-2} mol.

14.15 (a)

(b) After 20 minutes, the amount reacted is (20 min)(0.25 molecules/min) = 5 molecules.

14.17 Rates of elementary two-body reactions depend on concentration of each type of body: When a concentration is increased by some factor, the rate increases by the same factor.
(a) There are half as many red balls, so half as many pairs will form: 2;
(b) There are 1.5 times as many white balls, so 1.5 times as many pairs will form: 6;
(c) There are half as many red but twice as many white balls: Pairs = (1/2)(2)(4) = 4;
(d) There are twice as many of each type of balls: Pairs = (2)(2)(4) = 8;
(e) When the concentration of either type of ball changes, the frequency of collisions with other balls changes in the same proportion, so the rate of formation of pairs, which depends on collision rate, also changes proportionally.

14.19 (a) The rate law for an elementary step contains the product of the reactant concentrations: Rate = k[C][AB];
(b) The units of a rate constant have time in the denominator along with concentrations one power less than the number of reactants: units = $(conc.)^{-1} (time)^{-1}$;

(c) The steps of the mechanism must sum to give the observed stoichiometry for the reaction. Intermediate A can react with AB, then B and C can react:
C + AB → BC + A
A + AB → B + A_2
B + C → BC
Net: 2 C + 2 AB → 2 BC + A_2

14.21 (a) The rate law for an elementary step contains the product of the reactant concentrations: Rate = k[C][C][AB] = $k[C]^2[AB]$;

(b) The units of a rate constant have time in the denominator along with concentrations one power less than the number of reactants: units = $(conc.)^{-2} (time)^{-1}$;

(c) The steps of the mechanism must sum to give the observed stoichiometry for the reaction. A bimolecular reaction between the intermediate (AC) and AB is the simplest possibility:
2 C + AB → BC + AC
AC + AB → BC + A_2
Net: 2 C + 2 AB → 2 BC + A_2

14.23 According to the stated rate law, the rate of reaction is proportional to each concentration. A contains 6 molecules of each type, while B contains 10 NO and 2 O_3:
A: Rate = $k(6)(6) = 36\,k$;   B: Rate = $k(10)(2) = 20\,k$;
B will react slower by a factor of $20/36 = 0.56$.

14.25 (a) One way to treat experimental data is by plotting: For first-order behavior, ln c vs. t is a straight line, and for second-order behavior, $1/c$ vs. t is a straight line. Here, the plot of ln c vs. t gives a straight line, so this reaction is first-order.

(b) For a first-order reaction, $k = -$ slope of the graph

$$k = -(\text{slope}) = \frac{\Delta y}{\Delta x} = \pm\frac{(1.53)-(2.50)}{(800\text{ s})-(0\text{ s})} = 1.21 \times 10^{-3}\text{ s}^{-1}$$

(c) To find concentration at any particular time, use $\ln[A] = \ln[A]_o - kt$
$\ln[A] = \ln[2.50] - (1.21 \times 10^{-3}\text{ s}^{-1})(1600\text{ s}) = -1.0197$
$[A] = e^{-1.0197} = 0.361$ atm

(d) To find the time at which concentration reaches a particular value, use $kt = \ln([A]_o/[A])$

$$t = \frac{\ln([A]_o/[A])}{k} = \frac{\ln([2.50\text{ atm}]/[0.500\text{ atm}])}{(1.21 \times 10^{-3}\text{ s}^{-1})} = 1.33 \times 10^3\text{ s}$$

14.27 This is stated to be a second-order reaction, so Rate = $k[NOBr]^2$ and Equation 14-3 applies:

$\dfrac{1}{[A]} = \dfrac{1}{[A]_o} + kt$ and $kt = \dfrac{1}{[A]_o} - \dfrac{1}{[A]}$; The problem states that $k = 25\text{ M}^{-1}\text{ min}^{-1}$.

(a) $kt = \dfrac{1}{(0.025\text{ M})} - \dfrac{1}{(0.010\text{ M})} = (40 - 100)\text{ M}^{-1} = 60\text{ M}^{-1}$;
$t = (60\text{ M}^{-1}) \div (25\text{ M}^{-1}\text{ min}^{-1}) = 2.4$ min;

(b) $\dfrac{1}{[A]} = \dfrac{1}{(0.025\text{ M})} + (25\text{ M}^{-1}\text{ min}^{-1})(125\text{ min}) = (40 + 3125)\text{ M}^{-1} = 3165\text{ M}^{-1}$
$[A] = 3.2 \times 10^{-4}$ M.

14.29 Here, there are two concentration variables, [A] and [B]. In the first experiment, [B] >> [A], so A is isolated and the data can be analyzed to determine the order with respect to A. The data could be plotted, but half-lives can be recognized. It takes 30 s for the concentration to fall from 0.050 M to 0.025 M, 30 s to fall from 0.040 to 0.020 M, and 30 s to fall from 0.032 to 0.016 M. Thus the half-life is a constant value, 30 s, and the reaction is first-order in A. In the second experiment, the rate of disappearance of A does not change even though the initial concentration of B is increased. Consequently, the rate law does not depend on [B].
(a) Rate = k[A];
(b) The half-life of a first-order reaction provides a convenient way to calculate a rate constant. Use Equation 14-4: $t_{1/2}$ = (ln 2)/k, rearranged to give k = (ln 2)/$t_{1/2}$;
k = (0.693) ÷ (30 s) = 2.3 x 10^{-2} s^{-1}

14.31 To determine the order of a reaction from a set of experimental data, prepare plots of ln c vs. t and 1/c vs. t:

The first-order plot is linear, while the second-order plot is not. This reaction is first-order. Determine the rate constant from k = – (Slope) = 2.55 x 10^{-3} s^{-1}.

14.33 (a) $Cl_2 \rightleftarrows Cl + Cl$, followed by $\quad NO + Cl \rightarrow NOCl$
The rate law is determined by the rate-determining step: Rate = k_2[NO][Cl]. This is not satisfactory, however, because it contains the concentration of an intermediate. Set the rates equal for the forward and reverse first step: $\quad k_1[Cl_2] = k_{-1}[Cl]^2$;

Solve this equality for [Cl]: [Cl] = $(k_1/k_{-1})^{1/2}[Cl_2]^{1/2}$;

Substitute into the rate expression: Rate = $k_2(k_1/k_{-1})^{1/2}$[NO]$[Cl_2]^{1/2}$
(b) $NO + Cl_2 \rightleftarrows NOCl + Cl$, followed by $\quad NO + Cl \rightarrow NOCl$
The rate law is determined by the rate-determining step: Rate = k_2[NO][Cl]. This is not satisfactory, however, because it contains the concentration of an intermediate. Set the rates equal for the forward and reverse first step: $\quad k_1$[NO][Cl_2] = k_{-1}[NOCl][Cl];

Solve this equality for [Cl]: [Cl] = $\left(\dfrac{k_1}{k_{-1}}\right)\dfrac{[NO][Cl_2]}{[NOCl]}$;

Substitute into the rate expression: Rate = $k_2\left(\dfrac{k_1}{k_{-1}}\right)\dfrac{[NO]^2[Cl_2]}{[NOCl]}$

Chapter 14: Mechanisms of Chemical Reactions

14.35 (a) The steps of a mechanism must sum to give the observed overall stoichiometry of the reaction. For ozone decomposition, this is $2\,O_3 \rightarrow 3\,O_2$. The two steps proposed by the student consume $1\,O_3$, produce $1\,O_2$, and generate an O atom which must be consumed:
$$O_3 + O \rightarrow 2\,O_2$$

(b) The rate law is determined by the rate-determining step: Rate = $k_2[O_5]$. This is not satisfactory, however, because it contains the concentration of an intermediate. Set the rates equal for the forward and reverse first step: $\quad k_1[O_3][O_2] = k_{-1}[O_5]$;

Solve this equality for $[O_5]$: $[O_5] = \dfrac{k_1}{k_{-1}}[O_3][O_2]$;

Substitute into the rate expression:
$$\text{Rate} = k_2 \dfrac{k_1}{k_{-1}}[O_3][O_2]$$

14.37 (a) The rate law is determined by the rate-determining step: Rate = $k_2[N_2O_2][O_2]$. This is not satisfactory, however, because it contains the concentration of an intermediate. Set the rates equal for the forward and reverse first step:
$$k_1[NO]^2 = k_{-1}[N_2O_2];$$

Solve this equality for $[N_2O_2]$: $[N_2O_2] = \left(\dfrac{k_1}{k_{-1}}\right)[NO]^2$;

Substitute into the rate expression:
$$\text{Rate} = k_2\left(\dfrac{k_1}{k_{-1}}\right)[NO]^2[O_2]$$

(b) This rate expression has an overall order of $(2 + 1) = 3$, so the mechanism is consistent with third-order behavior.

(c) The intermediate species is N_2O_2. Two NO molecules could bind together in several ways:

○ = O
● = N

In the second step, O_2 collides with the intermediate and reacts to form two NO_2 molecules. The ONNO arrangement is the only one for which this rearrangement of bonds can easily occur:

14.39 The rate constant for a reaction depends on temperature according to the Arrhenius equation (Equation 14-6): $k = Ae^{-E_a/RT}$. When $E_a = 0$, the exponent is $e^0 = 1$ and k is independent of temperature. A zero activation energy exists when a reaction can occur without first breaking any chemical bonds. The most common example is the combination of two free radicals, such as $H_3C\bullet + \bullet CH_3 \rightarrow H_3C-CH_3$.

14.41 The only true statements are (d) and (f). Statement (a) is false because $\Delta E_{rxn} = C - A$. Here is an energy diagram showing the quantities involved:

14.43 An exothermic reaction is "downhill" from reactants to products, and the activation energy plot should show this:

14.45 Activation energies are calculated from the Arrhenius equation using Equation 14-8:

$$E_a = R \ln\left(\frac{k_2}{k_1}\right) \div \left(\frac{1}{T_1} - \frac{1}{T_2}\right)$$

Development time is inversely proportional to rate constant, so $\frac{k_2}{k_1} = \frac{t_1}{t_2}$.

(a) $(t_1/t_2) = 2$; $T_1 = 20 + 273 = 293$ K; $T_2 = 293 + 10 = 303$ K;
$\left(\frac{1}{T_1} - \frac{1}{T_2}\right) = (3.413 \times 10^{-3}) - (3.300 \times 10^{-3}) = 1.13 \times 10^{-4}$ /K;
$E_a = (8.314$ J/mol K$)(10^{-3}$ kJ/J$)(\ln 2) \div (1.13 \times 10^{-4}$ /K$) = 51$ kJ/mol.

(b) Use $\ln(t_1/t_2) = \frac{E_a}{R}\left(\frac{1}{T_1} - \frac{1}{T_2}\right)$.

$E_a = 51$ kJ/mol; $T_1 = 20 + 273 = 293$ K, $t_1 = 10$ min; $T_2 = 25 + 273 = 298$ K;
$\left(\frac{1}{T_1} - \frac{1}{T_2}\right) = (3.413 \times 10^{-3}) - (3.356 \times 10^{-3}) = 5.7 \times 10^{-5}$ /K;
$\ln(10$ min$/t_2) = (51$ kJ/mol$)(5.7 \times 10^{-5}$ /K$) \div (8.314 \times 10^{-3}$ kJ/mol K$) = 0.351$;
10 min$/t_2 = 1.42$; $t_2 = (10$ min$) \div (1.42) = 7.0$ min.

14.47 (a) Pd metal is present at the start of the reaction but does not appear in the net stoichiometry, so it is a catalyst. It is solid while the reaction is between gases, so it is a heterogeneous catalyst.

(b) Here is a molecular picture showing the sequence of steps:

(1) H_2 adsorbs on Pd surface; (2) H_2 dissociates to atoms; (3) C_2H_4 collides and adds 2 H atoms; (4) C_2H_6 dissociates from surface.

14.49 Your explanation should include the main features of the induced-fit model: a cavity into which the molecule fits, that adjusts its shape to the target molecule (square), then distorts to catalyze decomposition of the square:

Cavity can accept either "square" or "triangle"

"Square" locks in place, while "triangle" does not

After bond breakage, fragments leave the cavity

14.51 When the concentration of a reactant increases by a factor of three (triples), the rate of reaction changes by 3^n, where n is the order with respect to that concentration.
(a) nine-fold increase; (b) no change; (c) rate increases by 5.2 times.

14.53 (a) The information provided suggests that transfer of an O atom between NO_2 and O_3, giving NO_3, is a reasonable first step. Then NO_3 could react with NO_2, giving the correct overall stoichiometry: If the first reaction is rate-determining, the observed rate law results:

$NO_2 + O_3 \rightarrow NO_3 + O_2$ (slow, rate determining)
$NO_3 + NO_2 \rightarrow N_2O_5$ (rapid)

(b) A molecular picture of a reaction shows reactants, reactive intermediate, and products:

(c) The activation energy diagram should indicate the exothermicity and activation energy of the overall process:

Reaction coordinate

14.55 (a) A rate expression relates rate to Δc: Rate = $\Delta[C_6H_6]/\Delta t = -(1/3)(\Delta[C_2H_2]/\Delta t)$
(b) Rate laws must always be determined experimentally. Thus there is insufficient information to write the rate law. Experiments would have to be carried out measuring the rate as a function of $[C_2H_2]$ and the data analyzed using techniques described in your textbook.

14.57 The essential information needed to construct an activation energy diagram are the energy change and activation energy. Here, the reaction is a formation reaction and the change in moles of gas is zero during the reaction, so $\Delta E \cong \Delta H_f^o$:

Reaction coordinate

14.59 Reaction times for first-order reactions can be calculated using Equation 14-3, suitably rearranged: $t = \dfrac{\ln([A]_o/[A])}{k}$:

for 10.0%, $[A] = 0.900[A]_o$ and $\ln([A]_o/[A]) = \ln(1.0 \div 0.900) = 0.105$;
$t = (0.105) \div (5.5 \times 10^{-4}\ s^{-1}) = 1.9 \times 10^2$ s;

for 50.0%, $[A] = 0.500[A]_o$ and $\ln([A]_o/[A]) = \ln(1.0 \div 0.500) = 0.693$;
$t = (0.693) \div (5.5 \times 10^{-4}\ s^{-1}) = 1.3 \times 10^3$ s;

for 99.9%, $[A] = 0.001[A]_o$ and $\ln([A]_o/[A]) = \ln(1.0 \div 0.001) = 6.91$;
$t = (6.91) \div (5.5 \times 10^{-4}\ s^{-1}) = 1.3 \times 10^4$ s.

Chapter 14: Mechanisms of Chemical Reactions

14.61 (a) False (overall reaction would give fourth-order kinetics);
(b) False (rate constants must be measured at two or more temperatures to calculate E_a);
(c) True (rates of reaction increase with increasing temperature);
(d) True (a unimolecular step has first-order kinetics).

14.63 The mechanism has the rate law, Rate = $k[X_2]$, first-order in X_2 and independent of Y. A contains 5 X_2 and 8 Y, while B contains 10 X_2 and 8 Y. The rate for B will be twice that for A, because the concentration of X_2 is twice as great.

14.65 Reaction orders are given by the exponents on the concentrations appearing in the rate law. Overall order is the sum of those exponents.
(a) first-order in N_2O_5 and first-order overall;
(b) second-order in N_2, first-order in H_2, and third-order overall;
(c) first-order in enzyme and first-order overall.

14.67 (a) Obtain the overall stoichiometry by adding the three steps, ignoring the reverse reaction of step one which leads to no net change: $2\,NO + 2\,H_2 \rightarrow N_2 + 2\,H_2O$;
(b) The rate law is determined by the rate-determining step: Rate = $k_2[N_2O_2][H_2]$. This is not satisfactory, however, because it contains the concentration of an intermediate. Set the rates equal for the forward and reverse first step:
$$k_1[NO]^2 = k_{-1}[N_2O_2];$$
Solve this equality for $[N_2O_2]$: $\quad [N_2O_2] = \dfrac{k_1}{k_{-1}}[NO]^2;$

Substitute into the rate expression: $\quad$ Rate = $k_2 \dfrac{k_1}{k_{-1}}[NO]^2[H_2]$

14.69 Ultraviolet light causes chlorofluorocarbons to fragment, producing Cl atoms which catalyze the destruction of ozone:

$$CF_2Cl_2 \xrightarrow{h\upsilon} CF_2Cl + Cl$$
$$O_3 \xrightarrow{h\upsilon} O_2 + O$$
$$Cl + O_3 \rightarrow ClO + O_2$$
$$ClO + O \rightarrow Cl + O_2$$

Because the fourth reaction regenerates a Cl atom, the 2nd-4th reactions occur many hundreds of times for every CF_2Cl_2 molecule that fragments.

14.71 (a) The half-life is not constant, so test for second-order behavior by plotting $1/c$ vs. t:

t	hr	0	1.0	2.0	3.0	5.0	7.0	9.0
c	M	0.500	0.375	0.300	0.250	0.188	0.150	0.125
$1/c$	M^{-1}	2.00	2.67	3.33	4.00	5.32	6.67	8.00

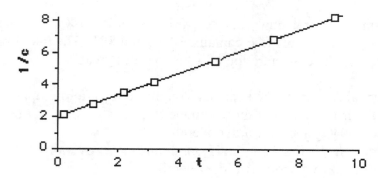

The second-order plot is a straight line, so Rate = $k[NH_4NCO]^2$;
(b) k = (Slope) = [(8.00 M^{-1}) – (2.00 M^{-1})] ÷ (9.0 hr – 0 hr) = 0.67 M^{-1} hr^{-1};
(c) At 50 °C, it takes 2.0 hr for concentration to fall from 0.500 to 0.300 M, while at 25 °C, it takes 6.0 hr for the same change in concentration to occur. The reaction is three times as fast at 50 °C as at 25 °C, so k_2/k_1 = 3.0; Use Equation 14-8 to calculate E_a:

$$E_a = \frac{R\ln(k_2/k_1)}{[(1/T_1) \pm (1/T_2)]}. \text{ with } T_2 = 50 + 273 = 323 \text{ K and } T_1 = 25 + 273 = 298 \text{ K};$$

E_a = (8.314 J/mol K)(10^{-3} kJ/J)[ln(3.0)] ÷ (2.6 x 10^{-4} K^{-1}) = 36 kJ/mol.

14.73 (a) At the molecular level, a catalyst binds to one or more of the reactants in a way that weakens chemical bonds and makes it easier for bonds to rearrange to form the products.
(b) When temperature increases, the average kinetic energies of the molecules increases, with the result that enough energy is present for reaction to occur in a larger fraction of molecular collisions.
(c) When concentration increases, the molecular density increases. There are more molecules to react. In addition, there is a higher rate of molecular collisions. Both factors contribute to a greater rate of reaction.

14.75 Flask 1 contains 5 molecules, and Flask 2 contains 10 molecules, so the concentration is doubled. The problem states that Flask 2 reacts four times as fast. This is (2)^2, so the rate law is Rate = $k[A]^2$. At the molecular level, the rate-determining step could be a reaction between two A molecules. For this process, the rate increases because of higher concentration and also because of a higher rate of collisions.

14.77 Plots of ln c vs. t and $1/c$ vs. t are not linear, showing that the reaction is neither first-order or second-order. Further examination of the data shows that the concentration change per unit time is independent of the concentration:
0 – 10 s, $\Delta c/\Delta t$ = (0.12 M) ÷ (10 s) = 0.012 M/s;
10 – 20 s, $\Delta c/\Delta t$ = (0.12 M) ÷ (10 s) = 0.012 M/s;
20 – 30 s, $\Delta c/\Delta t$ = (0.12 M) ÷ (10 s) = 0.012 M/s;
A rate that is totally independent of concentration indicates zero-order behavior: Rate = k.

Chapter 14: Mechanisms of Chemical Reactions

14.79 The hint for this problem suggests using the Arrhenius equation, $k = Ae^{-E_a/RT}$. Evaluate E_a/RT for the catalyzed and uncatalyzed situations; $T = 21 + 273 = 294$ K:

$$(E_a/RT)_{uncat} = \frac{(125 \text{ kJ/mol})}{(8.314 \times 10^{-3} \text{ kJ/mol K})(294 \text{ K})} = 51.1;$$

$$(E_a/RT)_{cat} = \frac{(46 \text{ kJ/mol})}{(8.314 \times 10^{-3} \text{ kJ/mol K})(294 \text{ K})} = 18.8;$$

$k_{uncat} = Ae^{-51.1}$ and $k_{cat} = Ae^{-18.8}$;

Divide one of these by the other to eliminate A and find the ratio of rate constants:
$k_{cat}/k_{uncat} = e^{(51.1 - 18.8)} = e^{32.3} = 1.1 \times 10^{14}$.

14.81 In any reacting system, concentrations of reactants and products are changing with time, even though these changes may be slow for some species. A changing concentration requires a non-zero net rate for each process involved in the mechanism of the reaction. Thus, for a reacting system, forward and reverse rates can never be exactly equal. (As discussed in Chapter 15, when forward and reverse rates are exactly equal the system is at equilibrium).

14.83 Figure 14-25 shows that while the activation energy for the uncatalyzed forward reaction, $O_3 + O$, is about 19 kJ/mol, the activation energy for the uncatalyzed reverse reaction, $O_2 + O_2$, is about 410 kJ/mol. Adding the catalyst reduces both values by about 15 kJ/mol, so the forward reaction has an activation energy of only about 4 kJ/mol, but the reverse reaction still has an immense activation energy, about 395 kJ/mol. The rate of the forward reaction is increased immensely by the catalyst, while the reverse reaction has essentially zero rate whether catalyzed or uncatalyzed.

14.85 The rate of reaction depends on the concentration of the enzyme-reactant complex. At low concentrations of reactant, the concentration of this complex increases linearly with the concentration of reactant, so the reaction rate increases linearly with concentration of reactant. At high concentrations of reactant, all the enzyme is in the form of enzyme-reactant complex, so the rate of reaction reaches an upper limit where it is independent of reactant concentration.

14.87 This reaction is 1/2 order in one reactant and first-order in the other reactant. There is more than one way to increase the concentrations so that the rate doubles. The simplest is to double the concentration of the reactant for which there is first-order behavior. The illustration shows three diatomic molecules and six atoms, so the new view should show three diatomic molecules and twelve atoms:

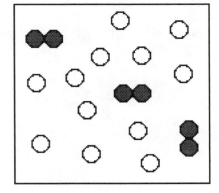

14.89 The decomposition of ozone described in Box 14-2 involves only three species: O_3, O_2, and O. O_3 is the only reactant and O_2 is the only product, so O is the only intermediate.

14.91 Neither intermediates nor catalysts appear in the overall stoichiometry of the reaction, so any species that appears in the mechanism but not in the overall stoichiometry is either an intermediate or a catalyst. Catalysts are consumed in early steps and regenerated in later steps, while intermediates are produced in early steps and consumed in later steps.

14.93 To determine which set of concentrations leads to the fastest rate, determine the product of concentrations for each set:

(a) $\left(\dfrac{0.5 \text{ mol}}{2.0\text{L}}\right)\left(\dfrac{0.5 \text{ mol}}{2.0\text{L}}\right) = 0.06 \text{ mol}^2/\text{L}^2$; (b) $\left(\dfrac{0.5 \text{ mol}}{1.0\text{L}}\right)\left(\dfrac{0.5 \text{ mol}}{1.0\text{L}}\right) = 0.25 \text{ mol}^2/\text{L}^2$;

(c) $\left(\dfrac{0.1 \text{ mol}}{1.0\text{L}}\right)\left(\dfrac{2.0 \text{ mol}}{1.0\text{L}}\right) = 0.20 \text{ mol}^2/\text{L}^2$;

thus (b), with the highest product, has the fastest rate.

Chapter 15 Principles of Chemical Equilibrium

15.1 Dynamic Equilibrium

QUESTIONS TO ANSWER, SKILLS TO LEARN
1. **What is chemical equilibrium and what are its characteristics?**
2. **What is the equilibrium constant, K_{eq}?**
3. **What are the characteristics of K_{eq}?**

Conversion of the iron oxide, hematite Fe_2O_3, to iron metal by reaction with carbon monoxide has been important in the iron manufacturing industry since the 1800's.

$$Fe_2O_{3\,(s)} + 3\,CO_{(g)} \rightarrow 2\,Fe_{(l)} + 3\,CO_{2\,(g)}$$

However, under the conditions used for the process, not all the CO was consumed, so the engineers at that time thought there was insufficient reaction time. The remedy proposed was to increase the time for the reaction by building a larger furnace (at quite a high cost) so that the gas would be in contact with the molten ore for a longer period and therefore have more time to react. However, upon running the new furnace, it was discovered that the proportion of CO in the gases released was the same as that in smaller furnaces. Therefore, the incomplete reaction was not due to lack of time. The nature of the reaction was such that not all the reactants were converted to products. This is an example of a reaction at *chemical equilibrium* with significant amounts of products and reactants present.

Chemical equilibrium is a curious phenomenon that requires two apparently contradictory observations:

- There is no observable change at the *macroscopic* level; measurable concentrations, pressures and other properties are constant.
- However, there is constant activity at the *microscopic* level where reactants are being converted to products at exactly the same rate as products are converted to reactants.

Consider the reduction of hematite to iron by CO described above by the chemical equation of reaction and below by a molecular picture as an example of an equilibrium process.

Note that in the picture, all four substances in the reaction are present in various amounts For the system to be at equilibrium, the partial pressures of CO and CO_2 in the system must be constant and the amounts of iron and hematite also must be constant. However, if the system is truly at equilibrium, there is chemical change occurring. For example, if we were to add some CO_2 which has been labeled by using a different isotope, ^{13}C, to the reaction chamber (the C* in the figure below), soon some of the ^{13}C label would be found in CO, not just CO_2. This shows that at equilibrium, not only are reactants being converted to products, but products are being converted to reactants. Chemical equilibrium is a *dynamic* process, where both the forward process and its reverse are occurring simultaneously and at the same rate.

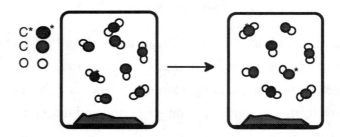

Exercise 15.1 Which of these processes is likely to be at chemical equilibrium?
(a) A balloon filled with hydrogen and oxygen ($2 H_2 + O_2 \rightarrow 2 H_2O$)
(b) A flask containing solid sodium chloride and a saturated solution of sodium chloride ($NaCl_{(s)} \rightarrow Na^+_{(aq)} + Cl^-_{(aq)}$)
(c) A flask containing a solution of acetic acid, acetate ions and hydrogen ions where the concentration of hydrogen ions is constant. ($CH_3COOH_{(aq)} \rightarrow H_3O^+_{(aq)} + CH_3COO^-_{(aq)}$)

Steps to Solution: *If a system is at chemical equilibrium, there must be no macroscopic changes; at the microscopic/molecular level, chemical processes in both the forward and reverse directions of the reaction should be proceeding. For the forward and reverse processes to be occurring simultaneously, all the substances involved in the process must be present (although their concentration may be very low).*

(a) *This reaction is not at equilibrium because there is no water present. Also, experience tells us that for this reaction, equilibrium consists of mainly water because the two gases, hydrogen and oxygen, readily react to form water. Therefore this system is metastable; there is no chemical reaction occurring, but it is not at equilibrium with the more stable product, water.*

(b) *This system has several features that suggest it is at equilibrium. First, all the products and reactants are present. Second, because the solution is saturated, the concentration of sodium and chloride ions will not change. As discussed previously, systems composed of ionic solids and solutions of their ions are in dynamic equilibrium when the process of sodium chloride dissolving is equal to the rate of sodium chloride precipitating.*

Chapter 15: Principles of Chemical Equilibrium

$$NaCl_{(s)} \rightarrow Na^+_{(aq)} + Cl^-_{(aq)}$$

$$Na^+_{(aq)} + Cl^-_{(aq)} \rightarrow NaCl_{(s)}$$

or to use a more compact notation, a double arrow is used to indicate the reversibility of the reaction:

$$Na^+_{(aq)} + Cl^-_{(aq)} \rightleftarrows NaCl_{(s)}$$

(c) *As in (b), this system has several features that suggest it is at equilibrium. All substances involved in the reaction are present, and one of the observable concentrations ($[H_3O^+]$) is constant. In the next chapter, we will see that in aqueous solution, proton transfer reactions are fast and reversible. Therefore, the two reactions:*

$$CH_3COOH_{(aq)} \rightarrow H_3O^+_{(aq)} + CH_3COO^-_{(aq)}$$

$$H_3O^+_{(aq)} + CH_3COO^-_{(aq)} \rightarrow CH_3COOH_{(aq)}$$

are proceeding at equal rates and the system is at equilibrium. For a reaction at equilibrium, the reaction will often be written with the double arrow, indicating the reversibility of the process:

$$CH_3COOH_{(aq)} \rightleftarrows H_3O^+_{(aq)} + CH_3COO^-_{(aq)}$$

Because at equilibrium, the rates of the forward reaction (reactants to products or left to right) and backward reaction (products to reactants) are the same, we can equate them mathematically:

$$\text{rate}_{forward} = \text{rate}_{back}$$

For the reaction in Exercise 15.1 (c), the rates of the forward and back reactions are:

$$\text{rate}_{forward} = k_1[CH_3COOH] \quad \text{and} \quad \text{rate}_{back} = k_{-1}[H_3O^+][CH_3COO^-]$$

Equating them gives (the subscript "eq" indicates that these are concentrations at equilibrium):

$$k_1[CH_3COOH]_{eq} = k_{-1}[H_3O^+]_{eq}[CH_3COO^-]_{eq}$$

$$\frac{k_1}{k_{-1}} = \frac{[H_3O^+]_{eq}[CH_3COO^-]_{eq}}{[CH_3COOH]_{eq}}$$

The ratio to the left of the equal sign is a constant that is called the **equilibrium constant, K_{eq}**.

Chapter 15: Principles of Chemical Equilibrium

$$K_{eq} = \frac{k_1}{k_{-1}} = \frac{[H_3O^+]_{eq}[CH_3COO^-]_{eq}}{[CH_3COOH]_{eq}}$$

Because K_{eq} is a constant (at a given temperature), this is a very powerful expression. It can be used to calculate concentrations of reactants and products at equilibrium given initial conditions.

We do not need to know the mechanism of a reaction to write the expression for the equilibrium constant; the value of the equilibrium constant depends only on the reactants and products and their states. For a general reaction, K_{eq} is related to the concentrations and stoichiometric coefficients:

$$aA + bB \rightleftharpoons cC + dD \qquad K_{eq} = \frac{[C]_{eq}^c[D]_{eq}^d}{[A]_{eq}^a[B]_{eq}^b}$$

This expression is sometimes called the mass action law. Several features regarding this expression should be recognized. Some of these features are illustrated in Exercise 15.2.

1. K_{eq} is a constant at a given temperature. However, only at equilibrium will the ratio of concentrations be equal to the equilibrium constant.

2. K_{eq} is independent of initial concentrations and conditions. Regardless of initial concentrations, when equilibrium is reached, the concentrations will be such that the equilibrium constant expression equality will be satisfied.

3. The value of K_{eq} depends upon the stoichiometry of the balanced equation of reaction.

4. The value of K_{eq} does change with temperature.

Exercise 15.2 Write the equilibrium constant expressions for the following reactions.
(a) $N_{2\,(g)} + O_{2\,(g)} \rightleftharpoons 2\,NO_{(g)}$
(b) $2\,N_{2\,(g)} + 2O_{2\,(g)} \rightleftharpoons 4\,NO_{(g)}$
(c) $2\,NO_{(g)} \rightleftharpoons N_{2\,(g)} + O_{2\,(g)}$
(d) $C_2H_{4\,(g)} + 2HCl_{(g)} + \frac{1}{2}O_{2\,(g)} \rightleftharpoons CH_2ClCH_2Cl_{(g)} + H_2O_{(g)}$

Steps to Solution: To write the equilibrium constant expression one should remember the rule "products over reactants" and remember that the concentrations of the products and reactants are raised to the power of the stoichiometric coefficients.

(a) Remember, products over reactants. Also, because the stoichiometric coefficient of NO is 2, the concentration of NO will be raised to the power of 2. $\qquad K = \dfrac{[NO]^2}{[N_2][O_2]}$

556

(b) In this case the stoichiometric coefficients of N_2 and O_2 are 2 so their concentrations will be raised to the second power. The stoichiometric coefficient of NO is 4; therefore its concentration will be raised to the power of 4

$$K' = \frac{[NO]^4}{[N_2]^2[O_2]^2}$$

(c) This reaction is the same as that in part (a) except that the products and reactants have been interchanged. Therefore the equilibrium constant expression is the reciprocal of that in part (a).

$$K'' = \frac{[N_2][O_2]}{[NO]^2}$$

(d) This reaction is a bit more complicated but the same form, products over reactants, is used. In this reaction, the stoichiometric coefficient of HCl is 2; its concentration will be raised to the power 2. Similarly, the coefficient of O_2 is $\frac{1}{2}$; the concentration of oxygen will be raised to the $\frac{1}{2}$ power.

$$K = \frac{[CH_2ClCH_2Cl][H_2O]}{[C_2H_4][HCl]^2[O_2]^{1/2}}$$

Notice that in parts (a), (b) and (c), the same reaction is written in different ways, varying the coefficients or the direction. For example, in parts (b), the coefficients are two times those of the coefficients in part (a). At equilibrium,

$$K'' = (K')^2$$

Increasing all the stoichiometric coefficients of a reaction with the equilibrium constant K by a multiplicative factor changes the numerical value of the equilibrium constant by raising K to the power of that multiplicative factor.

15.2 Properties of Equilibrium Constants

QUESTIONS TO ANSWER, SKILLS TO LEARN
1. **What concentration units are used for calculating equilibrium constants?**
2. **Writing the expression for K_{eq}**
3. **How the form of the equation of reaction affects K_{eq}**

The equilibrium constants dealt with above have all of the reactants and products in the same phase. However, in many reactions and processes, the reactants and products may be in several phases, so we need a set of standard concentration units used for the different phases.

Chapter 15: Principles of Chemical Equilibrium

Phase	Concentration Unit
Solid or liquid	Mole fraction, X
Gas	Partial Pressure, atm
Solutes	Molarity, M
Solvent	Mole fraction, X

Pure solids or liquids do not appear in the equilibrium constant expression because they have a mole fraction of 1.

Concentrations of solvents do not appear in the equilibrium constant expression because the mole fraction of the solvent is very nearly 1.

These conventions simplify the form of the equilibrium constant as shown in the following exercises.

Exercise 15.3 Write the equilibrium constant expressions for the following processes in which reactants and products are in different phases.
(a) $AgCl_{(s)} \rightleftarrows Ag^+_{(aq)} + Cl^-_{(aq)}$
(b) $Fe_2O_3{}_{(s)} + 3\,CO_{(g)} \rightleftarrows 2\,Fe\,(l) + 3\,CO_2{}_{(g)}$
(c) $CO_2{}_{(s)} \rightleftarrows CO_2{}_{(g)}$
(d) $Cl_2{}_{(g)} + H_2O\,(l) \rightleftarrows HOCl_{(aq)} + Cl^-_{(aq)} + H_3O^+_{(aq)}$

Steps to Solution: For each of the parts we must first set up the equilibrium constant expression and then consider the phase of the substance. After determining the phase of the substance, we choose the appropriate concentration units. The choice of units listed in the table can lead to considerable simplification of the equilibrium constant expression.

(a) The equilibrium constant expression for this reaction and the phases of the different chemical species involved and concentration units are given in the table below.

$AgCl_{(s)} \rightleftarrows Ag^+_{(aq)} + Cl^-_{(aq)}$

$$K = \frac{[Ag^+][Cl^-]}{[AgCl]}$$

Species	Phase	Concentration Unit
$AgCl_{(s)}$	pure solid	mole fraction
$Ag^+_{(aq)}$	solute	molarity, M
$Cl^-_{(aq)}$	solute	molarity, M

The mole fraction of the pure solid AgCl is 1; therefore we do not have to include the concentration of AgCl in the equilibrium constant expression. The other two chemical species are solutes whose concentrations are expressed in units of molarity, which are denoted by the square brackets. The final expression is written:

$$\boxed{K = [Ag^+][Cl^-] = M \cdot M = M^2}$$

(b) The equilibrium constant expression for this reaction and the phases of the different chemical species involved and concentration units are given in the table below.

$$Fe_2O_{3\,(s)} + 3\,CO_{(g)} \rightleftarrows 2\,Fe\,(l) + 3\,CO_{2\,(g)}$$

Species	Phase	Concentration Unit
$Fe_{(l)}$	pure liquid	mole fraction
$CO_{2\,(g)}$	gas	pressure, atm
$Fe_2O_{3\,(s)}$	solid	mole fraction
$CO_{(g)}$	gas	pressure, atm

$$K = \frac{[Fe]^2[CO_2]^3}{[Fe_2O_3][CO]^3}$$

The mole fractions of the pure solid Fe_2O_3 and the liquid Fe are 1; therefore the concentrations of these two components of the system are not written in the final equilibrium constant expression. The two gases will have their concentrations expressed in units of pressure, atm. Therefore the equilibrium constant expression can be written as:

$$K = \frac{p_{CO_2}^3}{p_{CO}^3} = \frac{atm^3}{atm^3} = 1 \quad \text{(this equilibrium constant will have no units!)}$$

(c) The table below gives the equilibrium constant expression for this reaction as well as the phases of the different chemical species involved and their concentration units.

$$CO_{2\,(s)} \rightleftarrows CO_{2\,(g)}$$

Species	Phase	Concentration Unit
$CO_{2(s)}$	solid	mole fraction
$CO_{2\,(g)}$	gas	pressure, atm

Because the mole fraction of solid CO_2 is unitless and 1 and the concentration of the gas is to be expressed in atm, the units of the equilibrium constant for the sublimation of CO_2 are:

$$K = \frac{p_{CO_2}}{X_{CO_2}} = p_{CO_2} = \frac{atm}{1} \quad \text{(units of } K \text{ are atm)}$$

(d) The equilibrium constant expression for this reaction and the phases of the different chemical species involved and concentration units are given in the table below.

$$Cl_{2\,(g)} + H_2O\,(l) \rightleftarrows HOCl_{(aq)} + Cl^-_{(aq)} + H_3O^+_{(aq)}$$

$$K = \frac{[H_3O^+][Cl^-][HOCl]}{[Cl_2][H_2O]}$$

Species	Phase	Concentration Unit
$Cl_{2(g)}$	gas	pressure, atm
$H_2O_{(l)}$	solvent (liquid)	mole fraction
$HOCl$	solute	molarity, M
Cl^-	solute	molarity, M
H_3O^+	solute	molarity, M

The solvent, water, has a mole fraction that is very near 1; therefore, as an approximation, the mole fraction of the solvent in this (and in nearly all the other examples in this text) wil be considered to be 1 and therefore will not appear in the equilibrium constant expression. The rest of the units are standard; omitting the solvent term the equilibrium constant expression is:

$$K = \frac{[H_3O^+][Cl^-][HOCl]}{p_{Cl_2}} = \frac{M \cdot M \cdot M}{atm} = \frac{M^3}{atm}$$

Magnitudes of Equilibrium Constants

The size of the equilibrium constant of a reaction is an indication of the composition of the mixture of substances present at equilibrium. Equilibrium constants expressed in terms of molarity and/or atm that are greater than 10^4 are usually considered large, and those that are smaller than 10^{-4} are usually considered small. Reactions that have small equilibrium constants have more reactants than products present at equilibrium; conversely, those with large equilibrium constants have much higher product concentrations than reactants at equilibrium. Equilibrium constants with values in between these guidelines tend to have substantial amounts of both products and reactants present at equilibrium.

Reactions used in the industrial manufacture of chemicals tend to have large equilibrium constants so that most of the reactants are converted to products. If not, much effort is expended to devise clever methods to separate the products from the reactants at minimal cost.

15.3 Thermodynamics and Equilibrium

QUESTIONS TO ANSWER, SKILLS TO LEARN
1. **Using the concentration quotient to predict the direction a reaction will proceed to reach equilibrium**
2. **Using $\Delta G°$ to estimate equilibrium constants**
3. **What is the relationship between ΔG, $\Delta G°$, Q and K?**

In Chapter 13, the notion of the *concentration quotient* was developed for determining the spontaneity of a reaction. For the general reaction,

$$aA + bB \rightarrow cC + dD$$

the concentration quotient is:

$$Q = \frac{[C]^c[D]^d}{[A]^a[B]^b} \text{ (for solutions)} \quad \text{or} \quad Q = \frac{p_C^c \, p_D^d}{p_A^a \, p_B^b} \text{ (for gases)}$$

The values for Q are unlimited as Q can represent *any* set of concentrations or pressures (nonequilibrium conditions) at some temperature, unlike the equilibrium constant for which the equilibrium concentrations (or pressures) must satisfy the mass action law.

Chapter 15: Principles of Chemical Equilibrium

$$K = \frac{[C]_{eq}^c [D]_{eq}^d}{[A]_{eq}^a [B]_{eq}^b}$$

Consider what it means if Q is greater than K for any reaction. Because Q represents nonequilibrium conditions, the pressures of reactants and products will change so that Q decreases until equilibrium is reached and $Q = K$. Q will decrease if the concentration(s) of product(s) decrease(s) and the concentration(s) of reactant(s) increase(s). Products will be converted to reactants; the reaction will *shift* to the right.

$$aA + bB \rightleftarrows cC + dD \qquad \Delta G = \Delta G° + RT\ln Q$$

Remember when $Q = K$, $\Delta G = 0$:

$$\Delta G = 0 = \Delta G° + RT\ln K \qquad \Delta G° = -RT\ln K \text{ or } K = \exp\left(-\frac{\Delta G°}{RT}\right)$$

where $\Delta G°$ is expressed in Joules. The following table gives the relation between ΔG and the relative sizes of Q and K.

Q compared to K	ΔG	Reaction proceeds to:
$Q > K$	>0	form reactants from products (shift to right)
$Q < K$	<0	form products from reactants (shift to left)
$Q = K$	= 0	no change in concentrations at equilibrium

Because we can calculate $\Delta G°$ at different temperatures with these quantities, we can also estimate the equilibrium constant at different temperatures, as shown below in Exercise 15.4.

Exercise 15.4 The "water gas shift reaction" is shown below. Calculate K at (a) 298 K and (b) 1000 K and (c) find the temperature at which the equilibrium constant is 1.

$$CO_{(g)} + H_2O_{(g)} \rightleftarrows CO_{2(g)} + H_{2(g)} \qquad K = \frac{p_{CO_2} p_{H_2}}{p_{CO} p_{H_2O}}$$

Steps to Solution: We must first find the enthalpy and entropy changes for the reaction. Then we must calculate $\Delta G°$ at the temperatures of interest in parts (a) and (b). In part (c) the answer requires analysis of the expression for ΔG in terms of $\Delta G°$ and K.

(a) *Using a table of thermodynamic data, we find the following values for standard enthalpies of formation and standard entropies for the reactants and products:*

561

Chapter 15: Principles of Chemical Equilibrium

	$CO_{(g)}$	$H_2O_{(g)}$	$CO_{2(g)}$	$H_{2(g)}$	Δ
$\Delta H_f°$ (kJ/mol)	-110.523	-241.83	0	-393.513	$\Delta H° = -41.16$ kJ
$S°$ (J/K•mol)	197.91	188.72	130.59	213.64	$\Delta S° = -42.4$ J/K

The enthalpy and entropy changes are calculated by the methods in Chapter 13. At 298 K,

$$\Delta G° = -41.16 \text{ kJ} - (298 \text{ K})\left(-42.4 \text{ J} \cdot \frac{1 \text{ kJ}}{1000 \text{ J}}\right) = -28.52 \text{ kJ}$$

$$K = \exp\left(-\frac{-28.52 \text{ kJ} \cdot 1000 \text{ J/kJ}}{8.314 \text{ J/mole} \cdot \text{K} \cdot 298 \text{ K}}\right) = 11.51 = \boxed{1.0 \times 10^5}$$

(Note: the antilog of a number is obtained on most calculators by entering the number and using the *exp* key or using the inverse (*INV*) of the *ln* function)

The reaction favors products because K is large.

(b) *Now we calculate K at 1000 K; the same values of $\Delta H°$ and $\Delta S°$ may be used because an estimate is being sought.*

$$\Delta G° = -41.16 \text{ kJ} - (1000 \text{ K})\left(-42.4 \text{ J/K} \cdot \frac{1 \text{ kJ}}{1000 \text{ J}}\right) = 1.24 \text{ kJ}$$

$$K = \exp\left(-\frac{1.24 \text{ kJ} \cdot 1000 \text{ J/kJ}}{8.314 \text{ J/mol} \cdot \text{K} \cdot 1000 \text{ K}}\right) = e^{-0.149} = \boxed{0.86}$$

Now K is less than 1 and therefore reactants are favored at this temperature (although not by much). It is customary to state that a reaction becomes spontaneous when K > 1. Note that this temperature must lie between 298 and 1000 K for the water gas shift reaction as K < 1 at 1000 and K > 1 at 298 K.

(c) *At what temperature does K change from being spontaneous to not being spontaneous (or at what temperature is K = 1)?* Because we are seeking an equilibrium condition, the following expression will be useful as it includes the free energy and the equilibrium constant. Because the reaction is at equilibrium, $\Delta G = 0$ and so the following expression is obtained:

$$\Delta G = 0 = \Delta G° + RT\ln K \qquad \Delta G° = -RT\ln K = \Delta H° - T\Delta S°$$

Because K = 1, lnK = ln(1) = 0: $\quad 0 = \Delta H° - T\Delta S° \qquad$ so $\qquad \Delta H° = T\Delta S°$

Chapter 15: Principles of Chemical Equilibrium

which gives: $\quad\quad\quad T = \dfrac{\Delta H°}{\Delta S°}$ for $K = 1$

Substituting the values for the enthalpy change and entropy change obtained in part (a) gives:

$$T = \dfrac{41.16 \times 10^3 \text{ J}}{42.4 \text{ J/K}} = \boxed{9.71 \times 10^2 \text{ K}}$$

Because ΔH and ΔS vary to some degree with changes from standard conditions, this value is approximate (remember we used the standard enthalpies of formation and standard entropies to find ΔH and $\Delta S°$); the calculated temperature is probably precise to about one significant figure in view of these approximations. K will be 1 at about 1000 K.

15.4 Shifts in Equilibrium

QUESTIONS TO ANSWER, SKILLS TO LEARN
1. What is meant by "shifting equilibrium"?
2. What is Le Chatelier's principle?
3. Predicting the direction of the shift upon changing concentrations of reagents, pressure or temperature

A reaction at equilibrium will be affected by nearly any change such that the concentrations or pressures of the reactants or products will change. If more reactants are produced as a result of the change, it is commonly said that the reaction *shifts toward reactants*. If more products are formed, then the reaction *shifts toward products*.

The changes of interest are:
- changes in amounts (concentrations or partial pressures of reactants or products)
- changes in temperature

A simple statement provides qualitative guides to which direction the reaction will shift. Le Chatelier's principle states:

> **When a change is introduced into a chemical system at equilibrium, the system will shift in the direction that counteracts that change.**

Consider the Haber process of ammonia production as an example of how the above changes can shift equilibrium.

Chapter 15: Principles of Chemical Equilibrium

$$N_{2(g)} + 3\,H_{2(g)} \rightleftarrows 2\,NH_{3(g)} \qquad \Delta H^\circ = -92.38\text{ kJ} \qquad K = \frac{p_{NH_3}^2}{p_{N_2} p_{H_2}^3}$$

Changes in Amounts

Addition of any substance that is involved in the chemical reaction will have an effect because the system will no longer be at equilibrium. We can best understand this effect by examining the effect of that substance on the value of the concentration quotient, Q, for the system. If $Q > K$, then the reaction will shift toward reactants; if $Q < K$, the reaction will shift toward products.

$$Q = \frac{p_{NH_3}^2}{p_{N_2} p_{H_2}^3}$$

The effects of changes in amounts in this reaction are examined in Exercise 15.5.

Exercise 15.5 A system containing nitrogen, ammonia and hydrogen is at equilibrium. In which direction would the reaction shift if: (a) H_2 is added? (b) NH_3 is added? (c) Ar is added? (d) NH_3 is removed?

Steps to Solution: The effects of adding a substance to the system are best understood by examining the effect that addition has on Q. The system is at equilibrium before the addition; therefore before addition, Q = K. Addition of a substance will increase its pressure. If the substance appears in the numerator of Q, Q will increase and be greater than K. If the substance is a reactant (appears in the denominator), an increase in its size will decrease Q so Q < K, and the reaction will shift toward products.

(a) *If H_2 is added to the system at equilibrium, the pressure of H_2 will increase, therefore making Q < K (p_{H_2} is in the denominator, so we are dividing by a larger number!).*

$$Q = \frac{p_{NH_3}^2 (eq)}{p_{N_2}(eq)\, p_{H_2}^3} < K \text{ for } p_{H_2} > p_{H_2}(eq)$$

For Q < K, the reaction will proceed to form products, or the reaction shifts toward products.

(b) *Addition of NH_3 to the system at equilibrium will increase the ammonia pressure, making Q > K. For Q > K, the reaction will shift toward reactants, using up ammonia to decrease its pressure and increasing the pressure of the reactants until Q = K.*

(c) *Addition of Ar gas will have no effect on the position of equilibrium (pressures of nitrogen, hydrogen and ammonia) because argon is not involved in the equilibrium reaction.*

(d) *Removal of a substance from a system at equilibrium will decrease its pressure (or concentration if a dissolved species). We can determine the effect of such a change by examining the effect on Q of the decrease. Removing ammonia gas, a product, will decrease its partial pressure and also decrease Q.*

What is the effect of adding a *catalyst* to a reaction? Using the above criterion for Q, it is clear that there should be no effect because the catalyst will not appear in the concentration quotient. In addition, remember that the only purpose a catalyst serves is to change the rate at which a reaction occurs by changing the mechanism of the reaction. The equilibrium constant is a thermodynamic quantity and depends on the enthalpy (bond energies) and entropy changes in the reaction, which are properties of the reactant and product molecules.

Effect of Temperature Changes

The effect of temperature changes can be analyzed by understanding the mathematical relationship between K and T.

$$\ln K = -\frac{\Delta H°}{RT} + \frac{\Delta S°}{R}$$

If $\Delta H > 0$, then the $-\frac{\Delta H°}{RT}$ term is always negative. Because K is proportional to $\exp\left(-\frac{\Delta H°}{RT}\right)$, K is less than 1 for terms where $-\frac{\Delta H°}{RT} < 0$. As T increases, this term becomes smaller in magnitude and therefore $\ln K$ (and K) grows larger, especially if $\Delta S° > 0$. Conversely, if $\Delta H° < 0$, then the $-\frac{\Delta H°}{RT}$ term is always positive. As T increases, the magnitude of this number grows smaller and so does $\ln K$.

K decreases with increasing temperature for exothermic processes.
K increases with temperature for endothermic processes.

Another way to think about this relationship is to remember that in an exothermic reaction, heat is a product of the reaction. Addition of heat (increasing the temperature) drives equilibrium toward reactants (lowers K). For an endothermic process, heat is a reactant, so addition of heat drives the reaction toward products (K increases with increasing temperature).

Exercise 15.6 What will be the shift in equilibrium for each of the following reactions upon increasing temperature?
(a) $N_{2\,(g)} + 3\,H_{2\,(g)} \rightleftarrows 2\,NH_{3\,(g)}$ $\Delta H° = -92.38$ kJ
(b) $N_{2\,(g)} + O_{2\,(g)} \rightleftarrows 2\,NO_{(g)}$ $\Delta H° = 90.3$ kJ
(c) $CO_{(g)} + H_2O_{(g)} \rightleftarrows CO_{2\,(g)} + H_{2\,(g)}$ $\Delta H° = -41.16$ kJ

Steps to Solution: *We can determine the shift in the equilibrium constant by considering the enthalpy change. For $\Delta H° < 0$, K decreases with increasing temperature.*

(a) *Because the reaction is exothermic, K decreases with increasing temperature.*

(b) *Because the reaction is endothermic, K increases with increasing temperature. This reaction was once tried for the commercial production of nitric acid, but was deemed not feasible because of the high energy costs in reaching high enough temperatures for significant conversion to products. The reaction does occur to a small extent in the burning of fuels in internal combustion engines.*

(c) *Because the reaction is exothermic, K decreases with increasing temperature. Two values of K for this reaction were calculated above in Exercise 15.8 and showed that K decreases with increasing temperature.*

15.5 Types of Equilibria

QUESTIONS TO ANSWER, SKILLS TO LEARN
1. **What are common physical changes described by K_{eq}?**
2. **What are common chemical changes that are described by K_{eq}?**

Physical Phenomena

In previous chapters physical phenomena have been described as equilibrium processes. These processes include phase changes and the solubility of gases. In the following pages these equilibria and special symbols that have been assigned to these processes are described.

Vaporization

Vaporization equilibria involve both liquid-vapor and solid-vapor equilibria.

$$C_6H_{6\,(l)} \rightleftarrows C_6H_{6\,(g)} \qquad\qquad C_{10}H_{8\,(s)} \rightleftarrows C_{10}H_{8\,(g)}$$

(liquid-vapor) (solid-vapor)

The equilibrium constant expressions for these two processes involve only the partial pressure of the vapor because the mole fraction of the liquid (and the solid) is 1:

$$K = \frac{p_{C_6H_6}}{X_{C_6H_6(l)}} = \frac{p_{C_6H_6}}{1} = p_{C_6H_6} \qquad\qquad K = \frac{p_{C_{10}H_8}}{X_{C_{10}H_8(s)}} = \frac{p_{C_{10}H_8}}{1} = p_{C_{10}H_8}$$

Chapter 15: Principles of Chemical Equilibrium

Gas Solubility

Gases have a measurable solubility in most solvents, which we described using Henry's Law in Chapter 12. For example, carbon dioxide dissolves in water as described by the chemical equation of reaction and the equilibrium constant expression below.

$$CO_{2\,(g)} \rightleftarrows CO_{2\,(aq)} \qquad K = \frac{[CO_2(aq)]}{p_{CO_2}}$$

Rearranging the equilibrium constant expression gives the equation recognized as Henry's Law and also identifies the equilibrium constant as the Henry's Law constant, K_h:

$$[CO_{2\,(aq)}] = K \cdot p_{CO_2} = K_h \cdot p_{CO_2} \text{ therefore } K = K_h$$

Chemical Processes

All chemical processes have equilibrium constants that can be calculated for them; there are several general classes of reactions for which the equilibria are particularly common and are sufficiently important that they have been assigned special symbols.

Solubility of Salts: K_{sp}

The process of a salt dissolving in water is an equilibrium. The addition of mercury(I) chloride to water leads to dissolving of the salt until no more of the salt will dissolve and the solution is saturated.

$$Hg_2Cl_{2\,(s)} \rightleftarrows Hg_2^{2+}{}_{(aq)} + 2\,Cl^-{}_{(aq)}$$

At saturation, the rate of the solid dissolving is exactly balanced by the rate of the aqueous ions crystallizing. The equilibrium constant expression for this process is:

$$K = \frac{[Hg_2^{2+}][Cl^-]^2}{X_{Hg_2Cl_2(s)}} = K_{sp} = [Hg_2^{2+}][Cl^-]^2$$

Because the mole fraction of the solid salt is 1, it is usually not written in the equilibrium constant expression. Solubility equilibria are encountered often enough to merit their own symbol, K_{sp}, and the value of the equilibrium constant is called the **solubility product.**

Water Self-Ionization: K_w

Water is a substance that has very polar bonds between O and H and also has lone pairs. Because of the polarity of the O-H bonds, it is not surprising that water can act as a proton donor: the lone pairs allow water molecules to also act as a proton acceptor. What is perhaps more difficult to comprehend is that in pure water, some of the water molecules undergo proton transfer to other water molecules; this important equilibrium is called the **auto-ionization of water** and the equilibrium constant is given the symbol K_w.

$$2\,H_2O_{(aq)} \rightleftarrows H_3O^+_{(aq)} + OH^-_{(aq)} \qquad K_w = [H_3O^+][OH^-] = 1.00 \times 10^{-14}\,M^2$$

The square brackets in the picture are to remind you that the chemical species are aqueous ions that are interacting with several water molecules that are not shown.

Acid-Base (Proton Transfer) Equilibria: K_a and K_b

One class of reactions whose equilibria are of great importance is the proton transfer reaction from an acid to a base. In an aqueous solution of a substance, water can accept or release protons depending upon how strongly the solute binds protons. Some substances transfer protons to water such as hydrochloric acid (HCl) or formic acid (H-(C=O)-OH). The equilibrium constant for the extent to which the substance is converted to ions and aqueous hydrogen ions is called the **acid dissociation constant, K_a**.

$$HCl_{(aq)} + H_2O_{(l)} \rightleftarrows H_3O^+_{(aq)} + Cl^-_{(aq)} \qquad K_a = \frac{[H_3O^+][Cl^-]}{[HCl]} = 1 \times 10^3$$

$$H(C=O)OH_{(aq)} + H_2O_{(l)} \rightleftarrows H_3O^+_{(aq)} + H(C=O)O^-_{(aq)}$$

$$K_a = \frac{[H_3O^+][H(C=O)O^-]}{[H(C=O)OH]} = 1.8 \times 10^{-4}$$

Gas Phase Equilibria

Many reactions occur in the gas phase and do not fit any of the above reaction types. However, some reactions important in the industrial manufacture of substances like nitric and sulfuric acids do occur in the gas phase. Reactions that occur with the reactants and products in the gas phase will be described simply as gas phase equilibria.

15.6 Working with Equilibria

QUESTIONS TO ANSWER, SKILLS TO LEARN
1. **Using the molecular perspective to analyze equilibrium problems**
2. **Using five steps to analyze and solve equilibrium problems**
3. **Calculating equilibrium constants from equilibrium concentrations**
4. **Estimating concentrations at equilibrium given initial, nonequilibrium concentrations**
5. **How is the "position of equilibrium" related to the magnitude of K_{eq}?**
6. **Using the position of equilibrium to simplify equilibrium calculations**

Probably one of the most important (and common) types of calculations involving equilibria is the calculation of concentrations (or pressures) of substances at equilibrium given initial conditions (concentrations or pressures of reactants and/or products). These problems may be solved using some fundamental algebra and a molecular perspective to keep track of the stoichiometry. Remember, perfect algebra with the wrong assumptions about stoichiometry gives the wrong answer. The following five steps are part of solving every equilibrium problem:

1. Identify all the chemical species that are present before reaction occurs. This allows identification of the types(s) of reactions that may occur.

2. Write the net chemical reaction (or the sum of the reactions) that the substances can undergo.

3. Write the equilibrium constant expression(s) for the reaction(s) and see if the values for the equilibrium constant are to be solved for or are provided.

4. Perform a stoichiometric analysis of the reactants and products to determine mathematical expressions (formulas/equations) for the equilibrium concentrations. One *may* be able to solve these equations to give values for the equilibrium concentrations.

5. Substitute the equilibrium concentration expressions found in step 4 into the equilibrium constant expression and solve for the unknown.

Calculation of Equilibrium Constants

One problem involves the calculation of the equilibrium constant given information regarding the concentrations of the substances present. In these cases the crux of the problem is determining the stoichiometric relationships between the different reactants and products. The five step method helps to organize your attack of the problem.

Exercise 15.7 The conversion of nitrogen and hydrogen to ammonia is an important industrial reaction:

Chapter 15: Principles of Chemical Equilibrium

$$N_{2\ (g)} + 3\ H_{2\ (g)} \rightleftarrows 2\ NH_{3\ (g)}$$

If a tank initially containing only nitrogen at 1.0 atm and hydrogen at 3.0 atm converts 13.0 % of the nitrogen to ammonia, what is the value of the equilibrium constant for the reaction as written above?

Steps to Solution: This question asks us to find the value of the equilibrium constant which requires knowledge of the equilibrium concentrations. We will use the five-step method to deduce the equilibrium pressures and calculate K.

1. The chemical species initially present are nitrogen and hydrogen at pressures of 1.0 and 3.0 atm, respectively.
2. The reaction that the substances can undergo is the formation of ammonia, given in the problem.
3. The equilibrium constant expression is:

$$K = \frac{p_{NH_3}^2}{p_{N_2}\ p_{H_2}^3}$$

4. Perform stoichiometric analysis to determine the partial pressures of the reactants and products at equilibrium. This problem gives initial conditions and the percent of a starting material converted to products at equilibrium. It may be useful to construct a table that summarizes the knowns and unknowns for the problem.

Compound	Initial Pressure	Change	Equilibrium Pressure
N_2	1.0 atm	13 % consumed	
H_2	3.0 atm		
NH_3	0		

In this problem, the gases will be treated as ideal gases. One property of ideal gases is that pressure is independent of the type of molecule in the gas, and depends only on the number of molecules of gas. Therefore, because 13% of the N_2 is consumed, the pressure of N_2 will drop by 13% of its original value.

$$\text{decrease in } N_2 \text{ pressure} = \frac{13\%}{100}(1\text{ atm}) = 0.13\text{ atm}$$

Because this is a decrease in pressure, the value in the table will be negative. The stoichiometry requires 3 H_2 per N_2 the change in the H_2 pressure is:

$$\text{decrease in } H_2 \text{ pressure} = \frac{3\ H_2}{1\ N_2} \bullet (\text{change in } N_2) = 3 \bullet (0.13\text{ atm}) = 0.39\text{ atm}$$

Chapter 15: Principles of Chemical Equilibrium

This is a decrease in pressure; the value in the table will be negative. From the stoichiometry, there are 2 mols of NH_3 produced for every mole N_2 consumed. The increase in NH_3 pressure is:

$$\text{increase in } NH_3 \text{ pressure} = \frac{2\ NH_3}{1\ N_2} (\text{change in } N_2) = 2(0.13\ \text{atm}) = 0.26\ \text{atm}$$

Placing these values in the "change" column of the table allows us to calculate the equilibrium pressures by adding the initial pressure and the change as done below.

Compound	Initial Pressure	+ Change	= Equilibrium Pressure
N_2	1.0 atm	- 0.13 atm (decrease)	0.87 atm
H_2	3.0 atm	- 0.39 atm (decrease)	2.61 atm
NH_3	0	+ 0.26 atm (increase)	0.26 atm

5. Substituting the equilibrium pressures into the equilibrium constant expression gives:

$$K = \frac{p_{NH_3}^2}{p_{N_2}\, p_{H_2}^3} = \frac{(0.26\ \text{atm})^2}{0.87\ \text{atm}\ (2.61\ \text{atm})^3} = \boxed{4.4 \times 10^{-3}\ \text{atm}^{-2}}$$

Calculating Equilibrium Concentrations from Initial Conditions

One of the more common uses of equilibrium constants is the prediction of concentrations or pressures at equilibrium given initial concentrations. In this type of problem, the equilibrium constant and some initial concentrations or pressures are supplied, and the task is to calculate the change in the concentrations. The concentration tables introduced above are very useful in working these types of calculations. Substitution of the mathematical expression for the equilibrium concentration into the equilibrium constant expression leads to an equation that can be solved for the change in concentration. This process is illustrated in Exercise 15.7.

Exercise 15.7 The water gas shift reaction is used to remove CO from the mixture of gases that contain the hydrogen in ammonia production:

$$CO_{(g)} + H_2O_{(g)} \rightleftarrows CO_{2\,(g)} + H_{2\,(g)}$$

At 300 °C, the equilibrium constant for this reaction is $K = 36$. If a tank with a volume of 30.0 L is charged with 1.0 mole of CO and 2.2 mols of H_2O and the temperature brought to 300 °C, what are the equilibrium pressures of CO, H_2, H_2O and CO_2 in the tank?

Chapter 15: Principles of Chemical Equilibrium

Steps to Solution: This problem gives initial conditions and an equilibrium constant, and the answers desired are the equilibrium partial pressures of the reactants and products. Using the five-step method and a table of concentrations will organize the data and help solve the problem.

1. The system initially contains 1.0 mole of CO and 2.2 mols of H_2O in a 30.0 L tank at 300 °C.
2. The reaction is given: $CO_{(g)} + H_2O_{(g)} \rightleftharpoons CO_{2(g)} + H_{2(g)}$
3. The equilibrium expression is:

$$K = \frac{p_{CO_2} p_{H_2}}{p_{CO} p_{H_2O}}$$

4. Use of a concentration table (actually a pressure table in this case) will organize the data. First we must convert the initial amounts of CO and H_2O to units of atmospheres. The pressure of CO is initially:

$$p_{CO} = \frac{nRT}{V} = \frac{1.0 \text{ mol } 0.0821 \text{L} \cdot \text{atm/mol} \cdot \text{K}(300+273\text{K})}{30.0\text{L}} = 1.57 \text{ atm}$$

(Note that an extra significant figure is being carried; it will be rounded off at the end.)

similarly, $p_{H_2O} = 3.45$ atm

As the reaction proceeds, some of the CO and water will be consumed and some H_2 and CO_2 will be produced. We cannot readily solve directly for the equilibrium pressures of the substance, but the changes in partial pressure of the substances are all related by stoichiometry. If the pressure of CO decreases by some amount, x atm, the decrease in pressure of H_2O is:

$$p_{H_2O} = p_{initial} - \frac{1 \text{ mol } H_2O}{1 \text{ mol CO}} \cdot x = p_{initial} - x$$

The partial pressure of CO_2 will increase by x times the stoichiometric ratio:

$$p_{CO_2} = p_{initial} + \frac{1 \text{ mole } CO_2}{1 \text{ mole CO}} \cdot x = p_{initial_{CO_2}} + x$$

Now we can complete the table:

Compound	Initial Pressure	+ Change	= Equilibrium Pressure
CO	1.57 atm	-x	1.57 - x
H_2O	3.45 atm	-x	3.45 - x
CO_2	0	+x	x
H_2	0	+x	x

5. Substitute into the equilibrium expression and solve for x. Then we can use the value of x to solve for the equilibrium pressures in the table.

$$K = \frac{p_{CO_2} p_{H_2}}{p_{CO} p_{H_2O}} = \frac{x \cdot x}{(1.57 - x)(3.45 - x)} = \frac{x^2}{(5.42 - 5.02x + x^2)}$$

Gathering all the terms together gives a quadratic equation:

$$(5.42 - 5.02x + x^2)K = x^2 \text{ which gives: } (K - 1)x^2 - 5.02(K)x + 5.42K = 0$$

A quadratic equation has the form: $(ax^2 + bx + c = 0)$ where:

$a = K - 1 = 36 - 1 = 35$ $b = -5.02(K) = -5.02 \cdot 36$ $c = 5.42(K) = 5.42 \cdot 36$
$\qquad\qquad\qquad\qquad\qquad b = -181 \qquad\qquad\qquad c = 195$

The two values of x which will satisfy the quadratic equation are found using:

$$x = \frac{-b \pm \sqrt{b^2 - 4ac}}{2a}$$

Substituting these values into the equation for x gives:

$$x = \frac{-(-181) + \sqrt{(-181)^2 - 4(35)(195)}}{2(35)} = 3.63 \qquad x = \frac{-(-181) - \sqrt{(-181)^2 - 4(35)(195)}}{2(35)} = 1.53$$

Which of these two values of x is the correct one? The first value gives pressures of reactants at equilibrium that are less than zero, which are physically unreasonable. The second root of the equation gives physically reasonable pressures:

$p_{CO} = 1.568 - 1.534$ atm $= 0.034$ atm $\qquad p_{H_2O} = 3.45 - 1.53 = 1.91$ atm

$$p_{CO_2} = x = p_{H_2} = 1.53 \text{ atm}$$

To verify that no errors have been made, one can substitute these values into the equilibrium expression and see if the value of the equilibrium constant is obtained.

$$K = \frac{(1.53)(1.53)}{(0.034)(1.91)} = 36$$

Chapter 15: Principles of Chemical Equilibrium

The equilibrium pressures are:

$$p_{CO} = 0.034 \text{ atm} \quad p_{H_2O} = 1.91 \text{ atm} \quad p_{CO_2} = 1.53 \text{ atm} \quad p_{H_2} = 1.53 \text{ atm}$$

Although the quadratic equation seems messy, there are many equilibrium constants that give higher-order (cubic, x^3 and quartic, x^4) equations that have no such simple solution. Those equations will not be encountered in this text unless some of the simplifying assumptions discussed in the next section are applicable.

Using the Position of Equilibrium to Simplify Equilibrium Problems: Small K

In the preceding example, K was of an intermediate size; as a result, significant amounts of products and reactants are present at equilibrium. However, if K is quite small (10^{-5} or smaller) only a very small amount of reactants is converted to products. Therefore it may be possible to simplify the problem by comparing the anticipated value of x to the pressure or concentration it is being subtracted from (or added to). This simplification is illustrated in Exercise 15.8.

Exercise 15.8 Solid CaF_2 is added to a 0.10 M solution of $CaCl_2$ until no more will dissolve. If the K_{sp} of CaF_2 is 3.9×10^{-11} M³, what are the concentrations of $Ca^{2+}{}_{(aq)}$ and $F^-{}_{(aq)}$ at equilibrium?

Steps to Solution: Use the five-step method to examine this system and be aware of situations where the change in concentration is quite small compared to the initial concentration.

1. *Initially, the system can be considered to consist of solid CaF_2 and aqueous Ca^{2+} (0.10 M) and Cl^- (0.20 M) ions.*

2. *The reaction of interest is dissolving CaF_2:*

$$CaF_{2\,(s)} \rightleftarrows Ca^{2+}{}_{(aq)} + 2\,F^-{}_{(aq)}$$

   $CaCl_2$ is a soluble salt, so its solubility equilibrium will not come into play with this reaction.

3. *The equilibrium expression is: $K_{sp} = [Ca^{2+}][F^-]^2$*

4. *These data are readily organized into a concentration table of the chemical species active in the equilibrium to proceed with the solution of the problem. The concentration of Ca^{2+} initially is 0.10 M because of the calcium ion from the calcium chloride solution.*

Chapter 15: Principles of Chemical Equilibrium

Compound	Initial Concentration	+ Change	= Equilibrium Concentration
[Ca^{2+}]	0.10 M		
[F$^-$]	0.0 M		

The change in concentration of calcium ion will be called x. The change in F$^-$ concentration will be related to x by the composition of the salt:

$$[F^-] = \frac{2 \text{ mole F}^-}{1 \text{ mole CaF}_2 \text{ dissolved}}(x) = 2x$$

The concentration table can be completed with this information.

Compound	Initial Concentration	+ Change	= Equilibrium Concentration
[Ca^{2+}]	0.10 M	x	0.10 + x
[F$^-$]	0.0 M	2x	2x

Now we can proceed with step 5: substitute and solve for x.

$$K_{sp} = [Ca^{2+}][F^-]^2 = (0.10 + x)(2x)^2 = 3.9 \times 10^{-11}$$

This will result in a cubic equation. However, note that K_{sp} is small and the increase in calcium ion is expected to be small compared to the initial concentration of 0.10 M. If we make the approximation that:

$$0.10 + x \approx 0.10 \quad \text{only if x is much smaller than 0.10}$$

then the substituted equilibrium expression becomes simpler:

$$K_{sp} = (0.10)(2x)^2$$

$$x = \sqrt{\frac{K_{sp}}{4(0.10)}} = \sqrt{\frac{3.9 \times 10^{-11}}{4(0.10)}} = 9.9 \times 10^{-6} \text{ M}$$

Note that one cannot neglect the x in the [F$^-$] term; either one obtains an inequality (if x is allowed to be 0) or another common error is to let x = 1 so it "disappears". Neither of these methods is valid!

Chapter 15: Principles of Chemical Equilibrium

> **Common Pitfall:** "Neglecting" x when it does *not* appear in a sum or difference by letting it be 1. **DON'T DO IT!**
> An example of this fatal mistake would be to convert
> $K_{sp} = (0.10 + x)(2x)^2$ to $K_{sp} = (0.10 + x)$

Therefore, the equilibrium concentrations of calcium ion and fluoride ion are:

$$[Ca^{2+}] = 0.10 \text{ M} + 9.9 \times 10^{-6} = 0.10 \text{ M} \qquad [F^-] = 2(9.9 \times 10^{-6} \text{ M}) = 2.0 \times 10^{-5} \text{ M}$$

Note that the concentration of calcium ion is not significantly changed by the addition of CaF_2. However, remember that x is not neglected; it is just too small to affect the precision of the initial value of the concentration. The approximation is valid only when x is less than 5% of the value it is compared to. To test this validation:

$$\text{test} = \frac{x}{[Ca^{2+}(\text{initial})]} 100\% = \frac{9.9 \times 10^{-6}}{0.10} \cdot 100\% = 9.9 \times 10^{-3} \% \text{ (less than 5\%)}$$

A final test is to substitute the equilibrium concentrations into the equilibrium expression to see if the value of the equilibrium constant is obtained:

$$K = ? = (0.10)(2.0 \times 10^{-5})^2 = 4.0 \times 10^{-11} \approx 3.9 \times 10^{-11}$$

Because 3.9×10^{-11} is nearly the same as 4.0×10^{-11}, the calculation is correct.

Large K

In Exercise 15.6, the reaction did not proceed very far towards the products because the reaction had a small equilibrium constant. In other cases, the reaction may have a large equilibrium constant and only relatively small amounts of reactants will remain at equilibrium. In Exercise 15.6, it was also shown that small concentrations of reagents may sometimes be ignored to simplify the calculations. To exploit this simplifying tool in reactions with large equilibrium constants, we use a property of equilibria that seemed unimportant before:

The position of equilibrium is the same regardless of the direction it is approached from, reactants or products.

This means that in reactions with large equilibrium constants, the problem can be treated as if the reaction goes to completion, using all of the limiting reagent. Equilibrium is then approached from the direction of products, with the formation of small concentrations (or pressures) of reactants.

Exercise 15.8 A sample of 0.134 g of $CuCl_2 \cdot 2H_2O$ is added to 0.10 L of an 1.0 M ammonia solution. It completely dissolves to give aqueous Cu^{2+} ions which react with the ammonia to form the coppertetraammine complex ion:

Chapter 15: Principles of Chemical Equilibrium

$$Cu^{2+}_{(aq)} + 4\ NH_{3(aq)} \rightleftarrows Cu(NH_3)_4^{2+}_{(aq)} \quad K_f = 5.0 \times 10^{13}$$

What are the concentrations of copper(II), ammonia and the complex ion, $Cu(NH_3)_4^{2+}$, at equilibrium?

Steps to Solution: Again, as with equilibrium problems in general, we must apply the five step method, and generate a concentration table to find the concentrations in terms of initial conditions and the amount of some reagent, x, which is related to the concentrations of the other substances by the stoichiometric ratios.

1. *The substances and amounts initially present are:*

   NH_3 : 0.10 L of 1.0 M solution *or* 1.0 mol/L (0.10 L) = 1.0×10^{-1} mole NH_3

   $CuCl_2 \cdot 2H_2O$: 0.134 g *or* $\dfrac{0.134\ g}{170.5\ g/mole}$ = 7.86×10^{-4} mole $CuCl_2 \cdot 2\ H_2O$

2. *The chemical reaction that occurs is the formation of the complex ion*:

   $$Cu^{2+}_{(aq)} + 4\ NH_{3(aq)} \rightleftarrows Cu(NH_3)_4^{2+}_{(aq)}$$

3. *The equilibrium constant expression is*:

   $$K = 5.0 \times 10^{13} = \dfrac{[Cu(NH_3)_4^{2+}]}{[Cu^{2+}][NH_3]^4}$$

4. *Now set up the concentration table. However, because K is quite large, we will use the trick of "letting the reaction go to completion". Then the products will be "allowed" to decompose to reactants to reach equilibrium concentrations.*

REALITY CHECK: THIS IS JUST A DEVICE TO MAKE CALCULATIONS EASIER.

You should readily see that copper(II) chloride dihydrate is the limiting reagent. At completion,

$$\text{mols } Cu(NH_3)_4^{2+} = \text{mols } CuCl_2 \cdot 2H_2O = 7.9 \times 10^{-4}\ \text{mol}$$

$$\text{mole } NH_3 = 1.0 \times 10^{-1}\ \text{mole} - \dfrac{4\ \text{mol } NH_3}{1\ \text{mol } Cu(NH_3)_4^{2+}}(7.9 \times 10^{-4}\ \text{mol } Cu(NH_3)_4^{2+})$$

$$\text{mole } NH_3 = 9.7 \times 10^{-2}\ \text{mole } NH_3$$

mole $CuCl_2 \cdot 2\ H_2O$ = 0 (it was consumed to make $Cu(NH_3)_4^{2+}$)

Now we can tackle the equilibrium calculation from the products side where the changes in concentration will be small. Let the decrease in concentration of

$Cu(NH_3)_4^{2+}$ that happens from loss of NH_3 to give Cu^{2+} be x. The change in NH_3 concentration will be:

$$\text{mole } NH_3 \text{ produced} = x \text{ mol } Cu(NH_3)_4^{2+}\left(\frac{4 \text{mol } NH_3}{\text{mol } Cu(NH_3)_4^{2+}}\right) = 4x$$

Compound	Initial Concentration	+ Change	= Equilibrium Concentration
$Cu(NH_3)_4^{2+}$	$\frac{7.9 \times 10^{-4} \text{ mole Cu(II)}}{0.10 L} = 7.9 \times 10^{-3}$ M	$-x$	$7.9 \times 10^{-3} - x$
NH_3	$\frac{9.7 \times 10^{-2} \text{ mole } NH_3}{0.10 L} = 9.7 \times 10^{-1}$ M	$+4x$	$9.7 \times 10^{-1} + 4x$
Cu^{2+}	0	$+x$	x

Now substitute and solve (Step 5). Using the expression for K_f:

$$K_f = \frac{[Cu(NH_3)_4^{2+}]}{[Cu^{2+}][NH_3]^4} = \frac{7.9 \times 10^{-3} - x}{x(9.7 \times 10^{-1} + 4x)^4} = 5.0 \times 10^{13} \text{ M}^{-4}$$

Because K_f is large, the following assumption will be tried:

$$\text{let } x \ll 7.9 \times 10^{-3} \text{ M so that}$$

$7.9 \times 10^{-3} - x \approx 7.9 \times 10^{-3}$ M and $9.7 \times 10^{-1} + 4x \approx 9.7 \times 10^{-1}$ M

Then:
$$K_f = \frac{7.9 \times 10^{-3}}{x(9.7 \times 10^{-1})^4} = 5.0 \times 10^{13}$$

and:
$$x = \frac{7.9 \times 10^{-3}}{(9.7 \times 10^{-1})^4} \cdot \frac{1}{5.0 \times 10^{13}} = 1.8 \times 10^{-16} \text{ M}$$

The assumption that x is much smaller than 7.9×10^{-3} is valid. Now the concentrations of the substances at equilibrium are found:

$[Cu^{2+}{}_{(aq)}] = x = \boxed{1.8 \times 10^{-16} \text{ M}}$ $[NH_3] = 9.7 \times 10^{-1} + 4(1.8 \times 10^{-16}) = \boxed{9.7 \times 10^{-1} \text{ M}}$

$[Cu(NH_3)_4^{2+}] = 7.9 \times 10^{-3} - 1.8 \times 10^{-16} = \boxed{7.9 \times 10^{-3} \text{ M}}$

Use these concentrations to calculate the equilibrium constant to show that there is no error in the calculations.

Test Yourself

1. Write the equilibrium constant expressions for the following reactions in terms of concentrations:
   (a) $2\, NO_{2\,(g)} + 7\, H_{2\,(g)} \rightleftarrows 2\, NH_{3\,(g)} + 4\, H_2O_{(g)}$
   (b) $C_{(s)} + CO_{2\,(g)} \rightleftarrows 2\, CO_{(g)}$
   (c) $2\, PbS_{(s)} + 3\, O_2 \rightleftarrows 2\, PbO_{(s)} + 2\, SO_{2\,(g)}$

2. Predict whether each equilibrium reaction will shift toward products or reactants with a decrease in temperature.
   (a) $CH_4 + H_2O \rightleftarrows CO + 3\, H_2$ $\Delta H° = 206$ kJ
   (b) $2\, SO_2 + O_2 \rightleftarrows 2\, SO_3$ $\Delta H° = -198$ kJ

3. A sample of 3.00×10^{-1} mole of pure phosgene gas, $COCl_2$, was placed in 15.0 L container and heated to 800 K. At equilibrium, the partial pressure of CO was found to be 0.497 atm. Calculate the equilibrium constant for the following reaction:

$$COCl_{2\,(g)} \rightleftarrows CO_{(g)} + Cl_{2\,(g)}$$

   in units of pressure (atm).

4. At 1033 K, $K = 33.3$ M for the equilibrium reaction:

$$PCl_{5\,(g)} \rightleftarrows PCl_{3\,(g)} + Cl_{2\,(g)}$$

   If a mixture of 0.100 mole of PCl_5 and 0.300 mole of PCl_3 is placed in a 2.00 L reaction vessel and heated to 1033 K, what are the numbers of mols of each component at equilibrium?

5. The K_{sp} of $PbBr_2$ is 8.9×10^{-6}. Determine the equilibrium concentration of Pb^{2+} in pure water and in (b) 0.20 M KBr.

Answers

1 (a) $K = \dfrac{[NH_3]^2[H_2O]^4}{[NO_2]^2[H_2]^7}$ (b) $K = \dfrac{[CO]^2}{[CO_2]}$ (c) $K = \dfrac{[SO_2]^2}{[O_2]^3}$

2. (a) towards reactants (b) towards products

3. 0.304 atm

4. mols of PCl_5 = 4.0 x 10^{-4} mol; mols of PCl_3 = 0.398 mole ; mols of Cl_2 = 9.94 x 10^{-4} mol

5. In pure water, $[Pb^{2+}]_{eq}$ = 0.013 M; in 0.20 M KBr, $[Pb^{2+}]_{eq}$ = 2.2 x 10^{-4} M

Solutions to Chapter Problems

15.1 The reaction proceeds, converting *cis*-butene into *trans*-butene, but as the concentration of *trans*-butene builds up, the reverse reaction becomes increasingly important until, when the concentration ratio is 3, the rates are equal. Then there is no net change. A plot of concentration vs. time appears similar to Figure 15-1, with 0.75 atm for the final pressure of *trans*-butene:

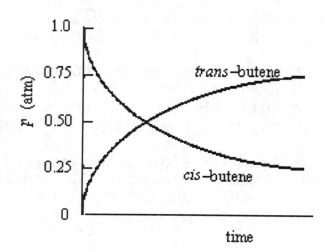

15.3 (a) According to the principle of reversibility, every elementary reaction can run in either direction, so as the system approaches equilibrium, the reverse reactions become important:

$$Cl^-_{(aq)} + ClO_2^-_{(aq)} \xrightarrow{k_{-1}} ClO^-_{(aq)} + ClO^-_{(aq)}$$

$$Cl^-_{(aq)} + ClO_3^-_{(aq)} \xrightarrow{k_{-2}} ClO^-_{(aq)} + ClO_2^-_{(aq)}$$

(b) The net reaction is 3 $ClO^-_{(aq)} \rightleftharpoons$ 2 $Cl^-_{(aq)}$ + $ClO_3^-_{(aq)}$.

By inspection, the equilibrium constant expression is $K_{eq} = \dfrac{[ClO_3^-]_{eq}[Cl^-]^2_{eq}}{[ClO^-]^3_{eq}}$

(c) The net reaction is the sum of the two elementary steps, so the equilibrium constant is the product of the rate ratios of the two steps: $K_{eq} = \dfrac{k_1 k_2}{k_{-1} k_{-2}}$.

15.5 A molecular picture of an elementary reaction shows the reactants, the products, and (if necessary) the intermediate collision complex.

15.7 To test the reversibility of a reaction, set up a system containing the products and observe whether or not reactants form. Here, a solution containing Cl⁻ and ClO₃⁻ ions should react to form some ClO⁻ ions. (If the reaction of ClO⁻ to form Cl⁻ and ClO₃⁻ goes virtually to completion, this experiment may not succeed.)

15.9 Equilibrium constant expressions can be written by inspection of the overall stoichiometry, remembering that pure liquids and solids do not appear in the expression:

(a) $K_{eq} = \dfrac{[IF_5]_{eq}^2}{[F_2]_{eq}^5}$; (b) $K_{eq} = \dfrac{1}{[O_2]_{eq}^5}$; (c) $K_{eq} = [CO]_{eq}^2$;

(d) $K_{eq} = \dfrac{1}{[CO]_{eq}[H_2]_{eq}^2}$; (e) $K_{eq} = \dfrac{[H_3O^+]_{eq}[PO_4^{3-}]_{eq}^3}{[H_3PO_4]_{eq}}$.

15.11 Equilibrium constant expressions can be written by inspection of the overall stoichiometry, remembering that pure liquids and solids do not appear in the expression:

(a) $2IF_{5\,(g)} \rightleftharpoons I_{2\,(g)} + 5\,F_{2\,(g)}$ $\qquad K_{eq} = \dfrac{[F_2]_{eq}^5}{[IF_5]_{eq}^2}$;

(b) $P_4O_{10\,(s)} \rightleftharpoons P_{4\,(s)} + 5\,O_{2\,(g)}$ $\qquad K_{eq} = [O_2]_{eq}^5$;

(c) $BaO_{(s)} + 2\,CO_{(g)} \rightleftharpoons BaCO_{3\,(s)} + C_{(s)}$ $\quad K_{eq} = 1/[CO]_{eq}^2$;

(d) $CH_3OH_{(l)} \rightleftharpoons CO_{(g)} + 2\,H_{2\,(g)}$ $\qquad K_{eq} = [CO]_{eq}[H_2]_{eq}^2$;

(e) $PO_4^{3-}{}_{(aq)} + 3\,H_3O^+{}_{(aq)} \rightleftharpoons H_3PO_{4\,(aq)} + 3\,H_2O_{(l)}$ $\quad K_{eq} = \dfrac{[H_3PO_4]_{eq}}{[H_3O^+]_{eq}[PO_4^{3-}]_{eq}^3}$.

15.13 The standard states of gases are gases at 1 atm, the standard states of solutes are solutions at
1 M, and the standard states of solvents and pure liquids and solids are mole fraction = 1.
(a) $p = 1$ atm for F_2 and IF_5, $X = 1$ for I_2;
(b) $p = 1$ atm for O_2, $X = 1$ for P_4 and P_4O_{10};
(c) $p = 1$ atm for CO, $X = 1$ for others;
(d) $p = 1$ atm for CO and H_2, $X = 1$ for CH_3OH;
(e) $c = 1$ M for H_3PO_4, H_3O^+, and PO_4^{3-}, $X = 1$ for H_2O.

15.15 To determine if all solid dissolves, evaluate Q for all solid dissolving and compare its value with the value for K_{eq}. If $Q < K_{eq}$, the spontaneous direction of reaction is to the right even after all solid dissolves, so all solid dissolves, while if $Q > K_{eq}$, not all solid dissolves.

The reaction is $PbCl_{2\,(s)} \rightleftharpoons Pb^{2+}_{(aq)} + 2\,Cl^-_{(aq)}$,
$K_{eq} = K_{sp} = 2 \times 10^{-5}$ and $Q = [Pb^{2+}][Cl^-]^2$;

Use mole-mass conversions to determine concentrations:
$n = m/MM = (0.50\text{ g}) \div (278.1\text{ g/mol}) = 1.80 \times 10^{-3}$ mol;

$$[Pb^{2+}] = \frac{(1.80 \times 10^{-3}\text{ mol})}{(300\text{ mL})(10^{-3}\text{L/mL})} = 6.0 \times 10^{-3}\text{ M};\quad [Cl^-] = 2[Pb^{2+}] = 1.2 \times 10^{-2}\text{ M};$$

$Q = (6.0 \times 10^{-3})(1.2 \times 10^{-2})^2 = 8.6 \times 10^{-7}$; $Q < K_{eq}$, so all the salt dissolves.

15.17 To determine an equilibrium constant at standard temperature from thermodynamic tables, calculate $\Delta G°_{rxn}$ from tabulated values for $\Delta G°_f$ and then use Equation 15-3:
$$\Delta G° = -RT \ln K_{eq}$$

(a) $\Delta G°_{rxn} = 1(0) + 1(-394.359) - 1(-137.168) - 1(-237.129) = -20.062$ kJ/mol;

$$\ln K_{eq} = -\frac{(-20.062\text{ kJ/mol})(10^3\text{J/kJ})}{(8.314\text{ J/mol K})(298\text{ K})} = 8.10;\quad K_{eq} = e^{8.10} = 3.3 \times 10^3;$$

(b) $\Delta G°_{rxn} = 2(-394.359) - 1(0) - 2(-137.168) = -514.38$ kJ/mol;

$$\ln K_{eq} = -\frac{(-514.38\text{ kJ/mol})(10^3\text{J/kJ})}{(8.314\text{ J/mol K})(298\text{ K})} = 207.62;\quad K_{eq} = e^{207.62} = 1.5 \times 10^{90};$$

(c) $\Delta G°_{rxn} = 1(-525.1) - 2(-137.168) - 1(0) - 1(-1137.6) = 338.16$ kJ/mol;

$$\ln K_{eq} = -\frac{(338.16\text{ kJ/mol})(10^3\text{J/kJ})}{(8.314\text{ J/mol K})(298\text{ K})} = -136.49;\quad K_{eq} = e^{-136.49} = 5.3 \times 10^{-60};$$

(d) $\Delta G°_{rxn} = 3(-237.129) + 1(70.62) - 6(0) - 3(-137.168) = -229.263$ kJ/mol;

$$\ln K_{eq} = -\frac{(-229.263\text{ kJ/mol})(10^3\text{J/kJ})}{(8.314\text{ J/mol K})(298\text{ K})} = 92.54;\quad K_{eq} = e^{92.54} = 1.5 \times 10^{40}.$$

15.19 To estimate the equilibrium constant at a temperature different from 298 K, calculate ΔH^o_{rxn} and ΔS^o_{rxn} at 298 K and then use Equations 13-9 and 15-3:
$$\Delta G^o = \Delta H^o - T\Delta S^o \qquad \Delta G^o = -RT \ln K_{eq}$$

15.17 (b) $\Delta H^o_{rxn} = 2(-393.509) - 1(0) - 2(-110.525) = -565.968$ kJ/mol;

$\Delta S^o_{rxn} = 2(213.74) - 1(205.138) - 2(197.674) = -173.01$ J/mol K;

$\Delta G^o_{rxn,\ 250\ K} = (-565.968$ kJ/mol$) - (250$ K$)(-173.01$ J/mol K$)(10^{-3}$ kJ/J$) = -522.7$ kJ/mol

$\ln K_{eq} = -\dfrac{(-522.7 \text{ kJ/mol})(10^3 \text{ J/kJ})}{(8.314 \text{ J/mol K})(250 \text{ K})} = 251.5; \quad K_{eq} = e^{251.5} = 1.6 \times 10^{109};$

15.17 (c) $\Delta H^o_{rxn} = 1(-553.5) + 2(-110.525) - 1(0) - 1(-1216.3) = 441.75$ kJ/mol;

$\Delta S^o_{rxn} = 2(197.674) + 1(70.42) - 1(5.740) - 1(112.1) = 347.93$ J/mol K;

$\Delta G^o_{rxn,\ 250\ K} = (441.75$ kJ/mol$) - (250$ K$)(347.93$ J/mol K$)(10^{-3}$ kJ/J$) = 354.76$ kJ/mol

$\ln K_{eq} = -\dfrac{(354.76 \text{ kJ/mol})(10^3 \text{ J/kJ})}{(8.314 \text{ J/mol K})(250 \text{ K})} = -170.68; \quad K_{eq} = e^{-170.68} = 7.5 \times 10^{-75}.$

15.21 To determine an equilibrium constant at standard temperature from thermodynamic tables, calculate ΔG^o_{rxn} from tabulated values for ΔG^o_f and then use Equation 15-3:
$$\Delta G^o = -RT \ln K_{eq}$$

15.17 (a) $\Delta G^o_{rxn} = 1(0) + 1(-394.359) - 1(-137.168) - 1(-228.72) = -28.47$ kJ/mol;

$\ln K_{eq} = -\dfrac{(-28.47 \text{ kJ/mol})(10^3 \text{ J/kJ})}{(8.314 \text{ J/mol K})(298 \text{ K})} = 11.49; \quad K_{eq} = e^{11.49} = 9.8 \times 10^4;$

15.17 (d) $\Delta G^o_{rxn} = 3(-228.72) + 1(70.62) - 6(0) - 3(-137.168) = -204.04$ kJ/mol;

$\ln K_{eq} = -\dfrac{(-204.04 \text{ kJ/mol})(10^3 \text{ J/kJ})}{(8.314 \text{ J/mol K})(298 \text{ K})} = 82.35; \quad K_{eq} = e^{82.35} = 5.8 \times 10^{35}.$

15.23 To determine an equilibrium constant at standard temperature from thermodynamic tables, calculate ΔG^o_{rxn} from tabulated values for ΔG^o_f; to estimate the equilibrium constant at a temperature different from 298 K, calculate ΔH^o_{rxn} and ΔS^o_{rxn} at 298 K and then use Equations 13-9 and 15-3:
$$\Delta G^o = \Delta H^o - T\Delta S^o \qquad \Delta G^o = -RT \ln K_{eq}$$

$\Delta G^o_{rxn} = 2(51.3) - 2(86.55) - 1(0) = -70.48$ kJ/mol;

$$\ln K_{eq} = -\frac{(-70.48 \text{ kJ/mol})(10^3 \text{ J/kJ})}{(8.314 \text{ J/mol K})(298 \text{ K})} = 28.45; \quad K_{eq} = e^{28.45} = 2.3 \times 10^{12};$$

$$\Delta H^\circ_{rxn} = 2(33.18) - 2(90.25) - 1(0) = -114.14 \text{ kJ/mol};$$

$$\Delta S^\circ_{rxn} = 2(240.06) - 2(210.761) - 1(205.138) = -146.54 \text{ J/mol K};$$

$$\Delta G^\circ_{rxn, 950 K} = (-114.14 \text{ kJ/mol}) - (950 \text{ K})(-146.54 \text{ J/mol K})(10^{-3} \text{ kJ/J}) = 25.07 \text{ kJ/mol}$$

$$\ln K_{eq} = -\frac{(25.07 \text{ kJ/mol})(10^3 \text{ J/kJ})}{(8.314 \text{ J/mol K})(950 \text{ K})} = -3.17; \quad K_{eq} = e^{-3.17} = 4.2 \times 10^{-2}.$$

15.25 Use Le Chatelier's principle to determine the effect on an equilibrium position caused by adding one reagent:
In reactions (a), (b), and (d), CO is a reactant, so adding CO causes the reaction to go to the right, forming products; in reaction (c), CO is a product, so adding CO causes the reaction to proceed to the left, forming reactants.

15.27 Use Le Chatelier's principle to determine the effect on an equilibrium position caused by a change in conditions:
(a) Solid reactants do not appear in the equilibrium constant expression, so adding $BaCl_2$ has no effect;
(b) Addition of water dilutes the solution, reducing the concentrations of the product ions, so more of the solid will dissolve;
(c) Addition of NaCl increases the concentration of one of the product ions, Cl^-, so some solid will precipitate;
(d) Addition of KNO_3 does not change the concentrations in the equilibrium constant expression, so adding this solid has no effect.

15.29 Use Le Chatelier's principle to determine what changes in conditions will drive an equilibrium in any given direction. This reaction can be driven to the left by removing SO_2, removing Cl_2, adding SO_2Cl_2, or increasing the temperature.

15.31 The standard types of equilibria are phase change (melting, vaporization, or sublimation), solubility (gas or solid), acid-base, and gas reactions.

(a) reverse of sublimation: $K_{eq} = \dfrac{1}{(p_{I_2})_{eq}} = 1/K_{subl}$;

(b) gas solubility: $K_{eq} = \dfrac{[I_2]_{eq}}{(p_{I_2})_{eq}} = K_H$;

(c) reverse of solid solubility: $K_{eq} = \dfrac{1}{[Pb^{2+}]_{eq}[I^-]^2_{eq}} = 1/K_{sp}$;

(d) proton transfer to water: $K_{eq} = \dfrac{[IO^-]_{eq}[H_3O^+]_{eq}}{[HIO]_{eq}} = K_a$.

15.33 A molecular picture of an equilibrium between phases shows the concentrations of the species and arrows indicating transfers between phases:

15.35 Equilibrium exists when the rates of evaporation and condensation are equal. A test tube has a smaller surface area than a petri dish, so fewer molecules escape per unit time from the test tube. However, the rate of escape per unit surface area is the same for both samples, so equilibrium is established at the same pressure (molecules colliding per unit surface area) for both samples.

15.37 Table 15-1 indicates that the solubility of O_2 in water increases as the temperature falls from 25 °C to 0 °C. Use Henry's law to calculate the concentration at each temperature:

$$\frac{[O_{2(aq)}]_{eq}}{(p_{O_2})_{eq}} = K_H, \quad \text{so } [O_{2(aq)}]_{eq} = (p_{O_2})_{eq} K_H$$

The atmosphere is 22 % O_2, so the partial pressure of O_2 is 0.22 atm.

At 25 °C, $[O_{2(aq)}]_{eq}$ = (0.22 atm)(1.3 x 10^{-3}) = 0.286 x 10^{-3} M;

At 0 °C, $[O_{2(aq)}]_{eq}$ = (0.22 atm)(2.5 x 10^{-3}) = 0.55 x 10^{-3} M;

The percentage change is $(100\ \%) \dfrac{(0.55 \times 10^{-3} M) - (0.286 \times 10^{-3} M)}{(0.55 \times 10^{-3} M)} = 48\ \%$.

15.39 To calculate an equilibrium constant when amounts are available, convert amounts data into concentration using stoichiometric reasoning, then complete an amounts table and substitute into the equilibrium constant expression. Generally, pressures in atmospheres are required, but in this problem, the equilibrium constant expression contains p^2 in the numerator and denominator, so units cancel and amounts can be used directly. Set up an amounts table, using stoichiometric reasoning and the fact that the change is 0.49 mol for CO:

Species:	H_2	+	CO_2	⇌	H_2O	+	CO
Initial amount (mol):	1.00		1.00		0		0
Change to equilib (mol)	−0.49		−0.49		0.49		0.49
Equilib amount (mol)	0.51		0.51		0.49		0.49

Now substitute into the equilibrium constant expression and evaluate K:

$$K_{eq} = \frac{(p_{H_2O})_{eq}(p_{CO})_{eq}}{(p_{H_2})_{eq}(p_{CO_2})_{eq}} = \frac{(n_{H_2O})_{eq}(n_{CO})_{eq}}{(n_{H_2})_{eq}(n_{CO_2})_{eq}} = \frac{(0.49)^2}{(0.51)^2} = 0.92$$

Chapter 15: Principles of Chemical Equilibrium

15.41 To calculate an equilibrium constant when experimental data concerning amounts are available, convert amounts data into concentration using stoichiometric reasoning, identify the reaction, then complete an amounts table and substitute into the equilibrium constant expression. For solutes, concentrations must be in mol/L.

$[C_2H_5CO_2H]_{initial} = (0.0500 \text{ mol}) \div (0.500 \text{ L}) = 0.100 \text{ M}$

To complete the amounts table, use stoichiometric reasoning and the fact that the change is 1.15×10^{-3} M for H_3O^+:

Species:	H_2O +	$C_2H_5CO_2H$	⇌	$C_2H_5CO_2^-$	+	H_3O^+
Initial conc. (M):		0.100		0		0
Change (M)		-1.15×10^{-3}		1.15×10^{-3}		1.15×10^{-3}
Equilib conc. (M)		0.099		1.15×10^{-3}		1.15×10^{-3}

Now substitute into the equilibrium constant expression and evaluate K:

$$K_a = \frac{[C_2H_5CO_2^-]_{eq}[H_3O^+]_{eq}}{[C_2H_5CO_2H]_{eq}} = \frac{(1.15 \times 10^{-3})^2}{(0.099)} = 1.3 \times 10^{-5}$$

15.43 To calculate an equilibrium constant when experimental data concerning concentrations are available, identify the reaction, then complete an amounts table and substitute into the equilibrium constant expression.

To complete the amounts table, use stoichiometric reasoning and the fact that the change is 1.9×10^{-3} M for Pb^{2+}:

Species:	PbF_2	⇌	Pb^{2+}	+	$2 F^-$
Initial conc. (M):			0		0
Change (M)			1.9×10^{-3}		3.8×10^{-3}
Equilib conc. (M)			1.9×10^{-3}		3.8×10^{-3}

Now substitute into the equilibrium constant expression and evaluate K:

$$K_{sp} = [Pb^{2+}]_{eq}[F^-]_{eq}^2 = (1.9 \times 10^{-3})(3.8 \times 10^{-3})^2 = 2.7 \times 10^{-8}.$$

15.45 To calculate an equilibrium constant when experimental data concerning concentrations are available, identify the reaction, then complete an amounts table and substitute into the equilibrium constant expression.

The reaction is $Ca_3(PO_4)_2 \rightleftharpoons 3\,Ca^{2+}_{(aq)} + 2\,PO_4^{3-}_{(aq)}$

The solubility in mol/L is $(8.1 \times 10^{-4} \text{ g/L}) \div (310.174 \text{ g/mol}) = 2.61 \times 10^{-6}$ M.

Species:	$Ca_3(PO_4)_2$	⇌	$3\,Ca^{2+}$	+	$2\,PO_4^{3-}$
Initial conc. (M):			0		0
Change (M)			$3(2.61 \times 10^{-6})$		$2(2.61 \times 10^{-6})$
Equilib conc. (M)			7.83×10^{-6}		5.22×10^{-6}

Now substitute into the equilibrium constant expression and evaluate K:

$$K_{sp} = [Ca^{2+}]_{eq}^3[PO_4^{3-}]_{eq}^2 = (7.83 \times 10^{-6})^3(5.22 \times 10^{-6})^2 = 1.3 \times 10^{-26}.$$

Chapter 15: Principles of Chemical Equilibrium

15.47 To calculate concentrations at equilibrium from initial conditions, set up a concentration table. For a gas phase reaction, concentrations must be expressed in atm. Let x = change in [H_2]:

Species:	H_2	+	Br_2	⇌	2 HBr
Initial conc. (atm):	0		0		10.0
Change (atm)	+x		+x		−2x
Equilib conc. (atm)	x		x		10.0 − 2x

Now substitute into the equilibrium constant expression and solve for x:

$$K_{eq} = \frac{(p_{HBr})^2_{eq}}{(p_{H_2})_{eq}(p_{Br_2})_{eq}} = \frac{(10.0 - 2x)^2}{(x)(x)} = 1.6 \times 10^5$$

To simplify, assume that 2x << 10.0:

$$1.6 \times 10^5 = \frac{(10.0)^2}{(x)^2} \quad \text{so} \quad x^2 = (100) \div (1.6 \times 10^5) = 6.25 \times 10^{-4}$$

$x = 2.5 \times 10^{-2} = (p_{H_2})_{eq} = (p_{Br_2})_{eq}$; $(p_{HBr})_{eq} = 10.0 - 2(2.5 \times 10^{-2}) = 10.0$ atm

$2(2.5 \times 10^{-2}) = 0.050 << 10.0$, so the approximation is valid.

15.49 To calculate concentrations at equilibrium from initial conditions, set up a concentration table. For a gas phase reaction, concentrations must be expressed in atm. Let x = change in [CO_2]:

Species:	$FeO_{(s)}$ +	$CO_{(g)}$	⇌	$CO_{2(g)}$ +	$Fe_{(s)}$
Initial conc. (atm):	−	5.0		0	−
Change (atm)	−	−x		+x	−
Equilib conc. (atm)	−	5.0 − x		x	−

Now substitute into the equilibrium constant expression and solve for x:

$$K_{eq} = \frac{(p_{CO_2})_{eq}}{(p_{CO})_{eq}} = \frac{(x)}{(5.0 - x)} = 0.403$$

$x = (0.403)(5.0 - x) = 2.015 - 0.403 x$ so $1.403 x = 2.015$ and $x = 1.436$

$(p_{CO_2})_{eq} = 1.4$ atm and $(p_{CO})_{eq} = 3.6$ atm.

15.51 Equilibrium constant expressions are determined by the stoichiometry of the overall reaction, with pure liquids, solids, and solvent omitted from the expression:

(a) $K_{eq} = [CO_2]_{eq}[H_2O]_{eq}$; (b) $K_{eq} = \dfrac{[NH_3]^4_{eq}[O_2]^3_{eq}}{[N_2]^2_{eq}}$; (c) $K_{eq} = \dfrac{[CH_3CHO]^2_{eq}}{[C_2H_4]^2_{eq}[O_2]_{eq}}$;

(d) $K_{eq} = [Ag^+]^2_{eq}[SO_4^{2-}]_{eq}$; (e) $K_{eq} = [H_2S]^2_{eq}[NH_3]_{eq}$.

15.53 To calculate concentrations at equilibrium from initial conditions, set up a concentration table. For a gas phase reaction, concentrations must be expressed in atm. Let $-x =$ change in $[H_2]$:

Species:	H_2	$+$	F_2	$\rightleftarrows$	$2\,HF$
Initial conc. (atm):	3.0		3.0		0
Change (atm)	$-x$		$-x$		$+2x$
Equilib conc. (atm)	$3.0-x$		$3.0-x$		$2x$

Now substitute into the equilibrium constant expression and solve for x:

$$K_{eq} = 115 = \frac{(p_{HF})^2_{eq}}{(p_{H_2})_{eq}(p_{F_2})_{eq}} = \frac{(2x)^2}{(3.0-x)(3.0-x)} = \frac{4x^2}{(9.0-6.0x+x^2)}$$

$4x^2 = (115)(9.0 - 6.0x + x^2) = 1035 - 690x + 115x^2$
$0 = 111x^2 - 690x + 1035$

$$\frac{-b \pm \sqrt{b^2 - 4ac}}{2a} = \frac{690 \pm \sqrt{(690)^2 - 4(111)(1035)}}{2(111)} = \frac{690 \pm \sqrt{16560}}{222} = 3.7 \text{ or } 2.5$$

The initial pressures are 3.0 atm, so a change of 3.7 atm is impossible; thus, x = 2.5
$(p_{H_2})_{eq} = (p_{F_2})_{eq} = 3.0 - 2.5 = 0.5$ atm; $(p_{HF})_{eq} = 2(2.5) = 5.0$ atm.

15.55 To determine concentrations of ions at equilibrium, follow the five-step procedure for working equilibrium problems:
1.) Species present initially are Pb^{2+}, NO_3^-, and $PbCl_2$.
2.) Reaction is a solubility process: $PbCl_{2\,(s)} \rightleftarrows Pb^{2+}{}_{(aq)} + 2\,Cl^-{}_{(aq)}$
3.) $K_{eq} = K_{sp} = [Pb^{2+}]_{eq}[Cl^-]^2_{eq} = 1.58 \times 10^{-5}$;
4.) Set up and complete a concentration table, letting x be the increase in $[Pb^{2+}]$;

Species:	$PbCl_{2\,(s)}$	$\rightleftarrows$	$Pb^{2+}{}_{(aq)}$	$+$	$2\,Cl^-{}_{(aq)}$
Initial conc. (M):			0.650		0
Change (M)			x		2x
Equilib conc. (M)			0.650 + x		2x

5.) Substitute into the equilibrium constant expression and solve for x:
$1.58 \times 10^{-5} = (0.650 + x)(2x)^2$; Assume $x \ll 0.650$; then $4x^2 = (1.58 \times 10^{-5}) \div (0.650)$
$x^2 = 6.08 \times 10^{-6}$ and $x = 2.47 \times 10^{-3}$; increase in $[Pb^{2+}] = 2.47 \times 10^{-3}$ M;
To convert to moles Pb^{2+} that dissolve, divide by the solution volume:
$n = M/V = (2.47 \times 10^{-3}\,M) \div (0.750\,L) = 3.29 \times 10^{-3}$ mol;
The amount of $PbCl_2$ that dissolves equals the amount of increase in Pb^{2+}; convert to mass by multiplying by the molar mass of $PbCl_2$:
$m = n\,MM = (3.29 \times 10^{-3}\,mol)(278.1\,g/mol) = 0.514$ g;
$2.47 \times 10^{-3} \ll 0.650$, so the approximation is valid.

Chapter 15: Principles of Chemical Equilibrium

15.57 To predict effects of changes on equilibrium position, apply le Chatelier's principle: The system will respond in the direction that reduces the effect of the change. The reaction in question is $2\text{ NO}_{(g)} + \text{O}_{2(g)} \rightleftharpoons 2\text{ NO}_{2(g)}$, $\Delta H^o_{rxn} = -114.14$ kJ/mol (from Prob. 15.23);

(a) NO_2 is a product, so reducing its pressure causes the reaction to proceed to the right;
(b) The reaction is exothermic, so decreasing the temperature causes it to proceed to the right;
(c) Adding Ar does not change any of the pressures in the equilibrium expression, so this change has no effect.

15.59 To calculate an equilibrium constant from initial and equilibrium conditions, set up a concentration table:

Species:	$C_{(s)}$ +	$CO_{2(g)}$ $\rightleftharpoons$	$2\text{ CO}_{(g)}$
Initial concentration (atm)		0.464	0
Change to equilibrium (atm)		– x	+ 2x
Equilib concentration (atm)		0.464 – x	2 x

The problem gives the equilibrium pressure, which is the sum of partial pressures:
0.746 atm = (0.464 – x) + 2 x = 0.464 + x, from which x = (0.746 – 0.464) = 0.282;
Substitute into the equilibrium constant expression and calculate K:

$$K_{eq} = \frac{[CO]^2_{eq}}{[CO_2]_{eq}} = \frac{[(2)(0.282)]^2}{(0.464 - 0.282)} = (0.564)^2 \div (0.182) = 1.75$$

15.61 At equilibrium, a molecular picture should show the presence of both reactants and products, in relative amounts that are determined by the value of the equilibrium constant. Set up a "concentration" table to determine how many of each species are present at equilibrium:

Species:				
Initial number:	12	12	0	0
Change in number	– x	– x	+ x	+ x
Equilibrium number	12 – x	12 – x	x	x

Substitute in the equilibrium constant expression and solve for x:

$K_{eq} = 25 = \frac{(x)^2}{(12-x)^2}$; Taking the square root of each side gives $5 = \frac{(x)}{(12-x)}$;

(60 – 5 x) = x, so 6x = 60 and x = 10

The molecular picture should show (12 – 10) = 2 each of reactants and 10 each of products:

Chapter 15: Principles of Chemical Equilibrium

15.63 When the volume of a gas system changes, all partial pressures change proportionately. This generates a new set of "initial" conditions that form the starting point for a new concentration table. When volume is cut by a factor of three, pressure increases three-fold:
$[CO_2] = 3(0.182) = 0.546$ atm; $[CO] = 3 (0.564) = 1.692$ atm. To return to equilibrium, the system responds in the direction that reduces total pressure, so [CO] decreases.

Species:	$C_{(s)}$ +	$CO_{2\,(g)}$ ⇌	$2\,CO_{(g)}$
Initial concentration (atm)		0.546	1.692
Change to equilibrium (atm)		+ x	– 2x
Equilib concentration (atm)		0.546 + x	1.692 – 2 x

Substitute into the equilibrium constant expression and solve for x:

$$1.75 = \frac{(1.692 \pm 2x)^2}{(0.546 + x)}, \text{ from which } (1.75)(0.546 + x) = (1.692 - 2x)^2$$

$0.9555 + 1.75\,x = 2.862 - 6.768\,x + 4\,x^2,$ or $4\,x^2 - 8.518\,x + 1.9065 = 0$

$$x = \frac{-b \pm \sqrt{b^2 - 4ac}}{2a} = \frac{8.518 \pm \sqrt{(-8.518)^2 - 4(4)(1.9065)}}{2(4)} = \frac{8.518 \pm \sqrt{72.556 - 30.504}}{8};$$

$x = 0.254$; $[CO_2]_{eq} = (0.546 + 0.254) = 0.800$ atm; $[CO]_{eq} = 1.692 - 2(0.254)] = 1.18$ atm.

15.65 To calculate equilibrium concentrations, set up a concentration table. In this problem, the table can be short, because two of the three concentrations are provided:

Species:	$Br_{2\,(g)}$ +	$I_{2\,(g)}$ ⇌	$2\,IBr_{(g)}$
Equilibrium conc. (atm)	0.512	0.327	x

Substitute into the equilibrium constant expression and solve for x:

$$322 = \frac{[IBr]_{eq}^2}{[Br_2]_{eq}[I_2]_{eq}} = \frac{x^2}{(0.512)(0.327)}, \text{ from which } x^2 = (322)(0.512)(0.327) = 53.91;$$

$x = [IBr]_{eq} = 7.34$ atm.

15.67 To calculate equilibrium concentrations, set up a concentration table:

Species:	$Sn_{(s)}$ +	$2\,H_{2\,(g)}$ ⇌	$SnH_{4\,(g)}$
Initial concentration (atm)		200	0
Change (atm)		– 2 x	x
Equilib conc. (atm)		200 – 2x	x

Because K_{eq} is very small, we assume that $2x \ll 200$:

$$1.07 \times 10^{-33} = \frac{[SnH_4]_{eq}}{[H_2]_{eq}^2} = \frac{x}{(200)^2}, \text{ from which } x = (1.07 \times 10^{-33})(200)^2 = 4.28 \times 10^{-29}$$

Thus $[SnH_4]_{eq} = 4.28 \times 10^{-29}$ atm (a very small pressure).
To calculate the volume that would be expected to contain a single molecule, use the ideal gas equation, rearranged to solve for V: $V = nRT/p$, with $n = 1/N_A$:

$$V = \frac{RT}{N_A p} = \frac{(0.0821 \text{ L atm / mol K})(298 \text{ K})}{(6.022 \times 10^{23} / \text{mol})(4.28 \times 10^{-29} \text{ atm})} = 9.49 \times 10^5 \text{ L.}$$

Chapter 15: Principles of Chemical Equilibrium

15.69 To determine an equilibrium constant at standard temperature from thermodynamic tables, calculate ΔG^o_{rxn} from tabulated values for ΔG^o_f; calculate ΔH^o_{rxn} and ΔS^o_{rxn} at 298 K and then use Equations 13-9 and 15-3 to estimate the equilibrium constant at a temperature different from 298 K:

$$\Delta G^o = \Delta H^o - T\Delta S^o \qquad \Delta G^o = -RT \ln K_{eq}$$

$\Delta G^o_{rxn} = 1(97.89) - 2(51.31) = -4.73$ kJ/mol;

$\ln K_{eq} = -\dfrac{(-4.73 \text{ kJ / mol})(10^3 \text{ J / kJ})}{(8.314 \text{ J / mol K})(298 \text{ K})} = 1.91$; $K_{eq} = e^{1.91} = 6.75$;

$\Delta H^o_{rxn} = 1(9.16) - 2(33.18) = -57.20$ kJ/mol;

$\Delta S^o_{rxn} = 1(304.29) - 2(240.06) = -175.83$ J/mol K;

$\Delta G^o_{rxn, 525 K} = (-57.20 \text{ kJ/mol}) - (525 \text{ K})(-175.83 \text{ J/mol K})(10^{-3} \text{ kJ/J}) = 35.11$ kJ/mol

$\ln K_{eq} = -\dfrac{(35.11 \text{ kJ / mol})(10^3 \text{ J / kJ})}{(8.314 \text{ J / mol K})(525 \text{ K})} = -8.04$; $K_{eq} = e^{-8.04} = 3.2 \times 10^{-4}$.

15.71 (a) The general weak base reaction is $B + H_2O \rightleftharpoons BH^+ + OH^-$, and $B = (CH_3)_3N$:

$(CH_3)_3N + H_2O \rightleftharpoons (CH_3)_3NH^+ + OH^-$, $K_b = \dfrac{[(CH_3)_3NH^+]_{eq}[OH^-]_{eq}}{[(CH_3)_3N]_{eq}}$;

(b) The general weak acid reaction is $HA + H_2O \rightleftharpoons A^- + H_3O^+$, and $HA = HF$:

$HF + H_2O \rightleftharpoons F^- + H_3O^+$, $K_a = \dfrac{[F^-]_{eq}[H_3O^+]_{eq}}{[HF]_{eq}}$;

(c) A solubility reaction involves solid dissolving to produce ions:

$CaSO_{4(s)} \rightleftharpoons Ca^{2+}_{(aq)} + SO_4^{2-}_{(aq)}$, $K_{sp} = [Ca^{2+}]_{eq}[SO_4^{2-}]_{eq}$;

(d) The water reaction is $2 H_2O \rightleftharpoons H_3O^+_{(aq)} + OH^-_{(aq)}$, $K_w = [H_3O^+]_{eq}[OH^-]_{eq}$.

Chapter 15: Principles of Chemical Equilibrium

15.73 "Saturated aqueous solution" identifies this as a solubility equilibrium.

(a) Reaction is $MgF_{2\,(s)} \rightleftharpoons Mg^{2+}_{(aq)} + 2F^-_{(aq)}$, $K_{sp} = [Mg^{2+}]_{eq}[F^-]^2_{eq}$;

(b) If $[Mg^{2+}]_{eq} = 1.14 \times 10^{-3}$ M, by stoichiometry $[F^-]_{eq} = 2(1.14 \times 10^{-3}$ M);
$K_{sp} = (1.14 \times 10^{-3})[2(1.14 \times 10^{-3})]^2 = 5.93 \times 10^{-9}$;

(c) To estimate the equilibrium constant at a temperature different from 298 K, calculate ΔH^o_{rxn} and ΔS^o_{rxn} at 298 K and then use Equations 13-9 and 15-3:

$$\Delta G^o = \Delta H^o - T\Delta S^o \qquad \Delta G^o = -RT \ln K_{eq}$$

$\Delta H^o_{rxn} = 1(-466.85) + 2(-332.63) - 1(-1123.4) = -8.71$ kJ/mol;

$\Delta S^o_{rxn} = 1(-138.1) + 2(-13.8) - 1(57.24) = = -222.94$ J/mol K;

$\Delta G^o_{rxn,\,373\,K} = (-8.71 \text{ kJ/mol}) - (373 \text{ K})(-222.94 \text{ J/mol K})(10^{-3} \text{ kJ/J}) = 74.44$ kJ/mol

$\ln K_{eq} = -\dfrac{(74.44 \text{ kJ/mol})(10^3 \text{ J/kJ})}{(8.314 \text{ J/mol K})(373 \text{ K})} = -24.00$; $K_{eq} = e^{-24.00} = 3.8 \times 10^{-11}$.

15.75 When a precipitate forms upon mixing aqueous solutions, the equilibrium reaction is the reverse of the solubility reaction. Follow the standard procedure for a reaction with a large equilibrium constant.

Reaction is $Mn^{2+} + CO_3^{2-} \rightleftharpoons MnCO_{3\,(s)}$, $K_{eq} = 1/K_{sp}$

"Initial" concentrations are after mixing but before reaction:

$[Mn^{2+}] = \dfrac{(0.750 \text{ L})}{(0.750 + 0.150 \text{ L})}(0.0250 \text{ M}) = 0.0208$ M

$[CO_3^{2-}] = \dfrac{(0.150 \text{ L})}{(0.750 + 0.150 \text{ L})}(0.500 \text{ M}) = 0.0833$ M

Species:	Mn^{2+} +	CO_3^{2-} $\rightleftharpoons$	$MnCO_{3\,(s)}$
Initial concentration (M)	0.0208	0.0833	
Change to completion (M)	−0.0208	−0.0208	
Conc. at completion (M)	0	0.0625	
Change to equilibrium (M)	y	y	
Conc. at equilibrium (M)	y	0.0625 + y	

The reaction goes nearly to completion, so assume that $y \ll 0.0625$
$K_{sp} = 1.82 \times 10^{-11} = (y)(0.0625)$, from which $y = (1.82 \times 10^{-11}) \div (0.0625) = 2.91 \times 10^{-10}$

Thus $[Mn^{2+}] = 2.91 \times 10^{-10}$

To calculate the mass of precipitate, use standard stoichiometric methods. The reaction goes virtually to completion, so

$n_{MnCO_3} = n_{Mn^{2+}} = MV = (0.0250 \text{ mol/L})(0.750 \text{ L}) = 0.01875$ mol;
$m = n\,MM = (0.01875 \text{ mol})(114.946 \text{ g/mol}) = 2.16$ g.

15.77 (a) The equilibrium constant expression for a reaction can be written by inspection of the stoichiometry of the reaction, omitting pure liquids, solids, and solvents:

$$K_{eq} = \frac{[CO_3^{2-}]_{eq}}{(p_{CO_2})_{eq}[OH^-]_{eq}^2}$$

(b) Use le Chatelier's Principle to predict the effect of changes on a system at equilibrium. Dissolving Na_2CO_3 leads to an increase in the concentration of CO_3^{2-}, so the equilibrium shifts to the left and the pressure of CO_2 increases.

(c) At first glance, it may appear that HCl will not affect this equilibrium, but recall that HCl is a strong acid which generates H_3O^+ in solution. This, in turn, will react with OH^-, reducing the concentration of a reactant. Again the equilibrium shifts to the left and the pressure of CO_2 increases.

15.79 To determine an equilibrium constant at standard temperature from thermodynamic tables, calculate ΔG^o_{rxn} from tabulated values for ΔG^o_f; to estimate the equilibrium constant at a temperature different from 298 K, calculate ΔH^o_{rxn} and ΔS^o_{rxn} at 298 K and then use Equations 13-9 and 15-3:

$$\Delta G^o = \Delta H^o - T\Delta S^o \qquad \Delta G^o = -RT \ln K_{eq}$$

(a) $\Delta G^o_{rxn} = 1(-210.745) - 1(31.82) - 1(-178.6) = -63.97$ kJ/mol;

$$\ln K_{eq} = -\frac{(-63.97 \text{ kJ/mol})(10^3 \text{ J/kJ})}{(8.314 \text{ J/mol K})(298 \text{ K})} = 25.82; \quad K_{eq} = e^{25.82} = 1.63 \times 10^{11};$$

(b) When the equilibrium pressure is 1.00 atm, $K_{eq} = 1$, $\ln(K_{eq}) = 0$ and $\Delta G^o_{rxn} = 0$:

$0 = \Delta H^o - T\Delta S^o$, so $T\Delta S^o = \Delta H^o$ and $T = \Delta H^o/\Delta S^o$

$\Delta H^o_{rxn} = 1(-265.22) - 1(61.32) - 1(-224.3) = -102.24$ kJ/mol;

$\Delta S^o_{rxn} = 1(192.5) - 1(174.96) - 1(146.0) = = -126.46$ J/mol K;

$T = (-102.24 \text{ kJ/mol})(10^3 \text{ J/kJ}) \div (-126.46 \text{ J/mol K}) = 796$ K

(c) $\Delta G^o_{rxn, 1050 K} = (-102.24 \text{ kJ/mol}) - (1050 \text{ K})(-126.46 \text{ J/mol K})(10^{-3} \text{ kJ/J}) = 30.54$ kJ/mol

$$\ln K_{eq} = -\frac{(30.54 \text{ kJ/mol})(10^3 \text{ J/kJ})}{(8.314 \text{ J/mol K})(1050 \text{ K})} = -3.50; \quad K_{eq} = e^{-3.50} = 3.02 \times 10^{-2}.$$

15.81 The value of an equilibrium constant changes with temperature but is not affected by catalysts, because although catalysts increase the rate of reaction, they do not change the thermodynamic properties of the reaction, and it is the thermodynamic properties that determine the value of the equilibrium constant. Therefore no matter how effective a catalyst is developed, the equilibrium position remains unfavorable ($K_{eq} = 0.38$ at 1300 K).

Chapter 15: Principles of Chemical Equilibrium

15.83 Modern society depends heavily on intensive agriculture to feed its highly urbanized population. Agriculture removes nutrients from the soil which must be replenished to maintain fertility and crop yields. The main component of fertilizers is "fixed" nitrogen, which must be obtained from N_2. The only process that does this on the scale needed for intensive agriculture is the Haber process, in which NH_3 is formed from N_2 and H_2. The ammonia made in this process is in turn used to make nitric acid, urea, and ammonium nitrate, three top-20 chemicals used primarily for fertilizers.

15.85 Consult your textbook for the sequence of reactions used for industrial processes:

(a) Haber synthesis: $CH_{4\,(g)} + H_2O_{(g)} \xrightleftharpoons{Ni,\ 800\ °C} CO_{(g)} + 3\ H_{2\,(g)}$

$CO_{(g)} + H_2O_{(g)} \xrightleftharpoons{Fe_2O_3,\ Cr_2O_3,\ 250\ °C} CO_{2\,(g)} + H_{2\,(g)}$

$N_{2\,(g)} + 3\ H_{2\,(g)} \xrightleftharpoons{Fe,\ K_2O,\ Al_2O_3,\ 450\ °C} 2\ NH_{3\,(g)}$

(b) Ostwald process: $4\ NH_{3\,(g)} + 5\ O_{2\,(g)} \xrightarrow{Pt\ gauze,\ 1200\ K} 4\ NO_{(g)} + 6\ H_2O_{(g)}$

$2\ NO_{(g)} + O_{2\,(g)} \rightleftharpoons 2\ NO_{2\,(g)}$

$3\ NO_2 + H_2O \rightarrow 2\ HNO_3 + NO$

(c) Contact process: $S_{(s)} + O_{2\,(g)} \rightarrow SO_{2\,(g)}$

$2\ SO_{2\,(g)} + O_{2\,(g)} \xrightleftharpoons{V_2O_5,\ 700\ K} 2\ SO_{3\,(g)}$

$SO_{3\,(g)} + H_2SO_{4\,(l)} \rightarrow H_2S_2O_{7\,(l)}$

$H_2S_2O_{7\,(l)} + H_2O_{(l)} \rightarrow 2\ H_2SO_{4\,(l)}.$

15.87 This molecular picture illustrates starting conditions and equilibrium conditions. The symbols indicate BG_3 as the starting material and G_2 and GB as products. Count numbers of symbols to determine initial and equilibrium concentrations. Set up a concentration table using numbers of symbols:

Species:	BG_3 ⇌	G_2 +	GB
Initial conc. (number):	15	0	0
Change (number)	− 12	+ 12	+ 12
Equilib conc. (number)	3	12	12

(a) From the amounts of change, the stoichiometry is 1:1, so the net reaction is
$BG_3 \rightleftharpoons G_2 + GB$;

(b) Use numbers of symbols at equilibrium to calculate the equilibrium constant:

$$K_{eq} = \frac{[GB]_{eq}[G_2]_{eq}}{[BG_3]_{eq}} = \frac{(12)(12)}{(3)} = 48.$$

Chapter 15: Principles of Chemical Equilibrium

15.89 To calculate an equilibrium constant from initial and equilibrium conditions, set up a concentration table:

Species:	2 Ef$_{(g)}$ +	3 N$_{2\,(g)}$ ⇌	2 EfN$_{3\,(g)}$
Initial concentration (atm)	0.75	1.00	0
Change to equilibrium (atm)	– 2 x	– 3 x	+ 2 x
Equilib concentration (atm)	0.75 – 2 x	1.00 – 3 x	2 x

The problem gives the equilibrium pressure, which is the sum of partial pressures:
0.85 = (0.75 – 2x) + (1.00 – 3x) + 2x = 1.75 – 3x, from which 3x = (1.75 – 0.85) = 0.90;
Thus x = 0.30;

Equilib concentration (atm)	0.15	0.10	0.60

Substitute into the equilibrium constant expression and calculate K:

$$K_{eq} = \frac{[EfN_3]^2_{eq}}{[Ef]^2_{eq}[N_2]^3_{eq}} = \frac{(0.60)^2}{(0.15)^2(0.10)^3} = 1.6 \times 10^4.$$

15.91 (a) Follow the five-step procedure for solving equilibrium problems:
1.) Species are acetic acid (A) and its dimer (A$_2$)
2.) The reaction is $\quad$ 2 A$_{(g)}$ ⇌ A$_{2(g)}$

3.) $K_{eq} = \dfrac{(p_{A_2})_{eq}}{(p_A)^2_{eq}} = 3.72;$

4.) The total pressure at equilibrium is given, P_{total} = 0.75 atm.
5.) Let x = p_{A_2} ; then p_A = 0.75 – x;

Substitute into the equilibrium constant expression and solve for x:

$$3.72 = \frac{x}{(0.75 - x)^2}$$

x = (3.72)(0.75 – x)^2 = (3.72)(x^2 – 1.50 x + 0.563) = 3.72 x^2 – 5.58 x + 2.093;
0 = 3.72 x^2 – 6.58 x + 2.093;

$$x = \frac{-b \pm \sqrt{b^2 - 4ac}}{2a} = \frac{6.58 \pm \sqrt{(6.58)^2 - 4(3.72)(2.093)}}{2(3.72)} = \frac{6.58 \pm \sqrt{43.296 - 31.144}}{7.44}$$

$x = \dfrac{6.58 \pm 3.49}{7.44}$ = 1.35 or 0.415; Rule out 1.35, which would give a negative pressure;

$(p_{A_2})_{eq}$ = 0.415 atm

(b) The reaction to form dimer is exothermic, so raising the temperature will shift the equilibrium position to the left. The equilibrium constant is lower at 200 °C.

15.93 Gas pressures and solubilities are connected through the equilibrium described by Henry's law: $K_H = \dfrac{[CO_{2(aq)}]_{eq}}{(p_{CO_2})_{eq}}$, from which $[CO_{2(aq)}]_{eq} = K_H (p_{CO_2})_{eq}$;

From Table 15-1, K_H (25 °C) = 3.4 x 10^{-2}; $[CO_{2(aq)}]_{eq}$ = (3.4 x 10^{-2})(1.10) = 0.0374 M (remember that in all equilibrium constant expressions, solute concentrations are molarities)

Complete the calculation using standard stoichiometric methods:
$n = MV$ = (0.0374 M)(225 mL)(10^{-3} L/mL) = 0.00842 mol;
$m = n\, MM$ = (0.00842 mol)(44.01 g/mol) = 0.37 g.

15.95 The standard types of equilibria are phase change (melting, vaporization, or sublimation), solubility (gas or solid), acid-base, and gas reactions.

(a) weak base equilibrium in reverse: $K_{eq} = \dfrac{[(CH_3)_2 NH]_{eq}}{[OH^-]_{eq}[(CH_3)_2 NH_2^+]_{eq}} = 1/K_b$;

(b) weak acid equilibrium: $K_{eq} = \dfrac{[CO_3^{2-}]_{eq}[H_3O^+]_{eq}}{[HCO_3^-]_{eq}} = K_a$;

(c) reverse of solid solubility: $K_{eq} = \dfrac{1}{[Zn^{2+}]_{eq}[OH^-]_{eq}^2} = 1/K_{sp}$;

(d) vaporization equilibrium: $K_{eq} = (p_{NH_3})_{eq} = vp$.

15.97 To determine the equilibrium concentrations when initial conditions are provided, set up a concentration table. Because K_{eq} is large, first take the reaction to completion:

Species:	2 $SO_{2(s)}$	+	$O_{2(g)}$ ⇌	2 $SO_{3(g)}$
Initial concentration (atm)	0.350		0.762	0
Change to completion (atm)	− 0.350		− 0.175	0.350
Concentration at completion (atm)	0		0.587	0.350
Change to equilibrium (atm)	+ 2 y		+ y	− 2 y
Equilib concentration (atm)	2 y		0.587 + y	0.350 − 2 y

Assume that y << 0.350, substitute into the equilibrium constant expression and solve for y:

$$5.60 \times 10^4 = \dfrac{(0.350)^2}{(2y)^2(0.587)}$$

$4y^2 = \dfrac{(0.587)}{(0.350)^2(5.60 \times 10^4)} = 0.856 \times 10^{-4}$, from which $y = 4.63 \times 10^{-3}$

$(p_{SO_2})_{eq} = 2(4.63 \times 10^{-3}) = 9.26 \times 10^{-3}$ atm; $(p_{O_2})_{eq} = 0.587 + 0.00463 = 0.592$ atm;

$(p_{SO_3})_{eq} = 0.350 - 2(0.00463) = 0.341$ atm.

Chapter 15: Principles of Chemical Equilibrium

15.99 To determine the effect of temperature, compare amounts of molecules at equilibrium for the two temperatures.

Equilibrium conditions at 1100 K: 4 O_2, 5 SO_3, and 5 SO_2 molecules.
Equilibrium conditions at 1300 K: 5 O_2, 3 SO_3, and 7 SO_2 molecules.

The equilibrium shifts to the left, forming additional reactant molecules, when temperature increases. According to le Chatelier's principle, the reaction shifts in the endothermic direction (absorbs heat) when temperature increases. Therefore the forward reaction is exothermic.

Chapter 16 Aqueous Equilibria

16.1 The Composition of Aqueous Solutions

QUESTIONS TO ANSWER, SKILLS TO LEARN
1. Identifying the "major" and "minor" components in a solution
2. What are the differences between ionic and molecular solutions?
3. What are spectator ions?

To simplify the discussion of equilibrium reactions in aqueous solution, the solutes are divided into two categories: those at relatively high concentrations are **major species** and those at relatively low concentrations are **minor species**.

The solvent of a solution is always a major species. The ions resulting from dissociation of a strong electrolyte (salt, strong acid, or strong base) are major species. Major species are solute molecules obtained from dissolving a nonelectrolyte or weak electrolyte. The ions resulting from the dissociation of a weak electrolyte are minor species.

Spectator ions are ions that do not participate in chemical reactions, but are the counterions of the ionic species that did participate in the reaction. Once the composition of a solution has been determined, the type(s) of chemical reaction(s) and the corresponding equilibrium(ia) may be determined and applied to solve the problem.

Exercise 16.1 Identify the major species and any spectator ions in each of the following systems:
  (a) Addition of solid $CaCl_2$ to water.
  (b) Addition of equal mols of NaCl *and* $AgNO_3$ to water.
  (c) Addition of liquid formic acid, HCOOH (a weak electrolyte), to water.
  (d) Addition of solid $Ba(OH)_2 \cdot 8H_2O$ to water.

Steps to Solution: To determine the major species, we must identify the nature of the substance that is being added to the solvent or solution to see if it will dissociate into ions. We do this by using the rules given in Chapter 4 to identify ionic substances. To identify spectator ions, we look for the types of reactions that might occur between ions in solution (precipitation or acid-base as described in Chapter 4) and identify the ions that do not participate in the reaction.

(a) *$CaCl_2$ is a soluble salt (it contains the alkaline earth cation Ca^{2+} and the anion Cl^-). Therefore the solution contains water, calcium ions and chloride ions as major species. Since no reaction occurs between ions in the solution, there are no spectator ions.*

(b) *This solution contains two ionic substances. Each of the substances is a soluble salt so we expect to obtain a solution containing the Na^+, Cl^-, Ag^+, and NO_3^- ions. However, examination of the solubility rules shows that AgCl is not soluble:*

$$Ag^+_{(aq)} + Cl^-_{(aq)} \rightarrow AgCl_{(s)}$$

The Na⁺ and NO₃⁻ ions do not participate in this reaction and therefore are spectator ions. They are also major species in the solution along with the solvent water. Most of the Ag⁺ and Cl⁻ have been precipitated, so they are not major species in the solution.

(c) *Formic acid is not a salt (it does not contain a metal and a polyatomic anion) or a strong acid or base. Therefore it will dissolve as molecules and the major species in the solution will be formic acid molecules (HCOOH) and water.*

(d) *$Ba(OH)_2 \cdot 8H_2O$ is a soluble strong base (it contains the alkaline earth Ba^{2+} ion and OH^- ions). Therefore it completely dissolves to give the Ba^{2+} ion and OH^- ions in water, the major species in solution.*

16.2 Solubility Equilibria

QUESTIONS TO ANSWER, SKILLS TO LEARN
1. **Using solubility equilibria (or precipitation equilibria)**

The equilibrium solubility of an ionic substance is governed by the solubility product, K_{sp}, as described in Chapter 1 The following criteria are used to classify substances as soluble, slightly soluble or insoluble.

K_{sp}	> 10^{-2}	$10^{-5} < K_{sp} < 10^{-2}$	< 10^{-5}
Solubility	soluble	slightly soluble	insoluble

The solubility of a substance is defined as the amount of substance that will be present in 1 L of a saturated solution. The concept of solubility can be used to determine how much substance will dissolve upon adding a solid to water or to determine how much substance will precipitate upon mixing two solutions containing ions that will form a precipitate.

Exercise 16.2 What is the concentration of Ca^{2+} in a saturated solution of calcium phosphate, $Ca_3(PO_4)_2$ ($K_{sp} = 2.07 \times 10^{-33}$)?

Steps to Solution: *In a saturated solution, the system is at equilibrium. The equilibrium conditions are determined by using the solubility product expression. Calcium phosphate dissolves to give calcium ions and phosphate ions:*

$$Ca_3(PO_4)_{2\,(s)} \rightleftarrows 3\,Ca^{2+}_{\,(aq)} + 2\,PO_4^{3-}_{\,(aq)} \qquad K_{sp} = [Ca^{2+}]^3[PO_4^{3-}]^2$$

Chapter 16: Aqueous Equilibria

If the number of moles of calcium phosphate that dissolves per liter is s, then the concentration table will be as shown below. Note that every mole of $Ca_3(PO_4)_2$ that dissolves gives 3 moles of Ca^{2+} and 2 moles of PO_4^{3-}.

Species	Initial Condition	+ Change	= Equilibrium
$Ca_3(PO_4)_{2\,(s)}$	solid present	solid dissolves	less $Ca_3(PO_4)_2$ is present
$Ca^{2+}_{\,(aq)}$	0	+ 3s	3s
$PO_4^{3-}_{\,(aq)}$	0	+2s	2s

Substitution of these values into the K_{sp} expression gives:

$$K_{sp} = [Ca^{2+}]^3[PO_4^{3-}]^2 = (3s)^3(2s)^2 = 27 \cdot 4 \cdot s^5 = 108\,s^5$$

Solving this equation for s gives:
$$s = \left(\frac{K_{sp}}{108}\right)^{\frac{1}{5}} = \left(\frac{2.07 \times 10^{-33} M^5}{108}\right)^{\frac{1}{5}} = 1.14 \times 10^{-7}\,M$$

(Use the y^x key on your calculator to calculate this quantity.) Note that the concentration of Ca^{2+} is 3s; therefore the concentration of Ca^{2+} is:

$$[Ca^{2+}] = 3s = 3(1.14 \times 10^{-7}\,M) = 3.42 \times 10^{-7}\,M$$

Exercise 16.3 A real world application of solubility products is found in softening "hard water" by removing calcium and magnesium ions present as the sulfate salts. We can decrease the amount of Ca^{2+} in solution by addition of sodium carbonate, a strong electrolyte:

$$Na_2CO_{3\,(s)} \rightarrow 2\,Na^+_{\,(aq)} + CO_3^{2-}_{\,(aq)}$$

By increasing the concentration of carbonate, we tend to drive the solubility equilibrium toward the reactant:

$$CaCO_{3\,(s)} \rightleftharpoons Ca^{2+}_{\,(aq)} + 3.42 \times 10^{-7}\,M \qquad K_{sp} = 4.7 \times 10^{-9}\,M^2$$

If the initial $[Ca^{2+}] = 5.0 \times 10^{-3}\,M$, what percent of the $[Ca^{2+}]$ will be removed if the carbonate concentration is maintained at $1.0 \times 10^{-3}\,M$?

<u>Steps to Solution:</u> This question asks for the percent of Ca^{2+} that will be removed by the precipitation reaction,

$$Ca^{2+}_{\,(aq)} + CO_3^{2-}_{\,(aq)} \rightleftharpoons CaCO_{3\,(s)} \qquad K = \frac{1}{K_{sp}}$$

The quantity being sought is the percentage of Ca^{2+} removed, which is 100% minus the percentage of Ca^{2+} remaining in solution.

$$\% \text{ removed} = 100\% - \% \text{ remaining} = 100\% - \frac{[Ca^{2+}]_{final}}{[Ca^{2+}]_{initial}} \cdot 100\%$$

The problem may be solved by finding the equilibrium concentration of Ca^{2+} and substituting into the above equation. Setting up a concentration table, we obtain the following values. Remember that the carbonate concentration is maintained at 1×10^{-3} M.

Species	Initial Condition	+ Change	= Equilibrium
$CaCO_{3(s)}$	none	solid precipitates	solid $CaCO_3$ formed
$Ca^{2+}_{(aq)}$	5.0×10^{-3}	$-x$	$5.0 \times 10^{-3} - x$
$CO_3^{2-}_{(aq)}$	1.0×10^{-3} M	0	1.0×10^{-3} M

However, because the equilibrium constant is large, it will be easier to consider this process by calculating the concentration of Ca^{2+} at equilibrium under these conditions; that is, let the reaction go to completion and treat it as if $CaCO_3$ is dissolving in the presence of 1×10^{-3} M carbonate (let $[CO_3^{2-}]$ be constant; if the carbonate concentration were not fixed, then the change would be +s).

Species	"Initial" Condition	+ Change	= Equilibrium
$CaCO_{3(s)}$	solid present	solid dissolves	less $CaCO_3$ present
$Ca^{2+}_{(aq)}$	0	+s	+s
$CO_3^{2-}_{(aq)}$	1.0×10^{-3} M	0 (constant)	1.0×10^{-3} M

At equilibrium,

$$[Ca^{2+}] = s \quad \text{and} \quad [CO_3^{2-}] = 1.0 \times 10^{-3} \text{ M}$$

$$4.7 \times 10^{-9} \text{ M}^2 = [Ca^{2+}][CO_3^{2-}] = s(1.0 \times 10^{-3} \text{ M})$$

$$s = \frac{4.7 \times 10^{-9} \text{ M}^2}{1.0 \times 10^{-3} \text{ M}} = 4.7 \times 10^{-6} \text{ M} = [Ca^{2+}] = 4.7 \times 10^{-6} \text{ M}$$

Chapter 16: Aqueous Equilibria

This is the equilibrium concentration of Ca^{2+} when the carbonate concentration is 1 x 10^{-3} M. With this equilibrium concentration and remembering that the initial concentration of Ca^{2+} was 3 x 10^{-3} M,

$$\% \text{ removed} = 100\% - \% \text{ remaining} = 100\% - \frac{4.7 \times 10^{-6} \text{ M}}{5.0 \times 10^{-3} \text{ M}} \cdot 100\%$$

$$\% \text{ removed} = 100\%(1 - 9.4 \times 10^{-4}) = \boxed{99.906 \%}$$

*Nearly all the calcium ion can be removed by the use of carbonate to precipitate $CaCO_3$. This is an example of the **common ion effect**, where the addition of an ion that participates in an equilbrium can be used to shift the position of equilibrium.*

16.3 Proton Transfer Equilibria

QUESTIONS TO ANSWER, SKILLS TO LEARN
1. **Identifying Brønsted-Lowry acids and bases and their conjugate bases and acids**
2. **Using the pH concept**
3. **Reactions of water with acids and bases**

One of the more general ways of defining acid-base reactions is as a reaction in which a proton is transferred from one chemical species to another. Thus the products differ from the reactants according to which species possesses the transferred proton. The chemical species in the products are called the **conjugate acid** and **conjugate base** of the corresponding base and acid in the products:

$$\text{acid} \quad + \quad \text{base} \quad \rightarrow \quad \text{conjugate acid} \quad + \quad \text{conjugate base}$$

$$\text{HA} \quad + \quad \text{B:} \quad \rightarrow \quad \underset{(protonated)}{\text{BH}^+} \quad + \quad \underset{(deprotonated)}{\text{A}^-}$$

Exercise 16.4 Identify the acids and bases and the conjugate acids and bases in the following acid-base reactions:
(a) $HCOOH_{(aq)} + NH_{3\,(aq)} \rightarrow HCOO^-_{\,(aq)} + NH_4^+_{\,(aq)}$
(b) $H_2O + H_2SO_4 \rightarrow H_3O^+ + HSO_4^-$
(c) $CH_3NH_2 + H_2CO_3 \rightarrow CH_3NH_3^+ + HCO_3^-$

Steps to Solution: This problem is addressed by examining the equation of reaction and determining which chemical species lose protons and which chemical species gain protons. The reactant that loses a proton is the acid; the reactant that gains the proton is the base. The products are similarly identified; the protonated base is the

Chapter 16: Aqueous Equilibria

conjugate acid of the reactant base; the deprotonated species is the conjugate base of the reactant acid.

(a) In this reaction, HCOOH loses a proton to become HCOO⁻; therefore HCOOH is the acid and HCOO⁻ is its conjugate base. The NH_3 molecule accepts the proton to become the ammonium ion; therefore NH_3 is the base and the ammonium ion is its conjugate acid.

(b) In this reaction, H_2SO_4 transfers a proton to H_2O; therefore sulfuric acid is the acid and water is the base. The product hydronium ion is the conjugate acid of the base, H_2O. HSO_4^- is the product of proton loss from sulfuric acid; therefore it is the conjugate base of sulfuric acid.

(c) In this reaction a proton is transferred from carbonic acid, H_2CO_3, to methyl amine, CH_3NH. Therefore H_2CO_3 is the acid and CH_3NH_2 is the base. Therefore $CH_3NH_3^+$ is the conjugate acid of methylamine and the hydrogen carbonate ion, HCO_3^-, is the conjugate base of carbonic acid.

It was noted earlier that water molecules undergo proton transfer reactions with other water molecules with a small, but very significant equilibrium constant:

$$2\,H_2O_{(l)} \rightleftarrows H_3O^+_{(aq)} + OH^-_{(aq)} \qquad K_w = [H_3O^+][OH^-] = 1.0 \times 10^{-14}\,M^2\;(298\,K)$$

The equilibrium relationship shows that the concentrations of hydronium and hydroxide ion are related: if one increases, the other must decrease for the equilibrium expression to be satisfied. The range of hydrogen and hydroxide ion concentrations in water and their relationship is explored in Exercises 16.5, 16.6 and 16.7.

Exercise 16.5 What are the hydrogen and hydroxide ion concentrations in the following solutions:
(a) 9.0 M HCl (b) 1.00 x10⁻⁵ M Ca(OH)₂ (c) 6.4 M NaOH.

Steps to Solution: In determining the hydrogen and hydroxide ion concentrations of these solutions, we must first identify the nature of the substance and its products upon dissolving in water. This step allows us to calculate the hydrogen (or hydroxide) ion concentration. Upon determining either the hydrogen or hydroxide ion concentration, we use the K_w expression to calculate the other concentration.

(a) *HCl is a strong acid that totally dissociates to give hydronium ions and chloride ions in water:*

$$HCl_{(aq)} \rightarrow H_3O^+_{(aq)} + Cl^-_{(aq)}$$

The $[H_3O^+]$ from the dissociation of water is insignificant compared to the $[H_3O^+]$ from HCl. Therefore $[H_3O^+] = [HCl] = 9.0\,M$. The concentration of hydroxide is calculated using the K_w expression:

$$K_w = [H_3O^+][OH^-]$$

$$[OH^-] = \frac{K_w}{[H_3O^+]} = \frac{1 \times 10^{-14}\ M^2}{9.0\ M} = \boxed{1.1 \times 10^{-15}\ M}$$

(b) Calcium hydroxide is a strong electrolyte which totally dissociates in water:

$$Ca(OH)_2 \rightarrow Ca^{2+}_{(aq)} + 2\ OH^-_{(aq)}$$

The concentration of hydroxide ion is:

$$[OH^-] = 1.00 \times 10^{-5}\ M\ Ca(OH)_2 \cdot \frac{2\ \text{mole}\ OH^-}{\text{mole}\ Ca(OH)_2} = \boxed{2.00 \times 10^{-5}\ M}$$

The concentration of hydronium is calculated using the K_w expression:

$$K_w = [H_3O^+][OH^-]$$

$$[H_3O^+] = \frac{K_w}{[OH^-]} = \frac{1 \times 10^{-14}\ M^2}{2.00 \times 10^{-5}\ M} = \boxed{5.00 \times 10^{-10}\ M}$$

(c) Sodium hydroxide is a strong electrolyte that totally dissociates in water: use the K_w expression to find $[H_3O^+] = \boxed{1.6 \times 10^{-15}\ M}$.

In Exercise 16.5, the $[H_3O^+]$ in parts (a) and (c) differ by a factor of nearly 10^{15}! An alternative way to express the concentration of hydrogen ion, as the (exponential) *power* of H^+, is called the pH.

$$p = -\log: \quad pH = -\log[H_3O^+] \quad pOH = -\log[OH^-] \quad pK = -\log K$$

$$pH + pOH = 14.0$$

If the pH is small, then the pOH is big (14 - pH) and the solution is acidic.

If the pH is big, then the pOH is small (14 - pH) and the solution is basic.

If pH = 7.0 at 298 K, the solution is neutral.

Exercise 16.6 Calculate the pH of the three solutions in Exercise 16.5.

Chapter 16: Aqueous Equilibria

Steps to Solution: To calculate the pH, we take the logarithm (log base 10, the **log** key, not the **ln** key on the calculator!) of $[H_3O^+]$ and then change the sign.

(a) The $[H_3O^+] = 9.0$ M, so

$$pH = -\log(9.0) = \boxed{-0.95}$$

Note that there are two significant figures in the concentration, so two significant figures are in the answer. In a logarithm, significant figures are counted to the right of the decimal point.

(b) The $[H_3O^+] = 5.00 \times 10^{-10}$ M, so

$$pH = -\log(5.00 \times 10^{-10}) = \boxed{9.301}$$

Note that the three significant figures in the concentration are reflected in the three places to the right of the decimal point.

(c) The pH = $\boxed{14.81}$

Exercise 16.7 What is the pH of an 800 mL solution prepared from 25 g of KOH and water?

Steps to Solution: First we must calculate the concentration of hydrogen ions, in this case by finding $[OH^-]$ and then using K_w to find $[H_3O^+]$. Then the pH can be calculated as above. First, we find M of KOH:

$$M = \frac{\frac{25g}{56 g/mol KOH}}{0.800 L} = 0.56 \text{ M}$$

Because KOH is a strong electrolyte, $[OH^-] = [KOH]$

$$[H_3O^+] = \frac{1 \times 10^{-14} M^2}{0.56 M} = 1.8 \times 10^{-14} \text{ M} \quad \text{and} \quad pH = -\log(1.8 \times 10^{-14}) = \boxed{13.74}$$

In the preceding exercises, the acids and bases have behaved conveniently, giving hydronium ion or hydroxide ion concentrations related directly to the acid or base concentration, respectively, by the stoichiometric coefficient. A large and important group of acids do not behave in this way; in these compounds, the acid and the conjugate base are in equilibrium with each other. These compounds are called **weak acids**.

Weak acids in aqueous solution are in equilibrium with small quantities of their conjugate base and hydronium ion.

Chapter 16: Aqueous Equilibria

Calculating pH's of solutions of weak acids is more complicated than working with strong acids because the hydronium ion concentration is not directly related to the acid concentration.

Exercise 16.8 A solution is made by adding 0.1000 mole of the weak acid, HF, to water, and then adding water until the volume of the solution is 1.000 L. Of the acid added, 8.5% is dissociated.
(a) What is K_a?
(b) What is the pH of the solution?

Steps to Solution: *The strategy is to identify the species in solution and the equilibria in which they participate. Then a concentration table can be drawn up to help determine the initial and equilibrium concentrations. Once we have found the equilibrium concentrations then we can calculate the pH and K.*

The solution contains HF, which is a weak acid. As a weak acid it reacts with water, the extent of which is described by K_a:

$$HF_{(aq)} + H_2O \rightleftharpoons H_3O^+_{(aq)} + F^-_{(aq)} \qquad K_a = \frac{[H_3O^+][F^-]}{[HF]}$$

We need to find the concentrations of the reagents to fill in the concentration table. The HF reacts to give F^- and H_3O^+ in equimolar amounts. We will assume that the amount of hydronium ion produced by the reaction of HF is much greater than that produced by the self-dissociation of water. $[F^-]$ is found by using the given percent dissociation.

$$[F^-] = \frac{[HF_{initial}](\% \text{ dissociation})}{100\%} = \frac{[0.1000](8.5\%)}{100\%} = 8.5 \times 10^{-3} \text{ M}$$

From the stoichiometry we know that

$$[H_3O^+] = [F^-] = 8.5 \times 10^{-3} \text{ M}$$

The concentration of HF started at 0.1000 M; however, some dissociated:

$$[HF] \text{ at equilibrium} = [HF]_{initial} - [F^-]_{equilibrium}$$

$$[HF] \text{ at equilibrium} = 0.1000 - 8.5 \times 10^{-3} = 0.0915 \text{ M}$$

Species	Initial Condition	+ Change	= Equilibrium
HF $_{(aq)}$	0.1000 M	-8.5×10^{-3} M	0.0915 M
H$_3$O$^+$ $_{(aq)}$	0	8.5×10^{-3} M	8.5×10^{-3} M
F$^-$ $_{(aq)}$	0	8.5×10^{-3} M	8.5×10^{-3} M

(a) Substituting the equilibrium concentrations into the K_a equation gives:

$$K_a = \frac{(8.5 \times 10^{-3} \text{ M})(8.5 \times 10^{-3} \text{ M})}{(0.0915 \text{ M})} = \boxed{7.9 \times 10^{-4} \text{ M}}$$

(b) The pH is calculated from [H$_3$O$^+$]:

$$\boxed{pH = -\log(8.5 \times 10^{-3}) = 2.07}$$

This exercise illustrated an important principle: addition of a weak acid increases the [H$_3$O$^+$] (decreasing the pH) of a solution by releasing its reactive proton to the water molecules, in much the same way that a strong acid decreases the pH of a solution. However, weak bases have a different mode of action than that of most of the strong bases we have discussed. A weak base reacts with water to abstract a proton, releasing a hydroxide ion in an equilibrium process with the equilibrium constant, K_b:

$$B: + H_2O \rightleftarrows BH^+ + OH^- \qquad K_b = \frac{[BH^+][OH^-]}{[B]}$$

This equation can be obtained from the chemical equation for the dissociation of the weak acid BH$^+$ and the K_w expression:

$$B: + H_3O^+ \rightleftarrows BH^+ + H_2O \qquad K = \frac{[BH^+]}{[B:][H_3O^+]} = \frac{1}{K_a}$$

$$+ \{2 H_2O \rightleftarrows H_3O^+ + OH^-\} \qquad K = K_w$$

$$\overline{B: + H_2O \rightleftarrows BH^+ + OH^- \qquad K_b = K_w \cdot \frac{1}{K_a}}$$

$$\boxed{K_w = K_a K_b \text{ (a memory bank equation!)}}$$

In Exercise 16.9 we will consider the reaction of the weak base, F$^-$, the conjugate base of HF, with water.

Chapter 16: Aqueous Equilibria

Exercise 16.9 What is the pH of a solution that is prepared by the addition of 15.0 g of NaF to sufficient water to make 1.2 L of solution?

Steps to Solution: The method to solve this and other equilibrium problems is the same: determine the chemical species present and then decide which equilibrium reactions will be important. Next, use a table of concentrations to determine the initial and equilibrium concentrations whether in numbers or in terms of some unknown amount. The chemical species present are the strong electrolyte, sodium fluoride, which dissociates to give sodium ions, which are spectator ions, and fluoride ions in aqueous solution. The weakly basic F^- ions can react with water to give HF and OH^-

$$F^- + H_2O \rightleftharpoons HF + OH^- \quad K = \frac{[OH^-][HF]}{[F^-]} = K_b$$

$$K_b = \frac{K_w}{K_a} = \frac{1.0 \times 10^{-14} \, M^2}{7.30 \times 10^{-4} \, M} = 1.37 \times 10^{-11} \, M$$

We must now determine the concentrations of the chemical species involved in the equilibrium reaction. The pre-equilibrium concentration of F^- is the same as the molarity of the NaF solution:

$$[F^-] = [NaF] = \frac{\text{mole NaF}}{\text{L solution}} = \frac{\frac{15.0 \, g}{41.99 \, g/\text{mole NaF}}}{1.20 \, L} = 0.298 \, M$$

We can assume that initially, $[HF] = 0 = [OH^-]$ (the hydroxide resulting from water self-reaction is neglected). Upon reaction with water, some F^- is converted to HF, resulting in a equimolar amount of OH^-. If the decrease in $[F^-]$ is x, the increase in $[HF]$ and $[OH^-]$ is also x. Now the concentration table may be completed.

Species	Initial Condition	+ Change	= Equilibrium
$F^-_{(aq)}$	0.298 M	-x	0.298 - x
$OH^-_{(aq)}$	0	+x	x
$HF_{(aq)}$	0	+x	x

Substitution into the equilibrium constant expression for K_b gives:

$$K_b = \frac{[OH^-][HF]}{[F^-]} = \frac{x \cdot x}{0.298 - x}$$

Since K is small, try the approximation where $0.298 - x \approx 0.298$ (x << 0.298)

$$K_b = \frac{x \cdot x}{0.298}$$

$$x = \sqrt{K_b \cdot 0.298} = 2.02 \times 10^{-6} \text{ M} = [OH^-] = [HF]$$

(Since $2.02 \times 10^{-6} \ll 0.298$, the approximation was valid.)

The $[H_3O^+]$ can now be calculated using:

$$[H_3O^+] = \frac{K_w}{[OH^-]} = \frac{1.00 \times 10^{-14} \text{ M}^2}{2.02 \times 10^{-6} \text{ M}} = 4.96 \times 10^{-9} \text{ M}$$

$$pH = -\log[H_3O^+] = -\log(4.96 \times 10^{-9}) = \boxed{8.305}$$

The classes of reactions shown in the table that follows will be important in the next sections; knowing the position of equilibrium from the size of K_a and K_b will help in understanding the chemical processes. For weak acids and bases, K_a and K_b are much smaller than 1 ($K_w = K_a \cdot K_b$).

Type of Reaction	General Equation	K and Its Size
weak acid-strong base	$HA + OH^- \rightarrow H_2O + A^-$	$K = \dfrac{1}{K_b}$ (small K_b, $K \gg 1$) reaction proceeds to near completion
strong acid-weak base	$A^- + H_3O^+ \rightarrow HA + H_2O$	$K = \dfrac{1}{K_a}$ (small K_a, $K \gg 1$) reaction proceeds to near completion

16.4 Acids and Bases

QUESTIONS TO ANSWER, SKILLS TO LEARN
1. **What structural features lead to acidity and basicity?**
2. **What are "polyprotic" acids?**
3. **Relating the strength of an acid to the strength of its conjugate base**

A Brønsted-Lowry acid is a proton donor to a base. Proton loss will be easier if the bond to hydrogen is polar with the positive charge on H.

$$\overset{\delta-}{X}-\overset{\delta+}{H}$$

For this kind of polar bond to be formed, the atom (or molecular fragment), X, must be more electronegative than hydrogen. With the exception of HCN, bonds between H and C will not be considered acidic in this text. Substances that are acidic in water can be divided into three general groups.

Binary Acids	Oxy Acids	Carboxylic Acids
These acids contain two elements, H and Group 6 or 7 elements. This group contains weak and strong acids.	These acids contain O atoms and acidic O-H groups bonded to a nonmetal or metal. There are weak and strong oxyacids.	These acids all contain the carboxyl functional group, $-CO_2H$. All carboxylic acids are weak acids.

Many molecules contain several different types of bonds to hydrogen. Determining which hydrogen atoms are acidic requires use of the above table and some deductions about the electronegativity differences leading to the different polarities of the bonds.

Exercise 16.10 Consider the following molecules and determine which if any of the H atoms would be acidic in water.
(a) 1-bromopropane
(b) H_3PO_3 (note: not all the H atoms are bound to O)
(c) Serine (see Chapter 11, Macromolecules).

Steps to Solution: First determine the structure using the methods described in Chapter 3. Then examine the structure to see if it belongs to one of the above three categories and identify the polar bonds to H that are expected to be acidic.

(a) The structure of 1-bromopropane is:	(b) The structure of H_3PO_3 is:	(c) The structure of serine is:
$Br-CH_2-CH_2-CH_3$ Examining the structure shows none of the three groups associated with acidic behavior nor any polar bonds to H. In accordance with these observations, the compound is not acidic.	(structure: H and two OH bonded to P, with P=O) This compound belongs to the second class of acids, the oxy acids. The acidic hydrogens are those bonded to the O atoms, as those bonds are more polar.	(structure: central C bonded to CH_2OH, COOH, NH_2, H) Serine contains the carboxylic acid group; therefore it is an acid, and the most acidic proton is that in the COOH group. The other bonds to H are not as polar and not as acidic.

Polyprotic Acids

A significant number of molecules have two or more essentially identical bonds to H that are acidic. Because more than one H atom is acidic, these molecules are called **polyprotic acids**. The molecule H_3PO_3 in Exercise 16.6.8(b) is an example of a polyprotic acid. The dissociation (or reaction) of the acidic hydrogen atoms proceeds in stages because loss of one proton makes the protons on the conjugate base less acidic, due in part to the increased negative charge on the conjugate base.

$$H_3PO_3 \rightleftarrows H_2PO_3^- + H^+ \qquad K_{a1} = 1.0 \times 10^{-2} \text{ M}$$

$$H_2PO_3^- \rightleftarrows HPO_3^{2-} + H^+ \qquad K_{a2} = 2.6 \times 10^{-7} \text{ M}$$

$$\frac{K_{a1}}{K_{a2}} = \frac{1.0 \times 10^{-2}}{2.6 \times 10^{-7}} = 3.8 \times 10^4$$

When working with polyprotic acids, we usually use only the equilibrium expression linking one acid-conjugate base pair because of the significant difference in acidity. The ratio of the K_a's reveals the difference in acidity; for example, H_3PO_3 is about 38,000 times more acidic than $H_2PO_3^-$. Therefore, in a system containing both H_3PO_3 and $H_2PO_3^-$ that reacts with a strong base, more H_3PO_3 will be consumed than $H_2PO_3^-$ because H_3PO_3 releases protons more readily to react with the base.

Bases

To act as a proton acceptor, a base must supply both electrons to form the bond to the proton. Therefore a base must have an atom with at least one lone pair. The three classes of bases in aqueous solution are:

Oxygen Bases	Nitrogen Bases	Other Bases
Water (a weak base) and hydroxide (a strong base) have a basic O atom.	Ammonia and amines contain a basic N atom.	The conjugate base of *any weak acid* contains a basic atom.

Exercise 16.11 Equal volumes of two solutions, one containing 0.23 M ammonia (pK_b = 4.74), and the other, 0.23 M HCl, are mixed. What is the pH of the resulting solution?

Steps to Solution: This problem involves the reaction of a strong acid and a weak base, so the final solution will require analysis of the equilibria that are occurring. First the chemical reactions that occur must be determined and the products of the reaction identified. We then analyze the equilibria involving these products using the methods developed in this and previous chapters. The reaction between a weak base and a strong acid will proceed essentially to completion; therefore this reaction will produce the salt ammonium chloride as the product:

$$NH_{3\,(aq)} + HCl_{(aq)} \rightarrow NH_4Cl_{(aq)}$$

Because the solutions mixed are of the same concentration and volume, equal numbers of moles of both reactants are mixed. Therefore the reaction will produce a solution containing the ammonium ion, water and the chloride ion as major species. The Cl^- ion is a spectator ion, but the NH_4^+ ion is a weak acid and the following equilibrium will determine the pH of the solution:

$$NH_4^+{}_{(aq)} \rightleftarrows NH_{3\,(aq)} + H_3O^+{}_{(aq)} \qquad K_a = \frac{1}{K_b} = 5.5 \times 10^{-10} = \frac{[NH_3][H_3O^+]}{[NH_4^+]}$$

A concentration table is now built as before. The concentration of the ammonium salt is initially $\frac{0.23 \text{ M}}{2} = 0.115$ M because the two solutions mixed are of equal volume. The concentration increase of NH_3 due to dissociation of NH_4^+ is x:

Species	Initial Condition	+ Change	= Equilibrium
$NH_4^+{}_{(aq)}$	0.115 M	-x	0.115 - x
$H_3O^+{}_{(aq)}$	0	+x	x
$NH_3{}_{(aq)}$	0	+x	x

Substitution into the K_a expression gives:

$$K_a = 5.5 \times 10^{-10} = \frac{[NH_3][H_3O^+]}{[NH_4^+]} = \frac{x^2}{0.115 - x}$$

let x << 0.115 as an approximation

$5.5 \times 10^{-10} = \dfrac{x^2}{0.115}$ gives $x = \sqrt{5.5 \times 10^{-10} (0.115)} = 7.9 \times 10^{-6}$ M = $[NH_3]$ = $[H_3O^+]$

The approximation is valid: pH = -log(7.9 x 10⁻⁶) = $\boxed{5.10}$

Addition of the conjugate acid of a weak base makes a solution acidic!

16.5 Buffer Solutions

QUESTIONS TO ANSWER, SKILLS TO LEARN
1. **What is a buffer solution?**
2. **Using the "buffer equation" to determine the pH of buffer solutions**
3. **What is the capacity of a buffer solution?**

A **buffer solution** is a solution that contains *both a weak acid and its conjugate base in similar concentrations*. In a buffer solution, the concentrations of the acid and its conjugate base are within a factor of 10 of each other. First we will calculate the pH of a solution containing a weak acid and its conjugate base.

Exercise 16.12 What is the pH of a solution made by the addition of 0.34 mole of Na_2HPO_4 and 0.65 mole of NaH_2PO_4 and sufficient water to give a total volume of 1.2 L?

Steps to Solution: This requires the usual equilibrium approach of finding the major species, finding the equilibria that involve these species and setting up a concentration table. At that point, the approximations may be used to simplify the calculations.

The major chemical species present are Na+ ions (which are spectators), HPO_4^{2-} and $H_2PO_4^-$ ions and water. The equilibrium reaction and expression involving the active chemical species are:

$$H_2PO_4^-{}_{(aq)} + H_2O \rightleftarrows H_3O^+{}_{(aq)} + HPO_4^{2-}{}_{(aq)} \quad K_a = \frac{[H_3O^+][HPO_4^{2-}]}{[H_2PO_4^-]} = 6.20 \times 10^{-8}$$

The initial concentrations of the phosphate species are calculated from the number of moles and the volume of the solution. The change will be for the acid to release protons to the conjugate base because there is more of the dihydrogen phosphate present.

Species	Initial Concentration	+ Change	= Equilibrium
$H_2PO_4^-{}_{(aq)}$	0.54 M	-x	0.54 - x
$H_3O^+{}_{(aq)}$	about 0	+ x	x
$HPO_4^{2-}{}_{(aq)}$	0.28 M	+ x	0.28 + x

Before substituting into the K_a expression, we will make the assumption that x << 0.28 M. The K_a expression now looks like:

$$K_a = \frac{[H_3O^+][HPO_4^{2-}]}{[H_2PO_4^-]} = 6.20 \times 10^{-8} \text{ M} = \frac{x(0.28 \text{ M})}{(0.54 \text{ M})}$$

$$x = 6.20 \times 10^{-8} \text{ M} \cdot \frac{0.54}{0.28} = 1.2 \times 10^{-7} = [H_3O^+]$$

Because 1.2×10^{-7} << 0.28, the approximation is valid

$$pH = -\log(1.2 \times 10^{-7}) = \boxed{6.93}$$

In a buffer solution, the pre-equilibrium concentrations of the acid and its conjugate base for practical purposes are the same as the equilibrium concentrations.

Chapter 16: Aqueous Equilibria

The defining property of a buffer solution is that addition of small amounts of a strong acid or base leads to only very small changes in pH compared to (1) pure water, (2) solutions containing strong acids or base or, (3) solutions containing only a weak acid or a weak base. In Exercise 16.13 we examine the difference in pH changes upon adding a strong base to the buffer in Exercise 16.12 and water.

Exercise 16.13 What is the pH upon adding 10 mL of 1.0 M NaOH to:
(a) 1.2 L of water and (b) the solution described above in Exercise 16.12?

Steps to Solution: *To calculate the pH of the solution in part (a), we first find [OH⁻] and then the pH by one of several routes; in this case we will convert [OH⁻] to pOH and then obtain the pH. In part (b), there is some chemistry going on; the strong base reacts essentially completely with the acid to decrease its concentration, leading to an increase in the concentration of the conjugate base:*

$$H_2PO_4^-{}_{(aq)} + OH^-{}_{(aq)} \rightarrow H_2O_{(l)} + HPO_4^{2-}{}_{(aq)}$$

If the change in concentration leaves the acid /conjugate base concentrations within a factor of 10, then we can utilize the above approximation of using the "initial condition" concentrations for the acid and the conjugate base.

(a) The major species are Na⁺ (a spectator ion) and OH⁻. The concentration of OH⁻ is found by using the definiton of molarity:

$$[OH^-] = \frac{\text{mole NaOH}}{\text{L solution}} = \frac{0.010 \text{ L} \cdot 1.0 \text{ mole/L}}{1.2 \text{ L} + 0.010 \text{ L}} = 8.3 \times 10^{-3} \text{ M}$$

$$pOH = -\log(8.3 \times 10^{-3}) = 2.08 \qquad pH = 14 - pOH = \boxed{11.92}$$

(b) The number of moles of NaOH added = $0.010 \text{ L} \cdot 1.0 \text{ mole/L} = 1.0 \times 10^{-2}$ mole. After addition of the base, the moles of $H_2PO_4^-$ will decrease by this amount:

$$\text{mole } H_2PO_4^- = 0.65 - 0.010 = 0.64 \text{ mole } H_2PO_4^-$$

and the moles of HPO_4^{2-} will increase :

$$\text{mole } HPO_4^{2-} = 0.34 + 0.010 = 0.35 \text{ mole } HPO_4^{2-}$$

The [H₃O⁺] can be found as in Exercise 16.12:

$$K_a = \frac{[H_3O^+][HPO_4^{2-}]}{[H_2PO_4^-]} = 6.20 \times 10^{-8} \text{ M} = \frac{x(0.35 \text{mole}/1.21 \text{ L})}{(0.64 \text{ mole}/1.21 \text{ L})}$$

$$x = 6.20 \times 10^{-8} \text{ M} \cdot \frac{0.53}{0.29} = 1.13 \times 10^{-7} \text{ M} = [\text{H}_3\text{O}^+] \qquad \text{pH} = -\log(1.13 \times 10^{-7}) = \boxed{6.94}$$

A very important point should be noted: upon the addition of identical amounts of base, the pH of water changed from 7.00 to 11.92 (nearly 5 pH units) whereas the pH of the buffer solution changed only 0.01 pH unit ! How does a buffer help a solution "resist" changes in pH?

How a Buffer Works

The diagrams that follow use a seesaw as an analogy to water and a buffered solution. In water, the pH-determining species are the minor species, the hydronium and hydroxide ions. Addition of strong acid decidedly "tips the balance" toward acid pH since all the hydronium ion added directly increases the hydronium ion concentration in the solution.

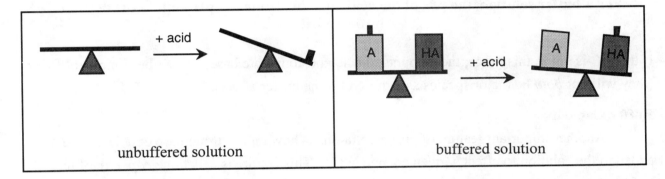

| unbuffered solution | buffered solution |

In a buffered solution, there are significant amounts of the weak acid and its conjugate base present. Addition of small amounts of strong acid leads to formation of more of the weak acid:

$$\text{H}_3\text{O}^+_{(aq)} + \text{A}^-_{(aq)} \rightarrow \text{HA}_{(aq)} + \text{H}_2\text{O}$$

which is not very effective at increasing hydronium ion concentration. In the seesaw analogy, the addition of small amounts of strong acid shifts a bit of the conjugate base to the acid side, but doesn't lead to much of a change in the position of equilibrium. Conversely, the addition of a strong base to a buffer solution converts some of the weak acid to its conjugate base, removing the hydroxide ion from solution. The buffer equation shows how the pH of a buffer solution depends on the concentrations of the acid and its conjugate base.

The Buffer Equation (also known as the Henderson-Hasselbach equation)

Because the quantity by which we measure acidity is pH, it is possible to rearrange the K_a expression so that pH is directly related to the pK_a and the concentrations of the acid and the conjugate base in a solution.

$$-\log\left(K_a = \frac{[\text{H}_3\text{O}^+][\text{A}^-]}{[\text{HA}]}\right) \qquad \text{gives} \qquad -\log K_a = -\log[\text{H}_3\text{O}^+] - \log\left(\frac{[\text{A}^-]}{[\text{HA}]}\right)$$

because $-\log K_a = pK_a$ and $-\log[H_3O^+] = pH$ we obtain $\quad pK_a = pH - \log\left(\dfrac{[A^-]}{[HA]}\right)$

Rearranging this equation, we obtain **the buffer equation**: $\quad pH = pK_a - \log\left(\dfrac{[HA]}{[A^-]}\right)$

The buffer equation is useful in the preparation of buffers as it helps in (1) the choice of the weak acid/conjugate base and (2) the ratio of their concentrations. Because the ratio of acid to conjugate base must be within the range of 0.1 to 10, this restricts the acid we may use in a buffer system (log 10 = 1, log 0.1 = -1).

For a buffer solution, the pK_a of the acid must be within ± 1 pH unit of the desired buffered pH

If the pK_a is not in this range, the amounts of acid and conjugate base will not be comparable (they will not *both* be major species as required for the buffer effect).

Buffer Capacity

Another important feature of buffer solutions is how much strong base or acid can be added to the solution to effect a given change in pH. This feature is called the *capacity* of the buffer; the more strong acid or base required, the greater the capacity of the buffer. Buffer capacity depends on the number of moles of the weak acid and its conjugate base present in the buffer:

$$pH = pK_a - \log\left(\dfrac{\text{mole HA}/V}{\text{mole A}^-/V}\right)$$

(V is the volume of the buffer solution)

Buffer capacity depends on:
- the concentration of the buffering agents (related to number of moles of HA and A⁻)
- the volume of the buffered solution (related to number of moles of HA and A⁻)
- the amount of pH variation that is tolerable (related to the pH sensitivity of the experiment)

The buffer changes pH as the ratio of acid to conjugate base changes with addition of strong acid and base. Remember that because both species are contained in the same solution, the volume, V, is the same and effectively cancels out of the expression:

$$pH = pK_a - \log\left(\dfrac{\text{mole HA}}{\text{mole A}^-}\right)$$

Chapter 16: Aqueous Equilibria

Exercise 16.14 Exercise 16.13 showed how the addition of a buffer resisted changes in pH with addition of some base. How many grams of NaOH would need to be added to increase the pH in the solution described in Exercise 16.12 by 0.2 pH units?

Steps to Solution: The point of this problem is to find how much strong base is required to increase the pH of a buffer by 0.2 pH units. The pH of a buffer is changed by varying the quotient of the moles of acid and conjugate base. First, we use the buffer equation to find the ratio of moles of acid to conjugate base. With this information we can execute the second step, determining the amount of base required for conversion of the acid to its conjugate base.

The important information from Exercise 16.12 is: initial pH = 6.93 and 0.34 mole Na_2HPO_4 and 0.65 mole NaH_2PO_4 in a total volume of 1.2 L. The target pH is 0.2 pH units higher: 6.93 + 0.2 = 7.1 The pK_a of the acid is 7.2. The buffer equation is used to find the ratio of acid to conjugate base at pH = 7.1

$$pH = pK_a - \log\left(\frac{\text{mole HA}}{\text{mole A}^-}\right)$$

$7.13 = 7.21 - \log\left(\frac{\text{mole HA}}{\text{mole A}^-}\right)$ gives $\log\left(\frac{\text{mole HA}}{\text{mole A}^-}\right) = 7.21 - 7.13 = 0.08$

$$\frac{\text{mole HA}}{\text{mole A}^-} = 10^{0.08} = 1.20$$

This is the ratio of the moles of acid and base that allows the number of moles of acid to be expressed in terms of the number of moles of conjugate base:

$$\text{mole HA} = 1.20 \text{ mole A}^-$$

The total number of moles of HA and A^- will be the same after addition of the base because nothing is being removed. The total number of moles of acid and conjugate base is found from the initial conditions:

$$\text{total moles} = \text{mole HA} + \text{mole A}^- = 0.65 + 0.34 = 0.99 \text{ mole}$$

At pH = 7.13, the sum of the moles of HA and A^- must be the same as it was before addition of base:

$$\text{total moles} = \text{mole HA} + \text{mole A}^- = 0.99 \text{ mole}$$

Since: $\text{mole HA} = 1.20 \text{ mole A}^-$

$$\text{total moles} = 1.20(\text{mole A}^-) + \text{mole A}^- = 0.99 \text{ mole} = 2.2 \text{ mole A}^-$$

619

$$\text{mole A}^- = \frac{0.99 \text{ mole}}{2.2} = 0.45 \text{ mole}$$

The number of moles of A⁻ has increased. The increase in moles of A⁻ is the same as the number of moles of NaOH added:

increase in mole A⁻ = mole NaOH = 0.45 - 0.34 = 0.11 mole NaOH required.

$$\text{g NaOH} = 0.11 \text{ mole}(40.00 \text{ g/mole NaOH}) = \boxed{4.4 \text{ g NaOH}}$$

It would require 4.4 g of NaOH to change the pH of the buffer solution by 0.2 pH units.

Buffer Preparation

There are several ways in which one can prepare a solution containing both a weak acid and its conjugate base. The more common ways are:

- mixing the acid and its conjugate base in solution (illustrated in Exercise 16.11)
- adding strong base to a solution of a weak acid (illustrated in Exercise 16.12)
- adding strong acid to a solution of a weak base

The choice of an acid-conjugate base pair is determined by the pH desired and the pK_a of the acid. The pK_a of the acid must be near the desired pH:

$$pK_a = pH \pm 1$$

The amounts of the various reagents used are determined by the *ratio of acid to conjugate base* obtained from the buffer equation and the *total number of moles* of acid and base. The total moles of acid and base is found from knowledge of the desired concentration and volume of the buffer to be prepared.

Exercise 16.15 In the selective precipitation of metal ions as sulfide salts, usable concentrations of the S^{2-} ions can be maintained by bubbling H_2S through a solution buffered at a pH of about 10.0. How many mL of 6 M HCl would need to be added to 3.5 mL of a solution 5.6 M in aqueous ammonia to form a solution buffered at pH 10.0?

Steps to Solution: In any buffer problem, the first step is to consider the chemistry that is occurring; in this reaction, the preparation of a buffer of a desired pH (10.0) is to be accomplished by the reaction of a weak base, ammonia, with HCl resulting in a solution containing both ammonia and the ammonium ion:

$$NH_{3\,(aq)} + HCl_{(aq)} \rightarrow NH_4^+{}_{(aq)} + Cl^-{}_{(aq)}$$

Chapter 16: Aqueous Equilibria

The second step is to determine the ratio of acid to conjugate base using the buffer equation:

$$pH = pK_a - \log\left(\frac{\text{mole HA}}{\text{mole A}^-}\right) \quad \text{gives} \quad \log\frac{\text{mole NH}_4^+}{\text{mole NH}_3} = pK_a - pH = 9.25 - 10.0$$

$$\log\frac{\text{mole NH}_4^+}{\text{mole NH}_3} = -0.75 \quad \text{gives} \quad \frac{\text{mole NH}_4^+}{\text{mole NH}_3} = 10^{-0.75} = 1.8 \times 10^{-1}$$

1 mole of NH_4^+ requires 1.8×10^{-1} mole of NH_3

or 1 mole of NH_3 requires 5.6 mole of NH_4^+.

We must determine the total moles of acid and conjugate base to be divided between the acid and conjugate base in the final buffer. In this problem the initial moles of ammonium (as a major species) is 0. The number of moles of NH_3 is:

$$\text{moles NH}_3 = M(V) = 5.6\,M(3.5 \times 10^{-3}\,L) = 2.0 \times 10^{-2} \text{ mole NH}_3$$

$$\text{total moles (NH}_3 \text{ and NH}_4^+) = 2.0 \times 10^{-2} + 0 = 2.0 \times 10^{-2} \text{ moles NH}_3 \text{ and NH}_4^+.$$

The moles of ammonium ion will be equal to the moles of HCl that need to be added to form the buffer. The equation will be solved in terms of moles of NH_4^+.

$$\text{total moles} = 2.0 \times 10^{-2} = \text{moles NH}_3 + \text{moles NH}_4^+$$

From the mole ratio; 1 mole NH_3 = 5.6 mole of NH_4^+

$$2.0 \times 10^{-2} = 5.6\,(\text{mole NH}_4^+) + (\text{mole NH}_4^+) = 6.6\,(\text{mole NH}_4^+)$$

$$\text{mole NH}_4^+ = \frac{2.0 \times 10^{-2}\,\text{mol}}{6.6} = 3.0 \times 10^{-3} \text{ mole} = \text{mole HCl required}$$

$$\text{mole} = M(V) \Rightarrow V = \frac{3.0 \times 10^{-3}\,\text{mol}}{6\,\text{mol/L}} = 5.0 \times 10^{-4}\,L = \boxed{5.0 \times 10^{-1}\,\text{mL}}$$

About 0.5 mL of acid will be needed.

16.6 Acid-Base Titrations

QUESTIONS TO ANSWER, SKILLS TO LEARN
1. Calculating the pH in a titration of acids and bases as the reactant (a strong base or strong acid, respectively) is added
2. What is the stoichiometric point in a titration?
3. Using the estimated pH at the stoichiometric point to choose a indicator to detect the end of the titration

In an acid-base titration, the number of moles of a substance in a sample is found by the addition of solution of a known concentration reactant (called the **titrant**) whose volume is measured to high precision. During the titration, the major species change three times so that four different stages of the titration can be delineated because of the different major species that are present. Calculations of the pH for acid-base titrations where either the acid or base is weak and the titrant is strong can be performed by sequentially considering these four different stages. The table on the following page summarizes these stages, describing the major species present at each stage and outlines the mathematical tools to solve for the pH. The amounts of substances are described in concentration or number of moles, depending on which simplifies the calculations. Exercise 16.16 illustrates these calculations.

Exercise 16.16 Calculate the pH in the titration of an 0.325 g sample of acetylsalicylic acid (see line drawing below) initially in 25.0 mL water with 0.102 M NaOH when (a) 0 mL (b) 6.23 mL (c) 8.84 mL (d) 17.68 mL and (e) 20.00 mL of titrant have been added.

(structure of acetylsalicylic acid)	HA = acetylsalicylic acid (aspirin) $C_9H_8O_4$ $K_a = 3.00 \times 10^{-4}$ M

Steps to Solution: This question is a series of acid-base equilibrium problems. Use the method of finding the major species in solution to select the appropriate equilibrium expressions to use.

(a) When no NaOH has been added the problem is a weak acid problem; we find the $[H_3O^+]$ using the K_a expression. The initial concentration of the acid must be determined to set up the table of concentrations.

Chapter 16: Aqueous Equilibria

Species	Initial Condition	+ Change	= Equilibrium
$HA_{(aq)}$	$[HA] = \dfrac{\dfrac{0.325 \text{ g}}{180.2 \text{ g/mol}}}{0.025 \text{ L}}$ $[HA] = 7.22 \times 10^{-2}$ M	$-x$ M	$(7.22 \times 10^{-2} - x)$ M
$H_3O^+_{(aq)}$	near 0	$+x$ M	x M
$A^-_{(aq)}$	0	$+x$ M	x M

Substitution gives:

$$K_a = \frac{x^2}{7.72 \times 10^{-2} - x} = 3.00 \times 10^{-4} \text{ M}$$

The usual approximation will be found to be invalid in this problem; the quadratic equation obtained by expanding this expression is:

$$x^2 + 3.00 \times 10^{-4} x - 2.165 \times 10^{-5} = 0$$

The physically reasonable root is $x = 4.50 \times 10^{-3}$ M $= [H_3O^+]$; $\boxed{pH = 2.35}$

(2 significant figures because the volume had only 2 significant figures)

(b) *When 6.23 mL of NaOH have been added, the strong base reacts nearly completely with the weak acid to give the conjugate base, so the major species in the solution are the weak acid, the conjugate base, Na^+, and water. In this situation, the buffer equation can be applied to calculate the pH.*

$$pH = pK_a - \log\left(\frac{\text{mole HA}}{\text{mole A}^-}\right)$$

mole A^- = mole OH^- added = 6.23×10^{-3} L$(0.102$ mole/L$) = 6.35 \times 10^{-4}$ mole

mole HA = initial mole HA − mole NaOH = $1.80 \times 10^{-3} - 6.35 \times 10^{-4}$
= 1.17×10^{-3} mole HA

$$pH = 3.523 - \log\left(\frac{1.17 \times 10^{-3} \text{ mole}}{6.35 \times 10^{-4} \text{ mole}}\right) = \boxed{3.258}$$

(c) *After 8.84 mL of NaOH solution have been added, the major species are the same as in part (b) so the problem may be solved in the same way.*

mole A^- = mole OH^- added = 8.84×10^{-3} L$(0.102$ mole/L$) = 9.02 \times 10^{-4}$ mole A^-

Chapter 16: Aqueous Equilibria

mole HA = initial mole HA - mole NaOH = $1.80 \times 10^{-3} - 9.02 \times 10^{-4} = 9.0 \times 10^{-3}$ mole HA

Note that the number of moles of acid and conjugate base are the same. This means that the titration is halfway to the stoichiometric point (sometimes called the half equivalence point)!

$$pH = 3.523 - \log\left(\frac{9.0 \times 10^{-4} \text{ mole}}{9.0 \times 10^{-4} \text{ mole}}\right) = \boxed{3.52} = pK_a$$

The pH at the half equivalence point in the titration of a weak acid is the pK_a of the acid!

(d) *First we need to determine the major species after 17.68 mL of NaOH solution have been added:*

mole A⁻ = mole OH⁻ added = 17.68×10^{-3} L(0.102 mole/L) = 1.80×10^{-3} mole

mole HA = initial mole HA - mole NaOH = $1.80 \times 10^{-3} - 1.80 \times 10^{-3}$ = "0 mole" HA

*At this point the number of moles of base added is equal to the number of moles initially present in the solution. This is the **stoichiometric point** (sometimes called the equivalence point) of the titration.*

Stoichiometric point: the volume at which the moles of titrant added react completely with the substance being titrated

Because the major species in solution are A⁻, Na⁺ (a spectator ion), and water, this problem is most easily solved by using the equilibrium of the weak base, A⁻, with water:

$$A^- + H_2O \rightleftharpoons HA + OH^- \qquad K_b = \frac{K_w}{K_a} = \frac{[HA][OH^-]}{[A^-]} = 3.33 \times 10^{-11} \text{ M}$$

Filling out the concentration table gives:

Species	Initial Condition	+ Change	= Equilibrium
A⁻ (aq)	$[A^-] = \dfrac{1.80 \times 10^{-3} \text{ mole}}{0.025 + 0.177 \text{ L}}$ $[A^-] = 4.23 \times 10^{-2}$ M	- x M	$(4.23 \times 10^{-2} - x)$ M
OH⁻ (aq)	near 0	+ x M	x M
HA (aq)	0	+ x M	x M

Substitution into the K_b equation gives:

$$3.33 \times 10^{-11} \text{ M} = \frac{[HA][OH^-]}{[A^-]} = \frac{x^2}{4.23 \times 10^{-2} - x}$$

for $x \ll 4.23 \times 10^{-2}$ M

$x = [OH^-] = \sqrt{3.33 \times 10^{-11} \text{ M} (4.23 \times 10^{-2} \text{ M})} = 1.2 \times 10^{-6}$ M (the approximation is valid)

([OH^-] is rounded to 2 significant figures because 2 significant figures are in the volume measurement (25 mL).)

$$pOH = 5.93 \quad \text{and} \quad pH = 14 - pOH = \boxed{8.07}$$

The pH at the stoichiometric point of a titration of a weak acid is not 7.00 because at the stoichiometric point a solution of a weak base has been formed.

(e) *At this stage in the titration, more strong base has been added than was required to neutralize the acetylsalicylic acid. The major species in solution are Na^+ (a spectator ion), A^- (a weak base) and OH^- (a strong base). The moles of strong base present will determine the pH of the solution.*

mole OH^- added = 20.00×10^{-3} L(0.102 mole/L) = 2.04×10^{-3} mole OH^- added

leftover OH^- = mole NaOH - initial mole HA = $2.04 \times 10^{-3} - 1.80 \times 10^{-3}$ mole
leftover OH^- = 2.4×10^{-4} mole OH^-

$$[OH^-] = \frac{2.4 \times 10^{-4} \text{ mole } OH^-}{0.025 + 0.020 \text{ L}} = 5.2 \times 10^{-3} \text{ M} \quad pOH = 2.28 \quad \boxed{pH = 11.72}$$

A qualitative plot of the pH vs. mL NaOH curve (titration curve) is shown below.

Indicators

A large number of compounds that are weak acids have the property that the weak acid and the conjugate base have distinctly different colors. These can be used to indicate when the pH of a solution is a certain value and are thus called **indicators**. For example, phenolphthalein is a weak acid with $pK_a = 9.0$. If a small quantity of phenolphthalein is placed in a solution (it is a minor species) containing much more of another weak acid, then the pH of the solution is determined by the major species. The relative amounts of the indicator, HIn, and its conjugate base, In⁻, give the solution its color and are determined by the pH of the solution.

$$HIn \rightleftharpoons H_3O^+ + In^-$$

$$\log\left(\frac{mol\,HIn}{mol\,In^-}\right) = pK_a - pH \quad \text{gives} \quad \left(\frac{mol\,HIn}{mol\,In^-}\right) = 10^{(pK_a - pH)}$$

The plot on page 454 shows how the mole fractions of In⁻ and HIn change with pH for phenolphthalein. At pH less than 7.5, nearly all the indicator is in the colorless acid form and the solution is colorless (mole fraction$_{In^-} \approx 0$, mole fraction$_{HIn} \approx 1$). As the pH increases, more of the intensely red-purple In⁻ form is present, which may be detected at a pH of about 8.0. As the pH grows higher yet, most of the phenolphthalein is present as the conjugate base.

Chapter 16: Aqueous Equilibria

Stage of titration	Weak Acid/Strong Base Titration	Weak Base/Strong Acid Titration
Major species: Initially (when no titrant is added)	weak acid $K_a = \dfrac{[H^+][A^-]}{[HA]}$ pH = $-\log[H^+]$	weak base $K_b = \dfrac{[OH^-][HA]}{[A^-]}$ pH = 14.00 − pOH
Major species: Buffer region (titrant added, but before equivalence point)	weak acid and conjugate base HA + OH⁻ → H₂O + A⁻ mole A⁻ = mole OH⁻ added mole HA = initial mole HA − mole OH⁻ added Solve using the buffer equation	weak base and conjugate acid A⁻ + H⁺ → HA mole HA = mole H⁺ added mole A⁻ = initial mole A⁻ − mole H⁺ added Solve using the buffer equation
Major species: major process Equivalence point	salt of weak acid hydrolysis by A⁻ $K_b = \dfrac{[OH^-][HA]}{[A^-]}$ mole A⁻ = initial mole HA	weak acid dissociation of HA $K_a = \dfrac{[H^+][A^-]}{[HA]}$ mole HA = initial mole A⁻
Major species: After equivalence point	salt of weak acid and excess strong base $[OH^-] = \dfrac{\text{mole excess OH}^-}{\text{total } V}$ excess OH⁻ = mole OH⁻ added − initial mole HA	weak acid and excess strong acid $[H^+] = \dfrac{\text{mole excess H}^+}{\text{total } V}$ excess H⁺ = mole OH⁻ − initial mole A⁻

This is the fundamental reason for the color change of an indicator: the relative amounts of the acid and the conjugate base change, leading to a change in the color of the solution. To signal the stoichiometric point of a titration, the pH at which the color change occurs must be near the pH of the stoichiometric point. Table 16-6 in your text lists a number of indicators and the pH range over which they change color.

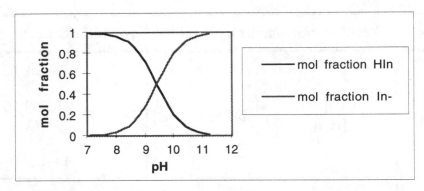

Exercise 16.17 From the list of indicators given in the text, choose two indicators that would be appropriate for detecting the stoichiometric point in a titration of salicylic acid (see Exercise 16.16).

Steps to Solution: An appropriate indicator is one that changes color at a pH near the stoichiometric point. The pH of the stoichiometric point was found to be 8.07 in part (d) of Exercise 16.16. Three indicators in the list change color near pH 8:

phenol red	6.4 (yellow) - 8.0 (red)
thymol blue	8.0 (yellow) - 9.6 (blue)
phenolphthalein	8.0 (colorless) - 10.0 (red-purple)

In this titration, the pH will be increasing, so the color change will be nearly complete at the stoichiometric point for phenol red; however, for thymol blue and phenolphthalein, the color change is just beginning. It will probably be easier to detect the color change for phenolphthalein and thymol blue than for phenol red.

The better indicators would be phenolphthalein and thymol blue.

Test Yourself

1. Predict the order of acidity of the following compounds: H_2O, H_2S, H_2Se.

2. The pH of a 0.30 M solution of a weak base, B, is 10.6 What is the K_b of the base?

3. The K_a of HNO_2 is 5.0×10^{-4}. What would be the pH of an 0.10 M solution of KNO_2?

4. A 40.0 mL sample of 0.10 M HNO_3 is added to 20 mL of 0.30 M aqueous NH_3. What is the pH of the resulting solution?

5. Consider the titration of 25.0 mL of an 0.11 M solution of lactic acid $CH_3CH(OH)COOH$ ($K_a = 8.4 \times 10^{-4}$ M) with 0.150 M NaOH.
   (a) Calculate the initial pH of the solution.
   (b) Calculate the pH after 10 mL of the NaOH solution has been added.
   (c) Calculate the pH of the solution at the stoichiometric point.
   (d) Calculate the pH of the solution at 10 mL of NaOH past the stoichiometric point.
   (e) Would phenol red or phenolphthalein be a better indicator for this titration?

Answers

1. $H_2O > H_2S > H_2Se$; the acidity parallels the electronegativity of the inner atom.

2. 7.1×10^{-7} M

3. pH = 8.15

4. pH = 8.95

5. 2.02 (b) 3.15 (c) 7.94 (d) 12.45 (e) phenol red; phenolphthalein would just be starting to change color after the stoichiometric point.

Solutions to Chapter Problems

16.1 To identify species in solution, first identify the nature of the solute. Strong acids, strong bases, and salts generate ions, while all other substances remain molecular:
(a) Weak acid, major species are H_2O and CH_3CO_2H; (b) salt, major species are H_2O, NH_4^+ and Cl^-; (c) salt, major species are H_2O, K^+ and Cl^-; (d) salt, major species are H_2O, Na^+ and $CH_3CO_2^-$; and (e) strong base, major species are H_2O, Na^+ and OH^-.

16.3 The equilibria among major species depend on the nature of the species. The proton transfer reaction of water always plays a role:
$H_2O_{(l)} + H_2O_{(l)} \rightleftharpoons H_3O^+_{(aq)} + OH^-_{(aq)}$ $K_w = 1.00 \times 10^{-14}$
(a) weak acid equilibrium:
$CH_3CO_2H_{(aq)} + H_2O_{(l)} \rightleftharpoons CH_3CO_2^-_{(aq)} + H_3O^+_{(aq)}$ $K_a = 1.8 \times 10^{-5}$, $K_a > K_w$
(b) ammonium ion is the conjugate base of NH_3, so is a weak acid:
$NH_4^+_{(aq)} + H_2O_{(l)} \rightleftharpoons NH_3_{(aq)} + H_3O^+_{(aq)}$ $K_a = 5.6 \times 10^{-10}$, $K_a > K_w$
(c) There are no equilibria other than the water equilibrium;
(d) acetate ion is the conjugate base of acetic acid, so is a weak base:
$CH_3CO_2^-_{(aq)} + H_2O_{(l)} \rightleftharpoons CH_3CO_2H_{(aq)} + OH^-_{(aq)}$ $K_b = 5.6 \times 10^{-10}$, $K_b > K_w$
(e) There are no equilibria other than the water equilibrium.

16.5 To identify species in solution, first identify the nature of the solute. Strong acids, strong bases, and salts generate ions, while all other substances remain molecular:
(a) major species are $(CH_3)_2CO$ (acetone) and H_2O; (b) salt, major species are H_2O, K^+, and Br^-; (c) strong base, major species are H_2O, Li^+, and OH^-; (d) strong acid, major species are H_2O, H_3O^+, and HSO_4^-.

Chapter 16: Aqueous Equilibria

16.7 To write a solubility product equilibrium expression, first determine the chemical formula and stoichiometry of the salt:

(a) $AgCl_{(s)} \rightleftharpoons Ag^+_{(aq)} + Cl^-_{(aq)}$ $\quad K_{sp} = [Ag^+]_{eq}[Cl^-]_{eq}$

(b) $BaSO_{4(s)} \rightleftharpoons Ba^{2+}_{(aq)} + SO_4^{2-}_{(aq)}$ $\quad K_{sp} = [Ba^{2+}]_{eq}[SO_4^{2-}]_{eq}$

(c) $Fe(OH)_{2(s)} \rightleftharpoons Fe^{2+}_{(aq)} + 2\ OH^-_{(aq)}$ $\quad K_{sp} = [Fe^{2+}]_{eq}[OH^-]^2_{eq}$

(d) $Ca_3(PO_4)_{2(s)} \rightleftharpoons 3\ Ca^{2+}_{(aq)} + 2\ PO_4^{3-}_{(aq)}$ $\quad K_{sp} = [Ca^{2+}]^3_{eq}[PO_4^{3-}]^2_{eq}$.

16.9 "Determine the mass that dissolves" means "calculate the amount present in solution at equilibrium". Initial amounts are zero, so it is easy to complete a concentration table. Calculate molarity from the solubility product expression, then convert to mass using standard stoichiometric methods. Let x be moles of salt dissolving in 1 L of solution.

(a) Species: $\quad AgCl_{(s)} \rightleftharpoons \quad Ag^+_{(aq)} + \quad Cl^-_{(aq)}$
Conc. at equilibrium (M): $\quad\quad\quad\quad\quad\quad\quad$ x $\quad\quad\quad$ x

$K_{sp} = 1.8 \times 10^{-10} = [Ag^+]_{eq}[Cl^-]_{eq} = x^2$, from which $x = 1.3 \times 10^{-5}$ M;
$m = MV\ MM = (1.3 \times 10^{-5}\ M)(0.475\ L)(143.4\ g/mol) = 8.8 \times 10^{-4}$ g;

(b) Species: $\quad BaSO_{4(s)} \rightleftharpoons \quad Ba^{2+}_{(aq)} + \quad SO_4^{2-}_{(aq)}$
Conc. at equilibrium (M): $\quad\quad\quad\quad\quad\quad\quad$ x $\quad\quad\quad$ x

$K_{sp} = 1.1 \times 10^{-10} = [Ba^{2+}]_{eq}[SO_4^{2-}]_{eq} = x^2$, from which $x = 1.0 \times 10^{-5}$ M;
$m = MV\ MM = (1.0 \times 10^{-5}\ M)(0.475\ L)(233.37\ g/mol) = 1.1 \times 10^{-3}$ g;

(c) Species: $\quad Fe(OH)_{2(s)} \rightleftharpoons \quad Fe^{2+}_{(aq)} + \quad 2\ OH^-_{(aq)}$
Conc. at equilibrium (M): $\quad\quad\quad\quad\quad\quad\quad$ x $\quad\quad\quad$ 2x

$K_{sp} = 7.9 \times 10^{-16} = [Fe^{2+}]_{eq}[OH^-]^2_{eq} = 4x^3$, from which $x = 5.8 \times 10^{-6}$ M;
$m = MV\ MM = (5.8 \times 10^{-6}\ M)(0.475\ L)(89.87\ g/mol) = 2.5 \times 10^{-4}$ g;

(d) Species: $\quad Ca_3(PO_4)_{2(s)} \rightleftharpoons \quad 3\ Ca^{2+}_{(aq)} + \quad 2\ PO_4^{3-}_{(aq)}$
Conc. at equilibrium (M): $\quad\quad\quad\quad\quad\quad\quad$ 3x $\quad\quad\quad$ 2x

$K_{sp} = 7.9 \times 10^{-16} = [Ca^{2+}]^3_{eq}[PO_4^{3-}]^2_{eq} = 108\ x^5$, from which $x = 7.1 \times 10^{-7}$ M;
$m = MV\ MM = (7.1 \times 10^{-7}\ M)(0.475\ L)(310.20\ g/mol) = 1.0 \times 10^{-4}$ g;

16.11 To calculate a solubility product from the mass that dissolves in solution, convert to molarity of ions and then evaluate the equilibrium constant expression:

$CaC_2O_{4(s)} \rightleftharpoons Ca^{2+}_{(aq)} + C_2O_4^{2-}_{(aq)}\quad K_{sp} = [Cu^{2+}]_{eq}[OH^-]_{eq}[Ca^{2+}]_{eq}[C_2O_4^{2-}]_{eq}$
$n = m/MM = (6.1\ mg)(10^{-3}\ g/mg) \div (128.1\ g/mol) = 4.76 \times 10^{-5}$ mol;
$M = n/V = (4.76 \times 10^{-5}\ mol) \div (1.0\ L) = 4.76 \times 10^{-5}$ mol/L;
$K_{sp} = [Cu^{2+}]_{eq}[OH^-]_{eq}(4.76 \times 10^{-5})^2 = 2.3 \times 10^{-9}$.

Chapter 16: Aqueous Equilibria

16.13 To determine the ion concentrations in a solution, follow the five step procedure for solving an equilibrium problem:
1.) The species in solution are Cu^{2+}, NO_3^-, K^+, OH^-, and H_2O.
2, 3.) Among these species, the only reaction is formation of $Cu(OH)_2$ precipitate:

$$Cu^{2+}_{(aq)} + 2\ OH^-_{(aq)} \rightleftharpoons Cu(OH)_{2\ (s)} \quad K_{eq} = 1/K_{sp} = 1/[Cu^{2+}]_{eq}[OH^-]^2_{eq}$$

4.) Initial concentrations are those present after mixing but before reaction occurs:

$$[Cu^{2+}] = (0.300\ M)\frac{(0.500\ L)}{(0.500\ L\ +\ 0.500\ L)} = 0.150\ M;\ [OH^-] = 0.200\ M;$$

Take the reaction to completion and then return to equilibrium:

Species:	$Cu^{2+}_{(aq)}\ +$	$2\ OH^-_{(aq)}$	$\rightleftharpoons Cu(OH)_{2\ (s)}$
Initial conc. (M):	0.150	0.200	
Change to completion (M):	-0.100	-0.200	
Conc. at completion (M):	0.050	0	
Change to equilib (M):	y	2 y	
Equilibrium conc. (M):	$0.050 + y$	2 y	

5.) Assume that $y \ll 0.050$, substitute into the solubility product expression, and solve:

$$K_{sp} = 2.2 \times 10^{-20} = [Cu^{2+}]_{eq}[OH^-]^2_{eq} = (0.050)(2\ y)^2$$

$y^2 = (2.2 \times 10^{-20}) \div (0.20) = 1.1 \times 10^{-19}$, from which $y = 3.3 \times 10^{-10}$;

$[Cu^{2+}]_{eq} = 0.050\ M$; $[OH^-]_{eq} = 2y = 6.6 \times 10^{-10}\ M$; $[NO_3^-] = 0.300\ M$; $[K^+] = 0.200\ M$.

16.15 Conversion from hydronium ion molarity to pH is accomplished by taking logarithm to base ten and changing sign, $pH = -(\log[H_3O^+])$: (a) -0.60; (b) 5.43; (c) 2.32; and (d) 11.13.

16.17 Use $pH + pOH = 14.00$ to convert pH to pOH. Then take 10^{-pOH} to convert pOH into hydroxide ion concentration: (a) $pOH = 13.34$, $[OH^-] = 4.6 \times 10^{-14}\ M$; (b) $pOH = 6.15$, $[OH^-] = 7.1 \times 10^{-7}\ M$; (c) $pOH = 10.32$, $[OH^-] = 4.8 \times 10^{-11}\ M$; (d) $pOH = -0.33$, $[OH^-] = 2.1\ M$.

16.19 Examine the species present to determine the nature of a solution: (a) weak acid; (b) Although NH_4Cl is a salt, NH_4^+ is the conjugate acid of NH_3, so the solution is a weak acid; (c) salt; (d) although $NaCH_3CO_2$ is a salt, $CH_3CO_2^-$ is the conjugate base of acetic acid, so the solution is a weak base; (e) strong base.

16.21 Examine the chemical formula to determine the nature of a compound: (a) weak base; (b) strong acid; (c) weak acid; (d) strong base.

16.23 Obtain the conjugate base of an acid by removing a proton; obtain the conjugate acid of a base by adding a proton: 16.1(a), CH_3CO_2H is a weak acid, $CH_3CO_2^-$ is the conjugate base; 16.1(b), NH_4^+ is a weak acid, NH_3 is the conjugate base; 16.1(d), $CH_3CO_2^-$ is a weak base, CH_3CO_2H is the conjugate acid; 16.21(a) NH_3 is a weak base, NH_4^+ is the conjugate acid; 16.21(c), $HClO$ is a weak acid, ClO^- is the conjugate base.

16.25 To calculate the pH of a solution, it is necessary to determine either the hydronium ion concentration or the hydroxide ion concentration.

(a) strong base, so $[OH^-] = 1.5$ M,
pOH = -0.18, pH = 14.18;

(b) weak base, carry out equilibrium calculation to determine $[OH^-]$:

Species:	H_2O +	C_5H_5N	⇌	$C_5H_5NH^+$	+	OH^-
Initial concentration (M):		1.5		0		0
Change to equilib (M)		$-x$		x		x
Equilib concentration (M)		$1.5 - x$		x		x

$pK_b = 8.72$, $K_b = 1.91 \times 10^{-9} = \dfrac{[C_5H_5NH^+]_{eq}[OH^-]_{eq}}{[C_5H_5N]_{eq}}$; Assume $x \ll 1.5$: $1.91 \times 10^{-9} = \dfrac{x^2}{1.5}$, so $x^2 = 2.86 \times 10^{-9}$ and $x = 5.35 \times 10^{-5}$; $[OH^-] = 5.35 \times 10^{-5}$ M, pOH = 4.27, and pH = 9.73;

(c) weak base, carry out equilibrium calculation to determine $[OH^-]$:

Species:	H_2O +	NH_2OH	⇌	NH_3OH^+	+	OH^-
Initial concentration (M):		1.5		0		0
Change to equilib (M)		$-x$		x		x
Equilib concentration (M)		$1.5 - x$		x		x

$pK_b = 7.96$, $K_b = 1.10 \times 10^{-8} = \dfrac{[NH_3OH^+]_{eq}[OH^-]_{eq}}{[NH_2OH]_{eq}}$; Assume $x \ll 1.5$:

$1.10 \times 10^{-8} = \dfrac{x^2}{1.5}$, from which $x^2 = 1.65 \times 10^{-8}$ and $x = 1.28 \times 10^{-4}$;
$[OH^-] = 1.28 \times 10^{-4}$ M, pOH = 3.89, and pH = 10.11;

(d) weak acid, carry out equilibrium calculation to determine $[H_3O^+]$:

Species:	H_2O +	HCO_2H	⇌	HCO_2^-	+	H_3O^+
Initial concentration (M):		1.5		0		0
Change to equilib (M)		$-x$		x		x
Equilib concentration (M)		$1.5 - x$		x		x

$pK_a = 3.74$, $K_a = 1.82 \times 10^{-4} = \dfrac{[HCO_2^-]_{eq}[H_3O^+]_{eq}}{[HCO_2H]_{eq}}$; Assume $x \ll 1.5$: $1.82 \times 10^{-4} = \dfrac{x^2}{1.5}$, from which $x^2 = 2.73 \times 10^{-4}$ and $x = 1.65 \times 10^{-2}$; $[H_3O^+] = 1.65 \times 10^{-2}$ M, and pH = 1.78.

Chapter 16: Aqueous Equilibria

16.27 Follow standard procedures for dealing with equilibrium problems:
(a) $N(CH_3)_3$ is a weak base. Major species: $N(CH_3)_3$, H_2O, Minor species: $HN(CH_3)_3^+$, OH^-;
(b) Carry out an equilibrium calculation to determine $[OH^-]$:

Species:	H_2O +	$N(CH_3)_3$	$\rightleftharpoons$	$HN(CH_3)_3^+$ +	OH^-
Initial concentration (M):		0.350		0	0
Change to equilib (M)		$-x$		x	x
Equilib concentration (M)		$0.350 - x$		x	x

$pK_b = 4.19$, $K_b = 6.46 \times 10^{-5} = \dfrac{[HN(CH_3)_3^+]_{eq}[OH^-]_{eq}}{[N(CH_3)_3]_{eq}}$; Assume $x \ll 0.350$:

$6.46 \times 10^{-5} = \dfrac{x^2}{0.350}$, from which $x^2 = 2.26 \times 10^{-5}$ and $x = 4.75 \times 10^{-3}$;

$[OH^-] = [HN(CH_3)_3^+] = 4.75 \times 10^{-3}$ M, $[N(CH_3)_3] = 0.350 - x = 0.345$ M;
(c) pOH = 3.32, and pH =11.68;
(d) The dominant equilibrium is proton transfer from water to trimethylamine:

16.29 Follow standard procedures for dealing with equilibrium problems:
(a) The compound is a salt, so major species are NH_4^+, NO_3^-, and H_2O;
(b) The species with acid-base properties are NH_4^+ (a weak acid) and H_2O, so the dominant equilibrium is: $H_2O + NH_4^+ \rightleftharpoons NH_3 + H_3O^+$
(c) Carry out equilibrium calculation to determine $[H_3O^+]$:

Species:	H_2O +	NH_4^+	$\rightleftharpoons$	NH_3 +	H_3O^+
Initial concentration (M):		0.0100		0	0
Change to equilib (M)		$-x$		x	x
Equilib concentration (M)		$0.0100 - x$		x	x

$pK_a = 9.25$, $K_a = 5.6 \times 10^{-10} = \dfrac{[NH_4^+]_{eq}[H_3O^+]_{eq}}{[NH_3]_{eq}}$; Assume $x \ll 0.0100$:

$5.6 \times 10^{-10} = \dfrac{x^2}{(0.0100)}$, from which $x^2 = 5.6 \times 10^{-12}$ and $x = 2.37 \times 10^{-6}$;

$[H_3O^+] = 2.37 \times 10^{-6}$ M, and pH = 5.63.

16.31 Tabulated equilibrium constants for acid-base reactions always refer to reactions in which H_2O is one of the reactants. The reaction in this problem is the reverse of a base reaction:
HPO_4^{2-} (aq) + OH^- (aq) $\rightleftharpoons$ PO_4^{3-} (aq) + H_2O (l)
Table 16-3 lists pK_a values for phosphoric acid:
HPO_4^{2-} (aq) + H_2O (l) $\rightleftharpoons$ PO_4^{3-} (aq) + H_3O^+ (aq) $pK_{a3} = 12.32$
K_a and K_b for a conjugate acid-base pair are related through $pK_a + pK_b = 14.00$:
$pK_b = 14.00 - 12.32 = 1.68$, and $K_b = 2.1 \times 10^{-2}$; thus $K_{eq} = 1/K_b = 48$

16.33 A buffer solution must contain both a weak acid and its conjugate weak base.
(a) The strong base reacts with NH$_4^+$, a weak acid, to generate the conjugate weak base. There is excess strong base (0.0375 mol compared with 0.025 mol), so no weak acid remains after mixing and this solution does not have buffer properties.

(b) The strong base reacts completely with the weak acid to generate the conjugate weak base.

Species:	NH$_4^+$ +	OH$^-$ ⇌	NH$_3$ +	H$_2$O
Initial amount (mol):	0.025	0.0125	0	
Change in amount (mol):	− 0.0125	− 0.0125	+ 0.0125	
Final amount (mol)	0.0125	0	0.0125	

This is an NH$_4^+$/NH$_3$ buffer. Use the buffer equation, Equation 16-5, with amounts:

$$pH = pK_a + \log\left\{\frac{(\text{mol } A^-)}{(\text{mol } HA)}\right\}, pK_a = 9.25, \text{ so } pH = 9.25 + \log\left\{\frac{(0.0125 \text{ mol})}{(0.0125 \text{ mol})}\right\} = 9.25;$$

(c) NH$_4^+$ is a weak acid and HCl is a strong acid. There is no conjugate base present in high concentration, so this solution does not have buffer properties;

(d) NH$_3$ is a weak base and HCl is a strong acid. The strong acid reacts completely with the weak base to generate the conjugate weak acid.

Species:	NH$_3$ +	H$_3$O$^+$ ⇌	NH$_4^+$ +	H$_2$O
Initial amount (mol):	0.025	0.0125	0	
Change in amount (mol):	− 0.0125	− 0.0125	+ 0.0125	
Final amount (mol)	0.0125	0	0.0125	

This is an NH$_4^+$/NH$_3$ buffer. Use the buffer equation, Equation 16-5, with amounts:

$$pH = pK_a + \log\left\{\frac{(\text{mol } A^-)}{(\text{mol } HA)}\right\} \quad pK_a = 9.25, \text{ so } pH = 9.25 + \log\left\{\frac{(0.0125 \text{ mol})}{(0.0125 \text{ mol})}\right\} = 9.25.$$

16.35 The two solutions that are buffered are (b) and (c). The concentration tables for these solutions show that they are identical in acid/base composition, despite being prepared differently. Thus only one calculation is required. The added strong acid reacts completely with the weak base of the buffer to generate the conjugate weak acid.

Species:	NH$_3$ +	H$_3$O$^+$ ⇌	NH$_4^+$ +	H$_2$O
Initial amount (mol):	0.0125	0.005	0.0125	
Change in amount (mol):	− 0.005	− 0.005	+ 0.005	
Final amount (mol)	0.0075	0	0.0175	

Use the buffer equation, Equation 16-5, with the new amounts:

$$pH = pK_a + \log\left\{\frac{(\text{mol } A^-)}{(\text{mol } HA)}\right\} = 9.25 + \log\left\{\frac{(0.0075 \text{ mol})}{(0.0175 \text{ mol})}\right\} = 9.25 - 0.37 = 8.88.$$

16.37 The two solutions that are buffered are (b) and (c). The concentration tables for these solutions show that they are identical in acid/base composition, despite being prepared differently. Thus only one calculation is required. Addition of strong base consumes weak acid, produces weak base, and increases the pH of the buffer. First use the buffer equation to calculate the acid/base ratio in a solution whose pH is greater by 0.10 unit than the initial buffer solution. Then set up a concentration ratio, using x = the amount of added base, and solve for x:

new pH = 9.25 + 0.10 = 9.35; [NH_3] = 0.0125 + x, [NH_4^+] = 0.0125 − x;

$9.35 = 9.25 + \log\left\{\frac{(0.0125 + x)}{(0.0125 - x)}\right\}$, $\log\left\{\frac{(0.0125 + x)}{(0.0125 - x)}\right\} = 0.10$, $\left\{\frac{(0.0125 + x)}{(0.0125 - x)}\right\} = 1.26$

(0.0125 + x) = 1.26(0.0125 − x) = 0.0157 − 1.26 x; 0.0032 = 2.26 x, and x = 1.2×10^{-3} mol.

16.39 A molecular picture of a buffer solution should show molecules/ions of the conjugate acid/base pair in the correct proportions, as determined by the pH of the buffer solution. For formic acid, pK_a = 3.74. Use the buffer equation to calculate the base/acid ratio of a formic acid buffer with pH = 4.04:

$\log\left\{\frac{(\text{mol } A^-)}{(\text{mol } HA)}\right\} = pH - pK_a = 4.04 - 3.74 = 0.30$; $\left\{\frac{(\text{mol } A^-)}{(\text{mol } HA)}\right\} = 2.0$.

The picture should show twice as many formate ions as formic acid molecules. The solution is acidic, so there should also be some hydronium ions, but the exact proportion depends on the total concentration of the buffer species. Here is a view showing 6 formate ions, 3 formic acid molecules, and 1 hydronium ion:

16.41 (a) Use the buffer equation to calculate the pH of a buffer solution:
The conjugate acid/base pair is $H_2PO_4^-$/HPO_4^{2-}, $pK_a = pK_{a2} = 7.21$;

$pH = pK_a + \log\left\{\frac{[A^-]}{[HA]}\right\} = 7.21 + \log\left\{\frac{(0.20\text{ M})}{(0.50\text{ M})}\right\} = 7.21 - 0.50 = 6.81$;

(b) Calculate the change in molarity due to the added base, then use the buffer equation to calculate the new pH. The added hydroxide reacts completely with the weak acid and generates weak base. $n_{\text{base}} = m/MM = (0.120\text{ g}) \div (40.0\text{ g/mol}) = 3.0 \times 10^{-3}$ mol;
$\Delta M = n_{\text{base}}/V = (3.0 \times 10^{-3}\text{ mol}) \div (0.15\text{ L}) = 0.020$ M;
[HPO_4^{2-}] = 0.20 + 0.020 = 0.22 M; [$H_2PO_4^-$] = 0.50 − 0.020 = 0.48 M;

$pH = pK_a + \log\left\{\frac{[A^-]}{[HA]}\right\} = 7.21 + \log\left\{\frac{(0.22\text{ M})}{(0.48\text{ M})}\right\} = 7.21 - 0.34 = 6.87$;

ΔpH = 6.87 − 6.81 = 0.06; (continued on next page)

16.41 (c) Adding acid reduces the concentration of weak base, increases the concentration of weak acid, and reduces the pH. Use the buffer equation to calculate the conjugate base/acid ratio at the limit of the pH range. Then calculate the change in molarity that this ratio represents, and convert to moles:

$$pH_{min} = 6.81 - 0.10 = 6.71 = 7.21 + \log\left\{\frac{[A^-]}{[HA]}\right\}; \log\left\{\frac{[A^-]}{[HA]}\right\} = 6.71 - 7.21 = -0.50;$$

$$\left\{\frac{[A^-]}{[HA]}\right\} = 0.32 = \left\{\frac{(0.20 - \Delta M)}{(0.50 + \Delta M)}\right\};$$

$(0.20 - \Delta M) = 0.32(0.50 + \Delta M) = 0.16 + 0.32\Delta M$, so $0.04 = 1.32\Delta M$;
$\Delta M = 3.0 \times 10^{-2}$ M; mol neutralized $= \Delta MV = (3.0 \times 10^{-2}$ M$)(0.25$ L$) = 7.5 \times 10^{-3}$ mol.

16.43 A buffer solution requires a conjugate acid/base pair whose pK_a is within 1 pH unit of the desired pH. For pH = 3.50, the HCO_2H/HCO_2^- system, $pK_a = 3.74$, would be most suitable. Sodium formate, $NaHCO_2$, could be used along with HCl solution which would protonate some formate anions to form the conjugate weak acid. For pH = 12.60, the HPO_4^{2-}/PO_4^{3-} system, $pK_a = 12.32$, would be most suitable. Potassium phosphate, K_3PO_4, could be used along with HCl solution which would protonate some phosphate anions to form the conjugate weak acid.

16.45 To determine the approximate pH at the stoichiometric point of an acid-base titration, examine the equilibrium that determines the pH in the vicinity of the stoichiometric point:
(a) This is a weak base-strong acid titration. At the stoichiometric point, major acid-base species are NH_4^+ and H_2O, and the dominant equilibrium is:
$NH_4^+_{(aq)} + H_2O_{(aq)} \rightleftharpoons NH_3_{(aq)} + H_3O^+_{(aq)}$
Hydronium ions are produced in this reaction, so pH < 7 at the stoichiometric point;
(b) This is a strong acid-strong base titration. The only acid-base species present at the stoichiometric point is H_2O, and the dominant equilibrium is:
$H_2O_{(aq)} + H_2O_{(aq)} \rightleftharpoons OH^-_{(aq)} + H_3O^+_{(aq)}$
Thus pH = 7 at the stoichiometric point.
(c) This is a weak acid-strong base titration. At the stoichiometric point, major acid-base species are $CH_3CO_2^-$ and H_2O, and the dominant equilibrium is:
$CH_3CO_2^-_{(aq)} + H_2O_{(aq)} \rightleftharpoons CH_3CO_2H_{(aq)} + OH^-_{(aq)}$
Hydroxide ions are produced in this reaction, so pH > 7 at the stoichiometric point.

16.47 The major species differ at various points during a titration, so there are different dominant equilibria which must be identified before doing an equilibrium calculation to determine pH.
(a) Before titration begins, the major species are a weak acid, aspirin (HA) and H_2O, and the dominant equilibrium is HA + H_2O $\rightleftharpoons$ A^- + H_3O^+, for which $K_a = 3.0 \times 10^{-4}$.

Species:	H_2O +	HA $\rightleftharpoons$	A^- +	H_3O^+
Initial conc. (M):		10^{-2}	0	0
Change (M)		$-x$	$+x$	$+x$
Equilib conc. (M)		$10^{-2} - x$	x	x

(continued on next page)

16.47 Now substitute into the equilibrium constant expression and solve for x:

$$K_a = 3.0 \times 10^{-4} = \frac{[A^-]_{eq}[H_3O^+]_{eq}}{[HA]_{eq}} = \frac{x^2}{(10^{-2} - x)};$$

$x^2 = (3.0 \times 10^{-4})(10^{-2} - x) = (3.0 \times 10^{-6}) - (3.0 \times 10^{-4})x;$
$x^2 + (3.0 \times 10^{-4})x - (3.0 \times 10^{-6}) = 0;$

$$x = \frac{-b \pm \sqrt{b^2 - 4ac}}{2a} = \frac{-(3.0 \times 10^{-4}) \pm \sqrt{(3.0 \times 10^{-4})^2 - 4(\pm 3.0 \times 10^{-6})}}{2} = 1.6 \times 10^{-3};$$

$[H_3O^+] = 1.6 \times 10^{-3}$ M, pH = 2.80;

(b) At the stoichiometric point, all the weak acid has been converted into its conjugate base, so the major acid-base species present are A^- and H_2O and the dominant equilibrium is:
$A^- + H_2O \rightleftharpoons HA + OH^-$, $K_{eq} = K_w/K_a = (1.0 \times 10^{-14}) \div (3.0 \times 10^{-4}) = 3.3 \times 10^{-11}$:

Species:	H_2O +	$A^- \rightleftharpoons$	HA +	OH^-
Initial conc. (M):		10^{-2}	0	0
Change (M)		$-x$	$+x$	$+x$
Equilib conc. (M)		$10^{-2} - x$	x	x

Now substitute into the equilibrium constant expression and solve for x:

$$K_a = 3.3 \times 10^{-11} = \frac{[HA]_{eq}[OH^-]_{eq}}{[A^-]_{eq}} = \frac{x^2}{(10^{-2} - x)}; \text{ Assume } x \ll 10^{-2};$$

$x^2 = (10^{-2})(3.3 \times 10^{-11}) = 3.3 \times 10^{-13}$, from which $x = 5.8 \times 10^{-7}$;
$[OH^-]_{eq} = 5.8 \times 10^{-7}$ M, pOH = 6.24, and pH = 14.00 − pOH = 14.00 − 6.24 = 7.76;

(c) At the midpoint of the titration, half of the weak acid has been converted into its conjugate base, and pH = pK_a = − log(3.0 × 10^{-4}) = 3.52.

16.49 At the second stoichiometric point, all the weak acid (H_2P) has been converted into its conjugate base (P^{2-}), so the major acid-base species present are P^{2-} and H_2O and the dominant equilibrium is:
$P^{2-} + H_2O \rightleftharpoons HP^- + OH^-$, $pK_{a2} = 5.41$, $K_{a2} = 3.9 \times 10^{-6}$;
$K_{eq} = K_w/K_{a2} = (1.0 \times 10^{-14}) \div (3.9 \times 10^{-6}) = 2.6 \times 10^{-9}$:

Species:	H_2O +	$P^{2-} \rightleftharpoons$	HP^- +	OH^-
Initial conc. (M):		0.015	0	0
Change (M)		$-x$	$+x$	$+x$
Equilib conc. (M)		$0.015 - x$	x	x

Now substitute into the equilibrium constant expression and solve for x:

$$K_{eq} = 2.6 \times 10^{-9} = \frac{[HP^-]_{eq}[OH^-]_{eq}}{[P^{2-}]_{eq}} = \frac{x^2}{(0.0150 - x)}; \text{ assume } x \ll 0.0150$$

$x^2 = (2.6 \times 10^{-9})(0.0150) = (3.9 \times 10^{-11})$; $x = 6.2 \times 10^{-6}$;
$[OH^-] = 6.2 \times 10^{-6}$ M, pOH = 5.20, pH = 14.00 − 5.20 = 8.80; a suitable indicator for this titration is thymol blue, $pK_{in} = 8.9$.

16.51 The best choice for a buffer solution generally is the weak acid whose pK_a is closest to the desired pH of the buffer solution. For a pH = 4.80 buffer solution, acetic acid/acetate, pK_a = 4.75, would be the best choice. To prepare the buffer solution, add the appropriate amounts of sodium acetate (NaAc) and acetic acid (HAc) solution to water and make up to 1.0 L with additional water. Calculate the acetate/acetic acid ratio using the buffer equation:

$$\log\left\{\frac{[Ac^-]}{[HAc]}\right\} = pH - pK_a = 4.80 - 4.75 = 0.05, \quad \left\{\frac{[Ac^-]}{[HAc]}\right\} = 10^{0.05} = 1.12;$$

Thus, $[Ac^-] = 1.12 [HAc]$; also, $[Ac^-] + [HAc] = 0.35$ M, so $[HAc] = 0.35$ M $- [Ac^-]$;
$[Ac^-] = 1.12(0.35$ M $- [Ac^-]) = 0.39 - 1.12 [Ac^-]$, so $2.12 [Ac^-] = 0.39$ and $[Ac^-] = 0.18$ M
$n_{Ac^-} = (0.18$ M$)(1.0$ L$) = 0.18$ mol; $m_{NaAc} = n\, MM = (0.18$ mol$)(82.0$ g/mol$) = 15$ g;
$[HAc] = 0.35 - 0.18 = 0.17$ M; $n_{HAc} = (0.17$ M$)(1.0$ L$) = 0.17$ mol;
$V_{HAc} = n/M = (0.17$ mol$) \div (1.00$ mol/L$) = 0.17$ L;
Mix together 15 g sodium acetate, 0.17 L 1.0 M acetic acid, and enough water to make 1.0 L.

16.53 "What mass will dissolve" means "calculate the amount present in solution at equilibrium." Initial amounts are zero, so it is easy to complete a concentration table. Calculate molarity from the solubility product expression, then convert to mass using standard stoichiometric methods. Let x be the number of moles of salt dissolving in 1 L of solution.

Species: $Zn(OH)_{2\,(s)} \rightleftharpoons$ $Zn^{2+}_{(aq)} +$ $2\, OH^-_{(aq)}$
Conc. at equilibrium (M):   x   2x

pK_{sp} (App. G) = 16.92, so $K_{sp} = 1.2 \times 10^{-17} = [Zn^{2+}]_{eq}[OH^-]^2_{eq} = 4x^3$ and $x = 1.4 \times 10^{-6}$;
$m = M\,V\,MM = (1.4 \times 10^{-6}$ M$)(1.00$ L$)(99.40$ g/mol$) = 1.4 \times 10^{-4}$ g.

16.55 (a) H_2SO_4 is a strong acid, so the major species in solution are H_2O, HSO_4^-, and H_3O^+. The hydrogensulfate ion is a weak acid, so the equilibrium reaction that determines pH is:
$H_2O + HSO_4^- \rightleftharpoons SO_4^{2-} + H_3O^+$;

(b) Na_2SO_4 is a salt, so the major species in solution are H_2O, SO_4^{2-}, and Na^+. The sulfate ion is the conjugate base of a weak acid, so the equilibrium reaction that determines pH is:
$SO_4^{2-} + H_2O \rightleftharpoons HSO_4^- + OH^-$;

(c) CO_2 associates with H_2O when it dissolves in water, so the major species in solution are CO_2, H_2CO_3, and H_2O. Carbonic acid is a weak acid, so the equilibrium reaction that determines pH is:
$H_2O + H_2CO_3 \rightleftharpoons HCO_3^- + H_3O^+$;

(d) NH_4Cl is a salt, so the major species in solution are H_2O, Cl^-, and NH_4^+. The ammonium ion is the conjugate acid of a weak base, so the equilibrium reaction that determines pH is: $H_2O + NH_4^+ \rightleftharpoons NH_3 + H_3O^+$.

16.57 Molecular pictures must show the correct relative numbers of the various species in the solution. From the starting condition (six molecules of oxalic acid), make appropriate changes and then draw new pictures:
(a) Hydroxide ions react with oxalic acid to form water and hydrogenoxalate: $H_2C_2O_4 + OH^- \rightleftharpoons H_2O + HC_2O_4^-$ The picture shows 2 molecules of oxalic acid and 4 each of water and hydrogenoxalate:

(b) When all oxalic acid has reacted, hydroxide ions react with hydrogenoxalate ions to form water and oxalate ions: $HC_2O_4^- + OH^- \rightleftharpoons H_2O + C_2O_4^{2-}$. The picture shows 4 hydrogenoxalate ions, 8 water molecules, and 2 oxalate ions:

(c) NH_3, a weak base, accepts a proton from oxalic acid, a weak acid:
$H_2C_2O_4 + NH_3 \rightleftharpoons NH_4^+ + HC_2O_4^-$
The picture shows 2 oxalic acid molecules and 4 each ammonium and hydrogenoxalate ions:

16.59 (a) An acid-base equilibrium reaction involves proton transfer, in this case from boric acid to water:

[Lewis structure diagram showing H₃BO₃ + H₂O ⇌ H₃BO₂⁻ (with extra OH) + H₃O⁺]

(b) $pK_a = -\log(K_a) = -\log(7.3 \times 10^{-10}) = 9.14$;
(c) To calculate the pH of a solution, follow the standard procedure for equilibrium calculations:

Species:	H_2O +	H_3BO_3	⇌	$H_3BO_2^-$	+	H_3O^+
Initial concentration (M):		0.050		0		0
Change to equilib (M)		$-x$		x		x
Equilib concentration (M)		$0.050 - x$		x		x

$K_a = 7.3 \times 10^{-10} = \dfrac{[H_3BO_2^-]_{eq}[H_3O^+]_{eq}}{[H_3BO_3]_{eq}}$; Assume $x \ll 0.050$:

$7.3 \times 10^{-10} = \dfrac{x^2}{0.050}$, from which $x^2 = 3.65 \times 10^{-11}$ and $x = 6.0 \times 10^{-6}$;

$[H_3O^+] = 6.0 \times 10^{-6}$ M, and pH = 5.22.

16.61 There are two amine groups, one at either end of the molecule, each of which can accept a proton from a water molecule. Convert the line structure to a Lewis structure using the standard procedures, then show the transfer of one proton to each N atom:

[Lewis structure diagrams showing the diamine reacting with water: first proton transfer to one nitrogen forming a monoprotonated species plus OH⁻, then second proton transfer to the other nitrogen forming a diprotonated species plus OH⁻]

640

Chapter 16: Aqueous Equilibria

16.63 To determine the pH of a solution, follow the standard procedure:
1.) Major species are H_2O, Na^+, and F^-;
2.) The dominant acid-base equilibrium is $H_2O + F^- \rightleftharpoons HF + OH^-$;
3.) $K_{eq} = K_b$; from Table 16-2, $pK_b = 10.55$, from which $K_b = 10^{-10.55} = 2.8 \times 10^{-11}$;
4.) Species: $\quad H_2O \quad + \quad F^- \quad \rightleftharpoons \quad HF \quad + \quad OH^-$
Initial concentration (M): $\qquad\qquad$ 0.250 $\qquad\qquad$ 0 $\qquad\qquad$ 0
Change to equilib (M): $\qquad\qquad$ $-x$ $\qquad\qquad$ x $\qquad\qquad$ x
Equilib concentration (M): $\qquad$ 0.250 − x $\qquad\qquad$ x $\qquad\qquad$ x

$K_b = 2.8 \times 10^{-11} = \dfrac{[HF]_{eq}[OH^-]_{eq}}{[F^-]_{eq}}$; Assume x << 0.250:

$2.8 \times 10^{-11} = \dfrac{x^2}{0.250}$, from which $x^2 = 7.0 \times 10^{-12}$ and $x = 2.7 \times 10^{-6}$;

$[OH^-] = 2.7 \times 10^{-6}$ M, pOH = 5.58, and pH = 14.00 − 5.58 = 8.42.

16.65 To determine concentrations of species in a solution, follow the standard procedure:
1.) This is a strong acid. Major species are H_2O, H_3O^+, and HSO_4^-;
2.) The dominant acid-base equilibrium is $H_2O + HSO_4^- \rightleftharpoons SO_4^{2-} + H_3O^+$;
3.) $K_{eq} = K_{a2}$; from Table 16-3, $pK_{a2} = 1.96$, from which $K_{a2} = 10^{-1.96} = 1.1 \times 10^{-2}$;
4.) In this solution, there is hydronium ion from the strong acid present initially:
Species: $\quad H_2O \quad + \quad HSO_4^- \quad \rightleftharpoons \quad SO_4^{2-} \quad + \quad H_3O^+$
Initial concentration (M): $\qquad\qquad$ 2.00 $\qquad\qquad$ 0 $\qquad\qquad$ 2.00
Change to equilib (M): $\qquad\qquad$ $-x$ $\qquad\qquad$ x $\qquad\qquad$ x
Equilib concentration (M): $\qquad$ 2.00 − x $\qquad\qquad$ x $\qquad\qquad$ 2.00 + x

$K_{eq} = 1.1 \times 10^{-2} = \dfrac{[SO_4^{2-}]_{eq}[H_3O^+]_{eq}}{[HSO_4^-]_{eq}}$; Assume x << 2.00:

$1.1 \times 10^{-2} = \dfrac{(2.00)(x)}{(2.00)}$, from which $x = 1.1 \times 10^{-2}$;

$[H_3O^+] = 2.00 + x = 2.01$ M; $[SO_4^{2-}] = 1.1 \times 10^{-2}$ M; $[HSO_4^-] = 2.00 - x = 1.99$ M.

16.67 To determine concentrations of species in a solution, follow the standard procedure for working equilibrium problems:
1.) This is a salt dissolving in a buffered solution. Major species are the buffer species, H_2O and the solid, $Cu(OH)_2$;
2.) The solubility equilibrium is $Cu(OH)_{2(s)} \rightleftharpoons Cu^{2+}_{(aq)} + 2\,OH^-_{(aq)}$;
3.) From Appendix H, $pK_{sp} = 19.66$, from which $K_{sp} = 2.2 \times 10^{-20}$
4.) Because the solution is buffered, $[OH^-]$ is held constant at its buffered value:
pOH = 14.00 − pH = 14.00 − 5.00 = 9.00, $[OH^-] = 1.0 \times 10^{-9}$ M;
Species: $\qquad Cu(OH)_{2(s)} \quad \rightleftharpoons \quad Cu^{2+}_{(aq)} \quad + \quad 2\,OH^-_{(aq)}$
Initial concentration (M): $\qquad\qquad\qquad\qquad$ 0 $\qquad\qquad$ 1.0×10^{-9}
Change to equilib (M): $\qquad\qquad\qquad\qquad$ x $\qquad\qquad$ 0
Equilib concentration (M): $\qquad\qquad\qquad\qquad$ x $\qquad\qquad$ 1.0×10^{-9}

$K_{sp} = 2.2 \times 10^{-20} = [Cu^{2+}]_{eq}[OH^-]^2_{eq}$; $= (1.0 \times 10^{-9})^2 x$, from which $x = 2.2 \times 10^{-2}$;

$[Cu^{2+}] = 2.2 \times 10^{-2}$ M.

Chapter 16: Aqueous Equilibria

16.69 There is much interesting chemical information provided in the statement of this problem, but the calculation is a straightforward equilibrium determination for a solution of a weak base:

Species:	H_2O +	LSD	$\rightleftharpoons$	$LSDH^+$	+	OH^-
Initial concentration (M):		0.55		0		0
Change to equilib (M)		$-x$		x		x
Equilib concentration (M)		$0.55 - x$		x		x

$pK_b = 6.12$, $K_b = 7.5 \times 10^{-7} = \dfrac{[LSDH^+]_{eq}[OH^-]_{eq}}{[LSD]_{eq}}$; Assume $x \ll 0.55$:

$7.5 \times 10^{-7} = \dfrac{x^2}{0.55}$, from which $x^2 = 4.2 \times 10^{-7}$ and $x = 6.5 \times 10^{-4}$;

$[OH^-] = 6.5 \times 10^{-4}$ M, pOH = 3.19, and pH = 14.00 − 3.19 = 11.81.

16.71 The pH of a saturated solution of a solid hydroxide can be calculated from the equilibrium concentration of hydroxide ions, which is found using a standard equilibrium calculation:

Species:	$Cd(OH)_{2\,(s)}$	$\rightleftharpoons$	$Cd^{2+}{}_{(aq)}$ +	$2\,OH^-{}_{(aq)}$
Conc. at equilibrium (M):			x	$2x$

pK_{sp} (App. G) = 13.60, from which $K_{sp} = 2.5 \times 10^{-14} = [Cd^{2+}]_{eq}[OH^-]^2_{eq} = 4x^3$

$x = 1.8 \times 10^{-5}$, from which $[OH^-] = 1.8 \times 10^{-5}$ M, pOH = 4.73, pH = 14.00 − 4.73 = 9.27.

16.73 The best choice for a buffer solution generally is the weak acid whose pK_a is closest to the desired pH of the buffer solution. For a pH = 10.00 buffer solution, hydrogencarbonate/carbonate, $pK_a = 10.32$, is the best choice.

(a) The buffer solution must contain both HCO_3^- and CO_3^{2-} as major species. There are three ways to prepare this buffer solution using the reagents available.
(1) Weigh appropriate masses of $NaHCO_3$ and Na_2CO_3 and dissolve in water;
(2) Weigh an appropriate mass of $NaHCO_3$, dissolve in water, and add NaOH;
(3) Weigh an appropriate mass of Na_2CO_3, dissolve in water, and add HCl;

(b) To determine amounts, first calculate the base/acid ratio using the buffer equation. Then use the stoichiometric information provided and standard stoichiometric methods.

$\log\left\{\dfrac{[CO_3^{2-}]}{[HCO_3^-]}\right\} = pH - pK_a = 10.00 - 10.32 = -0.32$, $\left\{\dfrac{[CO_3^{2-}]}{[HCO_3^-]}\right\} = 10^{-0.32} = 0.48$;

Thus, $[CO_3^{2-}] = 0.48\,[HCO_3^-]$; as stated in the problem, $[HCO_3^-] = 0.25$ M; $[CO_3^{2-}] = 0.48(0.25$ M$) = 0.12$ M

$n_{CO_3^{2-}} = (0.12$ M$)(0.500$ L$) = 0.060$ mol; $n_{HCO_3^-} = (0.25$ M$)(0.500$ L$) = 0.125$ mol;

Here is the procedure using sodium carbonate and HCl: The hydrogencarbonate ions form from reaction of hydronium ions with carbonate, so the total amount of carbonate required is the sum of the two amounts, $n_{total} = 0.060 + 0.125 = 0.185$ mol;

$m = n\,MM = (0.185$ mol$)(106.0$ g/mol$) = 19.6$ g Na_2CO_3 are required;

The buffer also requires 0.125 mol of acid;

$n = MV$, so $V = n/M = (0.125$ mol$) \div (6.0$ mol/L$) = 2.1 \times 10^{-2}$ L;

Mix together 19.6 g sodium carbonate, 21 mL 6.0 M HCl, and water to make 0.500 L.

Chapter 16: Aqueous Equilibria

16.75 The major species present are different at various points during a titration, so there are different dominant equilibria which must be identified before doing an equilibrium calculation to determine pH. (a) Before titration begins, the major species are a weak base, formate (HCO_2^-) and H_2O, and the dominant equilibrium is
$HCO_2^- + H_2O \rightleftharpoons HCO_2H + OH^-$, for which $pK_b = 10.26$ and $K_b = 5.5 \times 10^{-11}$.

Species:	H_2O +	$HCO_2^- \rightleftharpoons$	HCO_2H +	OH^-
Initial conc. (M):		0.200	0	0
Change (M)		$-x$	$+x$	$+x$
Equilib conc. (M)		$0.200 - x$	x	x

Now substitute into the equilibrium constant expression and solve for x:

$$K_b = 5.5 \times 10^{-11} = \frac{[HCO_2H]_{eq}[OH^-]_{eq}}{[HCO_2^-]_{eq}} = \frac{x^2}{(0.200-x)}; \text{ assume } x \ll 0.200$$

$x^2 = 1.1 \times 10^{-11}$, from which $x = 3.3 \times 10^{-6}$;
$[OH^-] = 3.3 \times 10^{-6}$ M, pOH = 5.48 and pH = 14.00 − 5.48 = 8.42;
(b) At the midpoint of the titration, the buffer equation applies and the concentrations of acid and conjugate base are equal, so pH = pK_a = 3.74;
(c) At the stoichiometric point, all the weak base has been converted into its conjugate acid, so the major acid-base species present are HCO_2H and H_2O and the dominant equilibrium is:
$HCO_2H + H_2O \rightleftharpoons HCO_2^- + H_3O^+$, for which $K_a = 3.0 \times 10^{-4}$.

Species:	H_2O +	$HCO_2H \rightleftharpoons$	HCO_2^- +	H_3O^+
Initial conc. (M):		0.200	0	0
Change (M)		$-x$	$+x$	$+x$
Equilib conc. (M)		$0.200 - x$	x	x

Now substitute into the equilibrium constant expression and solve for x:

$$K_a = 3.0 \times 10^{-4} = \frac{[HCO_2^-]_{eq}[H_3O^+]_{eq}}{[HCO_2H]_{eq}} = \frac{x^2}{(0.200-x)};$$

$x^2 = (0.200 - x)(3.0 \times 10^{-4}) = (6.0 \times 10^{-5}) - (3.0 \times 10^{-4})x;$
$x^2 + (3.0 \times 10^{-4})x - (6.0 \times 10^{-5}) = 0;$

$$x = \frac{-b \pm \sqrt{b^2 - 4ac}}{2a} = \frac{-(3.0 \times 10^{-4}) \pm \sqrt{(3.0 \times 10^{-4})^2 + 4(6.0 \times 10^{-5})}}{2} = 7.7 \times 10^{-3};$$

$[H_3O^+]_{eq} = 7.7 \times 10^{-3}$ M, and pH = 2.11;
(d) A suitable indicator for this titration must change color around pH = 2. Thymol blue, $pK_{In} = 1.75$, would be the best choice.

16.77 To identify an acid from the pH of its solution, use the pH to calculate the equilibrium constant of the acid. For pH = 2.71, $[H_3O^+]_{eq} = 1.94 \times 10^{-3}$ M = $[A^-]$;
$[HA]_{eq}$ = (0.060 M − 0.00194 M) = 0.058 M;

$$K_a = \frac{[A^-]_{eq}[H_3O^+]_{eq}}{[HA]_{eq}} = \frac{(1.94 \times 10^{-3})^2}{(0.058)} = 6.6 \times 10^{-5}; pK_a = 4.18;$$

The acid is benzoic acid, listed in Table 16-2, pK_a = 4.19.

16.79 In any titration of a weak acid, the pH at the midpoint of the titration equals the pK_a of the acid. Reading from the graph, the stoichiometric point is at $V = 8.6$ mL, so the midpoint is at $V = 4.3$ mL, where pH = 3.88. Thus pK_a = 3.88.

16.81 Consult Table 16-6 to find out what color various indicators have at pH = 8.5. An indicator with pK_{In} < 8.5 has its base color, while one with pK_{In} > 8.5 has its acid color: (a) methyl orange will be yellow; (b) phenol red will be red; (c) bromocresol green will be blue; (d) thymol blue will be changing from yellow to blue (probably will appear green).

16.83 Follow the standard procedure for dealing with equilibrium calculations:
The major species are Na^+, $C_6H_5O^-$, and H_2O, and the acid-base equilibrium is:
$C_6H_5O^-_{(aq)} + H_2O_{(l)} \rightleftharpoons C_6H_5OH_{(aq)} + OH^-_{(aq)}$
Set up a concentration table and use it to determine K_{eq}:
pOH = 14.00 − pH = 11.00; $[OH^-] = 10^{-11.00} = 1.00 \times 10^{-3}$ M;

Species:	H_2O +	$C_6H_5O^-$ $\rightleftharpoons$	C_6H_5OH	+	OH^-
Initial conc. (M):		0.0100	0		0
Change (M)		− 1.00 × 10⁻³	+ 1.00 × 10⁻³		+ 1.00 × 10⁻³
Equilib conc. (M)		0.0090	1.00 × 10⁻³		1.00 × 10⁻³

$K_{eq} = K_b = \dfrac{[C_6H_5OH]_{eq}[OH^-]_{eq}}{[C_6H_5O^-]_{eq}} = \dfrac{(1.00 \times 10^{-3})^2}{(0.0090)} = 1.1 \times 10^{-4}$; p$K_b$ = 3.96

pK_a = pK_w − pK_b = 14.00 − 3.96 = 10.04.

16.85 The acidic or basic nature of a solution depends on the dominant equilibrium, which is determined by the major species present:

(a) CH_3CO_2H is a weak acid, major species are H_2O and CH_3CO_2H; dominant equilibrium is $CH_3CO_2H + H_2O \rightleftharpoons CH_3CO_2^- + H_3O^+$, and the solution is acidic;
(b) NH_3 is a weak base, major species are H_2O and NH; dominant equilibrium is $NH_3 + H_2O \rightleftharpoons NH_4^+ + OH^-$, and the solution is basic;
(c) NH_4Cl is a salt, major species are H_2O, NH_4^+, and Cl^-; dominant equilibrium is $NH_4^+ + H_2O \rightleftharpoons NH_3 + H_3O^+$, and the solution is acidic;
(d) $NaCH_3CO_2$ is a salt, major species are H_2O, $CH_3CO_2^-$, and Na^+; dominant equilibrium is $CH_3CO_2^- + H_2O \rightleftharpoons CH_3CO_2H + OH^-$, and the solution is acidic;
(e) $NH_4CH_3CO_2$ is a salt, major species are H_2O, $CH_3CO_2^-$, and NH_4^+; dominant equilibrium is $CH_3CO_2^- + NH_4^+ \rightleftharpoons CH_3CO_2H + NH_3$, for which $K_{eq} = K_aK_b = 1$; thus this solution is neutral.

16.87 (a) The salt generates Na^+ and SO_3^{2-}, CH_3CO_2H is a weak acid, and H_2O is always major;
(b) The weak acid and weak base react: $CH_3CO_2H + SO_3^{2-} \rightleftharpoons CH_3CO_2^- + HSO_3^-$;
(c) CH_3CO_2H is the acid, SO_3^{2-} is the base, HSO_3^- is the conjugate acid, and $CH_3CO_2^-$ is the conjugate base.

Chapter 16: Aqueous Equilibria

16.89 To identify the precipitate, examine the major species in solution. The buffer is unspecified, but the other major species are H_2O, Zn^{2+}, Cl^-, and OH^- (because pH = 8, making this a basic solution). From Appendix G, pK_{sp} for $Zn(OH)_2$ is 16.92, $K_{sp} = 1.2 \times 10^{-17}$, so this salt is insoluble and is the precipitate that forms.

To determine the concentration remaining in solution, carry out an equilibrium calculation.

Because the solution is buffered, $[OH^-]$ is held constant at its buffered value:
pOH = 14.00 − 8.00 = 6.00, $[OH^-] = 1.00 \times 10^{-6}$ M;
Use this concentration in the K_{sp} expression and solve for $[Zn^{2+}]$:

$K_{sp} = 1.2 \times 10^{-17} = [Zn^{2+}]_{eq}[OH^-]^2_{eq} = (1.0 \times 10^{-6})^2[Zn^{2+}]$, so $[Zn^{2+}] = 1.2 \times 10^{-5}$ M.

16.91 To determine the pH of a solution, follow the standard procedure for equilibrium problems:
1.) Major species are H_2O, Na^+, and ClO^-;
2.) The dominant acid-base equilibrium is $H_2O + ClO^- \rightleftharpoons HClO + OH^-$;
3.) $K_{eq} = K_b$; from Table 16-2, $pK_b = 6.48$, from which $K_b = 10^{-6.48} = 3.3 \times 10^{-7}$;
4.) Species: H_2O + ClO^- $\rightleftharpoons$ $HClO$ + OH^-
Initial concentration (M):   0.0100   0   0
Change to equilib (M)   − x   x   x
Equilib concentration (M)   0.0100 − x   x   x

$K_b = 3.3 \times 10^{-7} = \dfrac{[HClO]_{eq}[OH^-]_{eq}}{[ClO^-]_{eq}}$; Assume x << 0.0100; $3.3 \times 10^{-7} = \dfrac{x^2}{0.0100}$,

from which $x^2 = 3.3 \times 10^{-9}$ and $x = 5.8 \times 10^{-5}$;
$[OH^-] = 5.8 \times 10^{-5}$ M, pOH = 4.24, and pH = 14.00 − 4.24 = 9.76.

16.93 The enzyme will lose its activity when the pH reaches 7.1. Use the buffer equation to calculate the $H_2PO_4^-$ / HPO_4^{2-} ratio corresponding to this pH, then calculate what the change in molarity is and convert to moles:

$\log\left\{\dfrac{[HPO_4^{2-}]}{[H_2PO_4^-]}\right\} = pH - pK_a = 7.1 - 7.21 = -0.1$, $\left\{\dfrac{[HPO_4^{2-}]}{[H_2PO_4^-]}\right\} = 10^{-0.1} = 0.8$;

Thus, $[HPO_4^{2-}] = 0.8\ [H_2PO_4^-]$;
also, $[HPO_4^{2-}] + [H_2PO_4^-] = 0.085$ M, so $[H_2PO_4^-] = 0.085$M − $[HPO_4^{2-}]$;
$[HPO_4^{2-}] = 0.8(0.085$ M − $[HPO_4^{2-}]) = 0.068 - 0.8\ [HPO_4^{2-}]$;
$1.8\ [HPO_4^{2-}] = 0.068$ and $[HPO_4^{2-}] = 0.0378$ M;
The change from initial concentration of the buffer is 0.0445 − 0.0378 = 0.007 M;
$n = MV = (0.007$ M$)(0.250$ L$) = 2 \times 10^{-3}$ mol. (This result has just 1 significant figure because the pH limit has only 1 decimal place).

16.95 (a) Use the buffer equation, Equation 16-5, with amounts:

$$pH = pK_a - \log\left\{\frac{\text{mol NH}_3}{\text{mol NH}_4^+}\right\} = (14.00 - 4.75) - \log\left[\frac{(0.360 \text{ L})(0.300 \text{ M})}{(0.640 \text{ L})(0.300 \text{ M})}\right]$$

$$pH = 9.25 - \log\left[\frac{(0.108 \text{ mol})}{(0.192 \text{ mol})}\right] = 9.25 - 0.25 = 9.00;$$

(b) HNO_3 is a strong acid, so the reaction is $NH_3 + H_3O^+ \rightleftharpoons NH_4^+ + H_2O$;

(c) The amount of added acid is n = (5.00 mL)(10^{-3} L/mL)(1.00 mol/L) = 0.005 mol; the amount of weak acid increases and of weak base decreases by this amount:

$$pH = 9.25 - \log\left[\frac{(0.108 - 0.005) \text{ mol}}{(0.192 + 0.005) \text{ mol}}\right] = 9.25 - 0.28 = 8.97;$$

(d) As long as there is weak acid present, the reaction is $NH_4^+ + OH^- \rightleftharpoons NH_3 + H_2O$;

(e) The amount of added base is n = (14.0 g) ÷ (56.11 g/mol) = 0.250 mol; this is more than the amount of weak acid present, so the solution no longer is buffered, and the pH is determined by the excess OH^-: n_{excess} = (0.250 − 0.192) = 0.058 mol; $[OH^-]$ = (0.058 mol) ÷ (1.00 L) = 0.058 M; pOH = 1.24 and pH = 14.00 − 1.24 = 12.76.

16.97 (a) The solubility equilibrium is $MgCO_{3(s)} \rightleftharpoons Mg^{2+}_{(aq)} + CO_3^{2-}_{(aq)}$, so the major species present in the saturated solution are H_2O, Mg^{2+}, and CO_3^{2-};

(b) pK_{sp} = 5.00, K_{sp} = 1.0 × 10^{-5} = $[Mg^{2+}][CO_3^{2-}]$; by stoichiometry, $[Mg^{2+}]$ = $[CO_3^{2-}]$; 1.0 × 10^{-5} = $[Mg^{2+}]^2$ and $[Mg^{2+}]$ = 3.2 × 10^{-3} M;

(c) The pH of this solution is determined by the presence of carbonate, which is a weak base:
$CO_3^{2-}{}_{(aq)} + H_2O_{(l)} \rightleftharpoons HCO_3^-{}_{(aq)} + OH^-{}_{(aq)}$, $pK_{eq} = pK_w - pK_{a2}$ = 14.00 − 10.32 ;
$K_{eq} = 10^{-3.68} = 1.8 \times 10^{-4}$;
Set up a concentration table and use it to determine $[OH^-]$:

Species:	H_2O +	CO_3^{2-} $\rightleftharpoons$	HCO_3^- +	OH^-
Initial conc. (M):		3.2 × 10^{-3}	0	0
Change (M)		− x	+ x	+ x
Equilib conc. (M)		3.2 × 10^{-3} − x	x	x

Now substitute into the equilibrium constant expression and solve for x:

$$K_{eq} = 1.8 \times 10^{-4} = \frac{[HCO_3^-]_{eq}[OH^-]_{eq}}{[CO_3^{2-}]_{eq}} = \frac{x^2}{(3.2 \times 10^{-3} - x)};$$

$x^2 = (3.2 \times 10^{-3} - x)(1.8 \times 10^{-4}) = (5.8 \times 10^{-7}) - (1.8 \times 10^{-4})x$;
$x^2 + (1.8 \times 10^{-4})x - (5.8 \times 10^{-7}) = 0;$ (continued on next page)

Chapter 16: Aqueous Equilibria

$$x = \frac{-b \pm \sqrt{b^2 - 4ac}}{2a} = \frac{-(1.8 \times 10^{-4}) \pm \sqrt{(1.8 \times 10^{-4})^2 + 4(5.8 \times 10^{-7})}}{2} = 6.7 \times 10^{-4};$$

$[OH^-]_{eq} = 6.7 \times 10^{-4}$ M, pOH = 3.17, and pH = 14.00 − pOH = 14.00 − 3.17 = 10.82;

(d) $CaCO_3$ will precipitate if the concentration of Ca^{2+} is high enough to exceed the solubility. $CaCO_{3(s)} \rightleftharpoons Ca^{2+}_{(aq)} + CO_3^{2-}_{(aq)}$, $pK_{sp} = 8.55$, $K_{sp} = 2.8 \times 10^{-9}$ M;
In a saturated solution of $MgCO_3$, $[CO_3^{2-}] = 3.2 \times 10^{-3} - 6.7 \times 10^{-4} = 2.5 \times 10^{-3}$ M;
Thus $[Ca^{2+}]_{max} = \dfrac{2.8 \times 10^{-9}}{2.5 \times 10^{-3}} = 1.1 \times 10^{-4}$ M.

16.99 Use the buffer equation to carry out calculations on buffer solutions:

(a) $pH = pK_a + \log\left\{\dfrac{[TRIS]}{[TRISH^+]}\right\} = (14.00 - 5.91) + \log\left[\dfrac{(0.30 \text{ M})}{(0.60 \text{ M})}\right] = 7.79;$

(b) Adding HCl generates H_3O^+ ions, which react quantitatively with TRIS to generate $TRISH^+$; the amount of added acid is $n = (5.0 \text{ mL})(10^{-3} \text{ L/mL})(12 \text{ M}) = 0.060$ mol; if this is added to 1.0 L of buffer, the change of concentrations is 0.060 M;

$$pH = (14.00 - 5.91) + \log\left[\dfrac{(0.30 - 0.06) \text{ M}}{(0.60 + 0.06) \text{ M}}\right] = 8.09 + \log\left[\dfrac{(0.24 \text{ M})}{(0.66 \text{ M})}\right] = 7.65.$$

16.101 Acid-base reactions are reversible, so a buffer that has been ruined by addition of excess acid can be regenerated by adding sufficient base to regenerate the weak base. In this case, addition of 0.42 mol of NaOH will regenerate the original buffer (apart from Na^+ and Cl^- spectator ions which were not originally present).

16.103 The proper indicator has pK_{In} close to the pH at the stoichiometric point of the titration. Calculate that pH for the acetic acid solution:

The equilibrium reaction at the stoichiometric point is
$CH_3CO_2^-_{(aq)} + H_2O_{(l)} \rightleftharpoons CH_3CO_2H_{(aq)} + OH^-_{(aq)}$ $K_b = 5.6 \times 10^{-10}$

Species:	H_2O +	$CH_3CO_2^- \rightleftharpoons$	CH_3CO_2H	+ OH^-
Initial concentration (M):		0.3	0	0
Change in conc. (M):		−x	+ x	+ x
Final concentration (M)		0.3 − x	x	x

Substitute into the equilibrium constant expression and solve for x

$$K_b = 5.6 \times 10^{-10} = \dfrac{[CH_3CO_2H]_{eq}[OH^-]_{eq}}{[CH_3CO_2^-]_{eq}} = \dfrac{x^2}{(0.3 - x)}; \text{ Assume } x \ll 0.3;$$

$x^2 = (0.3)(5.6 \times 10^{-10}) = 1.7 \times 10^{-10}$, from which $x = 1.3 \times 10^{-5}$;
$[OH^-]_{eq} = 1.3 \times 10^{-5}$ M, pOH = 4.89, and pH = 14.00 − pOH = 14.00 − 4.89 = 9.11;

The most suitable indicator is phenolphthalein, $pK_{In} = 9.4$.

Chapter 16: Aqueous Equilibria

16.105 Examples are hydroxides such as $Mg(OH)_2$, sulfates such as $BaSO_4$, sulfides such as ZnS, and carbonates such as $CoCO_3$. Salts whose ions have no acid-base properties have solubilities that are independent of pH. Examples are bromides such as CuBr, chlorides such as AgCl, and iodides such as PbI_2.

16.107 Calculate the concentration of ions in solution using the standard equilibrium procedures:

(a) Species: $Al(OH)_{3\,(s)} \rightleftharpoons Al^{3+}_{(aq)} + 3\,OH^-_{(aq)}$
Conc. at equilibrium (M): $\quad\quad\quad\quad\quad\quad\quad\quad\quad\quad x \quad\quad\quad 3x$

$K_{sp} = 1.8 \times 10^{-33} = [Al^{3+}]_{eq}[OH^-]^3_{eq} = 27\,x^4$, from which $x = 2.8 \times 10^{-9}$;
$[Al^{3+}]_{eq} = 2.8 \times 10^{-9}$ M;

(b) First determine the "initial" concentrations after mixing, then set up a concentration table and solve: $[OH^-]_{init} = (2.50\text{ M})\left(\dfrac{1.20 \times 10^3 \text{L}}{2.50 \times 10^3 \text{L}}\right) = 1.20$ M;

$[Al^{3+}]_{init} = (0.223\text{ M})\left(\dfrac{1.30 \times 10^3 \text{L}}{2.50 \times 10^3 \text{L}}\right) = 0.116$ M;

Species: $Al(OH)_{3\,(s)} \rightleftharpoons$	$Al^{3+}_{(aq)} +$	$3\,OH^-_{(aq)}$
Initial concentration (M)	0.116	1.20
Change to completion (M):	–0.116	–0.348
Conc. at completion (M):	0	0.85
Change to equilibrium (M):	$+y$	$+3y$
Conc. at equilibrium (M):	y	$0.85 + 3y$

$K_{sp} = 1.8 \times 10^{-33} = [Al^{3+}]_{eq}[OH^-]^3_{eq} = (y)(0.85 + 3y)^3$; assume $y \ll 0.85$;
$1.8 \times 10^{-33} = (y)(0.85)^3$, from which $y = 2.9 \times 10^{-33}$ M, and $[Al^{3+}]_{eq} = 2.9 \times 10^{-3}$ M;
All the Al precipitates, so $n = (0.223\text{ M})(1.30 \times 10^3 \text{ L}) = 2.90 \times 10^2$ mol;
$m = n\,MM = (2.90 \times 10^2 \text{ mol})(78.0 \text{ g/mol}) = 2.26 \times 10^4$ g or 22.6 kg.

16.109 At different points during a titration, different major species are present, because the titration reaction consumes formic acid and produces formate:
$$HCO_2H + OH^- \rightarrow HCO_2^- + H_2O$$
Point A is before titration begins, when the major species are H_2O and HCO_2H; point B is near the midpoint of the titration, when the major species are H_2O, HCO_2H, and HCO_2^-; point C is the stoichiometric point where all formic acid has been consumed, so the major species are H_2O and HCO_2^-; point D is beyond the stoichiometric point, so excess hydroxide is present and the major species are H_2O, OH^-, and HCO_2^-; the cation of the strong base will also be present as a major species at points B, C, and D.

Chapter 16: Aqueous Equilibria

16.111 Net ionic equations show only the reacting species. Remember that strong acids generate H_3O^+ in solution and react to completion with weak bases, and strong bases generate OH^- in solution and react to completion with weak acids:
(a) strong base reacting with weak acid: $OH^- + C_6H_5CO_2H \rightarrow H_2O + C_6H_5CO_2^-$;
(b) strong acid reacting with weak base: $H_3O^+ + (CH_3)_3N \rightarrow H_2O + (CH_3)_3NH^+$;
(c) weak base reacting with weak acid: $SO_4^{2-} + CH_3CO_2H \rightleftharpoons HSO_4^- + CH_3CO_2^-$; HSO_4^-, $pK_a = 1.96$; CH_3CO_2H, $pK_a = 4.75$; HSO_4^- is stronger, so this reaction proceeds to a small extent;
(d) strong base reacting with weak acid: $OH^- + NH_4^+ \rightarrow H_2O + NH_3$;
(e) weak base reacting with weak acid: $HPO_4^{2-} + NH_3 \rightleftharpoons PO_4^{3-} + NH_4^+$; HPO_4^{2-}, $pK_a = 12.32$; NH_4^+, $pK_a = 9.25$; NH_4^+ is stronger, so this reaction proceeds to a significant extent.

16.113 A molecular picture of a buffer solution should show molecules/ions of the conjugate acid/base pair in the correct proportions, as determined by the pH of the buffer solution. For acetic acid, $pK_a = 4.75$. Use the buffer equation to calculate the base/acid ratio of a formic acid buffer with pH = 4.27:

$$\log\left\{\frac{(\text{mol } A^-)}{(\text{mol } HA)}\right\} = pH - pK_a = 4.27 - 4.75 = 0.48; \quad \left\{\frac{(\text{mol } A^-)}{(\text{mol } HA)}\right\} = 0.33.$$

The picture should show three times as many acetic acid molecules as acetate ions. The solution is acidic, so there should also be some hydronium ions, but the exact proportion depends on the total concentration of the buffer species. Here is a view showing 6 acetic acid molecules, 2 acetate ions, and 2 hydronium ions:

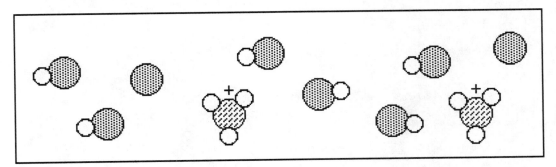

Chapter 17 Electron Transfer Reactions

17.1 Recognizing Redox Reactions

QUESTIONS TO ANSWER, SKILLS TO LEARN
1. **What is a redox reaction?**
2. **Assigning oxidation numbers to elements in compounds**
3. **Using oxidation numbers to recognize redox reactions**

Oxidation and reduction reactions were first discussed in Chapter 4. **Redox** is the contraction of *red*uction and *ox*idation. The mnemonic, OIL RIG, helps you remember the difference between oxidation and reduction.

oxidation: loss of electrons	Oxidation Is Loss
reduction: gain of electrons	Reduction Is Gain

A **redox reaction** is a chemical reaction in which one chemical species is oxidized and another is reduced. Because electrons have mass, one chemical species is reduced and another is oxidized in a redox reaction so that matter is conserved: the number of electrons lost by the oxidation process is the same as the number of electrons gained in the reduction process.

Oxidation and reduction must occur together since matter is conserved.

In the following redox reaction, copper gives up two electrons to two silver ions.

$$Cu + 2\,Ag^+ \rightarrow Cu^{2+} + 2\,Ag$$

One way to determine whether electrons are transferred in a reaction is by the use of oxidation numbers (a bookkeeping device for electrons). In a redox reaction, the oxidation numbers of some of the atoms change.

The oxidation number (ON) of an atom is the *apparent* or real charge that an atom has when all bonds between atoms of different elements are *assumed* to be ionic.

A set of rules that allow us to assign consistent oxidation numbers is helpful when dealing with redox reactions.

Rules for Determining Oxidation Numbers
1. The sum of oxidation numbers in the compound equals the charge on the chemical species (molecule or polyatomic ion). As a result:
   a. The oxidation number of an element in its elemental state (neutral) is zero.
   b. The oxidation number of a monoatomic ion is equal to the charge on the ion.

Chapter 17: Electron Transfer Reactions

2. When bonded to a nonmetal, the oxidation number of hydrogen is +1.

3. The most electronegative atom in a polyatomic species has an oxidation number equal to the number of valence electrons (VE) minus 8: ON = VE - 8.

When elements are chemically combined in a compound or polyatomic ion, we often can deduce their oxidation states from the groups in the periodic table to which they belong:

Group	ON	Examples	Exceptions
1	+1	Na_2S	
2	+2	$MgSO_4$	
17	-1	HCl	compounds with oxygen and other halogens
O	-2	H_2O	peroxides, RO-OR, ON = -1
H	+1	H_2O	M-H, ON = -1 (hydride)

Exercise 17.1 Assign oxidation numbers to all the elements in the following chemical species. (a) Na+ (b) HCO_2H (c) NO_2^- (d) titanium nitride, Ti_3N_4.

Steps to Solution: *The oxidation number in chemical species are determined by application of the rules/guidelines listed above, starting with the first rule and applying the others as necessary.*

(a) *In the sodium ion, application of rule 1 shows that the oxidation number is +1. The charge on Na+ = +1 so the oxidation number of sodium in the ion is +1.*

(b) *In HCO_2H (formic acid), the net charge is 0 (because it is a neutral molecule). The oxidation number on H is assumed to be +1 from rule 2. The most electronegative element is oxygen; its oxidation number is found from rule 3:*

$$ON (O) = VE - 8 = 6 - 8 = -2$$

The oxidation number of C is then found using rule 1 and the known oxidation numbers of H and O. The charge on formic acid is 0; therefore the sum of the oxidation numbers in the species is 0:

$$0 = 2(ON(H)) + 2(ON(O)) + ON(C)$$

$$0 = 2(+1) + 2(-2) + ON(C); \text{ therefore, } \boxed{ON (C) = +2}$$

(c) *In NO_2^- (nitrite ion) the net charge is -1. The most electronegative element is O; its oxidation number was found in part (b) to be -2. The sum of the oxidation numbers of the two O atoms and the N atom must be -1 from rule 1, so:*

$$-1 = 2(ON(O)) + ON(N); \text{ therefore, } \boxed{ON(N) = +3}$$

(d) *In Ti_3N_4 the net charge is 0. The most electronegative element is N; it has an oxidation number of:*

$$ON(N) = VE - 8 = 5 - 8 = -3$$

The sum of the oxidation numbers must be 0, giving:

$$0 = 3(ON(Ti)) + 4(ON(N)) = 3\,ON(Ti) + (-12); \text{ therefore, } \boxed{ON(Ti) = +4}$$

17.2 Balancing Redox Reactions

QUESTIONS TO ANSWER, SKILLS TO LEARN
1. Using oxidation numbers to balance redox reactions
2. What are half reactions?
3. How are half reactions used in balancing redox reactions?
4. What are the steps in balancing redox reactions?

Balancing redox reactions has all the requirements of balancing other types of chemical reactions; that is, mass must be conserved in the transformation. This requires that:

- Atoms of all elements must be conserved.
- Electrons must be conserved.
- Total electric charge must be conserved (this follows from electrons being conserved).
- All coefficients must be integers.

For charge to be conserved,

The number of electrons lost in the oxidation must equal the number of electrons gained in the reduction.

In balancing oxidation reduction reactions, we will use half reactions in which the "oxidation" and "reduction" processes are artificially separated to allow determination of the number of electrons transferred in the process. Several half reactions are shown below.

Oxidation half reactions (electrons are products)	Reduction half reactions (electrons are reactants)
$K \rightarrow K^+ + e^-$	$Au^{3+} + 3e^- \rightarrow Au$
$4OH^- \rightarrow O_2 + 2H_2O + 4e^-$	$K^+ + e^- \rightarrow K$
$H_2 \rightarrow 2H^+ + 2e^-$	$NO_3^- + 4H^+ + 3e^- \rightarrow NO + 2H_2O$

The half reaction method simplifies balancing redox reactions by breaking up one problem into several less complex ones:

Step 1 *Identify the species being oxidized and the species being reduced*

Step 2 *Write half reactions and balance them in both charge and matter (numbers of atoms).*
  a. In acidic solution, balance oxygen by addition of water and then hydrogen by addition of H^+.
  b. In basic solution, balance oxygen by addition of OH^- (remember to add twice as many OH^- as O atoms needed) and hydrogen by addition of H_2O.

Step 3 *Multiply each half reaction by a factor so that the number of electrons is the same in each half reaction.*

Step 4 *Add the two half reactions and clean up the resulting equation by canceling out molecules that appear in both products and reactants.*

Step 5 *Check the final net reaction to make sure that it is balanced in both charge and mass and that no reagent appears in both the products and reactants. Also, divide through by the lowest common denominator to give the simplest equation of reaction possible with integer coefficients. Check again.*

Exercise 17.2 Use the half reaction method to balance each of the following redox reactions:
  (a) $MnO_4^- + Sn^{2+} \rightarrow Sn^{4+} + Mn^{2+}$ (acidic solution)
  (b) $Cl_2 \rightarrow Cl^- + ClO^-$ (basic solution)
  (c) $I_2 + S_2O_3^{2-} \rightarrow I^- + S_2O_6^{2-}$ (basic solution)

<u>Steps to solution</u>: Follow the five steps described above for balancing redox reactions. These steps are referred to in the solutions shown below.

(a) $MnO_4^- + Sn^{2+} \rightarrow Sn^{4+} + Mn^{2+}$ (acidic solution)

The first step is to find the oxidation numbers of the substances in the redox reaction. The oxidation numbers for the monoatomic ions are the charges on the ions:

Chapter 17: Electron Transfer Reactions

Sn^{2+} ON = 2+ Sn^{4+} ON = 4+
Mn^{2+} ON = 2+

The ON of permanganate, MnO_4^-, is not as simple to find, and we must use the guidelines for calculation of oxidation numbers. We know that the sum of the oxidation numbers must equal the charge:

$$-1 = ON(Mn) + 4(ON(O))$$

Because ON (O) = -2 in this compound,

$$-1 = ON(Mn) + 4(ON(-2))$$

$$ON\ (Mn) = 8 - 1 = +7$$

In the oxidation half reaction, Sn^{2+} is being oxidized to Sn^{4+} (a change of 2 in oxidation number) and MnO_4^- is being reduced to Mn^{2+} (a change of 5 in oxidation number). The next step is to write and balance the half reactions.

Sn^{2+} is being oxidized to Sn^{4+} so electrons will appear as products in the half reaction:

$$Sn^{2+} \rightarrow Sn^{4+} + electrons$$

Because the change in oxidation numbers is 2, 2 electrons must be present as a product in the oxidation half reaction:

$$Sn^{2+} \rightarrow Sn^{4+} + 2\ e^-$$

This half reaction is balanced in charge and atoms.

In the reduction half reaction, the electrons are reactants, so we can write:

$$MnO_4^- + electrons \rightarrow Mn^{2+}$$

Mn is balanced; to balance the 4 O atoms, 4 H_2O molecules are added to the products. Eight H^+ are added to the reactants to balance the H atoms needed for the water molecules because the reaction is in acid solution:

$$8\ H^+ + MnO_4^- + electrons \rightarrow Mn^{2+} + 4\ H_2O$$

reactants' charge = 7   products' charge = 2

To balance this reaction in charge we must add 5 electrons to the reactants, the same as the change in oxidation number:

$$8\ H^+ + MnO_4^- + 5\ e^- \rightarrow Mn^{2+} + 4\ H_2O$$

This half reaction is balanced in charge and atoms.

The next step is to make sure that the same number of electrons are transferred in the reduction and oxidation half reactions. For this to be the case, the oxidation half reaction involving tin must be multiplied by 5 and the reduction half reaction by 2 so that 10 electrons are transferred in both. Then the two half reactions may be added.

Oxidation: $\quad\quad\quad\quad\quad\quad 5(Sn^{2+} \rightarrow Sn^{4+} + 2\ e^-)$
Reduction: $\quad\quad\quad 2(8\ H^+ + MnO_4^- + 5\ e^- \rightarrow Mn^{2+} + 4\ H_2O)$

Oxidation: $\quad\quad\quad\quad\quad 5\ Sn^{2+} \rightarrow 5\ Sn^{4+} + 10\ e^-$
Reduction: $\quad\quad 16\ H^+ + 2\ MnO_4^- + 10\ e^- \rightarrow 2\ Mn^{2+} + 8\ H_2O$

Sum: $\quad\quad 16\ H^+ + 2\ MnO_4^- + 5\ Sn^{2+} \rightarrow 2\ Mn^{2+} + 5\ Sn^{4+} + 8\ H_2O$

The reaction needs no cleanup and is balanced in charge and atoms. The final redox reaction is:

$$16\ H^+ + 2\ MnO_4^- + 5\ Sn^{2+} \rightarrow 2\ Mn^{2+} + 2\ Sn^{4+} + 8H_2O$$

(b) $Cl_2 \rightarrow Cl^- + ClO^-$ (in basic solution)

First we find the oxidation numbers of the three chemical species.
  Cl_2: ON = 0 $\quad\quad\quad\quad\quad\quad\quad\quad\quad\quad$ Cl^-: ON = -1
  ClO^-: ON(Cl) = x; $\;$ -1 = ON(Cl) + (2-) ; $\;$ ON(Cl) = +1

Therefore the oxidation half is Cl_2 being oxidized to ClO^-: $Cl_2 \rightarrow ClO^-$ and the reduction half is Cl_2 being reduced to Cl^-
[Cl(ON = 0) → Cl(ON = -1)]: $Cl_2 \rightarrow Cl^-$

We balance the oxidation half reaction by first balancing the Cl and O. A coefficient of 2 for ClO^- balances Cl:

$$Cl_2 \rightarrow 2\ ClO^-$$

To balance O in basic solution, we add twice as many OH^- as O atoms needed. The "extra" O and H atoms end up as water. In this reaction, 2 O atoms are needed so 4 OH^- ions are added to the reactants and 2 H_2O molecules to products:

$$Cl_2 + 4\ OH^- \rightarrow 2\ ClO^- + 2\ H_2O$$

Because 2 atoms of Cl are being oxidized by 1 electron each, 2 electrons must be products in this oxidation half reaction:

$$Cl_2 + 4\ OH^- \rightarrow 2\ ClO^- + 2\ H_2O + 2\ e^-$$

We balance the reduction half reaction by first balancing atoms. For chlorine to be conserved 2 atoms of Cl must appear in products:

$$Cl_2 \rightarrow 2\ Cl^-$$

Because 2 atoms Cl are being reduced by 1 electron each, 2 electrons must be reactants in this balanced reduction half reaction:

$$Cl_2 + 2e^- \rightarrow 2\ Cl^-$$

The third step is to multiply the reactions by a numerical factor so that the same number of electrons are transferred in the oxidation and reduction processes. However, in both these half reactions, the number of electrons is 2, so no such factor is required.

We proceed to the fourth step, adding the two half reactions and cleaning up:

$$Cl_2 + 4\ OH^- \rightarrow 2\ ClO^- + 2\ H_2O + 2\ e^-$$
$$Cl_2 + 2\ e^- \rightarrow 2\ Cl^-$$
$$\overline{2\ Cl_2 + 4\ OH^- \rightarrow 2\ ClO^- + 2\ Cl^- + 2\ H_2O}$$

All the coefficients are evenly divisible by 2. Cleanup requires that the coefficient be the smallest integer. The factor of 2 is removed to give the balanced redox reaction:

$$\boxed{Cl_2 + 2\ OH^- \rightarrow ClO^- + Cl^- + H_2O}$$

(c)
$$I_2 + S_2O_3^{2-} \rightarrow I^- + S_4O_6^{2-}$$

The first step is to identify the species being oxidized and reduced. We must find the oxidation numbers for the atoms in the substances in the chemical equation of reaction and compare those values in the reactants and products. In this reaction, O has ON = -2.

I_2 ON (I) = 0 $S_2O_3^{2-}$ ON (S) = 2
I^- ON (I) = -1

Calculating the oxidation number of S in $S_4O_6^{2-}$ uses the same method as in $S_2O_3^{2-}$ but gives an unusual result in that sulfur has a fractional oxidation number:

$$-2 = 4(ON(S)) + 6(O) = 4(ON(S)) + 6(-2): \quad ON(S) = \frac{10}{4} = 2.5$$

Iodine is being reduced and S is being oxidized.

The next step is to write and balance the half reactions.

We balance the oxidation half reaction by first balancing the S and O. A coefficient of 2 for $S_2O_3^{2-}$ balances both:

$$2\ S_2O_3^{2-} \rightarrow S_4O_6^{2-} + \text{electrons}$$

Because 4 atoms of S are being oxidized by $\frac{1}{2}$ electron each, 2 electrons must be products in this half reaction:

$$2\ S_2O_3^{2-} \rightarrow S_4O_6^{2-} + 2\ e^-$$

The oxidation half reaction is now balanced in charge and atoms. In the reduction half reaction, we first balance the atoms: for I to be conserved, 2 atoms of I must appear in products:

$$I_2 + \text{electrons} \rightarrow 2\ I^-$$

Because 2 atoms Cl are being reduced by 1 electron each, 2 electrons must be a reactant in this reaction:

$$I_2 + 2e^- \rightarrow 2\ I^-$$

The reduction half reaction is now balanced in charge and atoms. We now adjust the number of electrons transferred in the reduction and oxidation processes to be the same by multiplying one or both of the half reactions by some numerical factor. In this case, the number of electrons is the same (2) and no such factor is needed. In the fourth step, the two half reactions are added together to give the net redox reaction. If necessary, cleanup is performed by canceling out substances that appear in both products and reactants.

$$2\ S_2O_3^{2-} \rightarrow S_4O_6^{2-} + 2\ e^-$$
$$+\ I_2 + 2\ e^- \rightarrow 2\ I^-$$
$$\overline{2\ S_2O_3^{2-} + I_2 \rightarrow S_4O_6^{2-} + 2\ I^-}$$

Cleanup is not necessary.

17.3 Spontaneity of Redox Reactions

QUESTIONS TO ANSWER, SKILLS TO LEARN
1. **What are the two ways redox reactions are commonly performed?**
2. **How can half reactions be separated to harness electron transfer in redox reactions?**
3. **What is an electrochemical cell?**
4. **What are the parts and processes in an electrochemical cell?**

Chapter 17: Electron Transfer Reactions

The reaction of silver ions with copper is often used to deposit silver metal with the formation of copper ions:

$$2\,Ag^+_{(aq)} + Cu_{(s)} \rightarrow 2\,Ag_{(s)} + Cu^{2+}_{(aq)}$$

The standard free energy change for this reaction is:

$$\Delta G° = 2 \cdot \Delta G_f°(Ag_{(s)}) + \Delta G_f°(Cu^{2+}_{(aq)}) - \{2 \cdot \Delta G_f°(Ag^+_{(aq)}) + \Delta G_f°(Cu_{(s)})\}$$

$$\Delta G° = 2(0) + 64.98\,kJ - (2\,mole\,Ag^+\,(77.12\,kJ/mole\,Ag^+) + 0) = -89.27\,kJ$$

The reaction is predicted to be spontaneous because $\Delta G° < 0$. If you place a strip of copper metal in a 1 **M** solution of silver nitrate, the copper metal will become covered with silver metal and the solution will turn the characteristic blue color of aqueous copper(II) as the reaction proceeds. In the beaker, the electrons are being transferred to the silver ions from the atoms on the surface of the copper metal as silver ions collide with the surface. This is **direct electron transfer**. Direct electron transfer requires collisions between the species that are undergoing electron transfer.

Another type of electron transfer is possible. The free energy change is a state function; it only depends on the reactants' and products' initial and final states. It is possible to isolate the two reactants, silver ions and copper metal, in separate containers such that only reactants and products of each half reaction are present. In these cases **indirect electron transfer** occurs: the electrons are conducted through a wire (electron conductor) from one half reaction (the species being oxidized) to the other half reaction where another chemical species is being reduced. For indirect electron transfer to occur, there must be no buildup of electrical charge in the containers (or cells) where oxidation and reduction are occurring. Therefore cations must flow into the container (or cell) in which reduction is occurring and anions must flow into the cell where the oxidation half reaction is occurring. This transfer of ions can be accomplished by several techniques; one common method is to use a **salt bridge** that contains a solution of an electrolyte whose ions are spectators to the redox process.

Exercise 17.3 (a) Draw molecular pictures illustrating direct electron transfer in the reaction of silver(I) ions with copper metal.
(b) Draw a figure illustrating how a cell would be arranged for the same redox reaction but using indirect electron transfer and a salt bridge with KNO_3 solution. Indicate the direction of electron flow in the wire and the movement of ions in the salt bridge.

Steps to Solution: *This question asks us to illustrate (schematically) the differences between direct and indirect electron transfer. In part (a) we need to show that collisions must occur between the two reactants in direct transfer. In part (b), we need to show how the two half reactions are separated, the direction of electron movement and the direction of flow of anions and cations in the salt bridge.*

(a) *The following series of three illustrations shows the important features of direct electron transfer: approach of the reactants, collision, electron transfer and finally, the formation of products and release of the aqueous ion.*

(1)   (2)   (3)

Silver ions in aqueous solution near Cu metal (the water molecules are not shown).	Two silver ions collide with the copper metal and direct electron transfer occurs with the formation of a Cu^{2+} ion.	The Cu^{2+} ion is released into solution and the silver atoms form the beginning of a silver metal crystal.

(b) *Indirect electron transfer requires that (1) the half reactions occur separately, (2) electrons be transferred from one half reaction to the other by means of an electron conductor (a wire) and (3) ion transport takes place that will compensate for the change in electrical charge that occurs as a result of electron transfer. The two sketches below show one way to separate the half reactions so that direct electron transfer cannot occur: the product and reactant for each half reaction are placed in separate containers. Notice that there is a way to connect the electrical and chemical parts of the reactions: the metal strips in the solutions can react with the solutions and can conduct electrons. These are called* **electrodes**.

Oxidation   Reduction

(aqueous copper sulfate and copper)   (aqueous $AgNO_3$ and silver)

$Cu \rightarrow Cu^{2+} + 2\ e^-$ $2\ Ag + 2\ e^- \rightarrow 2\ Ag_{(s)}$

The next components needed are those required to let electrical charge flow from one container to another. We need something to conduct electrons: a metal strip or wire can conduct electrons (shown as "e^-" in the figure below) from the strip of metal in one container to the other. However, as electrons move from one container to the other, one container will build up an excess of negative charges (the container *containing the copper in this case) and the other is losing electrons and will build up a positive charge. This problem is overcome by insuring that there are ions in both solutions and a device that allows transfer of ions between the two containers, but does not allow direct electron transfer to occur at a significant rate. In the sketch below, a device called a **salt bridge** is shown. This salt bridge is a U-shaped tube filled with an electrolyte solution such as $KNO_{3\,(aq)}$.*

The figure above shows the motion of the various components of a redox reaction (electrons and ions) which harnesses the energy of a spontaneous chemical process to produce electrical energy: a battery or galvanic cell!

A cell contains all the components for a redox reaction to occur through indirect electron transfer.

- The anions of the electrolyte solution migrate toward the container where oxidation is occurring, because negative charges (electrons) are leaving that part of the cell.

    *An*ions migrate toward the *an*ode where they are *ox*idized.

- The cations in the electrolyte solution migrate toward the container where reduction occurs, because electrons are entering that container.

    *Cat*ions migrate toward the *cat*hode where they are *re*duced.

Critical components of an electrochemical cell are the **electrodes**, which allow the chemical part of the reaction (oxidation or reduction) to be linked to the electrical circuit (the wire through which electrons travel). Electrodes are of three general types:

Active: *a metal or other electronic conductor that is either a reactant or product in the redox reaction*

Passive: *a metal or other electronic conductor that does not participate in the redox reaction; it only serves to transport electrons to or from the species in solution.*

Reference: *an electrode that contains a half reaction under easily duplicated conditions so that one can measure the voltage of other half reactions compared to the reference.*

The electrodes in Exercise 17.3 are both active electrodes. Examples of passive electrodes are gold and platinum, which do not react readily and so can transfer electrons to or from species in the solution. Examples of a reference electrode are the *standard hydrogen electrode*, the *saturated calomel electrode (SCE)* and the *silver-silver chloride electrode*. Reference electrodes are discussed in Section 17.5.

17.4 Cell Potentials

QUESTIONS TO ANSWER, SKILLS TO LEARN
1. **What is cell potential and why can it be measured in volts?**
2. **Using standard cell voltages to calculate voltages of combinations of half reactions**

In the preceding sections we have seen that the free energy change in a redox reaction can be used to move electrons through an external circuit. When charged particles (such as electrons) move to a position of lower energy, the energy change may also be described in terms of electrical potential (volts). In a chemical system where a redox process can occur, the electrical potential difference between the two electrodes is the **electromotive force**, E, measured in volts, V. The electromotive force depends on the free energy change, as we will discuss in the next section. The energy that is released upon moving charged particles is found by multiplying the charge and the electrical potential:

$$\text{energy} = E \cdot q = (V)(C) \qquad 1\text{ J} = 1\text{ V} \cdot \text{C}$$

We can measure this voltage by using a voltmeter with the negative lead attached to the anode (where electrons emerge from the cell) and the positive lead to the cathode (at which electrode the electrons combine with chemical species). The source of the electromotive force is based on a number of factors, one of which is the relative affinity of two chemical species, the oxidizing agent and the reducing agent, for electrons. The E of a single half reaction cannot be determined because every redox reaction must have both an oxidizing agent and a reducing agent present. This presents a problem similar to measuring the elevation of a landmark. A standard must be chosen that will serve as the zero point; in elevation, the zero point is sea level. In redox chemistry, the standard is the reduction of aqueous hydronium ions to hydrogen under standard conditions of temperature and pressure (the **standard hydrogen electrode,** SHE) which has a potential of 0 V by definition.

Chapter 17: Electron Transfer Reactions

$$2 \text{ H}^+_{(aq)} + 2 \text{ e}^- \rightarrow \text{H}_{2\,(g)} \qquad E° = 0.00 \text{ V}$$

The potentials of half reactions as reductions are then determined by measuring the E of cells in which the SHE is the anode and the half reaction of interest is connected in the circuit as if it is the cathode. These values in tables of potentials are called **standard potentials** or **standard reduction potentials**. The standard potential is an intensive property: it depends on the relative attraction of the chemical species for electrons. Calculating the potential of a galvanic cell using the reaction of copper and silver ion illustrates the use of standard potentials.

In the redox reaction of silver ion and copper, the balanced equation is:

$$2 \text{ Ag}^+_{(aq)} + \text{Cu}_{(s)} \rightarrow 2 \text{ Ag}_{(s)} + \text{Cu}^{2+}_{(aq)}$$

The standard reduction potential for silver ion is:

$$\text{Ag}^+ + \text{e}^- \rightarrow \text{Ag} \qquad E° = E°_{cathode} = 0.7996 \text{ V}$$

Because silver is being reduced in the reaction as written, this is the reaction that will occur at the cathode. Copper is being oxidized in the reaction and is the anode. To obtain the complete, correct redox reaction, the copper half reaction must be written as an oxidation.

When a half reaction is reversed, the sign of the standard potential is changed.

$$\text{anode reaction} \qquad \text{Cu} \rightarrow \text{Cu}^{2+} + 2 \text{ e}^- \qquad E° = -0.3419 \text{ V}$$

The standard reduction potential for copper is:

$$\text{Cu}^{2+} + 2 \text{ e}^- \rightarrow \text{Cu} \qquad E° = E°_{anode} = 0.3419 \text{ V}$$

The *potential* (or voltage) that can be produced by the copper-silver ion redox reaction under standard conditions is the standard reduction potential of the reduction half reaction minus the standard reduction potential of the oxidation half reaction:

$$E°_{cell} = E°_{cathode} - E°_{anode} = 0.7996 - 0.3419 \text{ V} = 0.4577 \text{ V}$$

COMMON PITFALL: Multiplying cell potentials of half reactions by stoichiometric coefficients. Cell potentials are intensive properties. An example from above:
$E°_{cell} = E°_{cathode} - E°_{anode} = 2(0.7996) - 0.3419 \text{ V} = 1.2573 \text{ V}$: The **2** is **WRONG!**

Exercise 17.3 What would be the standard potential of voltaic cells that combine the following half reactions?
  (a) Pb to $PbSO_4$ and PbO_2 to $PbSO_4$ (acid solution)
  (b) The redox reaction $2 \text{ Na} + \text{S} \rightarrow \text{Na}_2\text{S}$ ($E°$ for $\text{S} + 2 \text{ e}^- = \text{S}^{2-}$ = -0.508 V)

Steps to Solution: First we must determine the half reactions that are occurring by finding the species being oxidized and reduced. Then the reduction half reactions (cathode reaction) and the oxidation half reaction (anode reaction) may be written and the standard potentials found in a table of reduction potentials.

(a) Pb to $PbSO_4$ and PbO_2 to $PbSO_4$ (acid solution)
This is the combination of reactions that occur in the automobile lead-acid battery. The species being oxidized is Pb: it is being converted to the salt, $PbSO_4$, in which lead has a 2+ charge. PbO_2 is being reduced to $PbSO_4$. The half reactions are listed in Table 17-1.

Oxidation (anode) half reaction: $Pb + SO_4^{2-} \rightarrow PbSO_4 + 2 e^-$ $E° = 0.3588$ V
Anode reaction as a reduction: $PbSO_4 + 2 e^- \rightarrow Pb + SO_4^{2-}$ $E° = -0.3588$ V
Cathode reduction: $PbO_2 + 2 H_2SO_4 + 2 e^- \rightarrow PbSO_4 + 2 H_2O + SO_4^{2-}$
$E°_{cathode} = 1.6913$ V

Net reaction: $Pb + PbO_2 + 2 H_2SO_4 \rightarrow 2 PbSO_4 + 2 H_2O$

$E°_{cell} = E°_{cathode} - E°_{anode} = 1.6913$ V $- (-0.3588$ V$) = \boxed{2.0501 \text{ V}}$

(b) $2 Na + S \rightarrow Na_2S$ ($E°$ for $S + 2 e^- = S_2^- = -0.508$ V)
This redox reaction is being considered as the basis of a high-energy battery. Sodium is being oxidized and sulfur is being reduced.

Oxidation (anode) half reaction $Na \rightarrow Na^+ + e^-$ $E° = 2.71$ V
Standard reduction reaction $Na^+ + e^- \rightarrow Na$ $E°_{anode} = -2.71$ V

Cathode half reaction $S + 2e^- \rightarrow S^{2-}$ $E° = -0.508$ V

The cell potential at standard conditions is:

$E°_{cell} = E°_{cathode} - E°_{anode} = -0.508 - (-2.71) = \boxed{2.20 \text{ V}}$

To find the standard potential for a redox reaction, all that is required is: (1) identify the half reactions and (2) find the standard reduction potentials of those half reactions.

Chapter 17: Electron Transfer Reactions

17.5 Free Energy and Electrochemistry

QUESTIONS TO ANSWER, SKILLS TO LEARN
1. What is the relationship between cell potential, E, and free energy, ΔG?
2. Using cell potentials to calculate equilibrium constants
3. Predicting how cell potential varies for nonstandard conditions
4. The Nernst equation
5. What is the relationship between electrical current, time and charge?
6. What are the units of electrical charge, and what is the charge on one mole of electrons?
7. Calculating the amount of electrically induced chemical change from the current passed and time

The free energy change and the cell potential correlate quite well for the prediction of a spontaneous reaction. The free energy and cell potential are related, as shown below.

$\Delta G° = -nFE°$ (for standard conditions) $\Delta G = -nFE$ (for nonstandard conditions)

The quantity, n, is the number of electrons transferred in the balanced redox reaction (the number of electrons in each half reaction before they are added) and F is the Faraday constant, 96485 C/mole. A summary of the relationships between reaction spontaneity, $\Delta G°$, K and $E°$ is shown below.

Standard state reaction is:	$\Delta G°$	K	$E°$
spontaneous	< 0	> 1	> 0
at equilibrium	0	1	0
not spontaneous	> 0	< 1	< 0

Exercise 17.4 Calculate the standard free energy changes for the redox reaction between silver ion and copper and the redox reaction described in Exercise 17.5, part (a).

Steps to Solution: This problem requires that we determine the standard potential, $E°$, for the redox process and also the balanced chemical equation of reaction, which is required to find the value of n for the equation:

$\Delta G° = -nFE°$ ($F = 96485$ C/mol)

Substitution of n and $E°$ into this equation gives the standard free energy change for the reaction.

(a) *The balanced chemical equation and $E°$ of the reaction are given below:*

$$2\ Ag^+_{(aq)} + Cu_{(s)} \rightarrow 2\ Ag_{(s)} + Cu^{2+}_{(aq)} \qquad E° = 0.4577\ V$$

Examining the half reactions, we find that the number of electrons transferred in the balanced redox reaction was 2; therefore n = 2. Substitution of the values of n and E° into the above equation gives:

$$\Delta G° = -nFE° = -2\ mole(96485\ C/mole)(0.4577\ V) = -8.832 \times 10^4\ C\cdot V = -8.832 \times 10^4\ J$$

$$\boxed{\Delta G° = -88.32\ kJ}$$

(b) *The balanced equation of reaction shows n = 2 and E° = 2.0501 V, as found in Exercise 17.5 (a). The free energy change is:*

$$\Delta G° = -nFE° = -2\ mole(96485\ C/mole)(2.0501\ V) = -3.956 \times 10^5\ C\cdot V = -3.956 \times 10^5\ J$$

$$\boxed{\Delta G° = -395.6\ kJ}$$

It is interesting to note that if the equation is multiplied by an integer factor (for example, each coefficient is multiplied by a factor of 2), $\Delta G°$ changes, but $E°$ is unchanged. This observation underlines the fact that $E°$ reflects the inherent reactivity, but the energy one can obtain from a chemical process depends on the inherent reactivity *and* the amount(s) of substance(s) that undergo reaction.

Standard Cell Potentials and Chemical Equilibrium

The free energy change is related to the equilibrium constant and to the cell potential. These mathematical expressions can be combined to calculate equilibrium constants from cell potentials.

$$\Delta G° = -RT\ln K = -nFE° \qquad\qquad E° = \frac{RT}{n}\ln K$$

This expression can be rewritten for $T = 298$ K and using log (base 10) instead of ln (because many calculations use pK's (-logK)) as:

$$E° = \left(\frac{5.916 \times 10^{-2}\ V}{n}\right)\log K \quad or \quad \log K = \frac{E° \cdot n}{5.916 \times 10^{-2}\ V}$$

Exercise 17.5 Calculate the equilibrium constants for the following redox reactions:
(a) $2\ Cu^{2+}_{(aq)} + Sn^{2+}_{(aq)} \rightarrow 2\ Cu^+_{(aq)} + Sn^{4+}_{(aq)}$
(b) $Fe^{3+}_{(aq)} + Cu^+_{(aq)} \rightarrow Fe^{2+}_{(aq)} + Cu^{2+}_{(aq)}$

Steps to Solution: To solve this problem, we must first write the half reactions involved and find the standard potentials for the half reactions, which allows us to calculate the cell potential for the redox reaction. Then the number of electrons, n, in the balanced redox reaction must be determined and then the values of n and E° calculated.

(a) The oxidation and reduction half reactions (adjusted to give the balanced redox reaction) and standard potentials are shown below.

Oxidation $Sn^{2+}_{(aq)} \rightarrow Sn^{4+}_{(aq)} + 2 e^-$ $E°_{anode} = 0.13$ V
Reduction $2 Cu^{2+}_{(aq)} + 2 e^- \rightarrow 2 Cu^+_{(aq)}$ $E°_{cathode} = 0.15$ V

$E°_{cell} = E°_{cathode} - E°_{anode} = 0.15 - 0.13$ V $= 0.02$ V

The value of n is 2; substituting into the equation for log K gives:

$$\log K = \frac{E° \cdot n}{5.916 \times 10^{-2} \text{ V}} = \frac{0.02 \text{ V} \cdot 2}{5.916 \times 10^{-2} \text{ V}} = 0.67$$

$K = 10^{0.67} = 4.7 \Rightarrow \boxed{K = 5}$ (rounded)

(b) The oxidation and reduction half reactions (adjusted to give the balanced redox reaction) and standard potentials are shown below.

Oxidation $Cu^+_{(aq)} \rightarrow Cu^{2+}_{(aq)} + e^-$ $E°_{anode} = 0.15$ V
Reduction $Fe^{3+}_{(aq)} + e^- \rightarrow 2 Fe^{2+}_{(aq)}$ $E°_{cathode} = 0.77$ V

$E°_{cell} = E°_{cathode} - E°_{anode} = 0.77 - 0.15$ V $= 0.62$ V

The value of n is 1; substituting into the equation for log K gives:

$$\log K = \frac{E° \cdot n}{5.916 \times 10^{-2} \text{ V}} = \frac{0.62 \text{ V} \cdot 1}{5.916 \times 10^{-2} \text{ V}} = 10.48$$

$K = 10^{10.48} = 3.02 \times 10^{10} \Rightarrow \boxed{K = 3 \times 10^{10}}$ (rounded)

Notice that a potential difference of little more than 0.5 V leads to a large equilibrium constant!

The Nernst Equation

Recently we found that the free energy change may be related to the concentrations of components when not in their standard state by the equation

$$\Delta G = \Delta G° + RT \ln Q$$

where Q is the reaction coefficient. Because

$$\Delta G = -nFE$$

substitution gives us an equation called the Nernst equation that allows us to calculate the potential, E, under nonstandard conditions:

$$E = E° - \frac{RT}{nF} \ln Q$$

For $T = 298$ K and using log (base 10) instead of ln (base e) gives:

$$E = E° - \left(\frac{5.916 \times 10^{-2} \text{ V}}{n}\right) \log Q$$

E is the measured cell potential; $E°$ is the standard cell potential

Exercise 17.6 Consider the Daniell cell where the cell reaction and standard potential are:

$$Zn_{(s)} + Cu^{2+}_{(aq)} \rightarrow Zn^{2+}_{(aq)} + Cu_{(s)} \qquad E° = 1.10 \text{ V}$$

If the cell is initially at standard conditions ($[Cu^{2+}] = [Zn^{2+}] = 1.00$ M), what are the concentrations of Cu^{2+} and Zn^{2+} when the cell potential has fallen to 1.06 V?

Steps to Solution: This problem asks for the concentrations of substances at a nonstandard potential, so the Nernst equation will be needed because it relates potential at nonstandard concentrations to the potential. First we need to find $E°$ and n to fill those two quantities into the Nernst equation.

$$E = E° - \left(\frac{5.916 \times 10^{-2} \text{ V}}{n}\right) \log Q$$

The pure metals Cu and Zn do not appear in Q, so

$$Q = \frac{[Zn^{2+}]}{[Cu^{2+}]}$$

It is the value of Q that will be needed to solve for the concentrations when e⁻ = 1.06 V. Substituting into the Nernst equation, and solving for Q:

$$E = 1.06 \text{ V} = 1.10 \text{ V} - \frac{0.0592}{2} \log \frac{[Zn^{2+}]}{[Cu^{2+}]} = 1.10 \text{ V} - 0.0296 \log \frac{[Zn^{2+}]}{[Cu^{2+}]} \text{ V}$$

Chapter 17: Electron Transfer Reactions

$$E = E° - 0.0286 \log \frac{[Zn^{2+}]}{[Cu^{2+}]}$$

$$\log Q = -\frac{(E - E°)}{0.0296} = -\frac{(1.06 - 1.10 \text{ V})}{0.0296 \text{V}} = 1.35$$

$$Q = 10^{(1.35)} = 22.5 = \frac{[Zn^{2+}]}{[Cu^{2+}]} \qquad [Zn^{2+}] = 22.5 \, [Cu^{2+}]$$

To find the concentrations, we must compute the concentrations of copper(II) and zinc(II). The initial concentrations of Cu(II) and Zn(II) were 1.00 M (standard conditions). Because the stoichiometry of the reaction is 1 mole Cu^{2+} reduced gives 1 mole Zn^{2+}, the sum of $[Cu^{2+}]$ and $[Zn^{2+}]$ will remain constant throughout the course of the reaction, so:

$$[Zn^{2+}] + [Cu^{2+}] = 2.00 \text{ M}$$

$$22.5 \, [Cu^{2+}] + [Cu^{2+}] = 2.00 \text{ M} \qquad [Cu^{2+}] = \frac{2.0 \text{ M}}{22.5 + 1} = 8.5 \times 10^{-2} \text{ M}$$

$$[Zn^{2+}] = 2.00 \text{ M} - [Cu^{2+}] = 2.00 \text{ M} - 8.5 \times 10^{-2} = 1.91 \text{M}$$

Notice that a difference in concentration of a factor of 20 changes the potential by only about 0.04 V!

Concentration Cells

Consider the effect of connecting two containers of equal volume by a small tube, one containing 2.0 M $AgNO_3$ and the other 1.00×10^{-4} M $AgNO_3$. We expect that the two solutions will mix and at equilibrium, the concentrations of silver ion will be the same in both compartments. This is a spontaneous process; therefore the free energy change must be < 0 and an electrical potential should exist when the concentrations of the solutions are different. If the electrochemical cell below is prepared with silver metal electrodes in both cells, the mixing process occurs by electron transfer through the external wire as indicated by the wire in the sketch. In the compartment to the left ($[Ag^+]$ = 2.0 M), the concentration of silver ion is lowered

Chapter 17: Electron Transfer Reactions

2.0 M Ag⁺   1 x10⁻⁴ M Ag⁺
(cathode)   (anode)

by reduction of silver ion:

$$Ag^+ + e^- \rightarrow Ag$$

In the right compartment, the silver ion concentration increases due to the oxidation of the silver metal electrode:

$$Ag \rightarrow Ag^+ + e$$

The potential can be calculated by the Nernst equation. The quantities required to find the potential with the Nernst equation are $E°$ and n, the number of electrons transferred. Because the half reactions for oxidation and reduction are the same, $E° = 0$ (and will be for *any* concentration cell). The number of electrons transferred is one, so $n = 1$. The reaction quotient, Q, is:

$$Q = \frac{[Ag^+]_{anode}}{[Ag^+]_{cathode}}$$

(the anode is the cell where the species in the concentration cell is at lower concentration)

$$E == E° - \left(\frac{5.916 \times 10^{-2} \text{ V}}{n}\right) \log Q = 0 - \frac{5.916 \times 10^{-2} \text{ V}}{1} \log \frac{1 \times 10^{-4} \text{ M}}{2.0 \text{ M}} = \boxed{0.25 \text{ V}}$$

Electrons possess electrical charge, so in a redox process, one can measure the amount of chemical change that occurs by measuring the charge that passes through the wire (external circuit). The SI unit for electrical charge is the coulomb, with the symbol C. The charge on an electron is 1.602×10^{-19} C. The charge on 1 mole of electrons is an important quantity named the **Faraday**, F.

$$1 F = 6.022 \times 10^{23} \text{ 1/mole } (1.602 \times 10^{-19} \text{ C}) = 96{,}484 \text{ C/mole e}$$

The rate at which charge flows in an electrical circuit is called the current, I (Amperes, A (C/s)), and is the charge, q (in C), that flows per unit time, t (in units of s):

$$I = \frac{q}{t} \qquad \text{(units: A, C/s)}$$

The charge that passes through an electrical circuit can be calculated from:

Chapter 17: Electron Transfer Reactions

$$q = I(t)$$

The number of moles of a substance produced or consumed in a redox reaction is related to the charge transferred in the reaction.

$$\text{moles electrons transferred, } n = \frac{q}{F}$$

Exercise 17.7 Aluminum is used in a battery in which the following reaction occurs:

$$4\ Al_{(s)} + 3\ O_{2\ (g)} + 4\ OH^-_{\ (aq)} + 6\ H_2O \rightarrow 4\ Al(OH)^{4-}_{\ (aq)}$$

If the battery must supply a current of 78 A for 4.0 hours, what mass of Al (in g) must be contained in the battery?

Steps to Solution: This problem can be started from several different directions; however, the words, "what mass of Al" tell us that this is a stoichiometry problem. The difference from previous stoichiometry problems is that the amount of product that we want to make is not described in moles of a product, but in the amount of charge that passes through a circuit. First, we must find the charge that passes through the circuit and then convert the charge from coulombs to moles of electrons. Then we must find the stoichiometric relationship between the moles of electrons and the moles of Al, which is found from the change in oxidation number. This will allow us to calculate the number of moles of Al required and thus calculate its mass.

First we calculate the charge. We must make sure that time is in units of s!

$$q = I(t) = 78 \frac{C}{s} (4.0\ hr) \cdot 60 \frac{min}{hr} \cdot 60 \frac{s}{min} = 1.12 \times 10^6\ C$$

The charge in C is now converted to moles of electrons, or Faradays:

$$\text{moles } e^- = \frac{q}{F} = \frac{1.12 \times 10^6\ C}{96485\ C/mol\ e} = 11.6\ \text{mole } e^-$$

The moles of electrons are related to the moles of aluminum by the number of electrons lost by Al when it is oxidized in the reaction. This is just the change in oxidation number. In the reactants, Al has ON = 0; in the products, Al has ON = +3. There are 3 moles of electrons transferred per mole of Al:

$$\text{mole ratio: } \frac{1\ \text{mole Al}}{3\ \text{mole electrons}}$$

The mass of aluminum is found by multiplying the molar mass of Al and the moles of Al:

$$mass_{Al} = moles_{Al} \cdot MM_{Al}$$

$$\text{mass}_{Al} = (11.6 \text{ mole e}^-) \frac{1 \text{ mole Al}}{3 \text{ mole e}^-} \cdot 26.98154 \text{ g/mole Al} = 1.05 \times 10^2$$

$$\boxed{\text{g Al} = 1.0 \times 10^2 \text{ g}}$$

17.6 Redox in Action

QUESTIONS TO ANSWER, SKILLS TO LEARN
1. **Converting chemical energy to electrical energy: understanding batteries**
2. **What is corrosion and how can it be minimized?**

Batteries

A **battery** is a device in which a spontaneous redox reaction is used to produce electrical energy (sometimes a battery is called a voltaic cell or a galvanic cell; more properly, a battery is a combination of two or more voltaic cells, such as a car battery). A large number of batteries have been designed to meet the requirements of portable electrical power, from automobiles to flashlights to batteries that are used in artillery shells and must be able to withstand the shock of being shot from a cannon. The requirements for a battery depend on its planned use(s). Useful batteries have the following two features in common:

- A large cell potential, $E°$.
- A large concentration of ions in the electrolyte are necessary because the flow of charge (ions) in the electrolyte must be equal to the charge flow of electrons in the external circuit.

Other features that are desirable in a battery are:

- Compact size and inexpensive materials and construction
- Rechargeable and using nontoxic materials (environmentally safe)

Examples of commercial batteries are described in the text.

Corrosion

Corrosion is the oxidation of materials that results in chemical change of the material, resulting in weakening or other deterioration of the structures made of the material. The most common oxidizing agent is the oxygen present in the atmosphere. The text shows that under atmospheric conditions, the reduction potential of $O_{2(g)}$ is 0.88 V; therefore, metals with standard reduction potentials less than 0.88 V undergo oxidation with $e^- > 0$ V (spontaneous reactions). Most metals used in construction have reduction potentials that make them susceptible to oxidation by air.

Chapter 17: Electron Transfer Reactions

There are three common methods of minimizing corrosion:

- *Protection*: A physical barrier such as paint keeps oxygen from being able to interact with the metal surface.

- *Passivation*: A layer of the oxidation product coats the metal and prevents further reaction with oxygen.

- *Sacrificial anode*: A substance is placed in electrical contact with the metal object that is to be protected. The substance must have a lower (more negative) reduction potential than the object to be protected so that it will be preferentially oxidized.

Exercise 17.8 An automobile manufacturer once used aluminum water pumps bolted onto iron engines. These water pumps were noted to rapidly corrode and fail. Explain this observation.

Steps to Solution: This problem describes a system that contains two metals, one of which preferentially undergoes corrosion. Examination of the reduction potentials of both metals will help to determine the relative susceptibilities to oxidation. Also, the fact that the engine and pump are bolted together tells us that they are in electrical contact. This information may also be important in understanding the phenomenon described in the problem.

The reduction potentials of iron and aluminum to their most accessible (lowest potential) oxidation numbers are shown below.

$$Fe^{2+} + 2\ e^- \rightarrow Fe_{(s)} \quad E° = -0.41\ V \qquad Al^{3+} + 3\ e^- \rightarrow Al_{(s)} \quad E° = -1.71\ V$$

Corrosion requires that the oxygen be reduced at a site on the metal surface ($O_2 + 2\ H_2O + 4\ e^- \rightarrow 4\ OH^-$) and a metal be oxidized elsewhere ($M \rightarrow M^{n+} + ne$). If two metals are in contact, electrons can flow from one metal to the other because metals are good conductors of electrons. Therefore the electrons required for the reduction of oxygen will be supplied by the metal that is most easily oxidized (has the most negative reduction potential). Aluminum is the more easily oxidized of the two metals; therefore aluminum will be oxidized. The problem with the water pumps was that they were acting as a sacrificial anode to protect the engine from corrosion. However, it only required loss of a small amount of aluminum to result in failure of the water pump!

17.7 Electrolysis

QUESTIONS TO ANSWER, SKILLS TO LEARN
1. What is electrolysis?
2. Calculating amounts of substances produced by electrolysis
3. Predicting which substances will be reduced and oxidized upon application of electrical energy

Electrolytic cells use electrical energy from an external source (such as a battery or other electrical power supply) to drive a chemical reaction that is not spontaneous.

Electrolysis is the use of external electrical energy to drive a nonspontaneous redox reaction.

A hydraulic analogy to this process is: water spontaneously flows downhill; however, use of a water pump will send water uphill. In the second case, external energy must be supplied to run the pump. In electrolytic reactions, the power supply serves to "pump" electrons into the cell to reduce chemical species and pump them out of the other "side" to oxidize the other chemical species. The following redox reaction is not spontaneous as written:

$$2\ Ag_{(s)} + Cu^{2+}_{(aq)} \rightarrow 2\ Ag^{+}_{(aq)} + Cu_{(s)} \quad E° = -0.4577\ V$$

However, connecting a power supply with a potential greater than 0.4577 V to such a cell changes the situation; if the negative lead is connected to the Cu electrode, it is pumping electrons in, making the Cu strip the cathode:

$$Cu^{2+} + 2e^- \rightarrow Cu$$

The positive lead is connected to Ag, making it the anode:

$$Ag \rightarrow Ag^+ + e^-$$

This situation is the opposite of the spontaneous reaction described in Sections 17.4 and 17.5. In practice, the potential required to induce electrolysis is greater than the cell potential found either by the standard potential or by the Nernst equation; the extra potential required for electrolysis to occur at a useful rate is called the **overvoltage** or **overpotential**. The amount of substance produced by electrolysis is found by the charge passed through the cell.

Exercise 17.9 If an electrolytic cell driving the following redox reaction has a current of 4.02 A passed through it for 2.32 hours, how much Ag will be dissolved and how much Cu will be deposited?

$$2\ Ag_{(s)} + Cu^{2+}_{(aq)} \rightarrow 2\ Ag^{+}_{(aq)} + Cu_{(s)}$$

Steps to Solution: This is an electrochemical stoichiometry problem; we must first find how many moles of electrons pass through the circuit and then use the redox half reactions to relate the number of moles of electrons to the number of moles of each metal.

The first step is to find the number of moles of electrons using the relationship between current and charge and the Faraday:

$$\text{moles electrons} = \frac{I(t)}{F} = \frac{4.02 \text{ C/s}(2.32 \text{ hours} \cdot 60 \text{ min/hour} \cdot 60 \text{ s/min})}{96485 \text{ C/mole}} = 0.348 \text{ mole e}^-$$

From the half reactions we obtain the mole relationship between electrons and the metals:

each mole of Ag gives 1 mole of Ag+ (and 1 e⁻)
each mole of Cu²⁺ requires 2 moles of e⁻ to give 1 mole of Cu

Using the mole relationships, we can calculate the mass of the metals:

$$\text{g Cu deposited} = 0.348 \text{ mole e}^- \cdot \frac{1 \text{ mole Cu}}{2 \text{ mole e}^-} \cdot 63.546 \text{ g/mole Cu} = \boxed{11.1 \text{ g Cu}}$$

$$\text{g Ag dissolved} = 0.348 \text{ mole e}^- \cdot \frac{1 \text{ mole Ag}}{1 \text{ mole e}^-} \cdot 107.868 \text{ g/mole Ag} = \boxed{37.5 \text{ g Ag}}$$

Exercise 17.10 For a brine electrolysis cell (see redox reaction below) operating at 60,000 amps, how many kg of NaOH and Cl₂ would be produced in 24.0 hours?

$$2 \text{ NaCl}_{(aq)} + 2 \text{ H}_2\text{O} \rightarrow 2 \text{ NaOH}_{(aq)} + \text{Cl}_{2(g)} + \text{H}_{2(g)}$$

Steps to Solution: This problem is solved in the same manner as Exercise 17.12, but it illustrates an important industrial process.

$$\text{mole e}^- = \frac{60000 \text{ coul/s} \cdot 60 \text{ s/min} \cdot 60 \text{ min} \cdot 24.0 \text{ hr}}{96485 \text{ coul/mole}} = 5.37 \times 10^4 \text{ moles e}^-$$

$$\text{kg NaOH} = \frac{1 \text{ mole NaOH}}{\text{mole e}^-} \cdot \frac{40.0 \text{ g/mole}}{1000 \text{ g/kg}} \cdot 5.37 \times 10^4 \text{ mole e}^- = \boxed{2.15 \times 10^3 \text{ kg NaOH}}$$

$$\text{kg Cl}_2 = \frac{1 \text{ mole Cl}_2}{2 \text{ mole e}^-} \cdot \frac{70.9 \text{ g/mole}}{1000 \text{ g/kg}} \cdot 5.37 \times 10^4 \text{ mole e}^- = \boxed{1.90 \times 10^3 \text{ kg Cl}_2}$$

Chapter 17: Electron Transfer Reactions

Test Yourself

1. Cells that utilize the reaction of oxygen with aluminum in basic conditions giving $Al(OH)_4^-$ are manufactured by a number of foreign companies. How much Al would be consumed if a current of 5.3 A was to be drawn from the device for 35. minutes?

2. An electrochemical cell is made by immersing a piece of Cd metal into a solution of 0.100 M $CdSO_4$ and a Zn electrode into a solution of 1.00 M $ZnSO_4$ and placing a salt bridge to allow ion flow between the two solutions. (a) What voltage will be produced by the cell and (b) what metal is the anode?
   $(Cd^{2+} + 2e^- \rightarrow Cd \quad E° = -0.402 \text{ V})$

3. The same charge of 1.07×10^4 C is passed through three solutions: one each of Au^{3+}, Cu^+ and Pb^{2+} with strips of the metals as cathodes. In which cell will the greatest mass of metal be reduced and what is the mass of that metal?

4. What are the values of the coefficients E, X, Y, Z and L in the balanced redox equation for the reaction of copper with nitric acid:

$$X\,Cu_{(s)} + Y\,NO_3^-{}_{(aq)} + E\,H^+ \rightarrow X\,Cu^{2+}{}_{(aq)} + Z\,NO_{(g)} + L\,H_2O$$

Answers

1. 1.0 g Al

2. (a) 0.331 V   (b) Zn is anode

3. Pb: 11.5 g

4. X = 3; Y = 2, E = 8; Z = 2; L = 4

Chapter 17: Electron Transfer Reactions

Solutions to Chapter Problems

17.1 Oxidation numbers are determined by applying the rules given in the textbook.
(a) O (most electronegative in this compound) is –2, H is +1, and Fe must be +3 to give overall neutrality; (b) F (most electronegative) is –1, so N must be +3 to give overall neutrality; (c) O is –2, H is +1, and C is –2 to give overall neutrality; (d) ionic compound, with K^+ = +1 and CO_3^{2-}; in the anion, O (more electronegative) is –2 and C is +4 to give the –2 net charge on the ion; (e) ionic compound; in NH_4^+, H is +1 and N is –3, and in NO_3^-, O is –2 and N is +5; (f) Cl (more electronegative) is –1, so Ti is +4 to give overall neutrality; (g) ionic, Pb must be +2 to balance -2 charge on sulfate; O is –2, so S is +6 to give the –2 charge on the anion; (h) P is 0.

17.3 Identify redox reactions by determining whether or not oxidation numbers change.
(a) no change in oxidation numbers (Br remains –1), so not redox;
(b) redox, because Fe changes from +2 to +3 (O also changes);
(c) no change in oxidation numbers (Fe remains +2), so not redox;
(d) redox, because O changes from 0 to –2 (C also changes);
(e) redox, because N changes from 0 to –3 and H changes from 0 to +1.

17.5 Oxidation numbers are determined by applying the rules given in the textbook.
(a) F is –1, so Cl is +5 to give overall neutrality;
(b) Cl is more electronegative, so it is –1;
(c) ionic, ClO_4^- has net charge of -1, and O is –2, so Cl is +7 to give net charge of –1;
(d) Cl is 0 because this is a pure element;
(e) ionic, ClO_2^- has net charge of -1, and O is –2, so Cl is +3 to give net charge of –1;
(f) ionic, ClO^- has net charge of -1, and O is –2, so Cl is +1 to give net charge of –1.

17.7 Determine half-reactions by inspection, making use of oxidation numbers if necessary. Balance each half-reaction following the standard steps in order (balance all but H and O by inspection, balance O by adding H_2O, balance H by adding H^+, balance charge by adding e^-):
(a) The reactants are Na and H_2O, and one product is H_2; Na^+ must be a second product:
Na → Na^+ + e^- and 2 $H^+_{(aq)}$ + 2 e^- → H_2;
(b) The reactants are Au and NO_3^-, and the products are $[AuCl_4]^-$ and NO; aqua regia also contains Cl^- anions:
Au + 4 Cl^- → $[AuCl_4]^-$ + 3 e^- and NO_3^- + 4 $H^+_{(aq)}$ + 3 e^- → NO + 2 H_2O;
(c) The reactants are MnO_4^- and $C_2O_4^{2-}$, and the products are Mn^{2+} and CO_2:
MnO_4^- + 8 $H^+_{(aq)}$ + 5 e^- → Mn^{2+} + 4 H_2O and $C_2O_4^{2-}$ → 2 CO_2 + 2 e^-.

17.9 Balance each half-reaction following the standard steps in order: 1.) balance all but H and O by inspection; 2.) balance O by adding H_2O; 3.) balance H by adding H^+; 3a.) if the solution is basic, neutralize H^+ with OH^-; 4.) balance charge by adding e^-.
(a) (1) Cu is balanced, (2) add H_2O on left, (3) add 2 H^+ on right, (4) add 1 e^- on right:
Cu^+ + H_2O → CuO + 2 $H^+_{(aq)}$ + e^-;
(b) (1) S is balanced, (2) no O present, (3) add 2 H^+ on left, (4) add 2 e^- on left:
S + 2 $H^+_{(aq)}$ + 2 e^- → H_2S;

(continued on next page)

Chapter 17: Electron Transfer Reactions

17.9 (c) (1) multiply Ag by 2, (2) add H_2O on right, (3) add 2 H^+ on left, (3a) add 2 OH^- to each side, (4) add 2 e^- on left: $Ag_2O + 2 H^+_{(aq)} + 2 OH^- + 2 e^- \rightarrow 2 Ag + H_2O + 2 OH^-$;
Combine and cancel: $Ag_2O + H_2O + 2 e^- \rightarrow 2 Ag + 2 OH^-$;

(d) (1) I is balanced, (2) add 3 H_2O on left, (3) add 6 H^+ on right, (3a) add 6 OH^- to each side, (4) add 6 e^- on right: $I^- + 3 H_2O + 6 OH^- \rightarrow IO_3^- + 6 H^+_{(aq)} + 6 OH^- + 6 e^-$;
Combine and cancel: $I^- + 6 OH^- \rightarrow IO_3^- + 3 H_2O + 6 e^-$;

(e) (1) I is balanced, (2) add 2 H_2O on right, (3) add 4 H^+ on left, (3a) add 4 OH^- to each side, (4) add 4 e^- on left: $IO_3^- + 4 H^+_{(aq)} + 4 OH^- + 4 e^- \rightarrow IO^- + 2 H_2O + 4 OH^-$;
Combine and cancel: $IO_3^- + 2 H_2O + 4 e^- \rightarrow IO^- + 4 OH^-$;

(f) (1) C is balanced, (2) add H_2O on left, (3) add 4 H^+ on right, (4) add 4 e^- on right:
$H_2CO + H_2O \rightarrow CO_2 + 4 H^+_{(aq)} + 4 e^-$.

17.11 To combine half-reactions into a net redox reaction, multiply by appropriate integers so that the electrons will cancel:
(a) $2[Cu^+ + H_2O \rightarrow CuO + 2 H^+_{(aq)} + e^-] + [S + 2 H^+_{(aq)} + 2 e^- \rightarrow H_2S]$;
$2 Cu^+ + 2 H_2O + S \rightarrow 2 CuO + 2 H^+_{(aq)} + H_2S$;
(b) $3[Ag_2O + H_2O + 2 e^- \rightarrow 2 Ag + 2 OH^-] + [I^- + 6 OH^- \rightarrow IO_3^- + 3 H_2O + 6 e^-]$;
$3 Ag_2O + I^- \rightarrow 6 Ag + IO_3^-$;
(c) $2[I^- + 6 OH^- \rightarrow IO_3^- + 3 H_2O + 6 e^-] + 3[IO_3^- + 2 H_2O + 4 e^- \rightarrow IO^- + 4 OH^-]$;
$2I^- + IO_3^- \rightarrow 3 IO^-$;
(d) $2[S + 2 H^+_{(aq)} + 2 e^- \rightarrow H_2S] + [H_2CO + H_2O \rightarrow CO_2 + 4 H^+_{(aq)} + 4 e^-]$
$2S + H_2CO + H_2O \rightarrow 2 H_2S + CO_2$.

17.13 The species that loses electrons is oxidized and acts as the reducing agent, while the species that gains electrons is reduced and acts as the oxidizing agent. In this reaction, Cl^- loses electrons to become Cl_2, and Mn in MnO_4^- gains electrons to become Mn^{2+}.

17.15 Balance redox reactions following the standard procedure. Break into half-reactions, balance each half-reaction using the stepwise technique, then multiply by integers to cancel electrons:
(a) $CN^- \rightarrow CNO^-$ (1) C and N already balanced, (2) add 1 H_2O on left, (3) add 2 H^+ on right, (3a) add 2 OH^- on each side, (4) add 2 e^- on right:
$CN^- + H_2O + 2 OH^- \rightarrow CNO^- + 2 H^+_{(aq)} + 2 OH^- + 2 e^-$;
Combine and cancel: $CN^- + 2 OH^- \rightarrow CNO^- + H_2O + 2 e^-$;
$MnO_4^- \rightarrow MnO_2$ (1) Mn already balanced, (2) add 2 H_2O on right, (3) add 4 H^+ on left, (3a) add 4 OH^- on each side, (4) add 3 e^- on left:
$MnO_4^- + 4 H^+_{(aq)} + 4 OH^- + 3 e^- \rightarrow MnO_2 + 2 H_2O + 4 OH^-$;
Combine and cancel: $MnO_4^- + 2 H_2O + 3 e^- \rightarrow MnO_2 + 4 OH^-$;
Multiply first reaction by 3, second by 2, and add:
$3 [CN^- + 2 OH^- \rightarrow CNO^- + H_2O + 2 e^-] + 2 [MnO_4^- + 2 H_2O + 3 e^- \rightarrow MnO_2 + 4 OH^-]$;
$3 CN^- + 2 MnO_4^- + H_2O \rightarrow 3 CNO^- + 2 MnO_2 + 2 OH^-$;

(b) As $\rightarrow HAsO_2$ (1) As already balanced; (2) add 2 H_2O on left, (3) add 3 H^+ on right, (4) add 3 e^- on right: $As + 2 H_2O \rightarrow HAsO_2 + 3 H^+_{(aq)} + 3 e^-$;
$O_2 \rightarrow H_2O$ (1) no elements other than O and H, (2) add 1 H_2O on right, (3) add 4 H^+ on left, (4) add 4 e^- on left: $O_2 + 4 H^+_{(aq)} + 4 e^- \rightarrow 2 H_2O$;
Multiply first reaction by 4, second by 3, and add:

(continued on next page)

Chapter 17: Electron Transfer Reactions

4 [As + 2 H$_2$O → HAsO$_2$ + 3 H$^+$$_{(aq)}$ + 3 e$^-$] + 3 [O$_2$ + 4 H$^+$$_{(aq)}$ + 4 e$^-$ → 2 H$_2$O];
4 As + 3 O$_2$ + 2 H$_2$O → 4 HAsO$_2$;

(c) Br$^-$ → BrO$_3^-$ (1) Br already balanced, (2) add 3 H$_2$O on left, (3) add 6 H$^+$ on right, (3a) add 6 OH$^-$ on each side, (4) add 6 e$^-$ on right:
Br$^-$ + 3 H$_2$O + 6 OH$^-$ → BrO$_3^-$ + 6 H$^+$$_{(aq)}$ + 6 OH$^-$ + 6 e$^-$;
Combine and cancel: Br$^-$ + 6 OH$^-$ → BrO$_3^-$ + 3 H$_2$O + 6 e$^-$;
MnO$_4^-$ → MnO$_2$, same as reaction in part (a): MnO$_4^-$ + 2 H$_2$O + 3 e$^-$ → MnO$_2$ + 4 OH$^-$;
Multiply second reaction by 2 and add:
[Br$^-$ + 6 OH$^-$ → BrO$_3^-$ + 3 H$_2$O + 6 e$^-$] + 2 [MnO$_4^-$ + 2 H$_2$O + 3 e$^-$ → MnO$_2$ + 4 OH$^-$];
Br$^-$ + 2 MnO$_4^-$ + H$_2$O → BrO$_3^-$ + 2 MnO$_2$ + 2 OH$^-$;

(d) NO$_2$ → NO$_3^-$ (1) N already balanced; (2) add 1 H$_2$O on left, (3) add 2 H$^+$ on right, (4) add 1 e$^-$ on right: NO$_2$ + H$_2$O → NO$_3^-$ + 2 H$^+$$_{(aq)}$ + 1 e$^-$;
NO$_2$ → NO   (1) N already balanced; (2) add 1 H$_2$O on right, (3) add 2 H$^+$ on left, (4) add 2 e$^-$ on left: NO$_2$ + 2 H$^+$$_{(aq)}$ + 2 e$^-$ → NO + H$_2$O;
Multiply first reaction by 2 and add:
2 [NO$_2$ + H$_2$O → NO$_3^-$ + 2 H$^+$$_{(aq)}$ + 1 e$^-$] + [NO$_2$ + 2 H$^+$$_{(aq)}$ + 2 e$^-$ → NO + H$_2$O];
3 NO$_2$ + H$_2$O → 2 NO$_3^-$ + NO + 2 H$^+$$_{(aq)}$;

(e) ClO$_4^-$ → ClO$^-$ (1) Cl already balanced; (2) add 3 H$_2$O on right, (3) add 6 H$^+$ on left, (4) add 6 e$^-$ on left: ClO$_4^-$ + 6 H$^+$$_{(aq)}$ + 6 e$^-$ → ClO$^-$ + 3 H$_2$O ;
Cl$^-$ → Cl$_2$ (1) multiply Cl$^-$ by 2; (2,3) no H or O present, (4) add 2 e$^-$ on right:
2 Cl$^-$ → Cl$_2$ + 2 e$^-$;
Multiply second reaction by 3 and add:
[ClO$_4^-$ + 6 H$^+$$_{(aq)}$ + 6 e$^-$ → ClO$^-$ + 3 H$_2$O] + 3 [2 Cl$^-$ → Cl$_2$ + 2 e$^-$];
ClO$_4^-$ + 6 Cl$^-$ + 6 H$^+$$_{(aq)}$ → ClO$^-$ + 3 H$_2$O + 3 Cl$_2$;

(f) AlH$_4^-$ → Al^{3+} (1) Al already balanced, (2) no O present, (3) add 4 H$^+$ on right, (3a) add 4 OH$^-$ on each side, combine with H$^+$, (4) add 8 e$^-$ on right:
AlH$_4^-$ + 4 OH$^-$ → Al^{3+} + 4 H$_2$O + 8 e$^-$
H$_2$CO → CH$_3$OH   (1) C already balanced, (2) O already balanced, (3) add 2 H$^+$ on left, (3a) add 2 OH$^-$ on each side, combine with H$^+$, (4) add 2 e$^-$ on left:
H$_2$CO + 2 H$_2$O + 2 e$^-$ → CH$_3$OH + 2 OH$^-$;
Multiply second reaction by 4 and add:
[AlH$_4^-$ + 4 OH$^-$ → Al^{3+} + 4 H$_2$O + 8 e$^-$] + 4 [H$_2$CO + 2 H$_2$O + 2 e$^-$ → CH$_3$OH + 2OH$^-$]
AlH$_4^-$ + 4 H$_2$CO + 4 H$_2$O → Al^{3+} + 4 CH$_3$OH + 4 OH$^-$

17.17 To determine spontaneity using standard thermodynamic values, calculate $\Delta G°_{rxn}$ from tabulated values for $\Delta G°_f$:

(a) $\Delta G°_{rxn}$ = 2(–129.7) = –259.4 kJ; spontaneous;

(b) $\Delta G°_{rxn}$ = 2(–58.539) = –117.078 kJ; spontaneous;

(c) $\Delta G°_{rxn}$ = 1(–300.2) – 1(–53.6) = –246.6 kJ; spontaneous;

(d) $\Delta G°_{rxn}$ = 2(51.31) = + 102.62 kJ; not spontaneous.

17.19 A molecular view of a process occurring at an electrode should show the species involved in charge transfer and indicate the direction of movement of electrons. See Figure 17-3 for an example. At a silver-silver chloride electrode undergoing reduction, solid AgCl gains an electron from the electrode to generate chloride anions in solution and solid Ag:

17.21 A passive electrode and a supply of gas are needed to study a redox reaction the includes gases. See Figure 17-6 for a sketch of a hydrogen electrode. A chlorine electrode is exactly analogous:

17.23 Active electrodes participate in redox reactions, while passive electrodes only provide or accept electrons. The Ag-AgCl electrode is active, while the Pt electrodes are passive.

17.25 Standard electrode potentials are calculated using Equation 17-1:

$$E^o{}_{cell} = E^o{}_{cathode} - E^o{}_{anode}$$

MnO_4^- undergoes reduction to Mn^{2+} (cathode), $\quad E^o{}_{cathode} = 1.507$ V;

Cl^- undergoes oxidation to Cl_2 (anode), $\quad E^o{}_{anode} = 1.35827$ V;

$E^o{}_{cell} = 1.507 - 1.35827 = 0.149$ V.

17.27 Balance redox reactions following the standard procedure. Break into half-reactions, balance each half-reaction using the stepwise technique, then multiply by appropriate integers so the electrons cancel:
$NO \rightarrow N_2O$ (1) multiply NO by 2, (2) add 1 H_2O on right, (3) add 2 H^+ on left, (4) add 2 e^- on left: $2 NO + 2 H^+_{(aq)} + 2 e^- \rightarrow N_2O + H_2O$;
$NO \rightarrow NO_3^-$ (1) N already balanced, (2) add 2 H_2O on left, (3) add 4 H^+ on right, (4) add 3 e^- on right: $NO + 2 H_2O \rightarrow NO_3^- + 4 H^+_{(aq)} + 3 e^-$;
Multiply first reaction by 3 and second reaction by 2, then add:
$3 [2 NO + 2 H^+_{(aq)} + 2 e^- \rightarrow N_2O + H_2O] + 2 [NO + 2 H_2O \rightarrow NO_3^- + 4 H^+_{(aq)} + 3 e^-]$;
$8 NO + H_2O \rightarrow 3 N_2O + 2 NO_3^- + 2 H^+_{(aq)}$;
Standard electrode potentials are calculated using Equation 17-1:

$$E°_{cell} = E°_{cathode} - E°_{anode}$$

NO undergoes reduction to N_2O (cathode), $E°_{cathode} = 1.591$ V;

NO undergoes oxidation to NO_3^- (anode), $E°_{anode} = 0.957$ V;

$E°_{cell} = 1.591 - 0.957 = 0.634$ V.

17.29 The direct way to measure a standard reduction potential is with a cell containing a standard hydrogen electrode as a reference. Because F^- is the conjugate base of a weak acid, the F_2/F^- portion of the cell must be separated from the H_2/H^+ portion, requiring a porous plate. Set up a standard hydrogen electrode on one side and a Pt electrode immersed in 1 M solution of NaF with F_2 bubbling over the electrode:

The standard reduction potential of F_2 is + 2.866 volts, so F_2 will be reduced in this cell and the H_2/H^+ electrode will be the anode.

17.31 To act as a reducing agent under standard conditions, a substance must have a more negative standard reduction potential than the substance that is to be reduced. According to Appendix H, the standard reduction potential of Be is − 1.847 V. Scanning the table, we find the following metals whose standard potentials are more negative than this: Ba(−2.912), Ca (−2.868), Cs (−2.92), Li (−3.0401), Mg (−2.37), K (−2.925), and Na (−2.714). All these metals lie in the s block of the periodic table and easily lose 1 or 2 electrons.

Chapter 17: Electron Transfer Reactions

17.33 The standard free energy change of a reaction is related to the standard reduction potential through Equation 17-3: $\Delta G° = -nFE°$. The standard potential for the reaction is calculated in Problem 17.25: $E°_{cell} = 1.507 - 1.35827 = 0.149$ V.
For this reaction, $n = 10$ (10 Cl⁻ anions lose 1 electron each to become 5 Cl_2):
$\Delta G° = -(10 \text{ mol})(9.6485 \times 10^4 \text{ C/mol})(0.149 \text{ V})(10^{-3} \text{ kJ/J}) = -144$ kJ.
(Remember that 1 J = 1 V C)

17.35 Operating potentials are related to standard cell potentials through Equation 17-6:
$$E = E° - \frac{RT}{nF} \ln Q$$
The reaction for the nickel-cadmium battery is:
$2 \text{ NiO(OH)}_{(s)} + 2 \text{ H}_2\text{O}_{(l)} + \text{Cd}_{(s)} \rightarrow 2 \text{ Ni(OH)}_{2(s)} + \text{Cd(OH)}_{2(s)}$
Here, all the substances are solids or pure liquids, so $Q = 1$, $\ln Q = 0$, and $E = E° = 1.35$ V.

17.37 The connection between electric current and chemical amounts is provided by Equation 17-7, $n = It/F$, where n is moles of electrons flowing.
$n = (5.9 \text{ A})(15 \text{ s}) \div (9.6485 \times 10^4 \text{ C/mol}) = 9.2 \times 10^{-4}$ mol;

The half-reactions in a lead storage battery are:
$\text{Pb}_{(s)} + \text{HSO}_4^-{}_{(aq)} \rightarrow \text{PbSO}_{4 (s)} + \text{H}^+{}_{(aq)} + 2 \text{ e}^-$

$\text{PbO}_{2 (s)} + \text{HSO}_4^-{}_{(aq)} + 3 \text{ H}^+{}_{(aq)} + 2 \text{ e}^- \rightarrow \text{PbSO}_{4 (s)} + 2 \text{ H}_2\text{O}_{(l)}$
Thus $n_{Pb} = n_{PbO_2} = (9.2 \times 10^{-4} \text{ mol}) \div 2 = 4.6 \times 10^{-4}$ mol; $m_{Pb} = (4.6 \times 10^{-4} \text{ mol})(207.2$ g/mol$) = 9.5 \times 10^{-2}$ g; $m_{PbO_2} = (4.6 \times 10^{-4} \text{ mol})(239.2$ g/mol$) = 0.11$ g.

17.39 The connection between electric current and chemical amounts is provided by Equation 17-7, $n = It/F$, where n is moles of electrons flowing. Thus $t = nF/I$;

The reaction being driven by the alternator is :
$\text{PbSO}_{4 (s)} + \text{H}^+{}_{(aq)} + 2 \text{ e}^- \rightarrow \rightarrow \text{Pb}_{(s)} + \text{HSO}_4^-{}_{(aq)}$

Thus $n = 2 \, n_{PbSO_4} = 2(0.850 \text{ g}) \div (303.3 \text{ g/mol}) = 5.6 \times 10^{-3}$ mol; $I = (1.750 \text{ A} - 1.350 \text{ A})$;
$t = (5.6 \times 10^{-3} \text{ mol})(9.6485 \times 10^4 \text{ C/mol}) \div (0.400 \text{ A}) = 1.35 \times 10^3$ s.

17.41 Balance redox reactions following the standard procedure. Break into half-reactions, balance each half-reaction using the stepwise technique, then multiply by integers to cancel electrons:
$\text{Cr}_2\text{O}_7^{2-} \rightarrow \text{Cr}^{3+}$ (1) multiply Cr^{3+} by 2, (2) add 7 H_2O on right, (3) add 14 H^+ on left, (4) add 6 e⁻ on left: $\text{Cr}_2\text{O}_7^{2-} + 14 \text{ H}^+{}_{(aq)} + 6 \text{ e}^- \rightarrow 2 \text{ Cr}^{3+} + 7 \text{ H}_2\text{O}$;

$\text{CH}_3\text{CHO} \rightarrow \text{CH}_3\text{CO}_2\text{H}$ (1) C already balanced, (2) add 1 H_2O on left, (3) add 2 H^+ on right, (4) add 2e⁻ on right: $\text{CH}_3\text{CHO} + \text{H}_2\text{O} \rightarrow \text{CH}_3\text{CO}_2\text{H} + 2 \text{ H}^+{}_{(aq)} + 2 \text{ e}^-$;

Multiply second reaction by 3 and add:
$[\text{Cr}_2\text{O}_7^{2-} + 14 \text{ H}^+{}_{(aq)} + 6 \text{ e}^- \rightarrow 2 \text{ Cr}^{3+} + 7 \text{ H}_2\text{O}]$
$+ 3[\text{CH}_3\text{CHO} + \text{H}_2\text{O} \rightarrow \text{CH}_3\text{CO}_2\text{H} + 2 \text{ H}^+{}_{(aq)} + 2 \text{ e}^-]$;
$\text{Cr}_2\text{O}_7^{2-} + 3 \text{ CH}_3\text{CHO} + 8 \text{ H}^+{}_{(aq)} \rightarrow 2 \text{ Cr}^{3+} + 3 \text{ CH}_3\text{CO}_2\text{H} + 4 \text{ H}_2\text{O}$;

It takes 2 mol of electrons to oxidize 1 mol of CH_3CHO;
$n = 2(1.00 \text{ g}) \div (44.0 \text{ g/mol}) = 0.0455$ mol;
Each mole of $Cr_2O_7^{2-}$ delivers 6 mol of electrons:
$n_{\text{dichromate}} = (0.0455 \text{ mol}) \div 6 = 7.6 \times 10^{-3}$ mol; $m = (7.6 \times 10^{-3} \text{ mol})(262 \text{ g/mol}) = 1.99$ g.

17.43 The Nernst equation is Equation 17-6, $E = E° - \dfrac{RT}{nF} \ln Q$. To determine Q, obtain the balanced redox equation for the mercury battery: $Zn_{(s)} + HgO_{(s)} \rightarrow ZnO_{(s)} + Hg_{(l)}$; All the substances participating in this reaction are pure solids or liquids, so their concentrations do not appear in Q. $Q = 1$, $\ln Q = 0$, and $E = E°$; the potential of the battery does not decrease with use.

17.45 When several metals are present, the one that corrodes first is the one with the more negative standard reduction potential. Here are the values for Cr and Fe, from Appendix H:
$Cr^{2+} + 2\text{ e}^- \rightarrow Cr$, $E° = -0.913$ V; $Fe^{2+} + 2\text{ e}^- \rightarrow Fe$, $E° = -0.447$ V;
Thus, Cr oxidizes more easily and will preferentially corrode.

17.47 Mercury batteries are characterized by a stable potential and compact size but limited current capacity, making them well suited for use where large currents are not needed, such as pacemakers and cameras. They cannot supply the large current needed to start an automobile engine; moreover, they are irreversible so would have to be replaced very frequently.

17.49 The connection between electric current and chemical amounts is provided by Equation 17-7, $n = It/F$, where n is moles of electrons flowing. Examine the balanced half-reaction to determine the stoichiometric relationship between moles of electrons and moles of chemical species. Here, the reaction is $2\text{ Cl}^- \rightarrow Cl_2 + 2\text{ e}^-$, so 1 mol of Cl_2 is formed by 2 mol of electrons:
$n = \dfrac{(4.50 \text{ A})(225 \text{ min})(60 \text{ s/min})}{(2)(9.6485 \times 10^4 \text{ C/mol})} = 0.315$ mol;
$m = n\, MM = (0.315 \text{ mol})(70.9 \text{ g/mol}) = 22.3$ g.

17.51 When an external potential is applied to a galvanic cell, the reduction that occurs is the one with the least negative reduction potential, and the oxidation that occurs is the one with the least positive reduction potential. The oxidation reaction is the reverse of the galvanic reduction, $Cu \rightarrow Cu^{2+} + 2\text{ e}^-$. However, instead of $Zn^{2+} + 2\text{ e}^- \rightarrow Zn$ ($E° = -0.7618$ V), the reduction of H_2O occurs: $2\text{ H}^+_{(aq)} + 2\text{ e}^- \rightarrow H_2$ ($E° = 0$ V). In neutral H_2O, the H_3O^+ ion concentration is only 10^{-7} M, but the resulting E is still less negative than $E°$ for Zn^{2+}/Zn.

17.53 (a) To determine where to attach the negative wire from the charger, examine the half-reactions of the battery, which must be run in the reverse direction to recharge:
Cathode: $NiO(OH)_{(s)} + H_2O_{(l)} + \text{e}^- \rightleftarrows Ni(OH)_2{}_{(s)} + OH^-_{(aq)}$
Anode: $Cd_{(s)} + 2\text{ OH}^-_{(aq)} \rightleftarrows Cd(OH)_2{}_{(s)} + 2\text{ e}^-$

To recharge, electrons must be supplied to the cadmium electrode, so this is the electrode to which to attach the negative wire from the charger should be attached. The reaction is: $Cd(OH)_{2\ (s)} + 2\ e^- \rightarrow Cd_{(s)} + 2\ OH^-_{(aq)}$.

(b) The connection between electric current and chemical amounts is provided by Equation 17-7, $n = It/F$, where n is moles of electrons flowing. Thus $t = nF/I$; $n_{Cd(OH)_2} = (1.55\ g) \div (146.4\ g/mol) = 0.0106\ mol$; $n = 2\ n_{Cd(OH)_2} = 0.0212\ mol$;

$$t = \frac{(0.0212\ mol)(9.6485 \times 10^4\ C/mol)}{(125\ mA)(10^{-3}\ A/mA)(60\ s/min)(60\ min/hr)} = 4.54\ hr.$$

17.55 Quantity of charge can be calculated once the amount of electron flow has been determined. Use the half-reaction to determine the relationship between amount of chemical change and amount of electron flow:
$Ag^+ + e^- \rightarrow Ag$, so $n = n_{Ag} = (12.89\ g - 10.77\ g) \div (107.9\ g/mol) = 0.0196\ mol$;
charge $= nF = (0.0196\ mol)(9.6485 \times 10^4\ C/mol) = 1.90 \times 10^3\ C$;

$$i = \text{charge}/t = \frac{(1.90 \times 10^3\ C)}{(15.0\ min)(60\ s/min)} = 2.11\ A.$$

17.57 A molecular picture must show the species undergoing redox reactions and the direction of electron flow. In a silver coulometer, both electrodes are silver. Ag is oxidized to Ag^+ at the anode, and Ag^+ is reduced to Ag at the cathode.

17.59 (a) Balance redox reactions following the standard procedure. Break into half-reactions, balance each half-reaction using the stepwise technique, then multiply by appropriate integers so the electrons cancel:
(a) $H_2 \rightarrow H_2O$ (1) no elements except O and H, (2) add 1 H_2O on left, (3) add 2 H^+ on right, (3a) add 2 OH^- on both sides, (4) add 2 e^- on right:
$H_2 + 2\ OH^- + H_2O \rightarrow H_2O + 2\ H^+_{(aq)} + 2\ OH^- + 2\ e^-$; Combine and cancel:
$H_2 + 2\ OH^- \rightarrow 2\ H_2O + 2\ e^-$;
$Cr(OH)_3 \rightarrow Cr$ (1) Cr already balanced, (2) add 3 H_2O on right, (3) add 3 H^+ on left, (3a) add 3 OH^- on both sides, (4) add 3 e^- on left:
$Cr(OH)_3 + 3\ H^+_{(aq)} + 3\ OH^- + 3\ e^- \rightarrow Cr + 3\ H_2O + 3\ OH^-$;
Combine and cancel: $Cr(OH)_3 + 3\ e^- \rightarrow Cr + 3\ OH^-$;
Multiply first reaction by 3 and second reaction by 2, then add:
$3\ [H_2 + 2\ OH^- \rightarrow 2\ H_2O + 2\ e^-] + 2\ [Cr(OH)_3 + 3\ e^- \rightarrow Cr + 3\ OH^-]$;
$3\ H_2 + 2\ Cr(OH)_3 \rightarrow 6\ H_2O + 2\ Cr$;
(b) Consult $E°$ values in Appendix H, and subtract $E°$ for the oxidation from $E°$ for the reduction: $H_2 + 2\ OH^- \rightarrow 2\ H_2O + 2\ e^-$, $E° = -0.828\ V$;
$Cr(OH)_3 + 3\ e^- \rightarrow Cr + 3\ OH^-$, $E° = -1.48\ V$; $E° = (-1.48\ V) - (-0.828\ V) = -0.65\ V$;
(c) $\Delta G° = -nFE° = -(6\ mol)(9.6485 \times 10^4\ C/mol)(-0.65\ V)(10^{-3}\ kJ/J) = 3.8 \times 10^2\ kJ$.

Chapter 17: Electron Transfer Reactions

17.61 To calculate a concentration from cell potential measurements, use the Nernst equation.

For a concentration cell, $E° = 0$ V, so $E = -\dfrac{0.0591}{n} \log Q$; and $\log Q = -\dfrac{nE}{(0.0591)}$;

The reactions in this cell are $AgCl + e^- \rightarrow Ag + Cl^-$, and the problem states that the 1.00 M KCl side of the cell is the anode, where oxidation occurs. Thus the net reaction is: $Cl^-_{(1.00\ M)} \rightarrow Cl^-_{(x\ M)}$, $n = 1$, and $Q = x \div (1.00\ M)$;

$\log\left(\dfrac{x}{1.00\ M}\right) = \dfrac{-(1)(0.0785\ V)}{(0.0591)} = -1.328$, $\dfrac{x}{1.00\ M} = 0.047$, $x = 0.047$ M.

17.63 Standard electrode potentials allow calculation of equilibrium constants if the half-reactions can be combined to give the equilibrium reaction whose constant is desired. In this case, subtract the second reaction from the first:

$SO_4^{2-}{}_{(aq)} + 4\ H^+{}_{(aq)} + 2\ e^- \rightleftharpoons SO_{2(g)} + 2\ H_2O$ $E° = 0.20$ V

$-\ [SO_4^{2-}{}_{(aq)} + 4\ H^+{}_{(aq)} + 2\ e^- \rightleftharpoons H_2SO_{3(aq)} + H_2O]$ $E° = 0.17$ V)

Net: $H_2SO_{3(aq)} \rightleftharpoons SO_{2(g)} + H_2O$ $E° = 0.20$ V $- 0.17$ V $= 0.03$ V;

Use Equation 17-5, rearranged: $\log K_{eq} = \left(\dfrac{nE°}{5.916 \times 10^{-2}\ V}\right) = \left(\dfrac{2(0.03\ V)}{5.916 \times 10^{-2}\ V}\right) = 1.014$;

$K_{eq} = 10.33 = \dfrac{(p_{SO_2})_{eq}}{[H_2SO_3]_{eq}}$; $(p_{SO_2})_{eq} = (10.33)(1.00) = 1 \times 10^1$ atm

17.65 To determine the redox chemistry taking place in an electrochemical cell, start with the species present in each compartment.

The first compartment contains an Ag/AgCl electrode and 0.500 M HCl, and the second compartment contains a Mg electrode and 1.00 M $MgCl_2$ solution.

Compartment 1: $AgCl + e^- \rightleftharpoons Ag + Cl^-$; Compartment 2: $Mg^{2+} + 2e^- \rightleftharpoons Mg$;

(a) Net reaction (Mg has a more negative $E°$): $Mg + 2\ AgCl \rightleftharpoons Mg^{2+} + 2\ Ag + 2\ Cl^-$;

(b) Use the Nernst equation to calculate the cell emf:

$E = E° - \dfrac{(0.0591)}{n} \log\{[Cl^-]^2[Mg^{2+}]\}$;

$E° = (0.22233\ V) - (-2.372\ V) = 2.612$ V;

$E = 1.612\ V - \dfrac{(0.0591)}{2} \log[(0.500)^2(1.00)] = 1.612 + 0.018\ V = 1.630$ V;

(c) A molecular picture of an electrochemical cell must show the species undergoing redox reactions and the direction of electron flow.

17.67 (a) Balance the half-reaction using the stepwise technique:
$Cr_2O_7^{2-} \rightarrow Cr^{3+}$ (1) multiply Cr^{3+} by 2, (2) add 7 H_2O on right, (3) add 14 H^+ on left, (4) add 6 e^- on left: $Cr_2O_7^{2-} + 14\ H^+_{(aq)} + 6\ e^- \rightarrow 2\ Cr^{3+} + 7\ H_2O$;

(b) It takes 2 mol K_2CrO_4 to generate 1 mol $Cr_2O_7^{2-}$, which in turn consumes 6 mol of e^-, so 1 mol K_2CrO_4 generates 3 mol e^- and (0.250 mol) ÷ 3 = 0.0833 mol are required; $m = n\ MM = (0.0833\ mol)(194\ g/mol) = 16.2\ g$.

17.69 Identify what is taking place in a galvanic cell by identifying the species present.

(a) On the left, $Ni^{2+} + 2e^- \rightleftharpoons Ni$, $E° = -0.257$ V;
on the right, $Fe^{2+} + 2e^- \rightleftharpoons Fe$, $E° = -0.447$ V;

(b) To generate a positive overall potential, the iron half-reaction operates as oxidation:
$Fe + Ni^{2+} \rightleftharpoons Ni + Fe^{2+}$, $E° = -0.257\ V - (-0.447\ V) = 0.190\ V$;

(c) oxidation occurs at the anode, so the Fe electrode is the anode and Ni is the cathode;

(d) A molecular picture of an electrochemical cell must show the species undergoing redox reactions and the direction of electron flow.

17.71 Use the Nernst equation to determine the concentration at which $E = 0$:

$$Fe + Ni^{2+} \rightleftharpoons Ni + Fe^{2+},\ E° = -0.257\ V - (-0.447\ V) = 0.190\ V;$$

$$E = 0 = E° - \frac{(0.0591)}{n} \log\left(\frac{[Fe^{2+}]}{[Ni^{2+}]}\right);\ \log\left(\frac{[Fe^{2+}]}{[Ni^{2+}]}\right) = \left(\frac{nE°}{5.916 \times 10^{-2}\ V}\right);$$

The concentration of reactant must be reduced to reach a potential of 0 V:

$$\log\left\{\frac{(1.00)}{[Ni^{2+}]}\right\} = \frac{(2)(0.190\ V)}{(5.916 \times 10^{-2}\ V)} = 6.42,\ [Ni^{2+}] = 10^{-6.42} = 3.8 \times 10^{-7}\ M.$$

Chapter 17: Electron Transfer Reactions

17.73 Balance redox reactions following the standard procedure. Break into half-reactions, balance each half-reaction using the stepwise technique, then multiply by appropriate integers so the electrons cancel. First balance the half-reaction involving manganese: $MnO_4^- \to Mn^{2+}$ (1) Mn is balanced, (2) add 4 H_2O on right, (3) add 8 H^+ on left, (4) add 5 e^- on left: $MnO_4^- + 8 H^+_{(aq)} + 5 e^- \to Mn^{2+} + 4 H_2O$;

(a) $H_2SO_3 \to HSO_4^-$ (1) S already balanced, (2) add 1 H_2O on left, (3) add 3 H^+ on right, (4) add 2e^- on right: $H_2SO_3 + H_2O \to HSO_4^- + 3 H^+_{(aq)} + 2 e^-$;
Multiply Mn reaction by 2, S reaction by 5, and add:
2 [$MnO_4^- + 8 H^+_{(aq)} + 5 e^- \to Mn^{2+} + 4 H_2O$]
    + 5 [$H_2SO_3 + H_2O \to HSO_4^- + 3 H^+_{(aq} + 2 e^-$]
$2 MnO_4^- + 5 H_2SO_3 + H^+_{(aq)} \to 2 Mn^{2+} + 5 HSO_4^- + 3 H_2O$;

(b) $SO_2 \to HSO_4^-$ (1) S already balanced, (2) add 2 H_2O on left, (3) add 3 H^+ on right, (4) add 2e^- on right: $SO_2 + 2 H_2O \to HSO_4^- + 3 H^+_{(aq)} + 2 e^-$;
Multiply Mn reaction by 2, S reaction by 5, and add:
2 [$MnO_4^- + 8 H^+_{(aq)} + 5 e^- \to Mn^{2+} + 4 H_2O$]
    + 5 [$SO_2 + 2 H_2O \to HSO_4^- + 3 H^+_{(aq)} + 2 e^-$]
$2 MnO_4^- + 5 SO_2 + 2 H_2O + H^+_{(aq)} \to 2 Mn^{2+} + 5 HSO_4^-$;

(c) $H_2S \to HSO_4^-$ (1) S already balanced, (2) add 4 H_2O on left, (3) add 9 H^+ on right, (4) add 8 e^- on right: $H_2S + 4 H_2O \to HSO_4^- + 9 H^+_{(aq)} + 8 e^-$;
Multiply Mn reaction by 8, S reaction by 5, and add:
8 [$MnO_4^- + 8 H^+_{(aq)} + 5 e^- \to Mn^{2+} + 4 H_2O$]
    + 5 [$H_2S + 4 H_2O \to HSO_4^- + 9 H^+_{(aq)} + 8 e^-$]
$8 MnO_4^- + 5 H_2S + 19 H^+_{(aq)} \to 8 Mn^{2+} + 5 HSO_4^- + 12 H_2O$;

(d) $H_2S_2O_3 \to HSO_4^-$ (1) Multiply HSO_4^- by 2, (2) add 5 H_2O on left, (3) add 10 H^+ on right, (4) add 8 e^- on right: $H_2S_2O_3 + 5 H_2O \to 2 HSO_4^- + 10 H^+_{(aq)} + 8 e^-$;
Multiply Mn reaction by 8, S reaction by 5, and add:
8 [$MnO_4^- + 8 H^+_{(aq)} + 5 e^- \to Mn^{2+} + 4 H_2O$]
    + 5 [$H_2S_2O_3 + 5 H_2O \to 2 HSO_4^- + 10 H^+_{(aq)} + 8 e^-$]
$8 MnO_4^- + 5 H_2S_2O_3 + 14 H^+_{(aq)} \to 8 Mn^{2+} + 10 HSO_4^- + 7 H_2O$.

17.75 The connection between electric current and chemical amounts is provided by Equation 17-7, $n = It/F$, where n is moles of electrons flowing. Thus $t = nF/I$;

The reaction is $Cu^{2+} + 2e^- \to Cu$, so $n = 2 n_{Cu}$
$n_{Cu} = (0.250 L)(0.245 mol/L) = 0.0613 mol$; $n = 2 n_{Cu} = 0.126 mol$;

$$t = \frac{(0.126 \text{ mol})(9.6485 \times 10^4 \text{C / mol})}{(2.45 \text{ A})(60 \text{ s / min})} = 80.4 \text{ min.}$$

17.77 The process in Problem 17.75 is electrodeposition, presumably with Cu serving as both electrodes. Reduction of Cu^{2+} to Cu occurs at the cathode, and oxidation of Cu to Cu^{2+} occurs at the anode:

17.79 An equilibrium constant can be calculated from standard reduction potentials provided the half-reactions can be combined to give the equilibrium reaction whose constant is desired.
Appendix H provides these standard potentials:

$Zn^{2+} + 2\,e^- \rightleftharpoons Zn$, $E° = -0.7618$ V;

$Zn[NH_3]_4^{2+} + 2\,e^- \rightleftharpoons Zn + 4\,NH_3$, $E° = -1.04$ V;

Subtract the second half-reaction from the first to give the desired equilibrium:

$E° = (-0.7618\text{ V}) - (-1.04\text{ V}) = +0.28$ V;

$$\log K_{eq} = \left(\frac{nE°}{5.916 \times 10^{-2}\text{ V}}\right) = \left(\frac{2(0.28\text{ V})}{5.916 \times 10^{-2}\text{ V}}\right) = 9.47; \quad K_{eq} = 10^{9.47} = 2.9 \times 10^9$$

17.81 The better oxidizing agent under standard conditions is the reactant having the more positive reduction potential. (a) MnO_4^- (1.695 vs. 1.232 V); (b) Cl_2 (1.3583 vs. 1.087 V); (c) Sn^{2+} (–0.137 vs. –0.447 V).

17.83 Although the anions are different in the two solutions, the redox process involves only Zn metal and Zn^{2+} cations, so this is a concentration cell.
(a) The molecular picture shows Zn^{2+} cations depositing on the electrode in the 1.25 M solution and dissolving off the electrode in the 0.250 M solution:

Chapter 17: Electron Transfer Reactions

17.83 (b) To calculate a cell potential from concentration measurements, use the Nernst equation. For a concentration cell, $E^\circ = 0$ V, so $E = -\dfrac{(0.05916)}{n} \log Q$;

The reactions are $Zn^{2+} + 2\,e^- \rightarrow Zn$, and the concentrations are 0.250 M and 1.25 M:

$Zn^{2+}{}_{(1.25\,M)} \rightarrow Zn^{2+}{}_{(0.250\,M)}$, $n = 2$, and $Q = \dfrac{0.250\,M}{1.25\,M}$;

$E = -\dfrac{(0.05916)}{2} \log \dfrac{0.250\,M}{1.25\,M} = 0.0207$ V

17.85 An equilibrium constant can be calculated from standard reduction potentials provided the half-reactions can be combined to give the equilibrium reaction whose constant is desired.
Appendix H provides these standard potentials:

$O_2 + 2\,H^+{}_{(aq)} + 2\,e^- \rightleftharpoons H_2O_2$, $E^\circ = +0.695$ V;

$O_2 + 4\,H^+{}_{(aq)} + 4\,e^- \rightleftharpoons 2\,H_2O$, $E^\circ = +1.229$ V;

Multiply the first half-reaction by 2 and subtract from the second to give the desired equilibrium: $2\,H_2O_2 \rightleftharpoons 2\,H_2O + O_2$: $E^\circ = (1.229\,V) - (0.695\,V) = +0.534$ V;

$\log K_{eq} = \left(\dfrac{nE^\circ}{5.916 \times 10^{-2}\,V}\right) = \left(\dfrac{4(0.534\,V)}{5.916 \times 10^{-2}\,V}\right) = 36.11$; $K_{eq} = 10^{36.11} = 1.3 \times 10^{36}$

Solutions of hydrogen peroxide are quite unstable and spontaneously decompose, releasing oxygen. Fortunately, this process is kinetically slow (partly because of the large number of electrons that must be transferred), but solutions of hydrogen peroxide deteriorate with time.

17.87 Recall that to determine a yield (efficiency), first it is necessary to determine the theoretical yield. For an electrochemical reaction, this requires calculation of moles of electrons.

$n = It/F = \dfrac{(5.8\,A)(5.0\,hr)(60\,s/min)(60\,min/hr)}{(9.6485 \times 10^4\,C/mol)} = 1.08$ mol;

The half-reaction is $Cu^{2+} + 2\,e^- \rightarrow Cu$, so the theoretical yield is $(1.08\,mol) \div 2 = 0.54$ mol;
Convert to mass: $(0.54\,mol)(63.55\,g/mol) = 34.4$ g;
Efficiency = (100%)(actual mass)/(theoretical mass) = (100%)(32 g) ÷ (34.4 g) = 93 %.

Chapter 17: Electron Transfer Reactions

17.89 An equilibrium constant can be calculated from standard reduction potentials provided the half-reactions can be combined to give the equilibrium reaction whose constant is desired.

Although standard reduction potentials of three metal hydroxides appear in Table 17-1, the additional reduction potential, for the corresponding metal cation appears only for Ca:

$Ca^{2+} + 2\,e^- \rightleftharpoons Ca$, $E° = -2.868$ V;

$Ca(OH)_2 + 2\,e^- \rightleftharpoons Ca + 2\,OH^-$, $E° = -3.02$ V;

Subtract the first half-reaction from the second to give the desired solubility equilibrium:

$Ca(OH)_2 \rightleftharpoons Ca^{2+} + 2\,OH^-$, $E° = (-3.02\text{ V}) - (-2.868\text{ V}) = -0.15$ V;

$\log K_{eq} = \left(\dfrac{nE°}{5.916 \times 10^{-2}\text{ V}}\right) = \left(\dfrac{2(\pm 0.15\text{ V})}{5.916 \times 10^{-2}\text{ V}}\right) = -5.07$; $K_{eq} = 10^{-5.07} = 8.5 \times 10^{-6}$.

17.91 Use the Nernst equation to determine a cell potential under non-standard conditions. First determine the net reaction in order to find n, $E°$, and Q.

The reactions are: $Tl^+ + e^- \rightleftharpoons Tl$, $E° = -0.34$ V;

$2\,H^+_{(aq)} + 2\,e^- \rightleftharpoons H_2$, $E° = 0$ V;

Multiply the first reaction by 2 and subtract it from the second:

$2\,Tl + 2\,H^+_{(aq)} \rightleftharpoons 2\,Tl^+ + H_2$, $E° = 0\text{ V} - (-0.34\text{ V}) = 0.34$ V;

$E = E° - \dfrac{(0.05916)}{n}\log Q = E° - \dfrac{(0.05916)}{n}\log\left\{\dfrac{[Tl^+]^2(p_{H_2})}{[H^+_{(aq)}]^2}\right\}$;

Now substitute values and calculate the potential:

$E = 0.34\text{ V} - \dfrac{(0.05916)}{2}\log\left\{\dfrac{(0.050)^2(0.90)}{(0.50)^2}\right\} = 0.34\text{ V} - (-0.061\text{ V}) = 0.40$ V.

17.93 The most negative half-cell potentials are around -3 V, so to generate a net cell potential greater than 5 V requires a half-reaction with a positive potential greater than $+2$ V. There is only one half-reaction in Table 17-1 that meets these requirements:

$F_2 + 2\,e^- \rightleftharpoons 2\,F^-$, $E° = 2.866$ V;

This half-reaction could be coupled with any of the following:

$Mg^{2+} + 2\,e^- \rightleftharpoons Mg$, $E° = -2.372$ V;
$Ca^{2+} + 2\,e^- \rightleftharpoons Ca$, $E° = -2.868$ V;
$Na^+ + e^- \rightleftharpoons Na$, $E° = -2.714$ V;
$Li^+ + e^- \rightleftharpoons Li$, $E° = -3.0401$ V;
$Ca(OH)_2 + 2\,e^- \rightleftharpoons Ca + 2\,OH^-$, $E° = -3.02$ V;

The problem with any of these batteries is that they would have to contain F_2 gas, which would require a large, heavy container. In addition, fluorine is corrosive and highly toxic.

17.95 The lead electrode acts as the anode, meaning that oxidation occurs at that electrode. Thus
$E°_{cell} = E°_{Ru} - E°_{Pb}$; $E°_{Ru} = E°_{cell} + E°_{Pb} = 0.745$ V $+ (-0.3588$ V$) = 0.386$ V.

17.97 The standard potential for a redox reaction is easily calculated from the equilibrium constant for the reaction by using Equation 17-5, $E° = \left(\dfrac{5.916 \times 10^{-2} \text{ V}}{n}\right) \log K_{eq}$:

$E° = 1/2(5.916 \times 10^{-2}$ V$) \log(2.69 \times 10^{12}) = (0.01958)(12.430) = 0.368$ V.

17.99 Before carrying out calculations on a cell reaction, it is necessary to have the net reaction and its standard cell potential. From Appendix H:

$I_2 + 2$ e$^- \rightleftharpoons 2$ I$^-$, $E° = 0.5355$ V
2 IO$_3^- + 12$ H$^+_{(aq)} + 10$ e$^- \rightleftharpoons I_2 + 6$ H$_2$O, $E° = 1.195$ V;
Multiply the first reaction by 5 and subtract from the second reaction:
2 IO$_3^- + 12$ H$^+_{(aq)} + 10$ e$^- \rightleftharpoons I_2 + 6$ H$_2$O $- 5[I_2 + 2$ e$^- \rightleftharpoons 2$ I$^-]$
2 IO$_3^- + 12$ H$^+_{(aq)} + 10$ I$^- \rightleftharpoons 6 I_2 + 6$ H$_2$O; simplify by dividing by 2:
IO$_3^- + 6$ H$^+_{(aq)} + 5$ I$^- \rightleftharpoons 3 I_2 + 3$ H$_2$O; $n = 5$, $E° = 1.195 - 0.5355 = 0.660$ V;

(a) To determine the spontaneous direction at pH $= 2.00$, use the Nernst equation to calculate E at this pH:

$E = E° - \dfrac{(0.05916)}{n} \log Q = 0.660$ V $- \dfrac{(0.05916)}{5} \log\left\{\dfrac{1}{[\text{IO}_3^-][\text{H}^+_{(aq)}]^6[\text{I}^-]^5}\right\}$

$E = 0.660$ V $+ (0.0118$ V$) \log[(0.100)(0.0100)^6(0.100)^5] = 0.660$ V $- 0.213$ V $= 0.447$ V.
The reaction is spontaneous in the direction written;

(b) Repeat the calculation at pH $= 11.00$:
$E = 0.660$ V $+ (0.0118$ V$) \log[(0.100)(10^{-11})^6(0.100)^5] = 0.660$ V $- 0.852$ V $= -0.192$ V.
The reaction is spontaneous in the opposite direction.

(c) Set $E = 0$ and solve for the concentration of hydronium ions:

$E = 0.660$ V $+ \dfrac{(0.05916)}{5} \log\{[\text{IO}_3^-][\text{H}^+_{(aq)}]^6[\text{I}^-]^5\}$;

$0 = 0.660$ V $+ (0.0118$ V$)(-6) + (0.0118$ V$) \log\{[\text{H}^+_{(aq)}]^6\}$;

$\log\{[\text{H}^+_{(aq)}]^6\} = -49.93$, $\log[\text{H}^+_{(aq)}] = (-49.93) \div 6 = -8.32$, pH $= 8.32$.

Chapter 18. The Transition Metals

18.1 Overview of the Transition Metals

QUESTIONS TO ANSWER, SKILLS TO LEARN
1. **What are the transition metals?**
2. **Using electronic configuration to predict physical properties of the transition metals.**

The transition metals are the elements that have valence electrons only in ns and $(n-1)d$ orbitals. These elements form the long rows of metals located in the between the group 2 and group 13 elements. Although the metal atoms' electronic configurations have valence electrons in both the ns and $(n-1)d$ orbitals, the ground state configurations of the the ions have valence electrons only in the $(n-1)d$ orbitals.

The metals have a variety of melting points. The trends in their melting points can be understood by remembering that the bonding in metals is delocalized. The transition metals have 9 valence orbitals (the ns, the three np and the 5 $(n-1)d$) which form bands. In the transition metals, the bands formed from the np orbitals can be ignored at this level. The first six valence electrons occupy bands that are strongly metal-metal bonding in character; subsequent electrons occupy bands that are antibonding.

The size of the metals follow some of the periodic trends described earlier in the text: atomic size increases going down a group, with the difference between the 1st and 2nd rows being greater than that between the second and third row transition metals. However, the atomic size of the metal going left to right first decreases until the group 8 elements, whereupon the antibonding electrons lead to an decrease in the metal-metal bond order and the bond length (and the measured atomic size) increases.

Exercise 18.1 What transition metals would be similar in size to ruthenium?

Steps to Solution *This question requires you to consider the periodic trends with some of the exceptions that have been noted above and in the text. We expect that the similar sized metals will be close to ruthenium in the periodic chart. Osmium would be somewhat larger, but not much since the radii of 2nd and 3rd row transition metals in the same group tend to be similar. Another transition metal that would of similar size is Mn. Manganese is in the group to the left, which would make it larger than iron (directly above ruthenium) but because it is one row higher (a first row transiton metal) it is smaller than techetium (immediately to the left of ruthenium). This is an example of a diagonal relationship.*

18.2 Coordination Complexes

QUESTIONS TO ANSWER, SKILLS TO LEARN
1. What features are common to coordination compounds?
2. What are ligands and what features are necessary for them to chelate a metal?
3. Learning the different geometries and isomers possible for coordination numbers 4 and 6
4. Learning to convert formulas to names and vice versa; nomenclature

Lewis acid-base adducts with metal ions are called **coordination complexes**. The word complex reflects the difficulty that early chemists had understanding the bonding in these systems. In many coordination complexes, six Lewis bases form bonds to the metal ion or atom (Exercise 18.1 (d)), in contrast to carbon compounds which formed no more than four bonds to any given carbon atom in a compound. The molecules that bind the metal are Lewis bases; Lewis bases in coordination complexes are often referred to as **ligands**.

Ligands **are the Lewis bases in coordination complexes.**

Ligands have quite a variety of structural types since any Lewis base can be a ligand. Many ligands have only one atom which bears an unshared pair, so the ligand may bind only through that atom. The atom that bears the lone pair is called the **donor atom** of the ligand. Molecules may have more than one donor atom (Lewis base site).

One classification of ligands is by the number of donor atoms with which they can bind the metal. If they have:

- **One** donor atom = **monodentate** ligand. Examples: Cl^-, OH^-, CN^-, NH_3, CO, $P(CH_3)_3$ and H_2O.

- **Two** donor atoms = **bidentate** ligand. These ligands can bind to the metal through two atoms. When a bidentate ligand binds a metal through both donor atoms it is called a **chelating ligand**. Examples of bidentate ligands are shown below. The donor atoms in the oxalate ligand are distinguished by the indicated lone pairs.

- **Three** donor atoms = **tridentate** ligand. These ligands may bind a metal with three atoms.

- **Four** donor atoms = **tetradentate** ligand. These ligands have four atoms that can bind the metal.

- **Six** donor atoms = **hexadentate** ligand. These ligands can bind a metal with six atoms. The most common hexadentate ligand is EDTA, $(^-O_2C\text{-}CH_2)_2N\text{-}CH_2\text{-}CH_2\text{-}N(CH_2CO_2^-)_2$, an ingredient in shampoos, soaps, food products and many other commercial products.

Structures of Coordination Complexes

The number of donor atoms bonded to the metal ion in coordination complexes is called the **coordination number**. The geometrical arrangement of ligands around the inner metal atom is described on the basis of the number of donor atoms bonded to the metal, which is called the **coordination geometry**. The common geometries for coordination numbers 4 and 6 are listed in the table below.

Coordination number	Geometry	Sketch of Structure (L = monodentate ligand)
4	tetrahedral (all L-M-L bond angles are 109.5°)	
4	square planar (all L-M-L bond angles are 90°)	
6	octahedral (all angles between adjacent bonds are 90°)	

If there is more than one type of ligand bonded to the metal atom/ion, then there may be different arrangements of ligands around the metal atom (isomers). Isomers of coordination complexes differ in the arrangement of the ligands about the metal atom. The best way to determine the number of isomers for a given geometry and different ligands is to systematically add one new ligand at a time and count the isomers obtained. To insure that you don't double count possible isomers, you should use a molecular model set to determine that the two arrangement of ligands around the metal are in fact different. After using models, and comparing the models to drawings, you will become much more accurate in your ability to use drawings to find all the isomers of coordination compounds with several different ligands.

> **COMMON PITFALL: Not using models.**
> To best determine whether or not two arrangements of ligands are isomers, make models of both and try to make them "overlap." Isomers can be superimposed only by taking apart one of the models!

Exercise 18.2. Find the number of isomers possible for:
    (a) a tetrahedral complex, $CoBr_2Cl_2^{2-}$
    (b) the square planar complex, $Pd(NH_3)_2Cl_2$
    (c) $Fe(bipy)(NH_3)_4^{2+}$

Steps to Solution: *To determine the number of isomers of coordination complexes, we draw the compound with one fewer ligand than the final number substituted. Then the other ligand is placed in the other possible positions. These possibilities are then compared to find out how many of these possibilities lead to truly different geometries which cannot be converted into each other by rotation of the molecule. Discarding the identical compounds, the number of isomers is determined. Use models if you are confused!*

(a) Tetrahedral $CoBr_2Cl_2^{2-}$
   The starting point for $CoBr_2Cl_2^{2-}$ could be either $CoBrCl_3^{2-}$ or $CoBr_3Cl^{2-}$. Using the first complex, $CoBrCl_3^{2-}$, as a starting point, we draw its structure (i) (remember that the solid wedge line indicates the bond is coming out of the plane of the paper). The second Br atom could replace any of the other three Cl atoms; possibilities are shown in ii and iii.

*The structures ii and iii initially look different, but models show that they are the same molecule, simply rotated so that your perspective is different. Therefore there is only **one isomer**, and structures ii and iii are just different views of its structure.*

(b) Square planar $Pd(NH_3)_2Cl_2$
   The starting point for $Pd(NH_3)_2Cl_2$ could be either $Pd(NH_3)_3Cl$ or $Pd(NH_3)Cl_3$. Using the first complex, $Pd(NH_3)_3Cl$, as a starting point, we draw its structure (i). The second Cl atom could replace either the NH_3 across from the lower NH_3 or the one adjacent to it. These possibilities are shown in ii and iii.

The structures in ii and iii look different, and models show that they are not the same molecule. Therefore there are two isomers; that shown in ii is called the trans *isomer (the like ligands are across from each other) and that in iii is the* cis *isomer (the like ligands are next to each other).*

(c) Octahedral Fe(bipy) $(NH_3)_4^{2+}$

We must remember that an octahedral complex has coordination number 6, with bond angles of 90° between the nearest ligand pairs. Next we must find the structure of the ligand whose abbreviation is bipy (see the above table). This bidentate ligand will occupy two of the octahedral coordination sites. Because the donor N atoms are separated by only three bonds (see i), this ligand can only reach adjacent, or cis, *coordination positions in the coordination complexes they form. The four NH_3 ligands will occupy the remaining four coordination sites, as shown in ii.*

Naming Coordination Compounds

The variety of coordination compounds requires systematic nomenclature for chemists to be able to deduce the structure of a complex from the name and vice versa. To translate a structure or formula to a written name, the following rules are used:

<u>Names of Complexes</u>

1. An ionic complex is formulated and written with the cation name followed by the anion name.

2. Complexes are named by listing the ligands and then the central metal atom (the reverse of the formula convention).

Chapter 18: The Transition Metals

3.a. Anionic ligands have names ending with the suffix -o.

Anion	Anion Name	Ligand Name
Cl^-	chlor*ide*	chlor*o*
OH^-	hydrox*ide*	hydrox*o*
CN^-	cyan*ide*	cyan*o*

3.b. When serving as ligands, anions whose names end in *-ate* have this ending replaced with *-ato*.

Anion	Anion Name	Ligand Name
SO_4^{2-}	sulf*ate*	sulf*ato*
$C_2O_4^{2-}$	oxal*ate*	oxal*ato*
SCN^-	thiocyan*ate*	thiocyan*ato*

3.c. Neutral ligands are named as the neutral molecule except for H_2O, the *aqua* ligand; NH_3, the *ammine* ligand; and CO, the *carbonyl* ligand.

4. The numbers of ligands in a complex are specified by using the Greek prefixes di-, tri-, tetra-, etc. The prefixes bis-, tris-, tetrakis-, etc., are used to eliminate confusion when the name of a ligand contains a Greek prefix, e.g. dichloro*bis* (ethylene*di*amine)cobalt(III) bromide.

5. The name of an anionic complex ends with *-ate*. This is added to the stem of the name of the metal. Chromium becomes chrom*ate*, aluminum becomes alumin*ate*, cupric becomes cupr*ate*, cobalt becomes cobalt*ate*, etc. Cationic and neutral complexes are identified with the English name of the metal and no suffix.

6. The oxidation state of the central metal atom is indicated with a Roman numeral (the Stock number) in parentheses at the end of the name of the metal. Also, since the oxidation number of the metal has been indicated, it is redundant to state the number of counterions. For example, potassium hexachloroplatinate(IV), hexamminecobalt(III) bromide.

Formulas of Coordination Complexes

1. The formula of a complex shows the central atom first, followed by the ligands. For example, $Cu(NH_3)_4^{2+}$, $CrCl_6^{3-}$

2. When the ligands in a complex are not all alike, they are listed alphabetically by name, ignoring prefixes.

Exercise 18.3. Write the missing name or formula for the following:
(a) $[Co(H_2O)_5Br]Cl_2$
(b) magnesium tetrachlorocobaltate(II)
(c) $K_2Ni(CN)_4$
(d) $[Cr(OH)_2(H_2O)_4]Br$

Chapter 18: The Transition Metals

Steps to Solution: When naming a compound from the formula, we identify the complex and the counterion(s) and then name the complex from the rules given above. It is important to work through the rules step by step, while you are still learning the rules. When writing the formula given the name of a coordination complex, one more step is required: finding the number of counterions given the oxidation number on the metal and the charges of the ligands bound to the metal.

(a) $[Co(H_2O)_5Br]Cl_2$
The cation in this compound is the complex ion, $Co(H_2O)_5Br_2^+$, with chlorides as the anionic counterions. The ligands are five water ligands (named aqua) and one bromide ligand (named bromo) and must be listed alphabetically by ligand name; pentaquabromo describes the ligands on the complex ion. The oxidation number is +3. The name of the coordination complex is pentaquabromocobalt(III). The name of the compound requires the name of the counterion: \X(pentaquabromocobalt(III) chloride) .

(b) Magnesium tetrachlorocobaltate(II)
In this compound, the complex is anionic, as indicated by the -ate ending on tetrachlorocobaltate; therefore magnesium is the counter ion. The ligands are four chlorides, as indicated by tetrachloro-. Because the oxidation number of Co is +2, the charge on the complex is -2 (+2 for Co; 4(-1) for the Cl^- ligands): one Mg^{2+} ion will be required to form the neutral salt. The cation is listed first when writing the formula of a salt; the ligands are listed alphabetically in the complex: $\boxed{Mg[CoCl_4]}$.

(c) $K_2Ni(CN)_4$
The complex is $Ni(CN)_4^{2-}$ because there are two K^+ counterions. There are four cyano ligands; they are named tetracyano. The oxidation number on Ni is +2 (charge on complex = -2; charge on CN^- ligands 4(-1) = -4); the complex is tetracyanonickelate(II). The complete name requires specifying the counterion, potassium. The complete name is $\boxed{\text{potassium tetracyanonickelate(II)}}$.

(d) $[Cr(OH)_2(H_2O)_4]Br$
In this compound, the counterion is bromide. Therefore the complex is $Cr(OH)_2(H_2O)_4^+$. There are two OH^- ligands (= dihydroxo) and four water ligands (= tetraqua) and the oxidation number on Cr is +3 (charge on ligands = -2; charge on complex = +1). The name of the complex ion is tetraquadihydroxchromium(III); the complete name with the counterion is $\boxed{\text{tetraquadihydroxchromium(III) bromide}}$.

18.3 Bonding in Coordination Complexes

QUESTIONS TO ANSWER, SKILLS TO LEARN
1. Using crystal field theory to describe bonding in coordination compounds with coordination number 4 and 6
2. What are Δ (d orbital splitting) and P (pairing energy)?
3. Understanding how the differences in Δ and P lead to different electron configurations for complexes of the same metal ion with different ligands

Crystal field theory (CFT) was developed by Hans Bethe in 1929 for ionic solids but is useful for understanding how the energies of the $5d$ orbitals of a metal ion change when in different coordination geometries. In the gas phase with no ligands, the $5d$ orbitals have the same energy. In examining the energies of the different orbitals, the only factor we will consider is the electron repulsion that would be experienced by an electron in a particular d orbital in a given geometrical arrangement of ligands (which have lone pairs of electrons pointing at the metal). Orbitals directed at the ligands are less stable than those pointing between the incoming ligands. The three arrangements of ligands we will consider are octahedral, tetrahedral and square planar. The drawings that follow show octahedral and tetrahedral coordination geometries and the Cartesian coordinates commonly used.

octahedral geometry

tetrahedral geometry

Energies of d Orbitals in Octahedral Complexes

In octahedral complexes the d_{xz}, d_{yz} and d_{xy} orbitals point between the x, y and z axes and the $d_{x^2-y^2}$ and d_{z^2} orbitals point along the x, y and z axes (mentally or physically draw them in the picture above). The three orbitals that point between the axes are of equal energy and are referred to as the t_{2g} **orbitals**; the two that point along the axes are both at a higher energy and are called the e_g **orbitals**. The energy difference between the t_{2g} and e_g orbitals is called Δ. The energy splitting can be pictured as shown below in the first part of the diagram.

Chapter 18: The Transition Metals

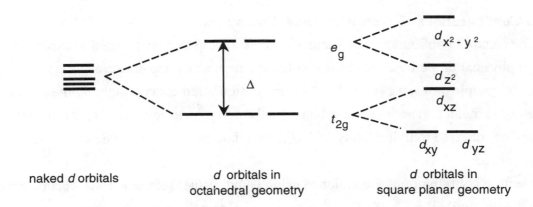

naked d orbitals d orbitals in octahedral geometry d orbitals in square planar geometry

Square Planar Geometry

Square planar geometry with coordination number 4 is related to octahedral geometry; removing the ligands on the +z and -z axis directions gives square planar geometry. The result of removing the "z" ligands is that orbitals that point in the +z and -z directions will be more stable, because there are no ligands to give rise to electron-electron repulsions. The $d_{x^2-y^2}$ orbital's lobes point directly at the ligands, and is highest in energy. The d_{xz} and d_{yz} orbitals point out of plane and between the ligands and thus are the most stable d orbitals.

Tetrahedral Geometry

In tetrahedral geometry the ligands can be envisioned as approaching the metal center from alternate corners of a cube (see above drawing). Orbitals that have electron density (point) along the x, y and z axes will be more stable than other orbitals because there are fewer electron-electron repulsions with the ligands in these directions than with orbitals which point in between the x, y and z axes. The orbitals split into two groups: the e_g and t_{2g}. The significant difference is that the t_{2g} orbitals are the less stable group in tetrahedral geometry.

Summary

Complex Geometry	Octahedral	Square Planar	Tetrahedral
Higher-energy orbitals	e_g ($d_{x^2-y^2}$, d_{z^2})	$d_{x^2-y^2}$	t_{2g} (d_{xz}, d_{yz}, d_{xy})
		d_{xy}	
Lower-energy orbitals	t_{2g} (d_{xz}, d_{yz}, d_{xy})	d_{z^2}	e_g ($d_{x^2-y^2}$, d_{z^2})
		d_{xz}, d_{yz}	

Chapter 18: The Transition Metals

Electron Configurations of Transition Metal Complexes

The electron configuration of a transition metal complex may be used to explain a number of physical properties, such as the color and number of unpaired electrons in the complex. In complex ions, we write the electron configuration concerning only the valence electrons of the metal that are in the d orbitals. If the complex is octahedral or tetrahedral, the d orbitals are not referred to individually (d_{xz}, d_{yz}, etc.), but the number of electrons in the group (t_{2g} or e_g) is used.

Electron configurations are assigned using Hund's Rules (electrons will occupy orbitals of the same energy until all are half filled). For octahedral transition metal complexes with three or fewer d electrons, the electron configurations are easily assigned.

Exercise 18.4 What is the electron configuration of the complex TiF_6^{4-} ?

Steps to Solution: We must find several pieces of information before answering this question: the oxidation number of Ti in the complex and then the number of d electrons present. The geometry must also be deduced so that the correct energy level diagram is used.

There are 6 donor atoms in this complex so the geometry is octahedral, and we use the octahedral energy level diagram to assign the electron configuration. The oxidation number of Ti in the complex is +2 (complex charge = -4 and the sum of charges on the ligands = 6(-1) = -6). The number of d electrons on Ti is two (Ti has an electron configuration of $4s^2 3d^2$: Ti^{2+} is $3d^2$). The lowest-energy orbitals in the complex are the t_{2g} orbitals; the two electrons occupy the t_{2g} orbitals. The electronic configuration is $\boxed{(t_{2g})^2}$.

When there are more than three d electrons, there is a choice of orbitals that the next electrons may occupy; they might occupy the e_g orbitals (and increase the number of unpaired electrons) or pair in the t_{2g} orbitals. It is energetically unfavorable to pair electrons because they repel each other; the energy required to overcome this repulsion is the **pairing energy, P.**

> The *pairing energy* is the amount of energy required to place two electrons in the same orbital.

In transition metal complexes, the splitting between the t_{2g} and e_g orbitals is nearly the same as the pairing energy. The size of Δ depends on the ligands; ligands that lead to larger values of Δ are called **strong field ligands**; those with smaller values of Δ are called **weak field ligands**.

If the splitting of the orbitals, Δ, is large enough, it will cost more energy for the electron to occupy an e_g orbital than to pair in a t_{2g} orbital. The fourth electron will occupy a t_{2g} orbital giving a t_{2g}^4 electron configuration. If the pairing energy is more than Δ, it will cost less energy to put the fourth electron in the e_g orbital, giving a configuration of $t_{2g}^3 e_g^1$. These two electron configurations are shown below using orbital energy diagrams.

Chapter 18: The Transition Metals

$$\Delta > P : t_{2g}^4 \qquad \Delta < P : t_{2g}^3 e_g^1$$

The different configurations give rise to different numbers of unpaired electrons: there are two unpaired electrons in the t_{2g}^4 configuration and four unpaired electrons in the $t_{2g}^3 e_g^1$ configuration. As discussed in Chapter 3, the number of unpaired electrons in paramagnetic compounds may be determined experimentally. Therefore these two configurations, which are obtained from the same number of valence d electrons, are readily distinguished from each other on the basis of their magnetic behavior. The one with the greater number of unpaired electrons is the **high spin** configuration and the other, with fewer unpaired electrons, is the **low spin** configuration.

Ligands that give $\Delta > P$ are strong field ligands and generally give low spin configurations.

Ligands that give $P > \Delta$ are weak field ligands and generally give high spin configurations.

A summary of the different electron configurations and the numbers of unpaired electrons is presented in the following table.

Number of d Electrons	P > Δ (High Spin) Configuration	# Unpaired e's	P < Δ (Low Spin) Configuration	# Unpaired e's
1	t_{2g}^1 1 unpaired electron (1 UPE)		same as high spin	
2	t_{2g}^2	2 UPE	same as high spin	
3	t_{2g}^3	3 UPE	same as high spin	
4	$t_{2g}^3 e_g^1$	4 UPE	t_{2g}^4	2 UPE
5	$t_{2g}^3 e_g^2$	5 UPE	t_{2g}^5	1 UPE
6	$t_{2g}^4 e_g^2$	4 UPE	t_{2g}^6	0 UPE
7	$t_{2g}^5 e_g^2$	3 UPE	$t_{2g}^6 e_g^1$	1 UPE
8	$t_{2g}^6 e_g^2$	2 UPE	$t_{2g}^6 e_g^2$	2 UPE
9	$t_{2g}^6 e_g^3$	1 UPE	same as high spin	

Exercise 18.5 The complex $Fe(C_2O_4)_3^{3-}$ has one unpaired electron. What is the electron configuration of this complex?

Steps to Solution: To determine the electron configuration, we must consider the possible ways that the d electrons can be placed in the d orbitals. To accomplish this, we must determine the number of d electrons for the iron, and the structure of the complex. From the structure, we will know the pattern of d orbital splitting and then we can decide between low spin and high spin possibilities by constructing both low spin and high spin configurations and comparing the number of unpaired electrons in those configurations to the experimental result of one unpaired electron.

The (C_2O_4) group is the oxalate ligand, $C_2O_4^{2-}$. This is a bidentate chelating ligand; because there are three oxalate ligands, the iron has a coordination number of 6, and octahedral geometry. The number of d electrons is found using the oxidation number, +3. Fe has 8 valence electrons; therefore Fe^{3+} has 8 - 3 = 5 d electrons. The two ways of arranging these electrons are shown in the following energy level diagrams:

low spin   high spin

The low spin case has one unpaired electron and the high spin case, five unpaired electrons. We were told the complex has one unpaired electron; therefore it is low spin and has the low spin d^5 electronic configuration, $\boxed{t_{2g}^5}$.

From the values of Δ determined for a large number of ligands, we can write a general ordering of ligands in terms of increasing Δ. This order is referred to as the **spectrochemical series**.

Spectrochemical Series

smaller Δ   larger Δ

$$I^- < Br^- < Cl^- < F^- < OH^- < H_2O < NH_3 < en < NO_2^- < CN^- < CO$$

Compounds with ligands that lie on the high end of the spectrochemical series tend to be low spin; complexes with ligands on the low end of the spectrochemical series tend to be high spin.

One of the more esthetically pleasing aspects of the different coordination compounds is the variety of colors they possess. The colors of these compounds arises from the fact that the energy separation between the occupied orbitals and the empty orbitals lies in the frequency range of visible light. When white light shines on the sample, some colors are absorbed and we observe those that are not absorbed (the **complementary color**). If the substance absorbs only ultraviolet light, the sample will appear colorless, or white.

The energy of the of the photon absorbed can be calculated to give the energy separation of the orbitals in the molecule energy because:

$$\Delta E = E_{photon} = h\nu = h\frac{c}{\lambda}$$

Exercise 18.6 Find the crystal field splitting for the following cobalt complexes in which only one ligand has changed. Compare the calculated values of Δ with the order of the ligands in the spectrochemical series.

Complex	$Co(NH_3)_5H_2O^+$	$Co(NH_3)_6^{3+}$	$Co(NCS)(NH_3)_5^{2+}$	$Co(NH_3)_5Cl^+$
λ_{max}	500 nm	430 nm	470 nm	530 nm
color observed	red	yellow	orange	purple
(absorbs)	blue/green	indigo	blue	green

Steps to Solution: The energy, Δ, is found by applying the formula,

$$\Delta E = h\frac{c}{\lambda}$$

After conversion of the wavelength from nm to m (illustrated for the $Co(NH_3)_5H_2O_3^+$ complex), the energy, Δ, is calculated to be:

$$\Delta E = 6.626 \times 10^{-34} \text{ J} \cdot \text{s} \frac{3.00 \times 10^8 \text{ m/s}}{500 \text{ nm} (10^{-9} \text{ m/nm})} = 3.97 \times 10^{-19} \text{ J}$$

The remaining values for Δ are calculated in an identical fashion.

Complex	$Co(NH_3)_5H_2O^+$	$Co(NH_3)_6^{3+}$	$Co(NCS)(NH_3)_5^{2+}$	$Co(NH_3)_5Cl^+$
Δ	3.97×10^{-19} J	4.62×10^{-19} J	4.23×10^{-19} J	3.75×10^{-19} J

Sorting these values of Δ for the different ligands from lowest to highest energy gives:

Ligand changed	Cl⁻	H₂O	NCS⁻	NH₃
Δ	3.75×10^{-19} J	3.97×10^{-19} J	4.23×10^{-19} J	4.62×10^{-19} J

Notice that the three ligands that are on the spectrochemical series supplied above have values of Δ consistent with their relative positions in the series: Cl⁻ has the lowest value of Δ and is the lowest of these ligands in the spectrochemical series. NCS⁻ is not in the above series, but the calculation suggests that it would fall between the aqua and ammine ligands in splitting the d orbitals in the complexes it forms.

18.4 Metallurgy

QUESTIONS TO ANSWER, SKILLS TO LEARN
1. **How are metals obtained from ores?**
2. **What are the fundamental steps in metal refining?**

Most metals are readily oxidized and so are found as ores combined with non-metal anions in nature. In addition to the fact that the metal is in an oxidized state, there are ususally other metal ions in the ore which have similar properties. To obtain the metal one must separate the ore from other mineral matter which is present, then the metal cation must be reduced to the metallic state. Finally, the mixture of metals which is obtained by this method must be further purified or refined.

The separation process can use physical properties such as density differences or chemical reactivity. Reduction relies on a cheap source of electrons, commonly elemental carbon (coke) which reacts cleanly with metal oxides to give the metal and carbon monoxide and carbon dioxide depending upon the stoichiometry. More expensive reduction techniques use electrolysis (copper and the main group metals) to reduce metal ions to the elemntal metal. Refining of the metal uses various techniques. Raw copper is purified by an electrolysis method, nickel is refined by formation of a volatile complex, $Ni(CO)_4$, which can be distilled.

Chapter 18: The Transition Metals

18.5 Applications of Transition Metals

QUESTIONS TO ANSWER, SKILLS TO LEARN
1. **Recognizing transition metals' uses.**

Transition metals have many applications, from structural materials to metallic catalysts to coordination compounds that are used to combat arthritus and cancer or to catalyze industrial processes.

Exercise 18.7 Read your text and then observe your surroundings. Then write down three applications of transition metals that you observe in daily life (not just structural).

_____

18.6 Transition Metals in Biology

QUESTIONS TO ANSWER, SKILLS TO LEARN
1. **What features make transition metals useful in biological systems?**

Transition metals are found in biological systems usually combinded with polypeptides, a class of molecules called **metalloproteins**. These metalloproteins are invovled in transport and storage of small molecules necessary for biolological functions, such as hemoglobin and myoglobin. Another role of metalloproteins is as catalysts, where they act as acids such as carboxyanhydrase (it catalyzes the conversion of CO_2 to monohydrogen carbonate ion) or as electron transfer agentsand the cytochromes.

_____

Test Yourself

1. How many isomers are there for square planar $Pt(NH_3)_2BrCl$?

2. Write the electron configuration for CoF_6^{3-} and state whether the complex is paramagnetic. If paramagnetic, give the number of unpaired electrons in the complex.

3. The complexes $Co(NH_3)_6^{3+}$ and $Mo(CO)_6$ are isoelectronic and diamagnetic. The first complex is orange and the second complex is white. What can you deduce about the value of Δ in both these complexes?

Chapter 18: The Transition Metals

Answers

1. Two

2. $t_{2g}^4 e_g^2$, 4 unpaired electrons

3. $Co(NH_3)_6^{3+}$ absorbs visible light, since it is orange, whereas $Mo(CO)_6$ does not absorb light in the visible spectrum. The complexes both have the t_{2g}^6 electron configuration. The ligands are both high field (the complexes are low spin); therefore the value of Δ in the Mo complex must be sufficiently large that no visible light is absorbed; therefore the light must be absorbed in the ultraviolet region of the spectrum, which is at higher energy than visible light. Therefore the value of Δ is larger in $Mo(CO)_6$ than in $Co(NH_3)_6^{3+}$.

Solutions to Chapter Problems

18.1 Oxidation states are determined by applying the rules given in Chapter 17. The procedure can be simplified when a polyatomic ion of known charge is present.
(a) CO_3^{2-} has a charge of –2, so Mn must be +2 to give overall neutrality;
(b) Cl (more electronegative) is –1, so Mo must be +5 to give overall neutrality;
(c) Na is +1, so VO_4 has overall charge –3; each O is –2, so V is +5 to give net charge of –3;
(d) SO_4^{2-} has a charge of –2, so Fe must be +3 to give overall neutrality;
(e) O is –2, so Au must be +3 to give overall neutrality.

18.3 Use the periodic table to locate and identify elements from their valence configurations.
(a) There are 6 valence electrons, so the element is in column 6; $3d$ orbital is filling: Cr;
(b) There are full s and d blocks, so the element is in column 12; $4d$ has just filled: Cd;
(c) This element is the fourth member of the $4f$ block: Nd;
(d) This element is the eighth member of the $5f$ block: Cm.

18.5 The properties of transition metals vary regularly with their valence configurations, so the starting point for predicting relative properties is location in the d block.
(a) Pd is in column 10, Cd is in column 12, both in the $n = 4$ row. Beyond the middle of the d block, melting point decreases with Z because electrons are placed in antibonding orbitals, so Pd melts at higher T;
(b) Cu and Au are both in column 11, but Au has higher Z, so Au has higher density;

(c) Cr is in column 6, Co is in column 9, both in the $n = 3$ row. IE_1 increases with Z across a row because Z_{eff} increases, so Co has the higher IE_1;

(d) V is in column 5, Zn is in column 12, both in the $n = 3$ row. EA increases with Z across a row because Z_{eff} increases, so Zn has the higher EA.

18.7 Use the charges of ligands and ions to determine the oxidation states of transition metals in coordination complexes. Because s electrons are always removed first, the count of d electrons is given by the number of valence electrons – the oxidation state.

(a) Each Cl is –1, and NH$_3$ is neutral, so Ru has oxidation state +2. Ru is in column 8 (8 valence electrons), giving d^6;

(b) Each I is –1, and en is neutral, so Cr has oxidation state +3. Cr is in column 6 (6 valence electrons), giving d^3;

(c) Each Cl is –1, and trimethylphosphine is neutral, so Pd has oxidation state +2. Pd is in column 10 (10 valence electrons), giving d^9;

(d) Each Cl is –1, and NH$_3$ is neutral, so Ir has oxidation state +3. Ir is in column 9 (9 valence electrons), giving d^6;

(e) CO is a neutral ligand, so Ni has oxidation state 0. Co is in column 9 (9 valence electrons), but here no electrons have been removed, so the valence configuration is $s^2 d^7$.

18.9 Compounds that contain coordination complexes are named following the six rules stated in your textbook: name cation first, name ligands in alphabetical order, name metal, add "o" for anions, use Greek prefixes, add "-ate" for anionic complexes, give oxidation number.
(a) hexaammineruthenium(II) chloride;
(b) *trans*-bis(ethylenediamine)diiodochromium(III) iodide;
(c) *cis*-dichlorobis(trimethylphosphine)palladium(II);
(d) *fac*-triamminetrichloroiridium(III);
(e) tetracarbonylnickel(0).

18.11 The structure of a metal complex usually is octahedral (6 ligands), tetrahedral (4 ligands), or square planar (4 ligands).

18.13 The name of a complex contains the information needed to determine its chemical formula. Determine the charge of the complex from charges on the ligands and the oxidation number. (a) cis-$[Co(NH_3)_4Cl(NO_2)]^+$; (b) $[Pt(NH_3)Cl_3]^-$; (c) trans-$[Cu(en)_2(H_2O)]^{2+}$; (d) $[FeCl_4]^-$.

18.15 The crystal field diagram for weak and strong octahedral fields always is the same, with the populations changing depending on how many d electrons must be accommodated. The valence configuration provides information about the number of d electrons: (a) Ti^{2+} (column 4 – 2 electrons) is d^2; (b) Cr^{3+} (column 6 – 3 electrons) is d^3. For these two ions, the low field and high field configurations are the same:

(c) Mn^{2+} (column 7 – 2 electrons) is d^5; (d) Fe^{3+} (column 8 – 3 electrons) is d^5. These two ions have identical diagrams, with low field differing from high field:

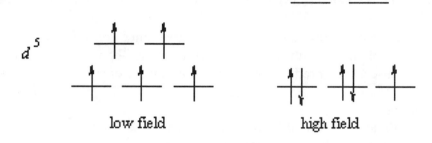

low field   high field

18.17 Complexes with four ligands can be square planar or tetrahedral. First-row transition metal complexes favor tetrahedral geometry, and Ni lies in the first row, so we predict tetrahedral geometry. Tetrahedral complexes have smaller splitting energies than octahedral complexes, so these complexes usually absorb in the infrared but not in the visible. Thus this complex is colorless. The complex is poisonous because it can release CO, a deadly poison gas.

18.19 The magnetic properties of a complex are determined by its number of d electrons and whether it is high field or low field.
(a) Ir^{3+} (column 9 – 3 electrons) is d^6, and NH_3 is a relatively high-field ligand, so all the electrons will be paired and the complex is diamagnetic;
(b) Cr^{3+} (column 6 – 3 electrons) is d^3, so regardless of field strength this complex is paramagnetic with three unpaired electrons;
(c) Pd has d^{10} configuration, so regardless of field strength this complex is diamagnetic;
(d) V^{3+} (column 5 – 3 electrons) is d^2, so regardless of field strength this complex is paramagnetic with two unpaired electrons.

18.21 The colors of transition metal complexes are generally determined by d-d transitions, but Zr^{4+} (column 4 – 4 electrons) has all its valence electrons removed, so there is no valence electron that can undergo a transition that would absorb visible light.

Chapter 18: The Transition Metals

18.23 Consult your textbook for the chemical reactions of various metallurgical processes.
(a) $CuFeS_2 + 2 O_2 \rightarrow CuO + FeO + 2 SO_2$;
(b) $Si + O_2 + CaO \rightarrow CaSiO_3$;
(c) $TiCl_4 + 4 Na \rightarrow 4 NaCl + Ti$.

18.25 This is a standard stoichiometry problem. First determine the balanced chemical equation, then do the stoichiometric calculation.
The reaction is $Cu_2S + O_2 \rightarrow 2 Cu + SO_2$.
$m_{Cu_2S} = (5.60 \times 10^4 \text{ kg})(10^3 \text{ g/kg})(2.37 \%)(10^{-2}/\%) = 1.327 \times 10^6$ g;
$n_{Cu_2S} = m/MM = (1.327 \times 10^6 \text{ g}) \div (159.2 \text{ g/mol}) = 8.34 \times 10^3$ mol;
$n_{Cu} = 2\ n_{Cu_2S} = 2(8.34 \times 10^3 \text{ mol}) = 1.668 \times 10^3$ mol;
$m = n\ MM = (1.668 \times 10^3 \text{ mol})(63.55 \text{ g/mol}) = 1.06 \times 10^6$ g;
$$V = \frac{nRT}{P} = \frac{(8.34 \times 10^3 \text{ mol})(0.08206 \text{ L atm / mol K})(273.2 + 23.5)\text{K}}{(755 \text{ torr})(1 \text{ atm / 760 torr})} = 2.05 \times 10^5 \text{ L}.$$

18.27 Standard free energy changes are calculated using standard free energies of formation found in Appendix E.
$ZnO_{(s)} + C_{(s)} \rightarrow Zn_{(s)} + CO_{(g)}$;
$\Delta G^o_{rxn} = 1(-137.168) + 1(0) - 1(-318.30) - 1(0) = 181.13$ kJ

$ZnO_{(s)} + CO_{(g)} \rightarrow Zn_{(s)} + CO_{2(g)}$;
$\Delta G^o_{rxn} = 1(-394.359) + 1(0) - 1(-318.30) - 1(-318.30) = 61.11$ kJ.

18.29 Consult your textbook, Figure 18.37, for the reactions that convert scheelite into tungsten:
$CaWO_4 + 2 H_3O^+ \rightarrow Ca^{2+} + H_2WO_4$; this reaction dissolves the ore;
$H_2WO_4 + 2 NH_3 \rightarrow (NH_4)_2WO_4 \xrightarrow{\text{heat}} WO_3 + 2 NH_3 + H_2O$; this reaction removes all elements except W and O;
$WO_3 + 3 H_2 \rightarrow W + 3 H_2O$; this reaction reduces the oxide to the pure metal.

18.31 The coinage metals are those that have been used since antiquity for coins: copper, silver, and gold. All are in column 11 of the periodic table. They are characterized by high electrical conductivity, good ductility, and low chemical reactivity, in particular resistance to oxidation. Hence they are used for money (a vanishing use in technologically advanced countries), for electrical wire, for jewelry, and for other decorative objects. See your text for special examples of uses for compounds of these elements.

18.33 Titanium is used as an engineering metal because of its relatively low density, high bond strength, resistance to corrosion, and ability to withstand high temperatures, all of which make it a favored structural material.

Chapter 18: The Transition Metals

18.35 The structural difference between hemoglobin and myoglobin is that the former has four subunits, while the later has just one. As a consequence, hemoglobin has much more complex cooperative chemical behavior than myoglobin does.

18.37 An iron ion bonded to four sulfur atoms from cysteines is in a tetrahedral environment, so the splitting pattern is the 2-3 pattern characteristic of tetrahedral complexes. The iron center loses one electron and is converted from Fe^{2+} to Fe^{3+}:

18.39 To determine electron configurations, start from the position of the element in the periodic table and remove s electrons preferentially.
(a) Cr is in column 6, configuration $[Ar]4s^1\,3d^5$; Cr^{2+} is $[Ar]\,3d^4$; Cr^{3+} is $[Ar]\,3d^3$;
(b) V is in column 5, configuration $[Ar]4s^2\,3d^3$; V^- is $[Ar]4s^2\,3d^4$; V^+ is $[Ar]4s^1\,3d^3$; V^{2+} is $[Ar]\,3d^3$; V^{3+} is $[Ar]\,3d^2$; V^{4+} is $[Ar]\,3d^1$; V^{5+} is $[Ar]$;
(c) Ti is in column 4, configuration $[Ar]4s^2\,3d^2$; Ti^{2+} is $[Ar]\,3d^2$; Ti^{4+} is $[Ar]$.

18.41 Compounds that contain coordination complexes are named following the six rules stated in your textbook: name cation first, name ligands in alphabetical order, name metal, add "o" for anions, use Greek prefixes, add "-ate" for anionic complexes, give oxidation number.
(a) *cis*-tetraaquadichlorochromium(III) chloride;
(b) bromopentacarbonylmanganese(I);
(c) *cis*-diamminedichloroplatinum(II).

18.43 The number of possible isomers of a complex is determined by its geometry and the number of ligands of each type.

18.45 Bidentate ligands form complexes with two links. Each Fe ion forms an octahedral complex with three oxalate anions:

18.47 Superoxide dismutase catalyzes the conversion of superoxide into molecular oxygen and hydrogen peroxide. This reaction occurs at a metal site that contains one Zn^{2+} ion and one Cu^{2+} ion, linked by a histidine ligand that bonds to both metal ions:

$$2\ O_2^- + 2\ H^+ \xrightarrow{SOD} O_2 + H_2O_2$$

18.49 Cu(II) in water forms an aqua complex. Addition of fluoride produces the insoluble green salt, CuF_2, while addition of chloride produces the tetrachlorocopper(II) complex, $[CuCl_4]^{2-}$.
$Cu^{2+}_{(aq)} + 2\ F^-_{(aq)} \rightarrow CuF_{2\ (s)}$
$Cu^{2+}_{(aq)} + 4\ Cl^-_{(aq)} \rightarrow [CuCl_4]^{2-}_{(aq)}$

18.51 Silver (column 11 of the periodic table) is difficult to oxidize, so it has good resistance to corrosion and is suitable for jewelry. Vanadium (column 5) is readily oxidized, so it corrodes rapidly and is unsuited to jewelry.

18.53 The *mer* isomer of an octahedral complex has three like ligands arranged in a meridian plane:

18.55 Silver tarnish is silver sulfide, Ag_2S, formed by reaction with trace amounts of H_2S in the atmosphere: $4\ Ag_{(s)} + 2\ H_2S_{(g)} + O_{2\ (g)} \rightarrow 2\ Ag_2S_{(s)} + H_2O_{(l)}$

18.57 The complex is octahedral, and the CO molecule is a high-field ligand, so the energy level splitting is expected to be large. All electrons are paired, filling the lower levels. The complex is colorless because the energy gap is sufficiently large that visible light has insufficient energy to be absorbed:

18.59 Brass is an alloy of zinc and copper, and superoxide dismutase contains zinc and copper ions in its reaction center.

18.61 The colors of substances depend on what wavelengths of light they absorb. According to Table 18-4, blue color results when a substance absorbs 610 nm light (196 kJ/mol), and red color results when a substance absorbs 500 nm light (239 kJ/mol). Thus deoxyhemoglobin has a smaller value for Δ than oxyhemoglobin, suggesting that O_2 is a higher-field ligand than H_2O (causes a larger splitting). The hypothesis could be tested by replacing the H_2O ligand in deoxyhemoglobin by a low-field ligand such as Cl^-, which should push the color toward the green. In addition, CO, the highest-field ligand, should lead to even larger splitting, pushing the color to red-orange or orange.

18.63 (a) CN^- is a high-field ligand, so we expect high splitting and a low-spin complex; Fe^{2+} is d^6:

(b) Cl^- is low-field, so we expect low splitting and a high-spin complex; Mn^{2+} is d^5:

(c) Coordination complexes of $4d$ elements are almost always low-spin, and Rh^{3+} is d^6:

(d) H_2O is medium-field, but lower oxidation states of $3d$ elements tend to be high-spin; Co^{2+} is d^7:

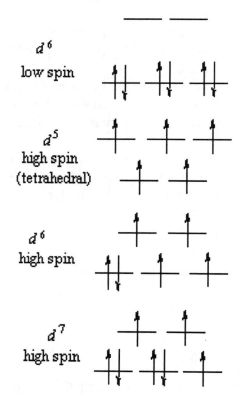

18.65 The chloride ions that form silver chloride precipitate are not bound to the complex and should not be included in the chemical formula of the complex ion:

PtCl$_4$•2 NH$_3$ is [PtCl$_4$(NH$_3$)$_2$], diamminetetrachloroplatinum(IV), *cis* and *trans*:

trans
```
        Cl
        |
  Cl,,,.|..NH3
      .Pt
  H3N/ | \Cl
        |
        Cl
```

cis
```
        Cl
        |
  H3N,,.|..Cl
      .Pt
   Cl/ | \Cl
        |
        NH3
```

PtCl$_4$•3 NH$_3$ is [PtCl$_3$(NH$_3$)$_3$], triamminetrichloroplatinum(IV), *mer* and *fac*:

mer
```
        Cl
        |
  Cl,,,.|..NH3
      .Pt
  H3N/ | \NH3
        |
        Cl
```

fac
```
        Cl
        |
  H3N,,.|..NH3
      .Pt
   Cl/ | \Cl
        |
        NH3
```

PtCl$_4$•4 NH$_3$ is [PtCl$_2$(NH$_3$)$_4$], tetraamminedichloroplatinum(IV), *cis* and *trans*:

trans
```
        Cl
        |
  H3N,,.|..NH3
      .Pt
  H3N/ | \NH3
        |
        Cl
```

cis
```
        Cl
        |
  H3N,,.|..NH3
      .Pt
  H3N/ | \Cl
        |
        NH3
```

PtCl$_4$•5 NH$_3$ is [PtCl(NH$_3$)$_5$], pentaamminechloroplatinum(IV), and PtCl$_4$•6 NH$_3$ is [Pt(NH$_3$)$_6$], hexaammineplatinum(IV):

```
        Cl
        |
  H3N,,.|..NH3
      .Pt
  H3N/ | \NH3
        |
        NH3
```

```
        NH3
        |
  H3N,,.|..NH3
      .Pt
  H3N/ | \NH3
        |
        NH3
```

18.67 Liquid mercury readily forms solutions with its similar neighbors, Au and Ag ("like dissolves like"), but other metals and compounds do not dissolve in this liquid metal. Because mercury boils at a much lower temperature than gold or silver, the pure metals can be obtained by boiling off the mercury.

18.69 When a surfactant molecule interacts with a mineral particle, the polar head groups are attracted to the ionic charges on the mineral particle, and the non-polar tails point outward, creating a mass that has a hydrophobic surface:

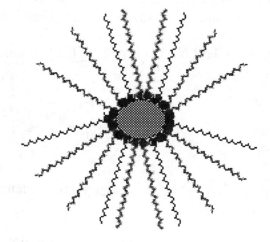

18.71 The dichromate anion has the chemical formula $Cr_2O_7^{2-}$. The tetrahedra about each Cr ion have one common O atom:

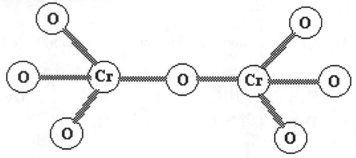

18.73 To determine the color of $[FeF_6]^{3-}$, prepare a table comparing the characteristics of the ligands in the complexes (see Table 18-4):

Ligand	Field strength	Color	Energy gap	Wavelength
CN^-	very high	deep red	239 kJ/mol	500 nm
H_2O	medium	pale green	166 kJ/mol	720 nm
F^-	very low	colorless	< 150 kJ/mol	> 750 nm

The energy gap in the fluoro complex is quite small, so the absorption lies in the infrared.

18.75 Your figure should incorporate the following features that are described in Section 18.6 of your textbook: hemoglobin transports O_2 throughout the body, myoglobin stores O_2 in cells, myoglobin has one heme and storage site, hemoglobin has four hemes and storage sites. You could include other details such as the nature of the binding sites, oxidation state of Fe, etc.

Chapter 19. The Main Group Elements

19.1 Lewis Acids and Bases

QUESTIONS TO ANSWER, SKILLS TO LEARN
1. **What are Lewis acids and bases?**

Molecular characteristics that lead to similar types of chemical reactions are very useful as they allow chemists to generalize and predict the products of new reactions. Earlier, acidity and basicity were defined in terms of the ability of a molecule or other chemical species to release a proton or to abstract a proton, respectively. A more general definition of acids and bases is the *Lewis definition*:

A Lewis acid is an electron pair acceptor.

A Lewis base is an electron pair donor.

A Lewis acid/base neutralization reaction between an acid, A, and a base, :B, can be written

$$A + :B \rightarrow A:B$$

The product, A:B, is a Lewis acid-base adduct. The atoms of A and B are joined by a *covalent* bond that is the same as the covalent bonds that have been previously described except that in this covalent bond, one of the species supplies both electrons to be shared in the bond. The Lewis definition includes Brønsted-Lowry acids and bases and another body of chemistry, that of coordination compounds, especially of transition metals.

Characteristics of Lewis Acids and Bases

Reactions of Lewis acids with Lewis bases involve the formation of a bond to the acid with the electron pair (or pairs) of the Lewis base (or bases). For this to be accomplished the acid must have an empty valence orbital (or several empty valence orbitals) in which to accommodate the electrons supplied by the base. The Lewis base must have one or more unshared pairs of electrons in its valence orbitals. What sort of chemical species fill these criteria?

Lewis Acids
- **Molecules with empty valence orbitals**: Molecules with vacant valence orbitals fall into two classes: (1) molecules that have fewer than 8 valence electrons and (2) those in the third and higher groups (valence electrons in orbitals of principal quantum number $n = 3$ and greater) with vacant nd orbitals. Examples of the first class are compounds such as $AlCl_3$ $_{(g)}$

and BCl_3 $_{(g)}$, which have only 6 valence electrons and have a vacant p orbital that may be used in forming the bond to the Lewis base. The second class is typified by molecules such as $SiCl_4$ and PF_5, which have 8 and 10 valence electrons, respectively.

- **Special Cases: Molecules with delocalized π orbitals involving O or F, such as CO_2, SO_2 and SO_3 and BF_3**: SO_2 and SO_3 are expected to be Lewis acids because we expect that the d orbitals of sulfur might be able to participate in bond formation. CO_2 is a Lewis acid because the C atom has two quite electronegative oxygen atoms removing electron density from it. Important to the acidic behavior in these molecules is the propensity of the double bond to O toward rearrangement upon attack by a base, as illustrated below for CO_2.

$$\ddot{\underset{..}{O}}=C=\ddot{\underset{..}{O}} \quad + \quad B: \quad \longrightarrow \quad :\ddot{\underset{..}{O}}-\underset{\underset{B}{|}}{C}=\ddot{\underset{..}{O}}$$

- **Metal cations and atoms**: Metal atoms and metal cations also possess empty valence orbitals. In main group metals, the empty orbitals are mainly s and p orbitals; in transition metals, the empty orbitals are s, p and d orbitals. The Lewis bases donate electron density from their lone pairs to these empty orbitals to form coordination complexes.

Lewis Bases

- **Molecules with unshared pairs in valence orbitals**: This class of Lewis bases includes nitrogen bases as described in Chapter 16 and other bases such as water, hydrogen sulfide (H_2S) and phosphine (PH_3).
- **Anions with unshared pairs in valence orbitals**: These Lewis bases include the halide anions (F^-, Cl^-, Br^-, I^-) and others such as HS^- and S^{2-}. The common feature of these anions is the presence of unshared paired that may be donated to a Lewis acid to form bonds.

Exercise 19.1. Identify the Lewis acids and bases in each of the following equations of reaction.
(a) H_3O^+ $_{(aq)}$ + OH^- $_{(aq)}$ → 2 H_2O $_{(l)}$
(b) BCl_3 $_{(g)}$ + $N(CH_3)_3$ $_{(g)}$ → $Cl_3B:N(CH_3)_3$ $_{(g)}$
(c) Al^{3+} $_{(aq)}$ + 4 OH^- $_{(aq)}$ → $Al(OH)_4^-$ $_{(aq)}$
(d) Fe^{3+} $_{(aq)}$ + 6 NH_3 $_{(aq)}$ → $Fe(NH_3)_6^{3+}$ $_{(aq)}$

Steps to Solution: To determine which substance is a Lewis acid in a reaction in which an adduct is formed requires examination of its Lewis structure. If the substance falls into one of the three categories described above, the substance is a Lewis acid. The substance that supplies the unshared pair of electrons to form the bond in the adduct is the Lewis base. In the following exercises one lone pair is shown (:) on the Lewis base in each reaction.

(a) $H_3O^+_{(aq)} + :OH^-_{(aq)} \rightarrow 2\,H_2O_{(l)}$
In this reaction, OH^- is identified as the Lewis base because it has three unshared pairs of electrons, one of which is used to form the bond to the H^+ to give water as a product. This Lewis acid-base reaction is also an acid-base reaction in the Brønsted-Lowry sense.

(b) $BCl_{3\,(g)} + :N(CH_3)_{3\,(g)} \rightarrow Cl_3B:N(CH_3)_{3\,(g)}$
In this reaction, BCl_3 is electron deficient because it only possesses 6 valence electrons at the B atom; therefore BCl_3 is the Lewis acid. A lone pair is present on the N atom in $N(CH_3)_3$ and can donate electron density to the B atom in BCl_3 to form the adduct. Trimethyl amine is the Lewis base.

(c) $Al^{3+}_{(aq)} + 4\,OH^-_{(aq)} \rightarrow Al(OH)_4^-{}_{(aq)}$
In this reaction, Al^{3+} is a metal ion, identifying it as the Lewis acid. The OH^- ions bear unshared electron pairs which are used to form bonds to the Al ion in the adduct; therefore the OH^- ions are the Lewis bases.

(d) $Fe^{3+}_{(aq)} + 6\,NH_{3\,(aq)} \rightarrow Fe(NH_3)_6^{3+}{}_{(aq)}$
Use the above methods to show that Fe^{3+} is the Lewis acid and the ammonia molecules are the Lewis bases.

19.2 Hard and Soft Acids and Bases

QUESTIONS TO ANSWER, SKILLS TO LEARN
1. How does polarizability change in atoms, ions and molecules?
2. What is the hard-soft acid-base principle (HSAB) and why does polarizability matter?

Lewis acids and bases are common in our natural world, but some acid-base adducts tend to be formed preferentially when the acceptor atoms(s) of the acid and the donor atom(s) of the base have similar polarizabilities. The polarizability of an atom depends on how tightly it binds its valence electrons: the tighter the valence electrons are bound, the less polarizable the atom is. In Chapter 10, the polarizability was used to describe the different magnitudes of dispersion forces. In this section, the polarizability is used to determine whether an acid or base is hard (not very polarizable) or soft (more polarizable).

Hard bases tend to be in the first two rows of the periodic table, because of the small size and large electronegativity of the Lewis base atoms in these groups. Soft bases tend to be in the lower rows of the periodic table, because these atoms are larger and the electrons are further away from the nucleus and not as tightly bound.

Chapter 19: The Main Group Elements

Most acids are hard because metal ions with positive charges tend to bind electrons tightly. Some of the lowest row metals with low oxidation numbers are classified as soft acids. Nonmetal acids also can be soft or hard. BCl_3 is a hard acid due to the small size and electron-withdrawing Cl groups bonded to it.

In addition, many Lewis acids and bases are not distinctly hard or soft and fall into a borderline category. Some examples of hard, soft and borderline acids and bases are shown below.

	Hard	Borderline	Soft
a c i d s	H^+, Al^{3+} Na^+, K^+ Mg^{2+} BF_3 CO_3^+, Fe^{3+}	Cu^+ Fe^{2+} CO_2	Hg^{2+} Hg_2^{2+} Ag^+
b a s e s	F^- NH_3, H_2O OH^-, Cl^-	Br^- SO_3^{2-}	$P(CH_3)_3$ CO, I^- HS^-

Exercise 19.3 Arrange the following groups of Lewis acids or bases in order of *increasing* hardness.
(a) Cd^{2+}, Zn^{2+}, Ag^+ (b) $N(CH_3)_3$, NH_3, $P(CH_3)_3$ (c) Mn^{2+}, Mn^{3+}, Mo.

Steps to Solution: *The relative hardness of a series of acids or bases can be deduced by estimating their relative polarizability. Two factors that lead to low polarizability (higher hardness) are small size and positive charge. Polarizability and softness increase with increasing size (descending a column) and decreasing positive charge. We may recognize that one of the substances is harder than the others, and then we can judge the softness or hardness relative to that compound. In judging relative softness or hardness, we compare the factors affecting polarizability in examining the series of substances listed in each part to determine which substances are harder than the others.*

(a) *Of these three Lewis acids, Zn^{2+} is the smallest (fourth row metal ion, while Cd^{2+} and Ag^+ are fifth row). Zn^{2+} also has a high positive charge it is the hardest of the three acids. Cd^{2+} and Ag^+ are isoelectronic, but Cd^{2+} has a higher positive charge. The higher positive charge would be expected to bind electrons more tightly with the* result that Cd^{2+} will be less polarizable and harder than Ag^+. The order in increasing hardness is: $Ag^+ < Cd^{2+} < Zn^{2+}$.

(b) In this group of Lewis bases, $P(CH_3)_3$ is the softest base, because the donor atom, P, is larger than the N donor atom in the other two substances. Comparing NH_3 and $N(CH_3)_3$ in terms of softness is more difficult. We must consider the effect of the groups bound to the N atom to determine relative hardness. In NH_3, the only bonds are N-H bonds, whereas in $N(CH_3)_3$, there are more electron-rich N-C bonds from which the N atom can draw electron density. Because the groups on $N(CH_3)_3$ can supply more electron density to the electronegative N atom, the lone pair is less strongly bound to the N atom. We expect that $N(CH_3)_3$, will be softer than NH_3. The order of the bases in increasing hardness is: $P(CH_3)_3 < N(CH_3)_3 < NH_3$.

(c) Of this group of Lewis acids, there are two different metals, one in the fifth row and the other in the fourth row. The 5th row metal, Mo has oxidation number 0; in the other choices, Mn has +2 and +3 oxidation numbers. Therefore Mo(0) is the softest acid because of its higher **n** and lower charge. Of the other two metals, Mn^{2+} has a lower positive charge and therefore will bind its electrons less strongly than Mn^{3+}. Therefore Mn^{3+} is harder than Mn^{2+}. The order of these acids in increasing hardness is: $Mo(0) < Mn^{2+} < Mn^{3+}$.

Hard and Soft Acid and Base (HSAB) Principle

The hard and soft acid and base principle states that:

Hard Lewis acids tend to combine with hard Lewis bases.
Soft Lewis acids tend to combine with soft Lewis bases.

These statements allow fairly accurate prediction of the types of compounds that will be formed when different Lewis acids and different Lewis bases of similar strengths are present in a reaction mixture.

Exercise 19.4 Group 13 elements. The reactions of Group 13 chlorides (BCl_3, $AlCl_3$, $GaCl_3$, $InCl_3$) with bases are predicted well by the HSAB principle.
(a) Predict the order of reactivity (completeness of adduct formation) of these compounds toward $N(CH_3)_3$.
(b) Predict the order of reactivity (completeness of adduct formation) of these compounds toward $S(CH_3)_2$.

Steps to Solution: The compounds listed are all Lewis acids, and we are asked to determine their relative reactivity towards a Lewis base. We can make this prediction using the HSAB principle. We should determine the relative order of hardness of the acids and then determine if the base is hard or soft. The HSAB then can be used to determine the order of reactivity; hard bases will bind best with hard acids, and less strongly to soft acids.

(a) The base $N(CH_3)_3$ is a hard base (the donor atom is from the second row); we expect that it will react best with hard acids. Because hardness decreases going down a column in analogous compounds (such as the one we are given), $N(CH_3)_3$

will react most completely (have the largest K) with the hard acid BCl_3 and react less completely with the acids lying lower in the column.

Most reactive: $BCl_3 > AlCl_3 > GaCl_3 > InCl_3$: least reactive

(b) *The base in this part is $S(CH_3)_2$, a soft base (the donor atom is from the third row of the periodic table). Therefore the softer acids will be expected to bind more strongly to it. The order of reactivity predicted by the HSAB will be the order of increasing softness.*

Most reactive: $InCl_3 > GaCl_3 > AlCl_3 > BCl_3$: least reactive

19.3 The Main Group Metals

QUESTIONS TO ANSWER, SKILLS TO LEARN
1. **What's unusual about aluminum?**
2. **Facts about aluminum production and uses.**
3. **What are important facts about the other main group metals, tin and lead?**

The main group metal with the highest visibility is aluminum due to its use in containers and construction. It is a reactive metal and has a very negative reduction potential, so very few substances are able to reduce aluminum oxides. As a result, aluminum has been in use for a relatively short time compared to metals such as iron or tin. The text outlines the difficulty in obtaining the metal: Exercise 19.5 asks you to review those steps.

Exercise 19.5: The following block diagram outlines the steps in manufacturing aluminum from ores. Find the information that is symbolized by the empty boxes.

Steps to Solution: Refer to your text for specifics. Identify the ore, all the reactions listed for purification of the ore and the specific process needed to convert the ore to the metal.

Exercise 19.6 What purposes does the cryolite and calcium fluoride serve in the refining of aluminum?

Steps to Solution: The question asks what purposes does the presence of cryolite and CaF_2 serve in the refining of aluminum. After reading the text, you should note that it

serves two purposes: (1) to make electrolysis more effective (by converting the aluminum oxide to ionic species) and (2) to save energy by lowering the melting point of the mixture.

Tin and Lead

Lead and tin metals have been used for much longer than aluminum, due to the fact that they are much more easily reduced to the metals from their ores. The principle ores for tin and lead are cassiterite (SnO_2) and galena (PbS), respectively.

Exercise 19.7 Compare the free energies of formation of galena and cassiterite to that of aluminum oxide. How do these figures explain the relatively late discovery of aluminum?

Steps to Solution The free energies of formation of the three ore materials follow: galena ($\Delta G_f° = -92.68$ kJ/mol); cassiterite ($\Delta G_f° = -519.6$ kJ/mol); aluminum oxide ($\Delta G_f° = -1576.4$ kJ/mol). It is clear that the free energy of formation of aluminum oxide is much more favorable (large and negative) than either lead sulfide or tin(IV) oxide. Therefore it will be much more difficult to convert aluminum oxide back to the metal than the other two main group metals.

19.4 The Metalloids

QUESTIONS TO ANSWER, SKILLS TO LEARN
1. **Important facts about boron and silicon**
2. **What are uses and sources of the rest of the metalloids?**

The metalloids are B, Si, Ge, As, Sb and Te. The more important of the metalloids (in terms of the amount produced) are boron and silicon. The more important products made from silicon include silicon semiconductors and the polymeric silicones. The other metalloids are produced in smaller quantities.

Exercise 19.8 Consult the periodic chart and/or the text and write the formulas for the common binary compounds of boron, arsenic, tellurium and silicon with bromine.

Steps to Solution First consult the periodic table to determine in which group the elements are and to find the number of valence electrons. Then, since bromine generally forms one bond to elements other than oxygen, add the appropriate number of bromine atoms. You should obtain the following formulas: BBr_3, $AsBr_3$, $TeBr_2$ and $SiBr_4$.

19.5 Phosphorus

QUESTIONS TO ANSWER, SKILLS TO LEARN
1. How is elemental phosphorus made?
2. How are phosphoric acid and fertilizers produced?
3. What are organophosphorus compounds used for?

Phosphorus is one of the more important elements as it has essential roles in biological systems and is also a major industrial commodity, especially as phosphoric acid and its derivatives. Elemental phosphorus is produced as a starting material for organophosphorus compounds and high purity phosphorus compounds. The production methods and reaction types are reviewed in Exercises 19.9 and 19.10.

Exercise 19.9 The production of phosphoric acid by the "wet" process involves the reaction of fluoroapetite with sulfuric acid. Write the reaction and describe why this acid-base reaction between two fairly strong acids proceeds to a high degree of completion.

Steps to Solution The reaction for this process can be found in your text:

$$Ca_5(PO_4)_3F_{(s)} + 5\ H_2SO_{4\ (aq)} \rightarrow 3\ H_3PO_{4\ (aq)} + HF_{(aq)} + 5\ CaSO_{4\ (s)}$$

The reason for this reaction proceeding to a high degree of completion is twofold; (1) sulfuric acid is a somewhat stronger acid that phosphoric acid (check the K_a values for the acids) but more importantly, (2) one of the products is a solid and precipitates, thus driving the reaction by le Chatelier's principle. Because one of the products is removed from the reaction mixture, more reactants are consumed to form products.

Exercise 19.10 In the manufacture of elemental phosphorus, what substance is being used as the reducing agent?

Steps to Solution The first step is to write the chemical equation of reaction for the production of phosphorus:

$$2\ Ca_3(PO_4)_{2\ (s)} + 6\ SiO_{2\ (s)} + 10\ C_{(s)} \rightarrow P_{4\ (g)} + 6\ CaSiO_{3\ (l)} + 10\ CO_{(g)}$$

Examination of this reaction and determination of the oxidation numbers of the elements participating in the reaction shows that carbon is oxidized from the 0 oxidation state to the +2 oxidation state.

19.6 Other Non-Metals

QUESTIONS TO ANSWER, SKILLS TO LEARN
1. What is important about sulfur?
2. What is important about chlorine and the other halogens?

Sulfur is one of the more important non-metals because it is the starting material for the industrial chemical, H_2SO_4, that is produced in the greatest quantity. In addition, it is often found as the counterion to metal ions in their ores or incorporated in fossil fuels, depending on its origin. Upon combustion of these fuels, the sulfur is converted to sulfur oxides, which are Lewis acids and react with water to give the Brønsted acids, sulfuric acid and sulfurous acid.

The halogens are also industrially important. Chlorine is the halogen produced in largest quantity and finds use as a source of chlorine and an oxidizing agent. Fluorine and bromine are produced in smaller quantities.

Exercise 19.11 Use your knowledge of intermolecular forces, polarizability and atomic size to explain why the fluorinated polymers, the Teflons, are so chemically inert.
Steps to Solution These compounds, $[CF_2CF_2]_n$ are inert because they are hydrophobic and have strong C-F bonds that are difficult to break. One reason that the C-F bonds are difficult to break is that the small fluorine atoms fit quite closely to the central carbon atom so that other molecules or ions cannot get close enough to the C atom to form a new bond.

Test Yourself

1. Why is sulfuric acid produced in such high volume.

2. Why is aluminum so difficult to produce by chemical reduction?

Answers

1. Among other substances, it is used in the manufacture of phosphoric acid, an important component in fertilizers.

2. The common ores for aluminum are quite stable (have large, negative free energies of formation) and also have quite high melting points. Therefore, they are difficult to melt (liquids react more readily than solids) and even when molten, have unfavorable free energy changes towards reduction.

Chapter 19: The Main Group Elements

Solutions to Chapter Problems

19.1 Lewis acids are electron pair acceptors, and Lewis bases are electron pair donors. Thus Lewis acids are electron-deficient, while Lewis bases have lone pairs to donate.
(a) Lewis acid is Ni, Lewis base is CO;
(b) Lewis acid is Sb, Lewis base is Cl^-;
(c) Lewis acid is Al, Lewis base is P;
(d) Lewis acid is B, Lewis base is F^-.

19.3 Construct Lewis structures and determine steric numbers to identify the three-dimensional structure of a molecule.
(a) The Lewis structure shows SN = 3 for Al, so the molecule has trigonal planar geometry and there is a vacant $3p$ orbital perpendicular to the molecular plane:

(b) The Lewis structure shows SN = 5 for Sb, so the molecule has trigonal bipyramidal geometry. There are vacant $3d$ orbitals that do not participate in hybridization:

(c) The Lewis structure shows SN = 3 for S, so the molecule is trigonal planar with a lone pair (bent) and there is a delocalized π orbital to which an electron pair can add:

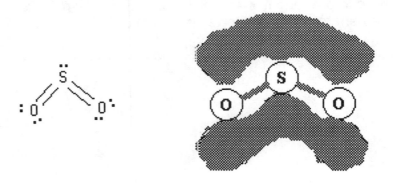

Chapter 19: The Main Group Elements

19.5 In the first step, an electron pair from the O atom of H$_2$O displaces a π bond in Lewis acid-base adduct formation. Then a proton from H$_2$O migrates to a C–O oxygen atom:

[Diagram: adduct formation → proton transfer]

19.7 Polarizability of cations increases with the values of n and Z and decreases with the charge on the ion.

V^{3+} (lowest Z, high charge) < Fe^{3+} (high charge) < Fe^{2+} < Pb^{2+} (large n).

19.9 Hard acids have low polarizability, and soft acids have high polarizability.

(a) the hardest acid is BF$_3$ (both elements from row 2), then BCl$_3$, and AlCl$_3$ is softest (both elements from row 3);

(b) the hardest acid is Al^{3+} (row 3), then Tl^{3+} (row 6), and Tl$^+$ (low charge) is softest;

(c) Polarizability increases with n, so hardest is AlCl$_3$, then AlBr$_3$, and AlI$_3$ is softest.

19.11 When an electronegative O atom bonds to a less negative S atom, it withdraws electron density from S, decreasing the polarizability about S and increasing the hardness of the base.

19.13 A metathesis reaction will occur if exchange of partners couples the harder Lewis acid with the harder Lewis base and the softer Lewis acid with the softer Lewis base.

(a) Al^{3+} is harder than Na$^+$, and Cl$^-$ is harder than I$^-$, so metathesis occurs, giving AlCl$_3$ (hard-hard) and NaI (soft-soft);

(b) the Lewis acid is Ti^{4+} in each substance, so no reaction occurs;

(c) Ca^{2+} is softer than H$^+$, and S^{2-} is softer than O^{2-}, so metathesis occurs, giving H$_2$O (hard-hard) and CaS (soft-soft);

(d) C is a soft base, and Li is a hard acid, so metathesis occurs, giving (CH$_3$)$_3$P (soft-soft) and LiCl (hard-hard);

(e) Ag and I are both soft, so no reaction occurs.

Chapter 19: The Main Group Elements

19.15 Descriptions of bonding always begin with a Lewis structure. As your textbook describes, Al$_2$Cl$_6$ contains two "bridging" chlorine atoms. Standard procedures predict tetrahedral geometry about all inner atoms, but the Al–Cl–Al bond angles of 91° indicate that Cl uses p orbitals. Each Al atom can be described as using $sp3$ hybrids to form four σ bonds to four different Cl atoms.
In Lewis acid-base terms, the bridged molecule forms from two AlCl$_3$ units linking together in double adduct formation between the Al Lewis acid atoms and two Cl Lewis base atoms.

19.17 Thallium lies below indium and gallium, so its properties should be similar to those metals: valence of 3, soft Lewis acid. Like its neighbor, Pb, it is toxic.

19.19 A BN pair has the same number of valence electrons as a pair of C atoms, so the structure and bonding of B$_3$N$_3$H$_6$ is just like that of benzene, a six-membered planar ring with a delocalized set of π bonds. There are 3(3) + 3(5) + 6(1) = 30 valence electrons, all of which are involved in the bonding network:

The B and N atoms have bonding and geometry that can be described using sp^2 hybrid orbitals, resulting in 3 σ bonds around each ring atom. In addition, there is a delocalized π bonding network encompassing all six ring atoms and containing 6 electrons.

19.21 The band gap decreases from top to bottom of each column of the periodic table, so Ge has a smaller band gap than Si. In orbital terms, this is because the principal quantum number of the valence orbitals increases, and this makes the valence orbitals larger, leading to less effective overlap and smaller energy difference between bonding and antibonding orbitals.

19.23 Polyphosphates form by sequential condensation of PO_3^- units, so the tetraphosphate has chemical formula $P_4O_{13}^{6-}$:

19.25 Phosphoric acid undergoes condensation reactions with alcohols in which water molecules are eliminated:

19.27 Phosphoric acid is produced directly from apatite by reaction with sulfuric acid:

$$Ca_5(PO_4)_3F_{(s)} + 5\ H_2SO_{4\ (aq)} \rightarrow 3\ H_3PO_{4\ (aq)} + 5\ CaSO_{4\ (s)} + HF_{(aq)}$$

This is a Brønsted acid-base reaction in which protons are transferred from sulfuric acid to the phosphate anions. There are no redox reactions in this process.

19.29 There is one industrial reaction in Section 19.6 in which sulfuric acid acts as a Brønsted acid:

$$CaF_{2\ (s)} + H_2SO_{4\ (l)} \rightarrow 2\ HF_{(g)} + CaSO_{4\ (s)}$$

19.31 The repeat structure of polyvinylchloride, in common with all polyethylene-type polymers, has an all-carbon backbone. There is a chlorine atom on every other carbon atom:

19.33 The reaction forming $TiCl_4$ from TiO_2 is as follows:

$$TiO_{2\ (s)} + C_{(s)} + 2\ Cl_{2\ (g)} \longrightarrow TiCl_{4\ (l)} + CO_{2\ (g)}$$

C goes from zero to +4 oxidation state, so it is oxidized and serves as the reducing agent. Cl goes from zero to -1 oxidation state, so it is reduced and serves as the oxidizing agent.

Chapter 19: The Main Group Elements

19.35 This is a straightforward stoichiometry problem which can be worked by determining the percentage composition of bauxite. Bauxite is AlOOH, MM = 60.00 g/mol;
% Al = (100 %)(26.982 g/mol) ÷ (60.00 g/mol) = 44.98 %;
1 kg of bauxite rock contains (1.00 kg)(85 %)(10^{-2}/%)(44.98 %)(10^{-2}/%) = 0.382 kg Al;
If the processing is 75 % efficient, each 1.00 kg of bauxite rock yields
(0.382 kg)(0.75) = 0.287 kg Al.
To produce 2500 kg of Al requires
(2500 kg Al) ÷ (0.287 kg Al/kg bauxite rock) = 8.7 x 10^3 kg bauxite rock.

19.37 Either position about phosphate detergents can be defended. Phosphates lead to eutrophication when used in excess, but other cleaning agents also have undesirable environmental consequences and phosphates are biodegradable.

19.39 Lewis acids are electron pair acceptors, and Lewis bases are electron pair donors. The As atom in $AsCl_3$ has a lone pair that it can donate, giving this compound Lewis base character. The As atoms also can accommodate additional electron pairs by using valence d orbitals, giving this compound Lewis acid character.
$AsCl_3 + BF_3 \rightarrow Cl_3As-BF_3$ (As acts as Lewis base)
$AsCl_3 + Cl^- \rightarrow AsCl_4^-$ (As acts as Lewis acid).

19.41 Balance a redox half-reaction following the standard steps (balance all but H and O by inspection, balance O by adding H_2O, balance H by adding H^+, balance charge by adding e^-):
$Al + 2 H_2O \rightarrow AlO_2^- + 4 H^+ + 3 e^-$; add 4 OH^- on each side to neutralize H^+:
$Al + 2 H_2O + 4 OH^- \rightarrow AlO_2^- + 4 H_2O + 3 e^-$; cancel 2 H_2O on each side:
$Al + 4 OH^- \rightarrow AlO_2^- + 2 H_2O + 3 e^-$;
(a) $NO_3^- + 9 H^+ + 8 e^- \rightarrow NH_3 + 3 H_2O$; add 9 OH^- on each side to neutralize H^+:
$NO_3^- + 9 H_2O + 8 e^- \rightarrow NH_3 + 3 H_2O + 9 OH^-$; cancel 3 H_2O on each side:
$NO_3^- + 6 H_2O + 8 e^- \rightarrow NH_3 + 9 OH^-$; multiply this half-reaction by 3 and the Al half-reaction by 8, and add:
$8 Al + 32 OH^- + 3 NO_3^- + 18 H_2O + 24 e^- \rightarrow 3 NH_3 + 27 OH^- + 8 AlO_2^- + 16 H_2O + 24 e^-$
Cancel duplicated species: $8 Al + 5 OH^- + 3 NO_3^- + 2 H_2O \rightarrow 3 NH_3 + 8 AlO_2^-$;
(b) $2 H_2O + 2 e^- \rightarrow H_2 + 2 OH^-$; multiply this half-reaction by 3 and the Al half-reaction by 2, and add: $2 Al + 8 OH^- + 6 H_2O + 6 e^- \rightarrow 3 H_2 + 6 OH^- + 2 AlO_2^- + 4 H_2O + 6 e^-$;
Cancel duplicated species: $2 Al + 2 OH^- + 2 H_2O \rightarrow 3 H_2 + 2 AlO_2^-$;
(c) $SnO_3^{2-} + 6 H^+ + 4 e^- \rightarrow Sn + 3 H_2O$; add 6 OH^- on each side to neutralize H^+:
$SnO_3^{2-} + 6 H_2O + 4 e^- \rightarrow Sn + 3 H_2O + 6 OH^-$; cancel 3 H_2O on each side:
$SnO_3^{2-} + 3 H_2O + 4 e^- \rightarrow Sn + 6 OH^-$; multiply this half-reaction by 3 and the Al half-reaction by 4, and add:
$4 Al + 16 OH^- + 3 SnO_3^{2-} + 9 H_2O + 12 e^- \rightarrow 3 Sn + 18 OH^- + 4 AlO_2^- + 8 H_2O + 12 e^-$
Cancel duplicated species: $4 Al + 3 SnO_3^{2-} + H_2O \rightarrow 3 Sn + 2 OH^- + 4 AlO_2^-$.

19.43 Nitrogen, at the top of Group 15, is a non-metal showing a valence of three that forms polar bonds most readily with other non-metals (B, C, O, the halogens). Phosphorus has a similar pattern of reactivity but can also involve d orbitals in its bonding. Arsenic and antimony are metalloids with useful semiconductor properties, and Bismuth is metallic.

Chapter 19: The Main Group Elements

19.45 The production of aluminum from its ore is described in Section 19.3. First, bauxite ore is treated with strong base to produce soluble $[Al(OH)_4]^-$:
$$Al(O)OH_{(s)} + OH^-_{(aq)} + H_2O \rightarrow [Al(OH)_4]^-_{(aq)}$$

When the solution is diluted with water, aluminum hydroxide precipitates:
$$[Al(OH)_4]^-_{(aq)} \rightarrow Al(OH)_{3\,(s)} + OH^-_{(aq)}$$

Strong heating drives off water, leaving aluminum oxide:
$$2\,Al(OH)_3 \xrightarrow{1250\,°C} Al_2O_{3\,(s)} + 3\,H_2O_{(g)}$$

Finally, electrolysis of aluminum oxide dissolved in cryolite reduces Al to pure metal:
$$2Al_2O_3 + 3\,C \rightarrow 3\,CO_{2\,(g)} + 4\,Al_{(l)}$$

19.47 Silicon dioxide is first converted into silicon tetrachloride by reaction with molecular chlorine:
$$SiO_{2\,(s)} + 2\,C_{(s)} + 2\,Cl_{2\,(g)} \xrightarrow{high\,T} SiCl_{4\,(g)} + 2\,CO_{(g)}$$
The $SiCl_4$ is purified by distillation and then reduced by reaction with Mg metal:
$$SiCl_4 + 2\,Mg \rightarrow Si + 2\,MgCl_2$$

19.49 Sulfur has d orbitals available for bond formation, allowing the formation of SF_6, in which the bonding can be described using d^2sp^3 hybrid orbitals on the S atom. Oxygen has no valence d orbitals available. In principle, SBr_6 could also form, but the Br atom is too large for six Br atoms to be accommodated around a central S atom.

19.51 This is a straightforward stoichiometry calculation. The reaction between H_2S and Cl_2 is a redox reaction, but it can be balanced easily by inspection: $8\,H_2S + 8\,Cl_2 \rightarrow S_8 + 16\,HCl$. To determine the amount of Cl_2 required, first calculate the H_2S content in the water:
mass H_2S = (25 ppm)(10^{-6}/ppm)(6.0 x 10^3 L)(1.00 g/mL)(10^3 mL/L) = 1.5 x 10^2 g;
Next, do appropriate mole-mass conversions:

$$m_{Cl_2} = m_{H_2S}\frac{MM_{Cl_2}}{MM_{H_2S}} = (1.5 \times 10^2\,g)(70.91\,g/mol) \div (35.08\,g/mol) = 3.0 \times 10^2\,g.$$

19.53 The reaction of aluminum sulfate with water to produce aluminum hydroxide is Brønsted acid-base reaction of the sulfate anion with H_2O:

$$Al_2(SO_4)_{3(s)} + 6\,H_2O \rightarrow 2\,Al(OH)_{3(s)} + 3\,HSO_4^-{}_{(aq)} + 3\,H^+{}_{(aq)}.$$

19.55 The reaction can be balanced by recognizing the chemical formulas of the substances involved. Calcium dihydrogenphosphate is $Ca(H_2PO_4)_2$:
$$Ca_5(PO_4)_3F_{(s)} + 7\,H_3PO_4\,_{(aq)} \rightarrow\ + 5\,Ca(H_2PO_4)_2\,_{(s)} + HF_{(aq)}$$
Calculate the mass percent phosphorus using standard stoichiometric techniques:
$MM_{compound} = \{40.078 + 2[(2)(1.008) + 30.974 + (4)(15.999)]\} = 234.05$ g/mol

$$\%\,P = (100\%)\,\frac{2(MM_P)}{MM_{compound}} = \frac{(100\%)(2)(30.974\text{ g/mol})}{(234.05\text{ g/mol})} = 26.47\,\%.$$

19.57 Gaseous Al_2Cl_6 is in equilibrium with gaseous $AlCl_3$: $Al_2Cl_6 \rightleftharpoons 2\,AlCl_3$.

Le Chatelier's principle predicts that because the forward reaction is endothermic (bonds must be broken), an increase in temperature shifts the position of the equilibrium to the right. Thus as temperature increases, the number of moles of gaseous substance increases, so the pressure increases faster than would be predicted by the ideal gas equation.

19.59 Soft metal ions are toxic because they are very soft Lewis acids which react readily with soft Lewis bases such as the sulfur atoms that hold essential enzymes in the conformations that those enzymes require to function properly. When a soft metal ion reacts with an S–S link in an enzyme, the conformation of the enzyme becomes distorted or destroyed and the enzyme can no longer function.

19.61 Mercury is a very soft Lewis acid, while zinc is intermediate in hard/soft character. Thus mercury preferentially associates with sulfur, a soft Lewis base, while zinc forms stable bonds with both soft (sulfur-containing) and hard (oxygen-containing) Lewis bases.

19.63 Polarizability increases with atomic size, and greater polarizability leads to larger dispersion forces. Thus the larger the atoms in a molecule, the larger the intermolecular forces and the easier it is to condense the substance. In the sequence BCl_3, BBr_3, BI_3, the halogens are increasing in size, accounting for the different stable phases at room temperature.

19.65 The Lewis acid-base adduct between tetrahydrofuran and BF_3 requires that the boron atom approach the oxygen atom closely. As structural formulas show, this is much easier to accomplish in the case of tetrahydrofuran than in the case of dimethyltetrahydrofuran, because methyl groups are much bulkier than hydrogen atoms:

19.67 The line structures of dioxin and DDT reveal their similarities and differences. Both contain benzene rings with chlorine atoms attached. Dioxin is a planar three-ring molecule with two O atoms bridging between two benzene rings, while the benzene rings in DDT are linked by a tetrahedral carbon atom, to which another tetrahedral carbon atom is attached:

dioxin   DDT

19.69 The reactions can be balanced by inspection:
$2 Al_2O_3 + 6 Cl_2 \rightarrow 4 AlCl_3 + 3 O_2$; $AlCl_3 + 3 Na \rightarrow Al + 3 NaCl$.

19.71 Compounds with Lewis acid-base properties will undergo metathesis reactions if a transfer of bonding partners associates harder acids with harder bases.

(a) Al^{3+} is quite hard due to its +3 charge, and Cl^- is softer than CH_3:
$AlCl_3 + 3 LiCH_3 \rightarrow Al(CH_3)_3 + 3 LiCl$;
(b) Sulfur is a Lewis acid as well as a Lewis base and forms an adduct with H_2O, a Lewis base: $SO_3 + H_2O \rightarrow H_2SO_4$ (aq)
(c) The antimony atom can bond to one additional F^- anion: $SbF_5 + LiF \rightarrow Li(SbF_6)$;
(d) S is harder than As, and F is harder than Cl, so there is no reaction.

19.73 If there is π bonding in BF_3, the B–F bonds in this compound should be shorter than bonds that are known to be single bonds. BF_3 and $[BF_4]^-$ provide illustrative data:

Species	B–F bond length	π bonding?
BF_3	131 pm	yes
$[BF_4]^-$	139 pm	no

Chapter 20 Nuclear Chemistry and Radiochemistry

20.1 Nuclear Stability

QUESTIONS TO ANSWER, SKILLS TO LEARN
1. How do nuclei differ, and how do we designate the different nuclides?
2. What is the "belt of stability," and what is its importance?
3. Using the relationship that equates mass and energy changes
4. What is the binding energy and how is it calculated?

The nucleus has been treated as a unchanging positive charge in the middle of a cloud of electrons; however, as we discussed much earlier in Chapter 2, the nucleus is composed of neutrons and protons. The following table summarizes the important physical properties of the three fundamental particles.

Particle	Symbol	Charge (e)	Mass (kg)	Molar Mass (g/mole)
Proton	$_1^1 p$	1	1.672633 x 10^{-27}	1.007276
Neutron	$_0^1 n$	0	1.674929 x 10^{-27}	1.008665
Electron	$_{-1}^0 e$	-1	0.000911 x 10^{-27}	5.487 x 10^{-4}

Isotopes of an element contain the same number of protons but different numbers of neutrons. In nuclear processes we are concerned about the nucleus; any particular combination of protons and neutrons is called a **nuclide**. A nuclide is described by stating its atomic number, Z (number of protons), and mass number, A (Z + number of neutrons, N). This notation for some nuclide, E, is:

$$\text{mass number} \atop \text{atomic number} \text{Elemental symbol} = {_Z^A}\text{E}$$

Exercise 20.1 Sulfur has four naturally occurring isotopes that have 16, 17, 18 and 20 neutrons. What are the symbols for the nuclei of these isotopes?

Steps to Solution: To write the symbols we must know the atomic number, the mass number and the elemental symbol. The atomic number is known because this element's name is given (S has Z = 16). The mass number, A, is the sum of Z and N, the number of neutrons. Because the number of neutrons is supplied in the problem, A is readily calculated.

16 neutrons: A = 16 + 16 = 32: The isotope is $_{16}^{32}$S.

17 neutrons: $A = 16 + 17 = 33$: The isotope is $^{33}_{16}\text{S}$.

The other two isotopes are: $^{34}_{16}\text{S}$ and $^{36}_{16}\text{S}$.

Nuclides *can* contain any combination of neutrons and protons, but not all of these combinations are stable.

Stable nuclei are those that do not decompose spontaneously to other nuclei and subatomic particles.

The factors that determine nuclear stability are well beyond the scope of this course, but observations of the known nuclei show a trend in Z and N. The positions of stable nuclei on a graph of their Z and N values show that the stable nuclei fall into a band of Z and N values called the **belt of stability**. The stable nuclides stop at about $Z = 82$. Nuclides with greater Z are unstable, although they may take a very long time to decompose.

Plot of N and Z of the Stable Nuclides

For low Z, the number of protons and neutrons is roughly equal; for higher Z, the ratio of neutrons to protons $\left(\dfrac{N}{Z}\right)$ is greater than 1, reaching about 1.4 to 1.5 for elements with Z around 80. This trend may be rationalized by considering the forces acting between the neutrons and protons in a nucleus. There are attractive forces between these particles, but there also strong repulsive forces acting between the positively charged protons. For small Z, about an equal number of neutrons helps to effectively dilute the positive charge so that the nuclide is stable. As Z increases, the number of neutrons compared to protons increases to reduce the proton-proton repulsions.

Exercise 20.2. Do you expect the following nuclides to be stable? Explain your reason if you believe it will not be stable. The isotope of :

Chapter 20: Nuclear Chemistry and Radiochemistry

(a) B with 6 neutrons

(b) $^{135}_{51}$Sb

(c) osmium with $A = 150$

(d) The element with $Z = 108$ and $A = 270$.

Steps to Solution: To determine whether a nuclide will be stable, you have the tools supplied in this chapter: the belt of stability, which is a summary of experimental data. If a nucleus lies near or on the belt, it will be expected to be stable; if it lies either well above, below or past the end of the belt of stability, it is not expected to be stable.

(a) This isotope of B has $Z = 5$ and $N = 6$. For smaller Z, $\left(\frac{N}{Z}\right)$ is near 1 for stable nuclei. In this case the ratio is near 1, so this nuclide is expected to be stable (this nuclide is the most common isotope of B).

(b) For this isotope, $N = 135 - 51 = 84$. The position of this isotope of Sb on the graph of N and Z of the stable nuclides is well above the belt of stability; therefore this nuclide is not expected to be stable.

(c) This nuclide has $Z = 76$; the value of $N = A - Z = 150 - 76 = 74$. The position of this nuclide on the graph of N and Z of the stable nuclides is below the belt of stability; therefore this nuclide is not expected to be stable.

(d) This nuclide has $Z = 108$ and is past the end of the belt of stability. In general, nuclides with $Z > 82$ are not stable. This nuclide is not expected to be stable.

Throughout this text, we have been operating on the assumption that mass is conserved; however, matter can be converted to energy and energy can be converted to matter. Einstein developed the following equation that relates energy and mass:

$$E = mc^2$$

Note that the proportionality constant between energy and mass is very large: the square of the speed of light! In a process in which energy is evolved or absorbed, the mass must change as well.

$$\Delta E = (\Delta m)c^2$$

(the energy change) = (the mass change) times c^2

Exercise 20.3. How much energy in kJ is released when 0.51 g of matter is converted to energy?

Steps to Solution: The energy change and the mass change are related by the equation

$$\Delta E = (\Delta m)c^2$$

However, the units of energy must be expressed in $J\left(\frac{kg\cdot m^2}{s^2}\right)$. The units of mass must be converted to kg before they can be be used in this equation. Finally, the energy in J must be converted to kJ.

$$\Delta E = (\Delta m)c^2 = (0.51\ g \cdot 10^{-3}\ kg/g)(2.998 \times 10^8\ m/s)^2 = 4.6 \times 10^{13}\ J = \boxed{4.6 \times 10^{10}\ kJ}$$

Exercise 20.4. To compare the difference in the size of energy changes of nuclear processes with chemical processes, calculate the mass of methane that would need to be combusted to give the same amount of energy in Exercise 20.3. ($\Delta H_{combustion}$ of CH_4 = -890 kJ/mole.)

Steps to Solution: This is a thermochemistry stoichiometry problem of the type you have solved before. The desired energy is divided by the enthalpy of combustion of methane to give the number of moles of methane required.

$$\text{moles } CH_4 = \frac{4.6 \times 10^{10}\ kJ}{890\ kJ/mole} = 5.2 \times 10^7 \text{ moles } CH_4$$

$$g\ CH_4 = 5.2 \times 10^7 \text{ moles } CH_4\ (16.04276\ g/mole) = \boxed{8.3 \times 10^8\ g\ CH_4}$$

This is a large amount of methane. The reason that mass changes in chemical processes are not observed is that they are so small. This reaction is quite exothermic, but the mass corresponding to the energy evolution is:

$$\text{percent mass consumed} = \frac{0.51\ g}{8.3 \times 10^8\ g} \cdot 100\% = 6.2 \times 10^{-8}\ \%$$

This percent change cannot be detected.

Because the energies involved in nuclear processes are so large, they are often computed using mass changes. An important quantity is the stabilization energy of a nuclide, which is the energy of stabilization of that nuclide compared to the component subatomic particles, sort of a nuclear enthalpy of formation. The stabilization energy of a nucleus is calculated by finding the mass difference between the atom of that nuclide and the sum of the masses of the protons, electrons and neutrons it contains. This mass difference can then be converted to energy by using Einstein's relationship. This energy is called the **binding energy**.

Binding energy: The energy difference found by the difference of the mass of the protons and neutrons and the mass of the nuclide.

Chapter 20: Nuclear Chemistry and Radiochemistry

> **COMMON PITFALL: Using average masses.**
> In calculating the binding energy you must use the exact mass of the nuclide, not the average mass found in the periodic table. The binding energy is different for each nuclide.

Exercise 20.5. Calculate the binding energy for $^{32}_{16}S$ (nuclide exact mass = 31.97207 g/mole) and compare its binding energy with that for $^{4}_{2}He$ (-2.730 x10⁹ kJ/mole)

Steps to Solution: To calculate the binding energy, we must first find the mass difference between the atom of that nuclide and the mass of the component parts. Then this is converted to energy using the energy-mass relationship.

The mass of the parts of the $^{32}_{16}S$ nuclide is found below:
Mass protons =	16(1.007276 g/mole) =	16.116416 g/mole
Mass neutrons =	16(1.008665 g/mole) =	16.13864 g/mole
Mass electrons =	16(0.0005487 g/mole) =	0.0087792 g/mole
Sum:		32.263835 g/mole

Δm = mass(product - mass(reactants) = 31.97207 - 32.263835 g/mole = -0.29176 g/mole

$\Delta E = (\Delta m)c^2$ = -0.29176 g/mole(0.001 kg/g)(2.998 x10⁸ m/s)²(0.001 kJ/J)

$$\Delta E = \boxed{-2.62 \times 10^{10} \text{ kJ/mole}}$$

The binding energy is exothermic, as expected, and is significantly larger than that of $^{4}_{2}He$ (-2.730 x 10⁹ kJ/mole). This is not surprising because there are more particles in the $^{32}_{16}S$ nucleus than in that of $^{4}_{2}He$.

Although nuclear reactions release considerable amounts of energy, they require large amounts of energy to induce. For two nuclei to undergo reaction, they must collide, an event that brings small positively charged particles together. The Coulomb repulsive forces are quite strong at such small distances, so nuclear reactions tend to have very large activation energies. The main exception to this rule is the reaction of the uncharged neutron with nuclei as neutrons have no charge and therefore experience no Coulomb repulsion.

Chapter 20: Nuclear Chemistry and Radiochemistry

20.2 Nuclear Decay

QUESTIONS TO ANSWER, SKILLS TO LEARN
1. **What is nuclear decay?**
2. **Balancing equations of reactions for nuclear processes**
3. **What are the common modes of nuclear decay?**
4. **Using the kinetics of nuclear decay processes**

We have earlier discussed stable isotopes, which do not undergo spontaneous change. Unstable nuclei (usually called **radioactive nuclei**) will undergo spontaneous change to other nuclides and subatomic particles, and sometimes emission of high-energy photons.

Nuclear decay: Spontaneous conversion of nuclides to other nuclides and other particles

The emitted particles are often described collectively as **radiation**. However, there are several types of radiation processes commonly observed. We can predict the product nuclides of a type of decay by balancing the equation of the nuclear process. There are several natural laws that cannot be violated when balancing these equations:

- Mass and energy must be conserved (remember that mass and energy may be interconverted).

- Charge must be conserved.

- Mass number must be conserved.

The types of particles that appear in a decay process are listed in the following table. Remember that the subscripts and superscripts have the following meaning:

$$^{\text{mass number}}_{\text{charge number}} E \quad \text{(E is the symbol of the particle)}$$

Particle Name	Symbol	Particle Name	Symbol
Neutron	$^1_0 n$ or n	alpha	$^4_2 He$ or $^4_2 \alpha$
Proton	$^1_1 p$ or p	positron	β^+ or $^0_{+1} e$ or $^0_{+1} \beta$
Electron or beta	β or $^0_{-1} e$	gamma	γ (no charge number or mass number)

Therefore the sum of the superscripts and subscripts of the reactants must equal the sum of the superscripts and subscripts of the products.

Chapter 20: Nuclear Chemistry and Radiochemistry

Exercise 20.6. Identify X, the missing nuclide or particle (from the above table), in each of the following nuclear equations of reaction:

(a) $^{39}_{17}Cl \rightarrow X + ^{39}_{18}Ar$ (b) $^{232}_{90}Th \rightarrow X + ^{228}_{88}Ra$ (c) $^{170}_{71}Lu \rightarrow ^{0}_{+1}e + X$ (d) $^{170}_{71}Lu + ^{0}_{-1}e \rightarrow X$

Steps to Solution: *To solve this type of problem, we must remember that the sum of the subscripts and superscripts in the products must equal the sum of the subscripts and superscripts in the reactants. In this way one can determine the subscripts and superscripts for the product X. We can analyze the change in mass number and then the change in charge number to find the mass and charge numbers of X. If X is a nuclide, we use the charge number to identify the element so that the elemental symbol can be written. If X is a particle, the table can be used to identify it.*

(a) *Let's determine the mass number and then the charge number of X:*
Mass number conservation: 39 = mass number of (X) + 39; mass number X = 0
Charge number conservation: 17 = charge number of (X) + 18; charge number of X = -1

The only particle with mass number 0 and charge number -1 is the β particle or electron. Therefore X is an electron or β particle.

(b) *Find the mass number and then the charge number of X:*
Mass number conservation: 232 = mass number of (X) + 228; mass number X = 4
Charge number conservation: 90 = charge number of (X) + 88; charge number of X = 2

The particle with mass number 4 and charge number +2 is the α particle.

(c) *Find the mass number and then the charge number of X:*
Mass number conservation: 170 = mass number of (X) + 0; mass number X = 170
charge number conservation: 71 = charge number of (X) + 1; charge number of X = 70

The nuclide with charge number 70 is Yb. Because the mass number is 170,
$X = {}^{170}_{70}Yb$.

(d) *Use the methods demonstrated above to show that $X = {}^{170}_{72}Hf$.*

Common Decay Processes

Nuclear decay occurs so that the mass and charge number of the product nuclide lie closer to the belt of stability than the mass and charge number of the starting nuclide. Nuclides that lie above the belt of stability are neutron rich; decay processes that lead to greater stability must lower the neutron to proton ratio. Decay processes that lower the neutron to proton ratio are the following:

741

Chapter 20: Nuclear Chemistry and Radiochemistry

- β emission converts a neutron to a proton.
- α emission removes 4 neutrons and 2 protons (also reduces mass number significantly).

Nuclides that lie below the belt of stability are proton rich and thus their decay processes must reduce the number of protons; α emission will not help since more neutrons are lost than protons and the situation is made worse. Decay processes that increase the neutron to proton ratio are the following:

- Positron (β+) emission converts a proton to a neutron.
- Electron capture (EC) converts a proton to a neutron.

If the nucleus is still not near or in the belt of stability, further decay processes may occur to give a product nuclide in the belt of stability. The different decay processes that occur on the way to the ultimate products(s) can be listed in the order that they occur, either as reactions or simply by listing the particles that are emitted. A list of the particles in the order they are emitted is called the **decay sequence**. For example, the first seven steps in the decay sequence of $^{238}_{92}U$ are α emission followed by β emission, then another β step, then α emission, α emission, α emission and another α emission. The decay sequence would usually be written: α, β, β, α, α, α, α. The product nuclide of each step is deduced using methods such those shown in Exercise 20.6.

Exercise 20.7. The decay sequence for $^{90}_{38}Sr$, a waste product from nuclear reactors is β, β. Write the nuclear reactions and name the final product nuclide.

Steps to Solution: Solving this problem is similar to solving Exercise 20.4, except for two factors. First, we have to write the equations for the two events. However, we know what the emitted particle is in both reactions, as that information is supplied in the decay sequence. We must solve for the decay product from the first β emission and then see what nuclides results from β emission from that product.

The first reaction can be written as:

$$^{90}_{38}Sr \rightarrow X + ^{0}_{-1}e$$

We need to find the mass number and the charge number of X:

Mass number conservation: 90 = mass number (X) + 0; mass number X = 90
Charge number conservation: 38 = charge number (X) + (-1); charge number of X = 39

Therefore $X = ^{90}_{39}Y$.

The second β emission reaction is:

$$^{90}_{39}Y \rightarrow X + ^{0}_{-1}e$$

Again, we need to find the mass number and the charge number of X:

Mass number conservation: 90 = mass number (X) + 0; mass number X = 90
Charge number conservation: 39 = charge number (X) + (-1); charge number of X = 40

Therefore $X = {}^{90}_{40}Zr$.

Rates of Nuclear Decay

The decay of nuclei is accompanied by the emission of particles. When these particles are detected, the detector records the emission product as a **count**.

Count: The signal observed by a detector indicating that a radioactive nuclide has decayed.

The number of counts, N, is proportional to the number of nuclides present.

The number of counts per unit time tells us how many nuclei decay in that amount of time; this quantity is $\dfrac{\Delta N}{\Delta t}$ (the number of nuclei decaying per unit time or **decay rate**). The decay rate can be used to calculate the half life of a nuclide by measuring the number of decays per unit time of a known amount of radioactive substance with a known half life.

$$\frac{\Delta N}{\Delta t} = -N \frac{\ln 2}{t_{1/2}}$$

In this equation, N is the total number of nuclei and $t_{1/2}$ is the half life of the nuclide.

Exercise 20.8. The isotope of uranium, ${}^{238}_{92}U$, is used for some purposes, such as the keels of sailboats or in armor-piercing artillery, because of its high density. This isotope decays by α emission with a half life of 4.5×10^9 years. How many decay events would occur per second in a sailboat keel containing 2500 kg of ${}^{238}_{92}U$?

Steps to Solution: We need to find the decay rate (number of decay events) of 2500 kg of ${}^{238}_{92}U$ in terms of seconds. We can use the above equation will solve for this quantity if we know the total number of nuclei present and the half life. The half life is given in the problem; we will have to convert the mass of ${}^{238}_{92}U$ to the number of nuclei present. Finally, we will have to convert the decay rate from counts per year to counts per second.

The mass of uranium 238 is readily converted to atoms of ^{238}U:

$$N = \frac{2500 \times 10^3 \text{ g } {}^{238}_{92}U}{238.051 \text{ g/mole } {}^{238}_{92}U} \cdot 6.022 \times 10^{23} \text{ atoms/mole} = 6.32 \times 10^{27} \text{ atoms}$$

We can now substitute in the equation:

$$\frac{\Delta N}{\Delta t} = -N \frac{\ln 2}{t_{1/2}} = -6.32 \times 10^{27} \text{ atoms/mole} = -9.74 \times 10^{17} \text{ decays/year}$$

The negative sign tells that the number of uranium nuclei is decreasing. Now, we must convert the decay rate to units of seconds:

$$\frac{\Delta N}{\Delta t} = -9.74 \times 10^{17} \frac{\text{decays}}{\text{year}} \cdot \frac{1 \text{ year}}{365 \text{ days}} \cdot \frac{24 \text{ hours}}{1 \text{ day}} \cdot \frac{1 \text{ hour}}{3600 \text{ s}} = \boxed{3.09 \times 10^{10} \text{ counts/s}}$$

This number may seem large, but remember that quite a large amount of substance is being considered. In Section 20.8, we will find that this amount of radiation is inconsequential compared to other sources of radiation in our daily lives.

Nuclear decay is a first-order process. The equation for first-order kinetics used earlier in Chapter 14 in a slightly different form is applicable for calculating the amount of a nuclide as time passes:

$$\ln \frac{N_o}{N} = \frac{\ln 2}{t_{1/2}} t \qquad t_{1/2} = \frac{\ln 2}{k}$$

In this equation, N is the amount of radioactive nuclide with half life $t_{1/2}$ that remains after some time, t, has passed from an initial amount of that nuclide, N_o. N and N_o can be expressed in variety of units *as long as they both have the same units.* Common units used for N and N_o are counts, mass units and, less commonly, moles.

Exercise 20.9. The nuclide, $^{90}_{38}\text{Sr}$, is a radioactive waste byproduct of nuclear reactions of $^{235}_{92}\text{U}$ that must be stored. It has a half life of 29 years. How many years must pass until 95% of this isotope decays?

Steps to Solution: *This problem can be solved by using the equation:*

$$\ln \frac{N_o}{N} = \frac{\ln 2}{t_{1/2}} t$$

We are given $t_{1/2}$ and the amount of the isotope that must decay; to find the answer, we must find the ratio $\frac{N_o}{N}$ and then solve this equation for time, t.

First we find the ratio. We need to find the amount of time it takes for 95% of the strontium 90 to decay; at that time, 5% of the $^{90}_{38}\text{Sr}$ will remain. Therefore,

$$N = \frac{5\%}{100\%} N_o = 0.05\, N_o \quad \text{and} \quad \frac{N_o}{N} = \frac{1 N_o}{0.05 N_o} = 20$$

N_ow the equation must be rearranged to solve for time:

$$t = \ln\frac{N_o}{N} \cdot \frac{t_{1/2}}{\ln 2} = \ln(20) \cdot \frac{29 \text{ yrs}}{0.69315} = 125 \text{ yrs} = \boxed{1 \times 10^2 \text{ yrs}}$$

That is a pretty long time to wait for this waste to deactivate itself!

20.3 Induced Nuclear Reactions

QUESTIONS TO ANSWER, SKILLS TO LEARN
1. What "reagents" are used to react with stable nuclei?
2. Why are accelerators used in nuclear reactions of stable nuclei?

Nuclear reactions are responsible for the creation of the elements, often using very vigorous conditions such those found in stars, a topic that will be discussed in Section 20.7. However, nuclear reactions can occur under milder conditions, and do, since a number of unstable nuclei with half lives much shorter than the age of the earth are present. These unstable nuclei are produced by binuclear reactions between nuclei and other particles, such as neutrons, electrons, or α particles. These reactions are called **induced nuclear reactions**; sometimes the nuclei that are changed are said to have been transmuted.

The reagent that requires the least energy to penetrate the nucleus and induce a nuclear reaction is the neutron, because it has no charge and is not repelled by any electrical charge. Neutrons are present in the earth's atmosphere because they are emitted from the sun. Capture of these solar neutrons is responsible for a great deal of the natural transmutation of nuclides on earth. However, neutron capture leads to formation of nuclei with higher neutron to proton ratios (which may or may not decay).

To produce nuclei that have higher proton to neutron ratios, a nucleus must react with another nucleus that contains protons; this requires that two positive charges be brought very close to one another. A large amount of energy is needed to overcome the Coulomb forces acting between nuclei when they are brought together. Manmade devices called **nuclear accelerators** are used to accelerate protons, α particles, and other small nuclei such as the $^{12}_{6}C$ nucleus to velocities at which they possess the necessary energy to merge with other nuclei, called **target nuclei**. The initial product of the merging of the positively charged nucleus with

Chapter 20: Nuclear Chemistry and Radiochemistry

the target is a larger nucleus called a **compound nucleus** whose identity is determined by balancing the nuclear equation of reaction. The compound nucleus may not be stable and may subsequently decompose. It is by use of accelerators that synthetic elements ($Z > 92$ and others) have been made.

Exercise 20.10. Technetium is one of the lightest synthetic elements. Propose a method for preparing technetium from niobium, which has the advantage of being naturally present in only the isotope of mass number 93.

Steps to solution: *To solve this problem, we need to consider it as a synthetic problem (which it is). Set up the equation of reaction with the desired product, the known reactant and the unknown reactant.*

$$^{93}_{41}\text{Nb} + X \rightarrow ^{??}_{43}\text{Tc}$$

We now need to find a particle that will increase the charge number by 2. Then we can balance the rest of the equation and ascertain the identity of the product (mass number as well).

Examining the equation of reaction shows that the difference in charge number is 2; therefore X must have a charge number of 2. The α particle has charge number 2. Substituting the α particle for X gives the equation below. Because the α particle has mass number 4, the mass number of the product nucleus will have increased by 4 as well.

$$^{93}_{41}\text{Nb} + ^{4}_{2}\alpha \rightarrow ^{97}_{43}\text{Tc}$$

Therefore, an α particle is a good reactant and the product of reaction of an α particle with $^{93}_{41}\text{Nb}$ would be $\boxed{^{97}_{43}\text{Tc}}$. *The $^{97}_{43}\text{Tc}$ is also the compound nucleus in this reaction.*

20.4 Nuclear Fission

QUESTIONS TO ANSWER, SKILLS TO LEARN
1. What is nuclear fission?
2. What is a nuclear chain reaction?
3. Calculating the energy released in a fission reaction

Nuclear fission is the *neutron capture-induced* decomposition of heavier nuclei into lighter nuclei and free neutrons. The product nuclides have a greater binding energy per nuclide than the starting materials; therefore energy is released. The most common fissionable nuclei are

Chapter 20: Nuclear Chemistry and Radiochemistry

$^{235}_{92}$U (naturally occurring) and the synthetic nuclides, $^{239}_{94}$Pu and $^{233}_{92}$U. Characteristics of fission are:

1. Fission is induced by neutron capture by heavy nuclides.
2. Fission gives a range of product nuclides.
3. Fission events produce more than one free neutron per nuclide that undergoes fission.
4. Fission produces more stable nuclides; therefore large amounts of energy are released.

The second point leads to considerable problems in nuclear power generation. The number of product nuclides is quite large (for $^{235}_{92}$U fission, no single nuclide is more than 7% of the total products) and leads to significant difficulties since one of the steps in treatment of used nuclear fuel is separation of the various products of the fission reactions.

Any reaction that produces one of the reactants is called a **chain reaction** because as the reaction proceeds, reactants are produced that can make the reaction rate larger. The third point shows that nuclear reactions are chain reactions; theoretically each neutron produced can initiate another fission event. In reality, there are limitations; most important, fissionable nuclei must be present to absorb the neutrons. If there are insufficient fissionable nuclei present to capture these neutrons, they will be absorbed by other nonfissionable nuclei and no free neutrons will be available to initiate more fission events. In this case the neutron recapture ratio (number of neutrons absorbed by fissionable nuclei/fission event) is much less than 1. If there is sufficient fissionable material present so that the neutron recapture ratio is near but less than 1, the reaction is self sustaining, because the number of fission events resulting from neutron recapture is about equal to the number of fission events releasing the neutrons. This is a desirable state of affairs for nuclear power generation because the nuclear reaction rate is constant and the energy released is also constant. For neutron recapture ratios greater than 1, the number of fission events increases exponentially with time and the amount of energy released increases at a similar rate. The mass of fissionable material required to reach this state is called the **critical mass**. When this amount of fissionable material exceeds the critical mass, the rate of the nuclear chain reaction increases and an explosion may result (as in a fission atomic bomb).

Exercise 20.11. $^{235}_{92}$U undergoes fission to give a wide variety of nuclides. Let's focus on one set of products for this exercise:

$$^{235}_{92}\text{U} + ^{1}_{0}n \rightarrow ^{142}_{54}\text{Xe} + ^{90}_{38}\text{Sr} + X ^{1}_{0}n$$

(a) Find the number of neutrons produced in this reaction.
(b) How much energy is released through the fission of 1.0 g $^{235}_{92}$U by this reaction if the masses of $^{142}_{54}$Xe and $^{90}_{38}$Sr are 141.929630 and 89.907738 g/mole, respectively?

Steps to Solution: First we need to balance the reaction in much the same way we balanced earlier reactions, except this time the identity of the product is known, but the number of neutrons produced must be determined. In part (b), the mass defect must be calculated to determine the energy released.

(a) Because neutrons have no charge, we do not have to worry about balancing the charge number. We use mass number conservation to balance this reaction:

$$235 + 1 = 142 + 90 + X(1)$$

$$X = 236 - (142 + 90) = 4$$

The balanced equation is: $\boxed{{}^{235}_{92}U + {}^{1}_{0}n \rightarrow {}^{142}_{54}Xe + {}^{90}_{38}Sr + 4\,{}^{1}_{0}n}$

(b) The energy released per mole of ${}^{235}_{92}U$ that undergoes fission is found from the mass defect in the nuclear reaction:

$$\Delta m = m({}^{90}Sr + {}^{142}Xe + 4\,n) - m({}^{235}U + n)$$

$$\Delta m = 89.907738 + 141.929630 + 4(1.008665) - (235.043924 + 1.008665) \text{ g (per mole } {}^{235}U)$$

$$\Delta m = -0.180561 \text{ g}$$

$$\Delta E = -0.180561 \text{ g} \cdot 1 \times 10^{-3} \text{kG/g} \,(3.00 \times 10^8 \text{ m/s})^2 = -1.625 \times 10^{13} \text{ J per mole } {}^{235}U$$

Converting this to energy per g of ${}^{235}U$ requires division by the molar mass of that nuclide:

$$\Delta E = \frac{-1.625 \times 10^{13} \text{ J/mole}}{235.04 \text{ g/mole}} = \boxed{-6.914 \times 10^{10} \text{ J/g } {}^{235}U}$$

A total of 6.914×10^{10} J may be obtained from the fission of 1 gram of ${}^{235}U$!

20.5 Nuclear Fusion

QUESTIONS TO ANSWER, SKILLS TO LEARN
1. What is nuclear fusion?
2. Calculating the energy released in fusion events
3. How are the elements synthesized in stars?
4. What is the difference between first and second generation stars?

In nuclear fission, larger nuclei undergo decomposition to smaller nuclei. **Nuclear fusion** is the process through which two smaller nuclei "fuse" to form a single, larger nucleus.

$$\text{two lighter nuclei} \rightarrow \text{a heavier nucleus} + \text{byproduct}$$

Two examples of fusion reactions are shown below.

$$^{2}_{1}H + ^{2}_{1}H \rightarrow ^{3}_{2}He + ^{1}_{0}n$$

$$^{2}_{1}H + ^{3}_{2}He \rightarrow ^{4}_{2}He + ^{1}_{0}n$$

Unlike fission, which requires a minimum amount of substance for the process to begin, fusion requires that the nuclides involved in the process have some threshold kinetic energy to overcome the repulsive forces of the like-charged nuclei. This energy requirement is one of several differences between fusion and fission. The coulombic repulsion will change with nuclear charge: as the nuclei have greater Z, the repulsive forces increase, and higher temperatures (kinetic energies) are required to initiate fusion. The energy change in fusion reaction is calculated from the mass defect in the reaction.

Exercise 20.12. Compare the energy released from the fusion of 1.0 g of deuterium $\left(^{2}_{1}H\right)$ with that produced from the fission of 1.0 g of uranium-235 calculated in Exercise 20.7.

Steps to Solution: *The energy released in any nuclear transformation is found by calculating the mass defect and then converting the mass change to energy. In this reaction, the mass change is found by:*

$$^{2}_{1}H + ^{2}_{1}H \rightarrow ^{3}_{2}He + ^{1}_{0}n$$

$$\Delta m = m\left(^{3}_{2}He\right) + m\left(^{1}_{0}n\right) - 2\left(m\left(^{2}_{1}H\right)\right)$$

$\Delta m = 3.01603 + 1.008665 - 2(2.0140)$ g $= -0.00330$ g for 2 mole of $_1^2H$

At this point, there are several ways to proceed to the energy calculation. The method used here will find the energy released for the reaction written (2 moles of deuterium), and then the molar mass of deuterium will be used to obtain the energy released from 1 g.

$$\Delta E = \Delta mc^2 = -0.00330 \text{ g} \cdot \frac{10^{-3} \text{ kg}}{\text{g}} \cdot (3.00 \times 10^8 \text{ m/s})^2 = -2.97 \times 10^{11} \text{ J}$$

This is the energy for fusion of 2 moles of deuterium. We must now convert this to the energy for 1 g of deuterium:

$$\Delta E = \frac{-2.97 \times 10^{11} \text{ J}}{2 \text{ mole } _1^2H} \cdot \frac{1 \text{ mole } _1^2H}{2.0140 \text{ g}} = \boxed{-7.38 \times 10^{10} \text{ J/g}}$$

Fusion of 1 g of deuterium releases 7.38×10^{10} J compared to the 6.914×10^{10} J of energy released upon fission of 1 g of uranium. The fission reaction releases a slightly smaller amount of energy than that of the fusion reaction of deuterium nuclei.

The energy source in stars is nuclear fusion. Fusion requires high temperatures to be initiated; ironically, it is the force of gravity that leads to the initiation of fusion in stars. Stars are formed from dust and gases in space; the first stars formed (**first generation stars**), are believed to have been formed from clouds of protons and electrons. The nuclei produced in a first generation star are produced through fusion events involving hydrogen atoms to give helium nuclides and higher Z elements up to $Z = 56$. When a first generation star dies, it explodes and the remnants of the star become intergalactic dust. **Second generation stars** are born by gravitational accumulation of this dust; the matter from which a second generation star is born contains not only hydrogen but large amounts of higher Z nuclei. These stars are responsible for formation of the elements with $Z > 56$. The debris from explosions of dying second generation stars is incorporated into **third generation stars**, of which the earth's star, the sun, is a prime example, because the matter found on earth contains elements with Z up to 92.

20.6 Effects of Radiation

QUESTIONS TO ANSWER, SKILLS TO LEARN
1. **How does radiation (the various emissions for radioactive decay) damage living organisms?**
2. **What is background radiation?**
3. **What is appropriate shielding to stop the different kinds of radioactive emission?**

Chapter 20: Nuclear Chemistry and Radiochemistry

Radiation causes damage to life forms because nuclear decay products have high energies, and this energy is transferred to the molecules and atoms they encounter. The main result is **ionization** (electrons are removed from atoms). The positive ions generated in this fashion have high reactivity and usually react to give new molecules before they can recapture the electron(s). In matter such as metals used in the construction of nuclear power plants, the ionization leads to weakening of structures; however, the extent of damage to such structures can be predicted with some accuracy and compensated for. In living organisms, the ionization and molecular changes caused by radiation are much more dramatic; a small change in the structure of an enzyme can lead to its being inactive. The amount of damage to humans caused by radiation depends on the type of particle (α, β or γ), its energy (velocity of the particle or energy of a γ photon) and the total number of particles. The measure of the biological hazard of radiation is the **rem** (this is short for *roentgen equivalent, man*, where the roentgen is an absolute measure of radiation in terms of energy). The short-term effects of radiation are summed up below:

The effects are most severe for cells that undergo division most frequently, because there is less time for the biological repair mechanisms to fix the damage. Other, more subtle changes may occur, such as in the DNA code, that will lead to mutations either at the cellular level (cancer) or at the organism level (birth defects).

The fact that organisms have mechanisms to repair damage to cells indicates that radiation is part of the natural environment. In fact, there are many sources of low-level radioactivity in our environment, called **background radiation**. Some of the unavoidable sources of background radiation are nuclides such as ^{14}C, ^{40}K, ^{222}Rn and "cosmic radiation." Cosmic radiation is radiation from the stars, most significantly the sun, totaling about 0.1 rem per year. Other sources of radiation that are more controllable are X-ray images, which average about 0.2 rem per picture, and cosmic rays, which have higher intensity at high altitudes. This means that frequent jet passengers have a higher exposure to these cosmic rays.

How can we be protected against radiation? Different types of radiation have different abilities to penetrate matter. X-rays and γ-rays have strong penetrating power because they are uncharged and are not affected by nuclear charges. X-rays require several mm thickness of lead to be absorbed, whereas more energetic γ-rays require thick (many cm) lead shields. The charged particles require less shielding: α particles are stopped before they travel more than 1 mm in a solid; β particles may require up to 100 mm shield thickness. Radiation is insidious: our natural senses cannot detect it, and we must rely on radiation detectors (such as Geiger counters and other devices) and careful handling.

20.7 Applications of Radioactivity

QUESTIONS TO ANSWER, SKILLS TO LEARN
1. **Using radioactivity to determine the ages of articles**
2. **What is a radioactive tracer and how is one used?**
3. **What are some medical uses of radioactivity?**

Radioactive decay has a high activation energy; therefore the rate constant for decay is not significantly affected by climatic changes on earth. As a result, we may use the amount of radioactivity to estimate the ages of different things provided we know the initial amount of the isotope that was present, the half life of the isotope and the current amount using the equation:

$$\ln\frac{N_o}{N} = \frac{\ln 2}{t_{1/2}} t$$

Exercise 20.13. ^{206}Pb is a decay product of ^{238}U. What is the ratio $\frac{^{206}Pb}{^{238}U}$ in a rock that is 7.6 x 10^8 years old? Assume that all the ^{206}Pb results from decay of ^{238}U ($t_{1/2}$ = 4.5 x 10^9 years).

Steps to Solution: This problem requires us to use the equation shown above with a bit of extra work. We can readily "plug and chug" in the above equation to find the ratio of ^{238}U currently present to that initially present:

$$\ln\frac{N_o}{N} = \frac{\ln(2)}{t_{1/2}} t \quad \text{gives} \quad \frac{N}{N_o} = \exp\left(-\frac{\ln(2)}{t_{1/2}} t\right)$$

$$\ln\frac{N_o}{N} = \exp\left(-\frac{0.693}{4.5 \times 10^9 \text{ years}} 7.6 \times 10^8 \text{ years}\right) = 0.89$$

N is the number of ^{238}U nuclei present at this time. N_o is the number of ^{238}U nuclei initially present. Because one Pb nuclide is formed from each U nuclide, the initial number of ^{238}U nuclei was the sum of the current number of ^{238}U and the number of ^{206}Pb nuclei (N_{Pb}) present:

$$N_o = N + N_{Pb}$$

If we initially had 100 ^{238}U nuclei, then 89 would be present at this time. Therefore the number of ^{206}Pb nuclei formed would be:

$$N_{Pb} = N_o - N = 100 - 89 = 11$$

The ratio would be: $\quad \dfrac{^{206}Pb}{^{238}U} = \dfrac{11}{89} = 0.12$

One of the dangerous aspects of radioactivity is the high energy of the emitted rays; this is also an asset, because these high-energy rays are very easy to detect. An isotope shows essentially the same chemical behavior until it decays and transmutes. Therefore we can introduce a radioactive isotope into a substance and "follow" the isotope by measuring the radioactivity. Sometimes the tracers are called **labels**, because the identity of the nucleus can be determined as that which was introduced. This ability to follow or trace nuclei can be useful in medicine and in research, and for that reason, radioactive nuclides in such substances are called **tracers**. For example, using tracers can help reveal what drugs (or other substances) are metabolized into and how long it takes for drugs to be removed from the body by metabolic processes.

Another important use is medical imaging, where radioactive isotopes are used to generate images of organs to detect abnormalities. Some of the requirements for good imaging agents are:

- Relatively harmless emission products, usually γ rays of low energy

- The ability to incorporate the nuclide into compounds that selectively aggregate at the organ or area that is to be imaged

- Lifetimes that are short enough that one can obtain an image rapidly (more counts/s) but long enough so that the imaging agent and nuclide can be prepared and then get to the target.

Another medical use of radioactivity takes advantage of its lethal aspects to kill undesirable cells. Cancer is a collection of diseases that are characterized by unusually rapid and abnormal growth of cells. Radioactivity is particularly deadly to rapidly dividing cells, so radiation therapy has been exploited to treat cancer for years. The problem with such treatments is the effects on healthy tissue, so the size of doses of radiation must be carefully controlled. One alternative that is being explored is the introduction of certain nuclides that are not radioactive into substances that will specifically bind to cancer tumors. These nuclides have the

Chapter 20: Nuclear Chemistry and Radiochemistry

property of becoming radioactive upon absorption of neutrons (which have relatively low toxicity compared to some radiation therapies). Irradiation of the tumor region with neutrons will then generate a radioactive substance only at the tumor site, perhaps providing medicine with a "magic bullet" in its fight against disease.

Test Yourself

1. A sealed sample initially containing 45.0 mg of the radioactive isotope ^{32}P ($t_{1/2}$ = 14.3 days) is found to contain 10.0 mg of ^{32}P. How long had the sample been decaying?

2. If a star has a energy output of about 4 x 10^{26} J/s, calculate the mass of ^{4}He produced per second if the energy is produced primarily through the fusion reaction:

$$4\,^1H + 2\,^{\,\,0}_{-1}e \rightarrow {}^4He + 6\gamma \quad \Delta E = -2.5 \times 10^9 \text{ kJ}$$

3. Which of the following isotopes of Cs would you expect to be stable using the "belt of stability"?
^{114}Cs; ^{133}Cs; ^{148}Cs.

4. ^{137}Cs is a byproduct isotope from fission (isotope mass = 136.907073 g/mole) with a half life of 30.17 years. If this nuclide decays to ^{137}Ba (isotope mass = 136.905812 g/mole), what particle is emitted and how much energy (in kJ) is released per mole of ^{137}Cs that decays?

Answers

1. 31 days

2. 6. x 10^{11} kg

3. ^{133}Cs

4. β emission, 6.41 x10^7 kJ/mole ^{137}Cs

Chapter 20: Nuclear Chemistry and Radiochemistry

Solutions to Chapter Problems

20.1 The elemental symbol identifies the value of Z, and the left superscript is A. $Z + N = A$:

Part:	(a)	(b)	(c)	(d)	(e)	(f)
Z	3	20	92	52	10	82
A	6	43	238	130	20	205
N	3	23	146	78	10	123.

20.3 Atomic number symbols have the elemental symbol accompanied by a left superscript denoting A and a left subscript denoting Z:

(a) $^{4}_{2}He$; (b) $^{184}_{74}W$; (c) $^{26}_{12}Mg$; (d) $^{60}_{28}Ni$.

20.5 Nuclides in the "belt of stability" are stable. Instability occurs if a nuclide has too few neutrons, too many neutrons, or $Z > 83$. In addition, most odd-odd nuclides are unstable:
(a) unstable, too many neutrons; (b) unstable, $Z > 83$; (c) stable; (d) unstable, odd-odd.

20.7 Energy releases are calculated using Equation 20-3: $\Delta E = (\Delta m)(8.988 \times 10^{10}$ kJ/g):
$\Delta m = (1.00$ met ton$)(1000$ kg/met ton$)(1000$ g/kg$) = 1.00 \times 10^{6}$ g;
$\Delta E = (1.00 \times 10^{6}$ g$)(8.988 \times 10^{10}$ kJ/g$) = 8.99 \times 10^{16}$ kJ.

20.9 Binding energy is calculated by determining the mass defect (difference between the mass of the atom and the sum of the masses of its individual components) and converting to energy. When an element has only one stable nuclide, its molar mass is the mass of that nuclide:
$MM_{Cs} = 132.905$ g/mol, stable nuclide has 55 protons and electrons, $133 - 55 = 78$ neutrons;
$m_{components} = 55(1.007276 + 0.0005486) + 78(1.008665) = 134.106223$ g/mol;
$\Delta m = 132.905 - 134.106223 = -1.201$ g/mol;
$\Delta E = (-1.201$ g/mol$)(8.988 \times 10^{10}$ kJ/g$) = -1.079 \times 10^{11}$ kJ/mol;
ΔE (per nucleon) $= (-1.811 \times 10^{11}$ kJ/mol$) \div (133$ nucleons$) = -8.11 \times 10^{8}$ kJ/mol nucleon.

20.11 The coulombic barrier for fusion depends on the product of the nuclear charges and on the radii of the two nuclides. The stability of the product nuclide depends on its proton-neutron ratio: $^{1}H + ^{6}Li \rightarrow ^{7}Be$, charge product $= (1)(3) = 3$ and product has n/p $= 0.75$;
$^{4}He + ^{4}He \rightarrow ^{8}Be$, charge product $= (2)(2) = 4$ and product has n/p $= 1.00$;
Thus the coulombic barrier is greater for the second reaction, but the product nuclide is more stable.

20.13 Symbols and names for nuclear particles appear in Table 20-3 and should be memorized:

	Description	Symbol	Name
(a)	high energy photon	γ	gamma ray
(b)	mass number 4	α	alpha particle
(c)	positive particle with m_e	β^+	positron.

20.15 To identify the products of nuclear decay processes, make use of the principles of conservation of mass number and charge:
(a) no change in mass or charge, product is ^{152}Te;
(b) EC changes nuclear charge by –1, no change in mass, product is $^{123}_{51}$Sb;
(c) β decay changes nuclear charge by +1, no change in mass, product is $^{127}_{53}$I.

20.17 Equation 20-4 is used to determine half-lives from radioactive decay data.
$$\text{Rate} = \frac{\Delta N}{\Delta t} = \frac{-N \ln 2}{t_{1/2}}$$
Summarize the known data: rate = –242/s; $m = 1.33 \times 10^{-12}$ g;
Convert mass to number of nuclei, using $MM \approx A = 26 + 33 = 59$ g/mol:
$$N = \frac{(1.33 \times 10^{-12} \text{ g})(6.022 \times 10^{23} / \text{mol})}{(59 \text{ g/mol})} = 1.36 \times 10^{10};$$
$t_{1/2} = -(1.36 \times 10^{10})(\ln 2) \div (-242/\text{s}) = 3.9 \times 10^7$ s;
Convert to a more convenient time unit:
$(3.9 \times 10^7 \text{ s}) \div (60 \text{ s/min})(60 \text{ min/hr})(24 \text{ hr/day})(365 \text{ day/yr}) = 1.2$ yr.

20.19 Identify products of nuclear decay by noting that α-decay changes A by –4 and Z by –2, while β-decay changes Z by +1 while leaving A unchanged:
$^{232}_{90}\text{Th} \to \alpha + {}^{228}_{88}\text{Ra} \to \beta + {}^{228}_{89}\text{Ac} \to \beta + {}^{228}_{90}\text{Th} \to \alpha + {}^{224}_{88}\text{Ra} \to \alpha + {}^{220}_{86}\text{Rn} \to \alpha + {}^{216}_{84}\text{Po};$
$^{216}_{84}\text{Po} \to \beta + {}^{216}_{85}\text{At} \to \alpha + {}^{212}_{83}\text{Bi} \to \beta + {}^{212}_{84}\text{Po} \to \alpha + {}^{208}_{82}\text{Pb}.$

20.21 Identify the compound nucleus and the final products of a nuclear reaction by applying the principles of conservation of mass number and charge:
(a) $^{12}_{6}\text{C} + n \to \left[{}^{13}_{6}\text{C}\right] \to {}^{12}_{5}\text{B} + p;$
(b) $^{16}_{8}\text{O} + \alpha \to \left[{}^{20m}_{10}\text{Ne}\right] \to {}^{20}_{10}\text{Ne} + \gamma;$
(c) $^{247}_{96}\text{Cm} + {}^{11}_{5}\text{B} \to \left[{}^{258}_{101}\text{Md}\right] \to {}^{255}_{101}\text{Md} + 3\,n.$

20.23 Pictures of nuclear reactions should show each nuclide with the appropriate numbers of protons and neutrons. The reaction in this problem is:

$^{14}_{7}\text{N} + \alpha \to \left[{}^{18}_{9}\text{F}\right] \to {}^{18}_{8}\text{O} + \beta^+$

● = proton   ◉ = neutron

20.25 Stable elements with mass numbers around 135 have N/Z ratios in the range of 1.4. Use $A = N + Z$ to calculate what Z value is likely to be stable:

$N = A - Z$, $\dfrac{135 \pm Z}{Z} = 1.4$, $\quad 2.4\,Z = 135$, $\quad Z = 56$.

Thus the most likely high-mass elements resulting from U-235 fission are Cs, Ba, La, Ce.

20.27 Calculate the mass defect from the masses of the individual reactants and products, then convert to energy released:
$\Delta m = [80.9199 + 151.9233 + 3(1.0087)] - [235.0439 + 1.0087] = -0.1833$ g/mol;
$\Delta E = (8.988 \times 10^{10}$ kJ/mol$)(-0.1833$ g/mol$) = -1.648 \times 10^{10}$ kJ/mol; this result is somewhat less than the result of the general calculation of 20.4.1 for a net change of 2 neutrons.

20.29 Your description should feature the fact that the core of a nuclear reactor generates radiation that converts any material in its vicinity into radioactive substances. Thus the heat exchanger in immediate contact with the core becomes radioactive and must be separated from the turbine that generates electricity. The primary heat exchanger transfers energy to a secondary heat exchanger which does not become radioactive and can safely drive the turbine.

20.31 The amount of energy released in a fusion reaction can be calculated from the energy per event and the total amount of matter undergoing fusion. The reactions in this example are:
^{2}H + ^{3}H → ^{4}He + n, $\Delta E = -1.7 \times 10^9$ kJ/mol
^{6}Li + n → ^{4}He + ^{3}H, $\Delta E = -4.6 \times 10^8$ kJ/mol
$n_H = (2.50$ g$) \div (2.014$ g/mol$) = 1.241$ mol; ;
$\Delta E = (1.241$ mol$)(-2.16 \times 10^9$ kJ/mol$) = -2.68 \times 10^9$ kJ;
In addition, 0.6207 mol Li react, $\Delta E = (0.6207$ mol$)(-4.6 \times 10^8$ kJ/mol$) = -2.86 \times 10^8$ kJ;
$\Delta E_{\text{total}} = (-2.68 \times 10^9$ kJ$) + (-2.86 \times 10^8$ kJ$) = -2.97 \times 10^9$ kJ.

20.33 To determine the speed of a nucleus that can fuse with another, first determine the energy needed to surmount the coulombic barrier, as described by Equation 20-1:

$$E_{\text{coulomb}} = \dfrac{(1.389 \times 10^5 \text{ kJ pm/mol})(Z_1)(Z_2)}{d}$$

Then calculate the speed from the equation for kinetic energy, $kE = 1/2\, mu^2$,
$u = (2\, kE/m)^{1/2}$
Here are the data: The radius of each H nuclide is 1.4×10^{-3} pm; $Z = 1$ for both nuclides; and $m = 3.01605$ g/mol for tritium:

$$E = \dfrac{(1.389 \times 10^5 \text{ kJ pm/mol})(1)(1)}{2(1.4 \times 10^{-3} \text{ pm})} = 5.0 \times 10^7 \text{ kJ/mol};$$

$$u = \left[\dfrac{(2)(5.0 \times 10^7 \text{ kJ/mol})(10^3 \text{ J/kJ})}{(3.01605 \text{ g/mol})(10^{-3} \text{ kg/g})}\right]^{1/2} = 5.8 \times 10^6 \text{ m/s}.$$

(Remember that 1 J = 1 kg m^2/s^2, so the units in this calculation cancel to give m/s).

Chapter 20: Nuclear Chemistry and Radiochemistry

20.35 Your description should include the extremely high energies required to initiate fusion and the difficulties in containing the fusion components at the temperature required for nuclei to have these high energies.

20.37 The characteristics of first-generation stars are described in your text:

Stage	Temperature	Composition
H-burning	4×10^7 K	H, He, e$^-$
He-burning	10^8 K	He, C, O (H depleted)
C-burning	10^9 K	C, O, Na, Ne, etc. up to Fe (He depleted)

20.39 Your description should include the fact that elements beyond $Z = 26$ are less stable than Fe, so they cannot be generated by fusion of lighter elements.

20.41 The problem asks for the energy released by a radioactive isotope. To determine this, it is necessary to calculate how much material decays during the time period. Determine the amount of the radioactive nuclide that decays in one day using Equation 20-5:

$$\ln\left(\frac{N_o}{N}\right) = \frac{t \ln 2}{t_{1/2}}.$$

$\ln\left(\frac{N_o}{N}\right) = \frac{(1 \text{ day})(0.693)}{(8.07 \text{ day})} = 0.0859$, from which $\left(\frac{N_o}{N}\right) = e^{0.0859} = 1.0897$;

$N = (7.45 \text{ pg}) \div (1.0897) = 6.84$ pg, and $\Delta N = 7.45$ pg $- 6.84$ pg $= 0.61$ pg;

Now convert to moles and multiply by the energy released per mole to obtain the amount of energy captured by the gland:

$n = (0.61 \text{ pg})(10^{-12} \text{ g/pg}) \div (131 \text{ g/mol}) = 4.7 \times 10^{-15}$ mol;

$E = (4.7 \times 10^{-15} \text{ mol})(9.36 \times 10^7 \text{ kJ/mol})(10^3 \text{ J/kJ}) = 4.4 \times 10^{-4}$ J.

20.43 Use Equation 20-4 to determine the rate of emission of a radioactive nuclide:

$N = (7.45 \text{ pg})(10^{-12} \text{ g/pg})(6.022 \times 10^{23} \text{ nuclei/mol}) \div (131 \text{ g/mol}) = 3.425 \times 10^{10}$ nuclei;

$$\frac{\Delta N}{\Delta t} = \frac{-N \ln 2}{t_{1/2}} = \frac{(3.425 \times 10^{10})(0.693)}{(8.07 \text{ day})(24 \text{ hr/day})(60 \text{ min/hr})(60 \text{ s/min})} = -7.09 \times 10^4;$$

There are 7.09×10^4 decays/s.

20.45 Exposure to radiation results first in damage to those cells that reproduce most quickly, including the white blood cells that are responsible for fighting infection and the mucous membrane lining of the intestinal tract. Thus the early symptoms of radiation exposure include reduced resistance to infection and nausea due to disruption of the digestive tract.

20.47 Dating techniques using radioisotopes are based on Equation 20-5: $\ln\left(\frac{N_o}{N}\right) = \frac{t \ln 2}{t_{1/2}}$.

To obtain an age estimate, the ratio N_o/N must be determined. Convert the mass ratio into the desired numerical ratio. Because both isotopes have the same value for A, their mole ratio and number ratio are the same as their mass ratio:

$N_o = N_{Sr} + N_{Rb}$; $\quad \frac{N_o}{N} = \frac{N_{Sr} + N_{Rb}}{N_{Rb}} = 1 + \frac{N_{Sr}}{N_{Rb}} = 1 + 0.0050 = 1.0050$;

$t = \left(\frac{t_{1/2}}{\ln 2}\right) \ln\left(\frac{N_o}{N}\right) = \left(\frac{4.9 \times 10^{11} \text{ yr}}{0.693}\right) \ln(1.0050) = 3.5 \times 10^9$ yr.

Chapter 20: Nuclear Chemistry and Radiochemistry

20.49 The best radioisotopes for medical imaging have intermediate half-lives, long enough to prepare and administer but short enough to decay in a reasonable length of time, and they decay by processes that do the least damage to body tissue (γ and β^+). Among the isotopes of iodine listed, $A = 123$ would be best because of its 13.3-hr half-life and EC/γ mode of decay.

20.51 To identify the products of nuclear decay processes, make use of the principles of conservation of mass number and charge:

(a) alpha emission changes nuclear charge by –2, mass by –4, product is $^{234}_{90}$Th;

(b) (n,p) changes nuclear charge by –1, no change in mass, product is $^{60}_{27}$Co;

(c) process changes nuclear charge by +6, mass by (+12 – 3), product is $^{248}_{99}$Es;

(d) (p,α) changes nuclear charge by –1, mass by –3, product is $^{32}_{16}$S;

(e) β decay changes nuclear charge by +1, no change in mass, product is $^{60}_{28}$Ni..

20.53 Binding energy is calculated by determining the mass defect (difference between the mass of the atom and the sum of the masses of its individual components) and converting to energy. When an element has only one stable nuclide, its molar mass is the mass of that nuclide:
$MM_{Bi} = 208.980$ g/mol, nuclide has 83 protons and electrons, 209 – 83 = 126 neutrons;
$m_{components} = 83(1.007276 + 0.0005486) + 126(1.008665) = 210.741232$ g/mol;

$\Delta m = 208.980 - 210.741232 = -1.761$ g/mol;
$\Delta E = (-1.761$ g/mol$)(8.988 \times 10^{10}$ kJ/g$) = -1.583 \times 10^{11}$ kJ/mol;

ΔE (per nucleon) $= (-1.583 \times 10^{11}$ kJ/mol$) \div (209$ nucleons$) =$
-7.574×10^8 kJ/mol nucleon.

20.55 Information about nuclear decays can be obtained using Equations 20-4 and 20-5.

$$\ln\left(\frac{N_o}{N}\right) = \frac{t \ln 2}{t_{1/2}}, \text{ from which } \ln N = \ln N_o - \frac{t \ln 2}{t_{1/2}}$$

$N_o = 5.0$ mg, $t_{1/2} = 138.4$ days, $t = 365$ days;

$\ln N = \ln(5.0) - \dfrac{(365 \text{ days})(\ln 2)}{(138.4 \text{ days})} = -0.2182, \quad N = e^{-0.2182} = 0.80$ mg;

$\dfrac{\Delta N}{\Delta t} = \dfrac{-N \ln 2}{t_{1/2}}$; $N = (0.80$ mg$)(10^{-3}$ g/mg$)(6.022 \times 10^{23}$/mol$) \div (210$ g/mol$) = 2.3 \times 10^{18}$

$\dfrac{\Delta N}{\Delta t} = \dfrac{(2.3 \times 10^{18})(\ln 2)}{(138.4 \text{ days})(24 \text{ hr / day})(60 \text{ min / hr})(60 \text{ s / min})} = 1.3 \times 10^{11}$/s.

Chapter 20: Nuclear Chemistry and Radiochemistry

20.57 Determine the products of nuclear decay by applying conservation of charge and mass number: The other product of neutron decay must be a particle with –1 charge and 0 mass, which is an electron: n → p + e.

Calculate the decay energy from the mass defect between reactant and products, using masses found in Table 20-1:
Δm = (1.007276 g/mol) + (0.005486 g/mol) – (1.008665 g/mol) = – 0.0008404 g/mol;
ΔE = (8.988 x 10^{10} kJ/g)(– 0.0008404 g/mol)(10^3 J/kJ) = – 7.554 x 10^{10} J/mol;

$kE_{electron}$ = (7.554 x 10^{10} J/mol) ÷ (6.022 x 10^{23}/mol) = 1.25 x 10^{-13} J.

20.59 Calculations of decay times make use of Equation 20-5:
$$\ln\left(\frac{N_o}{N}\right) = \frac{t \ln 2}{t_{1/2}}, \text{ from which } t = \left(\frac{t_{1/2}}{\ln 2}\right)\ln\left(\frac{N_o}{N}\right)$$

When 1% has decayed, $t = \left(\frac{1622 \text{ yr}}{\ln 2}\right)\ln\left(\frac{N_o}{0.99 N_o}\right)$ = 24 yr;

When 1% remains, $t = \left(\frac{1622 \text{ yr}}{\ln 2}\right)\ln\left(\frac{N_o}{0.01 N_o}\right)$ = 1.1 x 10^4 yr.

20.61 The n/p ratio of an isotope determines where it lies with respect to the belt of stability. ^{11}C has 6 protons and (11 – 6) = 5 neutrons, n/p = 0.833; ^{15}O has 8 protons and (15 – 8) = 7 neutrons, n/p = 0.875. Both isotopes are neutron-deficient and lie below (to the right of) the belt of stability. Isotopes that are neutron-deficient decay by positron emission, which increases their n/p ratios: ^{11}C → β$^+$ + ^{11}B and ^{15}O → β$^+$ + ^{15}N.

20.63 Calculations of decay times make use of Equation 20-5: $\ln\left(\frac{N_o}{N}\right) = \frac{t \ln 2}{t_{1/2}}$,

from which $t = \left(\frac{t_{1/2}}{\ln 2}\right)\ln\left(\frac{N_o}{N}\right)$;

When 1 μg is left, $t = \left(\frac{15 \text{ hr}}{\ln 2}\right)\ln\left(\frac{25 \text{ μg}}{1 \text{ μg}}\right)$ = 70 hr.

20.65 Information about nuclear decays can be obtained using Equation 20-5.
$\ln\left(\frac{N_o}{N}\right) = \frac{t \ln 2}{t_{1/2}}$, from which $\ln N = \ln N_o - \frac{t \ln 2}{t_{1/2}}$

N_o = (10^5/s)(30 s) = 3.0 x 10^6; t = (5 hr)(60 min/hr)(60 s/min) = 1.8 x 10^4 s;

If $t_{1/2}$ = 1100 s, $\ln N$ = $\ln(3.0 \times 10^6) - \frac{(1.8 \times 10^4 \text{ s})(\ln 2)}{(1100 \text{ s})}$ = 3.57,

N = $e^{3.57}$ = 36;

If $t_{1/2}$ = 876 s, $\ln N$ = $\ln(3.0 \times 10^6) - \frac{(1.8 \times 10^4 \text{ s})(\ln 2)}{(876 \text{ s})}$ = 0.67,

N = $e^{0.67}$ = 2.

20.67 Determine the products of nuclear decay using conservation of charge number and mass number. Use mass-energy equivalence to determine the mass of a product, given the mass of a reactant and the energy given off in the process:

(a) β-particles have 0 mass number, –1 charge number, so $^{64}_{29}Cu \rightarrow {}^{64}_{30}Zn + \beta$;

β+-particles have 0 mass number, –1 charge number, so $^{64}_{29}Cu \rightarrow {}^{64}_{28}Ni + \beta^+$;

(b) $\Delta E = (8.988 \times 10^{10}\ kJ/g)\Delta m$, from which $\Delta m = \Delta E \div (8.988 \times 10^{10}\ kJ/g)$;

For β, $\Delta m = \dfrac{(\pm 9.3 \times 10^{-14}\ J/atom)(6.022 \times 10^{23}\ atom/mol)}{(8.988 \times 10^{10}\ kJ/mol)} = -0.000623\ g/mol$;

$m_{Zn} = -0.000623 + 63.92976 - 0.0005486 = 63.92859\ g/mol$;

For β+, $\Delta m = \dfrac{(\pm 1.04 \times 10^{-13}\ J/atom)(6.022 \times 10^{23}\ atom/mol)}{(8.988 \times 10^{10}\ kJ/mol)} = -0.000697\ g/mol$;

$m_{Ni} = -0.000697 + 63.92976 - 0.0005486 = 63.92851\ g/mol$.

20.69 Energy releases are calculated using Equation 20-3: $\Delta E = (\Delta m)(8.988 \times 10^{10}\ kJ/g)$:

$\Delta m = (147.9146) + (4.00260) - 151.9205 = -0.0033\ g/mol$;
$\Delta E = (-0.0033\ g/mol)(8.988 \times 10^{10}\ kJ/g) = -2.97 \times 10^8\ kJ/mol$;
ΔE (per event) $= (-2.97 \times 10^8\ kJ/mol)(10^3\ J/kJ) \div (6.022 \times 10^{23}/mol) = -4.9 \times 10^{-13}\ J$;

Fraction carried off by the α–particle $= (3.59 \times 10^{-13}\ J) \div (4.93 \times 10^{-13}\ J) = 0.73$.

20.71 Nuclides in the "belt of stability" are stable. Instability occurs if a nuclide has too few neutrons, too many neutrons, or $Z > 83$. In addition, most odd-odd nuclides are unstable:
(a) too many neutrons;
(b) $Z > 83$;
(c) odd-odd;
(d) too many neutrons.

20.73 The decay schemes for heavy nuclides always proceed in increments of $A = -4$, because this is the change accompanying α-particle emission, while β- and γ-particles have $A = 0$. Thus the decay scheme of ^{238}U must produce ^{206}Pb (238 – 206 = 32 = 8(4)), while the decay scheme of ^{232}Th must produce ^{208}Pb (232 – 208 = 24 = 6(4)). A sample of Pb, some of which came from ^{238}U, will have more ^{206}Pb and less ^{208}Pb than a sample, some of which came from ^{232}Th. Thus the sample to which ^{238}U contributed will have a lower molar mass. Here is one example of a decay scheme:

$^{232}_{90}Th \rightarrow \alpha + {}^{228}_{88}Ra \rightarrow \beta + {}^{228}_{89}Ac \rightarrow \beta + {}^{228}_{90}Th \rightarrow \alpha + {}^{224}_{88}Ra \rightarrow \alpha + {}^{220}_{86}Rn \rightarrow \alpha + {}^{216}_{84}Po$;

$^{216}_{84}Po \rightarrow \beta + {}^{216}_{85}At \rightarrow \alpha + {}^{212}_{83}Bi \rightarrow \beta + {}^{212}_{84}Po \rightarrow \alpha + {}^{208}_{82}Pb$.

Chapter 20: Nuclear Chemistry and Radiochemistry

20.75 Dating techniques using radioisotopes are based on Equation 20-5: $\ln\left(\dfrac{N_o}{N}\right) = \dfrac{t \ln 2}{t_{1/2}}$.

To obtain an age estimate, the ratio N_o/N must be determined. Start by converting the volume of He gas into an amount in moles:

$n = pV/RT = \dfrac{(1 \text{ atm})(6.0 \times 10^{-5} \text{ cm}^3)(10^{-6} \text{ L/cm}^3)}{(0.08206 \text{ L atm/mol K})(298 \text{ K})} = 2.45 \times 10^{-9}$ mol;

The decay scheme includes 8 α-particles per U nucleus, so the amount of ^{238}U that has decayed is $(2.45 \times 10^{-9} \text{ mol}) \div 8 = 3.06 \times 10^{-10}$ mol;

The amount of ^{238}U remaining is $(1.3 \times 10^{-7} \text{ g}) \div (238 \text{ g/mol}) = 5.46 \times 10^{-10}$ mol;

Thus, $N = 5.46 \times 10^{-10}$ mol and $N_o = 5.46 \times 10^{-10} + 3.06 \times 10^{-10} = 8.52 \times 10^{-10}$ mol;

$t = \left(\dfrac{t_{1/2}}{\ln 2}\right) \ln\left(\dfrac{N_o}{N}\right) = \left(\dfrac{4.5 \times 10^9 \text{ yr}}{0.693}\right) \ln\left(\dfrac{8.52}{5.46}\right) = 2.9 \times 10^9$ yr.

20.77 Determine the nuclear reaction by applying conservation of mass number and charge number:

$^{10}_{5}\text{B} + \text{n} \rightarrow\,^{11}_{5}\text{B} \rightarrow\,^{7}_{3}\text{Li} + \alpha$. This reaction does not pose a significant health hazard, because Li-7 is a stable isotope and the α-particles are easily stopped using an appropriate shield.

20.79 Figure 20-10 indicates that the most likely mass numbers are $A = 95$ and 138. Figure 20-6 shows the stable isotopes with these mass numbers:
$A = 95$, the only stable isotope is ^{95}; other elements with stable isotopes between $A = 90$ and $A = 100$ are Zr, Nb, Ru, and Rh;
$A = 138$, the stable isotopes are Ba, La, and Ce; other elements with stable isotopes between $A = 132$ and $A = 143$ are Xe, Cs, Pr, and Nd.

20.81 To calculate the useful lifetime of the dating technique use Equation 20-5:

$\ln\left(\dfrac{N_o}{N}\right) = \dfrac{t \ln 2}{t_{1/2}}$, from which $t = \dfrac{t_{1/2}}{\ln 2} \ln\left(\dfrac{N_o}{N}\right)$

$t = \left(\dfrac{5730 \text{ yr}}{0.693}\right) \ln\left(\dfrac{15.3 \text{ counts/g min}}{0.03 \text{ counts/g min}}\right) = 5.2 \times 10^4$ yr.

20.83 Dating techniques using radioisotopes are based on Equation 20-5:
$$\ln\left(\frac{N_o}{N}\right) = \frac{t \ln 2}{t_{1/2}}.$$
First, calculate the N_o/N ratio for the age found by using the constant cosmic radiation assumption:
$$\ln\left(\frac{N_o}{N}\right) = \frac{(3250 \text{ yr})(0.693)}{(5730 \text{ yr})} = 0.3931, \text{ from which } \left(\frac{N_o}{N}\right) = e^{0.3931} = 1.482;$$

If the level of cosmic radiation was 20% higher at the time the tree grew, then the corrected ratio is $(1.20)(1.482) = 1.778$. Use this ratio to recompute the lifetime of the object:

$$t = \left(\frac{t_{1/2}}{\ln 2}\right)\ln\left(\frac{N_o}{N}\right) = \left(\frac{5730 \text{ yr}}{0.693}\right)\ln(1.778) = 4760 \text{ yr}.$$

20.85 Advantageous properties of Tc-99 include its intermediate half-life of 6 hr, its mode of emission (γ-rays, which are easy to detect but not as damaging as β- or α-particles), ease of obtaining pure from the target from which it is formed, and its ability to react chemically with many biochemical substances that can then be targeted at specific organs.

20.87 Determine the expected product using conservation of mass number and charge number:
$${}^{27}_{13}\text{Al} + {}^{4}_{2}\text{He} \rightarrow \left({}^{31}_{15}\text{P}\right) \rightarrow {}^{30}_{15}\text{P} + \text{n};$$ This nuclide is odd-odd, so is unstable. In addition, its n/p ratio is 1.00, so to become stable it must reduce its proton count. One way to do so is by positron emission (EC is another possibility) leading to ${}^{30}_{14}\text{Si}$, which is stable.

20.89 Use the data provided to calculate the total amount of ^{235}U available, then multiply by the energy per fission event to obtain the total amount of energy available and divide by yearly consumption to find the time.

Data: total amount of U = 1.0×10^7 metric tons, fractional content ^{235}U = 0.72%, energy released = 2.9×10^{-11} J/atom, energy need = 2.0×10^{17} kJ/yr.

Mass of ^{235}U available = $(1.0 \times 10^7 \text{ ton})(10^6 \text{ g/ton})(0.72\%)(10^{-2}/\%) = 7.2 \times 10^{10}$ g;
Atoms of ^{235}U = $(7.2 \times 10^{10} \text{ g})(6.022 \times 10^{23} \text{ atoms/mol}) \div (235 \text{ g/mol}) = 1.8 \times 10^{32}$ atoms;
Energy available = $(1.8 \times 10^{32} \text{ atoms})(2.9 \times 10^{-11} \text{ J/atom})(10^{-3} \text{ kJ/J}) = 5.2 \times 10^{18}$ J;

Length of time = $(2.0 \times 10^{17} \text{ kJ/yr}) \div (5.2 \times 10^{18} \text{ J}) = 26$ yr. (That's not a long time, but even in countries like Japan and France that rely heavily on nuclear power, no more than about 30% of the annual energy supply comes from this source).

20.91 To determine where the oxygen atom in the water molecule comes from, prepare a sample of the alcohol that is enriched in radioactive ^{18}O and run the reaction using this sample. Separate the products and measure the radioactivity of the ester and the water. If the C–OH bond in the alcohol breaks during the condensation, the ^{18}O will appear in the water, while if the C–OH bond in the carboxylic acid breaks during the condensation, the ^{18}O will appear in the ester.

20.93 Irradiation of foods kills microscopic organisms such as molds that cause food to spoil. However, irradiation might also convert some of the substances in food into harmful substances such as carcinogens that would pose a human health risk.

20.95 Dating techniques using radioisotopes are based on Equation 20-5: $\ln\left(\dfrac{N_o}{N}\right) = \dfrac{t \ln 2}{t_{1/2}}$.

First, calculate the N_o/N ratio, which is the ratio of counts/g hr for the new and old samples:
Fresh sample: count rate = (18,400 counts) ÷ (1.00 g)(20 hr) = 920 counts/g hr;
Old sample: count rate = (1020 counts) ÷ (0.250 g)(24 hr) = 170 counts/g hr;

$N_o/N = (920) \div (170) = 5.4$

$t = \left(\dfrac{t_{1/2}}{\ln 2}\right)\ln\left(\dfrac{N_o}{N}\right) = \left(\dfrac{5730 \text{ yr}}{0.693}\right)\ln(5.4) = 1.4 \times 10^4$ yr.